Photosynthesis

VOLUME II

Development, Carbon Metabolism, and Plant Productivity

CELL BIOLOGY: A Series of Monographs

G. M. Padilla, G. L. Whitson, and I. L. Cameron (editors). THE CELL CYCLE: Gene-Enzyme Interactions, 1969

A. M. Zimmerman (editor). HIGH PRESSURE EFFECTS ON CELLULAR PROCESSES, 1970

I. L. Cameron and J. D. Thrasher (editors). CELLULAR AND MOLECULAR RENEWAL IN THE MAMMALIAN BODY, 1971

I. L. Cameron, G. M. Padilla, and A. M. Zimmerman (editors). DEVELOPMENTAL ASPECTS OF THE CELL CYCLE, 1971

P. F. Smith. The BIOLOGY OF MYCOPLASMAS, 1971

Gary L. Whitson (editor). CONCEPTS IN RADIATION CELL BIOLOGY, 1972

Donald L. Hill. THE BIOCHEMISTRY AND PHYSIOLOGY OF *TETRAHYMENA*, 1972

Kwang W. Jeon (editor). THE BIOLOGY OF AMOEBA, 1973

Dean F. Martin and George M. Padilla (editors). MARINE PHARMACOGNOSY: Action of Marine Biotoxins at the Cellular Level, 1973

Joseph A. Erwin (editor). LIPIDS AND BIOMEMBRANES OF EUKARYOTIC MICROORGANISMS, 1973

A. M. Zimmerman, G. M. Padilla, and I. L. Cameron (editors). DRUGS AND THE CELL CYCLE, 1973

The list of titles in this series continues on the last page of this volume.

Photosynthesis

VOLUME II

Development, Carbon Metabolism, and Plant Productivity

Edited by

GOVINDJEE

Departments of Botany
and Physiology and Biophysics
University of Illinois at Urbana–Champaign
Urbana, Illinois

ACADEMIC PRESS 1982

A Subsidiary of Harcourt Brace Jovanovich, Publishers

New York London
Paris San Diego San Francisco São Paulo Sydney Tokyo Toronto

ACADEMIC PRESS, INC.
111 Fifth Avenue, New York, New York 10003

United Kingdom Edition published by
ACADEMIC PRESS, INC. (LONDON) LTD.
24/28 Oval Road, London NW1 7DX

Library of Congress Cataloging in Publication Data
Main entry under title:

Photosynthesis: Development, carbon metabolism, and plant productivity.
v. 2. Cell biology.
Includes bibliographies and index.
1. Photosynthesis. 2. Crop yields. 3. Primary productivity (Biology) I. Govindjee, Date.
II. Series. III. Series: Cell biology.
QK882.P546 1982, vol. 2 581.1'3342s 82-8700
ISBN 0-12-294302-3 (v. 2) [581.1'3342] AACR2

PRINTED IN THE UNITED STATES OF AMERICA

82 83 84 85 9 8 7 6 5 4 3 2 1

I am especially indebted to my daughter
ANITA GOVINDJEE
for her technical and editorial assistance
during the preparation of this volume.

Contents

7

Photorespiration 191

8

Crassulacean Acid Metabolism (CAM) 231

9

Environmental Regulation of Photosynthesis 263

Special Topics

16

Bicarbonate or Carbon Dioxide as a Requirement for Efficient Electron Transport on the Acceptor Side of Photosystem II

WIM F. J. VERMAAS and GOVINDJEE

List of Contributors

Numbers in parentheses indicate the pages on which the authors' contributions begin.

JAMES A. BASSHAM (141), Laboratory of Chemical Biodynamics, Lawrence Berkeley Laboratory, University of California, Berkeley, California 94720

JOSEPH A. BERRY (263), Department of Plant Biology, Carnegie Institution of Washington, Stanford, California 94305

NORMAN I. BISHOP (xxi), Department of Botany and Plant Pathology, Oregon State University, Corvallis, Oregon 97331

BOB B. BUCHANAN (141), Department of Plant and Soil Biology, University of California at Berkeley, Berkeley, California 94720

DENNIS E. BUETOW (13, 43), Department of Physiology and Biophysics, University of Illinois at Urbana–Champaign, Urbana, Illinois 61801

RAYMOND CHOLLET (191), Department of Agricultural Biochemistry, University of Nebraska, Lincoln, Nebraska 68583

A. LAWRENCE CHRISTY (499), Research Department, Monsanto Agricultural Products Company, St. Louis, Missouri 63141

W. JOHN S. DOWNTON (263), Division of Horticultural Research, Commonwealth Scientific and Industrial Research Organization, Adelaide, S. A. 5001, Australia

GERHART DREWS (89), Lehrstuhl für Mikrobiologie, Biologisches Institut II Albert Ludwigs Universität, D-7800 Freiburg, West Germany

DONALD R. GEIGER (345), Department of Biology, University of Dayton, Dayton, Ohio 45469

ROBERT T. GIAQUINTA (345), Central Research and Development Department, Experimental Station, E. I. du Pont de Nemours & Company, Wilmington, Delaware 19801

ROGER M. GIFFORD (419, 459), Plant Physiology Section, Division of Plant Industry, Commonwealth Scientific and Industrial Research Organization, Canberra City, ACT 2601, Australia

GOVINDJEE (541), Departments of Botany, and Physiology and Biophysics, University of Illinois at Urbana–Champaign, Urbana, Illinois 61801

JOHN D. HESKETH (387), United States Department of Agriculture, Agricultural Research Service, University of Illinois at Urbana–Champaign, Urbana, Illinois 61801

COLIN L. D. JENKINS (419), Biochemistry Section, Division of Plant Industry, Commonwealth Scientific and Industrial Research Organization, Canberra City, ACT 2601, Australia

MARTIN D. KAMEN (xix), Professor Emeritus, Department of Chemistry, University of California, San Diego, La Jolla, California 92093

MANFRED KLUGE (231), Fachbereich Biologie (10), Institut für Botanik, Technische Hochschule, D-6100 Darmstadt, West Germany

WILLIAM L. OGREN (191), United States Department of Agriculture, Agricultural Research Service, Department of Agronomy, University of Illinois at Urbana–Champaign, Urbana, Illinois 61801

ITZHAK OHAD (89), Department of Biological Chemistry, Institute of Life Sciences, The Hebrew University of Jerusalem, Jerusalem 91904, Israel

DOYLE B. PETERS (387), United States Department of Agriculture, Agricultural Research Service, Department of Agronomy, University of Illinois at Urbana–Champaign, Urbana, Illinois 61801

CLARK A. PORTER (499), Research Department, Monsanto Agricultural Products Company, St. Louis, Missouri 63166

ARCHIE R. PORTIS, JR. (1), United States Department of Agriculture, Agricultural Research Service, University of Illinois at Urbana–Champaign, Urbana, Illinois 61801

VENETIA A. SAUNDERS (13, 17), Department of Biology, Liverpool Polytechnic, Liverpool L3 3AF, England

ALAN J. STEMLER (513), Department of Botany, University of California at Davis, Davis, California 95616

WIM F. J. VERMAAS (541), Vakgroep Plantenfysiologisch Onderzoek, Landbouwhogeschool, 6703 BW Wageningen, The Netherlands

JOSEPH T. WOOLLEY (387), United States Department of Agriculture, Agricultural Research Service, Department of Agronomy, University of Illinois at Urbana–Champaign, Urbana, Illinois 61801

Preface

Future generations face challenges that are only now becoming of interest to the present generation. Fossil fuels (gasoline or petroleum), the products of past photosynthesis, are disappearing at a fast rate, and the population of our world is increasing with alarming speed. Thus, food and energy per capita is expected to decrease. The ultimate source of our food is from those organic compounds that are produced by higher plants and photosynthetic microorganisms through the process of photosynthesis. In plant photosynthesis, CO_2, H_2O, and 8–10 photons of light are converted into O_2 and 1/6 ($C_6H_{12}O_6$). In photosynthetic bacteria, H_2O is replaced by other H_2 donors, and O_2 is not evolved. Attempts are being made to improve the productivity of existing crops and to introduce new crops for food and biomass. It is our belief that a basic understanding of photosynthesis is needed before we can manipulate it for improving the overall rate of photosynthesis of a single plant. Can such an improvement lead to an increase in crop productivity? Can we improve photosynthesis by manipulating the genetics and molecular biology of the system? By regulating reactions in CO_2 fixation pathways? By decreasing photorespiration? Or, by manipulating environmental parameters (water, temperature, light, atmosphere, etc.)? It is our belief that a fundamental understanding of photosynthesis is necessary before we can answer these questions.

It is with this idea that we present here chapters by A. Portis (Introduction, Chapter 1), V. A. Saunders and D. E. Buetow (Genetics and Molecular Biology, Chapters 2–4), I. Ohad and G. Drews (Biogenesis of the Photosynthetic Apparatus, Chapter 5), J. A. Bassham and B. B. Buchanan (CO_2 Fixation Pathways, Chapter 6), W. L. Ogren and R. Chollet (Photorespiration, Chapter 7), M. Kluge (Crassulacean Acid Metabolism, Chapter 8), and J. Berry and W. J. S. Downton (Environmental Regulation, Chapter 9). Plant productivity depends upon a variety of factors including translocation of photosynthates, which is discussed by D. R. Geiger and R. T. Giaquinta (Chapter 10). Prediction of leaf behavior and canopy behavior is covered by J. D. Hesketh, J. T. Woolley, and

D. B. Peters (Chapter 11). There are many opinions on the prospects of applying knowledge of photosynthesis to plant productivity (for example, see *The Biology of Crop Productivity*). R. M. Gifford and C. L. D. Jenkins discuss this topic in Chapter 12. This is followed by a discussion of global photosynthesis and its relation to our food and energy needs also by R. M. Gifford (Chapter 13). Two *Special Topics* are included in this book. First, A. L. Christy and C. A. Porter (Chapter 14) discuss the canopy photosynthesis and yield in a particular crop (soybean). This is one of best examples in which whole plant photosynthesis is related to the crop yield, and it fosters optimism that studies on photosynthesis are indeed relevant to crop productivity. Whether the same is true for other crops depends upon the future. Second, A. Stemler (Chapter 15) discusses a unique effect of CO_2 or bicarbonate on photosynthetic electron transport; it is not related directly to the subject matter of the present volume, but is presented here so that the readers may realize that CO_2 is not only fixed in photosynthesis and provides us with food, but has other important roles (e.g., stimulation of photophosphorylation and of electron transport following light reactions). Stemler emphasizes his personal view that CO_2 may play a significant role in the evolution of O_2 from H_2O (see accompanying volume; *Photosynthesis: Energy Conversion by Plants and Bacteria,* Vol. I). Owing to the controversial nature of this view and in order to provide a balanced picture of the role of CO_2 in photosynthetic electron flow, W. F. J. Vermaas and I (Chapter 16) have provided a brief critique of the phenomena that emphasizes that the major unique role of CO_2 is not in the O_2-evolution mechanism but in the electron flow between the two photoreactions of photosynthesis. A close relationship between the binding site of CO_2 and the herbicides seems to emerge. Volume I covers aspects of photosynthesis such as the absorption of light, conversion of this energy into redox energy and proton motive force (proton gradient and membrane potential), and the production of the reducing power in the form of reduced nicotinamide adenine dinucleotide phosphate (NADPH) in plants (or NADH in bacteria) and adenosine triphosphate (ATP); the use of the latter two in CO_2 fixation are dealt with in this volume.

Each chapter in the present volume is written by international authorities in the field. As much as possible, the chapters provide an integrated approach to both green plant and bacterial photosynthesis. Molecular biology (Chapters 2–4), biogenesis of the photosynthetic apparatus (Chapter 5), and CO_2 fixation pathways (Chapter 6) in both green plants and bacterial systems have been presented; the authors were chosen for their expertise in one of the two systems. The remaining chapters deal mostly with work on higher plants because they are of agronomic importance and because this is where a great deal of attention has been paid recently. It seems that in the future there will be a larger emphasis on biomass production by photosynthetic microorganisms where the possibilities of applying genetic engineering seem closer at hand.

Each chapter in this volume is a comprehensive review of the area chosen and is illustrated with diagrams and bibliographies. The aim of most of the authors was to include a review of the historical development of major concepts, a critical analysis of experimental approaches, and an exposition of recent findings. It is hoped that the individual chapters will serve as a reference work integrating experimental results and theoretical considerations represented in a large number of research publications in addition to those in the authors' laboratories. We expect the present book not only to be a reference source for researchers but also an introductory book for graduate students in agronomy, plant biochemistry, plant biology, plant biophysics, botany, cell biology, ecology, microbiology, comparative physiology, and plant physiology. Some of the chapters will have a much wider audience. I hope that this book will be read by scientific administrators and research planners who sit in judgment on national priorities and on the future of biology.

I am grateful to my wife (Rajni) and my children (Anita and Sanjay) for tolerating me during the preparation of the manuscripts. I am thankful to my present graduate students (Danny Blubaugh, William Coleman, James Fenton and Julian Eaton-Rye) for not complaining loudly about the time I did not spend with them when I was doing the clerical work related to the editing of this book. I am also thankful to Shubha Govind, Wim F. J. Vermaas, Christa Critchley, and Ion Baianu. Thanks are due to several colleagues (the photosynthesis group of the University of Illinois at Urbana–Champaign that includes Tony Crofts, Don DeVault, Bill Ogren, Don Ort, Archie Portis, Tino Rebeiz, John Whitmarsh, and Colin Wraight) who aided me in the initial planning of this book. Don Ort, Charlie Arntzen, and John Whitmarsh were especially helpful.

GOVINDJEE

Hans Gaffron (1902–1979): A Tribute

A germinal figure in photosynthesis research such as Hans Gaffron is not to be encapsulated in any brief tribute. The esteem in which he was and is held is attested by the distinguished group of contributors dedicating this volume to his memory.

In an illustrious career extending well over five decades, Hans Gaffron made basic, important observations and fashioned penetrating insights into phenomena centering around the photometabolism of higher plants, algae, and photosynthetic bacteria. He will be remembered for his role with K. Wohl in the mid 1930s in generating the concept of the "photosynthetic unit." In addition, he was a leader and innovator in furthering knowledge of basic processes in photoreduction, dark reduction of CO_2 coupled to chemosynthetic metabolism, and evocation of photohydrogenase activity in algae following dark anaerobic adaptation. (I recall how eagerly I read his classic paper* in the late 1940s.) Nor did his level of creativity lessen as the years went by—witness his findings with W. Kowallik in the 1960s of the unexpected effect of blue light on chlorophyll-independent respiration in algae,† and the general productivity of the group he founded in Tallahassee (Florida State University) beginning a new career after two decades of distinguished research at the University of Chicago. N. I. Bishop and, later, P. Homann helped to form the Tallahassee laboratory, and others (Wiessner, Kaltwasser, Kowallik, and Schmid) helped to establish it. Always, Gaffron's work was characterized by adherence to the highest standards of reliability and integrity—an accomplishment aided incalculably by his laboratory associate and devoted wife, Clara. In the controversies with O. Warburg on quantum efficiency and with M. Calvin and A. Benson on the nature of the initial product of CO_2 assimilation, or in the confrontation with C. B. van Niel on the ultimate fate of

*H. Gaffron (1944). "Photosynthesis, Photoreduction and Dark Reduction in Certain Algae," *Biol. Revs.* **19,** 1–19.

†W. Kowallik and H. Gaffron (1966). "Respiration Induced by Blue Light," *Planta (Berlin)* **69,** 92–95.

organic substrates, he was usually correct in his assessment of the facts. No one could doubt his honesty and sincerity whether subsequent events confirmed his stand, as in the debates with Warburg and van Niel, or did not, as in that with Calvin and Benson. His criticisms did much to keep the California group on course in the early phases of their work.

He was a major generalist at a time when one was needed, for as the study of photosynthesis developed in the half century from an initial preoccupation with metabolism in the 1920s to a wholly different concern with the fast reactions of photochemistry in the living cell, the tendency of the research effort to fragment into special areas of expertise threatened to blur the general character of photosynthesis as an integrated process. He was a much sought after speaker and essayist on its general significance.* In a very special sense, he also became the conscience of the research community in photosynthesis in mounting a watchful guard against meretricious and cheap sensationalism.†

In an illuminating account‡ of the Chicago he knew so well, there is a quotation attributed to a "roving city editor" for the old Chicago *Herald and Examiner,* to wit: "The boss says he is getting tired of old clichés . . . we gotta get a lotta new clichès!" So uncompromising an intellectual as Hans was, he, nevertheless, would have enjoyed the sentiment and probably agreed with me that old clichès persist because they are so apt. Thus, I find it appropriate to resort to one in conclusion: "He was a scholar and a gentleman." Truly, it was never better applied. It was my good fortune to know him, and it is my high privilege to write these few lines in tribute.

MARTIN D. KAMEN

*H. Gaffron (1960). "Energy Storage: Photosynthesis," In *Plant Physiology* Vol. 1B (F. C. Steward, ed.), pp. 3–277. Academic Press, New York; also see H. Gaffron (1965). "Photosynthesis," *Biol. Sci. Curriculum Studies Pamphlet 24* (U. Lanham, ed.)

†He led a protest against some particularly noxious and misleading reports in the press in 1955, claiming that recent researches had succeeded in demonstrating sunlight could make food without using green plants, paving the way for making food from air and water. In this effort, he enlisted the support of William Arnold, Allan H. Brown, Robert Bandursky, Martin Gibbs, David Goddard, Henry Linschitz, Eugene Rabinowich, Birgit Vennesland, Leo Vernon, and the writer.

‡R. J. Casey (1947). *More Interesting People,* Bobbs-Merrill Co., Inc. p. 51. The same character, one Sam Makaroff, is alleged to have issued the order: "Hold it to half a paragraph!"—quite sage advice on this occasion.

Hans Gaffron, 1902–1979

Hans Gaffron was born in Lima, Peru, on May 17, 1902. He was educated in Germany and received a Ph.D. in Chemistry at the University of Berlin in 1925. From 1925 to 1930, he did postdoctoral research

Hans Gaffron

at the Kaiser Wilhem Institute of Biology, and then from 1932 to 1937 at the Institute of Biochemistry. It is during this period that he was introduced to research in photobiology. In 1936, in the laboratory of Otto Warburg and in collaboration with K. Wohl, he published the famous article in which the concept of the photosynthetic unit was developed. He then moved to the Hopkins Marine Station (in California) which was followed by his long association (1941–1960) with James Franck in the Research Institute and the Department of Biochemistry at the University of Chicago. Several of his major contributions to the photometabolism of algae and photosynthetic bacteria, including his discoveries of hydrogenases in photosynthetic organisms, were made during this period. In 1960, Hans Gaffron transferred to the Institute of Molecular Biophysics, Florida State University, where he directed a laboratory dedicated to photosynthesis research until his retirement in 1973; he then became emeritus research professor at the same institute. Throughout his academic career, Gaffron maintained an association with the Marine Biological Laboratory, Woods Hole, Massachusetts.

In recognition of his contributions to the basic understanding of photosynthesis, he was presented the Kettering Award for Excellence in Photosynthesis by the American Society of Plant Physiology in 1965. Gaffron's scholarly interests were broad and included not only his dedication to basic research but also a strong involvement in music, the arts, and philosophical aspects of man and science. He was a member of the American Society for Plant Physiologists, the American Association for the Advancement of Science, the American Society of Biological Chemists, the Society of General Physiology, and the Biophysical Society of America. He authored or coauthored numerous scientific and philosophical papers on photosynthesis and its relation to world energy supplies, past and present.

Hans Gaffron, an internationally recognized scholar and researcher in photosynthesis, died August 18, 1979, in Falmouth, Massachusetts after suffering a heart attack. He is survived by his wife, Clara, whom he married in 1932, and a son, Peter.

NORMAN I. BISHOP

Contents of Volume I

1

Introduction to Photosynthesis: Carbon Assimilation and Plant Productivity

ARCHIE R. PORTIS, JR.

I. Introduction

Photosynthesis is the process by which green plants, cyanobacteria, and photosynthetic bacteria produce organic matter from CO_2 and hydrogen donors. In green plants, this process occurs in the chloroplasts. The reactions begin by the conversion of light energy, H_2O, adenosine diphosphate (ADP), phosphate and oxidized nicotinamide adenine dinucleotide phosphate ($NADP^+$) into O_2, adenosine triphosphate (ATP), and reduced nicotinamide adenine dinucleotide phosphate (NADPH). This aspect of photosynthesis often called the *light reactions* is discussed in another volume (Govindjee, 1982). The utilization of the ATP and NADPH for the conversion of CO_2 to carbohydrates often called the *dark reactions* is the subject of the present volume.

The photosynthate thus formed is translocated to various parts of the plant and utilized for its growth. In this introductory chapter, we shall introduce the reader to the various chapters dealing with both the question of CO_2 assimilation and plant productivity.

II. General Discussion

The continuing existence of life on earth ultimately depends on the utilization of solar energy by the process known as photosynthesis. In the higher plants, algae, and cyanobacteria, this process is usually summarized as

$$6H_2O + 6CO_2 \xrightarrow{48\,h\nu} C_6H_{12}O_6 + 6O_2 \quad \Delta G'_s = 686 \text{ kcal}$$

representing the oxidation of water to oxygen and the concomitant re-

Photosynthesis: Development, Carbon Metabolism, and Plant Productivity, Vol. II

ISBN: 0-12-294302-3

duction of carbon dioxide to carbohydrate by the use of light quanta ($h\nu$). Photosynthetic bacteria, however, do not use H_2O as the electron donor, but instead use other types of donors (for example, H_2S; see Govindjee, 1975, 1982). The abundant production of reduced carbon by photosynthesis permitted the concurrent evolution of nonphotosynthetic life forms that rely on the utilization of part of the energy stored in carbohydrate by its reoxidation to CO_2 for their growth. As a result, a relatively integrated symbiotic relationship between photosynthetic and nonphotosynthetic life forms developed.

It is clear that with this closed system, the utilization of carbohydrate for energy by nonphotosynthetic life cannot exceed its formation, present and past, by photosynthesis. At present, it is estimated that about 50% of all the organic matter on earth consists of cellulose, largely due to the massive amount of forestry. Unfortunately, this energy source is not readily accessible for growth purposes to most higher nonphotosynthetic life forms. Man, of course, has learned to recover some of the energy stored in cellulose by burning. An appreciation of the importance and magnitude of photosynthesis is also gained by considering that 20–30% of the protein content of plants consists of a single protein: ribulose bisphosphate (RuBP) carboxylase. This protein, which is the one catalyzing the assimilation of CO_2, is therefore the most abundant protein on earth.

Unfortunately, man's need for fuel and energy is rapidly overtaking that available from photosynthesis. This is due to (1) an explosive increase in the world's population; and (2) an increasing per capita use of energy mainly as a result of increases in the "quality of life" enjoyed by the more industrialized nations. The supplies of oil, coal, and gas, which actually are also the result of photosynthesis and which once seemed unlimited, are rapidly being depleted. Consequently, the concern about the continuance of a comfortable and secure existence has led to a dramatic increase in interest in photosynthesis. It is clear that an increased knowledge of photosynthesis leading to increases in solar energy capture, through increases in photosynthesis, cannot be the only answer to the problems we face. The establishment of a stable human population and the development of alternative energy resources such as nuclear fusion and geothermal power will also have to be employed. Increased photosynthesis and crop productivity will serve to allow us more time in order to achieve these goals. It will, however, be of continual benefit for the feeding of the human population globally. The contribution of photosynthesis toward providing for our future food and energy needs is thoroughly discussed in Chapter 13 of this volume.

What is the current status of our knowledge about photosynthesis and the prospects for utilizing such knowledge? The following chapters (see

Fig. 1) attempt to present a relatively integrated view of the photosynthetic process in organisms, ranging from unicellular algae to the higher plants. The details of the mechanisms and control of the light-capturing processes, which lead to the production of adenosine triphosphate (ATP) and the reduced form of nicotinamide adenine dinucleotide phosphate (NADPH) that are needed for the fixation of carbon dioxide in photosynthesis, are contained in Volume 1 (see Govindjee, 1982). Subsequent details of the CO_2 assimilation process and aspects of its control and integration with the growth of photosynthetic organisms are contained in the present volume. The aim of these chapters has been to outline the basic status of our knowledge and especially to point out the present deficiencies and limitations, such that the problems and immediate prospects for improving photosynthesis become evident.

In the following paragraphs, we will briefly introduce some of the more prominent aspects of photosynthesis discussed in the following

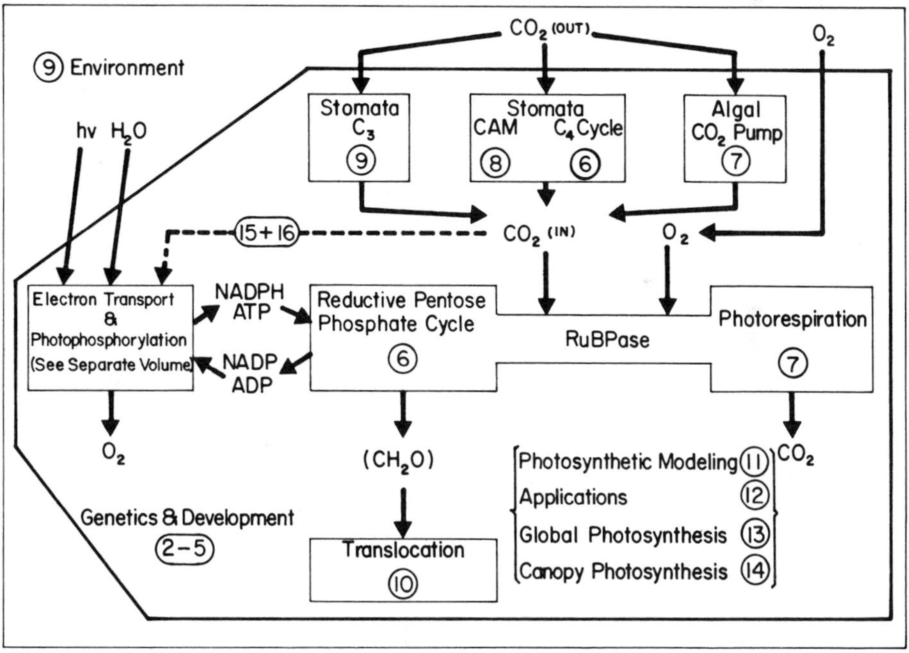

FIG. 1. Schematic diagram of some of the relationships between the various chapters of this book (see text). The circled numbers refer to chapter numbers. Abbreviations: ADP, ATP: Adenosine di(or tri) phosphate; C_3: plants in which the first carboxylation product is a three-carbon acid (e.g., phosphoglyceric acid); C_4: plants in which this product is a four-carbon acid (e.g., oxaloacetic); CAM: crassulacean acid metabolism; hv: light energy; NADP, NADPH: oxidized or reduced form of nicotinamide adenine dinucleotide phosphate; RuBPase: Ribulose bisphosphate carboxylase.

chapters of this volume, and in a few cases, attempt to offer some additional perspectives for the reader to consider.

Aspects of the genetics of photosynthesis in photosynthetic bacteria, cyanobacteria, and the chloroplasts of algae and higher plants are discussed in Chapters 2, 3, and 4 of this volume. The complexity of these systems, particularly in the higher plants, already appears to be enormous. Only a few of the most basic aspects are known, and much work needs to be done to even localize the genes involved in photosynthesis. As a result the early enthusiasm about the prospects for modifications in photosynthesis by "genetic engineering" techniques has been tempered by the realization of the enormous amount of work that remains to be done before the feasibility of many of the current popular proposals can even be ascertained. It would appear that the most simple modifications will likely be the most promising in the immediate future for the complex systems of higher plants. For example, changes in the amino acid composition of seed proteins for improved nutritional quality or simple changes in the amino acid sequence of particular proteins resulting in altered specific activities may be more successful than attempts to alter enzyme levels or to restructure and introduce new metabolic pathways, since so little is known about the regulatory aspects of genetic expression.

Some of the same considerations apply to the research into the developmental aspects of photosynthesis, which is outlined in Chapter 5 of this volume. Development consists of a regulated expression of the genetic potential of plants according to specific internal (age, the need to coordinate multiple activities, etc.) and external factors (light, CO_2, temperature, nutrition). Much more knowledge about the degree to which photosynthesis is controlled by developmental restrictions is needed, since *development* is the process by which the plant allocates available resources for particular goals in its struggle for survival and reproduction. Knowledge of whether relatively simple modifications of these processes can be achieved is extremely important in our attempts to increase crop productivity. In an evolutionary context, increased photosynthesis and the utilization of photosynthate for seed production has rather limited selective advantages under most circumstances, as witnessed by the increased yields of our agricultural crops achieved over the last century, largely by removing nutritional and other environmental limitations to plant growth (see Chapter 12, this volume; and Gifford and Evans, 1981).

The fundamental pathway utilized for CO_2 assimilation is the reductive pentose phosphate cycle, whose basic features are discussed in detail in Chapter 6 of this volume. Reflecting the rather conservative nature of the evolutionary process, this pathway is common to all oxygenic life. Although the pathway involves three unique enzymatic reactions, only RuBP carboxylase does not have any close parallel with other common enzymatic reactions. The phosphoribulokinase and sedoheptulose bisphosphatase are rather similar to other enzymes catalyzing phosphorylation and dephosphorylation reactions, such as phosphofructokinase and fructose bisphosphatase, respectively. The prospects for improving the absolute efficiency of these "dark reactions" appear rather slim since they are already about 90% efficient. In this respect, the light reactions (see Govindjee, 1975, 1982), which are only 32% efficient on the basis of photosynthetically active radiation, would appear to offer more room for improvement. However, as pointed out repeatedly in several chapters, not only the efficiency of a particular reaction or process needs to be considered, but also the capacity. A comparison between the light harvesting reactions and the dark reactions reveals that the turnover time of the light reaction is about 20 msec, whereas it is about 60 msec for RuBP carboxylase (for further details, see Govindjee, 1982; and Chapter 11, this volume). The difference necessitates that most plants invest 20–30% of their resources for protein synthesis in the production of RuBP carboxylase, as mentioned previously. The potential for improvement is obvious, but our present lack of the molecular details of this reaction and the structure of the active site (see Lorimer, 1981) do not permit a reasonable assessment of the prospects for improving the turnover of this enzyme. It should also be noted that we still do not completely understand the *in vivo* regulation of RuBP carboxylase and the coordination of the activity of the regenerative phase of the cycle with RuBP carboxylase activity and with the supply of ATP and NADPH supplied by the light reactions. Because of this, it is difficult to determine at this time the extent to which the capacity of the dark reactions could be increased.

In addition to the limited capacity of RuBP carboxylase, O_2 is an alternative substrate, resulting in the production of phosphoglycolate. In the more abundant C_3 plants (plants that initially incorporate CO_2 into the 3-carbon intermediate phosphoglyceric acid), a complex pathway has apparently developed to deal with this product. Various aspects of this process, termed *photorespiration,* are discussed in Chapter 7 of this volume. Some of the more interesting aspects of this process are (1) involvement of three different organelles in the higher plants; (2) release of NH_3 stoichiometrically with CO_2 and the need for it to be reassimilated; and (3) the complexity of the process, necessitating a considerable investment of resources. There is no known function for this pathway at present, except possibly to protect the plant under certain stress situations (see Heber and Krause, 1980). The potential for crop productivity increases is suggested by the 30% increases in yield of several crops upon CO_2 enrichment, a considerable portion of which is due to elimination of photorespiration. There would not be much incentive for elimination of photorespiration if CO_2 enrichment were generally economically feasible.

The considerable controversy over which enzyme(s) is the direct cause of photorespiration has now been resolved in favor of RuBP carboxylase as the primary source. Hopefully, this will result in even more interest in the molecular details of the enzymology of the active site of this enzyme. Such information will not only be useful in attempts to eliminate O_2 as a substrate, but also, as discussed earlier, to increase the turnover of the enzyme.

Photorespiration is also very limited in algae under normal conditions. In this case, it appears that an active transport of CO_2/HCO_3^- occurs. Details of this process and even its localization (plasmalemma or chloroplast envelope?) are not completely known at present, but this discovery has led to speculation about the possibility of incorporation of this activity into higher plant cells. The feasibility of this idea depends on more detailed knowledge about the biochemistry and physiology of the "CO_2 pump" in the algae. The unique cellular arrangement found in C_4 plants (plants that initially incorporate CO_2 into 4-carbon intermediates) might be indicative that a simple "CO_2 pump" localized in the plasmalemma of plant cells would not be very effective in practice.

Two variations of the reductive pentose phosphate pathway of higher plants have been elucidated. These are referred to as C_4 and crassulacean acid metabolism (CAM). C_4 pathways are discussed in Chapter 6 of this volume and CAM in Chapter 8. In each case, extensive changes in leaf structure and enzymology have occurred, which result in reduced photorespiration and increased water use efficiency.

In many cases, in CAM plants the whole physiology of the plant seems to be adapted for maximal water conservation, permitting plant growth and photosynthesis in extremely arid environments. These plants provide a particularly striking example of the sacrifices in absolute photosynthetic productivity that can be tolerated if growth and survival in otherwise totally unsuitable environments is possible. The increasing interest in utilizing arid land and in expanding our base of agronomically useful plants will continue to stimulate research into the photosynthesis of these plants.

Compared to CAM, considerably more interest has been focused on aspects of C_4 photosynthesis. This is not surprising considering the fact that most weeds, maize, sorghum, and other major crops are C_4 plants, whereas pineapple is one of the few important CAM plants. C_4 plants are so efficient at eliminating photorespiration that the extent of this process under normal conditions is still controversial. Our increasing knowledge of the biochemistry and physiology of these plants indicates that water use efficiency is probably also an equally important benefit of the differences observed between photosynthesis in these plants and C_3 plants. Both CAM and C_4 are probably evolutionary developments from C_3, but detailed genetic comparisons will be required to substantiate this idea (Smith and Robbins, 1974; Björkman, 1976). The increased cellular and spatial complexity of C_4 plants is certainly consistent with this conclusion. It is also clear that C_4 plants at least have the potential to photosynthetically outcompete C_3 plants in full sun, high temperature, and low water environments. However, C_3 plants, even with photorespiratory limitations, appear to be more productive in low light and cool environments. C_4 photosynthesis appears to have arisen independently several times, which would suggest that there is considerable selective pressure for this modification.

The advantages of C_4 photosynthesis have lead to attempts to breed the C_4 phenotype into C_3 plants within closely related species. These have not been very successful, although various C_4 features appeared to be present in the progeny (Björkman, 1976). This suggests that the C_4 phenotype requires both the necessary enzymes and the proper spatial and cellular organization. Therefore, without a rather complete knowledge of the genetics of both C_3 and C_4 plants, which is unlikely to be available in the near future, the feasibility of applying genetic engineering techniques to assist in this process is doubtful for the present.

Current agricultural practices rely heavily on the use of herbicides. With the exception of those that have little specificity between C_3 and C_4 plants, these herbicides act to inhibit electron transport (see Trebst, 1980). Our increasing knowledge of the differences between C_3 and C_4 photosynthesis will hopefully lead to the development of new herbicides (see Chapter 6, this volume; Edwards and Huber, 1979) exploiting the unique features of C_4 photosynthesis. The possibilities include inhibitors of the activity of (1) pyruvate phosphate dikinase; (2) phosphoenolpyruvate (PEP) carboxykinase; (3) nicotinamide adenine dinucleotide (NAD) malic enzyme or nicotinamide adenine dinucleotide phosphate (NADP) malic enzyme; or (4) the carbon transport systems in the chloroplast envelope, all of which appear to be significantly different in C_3 and C_4 plants. Much work remains to be done with the transport activities of C_4 chloroplasts and, except for the characterization of the enzymology unique to C_4 plants, much of the regulation of photosynthesis is presently assumed to be similar to that already found in C_3 photosynthesis.

Under most circumstances, plant productivity is not only limited by intrinsic factors in photosynthesis but also due to factors arising from the plant environment. For an extensive discussion of the environmental aspects of photosynthesis, see Chapter 9 of this volume. These environmental factors include temperature, light, water, CO_2, nutrients, parasites (diseases and pests), and, recently, pollutants. The constraints these factors place on photosynthesis have also been discussed by Good and Bell (1980). The environment can affect either the photosynthetic process itself or the capacity of the plant to utilize the available photosynthate (i.e., the capacity for plant growth). Much of the current research is directed at determining the degree to which the environment limits each of these processes, but it has been difficult to devise clear-cut experiments in order to separate the cause and the effect. More interest is also being directed at studies of plants that are adapted to particular environmental extremes, with the aim of understanding the biochemical and physiological nature of these successful adaptations and their cost to productivity. Knowledge of these plants is obviously of considerable importance in the attempts to either adapt current agricultural crops to less favorable environments or to find new crops suitable for use in such areas. The rapidly rising costs of removing some of the most common environmental limitations for increased photosynthesis and yield achieved by the practices of fertilization, irrigation, use of pesticides and herbicides will serve to further stimulate interest in maintaining or increasing photosynthesis under unfavorable environmental conditions.

The predominant intervening process between the production of photosynthate and its utilization for fruit and seed production is translocation. Aspects of translocation and its relationship with photosynthesis are discussed in Chapter 10 of this volume. Much of the research in translocation has dealt with source-sink effects on photosynthesis. In addition to a better characterization of the basic biochemistry of translocation, it is becoming clear that more knowledge about the developmental aspects of translocation and its operation at the whole plant level is needed.

Since the efficiency of the process appears to be relatively high at present (approximately 3% of the energy stored in sucrose is required for phloem loading), improved prospects for increasing yield are more likely to result from increased knowledge of the controlling factors that regulate photosynthetic utilization. For example, seed development requires a coordinated change from the utilization of photosynthate mainly for new leaf and stem production to its utilization mainly for seed growth. What factors control how and when this is accomplished, and can they be changed? It is also clear that many plants remobilize a portion of the carbon and nitrogen previously invested in leaf, root, and stem development for use in seed development. Can this process be increased and controlled for an agronomic benefit? These are very important areas for future research as our current knowledge of these areas is very limited.

Knowledge of the various aspects of photosynthesis is being integrated into a more comprehensive understanding of photosynthesis, growth, and plant productivity by a process termed *modeling*. Aspects of this type of research, the type of information required for such effort, and the type of information obtained are discussed in Chapter 11 of this volume. The potential importance of modeling is at least twofold. First, models not only attempt to summarize and integrate the multitude of interactions between various processes, as revealed by past and concurrent research, but the inadequacies and failures of the model tend to stimulate research into areas of particular ignorance. Second, the more successful models of photosynthesis will probably find practical applications in agricultural decision-making processes. Already, decisions about what crop varieties to plant are made on the basis of soil type, average rainfall, temperature, and soil fertility, and a considerable effort goes into weighing the economic benefits of fertilization and irrigation to increase yield. The development of more sophisticated photosynthesis models at all levels will be increasingly important in making decisions on how to allocate available resources more effectively in particular agricultural situations.

Some views on the prospects for applying knowledge of photosynthesis toward improving crop production are presented in Chapter 12 of this volume. The failures of past attempts to increase crop yield by apparent improvements in various individual subcomponents of crop photosynthesis are reviewed and some of the possible reasons for these failures are discussed. It is concluded that breeding for photosynthesis is unlikely to result in increased yield due to the hierarchical nature of photosynthesis. Instead, manipulations should be made "at a level of organization as close as possible to the level of organization where the problem appears to lie." Historically, breeding for yield itself with usually limited knowledge of the actual manipulations made at a biochemical level has been most successful. Recently, evidence has been obtained that improvement in photosynthetic rates can accompany increases in yield achieved in this manner (Larson *et al.*, 1981).

These past failures do not really indicate that increased knowledge of photosynthesis, even at very elementary levels, cannot result in dramatic increases in yield as a result of fundamental improvements in photosynthesis. Indeed, such knowledge may not be appropriate for breeding approaches, which are probably confined to rather high-level manipulations in the complex genetic systems of higher plants (e.g., genetic expression). The long-range potential of genetic engineering techniques is that they provide a complementary approach to traditional breeding approaches, and they may allow dramatic increases in the genetic potential for photosynthesis. However, this potential is dependent on a sufficient knowledge of photosynthesis and its interaction with other plant processes before we are able to specify what modifications would be most effective. Following such changes, breeding techniques might then be the most fruitful approach to achieve a maximal utilization of the increased genetic potential for photosynthesis and therefore, increased yield. Therefore, I believe that future prospects are not quite so limited as those discussed in Chapter 12 of this volume. It should also be apparent from the following chapters that a considerable amount of the current photosynthesis research is becoming less concerned with the basics of photosynthesis and more concerned with the interactions between the various components of the photosynthetic process and other aspects of plant growth and development. Such knowledge may be more suitable for achieving improvements in plant productivity by currently available technology, as discussed in Chapter 12 of this volume.

Photosynthesis should obviously determine plant productivity. It may therefore be surprising to some that it has been so difficult and that so much controversy has arisen over actually demonstrating a clear relationship between photosynthesis and productivity. Aspects of this problem are discussed in Chapter 14 of this volume and are also mentioned in several other chapters. The major difficulty is that photosynthesis, or what we could consider to be "primary productivity," is very physiologically removed from the economically productive parts of the plant, which is usually only the fruit or the seed. Another problem has been that the developmental aspects have been ignored, usually out of necessity, and attempts have focused on altering or measuring only one component of photosynthesis and thereby attempting to correlate it with yield. The ability to measure canopy photosynthesis throughout the whole growth period circumvents many of these problems and excellent correlations can now be obtained under well-defined conditions. Plants therefore have access to and actually can make use of a portion of the total accumulated pool of photosynthate. An important problem for future research is to elucidate how this pool is utilized and allocated under different conditions.

The problem of correlating photosynthesis with yield clearly illustrates that the life of a plant is directed toward acquisition and the utilization of all available resources, many of which are subject to rather rapid changes, to ensure both its immediate survival and the eventual production of the next generation. The current variety of the world's photosynthetic life forms and their multitude of developmental and reproductive strategies are evidence of this fact. The problems we will face in trying to promote the maximal acquisition of photosynthate—specifically, its maximal utilization for specific economical products—are less likely to be underestimated when the evolutionary and developmental aspects of plants are considered.

CO_2 appears to have an additional role in photosynthesis besides being the carbon source for carbohydrate. As outlined in Chapters 15 and 16 of this volume, photosystem II (PSII) activity [oxidation of water to O_2 and reduction of plastoquinone (PQ) to PQH_2] appears to have a CO_2 requirement. The localization of the CO_2 site is controversial (contrast Chapter 15 with Chapter 16 of this volume), with experiments showing effects on both the oxidizing (H_2O) and reducing (PQ) side of the reaction center. The majority of experiments show large effects on the reducing side, and direct effects on the water side have been questioned (see Chapter 16 of this volume). It remains to be proven that this CO_2 effect has any physiological role *in vivo.* (However, see Vermaas and Govindjee, 1981.)

III. Conclusions

We hope that this book will not only be a source of knowledge for the reader to learn about how green plants fix CO_2 into carbohydrates, how they translocate the photosynthate, and how they regulate these reactions, but that it will also stimulate the formation of new ideas about how plant productivity might be increased through photosynthesis research.

REFERENCES

Björkman, O. (1976). *In* "CO_2 Metabolism and Plant Productivity" (R. H. Burris and C. C. Black, eds.), pp. 287–309. University Park Press, Baltimore, Maryland.

Edwards, G. E., and Huber, S. C. (1979). *Encycl. Plant Physiol., New Ser.* **6,** 102–112.

Gifford, R. M., and Evans, L. T. (1981). *Annu. Rev. Plant Physiol.* **32,** 485–509.

Good, N., and Bell, D. H. (1980). *In* "The Biology of Crop Productivity" (P. S. Carlson, ed.), pp. 3–51. Academic Press, New York.

Govindjee, ed. (1975). "Bioenergetics of Photosynthesis." Academic Press, New York.

Govindjee, ed. (1982). "Photosynthesis: Energy Conversion by Plants and Bacteria," Vol. I. Academic Press, New York.

Heber, U., and Krause, G. H. (1980). *Trends Biochem. Sci.* **5,** 32–35.

Larson, E. M., Hesketh, J. D., Woolley, J. T., and Peters, D. B. (1981). *Photosynth. Res.* **2,** 3–20.

Lorimer, G. H. (1981). *Annu. Rev. Plant Physiol.* **32,** 349–383.

Smith, B. N., and Robbins, M. J. (1975). *Proc. Int. Congr. Photosynth., 3rd, 1974,* pp. 1579–1587.

Trebst, A. (1980). *In* "Methods in Enzymology" (A. San Pietro, ed.), Vol. 69, pp. 675–715. Academic Press, New York.

Vermaas, W. F. J., and Govindjee (1981). *Proc. Indian Natl. Sci. Acad. Part B,* 581–605.

2

Introduction to Genetics and Molecular Biology of Photosynthetic Bacteria, Cyanobacteria, and Chloroplasts

VENETIA A. SAUNDERS
DENNIS E. BUETOW

During recent years, research aimed at elucidating the mechanisms of and the controls over the process of photosynthesis has advanced at a rapidly increasing tempo. It is now clear that light has a dual role in photosynthesis. It functions as a substrate in supplying adenosine triphosphate (ATP) and nicotinamide adenine dinucleotide phosphate (NADPH) (Govindjee, 1982) and as a catalyst regulating enzymatic pathways that lead to the synthesis and to the degradation of energy-rich compounds (Buchanan, 1980; Bassham and Buchanan, Chapter 6, this volume). Light also is the ultimate regulator influencing the development of photosynthetic membranes in photosynthetic bacteria and of chloroplasts in eukaryotic algae and higher plants.

The last decade has witnessed a great increase in studies on the genetics and molecular biology of photosynthetic bacteria and of chloroplasts. Research on organelle heredity has been stimulated of course by the discovery that chloroplasts contain deoxyribonucleic acid (DNA). Therefore, "cytoplasmically inherited" genes influencing the phenotypes of chloroplasts are now known to reside within these organelles. A major goal in studying the genetics and molecular biology of chloroplasts and photosynthetic bacteria is to determine the arrangement and activity of those genes specifying the photosynthetic machinery and the modes of their regulation. The synthesis and assembly of the photosynthetic membrane in prokaryotes and of the chloroplast in eukaryotic algae and higher plants appear to be a morphogenetic process, which contains a coordinated and highly regulated sequence of events (see, e.g., Drews, 1978; Kaplan, 1978; Kirk and Tilney-Bassett, 1978). Valuable insights into this process are being gained by the utilization of biochemical, biophysical, and genetic technologies, the latter including the recently introduced techniques of DNA sequencing and cloning. Moreover, the discovery of genetic tools for manipulating photosynthetic bacteria and chloroplasts has led to the initiation of genetic and physical mapping

Photosynthesis: Development, Carbon Metabolism, and Plant Productivity, Vol. II

ISBN: 0-12-294302-3

studies. Knowledge of the genetic regulation of photosynthesis, however, remains in its infancy.

Phylogenetic relationships between photosynthetic prokaryotes and eukaryotes have been postulated often (e.g., Margulis, 1970; Grun, 1976). Indeed, close structural and functional similarities between chloroplasts and photosynthetic prokaryotes are found for photosynthetic membranes (e.g., Grun, 1976), ribosomes (Loening and Ingle, 1967), ribosomal ribonucleic acids (RNAs) (Pigott and Carr, 1972; Bonen and Doolittle, 1975; Zablen *et al.,* 1975; Dyer *et al.,* 1977), pathways of CO_2 fixation and photosynthetic electron flow (e.g., Taylor, 1970), and systems of protein biosynthesis (e.g., Ellis *et al.,* 1973). However, any scheme concerning the evolution of photosynthetic eukaryotes from photosynthetic prokaryotes ultimately must account for the partitioning of genes that control the development and function of chloroplasts between the nucleus and the organelles and for the existence in eukaryotic cytoplasms of ribosomes that are larger than those found in the organelles (e.g., Bell, 1970; Buetow, 1976).

The following two chapters consider recent data on the genetics and molecular biology of photosynthetic bacteria and chloroplasts and are best read in conjunction with each other. Chapter 3 explores the development of the molecular genetics of photosynthetic bacteria and the consequent contributions to our understanding of the mechanics of bacterial photosynthesis. It is designed to augment the wealth of biochemical and biophysical data concerning photosynthetic prokaryotes (e.g., Clayton, 1973; Parson, 1974; Jones, 1977; Clayton and Sistrom, 1978; Govindjee, 1982). Chapter 4 on the molecular biology of chloroplasts is intended to supplement and complement such thoughtful and extensive treatises as those of Sager (1972) and Gillham (1978) on chloroplast genetics and those of Margulis (1970) and Grun (1976) on chloroplast evolution.

REFERENCES

Bell, P. R. (1970). *Symp. Soc. Exp. Biol.* **24,** 109–127.

Bonen, L., and Doolittle, F. W. (1975). *Proc. Natl. Acad. Sci. U.S.A.* **72,** 2310–2314.

Buchanan, B. B. (1980). *Annu. Rev. Plant Physiol.* **31,** 341–374.

Buetow, D. E. (1976). *J. Protozool.* **23,** 41–47.

Clayton, R. K. (1973). *Annu. Rev. Biophys. Bioeng.* **2,** 131–156.

Clayton, R. K., and Sistrom, W. R., eds. (1978). "The Photosynthetic Bacteria." Plenum, New York.

Drews, G. (1978). *Curr. Top. Bioenerg.* **8,** 161–207.

Dyer, T. A., Bowman, C. M., and Payne, P. I. (1977). *NATO Adv. Study Inst. Ser., Ser. A* **A12,** 121–133.

Ellis, R. J., Blair, G. E., and Hartley, M. R. (1973). *Biochem. Soc. Symp.* **38,** 137–162.
Gillham, N. W. (1978). "Organelle Heredity." Raven, New York.
Govindjee, ed. (1982). "Photosynthesis: Energy Conversion by Plants and Bacteria," Vol. I. Academic Press, New York.
Grun, P. (1976). "Cytoplasmic Genetics and Evolution." Columbia Univ. Press, New York.
Jones, O. T. G. (1977). *Symp. Soc. Gen. Microbiol.* **27,** 151–183.
Kaplan, S. (1978). *In* "The Photosynthetic Bacteria" (R. K. Clayton and W. R. Sistrom, eds.), pp. 809–839. Plenum, New York.
Kirk, J. T. O., and Tilney-Bassett, R. A. E. (1978). "The Plastids," 2nd ed. Elsevier/North-Holland, Amsterdam.
Loening, U. E., and Ingle, J. (1967). *Nature (London)* **215,** 363–367.
Margulis, L. (1970). "Origin of Eukaryotic Cells." Yale Univ. Press, New Haven, Connecticut.
Parson, W. W. (1974). *Annu. Rev. Microbiol.* **28,** 41–59.
Pigott, G. H., and Carr, N. G. (1972). *Science* **175,** 1259–1261.
Sager, R. (1972). "Cytoplasmic Genes and Organelles." Academic Press, New York.
Taylor, D. L. (1970). *Int. Rev. Cytol.* **27,** 29–64.
Zablen, L., Kissil, M. S., Woese, C. R., and Buetow, D. E. (1975). *Proc. Natl. Acad. Sci. U.S.A.* **72,** 2418–2422.

3

Genetics and Molecular Biology of Photosynthetic Bacteria and Cyanobacteria

VENETIA A. SAUNDERS

ABBREVIATIONS

ATP	Adenosine triphosphate
D	Dalton(s)
DNA	Deoxyribonucleic acid
G + C content	Guanine plus cytosine content
GTA	Gene transfer agent
ppGpp	Guanosine 3′-diphosphate 5′-diphosphate
RNA	Ribonucleic acid
mRNA	Messenger RNA
rRNA	Ribosomal RNA
tRNA	Transfer RNA
Rp.	Rhodopseudomonas
Rs.	Rhodospirillum Svedberg unit

ABSTRACT

Photosynthetic prokaryotes provide attractive experimental systems for investigating photosynthesis, nitrogen fixation, and related phenomena. A limiting factor in such studies has been the inadequacy of operational genetic systems. This chapter describes the development of the genetics of photosynthetic prokaryotes and the strategies that might be adopted to dissect processes controlling photosynthesis.

Photosynthesis: Development, Carbon Metabolism, and Plant Productivity, Vol. II

ISBN: 0-12-294302-3

I. Introduction

Recent advances in the molecular biology and genetics of photosynthetic bacteria have highlighted the potential of these organisms for identifying the genetic mechanisms that govern photosynthesis. Research has been stimulated by the development of facilities for genetic manipulation within this biological group. Provision of gene transfer systems, notably for members of the Rhodospirillaceae (purple nonsulfur bacteria), now permits novel approaches to the study of photosynthesis and justifies expectations for resolution of fundamental processes that have hitherto remained intransigent.

In this chapter, progress towards elucidating the genetic control of photosynthesis and attendant phenomena is assessed. Particular attention is focused on the genetics of the Rhodospirillaceae, although genetic studies on other photosynthetic prokaryotes will be considered. Genetic developments within the Rhodospirillaceae are traced from their inception to the current era of recombinant DNA technology. In addition, the value of a combined biochemical and genetic approach to the resolution of photosynthetic mechanisms is discussed.

II. Genetic Organization of Photosynthetic Prokaryotes

A. Genomes

The deoxyribonucleic acid (DNA) base compositions of green and purple bacteria span a range from 45 to 72 mole % guanine plus cytosine (G + C), with the majority falling within the upper one-half of this scale (Mandel *et al.*, 1971). Among the Rhodospirillaceae, DNA base contents [as determined by thermal denaturation, density gradient centrifugation, and ultraviolet spectrophotometry (Hill, 1966; Pfennig, 1967; Mandel *et al.*, 1971; Silver *et al.*, 1971; de Bont *et al.*, 1981)] vary between 62 to 72 mole % G + C. The G + C content of DNA for members of the Chromatiaceae (purple sulfur bacteria) ranges from 45 to 70 mole %, whereas the value for the Chlorobiaceae (green and brown sulfur bacteria) is between 48.5 and 58 mole % (Hill, 1966; Pfennig, 1967; Mandel *et al.*, 1971).

There appears to be wider variation (from 35 to 71 mole % G + C) in DNA base composition among cyanobacteria (Edelman *et al.*, 1967; Stanier *et al.*, 1971; Herdman *et al.*, 1979a). Furthermore, in contrast to the purple and green bacteria, DNA base contents of the majority of cyanobacteria fall into the lower one-half of this G + C scale (Herdman *et al.*, 1979a; cf. Mandel *et al.*, 1971). Paradoxically, there is little correlation between DNA base composition and structural diversity among cyanobacterial species (Herdman *et al.*, 1979a).

The sizes of genomes of purple nonsulfur bacteria are estimated to be 2.9×10^9 daltons (D) for *Rhodopseudomonas* (*Rp.*) *capsulata* (Yen *et al.*, 1979), 2.1×10^9 D for *Rhodomicrobium vannielii* (Potts *et al.*, 1980) and 1.6×10^9 D for *Rp. sphaeroides* (Gibson and Niederman, 1970); these values are close to those of other bacteria (Gillis *et al.*, 1970; Wallace and Morowitz, 1973). By comparison cyanobacterial genomes lie in the range 1.6×10^9 to 8.6×10^9 D and fall into four discrete size classes (Herdman *et al.*, 1979b). There is a correlation between genome size and structural complexity of cyanobacteria (Herdman *et al.*, 1979b). A more extensive discussion of cyanobacterial genomes is provided in a review by Doolittle (1979).

B. Extrachromosomal Deoxyribonucleic Acid

Extrachromosomal genetic elements (plasmids) are prevalent among bacteria, where they specify a diversity of biological functions (see, for example, Falkow, 1975; Novick *et al.*, 1976; Bennett and Richmond, 1978; Broda, 1979). There has been a number of reports describing the presence of plasmid DNA in photosynthetic prokaryotes. Extrachromosomal DNA was first detected in photosynthetic bacteria by Suyama and Gibson (1966). Subsequent findings confirmed and extended these initial observations, and to date plasmid DNA has been identified in a number of species, including *Chromatium* (Suyama and Gibson, 1966), *Rp. sphaeroides* (Suyama and Gibson, 1966; Gibson and Niederman, 1970; Saunders *et al.*, 1976; Pemberton and Tucker, 1977; Tucker and Pemberton, 1978), *Rp. capsulata* (Hu and Marrs, 1979; V. A. Saunders and S. J. Scahill, unpublished data), and *Rhodospirillum* (*Rs.*) *rubum* (Kuhl and Yoch, 1981). In the majority of these cases, multiple classes of plasmid DNA have been detected. Molecular weights of the plasmids fall within a range from 5×10^6 to 100×10^6.

Extensive surveys of the extrachromosomal DNA of cyanobacteria have been undertaken (see, for example, Roberts and Koths, 1976; Simon, 1978a; Lau and Doolittle, 1979; Friedberg and Seijffers, 1979; van den Hondel *et al.*, 1979; Lau *et al.*, 1980; Reaston *et al.*, 1980). These studies indicate the relatively common occurrence of plasmid DNA in both filamentous and unicellular species of cyanobacteria. For certain unicellular cyanobacteria, restriction enzyme analysis and other criteria have revealed that a number of strains of *Synechococcus* (strains 6301, 6707, and 6908) harbor plasmids with high, if not identical, base sequence homology (Lau and Doolittle, 1979; van den Hondel *et al.*, 1979), even though their genomes differ in size and base composition (Herdman *et al.*, 1979a, b). This suggests that a natural mechanism exists for gene transfer between these organisms. In other unicellular cyanobacteria, Lau and co-workers (1980) identified different plasmids that nevertheless show discrete regions of sequence homology. This has prompted the suggestion that common sequence elements, possibly analogous to insertion sequences or transposons (Campbell *et al.*, 1979), may be distributed among different plasmids in different strains of cyanobacteria.

Restriction enzyme analyses have also identified potentially useful cloning vehicles among plasmids of unicellular cyanobacteria (Lau and Doolittle, 1979; van den Hondel *et al.*, 1979; Reaston *et al.*, 1982). In the absence of selectable phenotypic traits for the plasmids, transposable genetic elements could be inserted into them. Indeed, van den Hondel and co-workers (1980) have exploited the ability of antibiotic-resistance genes to transpose between replicons so as to construct plasmids of *Anacystis nidulans* R-2 carrying the transposon Tn 901, which specifies resistance to ampicillin (pUH24::Tn 901). Such plasmids have subsequently been tailored for specific cloning purposes. For instance, plasmid vectors have been constructed that carry two resistance markers for use in cloning procedures based on insertional inactivation. Bifunctional vectors for cloning in cyanobacteria and *Escherichia coli* are also being developed (Kuhlemeier *et al.*, 1981; van den Hondel *et al.*, 1981). Furthermore, cosmid vectors, which carry the cohesive ends of coliphage lambda (λ) enabling *in vitro* phage packaging (Hohn and Murray, 1977), have been derived from cyanobacterial plasmids (van den Hondel *et al.*, 1981). In addition to providing cloning vectors, the availability of genetically marked plasmids should expand prospects for determining a number of underlying biological phenomena, not the least of which is the role, if any, of such plasmids in genetic interplay.

To date, the identity of genetic determinants encoded by plasmids in photosynthetic prokaryotes remains largely unresolved. There has been some speculation that plasmid DNA may play a role in specifying the photosynthetic apparatus (Gibson and Niederman, 1970; Saunders *et al.*, 1976; Kuhl and Yoch, 1981). However, plasmids are generally dispensible (Novick, 1969; Novick *et al.*, 1976), which tends to reduce the credibility of this suggestion. Other possible functions for the plasmid DNA of photosynthetic prokaryotes include resistance to antibiotics and/or heavy metals, production of toxins, conjugative ability, and gas vacuolation. Some of these possibilities were discussed in detail elsewhere (Saunders *et al.*, 1976; Saunders, 1978).

There have been several reports of antibiotic resistance among photosynthetic prokaryotes (see, for example, Weaver *et al.,* 1975; Kushner and Breuil, 1977; Pemberton and Tucker, 1977; Saunders, 1978). An isolate of *Rp. capsulata,* strain SP108, is resistant to benzylpenicillin by virtue of the production and activity of a β-lactamase (Weaver *et al.,* 1975). Preliminary analyses of the plasmid content of this strain and a penicillin sensitive counterpart tend to suggest a plasmid location for the structural gene for β-lactamase (S. J. Scahill, unpublished data). β-Lactamase production has also been reported in the cyanobacteria *Cocchloris elabens* (strain 7003) and *Anabaena* sp. (strain 7120) (Kushner and Breuil, 1977). Furthermore, plasmids have been detected in strains of *C. elabens* (Lau *et al.,* 1980) and *Anabaena* spp. (Simon, 1978a), but as yet there has not been any correlation between the presence of plasmid DNA and β-lactamase production. The molecular sizes of some of the plasmids of photosynthetic prokaryotes are consistent with those of known conjugative plasmids, generally >20 megadaltons (MD). So far, however, unequivocal genetic exchange mediated by these indigenous plasmids has not been demonstrated. If such plasmids are shown to promote their own transfer by conjugation, they may be more effective at mobilizing chromosomal genes than foreign plasmids are, some of which are already capable of mediating chromosome transfer in members of the Rhodospirillaceae (see Section IV,D).

Reports suggest that the ability to produce gas vacuoles can be lost spontaneously from certain species of cyanobacteria (Das and Singh, 1976; Walsby, 1977). This led to the suggestion that the production of gas vacuole protein might be plasmid-specified in cyanobacteria. Indeed, there is some evidence to associate this phenotype with the presence of plasmid DNA in strains of *Halobacterium* (Simon, 1978b; Weidinger *et al.,* 1979).

It is likely that recombinant DNA techniques will be instrumental in resolving plasmid-encoded functions. Mazur and co-workers (1980) have ruled out the possibility that genes for nitrogen fixation (*nif*) are plasmid-located in *Anabaena* 7120 by utilizing a recombinant plasmid containing *Klebsiella nif* genes (Cannon *et al.,* 1977, 1979) as hybridization probe. By similar rationales, it should soon be possible to establish directly if genes for photosynthesis are carried by plasmids in photosynthetic prokaryotes.

Plasmids and other extrachromosomal elements are believed to be a potent driving force in bacterial evolution (see, for example, Reanney, 1976; Bennett and Richmond, 1978). Plasmids are instrumental in facilitating the shuffling of genetic material between bacterial species. Elucidation of the plasmid biology of photosynthetic prokaryotes might thus reveal natural routes of gene flow among these organisms.

C. Ribonucleic Acid

1. RIBOSOMAL RNA (rRNA) AND RIBOSOMES

Ribosomes of those photosynthetic bacteria examined have sedimentation coefficients close to 66 S, in line with those of most bacteria (Taylor and Storck, 1964). However, there is evidence to suggest that changes in ribosomal structure accompany growth of these organisms under different atmospheric environments. Differences in concentration of ribosomes and their r-protein composition have been reported for *Rp. palustris* during growth under aerobic and photosynthetic conditions (Mansour and Stachow, 1975). The significance of these findings with respect to the adaptive formation of photosynthetic membranes has yet to be determined.

Commonly, RNA molecules of three distinct molecular classes namely 1.1×10^6 D (23 S), 0.56×10^6 D (16 S) and 3.5×10^4 D (5 S) represent the stable RNA components of bacterial ribosomes (Osawa, 1968). Although the majority of members of the Rhodospirillaceae examined possess species of rRNA of about 1.1×10^6 D and 0.53×10^6 D, there are some exceptions (Gray, 1978). For instance, both *Rp. sphaeroides* and *Rp. capsulata* lack the component of 1.1×10^6 D and contain instead RNA of 0.4×10^6 D (14 S) as a stable component of their ribosomes (Lessie, 1965; Marrs and Kaplan, 1970; Robinson and Sykes, 1971). The rRNA complement of *Rp. viridis* is also atypical, with major species of 0.78×10^6 D and 0.4×10^6 D (Gray, 1978). In *Rp. sphaeroides,* there are two distinct RNA species of 0.53×10^6 D, one of which forms part of the 29 S ribosomal subunit and the other part of the 45 S ribosomal subunit, which also contains the component of 0.4×10^6 D. In addition to these higher molecular weight components, rRNA of 3.6×10^4 D has been demonstrated in *Rp. sphaeroides* (Marrs and Kaplan, 1970). The occurrence of these atypical rRNA species, which has been attributed to an unusual mode of processing during the maturation of rRNA (Marrs and Kaplan, 1970; Gray, 1978), may have implications for both ribosome biogenesis and ribosome function. Furthermore, it is noteworthy that cleavage of the 23 S rRNA of *Paracoccus denitrificans* to yield 14 S and 16 S rRNAs has been reported (Mackay *et al.*, 1979). Interestingly both 14 S and 16 S rRNAs of *P. denitrificans* and *Rp. sphaeroides* exhibit strong sequence homology (Gibson *et al.*, 1979; Mackay *et al.*, 1979), in line with the notion that *Rp. sphaeroides* and *P. denitrificans* are phylogenetically related (Dickerson *et al.*, 1976; Almassy and Dickerson, 1978; Dickerson, 1980). On the other hand, the Rhodospirillaceae as a group is apparently heterogeneous with respect to sequence homology of 16 S rRNAs (Gibson *et al.*, 1979; Woese *et al.*, 1980).

Novel species of RNA, presumably resulting from aberrant postmaturational degradation of rRNA species, have also been identified in cyanobacteria. Moreover many of the RNA processing and degradation events in cyanobacteria are triggered by light (Doolittle, 1973; Singer and Doolittle, 1974).

2. TRANSFER RNA (tRNA)

Qualitative and quantitative variations in isoaccepting species of tRNA have been observed in a number of biological systems, and it has been proposed that species of tRNA may be involved in the regulation of cellular processes (see, for example, Sueoka and Kano-Sueoka, 1970; Littauer and Inouye, 1973; Rich and Raj Bhandry, 1976). In *Rp. sphaeroides,* transition from aerobic to photosynthetic growth conditions elicits a pronounced variation in the isoaccepting species of phenylalanyl-tRNA ($tRNA^{Phe}$) and tryptophanyl-tRNA ($tRNA^{Trp}$) (DeJesus and Gray, 1971; Razel and Gray, 1978; Sheperd and Kaplan, 1978). This could reflect differential transcription of genes specifying these tRNAs. Alternatively, one isoaccepting species of the tRNA may be physiologically convertible to another. Indeed, this seems to be the case with $tRNA^{Phe}$. Razel and Gray (1978) proposed that the dominant species of $tRNA^{Phe}$, which are found during prolonged growth of *Rp. sphaeroides* under photosynthetic conditions, are formed by posttranscriptional modification of the species of $tRNA^{Phe}$ that predominates during aerobic growth. This conversion, which is not readily reversible, presumably involves enzyme activity formed only during photosynthetic growth (Razel and Gray, 1978). Such variation in isoaccepting tRNA could be indicative of a regulatory role for $tRNA^{Phe}$ in the morphogenesis of the photosynthetic apparatus of *Rp. sphaeroides* (DeJesus and Gray, 1971; Razel and Gray, 1978; Sheperd and Kaplan, 1978). For a discussion of the role of tRNA in the cellular regulation of cyanobacteria, see Doolittle (1979).

3. RNA SYNTHESIS AND THE PHOTOSYNTHETIC APPARATUS

Various lines of inquiry indicate that protein and RNA syntheses are necessary requisites for formation of the photosynthetic apparatus in the Rhodospirillaceae (Sistrom, 1962; Bull and Lascelles, 1963; Higuchi *et al.,* 1965; Gray, 1967). In facultatively photosynthetic bacteria, this morphogenetic process is influenced by a number of environmental stimuli, notably light intensity and oxygen tension (Cohen-Bazire *et al.,* 1957). RNA synthesized either during the transition from aerobic to photosynthetic conditions or following a shift down in light intensity apparently contains an increased proportion of mRNA (Cost and Gray, 1967;

Gray, 1967). This response is correlated in both cases with a stimulation of intracytoplasmic membrane formation, and it has been suggested that a transcriptional mode of regulation could be operative during adaptation of the photosynthetic membrane (Gray, 1978). However, no qualitative or quantitative differences could be detected between RNA species derived from aerobically and photosynthetically grown cells of *Rp. sphaeroides* or *Rs. rubrum* (Yamashita and Kamen, 1968; Witkin and Gibson, 1972a; Chow, 1976) on the basis of DNA:RNA hybridization analyses. Possibly, the limitations of this technique precluded the resolution of subtle differences that may exist. Provision of DNA hybridization probes specifically containing genes for photosynthesis (see Section IV,D) should add a further refinement to these studies. Furthermore, it may be informative to utilize RNA from cells that are actively adapting to photosynthetic conditions in addition to RNA from aerobically or photosynthetically grown cells.

The possibility of translational regulation of protein synthesis has been suggested for purple nonsulfur bacteria (Yamashita and Kamen, 1968; Witkin and Gibson, 1972a, b; Chow, 1976). Differences in stability of an RNA component appeared to be associated with different growth modes, possibly reflecting differential binding of ribosomes and RNA, thereby facilitating translation of selected RNA transcripts (Witkin and Gibson, 1972b).

The guanosine nucleotides, guanosine 3′-diphosphate 5′-diphosphate (ppGpp) and guanosine 3′-diphosphate 5′-triphosphate (pppGpp) have been implicated in the regulation of RNA synthesis in bacteria (Gallant *et al.*, 1971; Hochstadt-Ozer and Cashel, 1972; Cashel, 1975). In *Rp. sphaeroides,* a shiftdown in light intensity results in cessation of RNA accumulation concomitant with an abrupt increase in the cellular concentration of ppGpp (Eccleston and Gray, 1973). Moreover, upon transfer from anaerobic to aerobic conditions in the light, a decrease in the concentration of ppGpp is observed, which is reversed upon return to anaerobic conditions (Gray, 1978). Such findings would be consistent with a role for ppGpp in regulating synthesis of the photosynthetic apparatus.

ppGpp and pppGpp have also been detected in cyanobacteria, although there are contrasting reports as to the extent of guanosine nucleotide modulation under conditions affecting stable RNA synthesis (Mann *et al.*, 1975; Adams *et al.*, 1977; Rogerson *et al.*, 1978; Akinyanju and Smith, 1979). However, it is worth noting that in *A. nidulans* and *Anabaena variabilis* the rate of transcription per gene appears to be constant irrespective of the growth rate (Leach *et al.*, 1971; Mann and Carr, 1974). Thus a modulation of ppGpp concentration with RNA content would be inconsistent with a role for ppGpp as a regulator of transcription (Smith, 1979).

III. Mutations Affecting Photosynthesis in the Rhodospirillaceae

Mutants are proving invaluable as research tools for probing physiological processes of photosynthetic bacteria. A refined analysis of these processes is permitted by the coupling of genetic and biochemical concepts and techniques. For a general appraisal of the applicability of mutants to the resolution of photosynthetic mechanisms, the reader is referred to a review of Marrs (1978a).

The use of mutants of the Rhodospirillaceae (notably *Rp. capsulata, Rp. sphaeroides,* and *Rs. rubrum*) with lesions specifically affecting photosynthesis or respiration has enabled incisive analysis of respiratory and photosynthetic electron transfer systems (for example, Marrs and Gest, 1973a; del Valle-Tascón *et al.*, 1975, 1977; LaMonica and Marrs, 1976; Zannoni *et al.*, 1976a, b; Marrs, 1978a; Saunders, 1978; Michels and Haddock, 1980). In purple nonsulfur bacteria, development of the photosynthetic apparatus is associated with the fabrication of intracytoplasmic membranes, which contain the photosynthetic electron transfer components. However, it is not yet completely clear if these specialized membranes also accommodate respiratory components. In this regard, Zannoni and co-workers (1978) have examined the ability of certain respiratory mutants of *Rp. capsulata* (strains M6 and M7) to perform light-induced oxygen uptake. They have identified cytochrome b_{260} as a component of both aerobic and photosynthetic electron transfer systems. Their findings are consistent with the notion that photosynthetic and respiratory systems are interconnected and are probably located on the same membrane vesicles in *Rp. capsulata.* Other mutants, in particular those bearing temperature-sensitive mutations of the photosynthetic apparatus, may represent additional tools for investigating associations between the photosynthetic and respiratory electron transfer systems.

Photosynthetically incompetent strains that are impaired in bacteriochlorophyll synthesis have been exploited in reconstitution experiments. In essence, these studies have assessed the contribution of specific pigment moieties (reaction center and light-harvesting complexes) to the restoration of photosynthetic activities in bacteriochlorophyll-less membranes of *Rp. sphaeroides* (Jones and Plewis, 1974; Hunter and Jones, 1976, 1979a, b) and *Rp. capsulata* (Garcia *et al.*, 1974, 1975).

Hunter and Jones (1979b), monitoring the kinetics of light-induced electron flow in reconstituted membranes of *Rp. sphaeroides*, have shown that such membranes exhibit similar properties to chromatophores derived from fully competent photosynthetic cells. It has been suggested (Hunter and Jones, 1979b) that nonpigmented membranes from aerobically grown cells of *Rp. sphaeroides* are potentially capable of forming chromatophore membranes by insertion of the photochemical apparatus into the membrane infrastructure in the correct orientation.

Mutants with lesions affecting photopigment production have contributed substantially to the elucidation of mechanisms involved in the biosynthesis of bacteriochlorophyll and carotenoid (see Lascelles, 1975; Kaplan, 1978; Scolnik *et al.*, 1980; Rebeiz and Lascelles, 1982). Apparently enzymes involved in bacteriochlorophyll synthesis are subject to repression in the presence of oxygen (Lascelles, 1975). Mutants of *Rp. sphaeroides* have been isolated that are insensitive to this regulation in that they continue to synthesize bacteriochlorophyll and intracytoplasmic membrane under conditions of high aeration (Lascelles and Wertlieb, 1971; Brown *et al.*, 1972). This response might be attributable to derepression of synthesis of enzymes for bacteriochlorophyll production, implying the involvement of a control element that no longer responds normally to oxygen. Alternatively, a structural gene may have been altered by mutation such that the gene product is refractory to inhibition by oxygen.

A number of models have been proposed to account for the regulation of bacteriochlorophyll synthesis by oxygen and light intensity. Cohen-Bazire and co-workers (1957) advanced the "redox governor" hypothesis in which it was proposed that the oxidation–reduction state of a carrier in the electron transport chain mediated control of photopigment synthesis. Marrs and Gest (1973b) subsequently modified these proposals in line with their observations of mutants of *Rp. capsulata* that were defective in respiratory electron transport. These workers postulated that bacteriochlorophyll synthesis is regulated by an "effector" molecule that is inactivated by oxygen and reactivated by electrons siphoned from the electron transport chain, possibly at the level of cytochrome *c*. However, recent findings using mutants blocked in electron flow between cytochromes *b* and *c* (B. L. Marrs, personal communication) are irreconcilable with that part of the scheme relating to the role of cytochrome *c* in diverting electrons to the effector.

Kaplan (1978) formulated a provisional scheme pertaining to the regulation of intracytoplasmic membrane development by oxygen, light intensity, and other stimuli. Broadly, this model includes the involvement of a hypothetical co-repressor either alone or in combination with an aporepressor in the control of intracytoplasmic membrane synthesis.

The co-repressor could be generated via the direct and specific interaction of light with reaction-center bacteriochlorophyll or by the interaction of oxygen with the postulated "effector molecule" (Marrs and Gest, 1973b). This model makes a number of predictions, some of which could be tested by investigating the behavior of appropriate mutants in response to specific stimuli.

In this connection, a mutant of *Rp. capsulata,* strain Z1, which is resistant to arsenate (Lien *et al.*, 1971; Zilinsky *et al.*, 1971) has been examined. In comparison with the parental strain, this mutant has an increased amount of reaction-center bacteriochlorophyll (relative to light-harvesting bacteriochlorophyll) and exhibits a 25% reduction in intracytoplasmic membrane synthesis. This might be interpreted (Kaplan, 1978) as supporting the assumption that the interaction of light with reaction-center bacteriochlorophyll effects co-repressor synthesis and a consequent reduction in synthesis of photosynthetic membrane. Analyses of mutants defective in reaction-center bacteriochlorophyll might assist clarification of these proposals.

Mapping studies (Yen and Marrs, 1976; Marrs, 1981) have revealed that genes for bacteriochlorophyll and carotenoid synthesis are closely aligned on the genome of *Rp. capsulata* (see Section IV,E). This could imply co-regulation of these genes, thereby permitting synchronized production of photopigments. However, it appears that carotenoid synthesis is less dramatically suppressed by oxygen than bacteriochlorophyll synthesis in *Rp. capsulata* (B. L. Marrs, personal communication). This suggests that some form of differential control of carotenoid and chlorophyll syntheses might exist.

Drews and colleagues (1976) have used the gene transfer agent of *Rp. capsulata* (see Section IV,B) to transfer genes for bacteriochlorophyll synthesis to mutants unable to synthesize bacteriochlorophyll and specific chromatophore membrane proteins. Concomitant restoration of both these abilities could be demonstrated in various mutants, including those with single lesions affecting bacteriochlorophyll synthesis. Such results support the notion that cooperative regulatory events are operating within the cell during development of the photosynthetic apparatus.

Studies with glycerol auxotrophs of *Rp. capsulata* (Klein and Mindich, 1976) suggest an association between the synthesis of phospholipid and photopigments during assembly of the chromatophore membrane. More recently, studies on the effects of the antibiotic cerulenin on photosynthetic membrane formation in *Rp. sphaeroides* have reinforced such proposals, indicating a requirement for concomitant synthesis of phospholipid, protein and pigment during assembly of the chromatophore membrane (Broglie and Niederman, 1979).

IV. Genetic Exchange Systems and Genetic Mapping

Thus far genetic analyses of photosynthetic bacteria have been largely confined to two species, namely *Rp. capsulata* and *Rp. sphaeroides*. These are the only organisms for which operational gene transfer systems have been developed. With regard to the cyanobacteria, genetic studies have been handicapped to some extent by a lack of efficient gene transfer systems. To date, transformation remains the only well-defined genetic system available.

A. Transduction

There has been a number of reports describing the isolation and characterization of bacteriophages (including cyanophages) specific to particular photosynthetic prokaryotes (see Padan and Shilo, 1973; Safferman, 1973; Wolk, 1973; Marrs, 1978b; Sherman and Brown, 1978). However the value of these phages as agents of genetic exchange remains limited at present. Only in a few instances has phage-mediated gene transfer been demonstrated. A temperate phage RS-2 exhibiting low-level transducing activity for a variety of genetic markers in *Rp. sphaeroides* was isolated by Kaplan and colleagues (see Marrs, 1978b; Saunders, 1978). Perhaps after suitable modification to enhance transduction frequencies, this could become a useful system for fine structure genetic mapping in *Rp. sphaeroides.* In addition, Kaplan and co-workers have introduced the coliphage P1 into *Rp. sphaeroides* (see Marrs *et al.*, 1977). Phage P1 mediates generalized transduction between strains of *E. coli* and exists extrachromosomally in the prophage state (Ikeda and Tomizawa, 1968). The possibility thus emerges of P1-promoted gene transfer in *Rp. sphaeroides.*

Pemberton and Tucker (1977) demonstrated effective transduction of a penicillin-resistance determinant between strains of *Rp. sphaeroides* mediated by the viral R plasmid R∅6P. However, attempts to perform generalized transduction with R∅6P were abortive. R∅6P is a rather unusual biological entity combining properties of a virus and drug-resistance (R) plasmid. It has been proposed that R∅6P carries a transposable β-lactamase gene and exists as a dimeric plasmid in the prophage state (Tucker and Pemberton, 1978).

B. *The Gene Transfer Agent of* Rhodopseudomonas capsulata

The first genetic exchange system to be described for a photosynthetic bacterium was that mediated by the gene transfer agent (GTA) of *Rp.*

capsulata (Marrs, 1974). Of the wild type isolates of *Rp. capsulata* screened for intraspecific gene transfer, the majority produced nucleoprotein particles termed *gene transfer agents,* which could effect genetic exchange. Morphologically, these GTA particles resemble small, tailed bacterial viruses, albeit much smaller than any known phage of similar complexity (Marrs, 1978b). The nucleic acid of the GTA is linear duplex DNA of about 3×10^6 D (Solioz and Marrs, 1977; Yen *et al.,* 1979). GTA particles can apparently promote transfer of any region of the bacterial genome, including plasmid DNA, with equal probability and frequencies of recombination approaching 10^{-3} per recipient cell (Yen *et al.,* 1979). Accretion of genetic markers via GTA seems to involve homologous recombination events in recipient cells (Yen *et al.,* 1979). GTA-mediated genetic exchange, although superficially resembling generalized transduction, exhibits certain unusual features. There is no transfer of the capacity to produce GTA particles to recipient cells that receive genetic information from the GTA, and no overt viral activities have been associated with gene transfer (Marrs, 1978b). Moreover, it has been deduced from restriction endonuclease analyses and hybridization kinetics that DNA encoding the GTA particles is not amplified during GTA production. No GTA particles, or, at best, no more than a small proportion, contain exclusively phage-like DNA (Yen *et al.,* 1979). A plausible explanation for these observations is that the genetic determinants for GTA production are dispersed on the genome of *Rp. capsulata* in such a way that they fail to be packaged *en bloc* by GTA particles. This GTA-mediated mechanism of genetic exchange occurs commonly among wild strains of *Rp. capsulata* but is restricted to this species (Wall *et al.,* 1975). From an evolutionary standpoint, the GTA could be viewed as a defective phage that maintained an ability to transmit genetic material (Solioz *et al.,* 1975). Alternatively, the GTA could represent a progenitor of a contemporary bacterial virus evolving from bacterial genes concerned with genetic exchange (Marrs, 1978b; Yen *et al.,* 1979). In either case, this rather unusual biological agent would seem to be an effective mediator of genetic exchange and may be of significance in the genetic adaptation of wild populations of *Rp. capsulata* to changes in their environment.

C. *Transformation*

Transformation systems have proven to be of value in genetic analyses of various microorganisms (Portolés *et al.,* 1977; Glover and Butler, 1979). However, progress in developing effective transformation systems in photosynthetic bacteria has been relatively slow. Some headway has been made using spheroplasts of *Rp. capsulata* treated with divalent cations (B. L. Marrs, personal communication), although transformation in this case has not yet been demonstrated at frequencies that would provide a useful experimental system.

Factors that may influence transformability include the development of a state of competence for taking up DNA and the production and activity of both extracellular and intracellular nucleases that may inactivate the transforming DNA. In view of this, the utilization of liposomes as vectors for introduction of transforming DNA into recipient cells would seem to provide an attractive approach. Transfer of liposome-entrapped DNA to *E. coli* (Fraley *et al.*, 1979) and to *Rp. sphaeroides* (S. Kaplan, personal communication) has been achieved. The process is insensitive to added DNase. The precise mechanism of liposome uptake in these systems is unclear. It has been tentatively proposed (Fraley *et al.*, 1979) that the liposomes fuse with the outer membrane of the cells and release plasmid DNA into the cytoplasm or periplasmic space. It is tempting to speculate that this technique might ultimately be applicable for transforming other photosynthetic prokaryotes. In particular, it may serve to enhance transformation frequencies of those strains that are already transformable by more conventional techniques at low frequencies.

Tucker and Pemberton (1980) have achieved transformation of *Rp. sphaeroides* by DNA from the temperate bacteriophage R∅6P (see Section IV,A). R∅6P DNA is potentially an attractive transformation probe, providing lysogenized cells with the readily identifiable phenotype of penicillin resistance. Apparently, this transformation process relies directly upon concomitant infection of the recipient bacterium by a second closely related temperate phage R∅9. Penicillin-resistant transformants were obtained at a frequency of $\sim 2 \times 10^{-5}$ per survivor when cells of *Rp. sphaeroides* strain RS6143 infected with the temperate phage R∅9 were treated with R∅6P DNA. Transformation frequencies could be enhanced if strain RS6143 lysogenic for R∅9 (i.e., strain RS6579) replaced strain RS6143 as recipient. Furthermore, the frequency of transformation is crucially affected by the multiplicity of superinfection with R∅9. This transformation process is sensitive to DNase, but insensitive to RNase and pronase. The precise nature of such "helper" phage-mediated transformation is unclear. Nevertheless, it may provide a useful basis for a more generalized transformation system in *Rp. sphaeroides.*

Transformation systems are available for a few species of cyanobacteria (Delaney *et al.*, 1976). Transformation was first demonstrated in *A. nidulans* by Shestakov and Khyen (1970). Subsequently, intraspecific transformation systems have been described for *A. nidulans* (Herdman *et al.*, 1970; Herdman, 1973a,b; Mitronova *et al.*, 1973; Orkwiszewski and Kaney, 1974), *Aphanocapsa* 6714 (Astier and Espardellier, 1976), *Gloeocapsa alpicola* (Devilly and Houghton, 1977), and *Agmenellum quadruplicatum* (Stevens and Porter, 1980). In addition, there is some evidence for the transfer of drug-resistance markers from *A. nidulans* to *G. alpicola* and vice versa by transformation (Devilly and Houghton, 1977).

A topical use of transformation systems is in the cloning of DNA molecules (see Cohen and Chang, 1974). The availability of efficient transformation systems within photosynthetic prokaryotes should facilitate the application of recombinant DNA technology to the cloning of photosynthetic genes.

D. Conjugation

So far, there has been no definitive report describing an indigenous system of conjugative gene transfer within the photosynthetic prokaryotes. However, the introduction of promiscuous R plasmids into photosynthetic bacteria from unrelated organisms is proving a promising strategy in the development of conjugational systems for genetic analyses. Certain R plasmids, belonging principally to the P incompatibility (Inc) group, are transferrable to members of the Rhodospirillaceae, where they can promote chromosome transfer (presumably by conjugation). Olsen and Shipley (1973) described transfer of the R plasmid R1822 from *Pseudomonas aeruginosa* to *Rp. sphaeroides* and to *Rs. rubrum.* Although the R plasmid was not stably maintained in the absence of appropriate selection pressure, this demonstration of intergeneric transfer suggested that other related plasmids might provide useful genetic vehicles for manipulating genes within the Rhodospirillaceae. In this connection Sistrom (1977) has accomplished stable transfer of the R plasmid R68.45 from *P. aeruginosa* [strain PA0254 (R68.45)] to strains of *Rp. sphaeroides* and *Rp. gelatinosa.* Such transfer could only be achieved on solid medium (Sistrom, 1977; cf. Haas and Holloway, 1976). R68.45 is a derivative of the broad host range R plasmid R68 and has an enhanced ability to mobilize the chromosome of *P. aeruginosa* and of other bacteria (Holloway, 1979). Upon transfer to *Rp. sphaeroides,* mobilization of specific chromosomal genes (nutritional and antibiotic resistance markers) can be demonstrated, albeit at low transfer frequencies (10^{-6}–10^{-7} recombinants per recipient cell). However, it is worth noting that R68.45-mediated transfer of genes for photosynthesis was not reported. This could suggest that R68.45 promotes transfer of the chromosome of *Rp. sphaeroides* from only a limited number of sites. R68.45 can also be transferred to *Rp. capsulata* (Marrs *et al.*, 1977; Yu *et al.,* 1981) and *R. vannielii* (L. E. Potts and C. S. Dow, personal communication). It is, therefore, conceivable that other photosynthetic bacteria will accept this plasmid and that it will prove a useful genetic tool in developing novel conjugation systems in such organisms.

Miller and Kaplan (1978) have demonstrated the transfer of the Inc P plasmid RP4 from *P. aeruginosa* or *E. coli* to members of the Rhodospirillaceae, between members of this family and back to *E. coli.* However, the ability of RP4 to mediate transfer of chromosomal genes effectively within the Rhodospirillaceae has yet to be proven. Intraspecies transfer of RP4 in *Rp. sphaeroides,* like that of R68.45, occurs optimally on solid medium. However, transfer of RP4 is inhibited under anaerobic phototrophic conditions irrespective of whether donor or recipient has been grown anaerobically in the light or aerobically in darkness prior to the mating. This could reflect some innate physiological property of *Rp. sphaeroides,* which results in a failure to express those functions required for transfer and establishment of RP4 in recipient cells under phototrophic conditions (Miller and Kaplan, 1978).

A range of R plasmids from different incompatibility groups has been screened for the ability to transfer from *E. coli* to *Rp. sphaeroides* by Tucker and Pemberton (1979a). Only plasmids belonging to Inc P and Inc W groups were freely transmissible to *Rp. sphaeroides.* Certain plasmids of the Inc P group can mobilize nonconjugative plasmids to *Rp. sphaeroides* (D. J. S. Virk and V. A. Saunders, unpublished data). However Inc P and Inc W plasmids do not appear to be particularly proficient at mobilizing the chromosome of *Rp. sphaeroides* (Tucker and Pemberton, 1979a). Various R plasmids, including RH2, R751, and R751::Tn5, can be transmitted from *E. coli* to *Rs. rubrum* (Quivey *et al.,* 1981). It is relevant at this juncture to stress the importance of screening potential recipients for all selectable traits encoded by plasmids when performing mating experiments. Indeed some plasmid-encoded markers, notably carbenicillin resistance (bla^+) are not expressed or only transiently expressed in certain rhodopseudomonads (Sistrom, 1977; Miller and Kaplan, 1978; Pemberton, 1979a) and in other bacteria (Bruce and Warren, 1979).

A variant of the promiscuous plasmid RP1, denoted pBLM2, has been isolated by Marrs (1981). This plasmid has an enhanced ability (relative to the parental plasmid) to mobilize the chromosome of *Rp. capsulata.* Genetic markers including those for photopigment synthesis, cytochromes, reaction-center bacteriochlorophyll complex, and tryptophan synthetase can be mobilized and recombination frequencies greater than 10^{-4} per recipient cell have been attained. An apparent inability to transfer certain markers suggests that pBLM2 can promote transfer from one or a few unique sites on the chromosome of *Rp. capsulata.* The generation of R-prime plasmids containing segments of the bacterial chromosome covalently linked to plasmid pBLM2 indicates that pBLM2 can integrate into the genome of *Rp. capsulata* (Marrs, 1981). Indeed various R-prime plasmids carrying the photopigment region of *Rp. capsulata* have been isolated and subsequently introduced into a number of gram-negative bacteria. Genes for bacteriochlorophyll synthesis do not seem to function in any of these alternative hosts. These R-prime plasmids may be stably maintained in recombination-deficient (*rec A*) strains of *E. coli.* Apparently, those genes exclusive to photosynthesis are available as inserts carried by specific prime plasmids (B. L. Marrs, personal communication). Dissection of these plasmids by restriction endonuclease digestion led to the isolation and subsequent cloning of specific sections of the photopigment region of *Rp. capsulata* (Clark *et al.*, 1981).

The mechanism(s) by which plasmids mobilize the chromosomes of photosynthetic bacteria has yet to be established. It is possible that the acquired plasmid undergoes some transient or unstable association with the chromosome. On the other hand, some mechanism apparently not involving plasmid integration, like that implicated for mobilization of nonconjugative plasmids by conjugative plasmids in *E. coli.* (Falkow, 1975; Warren *et al.*, 1978) cannot be ruled out.

A possible explanation for the inability of particular Inc P plasmids to mobilize the chromosome of *Rp. sphaeroides* efficiently could be the lack of a specific region of genetic homology between the plasmid and host chromosome. Such homology might well be provided through the agency of insertion sequences or other transposable genetic elements (see Bukhari *et al.*, 1977). On this basis, Tucker and Pemberton (1979b) introduced a hybrid R plasmid, RP4::Mu*cts*62, into *Rp. sphaeroides.* Bateriophage Mu is capable of integrating efficiently at multiple sites in the chromosome of *E. coli* (Taylor, 1963; Bukhari and Zipser, 1972; Howe and Bade, 1975) providing a portable region of homology with plasmids carrying further copies of *Mu.* Dénarié and colleagues (1977) have assessed the virtues of RP4::Mu promoted chromosome mobilization in *E. coli* and other bacteria. Furthermore Mu-mediated transposition of chromosomal DNA segments into conjugative plasmids has been described and utilized for cloning the *nif* genes of *Klebsiella pneumoniae* in *E. coli* (see Dénarié *et al.*, 1977). Clearly expression of the Mu genome in *Rp. sphaeroides* should increase scope for *in vivo* genetic manipulation in this organism. Mu hybrid plasmids have been used to introduce transposons into rhodopseudomonads. The transposon Tn5 has been transferred to *Rp. sphaeroides* using plasmid pJB4J1 [=pPH1J1::Mu::Tn5

(Beringer *et al.*, 1978)] (Virk, D. J. S. and Saunders, V. A., unpublished data). Transfer of these transposable elements appears to be associated with the induction of pigment mutants among transconjugants.

E. Mapping Genes for Photosynthesis

The GTA has proven an exceptional genetic tool for fine structure analysis of the genome of *Rp. capsulata* (Marrs, 1978b). A map of the photopigment region has been constructed utilizing the GTA in conjunction with various mutants blocked in the biosynthesis of bacteriochlorophyll or carotenoid (Yen and Marrs, 1976). Map distances generated from co-transfer data have been shown to be additive. However, genetic analyses using the GTA are confined to short tracts (approximately 4–5 genes in length) of the genome of *Rp. capsulata.* Supplementary genetic mapping studies using plasmid pBLM2 have shown that genes for reaction-center bacteriochlorophyll complex and tryptophan synthetase are closely aligned with those for photopigment synthesis (Fig. 1). It is not yet clear whether genes for light-harvesting pigment proteins are close to those for bacteriochlorophyll and carotenoid syntheses. However, no linkage could be established between genes for cytochrome synthesis and those for photopigment synthesis (Marrs, 1981). It remains to be determined whether the cluster of genes encoding production of the photopigments represents a single functional operon. It has been tentatively proposed that the *bch D* gene (see Fig. 1) is a regulator gene, which governs expression either of those genes for chlorophyll synthesis alone or of those genes for photopigment synthesis *in toto* (Marrs, 1981). A genetic map of *Rp. sphaeroides* has been constructed using plasmid RP1::Tn501, which promotes a high frequency of chromosome transfer (Pemberton and Bowen, 1981). Interestingly, as with *Rp. capsulata* genes for photopigment synthesis are clustered on the genome of *Rp. sphaeroides.*

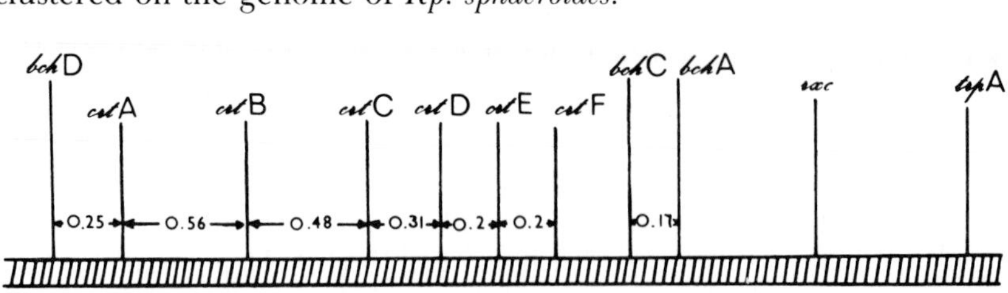

FIG. 1. **Genetic map of region for bacteriochlorophyll and carotenoid biosynthesis in *Rp. capsulata* modified from Yen and Marrs (1976) and Marrs (1981). The numbers refer to distances in map units between specific markers in each gene. *crt*, carotenoid synthesis; *bch*, bacteriochlorophyll synthesis; *rxc*, reaction-center synthesis; *trp*, tryptophan synthesis.**

Available data concerning synthesis and assembly of the photosynthetic membrane of purple nonsulfur bacteria indicates that this morphogenetic process entails a coordinated and strictly regulated sequence of events (see Drews, 1978; Kaplan, 1978; Ohad and Drews, Chapter 5, this volume). It is, therefore, likely that a network of regulatory mechanisms controls production of the photopigments and of other integral components of the photosynthetic apparatus. Precise details of these control systems should be revealed by the use of appropriate DNA cloning and sequencing techniques in conjunction with conventional genetic analyses.

V. Prospects for Applied Research

The recent upsurge of interest in the molecular biology of photosynthetic prokaryotes has prompted speculation that these organisms might be of value in applied research. For instance, it has been envisaged (Marrs, 1978a) that rhodopseudomonads could be genetically manipulated for the degradation of a diversity of carbon compounds that are recalcitrant to other bacteria. The feasibility of this proposition relies on the acquisition and maintenance by these photosynthetic organisms of "degradative" plasmids that specify the decomposition of a range of carbon compounds. An advantage of the use of photosynthetic bacteria over other bacteria would be that they have no absolute requirement for energy from the compounds degraded.

Other potential applications of photosynthetic prokaryotes include their use to produce hydrogen (ultimately with a view to its utilization as a fuel source) or to fix commercial amounts of nitrogen by the tapping of solar energy. In each of these cases, it may be possible to enhance existing biochemical abilities by the genetic manipulation of the organisms. In this context, mutants of *Rp. sphaeroides* have been isolated (Macler *et al.*, 1979) that can quantitatively convert glucose to hydrogen and carbon dioxide in the light. It should be possible to isolate and exploit other mutants capable of utilizing various reduced compounds, which are present for example in organic wastes, to produce hydrogen on an industrial scale.

Pilot studies have assessed the role of photosynthetic bacteria in the recycling of organic wastes (Crofts, 1971; Kobayashi *et al.*, 1971; Shipman *et al.*, 1977; Siefert *et al.*, 1978). It has been proposed that photosynthetic bacteria could be utilized in the purification of sewage, and the photosynthetic cell protein (a by-product of the process) could be used as a supplement for animal feeds (Kobayashi *et al.*, 1971). Furthermore, provisional schemes have been devised for the mass cultivation of photosynthetic bacteria with a view to the use of these bacteria as a source of human food (Shipman *et al.*, 1977). Genetic manipulation might play a role in improving growth efficiency, nutritional quality, and palatability of these photosynthetic organisms as food supplements.

The possible role of photosynthetic prokaryotes as hosts for *in vitro* recombinant DNA research warrants some attention in view of their ability to utilize solar energy for growth. It may ultimately prove more profitable to tailor these organisms to specific commercial needs rather than to use other bacteria. Recombinant DNA molecules are normally introduced into host cells by transformation (or transfection). Such genetic exchange systems are available for cyanobacteria and are currently being developed for photosynthetic bacteria. Moreover, suitable cloning vehicles may be provided from the extrachromosomal elements (plasmids and phages) available within this biological group. In the absence of suitable genetic transformation systems, *in vitro* packaging of recombinant DNA into appropriate bacteriophage vectors and subsequent infection of susceptible host cells may provide a viable alternative for the introduction of recombinant DNA into the host. Such techniques have previously been demonstrated in *E. coli* using bacteriophage lambda (Hohn and Murray, 1977).

VI. Concluding Remarks

The foundations for intensive genetic analysis of members of the Rhodospirillaceae have been laid. In particular, the development of conjugative systems by the introduction of promiscuous plasmids into photosynthetic bacteria is proving a fertile area for genetic manipulations. Conjugative plasmids have not only augmented GTA-mediated genetic studies of *Rp. capsulata* but have also extended genetic analysis to *Rp. sphäeroides*. Moreover, it is contemplated that such plasmids will provide effective genetic vehicles for conjugational analyses of other members of the Rhodospirillaceae. The potential of transposable genetic elements to promote *in vivo* genetic engineering of gram-negative bacteria (see Kleckner *et al.*, 1977) is gradually being realized for photosynthetic prokaryotes. Transposons have been grafted on to plasmids of these photosynthetic organisms where they can serve as genetic tags for otherwise cryptic plasmids. Furthermore, transposable elements can provide portable regions of homology with which other copies of the element will recombine. One consequence of this could be the construction of high frequency strains of photosynthetic organisms in which conjugative plasmids bearing transposable genetic elements might become integrated into the host chromosome at a site of homology afforded by a second copy of that element and promote high frequency chromosome transfer. Such chromosome transfer has been achieved in *Rp. sphaeroides* using RP1::Tn501. This has enabled formulation of the first comprehensive genetic map of the organism. It is feasible that similar approaches will allow compilation of a series of genetic maps of photosynthetic bacteria. Transposable elements could also be more extensively exploited to isolate mutants of photosynthetic organisms. This would be of particular advantage in those cases where the isolation of mutants by traditional procedures proves difficult.

The advent of recombinant DNA technology promises to extend the capability for genetic manipulation to organisms for which gene transfer systems are not readily available. The organization of genes for nitrogen fixation in *Anabaena* 7120 has already been investigated using this methodology (Mazur *et al.*, 1981). In addition, genes for photosynthesis in *Rp. capsulata* (Clark *et al.*, 1981) and for ribosomal RNA in *A. nidulans* (Tomioka *et al.*, 1981) are being characterized using cloning techniques. Indeed, there is every indication that the marriage between conventional genetics and *in vitro* recombinant DNA technology will rapidly accelerate the pace of our understanding of the molecular basis of gene expression in photosynthetic prokaryotes. In turn, this could expand prospects for the use of these organisms in applied research.

Acknowledgments

The author wishes to thank C. S. Dow, S. Kaplan, B. L. Marrs, and J. Pemberton for providing data prior to publication and to the Science Research Council (U.K.) for research grant GR/A 85797.

REFERENCES

Adams, D. G., Phillips, D. O., Nichols, J. M., and Carr, N. G. (1977). *FEBS Lett.* **81,** 48–52.

Akinyanju, J., and Smith, R. J. (1979). *FEBS Lett.* **107,** 173–176.

Almassy, R. J., and Dickerson, R. E. (1978). *Proc. Natl. Acad. Sci. U.S.A.* **75,** 2674–2678.

Astier, C., and Espardellier, F. (1976). *C.R. Hebd. Seances Acad. Sci., Ser. D* **282,** 795–797.

Bennett, P. M., and Richmond, M. H. (1978). *In* "The Bacteria" (I. C. Gunsalus, ed.), Vol. 6, pp. 1–69. Academic Press, New York.

Beringer, J. E., Beynon, J. L., Buchanan-Wollaston, A. V., and Johnston, A. W. B. (1978). *Nature (London)* **276,** 633–634.

Broda, P. (1979). "Plasmids." Freeman, San Francisco, California.

Broglie, R. M., and Niederman, R. A. (1979). *J. Bacteriol.* **138,** 788–798.

Brown, A. E., Eiserling, F. A., and Lascelles, J. (1972). *Plant Physiol.* **50,** 743–746.
Bruce, D. L., and Warren, R. A. J. (1979). *J. Bacteriol.* **139,** 811–816.
Bukhari, A. I., and Zipser, D. (1972). *Nature (London), New Biol.* **236,** 240–243.
Bukhari, A. I., Shapiro, J. A., and Adhya, S. L., eds. (1977). "DNA Insertion Elements, Plasmids and Episomes." Cold Spring Harbor Lab., Cold Spring Harbor, New York.
Bull, M. J., and Lascelles, J. (1963). *Biochem. J.* **87,** 15–28.
Campbell, A., Starlinger, P., Berg, D. E., Botstein, D., Lederberg, E. M., Novick, R. P., and Szybalski, W. (1979). *Plasmid* **2,** 466–473.
Cannon, F. C., Riedel, G. E., and Ausubel, F. M. (1977). *Proc. Natl. Acad. Sci. U.S.A.* **74,** 2963–2967.
Cannon, F. C., Riedel, G. E., and Ausubel, F. M. (1979). *Mol. Gen. Genet.* **174,** 59–66.
Cashel, M. (1975). *Annu. Rev. Microbiol.* **29,** 301–318.
Chow, C. T. (1976). *Can. J. Microbiol.* **22,** 228–236.
Clark, G., Taylor, D. P., Cohen, S. N., and Marrs, B. L. (1981). *81st Annu. Meet. Am. Soc. Microbiol.* Abstract No. H50.
Cohen-Bazire, G., Sistrom, W. R., and Stanier, R. Y. (1957). *J. Cell. Comp. Physiol.* **49,** 25–68.
Cohen, S. N., and Chang, A. C. Y. (1974). *In* "Microbiology—1974" (D. Schlessinger, ed.), pp. 66–75. Am. Soc. Microbiol., Washington, D.C.
Cost, H. B., and Gray, E. D. (1967). *Biochim. Biophys. Acta* **138,** 601–604.
Crofts, A. R. (1971). *Proc. R. Soc. London, Ser. B* **179,** 209–219.
Das, B., and Singh, P. K. (1976). *Arch. Microbiol.* **111,** 195–196.
de Bont, J. A. M., Scholten, A., and Hansen, T. A. (1981). *Arch. Microbiol.* **128,** 271–274.
DeJesus, T. G. S., and Gray, E. D. (1971). *Biochim. Biophys. Acta* **254,** 419–428.
Delaney, S. F., Herdman, M., and Carr, N. G. (1976). *In* "Genetics of the Algae" (R. A. Lewin, ed.), pp. 7–28. Blackwell, Oxford, England.
del Valle-Tascón, S., Giménez-Gallego, G., and Ramirez, J. M. (1975). *Biochem. Biophys. Res. Commun.* **66,** 514–519.
del Valle-Tascón, S., Giménez-Gallego, G., and Ramirez, J. M. (1977). *Biochim. Biophys. Acta* **459,** 76–87.
Dénarié, J., Rosenberg, C., Bergeron, B., Boucher, C., Michel, M., and Barate de Bertalmio, M. (1977). *In* "DNA Insertion Elements, Plasmids and Episomes" (A. I. Bukhari, J. A. Shapiro, and S. L. Adhya, eds.), pp. 507–520. Cold Spring Harbor Lab., Cold Spring Harbor, New York.
Devilly, C. I., and Houghton, J. A. (1977). *J. Gen. Microbiol.* **98,** 277–280.
Dickerson, R. E. (1980). *Nature (London)* **283,** 210–212.
Dickerson, R. E., Timkovich, R., and Almassy, R. J. (1976). *J. Mol. Biol.* **100,** 473–491.
Doolittle, W. F. (1973). *J. Bacteriol.* **113,** 1256–1263.
Doolittle, W. F. (1979). *Adv. Microb. Physiol.* **20,** 1–102.
Drews, G. (1978). *Curr. Top. Bioenerg.* **8,** 161–207.
Drews, G., Dierstein, R., and Schumacher, A. (1976). *FEBS Lett.* **68,** 132–136.
Eccleston, E. D., Jr., and Gray, E. D. (1973). *Biochem. Biophys. Res. Commun.* **54,** 1370–1376.
Edelman, M., Swinton, D., Schiff, J. A., Epstein, H. T., and Zelden, B. (1967). *Bacteriol. Rev.* **31,** 315–331.
Falkow, S. (1975). "Infectious Multiple Drug Resistance." Pion Ltd., London.
Fraley, R. T., Fornari, C. S., and Kaplan, S. (1979). *Proc. Natl. Acad. Sci. U.S.A.* **76,** 3348–3352.
Friedberg, D., and Seijffers, J. (1979). *FEBS Lett.* **107,** 165–168.
Gallant, J., Irr, J., and Cashel, M. (1971). *J. Biol. Chem.* **246,** 5812–5816.

Garcia, A. F., Drews, G., and Kamen, M. D. (1974). *Proc. Natl. Acad. Sci. U.S.A.* **71,** 4213–4216.

Garcia, A. F., Drews, G., and Kamen, M. D. (1975). *Biochim. Biophys. Acta* **387,** 129–134.

Gibson, J., Stackebrandt, E., Zablen, L. B., Gupta, R., and Woese, C. R. (1979). *Curr. Microbiol.* **3,** 59–64.

Gibson, K. D., and Niederman, R. A. (1970). *Arch. Biochem. Biophys.* **141,** 694–704.

Gillis, M., Deley, J., and De Cleene, M. (1970). *Eur. J. Biochem.* **12,** 143–153.

Glover, S. W., and Butler, L. O. (1979). "Transformation 1978." Cotswold Press, Oxford, England.

Gray, E. D. (1967). *Biochim. Biophys. Acta* **138,** 550–563.

Gray, E. D. (1978). *In* "The Photosynthetic Bacteria" (R. K. Clayton and W. R. Sistrom, eds.), pp. 885–897. Plenum, New York.

Haas, D., and Holloway, B. W. (1976). *Mol. Gen. Genet.* **144,** 243–251.

Herdman, M. (1973a). *In* "Bacterial Transformation" (L. J. Archer, ed.), pp. 369–386. Academic Press, New York.

Herdman, M. (1973b). *Mol. Gen. Genet.* **120,** 369–378.

Herdman, M., Faulkner, B. M., and Carr, N. G. (1970). *Arch. Microbiol.* **73,** 238–249.

Herdman, M., Janvier, M., Waterbury, J. B., Rippka, R., and Stanier, R. Y. (1979a). *J. Gen. Microbiol.* **111,** 63–71.

Herdman, M., Janvier, M., Rippka, R., and Stanier, R. Y. (1979b). *J. Gen. Microbiol.* **111,** 73–85.

Higuchi, M., Goto, K., Fujimoto, M., Namiki, O., and Kikuchi, G. (1965). *Biochim. Biophys. Acta* **95,** 94–110.

Hill, L. R. (1966). *J. Gen. Microbiol.* **44,** 419–437.

Hochstadt-Ozer, J., and Cashel, M. (1972). *J. Biol. Chem.* **247,** 7067–7072.

Hohn, B., and Murray, K. (1977). *Proc. Natl. Acad. Sci. U.S.A.* **74,** 3259–3263.

Holloway, B. W. (1979). *Plasmid* **2,** 1–19.

Howe, M. M., and Bade, E. G. (1975). *Science* **190,** 624–632.

Hu, N. T., and Marrs, B. L. (1979). *Arch. Microbiol.* **121,** 61–69.

Hunter, C. N., and Jones, O. T. G. (1976). *Biochem. Soc. Trans.* **4,** 669–670.

Hunter, C. N., and Jones, O. T. G. (1979a). *Biochim. Biophys. Acta* **545,** 325–338.

Hunter, C. N., and Jones, O. T. G. (1979b). *Biochim. Biophys. Acta* **545,** 339–351.

Ikeda, H., and Tomizawa, J. (1968). *Cold Spring Harbor Symp. Quant. Biol.* **33,** 791–798.

Jones, O. T. G., and Plewis, K. M. (1974). *Biochim. Biophys. Acta* **357,** 204–214.

Kaplan, S. (1978). *In* "The Photosynthetic Bacteria" (R. K. Clayton and W. R. Sistrom, eds.), pp. 809–839. Plenum, New York.

Kleckner, N., Roth, J., and Botstein, D. (1977). *J. Mol. Biol.* **116,** 125–159.

Klein, N. C., and Mindich, L. (1976). *J. Bacteriol.* **128,** 337–346.

Kobayashi, M., Kobayashi, M., and Nakamishi, H. (1971). *J. Ferment. Technol.* **49,** 817–825.

Kuhl, S. A., and Yoch, D. C. (1981). *81st Annu. Meet. Am. Soc. Microbiol.* Abstract No. H126.

Kuhlemeier, C. J., Borrias, W. E., van den Hondel, C. A. M. J. J., and van Arkel, G. A. (1981). *Mol. Gen. Genet.* **184,** 249–254.

Kushner, D. J., and Breuil, C. (1977). *Arch. Microbiol.* **112,** 219–223.

LaMonica, R. F., and Marrs, B. L. (1976). *Biochim. Biophys. Acta* **423,** 431–439.

Lascelles, J. (1975). *Ann. N.Y. Acad. Sci.* **244,** 334–347.

Lascelles, J., and Wertlieb, D. (1971). *Biochim. Biophys. Acta* **226,** 328–340.

Lau, R. H., and Doolittle, W. F. (1979). *J. Bacteriol.* **137,** 648–652.

Lau, R. H., Sapienza, C., and Doolittle, W. F. (1980). *Mol. Gen. Genet.* **178,** 203–211.

Leach, C. K., Old, J. M., and Carr, N. G. (1971). *J. Gen. Microbiol.* **68,** xiv.

Lessie, T. G. (1965). *J. Gen. Microbiol.* **39,** 311–320.

Lien, S., San Pietro, A., and Gest, H. (1971). *Proc. Natl. Acad. Sci. U.S.A.* **68,** 1912–1915.
Littauer, U. Z., and Inouye, H. (1973). *Annu. Rev. Biochem.* **42,** 439–470.
Mackay, R. M., Zablan, L. B., Woese, C. R., and Doolittle, W. F. (1979). *Arch. Microbiol.* **123,** 165–172.
Macler, B. A., Pelroy, R. A., and Bassham, J. A. (1979). *J. Bacteriol.* **138,** 446–452.
Mandel, M., Leadbetter, E. R., Pfennig, N., and Trüper, H. G. (1971). *Int. J. Syst. Bacteriol.* **21,** 222–230.
Mann, N., and Carr, N. G. (1974). *J. Gen. Microbiol.* **83,** 399–405.
Mann, N., Carr, N. G., and Midgley, J. E. M. (1975). *Biochim. Biophys. Acta* **402,** 41–50.
Mansour, J. D., and Stachow, C. S. (1975). *Biochem. Biophys. Res. Commun.* **62,** 276–281.
Marrs, B. (1974). *Proc. Natl. Acad. Sci. U.S.A.* **71,** 971–973.
Marrs, B. L. (1978a). *Curr. Top. Bioenerg.* **8,** 261–294.
Marrs, B. L. (1978b). *In* "The Photosynthetic Bacteria" (R. K. Clayton and W. R. Sistrom, eds.), pp. 873–883. Plenum, New York.
Marrs, B. (1981). *J. Bacteriol.* **146,** 1003–1012.
Marrs, B., and Gest, H. (1973a). *J. Bacteriol.* **114,** 1045–1051.
Marrs, B., and Gest, H. (1973b). *J. Bacteriol.* **114,** 1052–1057.
Marrs, B., and Kaplan, S. (1970). *J. Mol. Biol.* **49,** 297–317.
Marrs, B., Wall, J. D., and Gest, H. (1977). *Trends Biochem. Sci.* **2,** 105–108.
Mazur, B. J., Rice, D., and Haselkorn, R. (1980). *Proc. Natl. Acad Sci.* **77,** 186–190.
Michels, P. A. M., and Haddock, B. A. (1980). *FEMS Microbiol. Lett.* **7,** 327–331.
Miller, L., and Kaplan, S. (1978). *Arch. Biochem. Biophys.* **187,** 229–234.
Mitronova, T. N., Shestakov, S. V., and Zhevner, V. D. (1973). *Mikrobiologiya* **42,** 519–524.
Novick, R. P. (1969). *Bacteriol. Rev.* **33,** 210–263.
Novick, R. P., Clowes, R. C., Cohen, S. N., Curtiss, R., III, Datta, N., and Falkow, S. (1976). *Bacteriol. Rev.* **40,** 168–189.
Olsen, R. H., and Shipley, P. (1973). *J. Bacteriol.* **113,** 772–780.
Orkwiszewski, K. G., and Kaney, A. R. (1974). *Arch. Microbiol.* **98,** 31–37.
Osawa, S. (1968). *Annu. Rev. Biochem.* **37,** 109–130.
Padan, E., and Shilo, M. (1973). *Bacteriol. Rev.* **37,** 343–370.
Pemberton, J. M., and Bowen, A. R. St. G. (1981). *J. Bacteriol.* **147,** 110–117.
Pemberton, J. M., and Tucker, W. T. (1977). *Nature (London)* **266,** 50–51.
Pfennig, N. (1967). *Annu. Rev. Microbiol.* **21,** 285–324.
Portolés, A., Lopez, R., and Espinosa, M. (1977). "Modern Trends in Bacterial Transformation and Transfection." Elsevier/North-Holland Biomedical Press, Amsterdam.
Potts, L. E., Dow, C. S., and Avery, R. J. (1980). *J. Gen. Microbiol.* **117,** 501–507.
Quivey, R., Meyer, R. J., and Tabita, F. R. (1981). *81st Annu. Meet. Am. Soc. Microbiol.* Abstract No. H113.
Razel, A. J., and Gray, E. D. (1978). *J. Bacteriol.* **133,** 1175–1180.
Reanney, D. (1976). *Bacteriol. Rev.* **40,** 552–590.
Reaston, J., van den Hondel, C. A. M. J. J., van der Ende, A., van Arkel, G. A., Stewart, W. D. P., and Herdman, M. (1980). *FEMS Microbiol. Lett.* **9,** 185–188.
Reaston, J., van den Hondel, C. A. M. J. J., van Arkel, G. A., and Stewart, W. D. P. (1982). *Plasmid* **7,** 101–104.
Rebeiz, A., and Lascelles, J. (1982). *In* "Photosynthesis: Energy Conversion by Plants and Bacteria," (Govindjee, ed.), Vol. I, Academic Press, New York, 699–780.
Rich, A., and Raj Bhandry, V. L. (1976). *Annu. Rev. Biochem.* **45,** 805–860.
Roberts, T. M., and Koths, K. E. (1976). *Cell* **9,** 551–557.
Robinson, A., and Sykes, J. (1971). *Biochim. Biophys. Acta* **238,** 99–115.
Rogerson, A. C., Rowell, P., and Stewart, W. D. P. (1978). *FEMS Microbiol. Lett.* **3,** 299–303.

Safferman, R. S. (1973). *In* "The Biology of Blue-Green Algae" (N. G. Carr and B. A. Whitton, eds.), pp. 214–237. Blackwell, Oxford, England.

Saunders, V. A. (1978). *Microbiol. Rev.* **42,** 357–384.

Saunders, V. A., Saunders, J. R., and Bennett, P. M. (1976). *J. Bacteriol.* **125,** 1180–1187.

Scolnik, P. A., Walker, M. A., and Marrs, B. L. (1980). *J. Biol. Chem.* **255,** 2427–2432.

Sheperd, W. D., and Kaplan, S. (1978). *Arch. Microbiol.* **116,** 161–167.

Sherman, L. A., and Brown, R. M., Jr. (1978). *Compr. Virol.* **12,** 145–234.

Shestakov, S. V., and Khyen, N. T. (1970). *Mol. Gen. Genet.* **107,** 372–375.

Shipman, R. H., Fan, L. T., and Kao, I. C. (1977). *Adv. Appl. Microbiol.* **21,** 161–183.

Siefert, E., Irgens, R. L., and Pfennig, N. (1978). *Appl. Environ. Microbiol.* **35,** 38–44.

Silver, M., Friedman, S., Guay, R., Couture, J., and Tanguay, R. (1971). *J. Bacteriol.* **107,** 368–370.

Simon, R. D. (1978a). *J. Bacteriol.* **136,** 414–418.

Simon, R. D. (1978b). *Nature (London)* **273,** 314–317.

Singer, R. A., and Doolittle, W. F. (1974). *J. Bacteriol.* **118,** 351–357.

Sistrom, W. R. (1962). *J. Gen. Microbiol.* **28,** 607–616.

Sistrom, W. R. (1977). *J. Bacteriol.* **131,** 526–532.

Smith, R. J. (1979). *J. Gen. Microbiol.* **113,** 403–405.

Solioz, M., and Marrs, B. (1977). *Arch. Biochem. Biophys.* **181,** 300–307.

Solioz, M., Yen, H.-C., and Marrs, B. (1975). *J. Bacteriol.* **123,** 651–657.

Stanier, R. Y., Kunisawa, R., Mandel, M., and Cohen-Bazire, G. (1971). *Bacteriol. Rev.* **35,** 171–205.

Stevens, S. E., and Porter, R. D. (1980). *Proc. Natl. Acad. Sci. U.S.A.* **77,** 6052–6056.

Sueoka, N., and Kano-Sueoka, T. (1970). *Prog. Nucleic Acid Res. Mol. Biol.* **10,** 23–55.

Suyama, Y., and Gibson, J. (1966). *Biochem. Biophys. Res. Commun.* **24,** 549–553.

Taylor, A. L. (1963). *Proc. Natl. Acad. Sci. U.S.A.* **50,** 1043–1051.

Taylor, M. M., and Storck, R. (1964). *Proc. Natl. Acad. Sci. U.S.A.* **52,** 958–965.

Tomioka, N., Shinozaki, K., and Sugiura, M. (1981). *Mol. Gen. Genet.* **184,** 359–363.

Tucker, W. T., and Pemberton, J. M. (1978). *J. Bacteriol.* **135,** 207–214.

Tucker, W. T., and Pemberton, J. M. (1979a). *FEMS Microbiol. Lett.* **5,** 173–176.

Tucker, W. T., and Pemberton, J. M. (1979b). *FEMS Microbiol. Lett.* **5,** 215–217.

Tucker, W. T., and Pemberton, J. M. (1980). *J. Bacteriol.* **143,** 43–49.

van den Hondel, C. A. M. J. J., Keegstra, W., Borrias, W. E., and van Arkel, G. A. (1979). *Plasmid* **2,** 323–333.

van den Hondel, C. A. M. J. J., Verbeek, S., van der Ende, A., Weisbeck, P. J., Borrias, W. E., and van Arkel, G. A. (1980). *Proc. Natl. Acad. Sci. U.S.A.* **77,** 1570–1574.

van den Hondel, C. A. M. J. J., Kuhlemeier, C. J., Borrias, W. E., van Arkel, G. A., Tandeau de Marsac, N., and Castets, A. M. (1981). *Soc. Gen. Microbiol. Quart.* **8,** Abstract, No. P19, 138–139.

Wall, J. D., Weaver, P. F., and Gest, H. (1975). *Arch. Microbiol.* **105,** 217–224.

Wallace, D. C., and Morowitz, H. J. (1973). *Chromosoma* **40,** 121–126.

Walsby, A. E. (1977). *Arch. Microbiol.* **114,** 167–170.

Warren, G. J., Twigg, A. J., and Sherratt, D. J. (1978). *Nature (London)* **274,** 259–261.

Weaver, P. F., Wall, J. D., and Gest, H. (1975). *Arch. Microbiol.* **105,** 207–216.

Weidinger, G., Klotz, G., and Goebel, W. (1979). *Plasmid* **2,** 377–386.

Witkin, S. S., and Gibson, K. D. (1972a). *J. Bacteriol.* **110,** 677–683.

Witkin, S. S., and Gibson, K. D. (1972b). *J. Bacteriol.* **110,** 684–690.

Woese, C. R., Gibson, J., and Fox, G. E. (1980). *Nature (London)* **283,** 212–214.

Wolk, C. P. (1973). *Bacteriol. Rev.* **37,** 32–101.

Yamashita, J., and Kamen, M. D. (1968). *Biochim. Biophys. Acta* **161,** 162–169.

Yen, H.-C., and Marrs, B. L. (1976). *J. Bacteriol.* **126,** 619–629.
Yen, H.-C., Hu, N. T., and Marrs, B. L. (1979). *J. Mol. Biol.* **131,** 157–168.
Yu, P.-L., Cullum, J., and Drews, G. (1981). *Arch. Microbiol.* **128,** 390–393.
Zannoni, D., Melandri, B. A., and Baccarini-Melandri, A. (1976a). *Biochim. Biophys. Acta* **423,** 413–430.
Zannoni, D., Melandri, B. A., and Baccarini-Melandri, A. (1976b). *Biochim. Biophys. Acta* **449,** 386–400.
Zannoni, D., Jasper, P., and Marrs, B. (1978). *Arch. Biochem. Biophys.* **191,** 625–631.
Zilkinsky, J. W., Sojka, G. A., and Gest, H. (1971). *Biochem. Biophys. Res. Commun.* **42,** 955–961.

4

Molecular Biology of Chloroplasts

DENNIS E. BUETOW

ABBREVIATIONS

A	Adenine, adenosine
*Bam*HI	Restriction endonuclease from *Bacillus amyloliquefaciens* H
b.p.	Base pair
C	Cytosine, cytidine
D	Dalton(s)
DNA	Deoxyribonucleic acid
*Eco*RI	Restriction endonuclease from *Escherichia coli* RY13
G	Guanine, guanosine
LS	Large subunit of ribulosebisphosphate carboxylase
PSII	Photosystem II
Ψ	Pseudouridine
RNA	Ribonucleic acid
mRNA	Messenger ribonucleic acid
rRNA	Ribosomal ribonucleic acid
RNase	Ribonuclease
rrnA rRNA operon	One of three ribosomal ribonucleic acid operons of *Escherichia coli* that have genes for isoleucine and alanine transfer RNAs located in the spacer DNA region between the genes for 23 S and 16 S ribosomal ribonucleic acids

Photosynthesis: Development, Carbon Metabolism, and Plant Productivity, Vol. II

ISBN: 0-12-294302-3

S	Svedberg unit
T	Ribothymidine
tRNA	Transfer ribosomal ribonucleic acid
$tRNA^{Ala}$	Alanine transfer ribonucleic acid
$tRNA^{Asn}$	Asparagine transfer RNA
$tRNA^{Glu}$	Glutamic acid transfer ribonucleic acid
$tRNA^{His}$	Histidine transfer ribonucleic acid
$tRNA^{Ile}$	Isoleucine transfer ribonucleic acid
$tRNA_f^{Met}$	Formylated methionine transfer ribonucleic acid; the initiator transfer ribonucleic acid of chloroplasts, mitochondria, and prokaryotes
$tRNA_i^{Met}$	Nonformylated methionine transfer ribonucleic acid; the initiator transfer ribonucleic acid of eukaryotic cytoplasms
$tRNA^{Phe}$	Phenylalanine transfer ribonucleic acid
$tRNA^{Trp}$	Tryptophan transfer ribonucleic acid
$tRNA^{Val}$	Valine transfer ribonucleic acid
U	Uracil, uridine

ABSTRACT

Chloroplasts are semiautonomous structures that require transcripts from their own genome and that of the nucleus in order to develop and function. Chloroplasts of higher plants and algae contain DNA coding for ribosomal, transfer and messenger RNAs. The kinetic complexities of chloroplast DNAs from most higher plants and *Euglena gracilis* range from 82–96 × 10^6 daltons (D). At 56 × 10^6 D, the chloroplast genome of the green alga, *Codium fragile,* is the smallest known. *Acetabularia* has the largest chloroplast genome, i.e., 1.1–1.5 × 10^9 D. Chloroplast DNA molecules are circular in higher plants and in algae. In a given organism, the bulk of all the chloroplast DNA molecules are identical, although a small amount of heterogeneity has been detected. Chloroplasts are highly polyploid, each containing many copies of the circular genome.

Chloroplast DNA is not associated with histones, and it lacks the modified base 5-methylcytosine, which is characteristic of nuclear DNA. As is true of animal mitochondrial DNA, chloroplast DNA contains a small number of ribonucleotides. The significance of the individually inserted ribonucleotides is not known. Considerable base composition heterogeneity exists along the chloroplast DNA molecule, with the guanine plus cytosine content varying widely from one segment to another. Chloroplast DNA replicates semiconservatively. Both "Cairns" and "rolling circle" replicative intermediates have been observed, and a chloroplast-specific DNA polymerase has been demonstrated.

Chloroplast DNA is known to code for 16 S, 23 S, and 5 S ribosomal RNAs, at least 30 transfer RNAs, the large subunit of ribulosebisphosphate carboxylase, a 32,000-D polypeptide located in the thylakoid membrane (associated with photosystem II), and a small amount of polyadenylated RNA of unknown function. Higher plant chloroplast DNA also codes for a 4.5 S ribosomal RNA, which locates in the large subunit of the chloroplast ribosome. About 90% of the potential coding capacity of a single strand of the chloroplast DNA is not accounted for. Also, there is evidence that transcripts come from both strands of the DNA.

The chloroplast genomes of several higher plants and algae have been physically mapped with restriction endonucleases, and the genes for RNAs, tRNAs, and the two known polypeptides have been located. In some, the ribosomal RNA "operons" map as inverted repeats, in others as tandem repeats, and in still others, there is only one ribosomal RNA operon present. Some RNAs map in the DNA spacer region between the 16 S and 23 S rRNA genes, whereas others are distributed over the map. A chloroplast-specific RNA polymerase has been described. Much remains to be done to determine how the expression of chloroplast genes is regulated.

Chloroplast ribosomes show lower sedimentation values (~70 S) than do cytoplasmic ribosomes (~80 S). Polysomes have been isolated from the chloroplasts of algae and higher plants. The chloroplast rRNA operon of spinach at least is transcribed as a single large precursor RNA that is subsequently processed to the mature ribosomal RNAs. The mature 23 S rRNA of higher plants is labile.

Aminoacyl–tRNA synthetases specific to the chloroplast have been characterized. The synthetases and tRNAs from chloroplasts and from prokaryotes show some similarities because cross-aminoacylation reactions are possible. Several isoaccepting species of tRNA have been identified in chloroplasts. The chloroplast initiator tRNA is formylated like its prokaryotic counterpart.

Evolution of the chloroplast genome has been studied by restriction endonuclease and molecular hybridization analyses. A region comprising 10–20% of the chloroplast DNA molecule is more resistant to change than the rest of the molecule. The rRNA genes are within this highly conserved region. The structure of the chloroplast rRNA operon and the primary structure of the chloroplast 16 S rRNA demonstrate the phylogenetic relationship between chloroplasts and prokaryotes.

I. Introduction

Early on light microscopists discovered that chloroplasts proliferate by growth and division within the cells that contain them. Also, early genetic studies clearly showed that certain chloroplast defects are inherited in such a way that the mutated genes appear to reside in the organelle itself rather than in the nucleus. Results of such cytological and genetic observations led to the view that chloroplasts are autonomous, self-sufficient entities.

A series of discoveries in the 1940s and 1950s established that genetic information is stored in living organisms in the form of DNA (in certain viruses in the RNA). In the 1960s, chloroplasts were also shown to store genetic information as DNA (e.g., Tewari, 1971; Kung, 1977; Kirk and Tilney-Bassett, 1978). Genetic studies over the past 2 to 3 decades have shown that the inheritance of only some chloroplast proteins is controlled by the organelle, whereas the inheritance of the majority is controlled by the nucleus (e.g., Sager, 1972; Gillham, 1978). Therefore, chloroplasts are not autonomous bodies. Rather, they are semi-autonomous structures that require transcripts from their own genome and that of the nucleus (and possibly that of the mitochondrion) in order to develop and function.

Much research is now focused on the nature, the order, and the expression of genes in the chloroplast genome. Genetic analyses of the chloroplast have been well reviewed by others (e.g., Sager, 1972, 1977; Birkey, 1978; Gillham, 1978). Protein synthesis by chloroplasts *in vivo* and *in vitro* has also been well reviewed (e.g., Bryant, 1976; Ellis, 1977a,b; Kirk and Tilney-Bassett, 1978). The present chapter considers the physicochemical and coding properties of chloroplast DNA and the structure and function of the RNAs involved in the expression of the chloroplast genome.

II. The Chloroplast Genome

A. Analytic Complexity

Analytic complexity is the total amount of DNA per chloroplast. Higher plant chloroplasts are reported to contain in the range of 3–8 × 10^{-15} g DNA per plastid (Kirk and Tilney-Bassett, 1978). Algae show a wider range of DNA content per plastid, e.g., about 4.0 × 10^{-15} g per chloroplast in *Chlorella pyrenoidosa,* 10–12 × 10^{-15} g in *Euglena gracilis,* 17 × 10^{-15} g in the vegetative cell chloroplast of *Chlamydomonas reinhardtii* (Kirk and Tilney-Bassett, 1978), and 25 × 10^{-15} g DNA in the amyloplast of *Polytoma obtusum* (Siu *et al.*, 1975). However, these analytic complexities should be considered as representative only. The analytic complexity of a given chloroplast can vary. For example, the amount of DNA per chloroplast increases during development of the organelle in *Euglena* (Rawson and Boerma, 1976a; Chelm *et al.*, 1977b) and in higher plants (Kowallik and Herrmann, 1972; Herrmann and Possingham, 1980).

Each chloroplast in most algae and in higher plant cells contains DNA. However, the algae *Acetabularia* and *Polyphysa* appear to be exceptions, i.e., DNA has been detected in only 20–50% of their chloroplasts (Woodcock and Bogorad, 1970; Lüttke, 1981). The lack of DNA in the majority of chloroplasts in *Acetabularia* at a certain stage of the life cycle is confirmed by the use of the DNA-specific fluorochrome 4′, 6-diamidino-2-phenylindole (DAPI), which allows the structure of chloroplast DNA to be examined in situ with a fluorescent microscope (James and Jope, 1978; Coleman, 1979). Only one-quarter of the chloroplasts of vegetative-stage *Acetabularia* contain even one DNA particle (Coleman, 1979). However, a dramatic change occurs upon development. About 98% of the chloroplasts of *Acetabularia* at this stage of the life cycle contain up to nine DNA particles each.

B. Physicochemical Properties of Chloroplast Deoxyribonucleic Acid

1. BUOYANT DENSITY AND GUANINE-PLUS-CYTOSINE CONTENT

Large differences in the densities between nuclear DNA and chloroplast DNA of *Euglena* and *Chlamydomonas* permit the separation of these DNAs from each other by preparative density-gradient centrifugation (Sager and Ishida, 1963; Edelman *et al.*, 1964). In contrast, the buoyant densities of the nuclear and chloroplast DNAs of most higher plants are very similar (Table I). Fortunately, many higher plant chloroplasts can be obtained in a relatively intact condition. Contaminating mitochondria can be removed by washing with buffer, and contaminating nuclear DNA by treatment with DNase. Tewari (1979) has given a detailed description of a general method for isolating chloroplast DNA from many higher plants.

Since the buoyant density of chloroplast DNA from higher plants is very close to that of nuclear DNA, the purity of a chloroplast DNA preparation often cannot be determined reliably by analytical CsCl-gradient centrifugation. The most useful method for ascertaining purity is to do an equilibrium CsCl-gradient centrifugation with chloroplast DNA that has been denatured and renatured (Tewari, 1979). After denaturation, organelle DNA renatures under the appropriate conditions much more readily than nuclear DNA. Following renaturation of the chloroplast DNA, any contaminating nuclear DNA (which is only partially renatured) will band in a CsCl density gradient at a position relatively far-removed from the chloroplast DNA.

Kirk and Tilney-Bassett (1978) and Herrmann and Possingham (1980) have extensive compilations of the buoyant densities and guanine-plus-cytosine contents (G + C) of higher plant and algal chloroplast DNAs. The base composition of chloroplast DNA does not seem to bear any particular relationship to that of the nuclear DNA in the same plant. So far, all higher plant chloroplast DNAs have a base composition in the range 37.8 ± 1 moles % G + C and a buoyant density in

TABLE I
Buoyant Densities of Plant Chloroplast and Nuclear DNAs[a]

	Buoyant density ($g \cdot cm^{-3}$)	
Plant	**Nuclear DNA**	**Chloroplast DNA**
Pea	1.695	1.698
Spinach	1.694	1.697
Lettuce	1.694	1.697
Tobacco	1.695	1.697
Swiss chard	1.694	1.696
Broad and mung bean	1.695	1.697
Snapdragon	1.689	1.698
Onion	1.691	1.696
Wheat	1.702	1.697
Maize	1.702	1.697
Oat	1.701	1.697

[a]This table and Table II reprinted with permission from K. K. Tewari (1979) *in* "Nucleic Acids in Plants" (T. C. Hall and J. W. Davies, ed.), Vol. I, pp. 41–108. Copyright The Chemical Rubber Co., CRC Press, Inc.

CsCl of 1.697 ± 0.001 g·cm^{-3}. Algal plastid DNAs vary more widely in buoyant density and base composition. *Euglena* shows the lowest G + C content (24–28 moles%) and *Acetabularia* the highest buoyant density, (1.706 g·cm^{-3}).

2. KINETIC COMPLEXITY

In common with other double-stranded DNAs, chloroplast DNA denatures when heated to temperatures above its melting point. If heat-denatured chloroplast DNA is renatured, the separated strands reassociate to form double-stranded DNA again. When double-stranded DNA denatures, its buoyant density and its absorbance at 260 nm increase. When this DNA renatures, these parameters decrease again to near normal values. That chloroplast DNA follows this sequence of events was first shown by Tewari and Wildman (1966) with tobacco chloroplast DNA.

The kinetic complexity of a DNA can be determined from its kinetics of renaturation from the denatured state (Britten and Kohne, 1968; Wetmur and Davidson, 1968). Kinetic complexity is a measure of the total length of nonrepeated base sequence in the DNA, which in turn is a measure of its potential content of genetic information. The kinetic complexity of a plastid DNA was first measured by Wells and Birnstiel (1969), who used lettuce chloroplast DNA. The kinetic complexities of higher plant and algal chloroplast DNAs as determined in many studies have been tabulated by Ellis (1977a), Kirk and Tilney-Bassett (1978), Bedbrook and Kolodner (1979), and Tewari (1979). Chloroplast DNAs from most higher plants and *E. gracilis* show kinetic complexities in the range of 82–96 × 10^6 D, values similar to those derived by measuring the lengths of these chloroplast DNA molecules under electron microscopy (Table II). Somewhat larger genomes of about 100–120 × 10^6 D have been reported for *Spirodela* (van Ee *et al.*, 1980), tobacco (Seyer *et al.*, 1981), and mustard (Link *et al.*, 1981), whereas a somewhat smaller genome of about 79 × 10^6 D occurs in *Vicia faba* (Koller and Delius, 1980). The siphonous green alga, *Codium fragile,* has the smallest chloroplast genome known, i.e., 56 × 10^6 D (Hedberg *et al.*, 1981).

The chloroplast DNAs of various strains of *Chlorella* and *Chlamydomonas* show kinetic complexities ranging from 170 to 230 × 10^6 D. Lower values of 126–157 × 10^6 D, however, have been obtained from restriction endonuclease mapping (Section II, G) of the chloroplast DNAs of various *Chlamydomonas* species (Rochaix, 1978; Lemieux *et al.*, 1980). The values derived from kinetic complexity measurements on these two algae may be overestimates because, as discussed by Kirk and Tilney-Bassett (1978), renaturation rates were measured at the later stages of the reaction, where the rates may be low (Christiansen *et al.*, 1974). Under conditions that allow the early part of the reaction to be measured, the kinetic complexity of *Euglena* chloroplast DNA is 94×10^6 D (Slavik and Hershberger, 1975), a value considerably lower than that previously reported for this DNA (Stutz, 1970), but in accord with other physical measures (Table II; see also Section II,B,3).

Acetabularia chloroplast DNA appears to have an exceptionally high kinetic complexity at least 10 times higher than that of higher plants, i.e., $1.1-1.5 \times 10^9$ D for different species (Green *et al.*, 1977; Padmanabhan and Green, 1978). These results cannot be explained on the basis of contamination of this chloroplast DNA with bacterial or nuclear DNAs. The evolutionary meaning of this high kinetic complexity is not clear.

A rapidly renaturing component comprising less than 10% of the chloroplast DNA has been reported for several algae (Wells and Sager, 1971; Bayen and Rode, 1973; Siu *et al.*, 1975) and for lettuce (Wells and Birnstiel, 1969). Other investigators, using some of these as well as other chloroplast DNAs, do not find such rapidly renaturing components (Stutz, 1970; Tewari and Wildman, 1970; Bastia *et al.*, 1971; Kolodner and Tewari, 1972, 1975a,b,c; Howell and Walker, 1976; Tewari, 1979) and suggest that these rapidly renaturing sequences result from incomplete denaturation of chloroplast DNA prior to renaturation. However, cross-hybridization of different restriction endonuclease fragments (Section II,G) indicates that repeated sequences do constitute a small proportion of the chloroplast DNAs of maize (Bedbrook and Bogorad, 1976) and *C. reinhardtii* (Rochaix, 1972, 1978, 1981).

Nicked chloroplast DNAs have been denatured and examined by electron microscopy to detect any spontaneously renaturing sequences, i.e., inverted repeats. A large sequence repeated one time in an inverted orientation occurs in *Chlamydomonas* (Rochaix, 1978) and in some (Bedbrook and Bogorad, 1976; Whitfeld *et al.*, 1978a; Herrmann *et al.*, 1980; van Ee *et al.*, 1980; Palmer and Thompson, 1981; Seyer *et al.*, 1981; Link

TABLE II
Molecular Weights of Chloroplast DNA[a]

	Molecular weight ($\times 10^{-6}$)	
Organism	**Electron microscopy**	**Kinetic complexity**
Pea	95.0	86.9
Lettuce	103.2	90.9
Spinach	101.4	87.9
Maize	91.1	83.5
Euglena gracilis	97.1	94.0[b]

[a]Adapted from Bedbrook and Kolodner (1979; reproduced, with permission, from the *Annual Review of Plant Physiology*, Vol. 30. © 1979 by Annual Reviews Inc.) and Tewari (1979).
[b]From Slavik and Hershberger (1975).

et al., 1981) but not in all (Tewari, 1979; Palmer and Thompson, 1981) higher plants. In some cases at least, this inverted repeat contains the genes for chloroplast rRNA (Bedbrook *et al.*, 1977; Rochaix, 1978; Whitfeld *et al.*, 1978a; Jurgenson and Bourque, 1980; Fluhr and Edelman, 1981). *Chlamydomonas* chloroplast DNA also contains a set of duplex fragments 150 base pairs (b.p.) long accounting for about 5% of this DNA (Gelvin and Howell, 1979). The function of these latter inverted repeats is unknown.

In sum, chloroplasts do contain a small amount of repeated DNA but nowhere near the amount found in nuclear DNA (Britten and Kohne, 1968).

3. CIRCULAR MOLECULES

Manning *et al.* (1971) and Manning and Richards (1972a), who examined chloroplast DNA from lysed chloroplasts of *Euglena gracilis* under electron microscopy, were the first to show the existence of circular chloroplast DNA molecules. About 34% of the DNA existed as circular molecules with an average contour length of 44.5 μm. This length corresponds to a molecular weight close to that derived from kinetic complexity measurements (Table II). The circular molecules were stable when treated with detergent or pronase indicating the absence of protein links in the molecules.

As with *Euglena,* circular DNA molecules of 40–46 μm in contour length have been demonstrated in the chloroplasts of many higher plants as summarized by Kirk and Tilney-Bassett (1978) and Bedbrook and Kolodner (1979). Also as with *Euglena,* these lengths calculate to molecular weights corresponding to those obtained from kinetic complexity measurements (Table II). As much as 70–90% of higher plant chloroplast DNA molecules have been isolated as circles (Kolodner and Tewari, 1972, 1975a,b; Herrmann *et al.*, 1975). This suggests that large circular molecules are the major, and probably only, form of chloroplast DNA. When a high proportion of linear molecules is isolated, technical problems with the isolation procedure should be suspected. Tewari (1979) presented a method for isolating high yields of circular chloroplast DNA molecules from higher plants.

Circular chloroplast DNA molecules from different organisms need not be so uniform in contour length as the preceding discussion might imply. The chloroplast DNA of the green alga *Codium* is 27.3 μm in length (Hedberg *et al.*, 1981), whereas that of *Chlamydomonas* is 62 μm long (Behn and Herrmann, 1977). Also, small 4.2 μm circular molecules as well as linear molecules up to 200 μm in length are associated with chloroplasts from *Acetabularia* (Green and Burton, 1970; Green, 1976; Green *et al.*, 1977). These long linear molecules could be circular in situ. In any case, their existence suggests a rather large chloroplast genome for this organism.

Both circular dimers and catenated dimers have been reported in the chloroplast DNAs of higher plants (Kolodner and Tewari, 1972, 1975a,b, 1979; Herrmann *et al.*, 1975; Tewari, 1979; Herrmann and Possingham, 1980). The detection and structure of dimers is discussed in detail by Tewari (1979).

C. Molecular Size and Ploidy

The close agreement between the molecular size of circular chloroplast DNA molecules determined by electron microscopy and by renaturation kinetics (Table II) suggests that most of the circular chloroplast DNA molecules in a given organism have the same sequence. This suggestion is supported by restriction endonuclease mapping studies (Section II,G). When digested with a restriction endonuclease, the chloroplast DNAs from several higher plants (e.g., Bedbrook and Bogorad, 1976; Whitfeld *et al.*, 1978a; Tewari, 1979) and from algae (Gray and Hallick, 1977; Rochaix, 1977, 1978, 1981; Hallick, 1983) yield fragments whose molecular weights when summed are close to the size of the circular molecules determined by electron microscopy. Furthermore, in all cases a map of the endonuclease restriction fragments is circular. Also, the molecular weight of pea chloroplast DNA as determined by equilibrium sedimentation is 89×10^6 D (Kolodner *et al.*, 1976), a value similar to that determined by kinetic complexity measurements and by electron microscopy (Table II). In sum, all these data are consistent with the interpretation that, in a given organism, the bulk of all the chloroplast DNA molecules are identical. This conclusion plus the data on the amount of DNA per chloroplast (Section II,A) indicates that each chloroplast contains many copies of a circular genome. Thus, chloroplasts are highly polyploid.

A small amount of heterogeneity in the size of the chloroplast DNA molecules of a given species of higher plants (Bedbrook and Kolodner, 1979; Vacek and Bourque, 1980) and of various strains of *Euglena gracilis* (Jenni *et al.*, 1981; Wurtz and Buetow, 1981) has been reported. Whether this genome heterogeneity is due to inter- or intracellular heterogeneity or to inter- or intrachloroplast heterogeneity in the chloroplast DNA is not known.

D. Substructure

1. ABSENCE OF HISTONES AND 5-METHYLCYTOSINE

Microscopic studies showed that the chloroplasts of *Chlamydomonas* contain DNA fibrils similar to those in prokaryotes (Ris and Plaut, 1962). When released by gentle lysis of isolated chloroplasts, these fibrils are not associated with histone (Ellis, 1977a). The chloroplasts of higher plants and algae all contain such DNA fibrils (Kirk and Tilney-Bassett, 1978).

Although nuclear DNA always contains 5-methylcytosine, this modified base is not present in chloroplast DNA (Ellis, 1977a; Edelman, 1981). The absence of this base can be used as a criterion for determining the purity of a chloroplast DNA preparation (Whitfeld and Spencer, 1968; Ellis and Hartley, 1974).

2. PRESENCE OF RIBONUCLEOTIDES

Covalently closed circular chloroplast DNA is sensitive to alkali and to RNase. When sedimented in alkaline CsCl or when incubated in a mixture of pancreatic RNase and RNase T1, the covalently closed circular molecules are gradually converted to open circular molecules (Kolodner *et al.,* 1975; Tewari, 1979). Kinetic studies on the rate of nicking in the presence of NaOH indicated that higher plant chloroplast circular DNA molecules contain 12 ± 2–18 ± 2 ribonucleotides each, and electron microscopic mapping indicated 19 unique alkali-labile sites per molecule (Tewari, 1979). Therefore, one ribonucleotide is present per site. There are 9 sites on one strand of the molecule and 10 on the other strand. The significance of the individually inserted ribonucleotides in chloroplast DNA is not known. Interestingly, animal mitochondrial DNAs also contain ribonucleotides (Grossman *et al.,* 1973; Miyaki *et al.,* 1973; Wong-Staal *et al.,* 1973).

3. BASE COMPOSITION HETEROGENEITY ALONG THE CHLOROPLAST DNA MOLECULE

When *Euglena* chloroplast DNA is sheared and then centrifuged in CsCl, bands of different buoyant densities are resolved. The buoyant densities of the bands depend on the size to which the molecules are sheared. In addition to the main band of density 1.685 $g{\cdot}cm^{-3}$, a component of density 1.700 $g{\cdot}cm^{-3}$ is clearly detected when the DNA is sheared to an average of 5×10^6 D, is barely detectable in DNA sheared to 12.5×10^6 D, and is not detected in DNA sheared to $20–30 \times 10^6$ D (Stutz and Vandrey, 1971; Rawson and Haselkorn, 1973; Vandrey and Stutz, 1973; Stutz *et al.,* 1978; Gray and Hallick, 1979a). These results indicate that the complete *Euglena* chloroplast chromosome has an average G + C content of 25–26 moles% but contains within it a segment with an average G + C content of 33 moles%, and within the latter segment is a segment of 41–44 moles% G + C. That *Euglena* chloroplast DNA has a strong base compositional heterogeneity is also demonstrated by its oligophasic melting profile (Slavik and Hershberger, 1976; Crouse *et al.,* 1978) and by its degradation pattern with micrococcal nuclease as well as by the buoyant densities of its restriction enzyme fragments

(Crouse *et al.*, 1978). All these measures indicate multiple regions of different G + C content ranging from 12 to 60 moles%.

Base composition heterogeneity along the chloroplast chromosome has also been demonstrated by melting profiles of chloroplast DNAs from *Chlamydomonas* (Bastia *et al.*, 1971; Wells and Sager, 1971), *Chlorella* (Bayen and Rode, 1973; Dalmon and Bayen, 1975), *Spirodela* (van Ee *et al.*, 1980), and spinach (Crouse *et al.*, 1978; Schmitt *et al.*, 1981) by electron microscopic observation of pea-chloroplast band separation following partial denaturation with formamide (Kolodner and Tewari, 1975a) and by restriction endonuclease digestion of chloroplast DNA from many sources (Vedel *et al.*, 1976; Kirk and Tilney-Bassett, 1978; Bedbrook and Kolodner, 1979).

E. Replication of Chloroplast Deoxyribonucleic Acid

Studies with cultures synchronized for cell division showed that chloroplast DNA replicates at a time in the cell cycle distinct from that of nuclear DNA replication in *Euglena* (Cook, 1966), *Chlamydomonas* (Chiang and Sueoka, 1967; Keller and Ho, 1981), and *Chlorella* (Stange *et al.*, 1962; Senger and Bishop, 1966; Wanka *et al.*, 1970; Dalmon *et al.*, 1975) but at the same time as nuclear DNA in the alga *Dunaliella* (Marano, 1979). In a review of such studies, Buetow *et al.* (1980) postulated the existence of a chloroplast cycle (based on the time of synthesis of chloroplast DNA versus nuclear DNA) within the cell cycle of photosynthetic cells. Replication of chloroplast DNA in *Euglena* (Manning and Richards, 1972b), in spinach (Rose *et al.*, 1974), and during vegetative multiplication of *Chlamydomonas* (Chiang and Sueoka, 1967) is semiconservative. In dividing chloroplasts of *Ochromonas* (Gibbs and Poole, 1973) and of expanding spinach leaf disks (Rose *et al.*, 1974), there appeared to be an equal and ordered, rather than a random, distribution of chloroplast DNA to daughter chloroplasts.

1. TURNOVER AND REPLICATIVE INTERMEDIATES

In synchronized cultures of *Chlorella*, chloroplast DNA shows a rapid gain and loss of labeled DNA phosphorus throughout the cell cycle with no net DNA synthesis other than during a discrete period (Iwamura, 1955, 1960, 1962, 1966; Dalmon *et al.*, 1975). A continuous synthesis or "turnover" of chloroplast DNA also occurs in *Euglena* (Cook, 1966; Manning and Richards, 1972b; Richards and Manning, 1975; Lyman and Srinivas, 1978; Walfield and Hershberger, 1978), *Chlamydomonas* (Grant *et al.*, 1978), and higher plants (Sampson *et al.*, 1963; Hotta *et al.*, 1965; Šebesta *et al.*, 1965). In exponentially growing *Euglena*, the ratio of chloroplast DNA to nuclear DNA is always constant. Each time the nuclear DNA replicates once, however, the chloroplast DNA replicates 1.5 times

(Manning and Richards, 1972b). The nature of the over-synthesis of chloroplast DNA is not defined, but it may represent a high degree of repair synthesis or synthesis associated with the "turnover" of a significant fraction of the organelle DNA during each cell cycle. As pointed out by Bedbrook and Kolodner (1979), the over-synthesis could result from a "rolling circle" type of replication mechanism as discussed next.

Replicative intermediates of chloroplast DNA have been studied by electron microscopy. Replication of higher plant chloroplast DNA initiates (Kolodner and Tewari, 1975a,c; Tewari, 1979) by the formation of two d-loops located on opposite strands of the DNA. The d-loops elongate toward each other and fuse to form a "Cairns" replicative intermediate, which continues the DNA synthesis bidirectionally until one round of replication is completed. *Euglena gracilis* chloroplast DNA also replicates by Cairns intermediates (Richards and Manning, 1975). In higher plants, chloroplast DNA shows both Cairns and rolling-circle replicative intermediates, the latter apparently resulting from a continuation of a Cairns round of replication (Kolodner and Tewari, 1975c; Tewari, 1979). The significance of the rolling-circle replication is not known. It could relate, however, to the oversynthesis of chloroplast DNA (see preceding paragraph) because a rolling-circle intermediate allows for the synthesis of many copies of a DNA molecule, even though replication is initiated only once (Bedbrook and Kolodner, 1979).

2. DEOXYRIBONUCLEIC ACID POLYMERASE

A DNA polymerase has been demonstrated in isolated chloroplasts from spinach (Spencer and Whitfeld, 1967a,b, 1969), tobacco (Tewari and Wildman, 1967), *Euglena* (Scott *et al.*, 1968), and *Chlamydomonas* (Keller and Ho, 1981). The *Euglena* and *Chlamydomonas* enzymes have been partially purified (Keller *et al.*, 1973; Keller and Ho, 1981).

F. Genetic Information in Chloroplast Deoxyribonucleic Acid

If a particular RNA hybridizes to chloroplast DNA, it is reasonable to assume that the RNA in question is coded by chloroplast DNA. The first study of this kind on chloroplasts was that of Scott and Smillie (1967), who showed that *Euglena* chloroplast rRNA hybridizes to chloroplast DNA. As reviewed by Kirk and Tilney-Bassett (1978), a number of reports were published between 1967 and 1976, which showed that ribosomal cistrons constituted anywhere from 2 to 12% of the *Euglena* chloroplast genome. This variation resulted from the fact that a particular chromosome fragment of higher density than the average for the whole DNA was lost during preparation of the DNA in some cases. This fragment contains the cistrons for the rRNAs (Rawson and Haselkorn, 1973; Vandrey and Stutz, 1973; Rawson and Boerma, 1977). Also, it has been recognized only recently that different strains of *Euglena gracilis* contain on their chloroplast DNAs different numbers of complete rRNA cistrons (one or three), each of which contains the genes for the 5 S, 23 S and 16 S rRNAs (Wurtz and Buetow, 1981; Hallick, 1983; see also Section II,G,1). In addition, *Euglena* chloroplast DNA carries an extra gene for 16 S rRNA (Jenni and Stutz, 1979). In contrast to *Euglena,* the alga *Chlamydomonas reinhardtii* has two complete rRNA cistrons on its chloroplast DNA (Rochaix and Malnoe, 1978a).

The first measurements with higher plants indicated only a low degree of hybridization of chloroplast rRNA with chloroplast DNA (Tewari and Wildman, 1968, 1970; Ingle *et al.*, 1970). Later measurements, however, showed that there are two complete rRNA cistrons per chloroplast chromosome (Fig. 1) of many higher plants (Thomas and Tewari, 1974; Bedbrook and Bogorad, 1976; Herrmann *et al.*, 1976; Tewari, 1979; Herrmann and Possingham, 1980). The presence of two complete rRNA cistrons is not invariant in higher plants because the chloroplast chromosomes of *Vicia faba* and pea contain only one complete rRNA cistron (Koller and Delius, 1980; Palmer and Thompson, 1981).

Cross-hybridization and competition studies with pea, spinach, lettuce, oat, and maize chloroplast DNAs and rRNAs indicate that the structure of the rRNA cistrons is similar from one species to another (Thomas and Tewari, 1974). Cytoplasmic rRNA does not hybridize to any appreciable extent to chloroplast DNA (Tewari and Wildman, 1970).

Saturation hybridization measurements indicate about 40 tRNA genes on pea chloroplast DNA (Meeker and Tewari, 1980). Other higher plant and *Euglena* chloroplast DNAs hybridize chloroplast tRNAs (Fig. 1) to an extent allowing for the coding of about 30 tRNAs (Burkard *et al.*, 1970; Gruol and Haselkorn, 1976; Haff and Bogorad, 1976a; McCrea and Hershberger, 1976; Schwartzbach *et al.*, 1976; Driesel *et al.*, 1979; Weil, 1979; Herrmann and Possingham, 1980). This latter number of tRNAs is close to the figure of 32 needed to read all codons according to the wobble hypothesis (Crick, 1966).

Polyadenylated RNA hybridizes to maize chloroplast DNA (Haff and Bogorad, 1976b). When isolated from fully-green *Euglena,* polyadenylated RNA hybridizes 2.8% of the chloroplast DNA (Milner *et al.*, 1979). Up to 20% of the latter genome is reported to code for polyadenylated RNA in developing *Euglena* chloroplasts (Nigon *et al.*, 1978; Verdier, 1979a,b; Nigon and Verdier, 1983). Over 60%, however, of the chloroplast polyadenylated RNA fraction from fully green *Euglena* contains sequences whose hybridization to chloroplast DNA is competed by chloroplast rRNA (Milner *et al.*, 1979). Therefore, the value of up to 20% of the chloroplast genome coding for polyadenylated RNA in developing chloroplasts may be an overestimate.

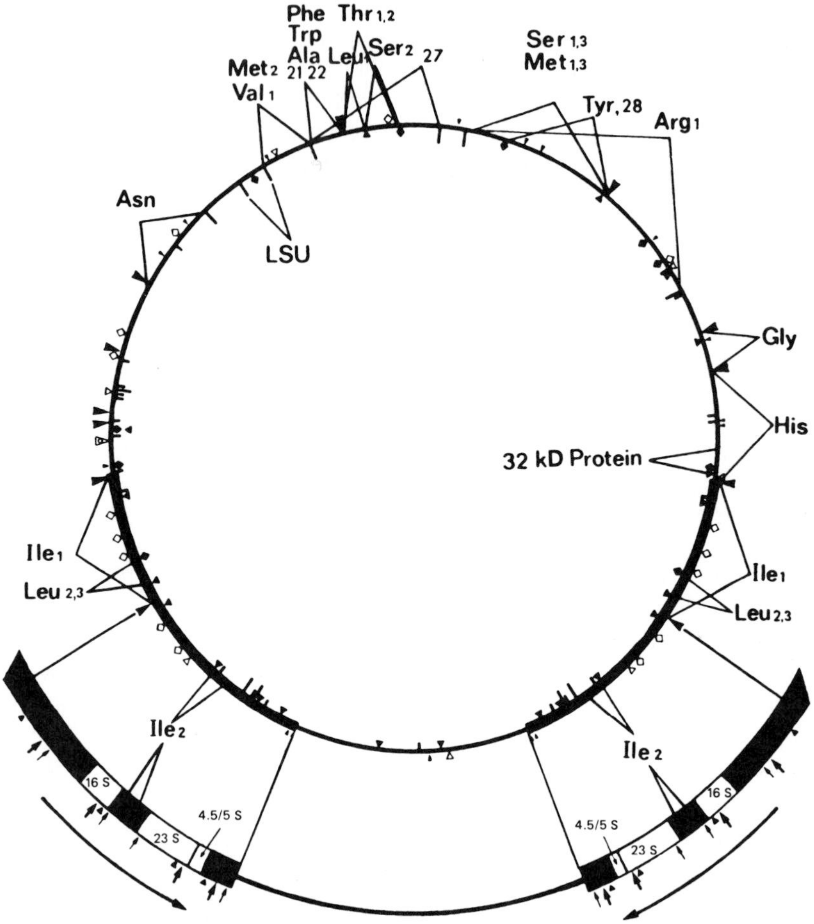

FIG. 1. The circular physical map of the *Spinacea oleracea* chloroplast chromosome. The chromosome accounts for about 95 × 10^6 D or 140 kb.p. of DNA. The map order of the various DNA fragments was determined by digestion with several restriction endonucleases. The relative cleavage sites of the endonucleases are drawn to scale and indicated as follows: *Sal*I = ▼, *Pst*I = ♦, *Kpn*I = ▌, *Xma*I = ▼, *Xho*I = ▌, *Bgl*I = ▽, *Pvu*I = ◇, *Sac*I = ▽. Restriction endonucleases are named according to Roberts (1977). The thick arc regions represent the inverted repeats of this chromosome. The location of the two sets of genes for rRNAs within the inverted repeats is shown in the expanded drawing with endonuclease cleavage sites indicated as follows: *Eco*RI = ⬇, *Bam*H1 = ↓, *Sma*I = ▼. The polarity of transcription in the ribosomal gene region proceeds from the gene for the 16 S rRNA towards the gene for the 23 S rRNA. The location of each tRNA or polypeptide gene is indicated by the smallest DNA segment to which each tRNA or mRNA (Bohnert *et al.*, 1980) is hybridized. LSU, large subunit of ribulosebisphosphate carboxylase; 32 kD protein = 32 kD protein of PSII. (From Herrmann and Possingham, 1980.)

At least some of the chloroplast-coded polyadenylated RNAs have been considered to be mRNAs (Haff and Bogorad, 1976b; Verdier, 1979a,b), as are the polyadenylated RNAs coded by the nucleus. In *Euglena* at least, the nature of the chloroplast-coded polyadenylated RNAs remains to be identified, because mature chloroplast mRNAs found in chloroplast polyribosomes are not polyadenylated (Milner *et al.*, 1979). In any case, the chloroplast genome of higher plants and algae is now known to code for the large subunit of ribulosebisphosphate carboxylase and for a membrane protein (Sections II,G,3 and 4).

Cistrons for rRNAs, tRNAs, polyadenylated RNAs, and the two known proteins account for about 5–15% of the potential information in a single strand of the chloroplast DNA of algae and higher plants. This leaves considerable room for the coding of more proteins. For example, about 34–53% of the chloroplast DNA of *Euglena* growing in the dark is transcribed (Chelm and Hallick, 1976; Rawson and Boerma, 1976b; Chelm *et al.*, 1978; Hallick, 1983), and the quantity and diversity of the transcripts increases when the cells are placed in the light (Brown and Haselkorn, 1971; Chelm and Hallick, 1976; Chelm *et al.*, 1977a,b, 1978; Heizmann *et al.*, 1978; Nigon *et al.*, 1978; Rawson and Boerma, 1979). Similar results have been obtained with *Chlamydomonas* (Howell and Walker, 1977; Matsuda and Surzycki, 1980). Also, the existence of transcripts from both strands of chloroplast DNA must be considered. Symmetrical transcription can be suspected in *Chlamydomonas,* because the overall complexity of the RNA, which hybridizes to chloroplast DNA is greater than the overall complexity of the DNA (Howell and Walker, 1977). In *Euglena,* three restriction endonuclease fragments of its chloroplast DNA appear to be transcribed symmetrically (Rawson *et al.*, 1981). In tobacco some tRNAs are coded by the chloroplast DNA strand opposite from that which codes the rRNAs (Kato *et al.*, 1981). Also, tobacco chloroplasts contain some double-stranded RNA (Ikegami and Fraenkel-Conrat, 1979), which is the expected product of symmetrical transcription. It should be noted, though, that this double-stranded RNA may be the product of an RNA-dependent RNA polymerase (Ikegami and Fraenkel-Conrat, 1979). In any case, rRNA genes on both strands of the chloroplast DNA in many higher plants and in *Chlamydomonas* (Section II,G,1) and other inverted genes in maize (Section II,G,5) clearly show that the chloroplast can use information on both strands of its DNA. It is clear that many of the functions of the chloroplast genome remain to be discovered.

G. Physical Mapping with Restriction Endonucleases

Recognition sequences for restriction endonucleases are very useful markers for the construction of physical maps of genomes (Roberts,

1976). Bedbrook and Kolodner (1979) reviewed the requirements for and the methods of physical mapping of chloroplast genomes. Physical maps based on the use of recognition sites for endonucleases as markers (Fig. 1) have been determined for the chloroplast DNA of *Euglena* (Gray and Hallick, 1977; Hallick, 1983), *Chlamydomonas* (Rochaix, 1977, 1978, 1981), maize (Bedbrook and Bogorad, 1976), spinach (Herrmann *et al.*, 1976, 1980, Crouse *et al.*, 1978; Steinmetz *et al.*, 1978; Whitfeld *et al.*, 1978a), pea (Bedbrook and Kolodner, 1979; Palmer and Thompson, 1981), tobacco (Jurgensen and Bourque, 1980; Takaiwa *et al.*, 1980; Fluhr and Edelman, 1981; Seyer *et al.*, 1981), mung bean (Palmer and Thompson, 1981), mustard (Link *et al.*, 1981), *Spirodela* (van Ee *et al.*, 1980), and *Oenothera* (Gordon *et al.*, 1981).

1. GENES FOR CHLOROPLAST RIBOSOMAL RIBONUCLEIC ACIDS

Physical mapping of chloroplast DNA molecules with restriction endonucleases shows two genes each for the 23 S and 16 S rRNAs of maize (Bedbrook and Bogorad, 1976; Bedbrook *et al.*, 1977), spinach (Whitfeld *et al.*, 1976, 1978a), tobacco (Jurgensen and Bourque, 1980; Fluhr and Edelman, 1981; Seyer *et al.*, 1981), and *Chlamydomonas* (Rochaix and Malnoe, 1978a,b) in confirmation of the results obtained by quantitative DNA–rRNA hybridization studies (Section II,F) and by electron microscopy (Kolodner and Tewari, 1979). Both two genes (Tewari, 1979) and one gene (Palmer and Thompson, 1981) have been reported for the pea. *Euglena gracilis* presents an interesting case in that the fine structure of the rRNA genes on its chloroplast genome is quite variable (Helling *et al.*, 1979; Wurtz and Buetow, 1981). *Euglena gracilis* strains Z and *bacillaris* both have three genes (Gray and Hallick, 1978; Jenni and Stutz, 1978; Rawson *et al.*, 1978; Helling *et al.*, 1979; Wurtz and Buetow, 1981), but differ in the size of the spacer region between the 16 S and 23 S rRNA genes. Also, another *Euglena gracilis* strain (designated Z-S) has only one rRNA cistron (Wurtz and Buetow, 1981).

rRNA genes physically map in an inverted repeat sequence (Section II,B,2; see also Fig. 1) found in the chloroplast DNA of maize (Bedbrook *et al.*, 1977), spinach (Whitfeld *et al.*, 1976; Crouse *et al.*, 1978), and tobacco (Jurgensen and Bourque, 1980; Fluhr and Edelman, 1981; Seyer *et al.*, 1981), mung bean (Palmer and Thompson, 1981), and *Chlamydomonas* (Rochaix and Malnoe, 1978a,b). As a consequence, each strand of the chloroplast DNA molecule contains a sequence complementary to rRNA (Bedbrook *et al.*, 1977). This location of rRNA genes has been confirmed in maize and spinach by the detection of displacement loops by electron microscopy following the hybridization of rRNA to its genes (Bedbrook *et al.*, 1977; Bedbrook and Kolodner, 1979). *Euglena*, pea, and broad bean chloroplast DNAs lack such inverted repeat sequences (Gray and Hallick, 1979a,b; Tewari, 1979; Koller and Delius, 1980; Palmer and Thompson, 1981). Therefore, the inverted arrangement for rRNA genes is not an absolute requirement for chloroplast DNA. The three chloroplast rRNA genes of *Euglena* strains Z and *bacillaris* are also arranged in head to tail repeats (Kopecka *et al.*, 1977; Gray and Hallick, 1978, 1979a,b; Jenni and Stutz, 1978; Knopf and Stutz, 1978; Rawson *et al.*, 1978; Helling *et al.*, 1979; Hallick, 1982a). *Euglena* strain Z-S (Wurtz and Buetow, 1981) and broad bean (Koller and Delius, 1980) contain only one rRNA gene on their chloroplast DNAs.

Physical maps (Fig. 1) show that the 5′ to 3′ order of the chloroplast rRNA genes is 16 S, 23 S, and 5 S rRNA in *Euglena* (Hallick, 1983), *Chlamydomonas* (Rochaix, 1981), maize (Bedbrook *et al.*, 1977), spinach (Whitfeld *et al.*, 1978a), lily (Sugiura and Hotta, 1980), tobacco (Fluhr and Edelman, 1981), and broad bean (Delius and Koller, 1980). The 16 S and 23 S genes are separated by a spacer region that is 2100 b.p. long in maize (Bedbrook *et al.*, 1977), 1680 b.p. in *Chlamydomonas reinhardtii* (Rochaix and Malnoe, 1978a), 1900–2200 b.p. in spinach (Whitfeld *et al.*, 1978a; Bohnert *et al.*, 1979) and 250–300 b.p. in the *Euglena* strain Z with three rRNA genes (Hallick *et al.*, 1978; Orozco *et al.*, 1980a). In *Euglena gracilis bacillaris,* the three sets of rRNA genes differ in length and sequence of their own spacers and with the spacers of strain Z (Helling *et al.*, 1979; Wurtz and Buetow, 1981).

In *Chlamydomonas reinhardtii,* the 23 S rRNA gene contains a 940 b.p. intron located about 300 b.p. from its 5′ end (Rochaix and Malnoe, 1978a). The rRNA gene regions flanking this intron have been sequenced (Allet and Rochaix, 1979; Rochaix, 1981). Insertions have not been detected in the rRNA genes of maize (Bedbrook *et al.*, 1977), peas (Tewari, 1979), spinach (Bedbrook and Kolodner, 1979), or *Euglena* (Gray and Hallick, 1979a), but there is some evidence from restriction enzyme mapping for an intron in the 23 S rRNA gene of tobacco chloroplasts (Jurgensen and Bourque, 1980).

The large subunit of higher plant chloroplast ribosomes contains a 4.5 S RNA (Section III,B,1) which is coded by the chloroplast genome (Hartley, 1979). The gene for this RNA (Fig. 1) maps between the 23 S and 5 S rRNA genes in spinach (Whitfeld *et al.*, 1978b), maize (Dyer and Bedbrook, 1980), lily (Sugiura and Hotta, 1980), broad bean (Sugiura, 1980), and tobacco (Fluhr and Edelman, 1981). A 4.5 S rRNA gene has not been detected in the chloroplast rRNA operons of *Chlamydomonas* (Rochaix, 1981) or *Euglena* (Hallick, 1982). *Chlamydomonas,* however, contains genes for a 3 S and a 7 S rRNA in the spacer between the 16 S and 23 S rRNA genes (Rochaix and Malnoe, 1978a,b). So far, 3 S and 7 S rRNAs have not been detected in other chloroplasts.

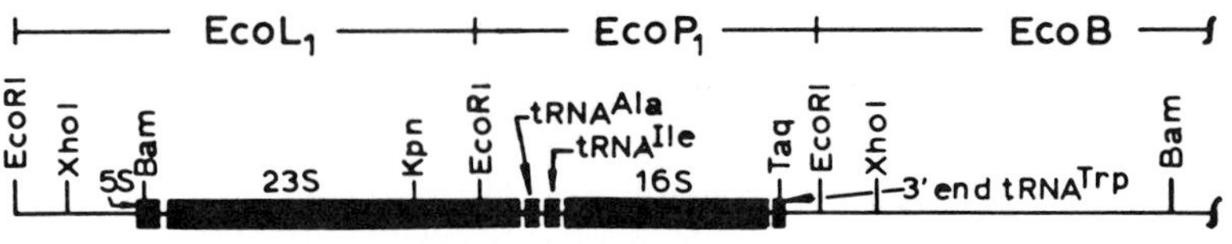

FIG. 2. The ribosomal RNA coding region of the chloroplast DNA of *Euglena gracilis* strain Z. One of the three tandemly repeated ribosomal RNA operons of this genome is shown. The 3′ end of the operon is to the left and the 5′ end to the right. Restriction endonucleases (Bam, EcoR1, Kpn, Taq, Xho1) and their cleavage sites are indicated. Genes for $tRNA^{Ala}$ and $tRNA^{Ile}$ map in the spacer region between the 16 S and 23 S rRNA genes and a partial (3′ end) gene for $tRNA^{Trp}$ maps proximal to the 5′ end of the 16 S rRNA gene. The 16 S rRNA gene is 1.5-kb.p. long, the $tRNA^{Ile}$ gene is 73 b.p., the $tRNA^{Ala}$ gene is 74 b.p., the 23 S rRNA gene is 2.9 kb.p., and the 5S rRNA gene is120 b.p. $EcoL_1$, $EcoP_1$, and EcoB are specific DNA fragments generated by EcoR1 restriction endonuclease. (From Hallick, 1983.)

2. GENES FOR CHLOROPLAST TRANSFER RIBONUCLEIC ACIDS

Most chloroplast tRNA genes map (Fig. 1) in the chloroplast DNA single copy region in spinach (Steinmetz *et al.*, 1978, 1980; Bohnert *et al.*, 1979; Driesel *et al.*, 1979), *Chlamydomonas* (Malnoe and Rochaix, 1978), and *Euglena* (Hallick *et al.*, 1978). In *Euglena,* the tRNA genes are distributed over the map including some of the adenine- and thymine-rich regions (Hallick, 1983). In spinach, different chloroplast genes code for the different isoaccepting species (Driesel *et al.*, 1979). In maize, the chloroplast gene for $tRNA^{His}$ overlaps the gene for an unidentified 1600 nucleotide-long RNA (Schwarz *et al.*, 1981).

Chloroplast tRNA genes also map in the DNA spacer region between the 16 S and 23 S rRNA genes of spinach (Steinmetz *et al.*, 1978, 1980; Bohnert *et al.*, 1979; Driesel *et al.*, 1979), maize (Koch *et al.*, 1981), *Chlamydomonas* (Malnoe and Rochaix, 1978), and *Euglena* (Hallick *et al.*, 1978; Keller *et al.*, 1980; Orozco *et al.*, 1980a,b). In maize, two tRNA genes, coding for $tRNA^{Ile}$ and $tRNA^{Ala}$, map in the 16 S–13 S spacer (Koch *et al.*, 1981). In spinach, a $tRNA^{Ile}$ gene (Fig. 1) maps in the spacer (Bohnert *et al.*, 1979; Steinmetz *et al.*, 1980). The chloroplast rRNA operon of spinach (Hartley and Ellis, 1973; Bohnert *et al.*, 1976; Hartley and Head, 1979) is transcribed as a single precursor RNA molecule (see also Section III,B,3). It is not known whether the chloroplast $tRNA^{Ile}$ gene shares a common promoter with the rRNA genes and is co-transcribed as part of a single precursor.

In *Euglena* strain Z, genes for $tRNA^{Ile}$ and $tRNA^{Ala}$ are found in the spacer region of all three rRNA operons (Fig. 2) and possibly also in the 3′ region, following the extra fourth 16 S rRNA gene (Keller *et al.*, 1980; Orozco *et al.*, 1980b). The 5′ region preceding the *Euglena* chloroplast 16 S rRNA gene contains a partial gene for $tRNA^{Trp}$ (Keller *et al.*, 1980;

Hallick, 1983), as well as a "pseudo" $tRNA^{Ile}$ gene which shows a 64% homology with the normal chloroplast $tRNA^{Ile}$ gene (Orozco *et al.*, 1980b). The partial gene (3′ end) for $tRNA^{Trp}$ is proximal to the 16 S rRNA gene (Fig. 2). The tRNA that hybridizes to this region is aminoacylated by *E. coli* aminoacyl tRNA synthetase (Keller *et al.*, 1980). Therefore, the 3′ end of the $tRNA^{Trp}$ gene may be part of a functional, split tRNA gene (Hallick, 1983). The $tRNA^{Ile}$ and $tRNA^{Ala}$ genes located in the 16 S–23 S rRNA spacer of maize chloroplasts are split genes (Koch *et al.*, 1981).

3. THE GENE FOR THE LARGE SUBUNIT OF RIBULOSEBISPHOSPHATE CARBOXYLASE

As reviewed by Gillham (1978), a large number of genetic and biochemical studies indicated that the large subunit (LS) of ribulosebisphosphate carboxylase is maternally inherited and coded for by chloroplast DNA. Evidence that the structural gene for the LS is located on chloroplast DNA has been obtained from several laboratories. The approaches used include (1) the *in vitro* transcription and translation of spinach total chloroplast DNA (Bottomley and Whitfeld, 1978, 1979); (2) the *in vitro* transcription and translation of a cloned chloroplast DNA fragment from maize (Coen *et al.*, 1977), *Chlamydomonas* (Rochaix and Malnoe, 1978a,b; Malnoe *et al.*, 1979), mustard (Link, 1981), tobacco (Seyer *et al;* 1981), and spinach (Whitfeld and Bottomley, 1980; Erion *et al.*, 1981); and (3) the hybridization of purified LS mRNA to a specific restriction endonuclease fragment from the chloroplast DNA of mustard (Link, 1981), *Chlamydomonas* (Gelvin *et al.*, 1977), and *Euglena* (Freyssinet *et al.*, 1981). The study of Gelvin *et al.* (1977), however, is open to question in that their claim that the LS gene of *Chlamydomonas* is located on the same DNA fragment as the rRNA genes has not been verified (Rochaix and Malnoe, 1978a). In the case of maize, the single structural gene for the LS has been localized on a 1600 b.p. sequence (Link and Bogorad, 1980), which in turn is localized on a 4350 b.p. long sequence (Bedbrook *et al.*, 1979). The latter sequence is located 30,000 b.p. from one of the rRNA repeats and about 71,000 b.p. from the other. The direction of transcription of the isolated maize LS gene has been established (Bogorad *et al.*, 1980; Link and Bogorad, 1980) and the entire gene sequenced (McIntosh *et al.*, 1980). The gene is not split, and it uses almost all possible codons.

4. CHLOROPLAST GENE FOR A 32,000-DALTON PHOTOSYSTEM II POLYPEPTIDE

A 34,500-D polypeptide is processed to 32,000 D in maize chloroplasts *in vivo*, but not *in vitro* (Grebanier *et al.*, 1978). The 32,000 D polypeptide is absent from the membranes of etioplasts, but it accumulates in chloroplast membranes during the light-dependent development of chloroplasts from etioplasts (Grebanier *et al.*, 1979). A similar, rapidly turning-over 32,000-D polypeptide processed from a larger precursor has also been identified in chloroplasts of *Spirodela* (Edelman and Reisfeld, 1980), mustard (Link, 1980), and spinach (Driesel *et al.*, 1980). The 32,000-D polypeptide was identified (Mattoo *et al.*, 1981; Mullet and Arntzen, 1981) as the previously hypothesized "proteinaceous shield" (Renger, 1976), which regulates electron flow through photosystem II (PSII) and mediates herbicide sensitivity.

The gene for the 32,000-D polypeptide is located on a 2200 b.p. portion of *Bam*Hl fragment 8 of maize chloroplast DNA (Bogorad *et al.*, 1980) and has been located on the physical map of the spinach chloroplast genome (Driesel *et al.*, 1980). Also, *Chlamydomonas* chloroplast DNA codes for a 35,000-D polypeptide (Malnoe *et al.*, 1979).

5. CHLOROPLAST GENE FOR A 2200-NUCLEOTIDE-LONG TRANSCRIPT

A 1400 b.p.-long, uninterrupted region that is co-linear with a chloroplast RNA occurs on *Bam*HI fragment 9 of maize chloroplast DNA (Bogorad *et al.*, 1980; Link and Bogorad, 1980). The transcript from this region is part of a 2200-nucleotide-long RNA whose gene is cut by *Bam*HI. Fragment 9 also contains the complete gene for the LS of ribulosebisphosphate carboxylase (Section II,G,3). The LS gene and the gene for the 2200 nucleotide transcript are close to one another, separated by an untranscribed intercistronic "gap" only about 330 b.p. long. The two genes are inverted on the chromosome, i.e., their 3′ termini are at opposite ends of the untranscribed gap, and they map on opposite strands. The function of the 2200 nucleotide transcript is not known.

H. Regulation of Chloroplast Genes

Though much progress has been made towards an understanding of gene expression in prokaryotes (Lewin, 1974), our understanding of the regulation of nuclear gene expression in eukaryotes lags far behind (Lewin, 1980). In the case of eukaryotic organelle genes, the regulation of their expression is just beginning to be studied.

A possible example of regulated synthesis of ribulosebisphosphate carboxylase occurs in maize. Bundle sheath cells contain the enzyme, whereas mesophyll cells do not (Hattersley *et al.*, 1977; see Bassham and Buchanan, chapter 6, this volume). The chloroplast genomes of the two cell types are identical by several physical criteria (Walbot, 1977), and both carry genes for the LS (Link *et al.*, 1978a,b). However, only bundle sheath cells contain LS mRNA. Though a rapid degradation of LS mRNA in mesophyll cells cannot be ruled out, these results suggest that one point of regulation of LS synthesis in the two cell types involves transcription.

The activity of ribulosebisphosphate carboxylase increases when dark-grown *Euglena* are placed in the light (Schiff, 1973; Nigon and Heizmann, 1978). With an antibody to the enzyme, Freyssinet *et al.* (1981) quantitated the amount of enzyme present in *Euglena* in the dark and in the light and determined the time course of an increase in amount of this enzyme during light-dependent chloroplast development. The increased activity of the enzyme found in the light was shown to be due to an increased synthesis of enzyme and not to an activation of enzyme already present. Since the enzyme is present in low amounts in dark-grown cells, light is not an absolute requirement for its synthesis, but light is required for an increase in its synthesis. Results suggest that light ultimately may control the transcription of the chloroplast gene for this enzyme.

In maize, a study on changes in the pools of chloroplast mRNA during light-dependent development of the organelles shows a 10-fold increase in the chloroplast-coded mRNA (Bedbrook *et al.*, 1978) for a 32,000 D PSII polypeptide (section II,G,4). There is a similar light-dependent regulation of the mRNA coding for the same polypeptide in *Spirodela* chloroplasts (Edelman and Reisfeld, 1978). Also, the pool of an mRNA for a possibly analogous protein in *Lemna* increases in the light (Tobin, 1978). As a whole, these results suggest that light ultimately controls the cellular level of the mRNA for the 32,000 D PSII polypeptide of chloroplasts.

An experimental result of potentially great significance for the study of the regulation of chloroplast genes was the isolation of a transcriptionally active chromosome, first from the chloroplasts of *Euglena* (Hallick *et al.*, 1976; Schiemann *et al.*, 1977) and subsequently from spinach chloroplasts (Briat *et al.*, 1979). These isolated chromosomes make possible a study of chloroplast transcription in a highly purified all-chloroplast system. These *in vitro* systems selectively initiate transcription at defined loci and elongate the products (Briat *et al.*, 1979; Rushlow *et al.*, 1980). Equally important has been another finding that maize chloroplast DNA-dependent RNA polymerase *in vitro* preferentially transcribed maize chloroplast DNA sequences incorporated in cloned bacterial plasmids (Jolly and Bogorad, 1980; Schwarz *et al.*, 1981). Preferential transcription is dependent on the presence of a 27,500 D polypeptide, which was purified from maize chloroplasts and also on the chloroplast DNA being in the supercoiled form. This system also has a high potential for further studies on chloroplast transcription.

III. Chloroplast Ribonucleic Acid

A. Ribonucleic Acid Polymerase

The finding that isolated chloroplasts from higher plants and *Euglena* incorporated ^{14}C-labeled nucleotides into RNA provided the first evidence for a chloroplast-associated RNA polymerase (Eisenstadt and Brawerman, 1964; Kirk, 1964a,b; Semal *et al.*, 1964; Shah and Lyman, 1966). The enzyme activity is tightly bound with DNA to chloroplast membranes (Spencer and Whitfeld, 1967b; Surzycki, 1969; Tewari and Wildman, 1969; Bottomley, 1970; Polya and Jagendorf, 1971; Wollgiehn and Munsche, 1972; Bogorad *et al.*, 1973), and as reviewed by Becker (1979) and Wollgiehn and Parthier (1980), an important factor in its solubilization is a low magnesium concentration. The first chloroplast RNA polymerases solubilized were those from maize (Bottomley *et al.*, 1971a) and wheat (Polya and Jagendorf, 1971). The chloroplast RNA polymerases from *Euglena* (Brandt and Wiessner, 1977) and from *Phaseolus* (Ness and Woolhouse, 1980) have also been solubilized. Chloroplast RNA polymerase complexed to chloroplast DNA has been isolated from *Euglena* and spinach (Section II,H).

The maize enzyme contains at least two polypeptides and possibly several more (Smith and Bogorad, 1974; Kidd and Bogorad, 1980). In pea chloroplasts, there appear to be two enzymes, one bound to membranes, the other in the stroma. These two differ in pH optima, template response, and G + C content of the reaction product (Joussaume, 1973). The *Euglena* chloroplast enzyme shows optimal activity at 28°–29°C and is inactive at 34°–35°C, whereas the nuclear RNA polymerases are optimally active at 32°–33°C and still active at 34°–35°C (Brandt and Wiessner, 1977). *Euglena* cultures are permanently bleached after about five population doublings at 34°–35°C (Mego and Buetow, 1967; Uzzo and Lyman, 1972). The inactivity of the chloroplast enzyme at this temperature apparently leads to bleaching (Brandt and Wiessner, 1977).

Unlike bacterial RNA polymerases, the isolated chloroplast enzyme of most higher plants and *Euglena* and the isolated chloroplast DNA–RNA polymerase complex from *Euglena* are not inhibited by rifamycins (Bottomley *et al.*, 1971a,b; Polya and Jagendorf, 1971; Hallick *et al.*, 1976; Wollgiehn and Parthier, 1979; 1980; Kidd and Bogorad, 1980). Chloroplast RNA synthesis *in vivo* in tobacco (Munsche and Wollgiehn, 1973) and *Chlorella* (Galling, 1971) is sensitive to rifamycin, but such *in vivo* sensitivities appear to be lost during enzyme purification (Wollgiehn and Parthier, 1980). The RNA polymerase of *Chlamydomonas* chloroplasts has been reported as both sensitive (Surzycki, 1969) and insensitive (Sirevag and Levine, 1972) to rifamycin SV (rifampicin).

When dark-grown seedings are placed in the light, a rapid increase in the activity of the chloroplast RNA polymerase occurs (Harel and Bogorad, 1973; Apel and Bogorad, 1976). The increased activity, however, is not paralleled by an increase in the amount of enzyme (Apel and Bogorad, 1976). Therefore, light apparently modulates the activity of the enzyme but not its synthesis. Surzycki and Shellenbarger (1976) have isolated a protein with properties like the bacterial sigma-factor, i.e., it stimulates the initiation of transcription by the chloroplast RNA polymerase of *Chlamydomonas.*

B. Ribosomal Ribonucleic Acid

1. RIBOSOMES

Lyttleton (1962) was the first to show that chloroplasts contain ribosomes that are distinct from those found in the cytoplasm of plant cells. Shortly thereafter, Clark (1964) and Clark *et al.* (1964) showed that the sedimentation value for chloroplast ribosomes of *Brassica* was 68 S and that for the cytoplasmic ribosomes was 86 S. That chloroplast ribosomes show lower sedimentation values (~70 S) than do cytoplasmic ribosomes (~80 S) has now been demonstrated in many plants and algae (Gillham, 1978).

Stutz and Noll (1967) first showed that chloroplast ribosomes contain RNA distinguishable from cytoplasmic ribosomal RNA. The major RNA species from chloroplast ribosomes have sedimentation values of 23 S and 16 S (Loening and Ingle, 1967), which correspond to MW of 1.1×10^6 and 0.56×10^6 D, respectively (Leaver and Ingle, 1971). The 16 S rRNA is a component of the small (30 S) subunit of chloroplast ribosomes. The large (50 S) subunit contains the 23 S rRNA (Dyer and Leech, 1968; Payne and Dyer, 1971; Gray and Hallick, 1979a; Leaver, 1979; Edelman, 1981). In higher plants at least, a 4.5 S RNA is also a component of the large subunit and is present in approximately equimolar amount as the 5 S rRNA (Dyer and Bowman, 1976; Bohnert *et al.*, 1977; Dyer *et al.*, 1977; Whitfeld *et al.*, 1978b; see also Section II,G,1). On the basis of their oligonucleotide patterns (Galling and Jordan, 1972; Dyer *et al.*, 1977; Bowman and Dyer, 1979) and nucleotide sequences (Dyer and Bowman, 1976; Dyer *et al.*, 1977; Takaiwa and Sugiura, 1980), both the 5 S and 4.5 S RNAs are distinct from the larger rRNAs and the tRNAs of the chloroplast, and the 5 S RNA is distinct from the similar RNA associated with the cytoplasmic ribosome.

The large and the small subunits of the chloroplast ribosome of *Nicotiana* contain 34–38 and 20–24 proteins, respectively (Bourque, 1977), of *Chlamydomonas reinhardtii* 21–34 and 22–25 (Hanson *et al.*, 1974; Brügger and Boschetti, 1975), and of *Euglena gracilis* 16–34 and 14–24 (Mendiola-Morgenthaler *et al.*, 1975; Freyssinet, 1977a,b). In each case, the higher numbers were obtained through the use of a high resolution two-dimensional gel electrophoretic technique and are probably more representative of the actual number of chloroplast ribosomal proteins than are the lower numbers.

2. POLYSOMES

Chloroplast polysomes are found both "free" in the stroma (Tao and Jagendorf, 1973; Alscher *et al.*, 1978) and "bound" to thylakoids (Falk, 1969; Chen and Wildman, 1970; Chua *et al.*, 1973; Tao and Jagendorf, 1973; Margulies and Michaels, 1974; Margulies and Weistrop, 1976; Margulies, 1980; Yamamoto *et al.*, 1981). Avadhani and Buetow (1972) isolated highly polymerized polysomes from the chloroplasts of *Euglena gracilis* through careful control of pH during isolation, which prevented activation of endogenous RNase, and by the use of RNase inhibitors during isolation of the chloroplasts and during all subsequent steps leading to the isolation of the polysomes. The isolated polysomes were active in protein synthesis without any added mRNA. The need for careful control of RNase activity during the isolation of chloroplast polysomes is further emphasized by the recent finding that thylakoids of *Chlamydomonas* contain RNase activity (Margulies, 1980). Chemical and physical parameters of *Euglena* chloroplast polysomes have been reviewed (Edelman, 1981; Avadhani and Freyssinet, 1983). Methods for the isolation of "free" and "bound" chloroplast polysomes from *Pisum* are given by Tao and Jagendorf (1973) and Alscher *et al.* (1978) and for bound chloroplast polysomes from *Chlamydomonas* by Margulies and Michaels (1974), Margulies (1980), and Bolli *et al.* (1981).

3. SYNTHESIS AND PROCESSING

In spinach, the chloroplast rRNA operon (Fig. 1) appears to be transcribed into a 2.7×10^6 D precursor molecule (Hartley and Ellis, 1973; Bohnert *et al.*, 1976), which is then processed to mature 1.1×10^6 D (23 S), 0.56×10^6 D (16 S) and 3.3×10^3 D (4.5 S) ribosomal RNAs (Bohnert *et al.*, 1977; Hartley *et al.*, 1977). The large precursor may also include within itself the mature 5 S rRNA (Edelman, 1981). Similar large primary transcripts and processing schemes have been indicated for the chloroplast rRNA of the higher plants *Phaseolus* (Grierson and Loening, 1974) and *Spirodela* (Posner and Rosner, 1975; Rosner *et al.*, 1977a,b).

No large primary transcript of the complete chloroplast rRNA operon has as yet been detected in algae (Heizmann, 1974; Miller and McMahon, 1974; Scott, 1976, 1977; Stempka and Richter, 1978; Wollgiehn and Parthier, 1979). An RNA precursor of $1.8–1.88 \times 10^6$ D has been detected in chloroplasts of *Chlorella* and *Cyanidium* (Stempka and Richter, 1978), *Euglena* (Wollgiehn and Parthier, 1979) and in the brown alga, *Pylaiella* (Loiseaux *et al.*, 1980). The nucleotide compositions of higher plant, *Euglena*, and *Chlorella* chloroplast rRNAs have been compiled and reviewed several times (Ellis and Hartley, 1974; Whitfeld, 1977; Gillham, 1978; Leaver, 1979; Edelman, 1981). The nucleotide sequence of chloroplast 5 S rRNAs appears to be highly conserved in higher plants (Delihas *et al.*, 1981). However, tobacco chloroplasts contain three species of 5 S rRNA with differing electrophoretic mobilities (Takaiwa and Sugiura, 1981).

4. LABILITY OF THE 23 S RIBOSOMAL RIBONUCLEIC ACID

Other than in the case of *Pinus,* the 23 S rRNA of higher plants is labile (Leaver, 1979; Edelman, 1981). In contrast, the 23 S rRNA of *Euglena* chloroplasts is stable (Avadhani and Buetow, 1972), but the 25 S rRNA of the cytoplasmic ribosome is labile (Heizmann, 1970; Avadhani and Buetow, 1972).

As reviewed by Leaver (1979), the lability of the chloroplast 23 S rRNA of higher plants first led to the suggestion that chloroplast ribosomes contained only one species of rRNA with a sedimentation value of about 16 S. The stability of the chloroplast large rRNA, however, has been found to depend on the conditions of extraction, the age of the leaf, and the species of plant (Leaver and Ingle, 1971). Also, the large subunit of the chloroplast ribosome clearly contains a 23 S rRNA (Leaver, 1979; Edelman, 1981). Degradation of the 23 S rRNA occurs in the ribosome but is not a random process. Rather, it produces discrete fragments that appear to be characteristic of the species involved (Leaver and Ingle, 1971; Rosner *et al.,* 1974). Also, newly synthesized 23 S rRNA is more stable than older 23 S rRNA (Ingle *et al.,* 1970; Rosner *et al.,* 1974; Grierson, 1974). The relationship between the cleavage of the 23 S rRNA and the function of the ribosome is not known.

C. *Transfer Ribonucleic Acid*

1. NUMBER AND ISOACCEPTORS

Chloroplast DNA codes for at least 30 species of tRNA (Section II,F; see also Fig. 1). Fractionation of chloroplast tRNAs by two-dimensional polyacrylamide gel electrophoresis yields 35 spots in the case of spinach and *Euglena* and 30 in the case of maize (Driesel *et al.,* 1979; Burkard *et al.,* 1980; Steinmetz *et al.,* 1980). Genes, corresponding to the tRNAs for 14 different amino acids, have been located on the restriction endonuclease map of spinach chloroplast DNA (Fig. 1). Several chloroplast isoaccepting species of tRNA have been identified and those for leucine, isoleucine, methionine, and serine are coded for by different genes (Williams *et al.,* 1973; Steinmetz *et al.,* 1978; Burkhard *et al.,* 1980). The leucine isoacceptors from bean chloroplasts differ greatly in their primary structure (Burkard *et al.,* 1980; Osorio-Almeida *et al.,* 1980), a result which verifies that they are coded by different genes. In some cases, different isoaccepting species apparently result from post-transcriptional modifications (Steinmetz and Weil, 1976). Weil (1979), Wollgiehn and Parthier (1980), and Mubumbila *et al.* (1980) have published extensive catalogs of the isoaccepting tRNA species found in the chloroplasts of a variety of higher plants and green algae. Weil (1979) also described techniques for the isolation, fractionation, and purification of plant tRNAs.

2. STRUCTURE OF CHLOROPLAST TRANSFER RIBONUCLEIC ACIDS

The primary sequence and secondary structure of several chloroplast tRNAs have been determined: $tRNA^{Asn}$ of tobacco (Kato *et al.*, 1981); $tRNA_3^{Leu}$, $tRNA^{Phe}$, $tRNA_3^{Thr}$, $tRNA^{Met}$, and $tRNA^{Val}$ of spinach (Kashdan *et al.*, 1980; Canaday *et al.*, 1980b; Pirtle *et al.*, 1981; Sprouse *et al.*, 1981); $tRNA^{Phe}$ (Guillemaut and Keith, 1977) and three $tRNA^{Leu}$ molecules (Osorio-Almeida *et al.*, 1980) of the bean *Phaseolus;* $tRNA^{Phe}$ of *Euglena* (Chang *et al.*, 1976); and initiator tRNAs of bean and spinach (Section III,C,3). The $tRNA_3^{Leu}$ molecules of the dicotyledons, spinach and bean, differ in only one position, thus showing a 99% homology. In contrast, the three chloroplast $tRNA^{Leu}$ molecules of the bean differ from each other in the total number of nucleotides and in much of their nucleotide sequences (Osorio-Almeida *et al.*, 1980). Spinach chloroplast $tRNA_3^{Thr}$, tobacco chloroplast $tRNA^{Asn}$, and maize chloroplast $tRNA^{His}$ all are equally homologous to both the respective eukaryotic cytoplasmic and prokaryotic tRNA molecules (Kashdan *et al.*, 1980; Kato *et al.*, 1981; Schwarz *et al.*, 1981). On the other hand, the $tRNA^{Met}$ is considerably more homologous (Pirtle *et al.*, 1981) to bacterial $tRNA^{Met}$ (67% homology) than to eukaryotic cytoplasmic $tRNA^{Met}$ molecules (50–55% homology).

$tRNA^{Phe}$ from spinach and bean (Guillemaut and Keith, 1977; Canaday *et al.*, 1980b) are 99% homologous in primary sequence. $tRNA^{Phe}$ from *Euglena* and bean (Chang *et al.*, 1976; Guillemaut and Keith, 1977) also are highly homologous, that is, 93%, but differ considerably in sequence from $tRNA^{Phe}$ molecules found in the cytoplasm of higher plants and algae (Hecker *et al.*, 1976; Barnett *et al.*, 1978). The similarity in structure between the higher plant and the algal organelle $tRNA^{Phe}$ molecules is an exception, however, and should not be taken as evidence that the structure of chloroplast tRNAs has been rigidly conserved in evolution (Burkard *et al.*, 1980). Heterologous hybridization reactions show that, though the total tRNAs of bean and spinach chloroplasts hybridize to the same extent to bean chloroplast DNA, the total tRNAs of maize chloroplasts hybridize to a lower level (65–75%), and the total tRNAs of *Euglena* chloroplasts hybridize only to a level of about 10–15% (Williams *et al.*, 1973; Burkard *et al.*, 1980; Mubumbila *et al.*, 1980). Furthermore, when 14 individual chloroplast tRNAs from *Euglena* were tested, only $tRNA^{Phe}$ hybridized to bean and spinach chloroplast DNA, whereas the others did not hybridize at all (Mubumbila *et al.*, 1980.

Like bacterial $tRNA^{Phe}$, the chloroplast $tRNA^{Phe}$ of *Euglena* and bean lack the fluorescent base "Y," a hypermodified guanine derivative located adjacent to the anticodon of eukaryotic cytoplasmic $tRNA^{Phe}$ molecules (Fairfield and Barnett, 1971; Guillemaut *et al.*, 1976; Hecker *et al.*, 1976; Delehanty *et al.*, 1983).

3. INITIATOR $tRNA_f^{Met}$

Unlike the cyotplasmic initiator $tRNA_i^{Met}$, but like the prokaryotic initiator, chloroplast initiator $tRNA_f^{Met}$ is formylated as demonstrated for *Phaseolus*, (Burkard *et al.*, 1969; Guillemaut *et al.*, 1973; Guillemaut and Weil, 1975), *Acetabularia* (Bachmeyer, 1970), wheat (Leis and Keller, 1970), spinach (Bianchetti *et al.*, 1971), and cotton (Merrick and Dure, 1971).

Chloroplast initiator $tRNA_f^{Met}$ molecules show prokaryotic structural features, e.g., the 5′ terminal nucleotide is not based-paired and the TΨCA sequence is present (Fig. 3). Bean (*Phaseolus*) and spinach chloroplast initiator $tRNA_f^{Met}$ molecules (Fig. 3) are highly similar to one another in sequence (92% homology) and quite similar to the bacterial $tRNA_f^{Met}$ sequence (77–80% homology). In contrast, the bean chloroplast and cytoplasmic initiator tRNAs are only 61% homologous to each other (Fig. 3). The bean chloroplast initiator $tRNA_f^{Met}$ is 73–81% homologous to various prokaryotic initiator $tRNA_f^{Met}$ molecules and the bean cytoplasmic initiator $tRNA_i^{Met}$ is 73–100% homologous to various eukaryotic cytoplasmic $tRNA_i^{Met}$ molecules (Canaday *et al.*, 1980a). Therefore, the nucleotide sequence of chloroplast initiator $tRNA_f^{Met}$ more strongly resembles the sequence of prokaryotic initiator tRNAs than the sequence of cytoplasmic initiator tRNAs (Fig. 3).

4. AMINOACYL–TRANSFER RIBONUCLEIC ACID SYNTHETASES

Chloroplast-specific tRNA synthetases have been characterized in a number of higher plants and *Euglena* (Lea and Norris, 1977; Parthier *et al.*, 1978; Weil, 1979; Wollgiehn and Parthier, 1980). The organelle synthetases differ from the corresponding cytoplasmic synthetases in their chromatographic behavior on hydroxyapatite or DEAE cellulose and in their specificity for tRNA substrates. The aminoacyl–tRNA synthetases and tRNAs from chloroplasts and from prokaryotes show some similarities because cross-aminoacylation reactions are possible (Parthier and Krauspe, 1974; Guillemaut and Weil, 1975; Guillemaut *et al.*, 1975; Jeannin *et al.*, 1976; Selsky, 1978). For example, cross-aminoacylation occurs between the $tRNA^{Leu}$ and $tRNA^{Ile}$ molecules and the corresponding tRNA synthetases of *Euglena* chloroplasts and of the cyanobacteria *Anacystis* (Parthier and Krauspe, 1974) and *Nostoc* (Selsky, 1978). In

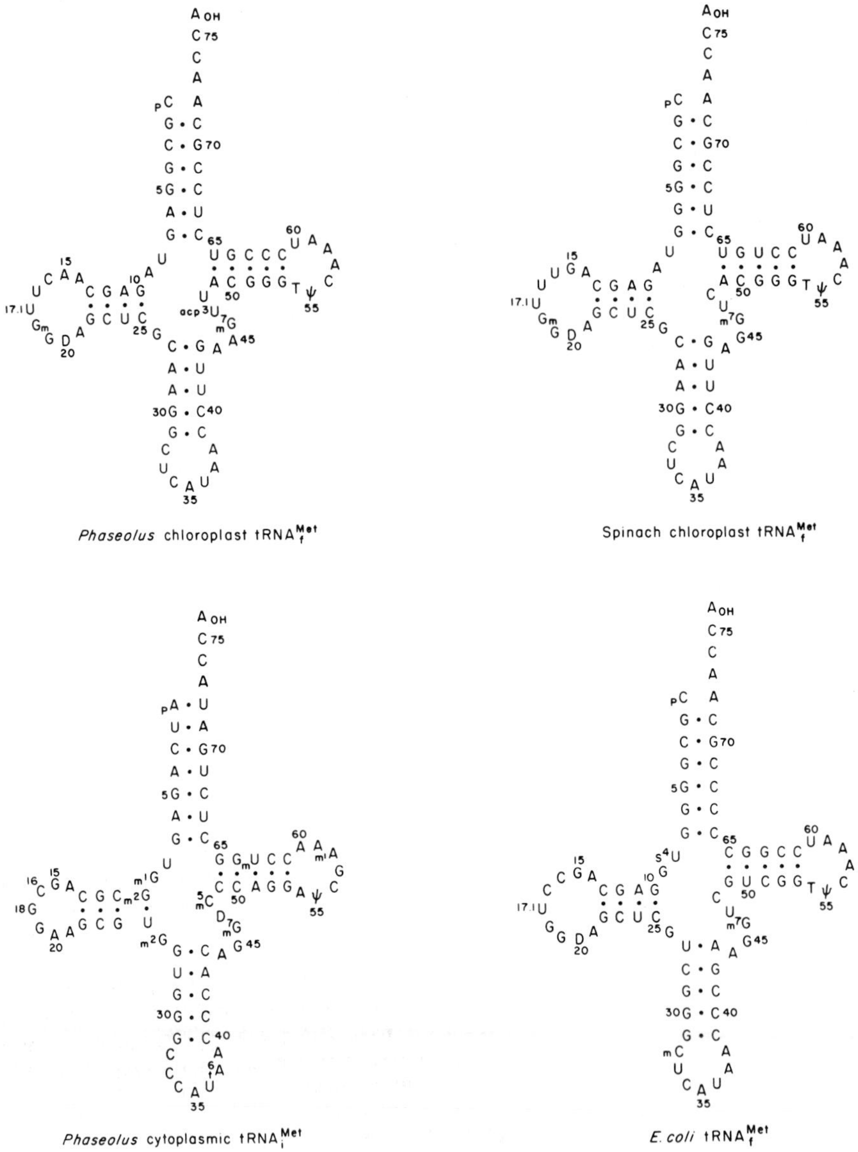

FIG 3. Nucleotide sequence of initiator tRNAs. *Phaseolus* chloroplast and cytoplasmic tRNA sequences from Canaday *et al.* (1980a); spinach chloroplast tRNA sequence from Calagan *et al.* (1980); *E. coli* tRNA sequence from Dube and Marcker (1969). (D = dihroduridine.) Numbered sequences and modified bases are designated according to Sprinzl *et al.* (1980) and Canaday *et al.* (1980a).

contrast, the chloroplast $tRNA^{Leu}$ is not aminoacylated by the cytoplasmic synthetase (Parthier and Krauspe, 1974). The high specificity shown by the *Euglena* chloroplast leucyl–tRNA synthetase, however, does not seem to hold true for other synthetases. Some cross-aminoacylation between chloroplast and cytoplasmic tRNAs and synthetases occurs in cotton cotyledons (Brantner and Dure, 1975), bean (Jeannin *et al.*, 1978), and in *Euglena* in the case of tRNAs other than $tRNA^{Leu}$ (Parthier *et al.*, 1978; Wollgiehn and Parthier, 1980).

Euglena chloroplast aminoacyl–tRNA synthetases are coded for by nuclear DNA, synthesized on cytoplasmic ribosomes, and transported into the chloroplasts (Reger *et al.*, 1970; Parthier, 1973; Hecker *et al.*, 1974; Barnett *et al.*, 1976; Parthier and Neumann, 1977; Weil, 1979). The chloroplast tyrosyl–tRNA synthetase from soybean (Locy and Cherry, 1978) and the chloroplast valyl–tRNA synthetase from *Euglena* (Sarantoglou *et al.*, 1978; Burkard *et al.*, 1980) have been purified. Following illumination of dark-grown *Euglena*, chloroplast leucyl–tRNA synthetase increases in activity (Parthier and Krauspe, 1974) as a result of de novo synthesis (Nover, 1976).

D. Messenger Ribonucleic Acids

The messenger RNA (mRNA) activity for the LS has been found only in the polyadenylate-lacking RNA fraction from chloroplasts of *Euglena* (Sagher *et al.*, 1976; Freyssinet *et al.*, 1981), spinach (Hartley *et al.*, 1975; Wheeler and Hartley, 1975), *Chlamydomonas* (Howell *et al.*, 1977), and *Spirodela* (Reisfeld *et al.*, 1978). The chloroplast mRNA for a 32,000-D PSII polypeptide (Section II,G,4) also is not polyadenylated (Rosner *et al.*, 1975, 1977b). These results raise the question of whether or not any chloroplast DNA-coded mRNA will be found to possess a polyadenylic acid sequence at its 3′ end as is found for nuclear-coded mRNAs (Sagher *et al.*, 1974). However, some polyadenylated RNA does hybridize to chloroplast DNA, though its relationship to mRNA is not clear (Section II,F). Also, Gray and Cashmore (1976) reported that some chloroplast polyadenylated RNA fractions do synthesize chloroplast proteins and Bartolf and Price (1979) reported that isolated spinach chloroplasts synthesize polyadenylated RNA. Contamination of chloroplast preparations with small amounts of cytoplasmic polyadenylated mRNAs is difficult to rule out, however. Possibly, chloroplast-coded mRNAs contain only short polyadenylic acid tracts that are not easily detected by the usual measure of binding to oligothymidylic acid [oligo(dT)]-cellulose or polyuridylic acid [poly(U)]-Sephadex (Haff and Bogorad, 1976b; Edelman *et al.*, 1977). Another possibility is that the polyadenylated segment is quite labile (Sano *et al.*, 1979).

The chloroplast LS mRNA shows a sedimentation value of 12–14 S in the case of *Chlamydomonas* (Howell *et al.*, 1977), 16 S in *Spirodela* (Rosner *et al.*, 1977b; Reisfeld *et al.*, 1978), and 13 S in *Euglena* (Freyssinet *et al.*, 1981). The chloroplast mRNA for the 32,000-D polypeptide from *Spirodela* (Rosner *et al.*, 1975, 1977b; Reisfeld *et al.*, 1978) and from spinach (Driesel *et al.*, 1980) sediments at 13–14 S. The 14 S mRNA fraction of spinach chloroplasts also contains mRNAs for unidentified polypeptides ranging from 12,600 to 45,000 D (Spiers and Grierson, 1978; Driesel *et al.*, 1980).

Hybridization of the spinach mRNAs for the LS and for the 32,000-D PSII polypeptide to the chloroplast DNA of *Oenothera* indicates a considerable homology between the relevant genes in these two closely related plant species (Bohnert *et al.*, 1980). In contrast, no homology could be detected between the chloroplast genes for these two polypeptides in spinach and *Euglena*. This latter result seems to indicate considerable divergence in chloroplast polypeptide genes in evolution.

Since only two chloroplast-coded mRNAs have been specifically identified so far, much work remains to be done to determine what other proteins are coded by the chloroplast genome (see also Section II,F).

IV. Evolution of the Chloroplast Genome

A. Restriction Enzyme and Molecular Hybridization Analyses

Some studies on chloroplast DNAs from a variety of higher plants indicated a high degree of similarity in these genomes, thus suggesting the possibility of conservation of chloroplast DNA during evolution. Examples in higher plants are the similar G + C content (Section II,B,1), kinetic complexity (Section II,B,2) and size (Section II,B,3) of their chloroplast DNAs, and the conservation of structure of their chloroplast rRNAs (Thomas and Tewari, 1974), some tRNAs (Section III,C,2 and 3), and the LS of ribulosebisphosphate carboxylase (Kwok and Wildman, 1974; Kung *et al.*, 1977). Experiments in which chloroplast DNAs from various higher plants were hybridized, however, indicate a 75–95% homology between the genomes of species of the same family, but only 15–57% homology between more distant species, e.g., between dicotyledons and monocotyledons (Walbot, 1977; Lamppa and Bendich, 1979; Bisaro and Siegel, 1980).

Restriction endonuclease enzymes have proven useful for studies on the evolution of the chloroplast genome. Because of the specificity of the cleavage site of a restriction enzyme, the fragmentation pattern produced is characteristic for a specific DNA molecule. This technique is more sensitive for detecting mutations than the technique of molecular hybridization (e.g., Thomspon *et al.*, 1974). Specific fragmentation patterns produced with *Eco*RI endonuclease, for example, show that chloroplast DNAs from different families or genera of higher plants have few, if any, fragments in common, but similarities do occur in some cases when species of the same genus are compared (Atchison *et al.*, 1976; Vedel *et al.*, 1976). There is, however, considerable variation in the fragments even between species within a genus, e.g., *Zea* (Timothy *et al.*, 1979) and *Oenothera* (Gordon *et al.*, 1981) and between isonuclear male-sterile lines of *Nicotiana* (Frankel *et al.*, 1979). Also, during the course of evolution of legumes, homologous sequences within their chloroplast genomes have been rearranged frequently (Palmer and Thompson, 1981).

The molecular hybridization and restriction enzyme studies together indicate the existence of a region of chloroplast DNA comprising about 10–20% of the total molecule that is more resistant to change than the rest of the molecule (e.g., Bisaro and Siegel, 1980). As discussed next, within this apparently highly conserved region are the rRNA genes.

B. Phylogenetic Relationships with Prokaryote Genomes

The primary structure of 16 S rRNA molecules is highly conserved in evolution (Woese *et al.*, 1975). Therefore, chloroplast rRNAs and rRNA operons have been used as indicators of evolutionary relationships between chloroplasts of different species and between chloroplasts and prokaryotes. Heteroduplex analysis of hybrids formed between *Vicia faba* chloroplast DNA and an *E. coli* DNA fragment containing rRNA genes indicates considerable homology between the genes on the two DNAs for both 16 S and 23 S rRNA (Delius and Koller, 1980). Spinach chloroplast rRNAs were hybridized to restriction endonuclease fragments of chloroplast DNA containing rRNA genes from spinach, *Oenothera* and *Euglena*, to *Acanthomoeba* mitochondrial DNA, and to an rRNA operon of *E. coli* (Bohnert *et al.*, 1980). The degree of homology is greatest for the 16 S rRNA gene (Table III). Greater than 90% homology occurs between the two higher plant genes, 80% homology to the algal (*Euglena*) gene and 60–70% to the bacterial gene (Table III). The latter value agrees well with the 75% homology reported between the maize chloroplast 16 S rRNA gene and the *E. coli* 16 S rRNA or its gene (Schwarz and Kössel, 1980). The 23 S and 5 S rRNA genes are also highly homologous between the two higher plants. The levels of their homology to the algal chloroplast and bacterial genes, however, indicate that these genes have diverged to a greater degree during evolution than has the 16 S rRNA gene. The mitochondrial rRNA genes show the least homology with the chloroplast genes (Table III).

Pigott and Carr (1972) and Phillips and Carr (1975) hybridized rRNAs of various prokaryotes with the chloroplast DNA of *Euglena* and

TABLE III
Percentage of Homology between rRNA Genes

Sources of rRNA genes	Percentage of homology with spinach chloroplast rRNA[a] 16 S	23 S	5 S
Spinach chloroplasts	100	100	100
Oenothera chloroplasts	>90	>90	>90
Euglena chloroplasts	80	50	50
Acanthamoeba mitochondria	20	5	0
Escherichia coli (rrnB rRNA operon)	60–70	40–50	40

Source of rRNA	Percentage of homology with *Euglena* chloroplast rRNA genes[b] Total rRNA genes[c]	5S rRNA genes[d]
Euglena chloroplasts	100	100
Euglena cytoplasm	1	19
Unicellular cyanobacteria		
Gloeocapsa alpicola	47	72
Anacystis nidulans	13	68
Chlorogloea fritschii	32	—
Filamentous cyanobacteria		
Nostoc muscorum	32	—
Nostoc sp.	—	52
Anabaena variabilis	21	46
Mastigocladus laminosus	12	—
Anabaena cylindrica	11	—
Photosynthetic bacteria		
Rhodomicrobium vanniellii	6	—
Rhodospirillum rubrum	4	—
Heterotrophic bacteria		
Klebsiella aerogenes	5	—
Serratia marcescens	4	—
Staphylococcus aureus	4	—
Escherichia coli	4	54
Pseudomonas sp.	2	—

[a]Determined by the degree of hybridization of rRNA genes from organelles and bacteria. (From Bohnert *et al.*, 1980.)

[b]Determined by the degree of hybridization of *Euglena* chloroplast rRNA genes with rRNA from prokaryotes.

[c]From Pigott, G. H., and Carr, N. G. (1972) *Science* **175,** 1259–1261. Copyright 1972 by the American Association for the Advancement of Science.

[d]From Phillips and Carr (1975).

showed a significant similarity between the genes for rRNA of the *Euglena* chloroplast and those of cyanobacteria (blue–green algae) and, in the case of 5 S rRNA genes, between the *Euglena* chloroplast and *E. coli* (Table III). By the use of oligonucleotide sequencing, the *Euglena* chloroplast 16 S rRNA was shown to be related phylogenetically to the chloroplast 16 S rRNA of the red alga, *Porphyridium,* and at least distantly to the 16 S rRNAs of cyanobacteria (Zablen *et al.,* 1975; Buetow, 1976; Buetow *et al.,* 1977).The 3′ terminal sequence of maize chloroplast 16 S rRNA is very similar to the 3′ terminal sequence of the 16 S rRNA from the cyanobacterium *Synechococcus* (Borbély and Simoncsits, 1981). Similarly, the 16 S rRNA of *Porphyridium* also is related to the 16 S rRNAs of cyanobacteria (Bonen and Doolittle, 1975, 1976). Furthermore, in similar studies, the chloroplast 5 S rRNAs of *Lemna* (Dyer *et al.,* 1977) and tobacco (Takaiwa and Sugiura, 1981) were shown to be related to that of the cyanobacterium *Anacystis nidulans.* Also, the maize chloroplast 23 S rRNA gene is 67% homologous with the *E. coli* 23 S rRNA gene (Edwards and Kossel, 1981) and the maize chloroplast 4.5 S rRNA appears to be the structural and functional equivalent of the 3′ terminal region of the bacterial 23 S rRNA (Edwards *et al.,* 1981). In contrast, in none of these studies were the chloroplast rRNAs shown to bear any obvious relationship to the similar nuclear-coded rRNAs found in the cytoplasms of eukaryotes. Fox *et al.* (1980) reviewed the oligonucleotide sequencing data used to draw phylogenetic relationships among the prokaryotes and concluded that the chloroplast of *Porphyridium* descended from the cyanobacteria, whereas the chloroplasts of *Euglena* and the duckweed, *Lemna,* are not direct descendents of but rather share a common ancestry with the cyanobacteria.

The phylogenetic relationship between chloroplasts and prokaryotes is best seen in the structure of their respective rRNA operons. The order and polarity of the 16 S, 23 S, and 5 S genes are the same on all the algal and higher plant chloroplast genomes examined so far (Section II,G,1) and also the same as on the *E. coli* genome (Kenerley *et al.,* 1977). Also both chloroplast and *E. coli* genomes have tRNA genes in the spacer region between the 16 S and the 23 S rRNA genes (Section II,G,2). The seven *E. coli* rRNA operons, however, fall into two groups in respect to the organization of tRNA genes in the spacer region, i.e., three with $tRNA^{Ile}$ and $tRNA^{Ala}$ genes in the spacer, designated as rrnA, D, and X operons, and four with a $tRNA^{Glu}$ gene in the spacer and designated as rrnB, C, E, and F operons (Nomura and Morgan, 1977). Heteroduplex analysis of *Vicia faba* chloroplast DNA (Delius and Koller, 1980) and the mapping of a $tRNA^{Ile}$ gene in the spacer of spinach and of the $tRNA^{Ile}$ and $tRNA^{Ala}$ genes in maize (Section II,G,2; see also Fig. 1) show that the chloroplast rRNA operons of these higher plants are more closely related to the *E. coli* rrnA, D, and X operons than to the other group of

E. coli operons. The *Euglena* chloroplast rRNA operon (Fig. 3) is highly analogous structurally (Graf *et al.*, 1980; Keller *et al.*, 1980; Orozco *et al.*, 1980b; Hallick, 1983) to the *E. coli* rrnA, D, and X operons. The *Euglena* operon has $tRNA^{Ile}$ and $tRNA^{Ala}$ genes in its 16 S–23 S rRNA gene spacer region in the same order, polarity, and with the same anticodons as in the *E. coli* rrnA, D, and X operons. Also, sequencing of the 5′ end of the *Euglena* chloroplast 16 S rRNA gene shows that the first 165 b.p. correspond, with a single 12-b.p. deletion, to the initial 177 b.p. of the *E. coli* 16 S rRNA gene. The overall homology is 72% (Orozco *et al.*, 1980b) and is consistent with the prediction of a common evolutionary origin for the coding regions of these operons.

V. Conclusions

The data reviewed here support the concept that in most, if not all, algae and higher plants the complete chloroplast genome is represented by the nucleotide sequence of a single circular DNA molecule. The chloroplast contains unique rRNAs and a unique, probably complete, set of tRNAs, which function in the expression of the organelle genome. So far, the chloroplast genome has been shown to contain the genes for these rRNAs and tRNAs as well as the genes for two chloroplast polypeptides. The size of the genome, about 90×10^6 D, suggests that it codes for other proteins also. The identity of at least some of these other proteins is suggested by genetic analysis of chloroplasts in *Chlamydomonas* (Sager, 1972, 1977; Gillham, 1978) and by studies of the products of protein synthesis in intact isolated chloroplasts (Ellis, 1977b). The latter system uniquely uses photosynthesis as a source of energy for protein synthesis, thus eliminating any contribution from contaminating cytoplasmic or mitochondrial protein-synthesizing systems. Genetic analyses, however, do not necessarily prove that a given protein is coded by the chloroplast genome, but may tell only that this genome is involved in the synthesis of the protein. Also, synthesis of a protein by isolated chloroplasts does not prove that the protein is coded by the organelle genome. The relevant mRNA could be coded by the nucleus and imported into the chloroplast for translation. Nevertheless, by combining information obtained from such genetic and biochemical analyses with physical mapping of the chloroplast genome by currently available techniques, further experiments can determine precisely which regulatory and structural genes are contained in this genome and where they are located.

Continued studies on the molecular biology of the chloroplast genome and of the RNAs involved in its expression will help clarify its role in the growth and development of the chloroplast itself as well as the nature of its interdependence with other genetic systems in the plant cell. Results of such studies also will help assess the feasibility of postulated "genetic engineering" approaches to improving the photosynthetic efficiency of important food, feed, and fiber plants (see, Day, 1977; Radmer and Kok, 1977; Andersen *et al.*, 1980).

NOTE ADDED IN PROOF

Since this chapter was completed, new relevant literature has appeared. A restriction enzyme map of the chloroplast DNA of *Chlorella ellipsoidea* has been reported [Yamada, T. (1982). *Plant Physiol.* **70,** 92–96]. Restriction enzyme fragments representing the entire chloroplast genomes from several higher plants have been stably cloned in *Escherichia coli* [Palmer, J. D., and Thompson, W. F. (1981). *Gene* **15,** 21–26]. The gene sequence in the rRNA gene cluster of tobacco chloroplast DNA follows the order: 16S, 23S, 4.5S, and 5S rRNA [Sugiura, M. *et al.* (1980). *In* "Genetics and Evolution of RNA Polymerase, tRNA and Ribosomes" (S. Ozawa *et al.*, eds.), pp. 437–449. Elsevier/North-Holland Biomedical Press, Amsterdam]; the 23S rRNA gene from maize chloroplasts (Edwards, K., and Kössel, H. (1981) *Nucleic Acids Res.* **9,** 2853–2869] and the 16S rRNA gene from tobacco chloroplasts [Tohdoh, N., and Sugiura, M. (1982). *Gene* **17,** 213–218] have been sequenced; and the chloroplast rRNA genes exist as inverted repeats in pearl millet [Rawson, J. R. Y. *et al.* (1981). *Gene* **16,** 11–19], petunia [Bovenberg, W. A. *et al.* (1981). *Nucleic Acids Res.* **9,** 503–517], and *Atriplex* and *Cucumis* [Palmer, J. D. (1982). *Nucleic Acids Res.* **10,** 1593–1605]. Chloroplast and nuclear rRNA genes of *Euglena gracilis* strain Z are 94% homologous in the regions that code for the 3′-ends of the respective small rRNAs and for the 5S rRNAs [Curtis, S. E., and Rawson, J. R. Y. (1982). *Plant Physiol.* **69,** 67–71]. The extra 16S rRNA gene in the *Euglena* chloroplast [Section II,F] has the same size and orientation as the same genes in the tandemly repeated rRNA operons and is flanked by nucleotide sequences as found between these operons [Koller, B., and Delius, H. (1982). *FEBS Lett.* **139,** 86–92 and **140,** 198–202]. Also, the 5′-region preceding the *Euglena* chloroplast rRNA operons contains an rRNA pseudogene cluster (Miyata, T. *et al.* (1982). *Nucleic Acids Res.* **10,** 1771–1780]. All the chloroplast tRNA genes have been mapped in *Spirodela* [Groot, G. S. P., and van Harten-Loosbroek, N. (1981). *Curr. Genet.* **4,** 187–190] and *E. gracilis* strain *bacillaris* (El-Gewely, M. R. *et al.* (1981). *Mol. Gen. Genet.* **181,** 296–305] and strain Z [Orozco, E. M., Jr., and Hallick, R. B. (1982). *J. Biol. Chem.* **257,** 3258–3264]. Also, a chloroplast tRNA gene cluster in strain Z has been sequenced [Orozco, E. M., Jr., and Hallick, R. B. (1982). *J. Biol. Chem.* **257,** 3265–3275]. Spinach chloroplast DNA contains about 42 tRNA genes and the structures of higher plant chloroplast tRNAs have diverged significantly [Meeker, R., and Tewari, K. K. (1982). *Biochim. Biophys. Acta* **696,** 66–75]. The genes for a spinach chloroplast threonine tRNA [Kashdan, M. A., and Dudock, B. S. (1982). *J. Biol. Chem.* **257,** 1114–1116] and for an isoleucine tRNA in both maize and spinach chloroplasts [Guillemaut, P., and Weil, J. H. (1982). *Nucleic Acids Res.* **10,** 1653–1659] have been sequenced. The latter maize gene contains an intron as suggested earlier [Koch, W. *et al.* (1981). *Cell* **25,** 203–213]. The spinach chloroplast LS gene has been sequenced [Zurawski, G. *et al.* (1981). *Nucleic Acids Res.* **9,** 3251–3270]. Wheat and maize chloroplast LS genes show extended stretches of homology interspersed with a fine pattern of non-homology [Koller, B. *et al.* (1982). *Eur. J. Biochem.* **122,** 17–23]. The chloroplast LS gene of *E. gracilis* strain and strain *bacillaris* is located on EcoR1 restriction fragment EcoA [Stiegler, G. L. *et al.* (1982). *Nucleic Acids Res.* **10,** 3427–3444] and not on fragment EcoC as first reported [Hallick, R. B. *et al.* (1979). *ICN-UCLA Symp. Mol. Cell Biol.* **15,** 127–141]. Also, the *Euglena* LS gene contains an intron. A 34 kD triazine herbicide-binding protein is coded by the chloroplast genome [Steinback, K. E. *et al.* (1981). *Proc. Natl. Acad. Sci. U.S.A.* **72,** 2418–2422]. An isolated spinach chloroplast chromosome transcribes *in vitro* rRNAs, tRNAs, and mRNAs with sizes equivalent to those transcribed *in vivo* [Blanc, M. *et al.* (1981). *Biochim. Biophys. Acta* **655,** 374–382].

Maize chloroplast rRNAs are more closely related in structure to bacterial than to eukaryotic cytoplasmic rRNAs [Küntzel, H., and Köchel, H. G. (1981). *Nature (London)* **293,** 751–755]. Tobacco chloroplast 4.5S rRNA is the structural equivalent of the 3′-end of bacterial 23S rRNA [MacKay, R. M. (1981). *FEBS Lett.* **123,** 17–18]. Chloroplast ribosomal proteins of spinach [Mache, R. *et al.* (1980). *Mol. Gen. Genet.* **184,** 484–488] and pea [Eneas-Filho, J. *et al.* (1981). *Mol. Gen. Genet.* **184,** 484–488] have been characterized by two-dimensional gel electrophoresis. The amino acid compositions and tryptic-digest maps of *Euglena* chloroplast and cytoplasm valyl- and leucyl-tRNA synthetases differ, indicating that these proteins are coded by different genes [Colas, B. *et al.* (1982). *Biochim. Biophys. Acta* **697,** 71–76].

REFERENCES

Allet, B., and Rochaix, J.-D. (1979). *Cell* **18,** 55–60.

Alscher, R., Patterson, R., and Jagendorf, A. T. (1978). *Plant Physiol.* **62,** 88–93.

Andersen, K., Shanmugam, K. T., Lim, S. T., Csonka, L. N., Tait, R., Hennecke, H., Scott, D. B., Hom, S. S. M., Haury, J. F., Valentine, A., and Valentine, R. C. (1980). *Trends Biochem. Sci.* **5,** 35–39.

Apel, K., and Bogorad, L. (1976). *Eur. J. Biochem.* **67,** 615–620.

Atchison, B. A., Whitfeld, P. R., and Bottomley, W. (1976). *Mol. Gen. Genet.* **148,** 263–269.

Avadhani, N. G., and Buetow, D. E. (1972). *Biochem. J.* **128,** 353–365.

Avadhani, N. G., and Freyssinet, G. (1983). *In* "The Biology of *Euglena*" (D. E. Buetow, ed.), Vol. 4. Academic Press, New York (in press).

Bachmeyer, H. (1970). *Biochim. Biophys. Acta* **209,** 584–586.

Barnett, W. E., Schwartzbach, S. D., Farelly, J. G., Schiff, J. A., and Hecker, L. I. (1976). *Arch. Microbiol.* **109,** 201–203.

Barnett, W. E., Schwartzbach, S. D., and Hecker, L. I. (1978). *Prog. Nucleic Acid Res. Mol. Biol.* **21,** 143–179.

Bartolf, M., and Price, C. A. (1979). *Biochemistry* **18,** 1677–1680.

Bastia, D., Chiang, K. S., Swift, H., and Siersma, P. (1971). *Proc. Natl. Acad. Sci. U.S.A.* **68,** 1157–1161.

Bayen, M., and Rode, A. (1973). *Eur. J. Biochem.* **39,** 413–420.

Becker, W. M. (1979). *In* "Nucleic Acids in Plants" (T. C. Hall and J. W. Davies, eds.), Vol. I, pp. 111–141. CRC Press, Boca Raton, Florida.

Bedbrook, J. R., and Bogorad, L. (1976). *Proc. Natl. Acad. Sci. U.S.A.* **73,** 4309–4313.

Bedbrook, J. R., and Kolodner, R. (1979). *Annu. Rev. Plant Physiol.* **30,** 593–620.

Bedbrook, J. R. Kolodner, R., and Bogorad, L. (1977). *Cell* **11,** 739–749.

Bedbrook, J. R., Link, G., Coen, D. M., Bogorad, L., and Rich, A. (1978). *Proc. Natl. Acad. Sci. U.S.A.* **75,** 3060–3064.

Bedbrook, J. R., Coen, D. M., Beaton, A. R., Bogorad, L., and Rich, A. (1979). *J. Biol. Chem.* **254,** 905–910.

Behn, W., and Herrmann, R. G. (1977). *Mol. Gen. Genet.* **157,** 25–30.

Bianchetti, R., Lucchini, G., and Sartirana, M. L. (1971). *Biochem. Biophys. Res. Commun.* **42,** 97–102.

Birkey, C. W., Jr. (1978). *Annu. Rev. Genet.* **12,** 471–512.

Bisaro, D., and Siegel, A. (1980). *Plant Physiol.* **65,** 234–237.

Bogorad, L., Mets, L. J., Mullinix, K. P., Smith, H. J., and Strain, G. C. (1973). *Biochem. Soc. Symp.* **38,** 17–41.

Bogorad, L., Jolly, S. O., Kidd, G., Link, G., and McIntosh, L. (1980). *In* "Genome Organization and Expression in Plants" (C. J. Leaver, ed.), pp. 291–311. Plenum, New York.
Bohnert, H.-J., Driesel, A. J., and Herrmann, R. G. (1976). *In* "Genetics and Biogenesis of Chloroplasts and Mitochondria" (T. Bücher, W. Neupert, W. Sebald, and S. Werner, eds.), pp. 629–636. Elsevier/North-Holland Biomedical Press, Amsterdam.
Bohnert, H.-J., Driesel, A. J., and Herrmann, R. G. (1977). *Colloq. Int. C.N.R.S.* **261,** 213–218.
Bohnert, H.-J., Driesel, A. J., Crouse, E. J., Gordon, K., Herrmann, R. G., Steinmetz, A., Mubumbila, M., Keller, M., Burkard, G., and Weil, J. H. (1979). *FEBS Lett.* **103,** 52–56.
Bohnert, H.-J., Gordon, K. H. J., and Crouse, E. J. (1980). *Mol. Gen. Genet.* **179,** 539–545.
Bolli, R., Mendiola-Morgenthaler, L., and Boschetti, A. (1981). *Biochim. Biophys. Acta* **653,** 276–284.
Bonen, L., and Doolittle, W. F. (1975). *Proc. Natl. Acad. Sci. U.S.A.* **72,** 2310–2314.
Bonen, L., and Doolittle, W. F. (1976). *Nature (London)* **261,** 669–673.
Borbély, G., and Simoncsits, A. (1981). *Biochem. Biophys. Res. Commun.* **101,** 846–852.
Bottomley, W. (1970). *Plant Physiol.* **46,** 437–441.
Bottomley, W., and Whitfeld, P. R. (1978). *In* "Chloroplast Development" (G. Akoyunoglou and J. H. Argyroudi-Akoyunoglou, eds.), pp. 657–662. Elsevier/North-Holland, Amsterdam.
Bottomley, W., and Whitfeld, P. R. (1979). *Eur. J. Biochem.* **93,** 31–39.
Bottomley, W., Smith, J. H., and Bogorad, L. (1971a). *Proc. Natl. Acad. Sci. U.S.A.* **68,** 2412–2416.
Bottomley, W., Spencer, D., and Whitfeld, P. R. (1971b). *Arch. Biochem. Biophys.* **143,** 269–275.
Bourque, D. P. (1977). *Colloq. Int. C.N.R.S.* **261,** 285–289.
Bowman, C. M., and Dyer, T. A. (1979). *Biochem. J.* **183,** 605–613.
Brandt, P., and Wiessner, W. (1977). *Z. Pflanzenphysiol.* **85,** 53–60.
Brantner, J. H., and Dure, L. S. (1975). *Biochim. Biophys. Acta* **414,** 99–114.
Briat, J.-F., Laulhere, J.-P., and Mache, R. (1979). *Eur. J. Biochem.* **98,** 285–292.
Britten, R. J., and Kohne, D. E. (1968). *Science* **161,** 529–540.
Brown, R. D., and Haselkorn, R. (1971). *Proc. Natl. Acad. Sci. U.S.A.* **68,** 2536–2539.
Brügger, M., and Boschetti, A. (1975). *Eur. J. Biochem.* **58,** 603–610.
Bryant, J. A. (1976). *In* "Molecular Aspects of Gene Expression in Plants" (J. A. Bryant, ed.), pp. 139–175. Academic Press, New York.
Buetow, D. E. (1976). *J. Protozool.* **23,** 41–47.
Buetow, D. E., Kissil, M. S., and Zablen, L. (1977). *Colloq. Int. C.N.R.S.* **261,** 227–233.
Buetow, D. E., Wurtz, E. A., and Gallagher, T. (1980). *In* "Nuclear-Cytoplasmic Interactions in the Cell Cycle" (G. L. Whitson, ed.), pp. 9–55. Academic Press, New York.
Burkard, G., Eclancher, B., and Weil, J. H. (1969). *FEBS Lett.* **4,** 285–287.
Burkard, G., Guillemaut, P., and Weil, J. H. (1970). *Biochim. Biophys. Acta* **224,** 184–198.
Burkard, G., Canaday, J., Crouse, E., Guillemaut, P., Imbault, P., Keith, G., Keller, M., Mubumbila, M., Osorio, L., Sarantaglou, V., Steinmetz, A., and Weil, J. H. (1980). *In* "Genome Organization and Expression in Plants" (C. J. Leaver, ed.), pp. 313–320. Plenum, New York.
Calagan, J. L., Pirtle, R. M., Pirtle, I. M., Kashdan, M. A., Vreman, H. J., and Dudock, B. S. (1980). *J. Biol. Chem.* **255,** 9981–9985.
Canaday, J., Guillemaut, P., and Weil, J. H. (1980a). *Nucleic Acids Res.* **8,** 999–1008.
Canaday, J., Guillemaut, P., Gloeckler, R., and Weil, J. H. (1980b). *Plant Sci. Lett.* **20,** 57–62.
Chang, S. H., Brum, C. K., Silberklang, M., RajBhandary, U. L., Hecker, L. I., and Barnett, W. E. (1976). *Cell* **9,** 717–724.
Chelm, B. K., and Hallick, R. B. (1976). *Biochemistry* **15,** 593–599.

Chelm, B. K., Hoben, P. J., and Hallick, R. B. (1977a). *Biochemistry* **16,** 776–781.
Chelm, B. K., Hoben, P. J., and Hallick, R. B. (1977b). *Biochemistry* **16,** 782–786.
Chelm, B. K., Gray, P. W., and Hallick, R. B. (1978). *Biochemistry* **17,** 4239–4244.
Chen, J. L., and Wildman, S. G. (1970). *Biochim. Biophys Acta* **209,** 207–219.
Chiang, K.-S., and Sueoka, N. (1967). *Proc. Natl. Acad. Sci. U.S.A.* **57,** 1506–1513.
Christiansen, C., Christiansen, G., and Bak, A. L. (1974). *J. Mol. Biol.* **84,** 65–82.
Chua, N.-H., Blobel, G., Siekevitz, P., and Palade, G. E. (1973). *Proc. Natl. Acad. Sci. U.S.A.* **70,** 1554–1558.
Clark, M. F. (1964). *Biochim. Biophys. Acta* **91,** 671–674.
Clark, M. F., Matthews, R. E. F., and Ralph, R. K. (1964). *Biochim. Biophys. Acta* **91,** 289–304.
Coen, D. M., Bedbrook, J. R., Bogorad, L., and Rich, A. (1977). *Proc. Natl. Acad. Sci. U.S.A.* **74,** 5487–5491.
Coleman, A. W. (1979). *J. Cell Biol.* **82,** 299–305.
Cook, J. R. (1966). *J. Cell Biol.* **29,** 369–373.
Crick, F. H. C. (1966). *J. Mol. Biol.* **19,** 548–555.
Crouse, E. J., Schmitt, J. M., Bohnert, H.-J., Gordon, K., Driesel, A. J., and Herrmann, R. G. (1978). *In* "Chloroplast Development" (G. Akoyunoglou and J. H. Argyroudi-Akoyunoglou, eds.), pp. 565–572. Elsevier/North-Holland, Amsterdam.
Dalmon, J., and Bayen, M. (1975). *Arch. Microbiol.* **103,** 57–61.
Dalmon, J., Bayen, M., and Gilet, R. (1975). *Colloq. Int. C.N.R.S.* **240,** 179–183.
Day, P. R. (1977). *Science* **197,** 1334–1339.
Delehanty, J., Farmerie, W. G., and Barnett, W. E. (1983). *In* "The Biology of *Euglena*" (D. E. Buetow, ed.), Vol. 4. Academic Press, New York (in press).
Delihas, N., Andersen, J., Sprouse, H. M., and Dudock, B. (1981). *Nucleic Acids Res.* **9,** 2801–2805.
Delius, H., and Koller, B. (1980). *J. Mol. Biol.* **142,** 247–261.
Driesel, A. J., Crouse, E. J., Gordon, K., Bohnert, H.-J., Herrmann, R. G., Steinmetz, A., Mubumbila, M., Keller, M., Burkard, G., and Weil, J. H. (1979). *Gene* **6,** 285–306.
Driesel, A. J., Speirs, J., and Bohnert, H. J. (1980). *Biochim. Biophys. Acta* **610,** 297–310.
Dube, S. K., and Marcker, K. A. (1969). *Eur. J. Biochem.* **8,** 256–262.
Dyer, T. A., and Bedbrook, J. R. (1980). *In* "Genome Organization and Expression in Plants" (C. J. Leaver, ed.), pp. 305–311. Plenum, New York.
Dyer, T. A., and Bowman, C. M. (1976). *In* "Genetics and Biogenesis of Chloroplasts and Mitochondria" (T. Bücher, W. Neupert, W. Sebald, and S. Werner, eds.), pp. 645–651. Elsevier/North-Holland Biomedical Press, Amsterdam.
Dyer, T. A., and Leech, R. M. (1968). *Biochem. J.* **106,** 689–698.
Dyer, T. A., Bowman, C. M., and Payne, P. I. (1977). *In* "Nucleic Acids and Protein Synthesis in Plants" (L. Bogorad and J. H. Weil, eds.), pp. 121–133. Plenum, New York.
Edelman, M. (1981). *In* "The Biochemistry of Plants" (A. Marcus, ed.), Vol. 6, pp. 249–301. Academic Press, New York.
Edelman, M., and Reisfeld, A. (1978). *In* "Chloroplast Development" (G. Akoyunoglou and J. H. Argyroudi-Akoyunoglou, eds.), pp. 641–652. Elsevier/North-Holland, Amsterdam.
Edelman, M., and Reisfeld, A. (1980). *In* "Genome Organization and Expression in Plants" (C. J. Leaver, ed.), pp. 353–362. Plenum, New York.
Edelman, M., Cowan, C. A. Epstein, H. T., and Schiff, J. A. (1964). *Proc. Natl. Acad. Sci. U.S.A.* **52,** 1214–1219.
Edelman, M., Sagher, D., and Reisfeld, A. (1977). *Colloq. Int. C.N.R.S.* **261,** 305–311.
Edwards, K., and Kössel, H. (1981). *Nucleic Acids Res.* **9,** 2853–2869.
Edwards, K., Bedbrook, J., Dyer, T., and Kössel, H. (1981). *Biochem. Internatl.* **2,** 533–538.
Eisenstadt, J. M., and Brawerman, G. (1964). *J. Mol. Biol.* **10,** 392–402.

Ellis, R. J. (1977a). *In* "The Molecular Biology of Plant Cells" (H. Smith, ed.), pp. 280–305. Blackwell, Oxford.
Ellis, R. J. (1977b). *Biochim. Biophys. Acta* **463,** 185–215.
Ellis, R. J., and Hartley, M. R. (1974). *MTP Int. Rev. Sci., Biochem.* **6,** 323–348.
Erion, J. L., Jarnowski, J., Weissbach, H., and Brot, N. (1981). *Proc. Natl. Acad. Sci. U.S.A.* **78,** 3459–3463.
Fairfield, S. A., and Barnett, W. E. (1971). *Proc. Natl. Acad. Sci. U.S.A.* **68,** 2972–2976.
Falk, H. (1969). *J. Cell Biol.* **42,** 582–587.
Fluhr, R., and Edelman, M. (1981). *Mol. Gen. Genet.* **181,** 484–490.
Fox, G. E., Stackebrandt, E., Hespell, R. B., Gibson, J., Maniloff, J., Dyer, T. A., Wolfe, R. S., Balch, W. E., Tanner, R. S., Magrum, L. J., Zablen, L. B., Blakemore, R., Gupta, R., Bonen, L., Lewis, B. J., Stahl, D. A., Luehrsen, K. R., Chen, K. N., and Woese, C. R. (1980). *Science* **209,** 457–463.
Frankel, R., Scowcroft, W. R., and Whitfeld, P. R. (1979). *Mol. Gen. Genet.* **169,** 129–135.
Freyssinet, G. (1977a). *Biochimie* **59,** 597–610.
Freyssinet, G. (1977b). *Physiol. Veg.* **15,** 519–550.
Freyssinet, G., Gallagher, T., Eichholz, R. L., Freyssinet, M., and Buetow, D. E. (1981). *Proc. Int. Cong. Photosynth., 5th,* 1980, **5,** 809–820.
Galling, G. (1971). *Planta* **98,** 50–62.
Galling, G., and Jordan, B. R. (1972). *Biochimie* **54,** 1257–1265.
Gelvin, S. R., and Howell, S. H. (1979). *Mol. Gen. Genet.* **173,** 315–322.
Gelvin, S. R., Heizmann, P., and Howell, S. H. (1977). *Proc. Natl. Acad. Sci. U.S.A.* **74,** 3193–3197.
Gibbs, S. P., and Poole, R. J. (1973). *J. Cell Biol.* **59,** 318–328.
Gillham, N. W. (1978). "Organelle Heredity." Raven, New York.
Gordon, K. H. J., Crouse, E. J., Bohnert, H. J., and Herrmann, R. G. (1981). *Theor. Appl. Genet.* **59,** 281–296.
Graf, L. Kössel, H., and Stutz, E. (1980). *Nature (London)* **286,** 908–910.
Grant, D., Swinton, D. C., and Chiang, K.-S. (1978). *Planta* **141,** 259–267.
Gray, P. W., and Hallick, R. B. (1977). *Biochemistry* **16,** 1665–1671.
Gray, P. W., and Hallick, R. B. (1978). *Biochemistry* **17,** 284–289.
Gray, P. W., and Hallick, R. B. (1979a). *Biochim. Biophys. Acta* **561,** 53–58.
Gray, P. W., and Hallick, R. B. (1979b). *Biochemistry* **18,** 1820–1825.
Gray, R. E., and Cashmore, A. R. (1976). *J. Mol. Biol.* **108,** 595–608.
Grebanier, A., Coen, D. M., Rich, A., and Bogorad, L. (1978). *J. Cell Biol.* **78,** 734–746.
Grebanier, A., Steinback, K., and Bogorad, L. (1979). *Plant Physiol.* **63,** 436–439.
Green, B. R. (1976). *Biochim. Biophys. Acta* **447,** 156–166.
Green, B. R., and Burton, H. (1970). *Science* **168,** 981–982
Green, B. R., Muir, B. L., and Padmanabhan, U. (1977). *In* "Progress in *Acetabularia* Research" (C. L. F. Woodcock, ed.), pp. 107–122. Academic Press, New York.
Grierson, D. (1974). *Eur. J. Biochem.* **44,** 509–515.
Grierson, D., and Loening, U. E. (1974). *Eur. J. Biochem.* **44,** 501–507.
Grossman, L. I., Watson, R., and Vinograd, J. (1973). *Proc. Natl. Acad. Sci. U.S.A.* **70,** 3339–3343.
Gruol, D. J., and Haselkorn, R. (1976). *Biochim. Biophys. Acta* **447,** 82–95.
Guillemaut, P., and Keith, G. (1977). *FEBS Lett.* **84,** 351–356.
Guillemaut, P., and Weil, J. H. (1975). *Biochim. Biophys. Acta* **407,** 240–248.
Guillemaut, P., Burkard, G., Steinmetz, A., and Weil, J. H. (1973). *Plant Sci. Lett.* **1,** 141–149.
Guillemaut, P., Steinmetz, A., Burkard, G., and Weil, J. H. (1975). *Biochim. Biophys. Acta* **378,** 64–72.
Guillemaut, P., Martin, R., and Weil, J. H. (1976). *FEBS Lett.* **63,** 273–277.

Haff, L. A., and Bogorad, L. (1976a). *Biochemistry* **15,** 4105–4109.

Haff, L. A., and Bogorad, L. (1976b). *Biochemistry* **15,** 4110–4115.

Hallick, R. B. (1983). *In* "The Biology of *Euglena*" (D. E. Buetow, ed.), Vol. 4. Academic Press, New York (in press).

Hallick, R. B., Lipper, C., Richards, O. C., and Rutter, W. J. (1976). *Biochemistry* **15,** 3039–3045.

Hallick, R. B., Gray, P. W., Chelm, B. K., Rushlow, K. E., and Orozco, E. M., Jr. (1978). *In* "Chloroplast Development" (G. Akoyunoglou and J. H. Argyroudi-Akoyunoglou, eds.), pp. 619–622. Elsevier/North-Holland, Amsterdam.

Hanson, M. R., Davidson, J. N., Mets, L. J., and Bogorad, L. (1974). *Mol. Gen. Genet.* **132,** 105–118.

Harel, E., and Bogorad, L. (1973). *Plant Physiol.* **51,** 10–16.

Hartley, M. R. (1979). *Eur. J. Biochem.* **96,** 311–320.

Hartley, M. R., and Ellis, R. J. (1973). *Biochem. J.* **134,** 249–262.

Hartley, M. R., and Head, C. (1979). *Eur. J. Biochem.* **96,** 301–309.

Hartley, M. R., Wheeler, A., and Ellis, R. J. (1975). *J. Mol. Biol.* **91,** 67–77.

Hartley, M. R., Head, C. W., and Gardiner, J. (1977). *Colloq. Int. C.N.R.S.* **261,** 419–423.

Hattersley, P. W., Watson, L., and Osmond, C. B. (1977). *Aust. J. Plant Physiol.* **4,** 523–539.

Hecker, L. I., Egan, J., Reynolds, R. J., Nix, C. E., Schiff, J. A., and Barnett, W. E. (1974). *Proc. Natl. Acad. Sci. U.S.A.* **71,** 1910–1914.

Hecker, L. I., Uziel, M., and Barnett, W. E. (1976). *Nucleic Acids Res.* **3,** 371–380.

Hedberg, M. F., Huang, Y.-S., and Hommersand, M. H. (1981). *Science* **213,** 445–447.

Heizmann, P. (1970). *Biochim. Biophys. Acta* **224,** 144–154.

Heizmann, P. (1974). *Biochem. Biophys. Res. Commun.* **56,** 112–118.

Heizmann, P., Verdier, G., and Younis, H. (1978). *In* "Chloroplast Development" (G. Akoyunoglou and J. H. Argyroudi-Akoyunoglou, eds.), pp. 623–628. Elsevier/North-Holland, Amsterdam.

Helling, R. B., El-Gewely, M. R., Lomax, M. I., Baumgartner, J. E., Schwartzbach, S. D., and Barnett, W. E. (1979). *Mol. Gen. Genet.* **174,** 1–10.

Herrmann, R. G., and Possingham, J. V. (1980). *In* "Chloroplasts" (J. Reinert, ed.), pp. 45–96. Springer-Verlag, Berlin and New York.

Herrmann, R. G., Bohnert, H.-J., Kowallik, K. V., and Schmitt, J. M. (1975). *Biochim. Biophys. Acta* **378,** 305–317.

Herrmann, R. G., Bohnert, H.-J., Driesel, A., and Hobom, G. (1976). *In* "Genetics and Biogenesis of Chloroplasts and Mitochondria" (T. Bücher, W. Neupert, W. Sebald, and S. Werner, eds.), pp. 351–359. Elsevier/North-Holland Biomedical Press, Amsterdam.

Herrmann, R. G., Whitfeld, P. R., and Bottomley, W. (1980). *Gene* **8,** 179–181.

Hotta, Y., Bassel, A., and Stern, H. (1965). *J. Cell Biol.* **27,** 451–457.

Howell, S. H., and Walker, L. L. (1976). *Biochim. Biophys. Acta* **418,** 249–256.

Howell, S. H., and Walker, L. L. (1977). *Dev. Biol.* **56,** 11–23.

Howell, S. H., Heizmann, P., Gelvin, S., and Walker, L. L. (1977). *Plant Physiol.* **59,** 464–470.

Ikegami, M., and Fraenkel-Conrat, H. (1979). *Proc. Natl. Acad. Sci. U.S.A.* **76,** 3637–3640.

Ingle, J., Possingham, J. V., Wells, R., Leaver, C. J., and Loening, U. E. (1970). *Symp. Soc. Exp. Biol.* **24,** 303–325.

Iwamura, T. (1955). *J. Biochem. (Tokyo)* **42,** 575–589.

Iwamura, T. (1960). *Biochim. Biophys. Acta* **42,** 161–163.

Iwamura, T. (1962). *Biochim. Biophys. Acta* **61,** 472–474.

Iwamura, T. (1966). *Prog. Nucleic Acid Res. Mol. Biol.* **5,** 133–155.

James, T. W., and Jope, C. (1978). *J. Cell Biol.* **79,** 623–630.

Jeannin, G., Burkard, G., and Weil, J. H. (1976). *Biochim. Biophys. Acta* **442,** 24–31.

Jeannin, G., Burkard, G., and Weil, J. H. (1978). *Plant Sci. Lett.* **13,** 75–81.

Jenni, B., and Stutz, E. (1978). *Eur. J. Biochem.* **88,** 127–134.
Jenni, B., and Stutz, E. (1979). *FEBS Lett.* **102,** 95–99.
Jenni, B., Fasnacht, M., and Stutz, E. (1981). *FEBS Lett.* **125,** 175–179.
Jolly, S. O., and Bogorad, L. (1980). *Proc. Natl. Acad. Sci. U.S.A.* **77,** 822–826.
Joussaume, M. (1973). *Physiol. Veg.* **11,** 69–82.
Jurgensen, J. E., and Bourque, D. P. (1980). *Nucleic Acids Res.* **8,** 3505–3516.
Kashdan, M. A., Pirtle, R. M., Pirtle, I. L., Calagan, J. L., Vreman, H. J., and Dudock, B. S. (1980). *J. Biol. Chem.* **255,** 8831–8835.
Kato, A., Shimada, H., Kusuda, M., and Sugiura, M. (1981). *Nucleic Acids Res.* **9,** 5601–5607.
Keller, M., Burkard, G., Bohnert, H. J., Mubumbila, M., Gordon, K., Steinmetz, A., Heiser, D., Crouse, E. J., and Weil, J. H. (1980). *Biochem. Biophys. Res. Commun.* **95,** 47–54.
Keller, S. J., and Ho, C. (1981). *Int. Rev. Cytol.* **69,** 157–190.
Keller, S. J., Biedenbach, S. A., and Meyer, R. R. (1973). *Biochem. Biophys. Res. Commun.* **50,** 620–628.
Kenerley, M. E., Morgan, E. A., Post, L., Lindahl, L., and Nomura, M. (1977). *J. Bacteriol.* **132,** 931–949.
Kidd, G., and Bogorad, L. (1980). *Biochim. Biophys. Acta* **609,** 14–30.
Kirk, J. T. O. (1964a). *Biochem. Biophys. Res. Commun.* **14,** 393–397.
Kirk, J. T. O. (1964b). *Biochem. Biophys. Res. Commun.* **16,** 233–238.
Kirk, J. T. O., and Tilney-Bassett, R. A. E. (1978). "The Plastids," 2nd ed. Elsevier/North-Holland, Amsterdam.
Knopf, U. C., and Stutz, E. (1978). *Mol. Gen. Genet.* **163,** 1–6.
Koch, W., Edwards, K., and Kössel, H. (1981). *Cell* **25,** 203–213.
Koller, B. and Delius, H. (1980). *Mol. Gen. Genet.* **178,** 261–269.
Kolodner, R., and Tewari, K. K. (1972). *J. Biol. Chem.* **247,** 6355–6364.
Kolodner, R., and Tewari, K. K. (1975a). *J. Biol. Chem.* **250,** 4888–4895.
Kolodner, R., and Tewari, K. K. (1975b). *Biochim. Biophys. Acta* **402,** 372–390.
Kolodner, R., and Tewari, K. K. (1975c). *Nature (London)* **256,** 708–711.
Kolodner, R., and Tewari, K. K. (1979). *Proc. Natl. Acad. Sci. U.S.A.* **76,** 41–45.
Kolodner, R., Warner, R. C., and Tewari, K. K. (1975). *J. Biol. Chem.* **250,** 7020–7026.
Kolodner, R., Tewari, K. K., and Warner, R. C. (1976). *Biochim. Biophys. Acta* **447,** 144–155.
Kopecka, H., Crouse, E. J., and Stutz, E. (1977). *Eur. J. Biochem.* **72,** 525–535.
Kowallik, K. V., and Herrmann, R. G. (1972). *J. Cell Sci.* **11,** 357–377.
Kung, S.-d. (1977). *Annu. Rev. Plant Physiol.* **28,** 401–437.
Kung, S.-d., Lee, C. L., Wood, D. D., and Moscarello, M. A. (1977). *Plant Physiol.* **60,** 89–94.
Kwok, S. Y., and Wildman, S. G. (1974). *J. Mol. Evol.* **3,** 103–108.
Lamppa, G. K., and Bendich, A. J. (1979). *Plant Physiol.* **63,** 660–668.
Lea, P. J., and Norris, R. D. (1977). *Prog. Phytochem.* **4,** 121–167.
Leaver, C. J. (1979). *In* "Nucleic Acids in Plants" (T. C. Hall and J. W. Davies, eds.), Vol. 1, pp. 193–215. CRC Press, Boca Raton, Florida.
Leaver, C. J., and Ingle, J. (1971). *Biochem. J.* **123,** 235–243.
Leis, J. P., and Keller, E. B. (1970). *Biochem. Biophys. Res. Commun.* **40,** 416–421.
Lemieux, C., Turmel, M., and Lee, R. W. (1980). *Curr. Genet.* **2,** 139–147.
Lewin, B. (1974). "Gene Expression," Vol. 1. Wiley, New York.
Lewin, B. (1980). "Gene Expression," 2nd ed., Vol. 2. Wiley (Interscience), New York.
Link, G. (1981). *Nucleic Acids Res.* **9,** 3681–3694.
Link, G., and Bogorad, L. (1980). *Proc. Natl. Acad. Sci. U.S.A.* **77,** 1832–1836.
Link, G., Coen, D. M., and Bogorad, L. (1978a). *In* "Chloroplast Development" (G.

Akoyunoglou and J. H. Argyroudi-Akoyunoglou, eds.), pp. 559–564. Elsevier/North-Holland, Amsterdam.
Link, G., Coen, D. M., and Bogorad, L. (1978b). *Cell* **15,** 725–732.
Link, G., Chambers, S. E., Thompson, J. A., and Falk, H. (1981). *Mol. Gen. Genet.* **181,** 454–457.
Locy, R. D., and Cherry, J. H. (1978). *Phytochemistry* **17,** 19–27.
Loening, U. E., and Ingle, J. (1967). *Nature (London)* **215,** 363–367.
Loiseaux, S., Rozier, S. C., and Dalmon, J. (1980). *Plant Sci. Lett.* **18,** 381–388.
Lüttke, A. (1981). *Exp. Cell Res.* **131,** 483–488.
Lyman, H., and Srinivas, U. K. (1978). *In* "Chloroplast Development" (G. Akoyunoglou and J. H. Argyroudi-Akoyunoglou, eds.), pp. 593–607. Elsevier/North-Holland, Amsterdam.
Lyttleton, J. W. (1962). *Exp. Cell Res.* **26,** 312–317.
McCrea, J. M., and Hershberger, C. L. (1976). *Nucleic Acids Res.* **3,** 2005–2018.
McIntosh, L., Poulsen, C., and Bogorad, L. (1980). *Nature (London)* **288,** 556–560.
Malnoe, P., and Rochaix, J.-D. (1978). *Mol. Gen. Genet.* **166,** 269–275.
Malnoe, P., Rochaix, J.-D., Chua, N. H., and Spahr, P. F. (1979). *J. Mol. Biol.* **133,** 417–434.
Manning, J. E., and Richards, O. C. (1972a). *Biochim. Biophys. Acta* **259,** 285–296.
Manning, J. E., and Richards, O. C. (1972b). *Biochemistry* **11,** 2036–2043.
Manning, J. E., Wolstenholme, D. R., Ryan, R. S., Hunter, J. A., and Richards, O. C. (1971). *Proc. Natl. Acad. Sci. U.S.A.* **68,** 1169–1173.
Marano, F. (1979). *Biol. Cell.* **36,** 65–70.
Margulies, M. M. (1980). *Biochim. Biophys. Acta* **606,** 13–19.
Margulies, M. M., and Michaels, A. (1974). *J. Cell Biol.* **60,** 65–77.
Margulies, M. M., and Weistrop, J. (1976). *In* "Genetics and Biogenesis of Chloroplasts and Mitochondria" (T. Bücher, W. Neupert, W. Sebald, and S. Werner, eds.), pp. 657–660. Elsevier/North-Holland Biomedical Press, Amsterdam.
Matsuda, Y., and Surzycki, S. J. (1980). *Mol. Gen. Genet.* **180,** 463–474.
Mattoo, A. K., Pick, U., Hoffman-Falk, H., and Edelman, M. (1981). *Proc. Natl. Acad. Sci. U.S.A.* **78,** 1572–1576.
Meeker, R., and Tewari, K. K. (1980). *Biochemistry* **19,** 5973–5981.
Mego, J. L., and Buetow, D. E. (1967). *In* "Le chloroplaste: Croissance et vieillissement" (C. Sironval, ed.), pp. 274–290. Masson, Paris.
Mendiola-Morgenthaler, L., Eikenberry, E. F., and Price, C. A. (1975). *Plant Cell Physiol.* **16,** 981–994.
Merrick, W. C., and Dure, L. S. (1971). *Proc. Natl. Acad. Sci. U.S.A.* **68,** 641–644.
Miller, M. J., and McMahon, D. (1974). *Biochim. Biophys. Acta* **366,** 35–44.
Milner, J. J., Hershberger, C. L., and Buetow, D. E. (1979). *Plant Physiol.* **64,** 818–821.
Miyaki, M., Koide, K., and Ono, T. (1973). *Biochem. Biophys. Res. Commun.* **50,** 252–258.
Mubumbila, M., Burkard, G., Keller, M., Steinmetz, A., Crouse, E., and Weil, J. H. (1980). *Biochim. Biophys. Acta* **609,** 31–39.
Mullet, J. E., and Arntzen, C. J. (1981). *Biochim. Biophys. Acta* **635,** 236–241.
Munsche, D., and Wollgiehn, R. (1973). *Biochim. Biophys. Acta* **294,** 106–117.
Ness, P. J., and Woolhouse, H. W. (1980). *J. Exp. Bot.* **31,** 223–234.
Nigon, V., and Heizmann, P. (1978). *Int. Rev. Cytol.* **53,** 211–290.
Nigon, V., and Verdier, G. (1983). *In* "The Biology of *Euglena*" (D. E. Buetow, ed.), Vol. 4. Academic Press, New York (in press).
Nigon, V., Verdier, G., Salvador, G., Heizmann, P., Ravel- Chapuis, P., and Freyssinet, G. (1978). *In* "Chloroplast Development" (G. Akoyunoglou and J. H. Argyroudi-Akoyunoglou, eds.), pp. 629–640. Elsevier/North-Holland, Amsterdam.
Nomura, M., and Morgan, E. A. (1977). *Annu. Rev. Genet.* **11,** 297–347.
Nover, L. (1976). *Plant Sci. Lett.* **7,** 403–407.

Orozco, E. M., Jr., Gray, P. W., and Hallick, R. B. (1980a). *J. Biol Chem.* **255,** 10991–10996.

Orozco, E. M., Jr., Rushlow, K. E., Dodd, J. R., and Hallick, R. B. (1980b). *J. Biol. Chem.* **255,** 10997–11003.

Osorio-Almeida, M. L., Guillemaut, P., Keith, G., Canaday, J., and Weil, J. H. (1980). *Biochem. Biophys. Res. Commun.* **92,** 102–108.

Padmanabhan, U., and Green, B. R. (1978). *Biochim. Biophys. Acta* **521,** 67–73.

Palmer, J. D., and Thompson, W. F. (1981). *Proc. Natl. Acad. Sci. U.S.A.* **78,** 5533–5537.

Parthier, B. (1973). *FEBS Lett.* **38,** 70–74.

Parthier, B., and Krauspe, R. (1974). *Biochem. Physiol. Pflanz.* **165,** 1–17.

Parthier, B., and Neumann, D. (1977). *Biochem. Physiol. Pflanz.* **171,** 547–562.

Parthier, B., Mueller-Uri, F., and Krauspe, R. (1978). *In* "Chloroplast Development" (G. Akoyunoglou and J. H. Argyroudi-Akoyunoglou, eds.), pp. 687–693. Elsevier/North-Holland, Amsterdam.

Payne, P. I., and Dyer, T. A. (1971). *Biochem. J.* **124,** 83–89.

Phillips, D. O., and Carr, N. G. (1975). *FEBS Lett.* **60,** 94–97.

Pigott, G. H., and Carr, N. G. (1972). *Science* **175,** 1259–1261.

Pirtle, R., Calagan, J., Pirtle, I., Kashdan, M., Vreman, H., and Dudock, B. (1981). *Nucleic Acids Res.* **9,** 183–188.

Polya, G. M., and Jagendorf, A. T. (1971). *Arch. Biochem. Biophys.* **146,** 649–657.

Posner, H. B., and Rosner, A. (1975). *Plant Cell Physiol.* **16,** 361–365.

Radmer, R., and Kok, B. (1977). *BioScience* **27,** 599–605.

Rawson, J. R. Y., and Boerma, C. L. (1976a). *Proc. Natl. Acad. Sci. U.S.A.* **73,** 2401–2404.

Rawson, J. R. Y., and Boerma, C. L. (1976b). *Biochemistry* **15,** 588–592.

Rawson, J. R. Y., and Boerma, C. L. (1977). *Biochem. Biophys. Res. Commun.* **74,** 912–918.

Rawson, J. R. Y., and Boerma, C. L. (1979). *Biochem. Biophys. Res. Commun.* **89,** 743–749.

Rawson, J. R. Y., and Haselkorn, R. (1973). *J. Mol. Biol.* **77,** 125–132.

Rawson, J. R. Y., Kushner, S. R., Vapnek, D., Alton, N. K., and Boerma, C. L. (1978). *Gene* **3,** 191–209.

Rawson, J. R. Y., Boerma, C. L., Andrews, W. H., and Wilkerson, C. G. (1981). *Biochemistry* **20,** 2639–2644.

Reger, B. J., Fairfield, S. A., Epler, J. L., and Barnett, W. E. (1970). *Proc. Natl. Acad. Sci. U.S.A.* **67,** 1207–1213.

Reisfeld, A., Jakob, K. M., and Edelman, M. (1978). *In* "Chloroplast Development" (G. Akoyunoglou and J. H. Argyroudi-Akoyunoglou, eds.), pp. 669–674. Elsevier/North-Holland, Amsterdam.

Renger, G. (1976). *Biochim. Biophys. Acta* **440,** 287–300.

Richards, O. C., and Manning, J. E. (1975). *Colloq. Int. C.N.R.S.* **240,** 213–221.

Ris, H., and Plaut, W. (1962). *J. Cell Biol.* **13,** 383–391.

Roberts, R. J. (1976). *CRC Crit. Rev. Biochem.* **4,** 123–164.

Roberts, R. J. (1977). *In* "DNA Insertion Elements, Plasmids and Episomes" (A. I. Bukhari, J. A. Shapiro, and S. L. Adhya, eds.), pp. 757–768. Cold Spring Harbor Lab., Cold Spring Harbor, New York.

Rochaix, J.-D. (1972). *Nature (London), New Biol.* **238,** 76–78.

Rochaix, J.-D. (1977). *Colloq. Int. C.N.R.S.* **261,** 77–83.

Rochaix, J.-D. (1978). *J. Mol. Biol.* **126,** 597–617.

Rochaix, J.-D. (1981). *Experientia* **37,** 323–332.

Rochaix, J.-D., and Malnoe, P. (1978a). *Cell* **15,** 661–670.

Rochaix, J.-D., and Malnoe, P. (1978b). *In* "Chloroplast Development" (G. Akoyunoglou and J. H. Argyroudi-Akoyunoglou, eds.), pp. 581–586. Elsevier/North-Holland, Amsterdam.

Rose, R. J., Cran, D. G., and Possingham, J. V. (1974). *Nature (London)* **251,** 641–642.

Rosner, A., Porath, D., and Gressel, J. (1974). *Plant Cell. Physiol.* **15,** 891–902.

Rosner, A., Jakob, K. M., Gressel, J., and Sagher, D. (1975). *Biochem. Biophys. Res. Commun.* **67,** 383–391.
Rosner, A., Gressel, J., and Jakob, K. M. (1977a). *Biochim. Biophys. Acta* **474,** 386–397.
Rosner, A., Reisfeld, A., Jakob, K. M., Gressel, J., and Edelman, M. (1977b). *Colloq. Int. C.N.R.S.* **261,** 561–568.
Rushlow, K. E., Orozco, E. M., Jr., Lipper, C., and Hallick, R. B. (1980). *J. Biol. Chem.* **255,** 3776–3792.
Sager, R. (1972). "Cytoplasmic Genes and Organelles." Academic Press, New York.
Sager, R. (1977). *Adv. Genet.* **19,** 287–340.
Sager, R., and Ishida, M. (1963). *Proc. Natl. Acad. Sci. U.S.A.* **50,** 725–730.
Sagher, D., Edelman, M., and Jakob, K. M. (1974). *Biochim. Biophys. Acta* **349,** 32–38.
Sagher, D., Grosfeld, H., and Edelman, M. (1976). *Proc. Natl. Acad. Sci. U.S.A.* **73,** 722–726.
Sampson, M., Katch, A., Hotta, Y., and Stern, H. (1963). *Proc. Natl. Acad. Sci. U.S.A.* **50,** 459–463.
Sano, H., Spaeth, E., and Burton, W. G. (1979). *Eur. J. Biochem.* **93,** 173–180.
Sarantoglou, V., Imbault, P., and Weil, J. H. (1978). *In* "Chloroplast Development" (G. Akoyunoglou and J. H. Argyroudi-Akoyunoglou, eds.), pp. 695–700. Elsevier/North-Holland, Amsterdam.
Schiemann, J., Wollgiehn, R., and Parthier, B. (1977). *Biochem. Physiol. Pflanz.* **171,** 474–478.
Schiff, J. A. (1973). *Adv. Morphog.* **10,** 265–312.
Schwartzbach, S. D., Hecker, L. I., and Barnett, W. E. (1976). *Proc. Natl. Acad. Sci. U.S.A.* **73,** 1984–1988.
Schwarz, Zs., and Kössel, H. (1980). *Nature (London)* **283,** 520–523.
Schwarz, Zs., Jolly, S. O., Steinmetz, A., and Bogorad, L. (1981). *Proc. Natl. Acad. Sci. U.S.A.* **78,** 3423–3427.
Scott, N. S. (1976). *Phytochemistry* 15, 1207–1213.
Scott, N. S. (1977). *Colloq. Int. C.N.R.S.* **261,** 431–434.
Scott, N. S., and Smillie, R. M. (1967). *Biochem. Biophys. Res. Commun.* **28,** 598–603.
Scott, N. S., Shah, V. C., and Smillie, R. M. (1968). *J. Cell Biol.* **38,** 151–157.
Šebesta, K., Bauveravá, J., and Šormová, Z. (1965). *Biochem. Biophys. Res. Commun.* **19,** 54–61.
Selsky, M. I. (1978). *Biochim. Biophys. Acta* **520,** 555–567.
Semal, J., Spencer, D., Kim. Y. T., and Wildman, S. G. (1964). *Biochim. Biophys. Acta* **91,** 205–216.
Senger, H., and Bishop, N. I. (1966). *Plant Cell Physiol.* **7,** 441–455.
Seyer, P., Kowallik, K. V., and Herrmann, R. G. (1981). *Curr. Genet.* **3,** 189–204.
Shah, V. C., and Lyman, H. (1966). *J. Cell Biol.* **29,** 174–176.
Sirevag, R., and Levine, R. P. (1972). *J. Biol. Chem.* **247,** 2586–2591.
Siu, C.-H., Chiang, K.-S., and Swift, H. (1975). *J. Mol. Biol.* **98,** 369–391.
Slavik, N. S., and Hershberger, C. L. (1975). *FEBS Lett.* **52,** 171–174.
Slavik, N. S., and Hershberger, C. L. (1976). *J. Mol. Biol.* **103,** 563–581.
Smith, H. J., and Bogorad, L. (1974). *Proc. Natl. Acad. Sci. U.S.A.* **71,** 4839–4842.
Spencer, D., and Whitfeld, P. R. (1967a). *Biochem. Biophys. Res. Commun.* **28,** 538–542.
Spencer, D., and Whitfeld, P. R. (1967b). *Arch. Biochem. Biophys.* **121,** 336–345.
Spencer, D., and Whitfeld, P. R. (1969). *Arch. Biochem. Biophys.* **132,** 477–488.
Spiers, J., and Grierson, D. (1978). *Biochim. Biophys. Acta* **521,** 619–633.
Sprinzl, M., Grueter, F., Spelzhaus, A., and Gauss, D. H. (1980). *Nucleic Acids Res.* **8,** r1–r22.

Sprouse, H. M., Kashdan, M., Otis, L., and Dudock, B. (1981). *Nucleic Acids Res.* **9,** 2543–2547.
Stange, L., Kirk, M., Bennett, E. L., and Calvin, M. (1962). *Biochim. Biophys. Acta* **61,** 681–695.
Steinmetz, A., and Weil, J. H. (1976). *Biochim. Biophys. Acta* **454,** 429–435.
Steinmetz, A., Mubumbila, V., Keller, M., Burkard, G., Weil, J. H., Driesel, A. J., Crouse, E. J., Gordon, K., Bohnert, H.-J., and Herrmann, R. G. (1978). *In* "Chloroplast Development" (G. Akoyunoglou and J. H. Argyroudi-Akoyunoglou, eds.), pp. 573–580. Elsevier/North-Holland, Amsterdam.
Steinmetz, A., Mubumbila, M., Keller, M., Burkard, G., and Weil, J. H. (1980). *In* "Transfer RNA: Biological Aspects" (D. Söll, J. N. Abelson, and P. R. Schimmel, eds.), pp. 281–286. Cold Spring Harbor Lab., Cold Spring Harbor, New York.
Stempka, R., and Richter, G. (1978). *Arch. Microbiol.* **119,** 187–196.
Stutz, E. (1970). *FEBS Lett.* **8,** 25–28.
Stutz, E., and Noll, E. (1967). *Proc. Natl. Acad. Sci. U.S.A.* **57,** 744–781.
Stutz, E., and Vandrey, J. P. (1971). *FEBS Lett.* **17,** 277–280.
Stutz, E., Jenni, B., Knopf, U. C., and Graf, L. (1978). *In* "Chloroplast Development" (G. Akoyunoglou and J. H. Argyroudi-Akoyunoglou, eds.), pp. 609–618. Elsevier/North-Holland, Amsterdam.
Sugiura, M. (1980). *Curr. Genet.* **2,** 95–98.
Sugiura, M., and Hotta, Y. (1980). *Plant Cell Physiol.* **21,** 1129–1132.
Surzycki, S. J. (1969). *Proc. Natl. Acad. Sci. U.S.A.* **63,** 1327–1334.
Surzycki, S. J., and Shellenbarger, D. L. (1976). *Proc. Natl. Acad. Sci. U.S.A.* **73,** 3961–3965.
Takaiwa, F., and Sugiura, M. (1980). *Nucleic Acids Res.* **8,** 4125–4129.
Takaiwa, F., and Sugiura, M. (1981). *Mol. Gen. Genet.* **182,** 385–389.
Takaiwa, F., Tohdoh, N., and Sugiura, M. (1980). *Jpn. J. Genet.* **55,** 121–125.
Tao, K.-L. J., and Jagendorf, A. T. (1973). *Biochim. Biophys. Acta* **324,** 518–532.
Tewari, K. K. (1971). *Annu. Rev. Plant Physiol.* **22,** 141–168.
Tewari, K. K. (1979). *In* "Nucleic Acids in Plants" (T. C. Hall and J. W. Davies, eds.), Vol. I, pp. 41–108. CRC Press, Boca Raton, Florida.
Tewari, K. K., and Wildman, S. G. (1966). *Science* **153,** 1269–1271.
Tewari, K. K., and Wildman, S. G. (1967). *Proc. Natl. Acad. Sci. U.S.A.* **58,** 689–696.
Tewari, K. K., and Wildman, S. G. (1968). *Proc. Natl. Acad. Sci. U.S.A.* **59,** 569–576.
Tewari, K. K., and Wildman, S. G. (1969). *Biochim. Biophys. Acta* **186,** 358–372.
Tewari, K. K., and Wildman, S. G. (1970). *Symp. Soc. Exp. Biol.* **24,** 147–179.
Thomas, J. R., and Tewari, K. K. (1974). *Proc. Natl. Sci. U.S.A.* **71,** 3147–3151.
Thompson, R., Hughes, S. G., and Broda, P. (1974). *Mol. Gen. Genet.* **133,** 144–149.
Timothy, D. H., Levings, C. S., III, Pring, D. R., Conde, M. F., and Kermicle, J. L. (1979). *Proc. Natl. Acad. Sci. U.S.A.* **76,** 4220–4224.
Tobin, E. M. (1978). *Proc. Natl. Acad. Sci. U.S.A.* **75,** 4749–4753.
Uzzo, A., and Lyman, H. (1972). *Photosynth., Two Centuries Its Discovery Joseph Priestley, Proc. Int. Congr. Photosynth. Res., 2nd, 1971* pp. 2585–2589.
Vacek, A. T., and Bourque, D. P. (1980). *Plasmid* **4,** 205–214.
Vandrey, J. P., and Stutz, E. (1973). *FEBS Lett.* **37,** 174–177.
van Ee, J. H., Vos, Y. J., and Planta, R. J. (1980). *Gene* **12,** 191–200.
Vedel, F., Quétier, F., and Bayne, M. (1976). *Nature (London)* **263,** 440–442.
Verdier, G. (1979a). *Eur. J. Biochem.* **93,** 573–580.
Verdier, G. (1979b). *Eur. J. Biochem.* **93,** 581–589.
Walbot, V. (1977). *Cell* **11,** 729–733.
Walfield, A. M., and Hershberger, C. L. (1978). *J. Bacteriol.* **133,** 1437–1443.
Wanka, F., Joosten, H. F. P., and deGrip, W. J. (1970). *Arch. Mikrobiol.* **75,** 25–36.

Weil, J. H. (1979). *In* "Nucleic Acids in Plants" (T. C. Hall and J. W. Davies, eds.), Vol. 1, pp. 143–192. CRC Press, Boca Raton, Florida.

Wells, R., and Birnstiel, M. (1969). *Biochem. J.* **112,** 777–786.

Wells, R., and Sager, R. (1971). *J. Mol. Biol.* **58,** 611–622.

Wetmur, J. G., and Davidson, N. (1968). *J. Mol. Biol.* **31,** 349–370.

Wheeler, A. M., and Hartley, M. R. (1975). *Nature (London)* **257,** 66–67.

Whitfeld, P. R. (1977). *In* "The Ribonucleic Acids" (P. R. Stewart and D. S. Letham, eds.), pp. 297–332. Springer-Verlag, Berlin and New York.

Whitfeld, P. R., and Bottomley, W. (1980). *Biochem. Int.* **1,** 172–178.

Whitfeld, P. R., and Spencer, D. (1968). *In* "Replication and Recombination of Genetic Material" (W. J. Peacock and R. D. Brock, eds.), pp. 74–85. Aust. Acad. Sci., Canberra.

Whitfeld, P. R., Atchison, B. A., Bottomley, W., and Leaver, C. J. (1976). *In* "Genetics and Biogenesis of Chloroplasts and Mitochondria" (T. Bücher, W. Neupert, W. Sebald, and S. Werner, eds.), pp. 361–368. Elsevier/North-Holland Biomedical Press, Amsterdam.

Whitfeld, P. R., Herrmann, R. G., and Bottomley, W. (1978a). *Nucleic Acids Res.* **5,** 1741–1751.

Whitfeld, P. R., Leaver, C. J., Bottomley, W., and Atchison, B. A. (1978b). *Biochem. J.* **175,** 1103–1112.

Williams, G. R., Williams, A. S., and George, S. A. (1973). *Proc. Natl. Acad. Sci. U.S.A.* **70,** 3498–3502.

Woese, C. R., Fox, G. E., Zablen, L. Uchida, T., Bonen, L., Pechman, K., Lewis, B. J., and Stahl, D. (1975). *Nature (London)* **254,** 83–86.

Wollgiehn, R., and Munsche, D. (1972). *Biochem. Physiol. Pflanz.* **163,** 137–155.

Wollgiehn, R., and Parthier, B. (1979). *Plant Sci. Lett.* **16,** 203–210.

Wollgiehn, R., and Parthier, B. (1980). *In* "Chloroplasts" (J. Reinert, ed.), pp. 97–145. Springer-Verlag, Berlin and New York.

Wong-Staal, F., Mendelsohn, J., and Goulian, M. (1973). *Biochem. Biophys. Res. Commun.* **53,** 140–148.

Woodcock, C. L. F., and Bogorad, L. (1970). *J. Cell Biol.* **44,** 361–375.

Wurtz, E. A., and Buetow, D. E. (1981). *Curr. Genet.* **3,** 181–187.

Yamamoto, T., Burke, J., Autz, G., and Jagendorf, A. T. (1981). *Plant Physiol.* **67,** 940–949.

Zablen, L., Kissil, M. S., Woese, C. R., and Buetow, D. E. (1975). *Proc. Natl. Acad. Sci. U.S.A.* **72,** 2418–2422.

5

Biogenesis of the Photosynthetic Apparatus in Prokaryotes and Eukaryotes

ITZHAK OHAD
GERHART DREWS

ABBREVIATIONS

ATP	Adenosine triphosphate
B800	A bacteriochlorophyll light-harvesting complex with an absorption band at 800nm
BChl	Bacteriochlorophyll
CD	Chloramphenicol-degreened cells
Chl	Chlorophyll
CP	A PS chlorophyll-protein complex (I–IV)
Cytoplast	Used to mean all the cell minus the plastid
D	Daltons
EF	Exoplasmic fracture face
LAC	Light-absorbing complex
LHC	Light-harvesting complex
LHCP	Light-harvesting chlorophyll-protein complex
PF	Protoplasmic fracture face
PS	Photosystem
Rp.	*Rhodopseudomonas*
Rs.	*Rhodospirillum*
SDS–PAGE	Sodium dodecyl sulfate-polyacrylamide gel electrophoresis

Photosynthesis: Development, Carbon Metabolism, and Plant Productivity, Vol. II

ISBN: 0-12-294302-3

ABSTRACT

In this chapter the development of the photosynthetic membrane in bacteria will be described separately from that of the chloroplast. The reasons for this separation are not entirely arbitrary. Many differences do indeed exist between the photosynthetic membranes of the prokaryotic as compared with that of the eukaryotic cell and in their development. Among the differences are the presence or the absence of components, such as the absence of an oxygen-evolving system and chlorophyll *a* and *b* and the presence of bacteriochlorophyll in bacteria. A major difference between the two systems is the fact that the photosynthetic membrane of the prokaryote is actually an extension of the plasma membrane of the cell and as such forms a continuum with the membrane carrying out oxidative electron flow. Here one deals with the differentiation of the same membrane—a result of the expression of the same genome having a dual potential. On the other hand, the eukaryotic photosynthetic membrane forms a separate entity enclosed within a specialized organelle, and its formation is the result of the expression of two different genomes and their controlled interactions.

Despite these basic differences, there are many common features in the developmental process of both membrane types. In both, development is a process of synthesis and assembly of various components via insertion into pre-existing structures, which expand to form a differentiated membrane. However, the sequence of addition of various complexes to the growing membrane might vary with the developmental conditions and type of organism.

In both systems, formation of reaction centers can occur independently of the formation of the antennae complexes. Variations in the order of assembly of various components and subcomplexes of the antennae system are also possible. At the same time, the assembly of chlorophyll–protein complexes proper—especially of those containing several chlorophyll molecules bound to polypeptide(s)—seem to be coordinated in time and require simultaneous synthesis of both the chlorophyll and protein components.

Specific growing sites might exist in the bacterial membrane, but it is not clear whether this situation also prevails in the eukaryotic one. In both cases, lateral diffusion and mixing of pre-existing and newly formed components might account for the redistribution and association of reaction centers, electron carriers, and antennae in order to form integrated functional units. As a result of the large degree of freedom in the sequence of the assembly process, wide variations are possible in the size of the photosynthetic units, integration of components of the electron carrier chain and collection and transfer of energy between the antennae and reaction centers.

In both pro- and eukaryotic membrane development, there is a response to a change in the environmental factors, and as such, it has an adaptive value. The exact features of the control mechanisms exerted by changes in the oxygen pressure and light intensity and quality on the developmental process are not yet well understood. In the eukaryotes, light appears to regulate both transcription and translation processes. Finally, the integration of chlorophyll and proteins within the membrane and the assembly of chlorophyll-protein complexes might play a stabilizing role reducing degradation and resulting in the accumulation of both chlorophyll and polypeptides. This process should be considered as playing a major part in the mechanism(s) controlling the membrane developmental processes.

I. Introduction

The phototrophic bacteria, the cyanobacteria, and the prochlorophyta are photosynthetically active organisms with a prokaryotic cellular organization. The chemotrophic bacterium *Halobacterium halobium* is able to produce adenosine triphosphate (ATP) by a light driven proton pump under specific conditions. The photosynthetic apparatuses of the phototrophic prokaryotes differ considerably in their composition, organization, and function. This variety of structure and functional organization supports the idea that these organisms followed separate lines of evolution. Composition and structure of the photosynthetic apparatus have been reviewed (Oelze and Drews, 1972; Pfennig and Trüper, 1974; Lewin, 1977; Stanier and Cohen-Bazire, 1977; Drews, 1978; Pfennig, 1978; Trüper and Pfennig, 1978; Carr and Whitton, 1981; Drews and Oelze, 1981; Kaplan and Arntzen, 1982). The photosynthetic apparatus of all phototrophic prokaryotic organisms is localized on cytoplasmic or intracytoplasmic membranes, which are not separated from the rest of the cell. Thus, prokaryotes have no organelles with their own genome as the chloroplasts of eukaryotic organisms have. Instead, the prokaryotic photosynthetic apparatus is the expression of only one genome, i.e., the bacterial chromosome. The localization of genes for the photosynthetic apparatus on plasmids or other extrachromosomal DNA cannot be excluded but seems to be unlikely (Marrs, 1978, Saunders, 1978). Composition and function of the cyanobacterial and the prochlorophytal photosynthetic apparatus are of the plant (eukaryotic) type (oxygenic photosynthesis). The phototrophic bacteria perform an anoxygenic photometabolism under anaerobic conditions and develop their photosynthetic apparatus only at low oxygen tensions. The main function of the bacterial photosynthetic apparatus is the production of ATP, or membrane energetization, by formation of an electric field and a proton gradient across membrane mediated by a cyclic electron transport system (see Govindjee, 1982).

Some of the phototrophic bacteria have been adapted to a chemotrophic mode of energy metabolism, which is an oxidative respiratory electron-transport chain phosphorylation (Pfennig, 1978). They are unique so far as the respiratory and the light-driven cyclic electron transport systems are bound to the same membrane system. Both electron transport chains interact with each other, and some of the redox carrier or even sections of the electron-transport chains share both electron transport paths (Keister, 1978; Kaplan and Arntzen, 1982). The coupling factor, bound to such a membrane system, produces ATP driven by the proton motive force generated either by respiratory or light-driven electron transport (Baccarini-Melandri and Melandri, 1978; D. Ort and B. A. Melandri, Chapter 12, Vol. I, edited by Govindjee, 1982).

The photosynthetic apparatus of all organisms consist functionally of four major parts. These are (a) the light-harvesting or antenna pigment complexes, which absorb light quanta and funnel light energy via exciton migration to the reaction centers; (b) the photochemical reaction center(s), where the exciton energy is transformed to excited states, finally leading to the formation of an electric field and a redox span across the membrane; (c) redox-carriers that catalyze electron and proton transport across the membrane; (d) the membrane-bound coupling factor ATPase, which catalyzes the formation of ATP and inorganic phosphate, energized by the membrane potential.

Prokaryotic, phototrophic organisms have been adapted during evolution to many ecological niches. But the major habitat for all phototrophic bacteria is the anaerobic and light-exposed zone of water bodies. External factors such as light intensity, oxygen partial pressure, temperature, and nutrition influence the development of the photosynthetic apparatus and membrane differentiation in different ways. Cyanobacteria live in many habitats that are exposed to light and air (Carr and Whitton, 1981).

The first part of this chapter will deal with a few members of the phototropic prokaryotes. The structural basis of development and differentiation has been discussed by Kaplan and Arntzen (1982).

Living organisms are highly organized and differentiated at the level of cells, tissues, and organs. Differentiation of structures in these organisms is a highly regulated process during which composition, molecular organization, biochemical activities, and morphology of the cellular structures are coordinatedly altered. Differentiation can occur in growing, multiplying, or resting cells.

The development of the photosynthetic apparatus in eukaryotic cells has been the object of wide interest for the last three decades. Studies have been focused on various aspects of this process, including development of structure and function and the subcellular origin, synthesis, and assembly of thylakoid components.

As opposed to the prokaryotes, the eukaryotic chloroplast membrane is a "hybrid" structure containing polypeptides coded by an eukaryotic (nuclear) and prokaryotic (chloroplast) genome and translated by their respective protein synthesis machinery. The interrelationship between these two genomes and the control of their expression during chloroplast development has also been studied extensively. The knowledge accumulated in these fields of research has been periodically reviewed (Bogorad, 1975; Ohad, 1975; Boardman, 1977; Boardman *et al.*, 1978), and treaties (Kirk and Tilney-Basset, 1978), as well as synopses of symposia dedicated to the biology, structure, and function of the chloroplast have been published (Hall *et al.*, 1977; Akoyunoglou and Argyroudi-Akoyunoglou, 1978).

As in prokaryotes, one of the major aspects of the eukaryotic chloroplast membrane architecture and function is that of the organization and interaction between the chlorophyll–protein complexes forming, in this case, two reaction centers; photosystem (PS) core I and II and the respective light harvesting antennae. In this chapter, we will try to focus attention specifically on the development of these membrane components. A detailed description of the composition and function of Chl–proteins can be found in several review articles (Boardman, 1977; Thornber and Alberte, 1978; Thornber *et al.*, 1979), as well as by Kaplan and Arntzen (1982). Thus, only a brief description of the chloroplast membrane polypeptides and chl–protein complexes will be presented here.

At present, the processes of chloroplast development in eukaryotic systems and of membrane differentiation in both eukaryotic and prokaryotic systems can be described and correlated with variations of external and internal factors that influence this process. But we are still far from a complete understanding of the molecular basis of development and differentiation of the photosynthetic apparatus.

II. Formation of the Photosynthetic Apparatus in Facultative Phototrophic Bacteria

A. Coordination between Pigment and Protein Synthesis

During membrane differentiation in cells of facultative phototrophic bacteria that are adapting from chemotrophic to phototrophic growth conditions, the incorporation of reaction center and light-harvesting pigment complexes into the membranes is the dominating process. It is accompanied by modifications in the composition and activities of the membrane-bound electron-transport chain. Reaction centers and light-harvesting antenna pigment complexes consist of BChl carotenoids, and polypeptides, which are specifically and stoichiometrically bound in a noncovalent mode to each other (Oelze and Drews, 1981; Kaplan and Arntzen, 1982).

An inhibition or mutational blockade in the biosynthesis of BChl or single proteins of the pigment complexes prevent the formation of a functional photosynthetic apparatus. Addition of chloramphenicol or puromycin inhibits formation and assembly of the specific polypeptides in the membrane and stops BChl synthesis, while formation of BChl precursors continue (Lascelles, 1959; Bull and Lascelles, 1963; Higuchi *et al.*, 1965; Drews, 1966; Biedermann *et al.*, 1967).

Cerulenin, an inhibitor of fatty acid synthesis, inhibits BChl synthesis and incorporation of the pigment binding polypeptides into the membrane, but it does not interrupt formation of other proteins (Broglie and Niederman, 1979). In contrast, inhibition of carotenoid formation does not disturb the synthesis of BChl or the formation of intracytoplasmic membranes (Maudinas *et al.*, 1973). However, mutants in carotenoid synthesis are frequently blocked in biosynthesis of the light-harvesting complex B800–B850 (Drews *et al.*, 1976; Marrs, 1978). Mutants, blocked in the last steps of BChl synthesis, excrete protein-bound BChl precursors; they do not incorporate polypeptides into the membranes that are specifically associated with BChl (Oelze *et al.*, 1970; Drews, 1974; Drews *et al.*, 1976).

B. The Formation of the Photosynthetic Apparatus: A Multistep Process

The coordinated synthesis of BChl and specific polypeptides and their assembly in the membrane can be observed in cells of facultative phototrophic bacteria precultivated under aerobic conditions in the dark and transferred to anaerobic light conditions or low aeration in the dark. After a lag phase, the rate of BChl synthesis increases exponentially. Later, the rate decreases, and the cellular level of BChl approaches a steady state concentration. The kinetics of BChl formation during the adaptation phase depend on the organism and the conditions of cultivation.

BChl and specific polypeptides are incorporated simultaneously into the membrane (Takemoto, 1974; Nieth and Drews, 1975; Niederman *et al.*, 1976) and aggregate immediately (Schumacher and Drews, 1978). Free BChl and precursors are only detectable very early and in small amounts during adaptation to phototrophic growth (Cellarius and Peters, 1968; Niederman *et al.*, 1976; Pradel *et al.*, 1978; H. Reidl, A. G. Garcia, and G. Drews, unpublished). Under conditions of growth-limited adaptation to phototrophic conditions, the pigment-associated polypeptides are the first and major proteins that are synthesized (Schumacher and Drews, 1978; Dierstein *et al.*, 1981). The autoradiography of membrane proteins, which are pulse-labeled after induction, shows the coordinated synthesis of the three reaction-center polypeptides. The light-harvesting complex B870 is formed simultaneously. But during the first hours of adaptation, the ratio B870–BChl to reaction center is lower as under steady state conditions (20–30 molecules BChl B870 per reaction center; Garcia *et al.*, 1981). in cells of *Rhodospirillum* (*Rs*) *rubrum*, *Rhodopseudomonas* (*Rp.*) *sphaeroides* and *Rp. capsulata* (Aagaard and Sistrom, 1972; Lien *et al.*, 1973; Nieth and Drews, 1975; Oelze and Pahlke, 1976; Schumacher and Drews, 1978, 1979). The size of the photosynthetic unit, i.e., the total amount of BChl per reaction center, decreases initially and then increases. This indicates that during the early stages of adaptation, reaction centers are preferentially synthesized.

In cells of *Rs. rubrum,* which produce only the B870 type of light-harvesting pigment complex, reaction center and the B870 complex are synthesized after the first period of adaptation in a constant ratio.

In cells of *Rp. sphaeroides* and *Rp. capsulata,* a relatively short phase of adaptation, characterized by domination of reaction center and light-harvesting B870 synthesis, is followed by a phase of increased formation of light-harvesting units B800–B850 (Takemoto, 1974; Nieth and Drews, 1975; Niederman *et al.*, 1976; Schumacher and Drews, 1978; Dierstein *et al.*, 1981; Garcia *et al.*, 1981). During the early stages of adaptation to phototrophy, the efficiency of energy transfer between light-harvesting pigment complexes and reaction center is low (Pradel *et al.*, 1978; Hunter *et al.*, 1979a), and the capacity for photophosphorylation is high, when calculated on a BChl basis, but it decreases as more light-harvesting BChl forms (Garcia and Drews, 1980). Redox components, for instance cytochrome *b,* do not reach a final level at the beginning of adaptation (Hunter *et al.*, 1979a,b; Garcia and Drews, 1980; Garcia *et al.*, 1981).

C. The Sites of Incorporation of Membrane Constituents

Newly synthesized constituents of the photosynthetic appraratus are incorporated and assembled into functional subunits at specific sites of the membrane system. In cells of *Rs. rubrum,* depleted of intracytoplasmic membranes by chemotrophic aerobic preculture, the first photosynthetic units form in the cytoplasmic membrane. An increase of the number of photosynthetic units above a threshold value is accompanied by invaginations of the cytoplasmic membrane (Oelze and Drews, 1969; Oelze *et al.*, 1969). The number of invagination sites increases parallel to the enlargement of the intracytoplasmic membrane system (Golecki and Oelze, 1975). As soon as the intracytoplasmic membrane forms, newly synthesized photosynthetic units assemble exclusively in this membrane (Oelze and Drews, 1972, 1981). The physical continuity of intracytoplasmic and cytoplasmic membranes (reviewed in Remsen, 1978; Oelze and Drews, 1972, 1981), the increase of invagination zones during adaptation to phototrophic growth, and the continuity of membrane constituents from the cytoplasmic to intracytoplasmic membranes (Oelze and Drews, 1969; Niederman *et al.*, 1979) support the idea that both membranes are morphogenetically linked.

Another proposed hypothesis that the intracytoplasmic membrane is formed de novo and can fuse in a secondary process with the cytoplasmic membrane (Kaplan, 1978). Cells of *Rp. capsulata* contain under aerobic conditions polar-arranged intracytoplasmic membranes that differ from the intracytoplasmic membranes of phototrophically grown cells by a lower density, a higher ratio of respiratory to photosynthetic electron transport, and a tubular instead of a vesicular form (Fig. 1a; Lampe *et al.*, 1972; Garcia *et al.*, 1981).

In cells of *Rp. capsulata* and *Rp. sphaeroides* newly synthesized, photosynthetic units are incorporated into a pink-colored membrane fraction much faster than into other membrane fractions (Dierstein, 1978; Hunter *et al.*, 1979b; Niederman *et al.*, 1979; Dierstein *et al.*, 1981; Garcia *et al.*, 1981). The pink fraction is enriched in reaction center and light-harvesting complex B875; it has a small photosynthetic unit of about 15 Chl, a low cytochrome *b* content, and a low efficiency of electron transport, but it has a high capacity for oxidative and light-driven phosphorylation (Hunter *et al.*, 1979a,b; Garcia and Drews, 1980; Garcia *et al.*, 1981). The B800–B850 complex, which can be found later in this fraction, is not energetically coupled to reaction centers in this fraction and does not contain succinate dehydrogenase and the cyanide-sensitive oxidase (Hunter *et al.*, 1979a; Garcia *et al.*, 1981).

In all three species (*Rs. rubrum, Rp. sphaeroides* and *Rp. capsulata*), the cytoplasmic membrane conserves respiratory activity and has, under all growth conditions, a low content of photosynthetic units (Throm *et al.*, 1970; Lampe and Drews, 1972; Parks and Niederman, 1978).

Cells of *Rs. tenue* and *Rp. gelatinosa* do not form intracytoplasmic membranes. Membrane differentiation is restricted to the cytoplasmic membrane. Upon adaptation to phototrophic growth, reaction center and light-harvesting pigment complexes are incorporated into the cytoplasmic membrane. Simultaneously, the number of intramembrane particles increases, especially of a large size class (Wakim *et al.*, 1978). Although the cells do not grow and the respiratory capacity remains constant, the number and the size of photosynthetic units increases strongly (Wakim and Oelze, 1980). The cell volume and the number of functional units per membrane area seems to increase during this mode of membrane differentiation.

Reaction center, electron transport system, and coupling factor of photosynthetic apparatus of green bacteria are bound to the cytoplasmic membrane. The antenna BChl-carotenoid complexes, however, are lo-

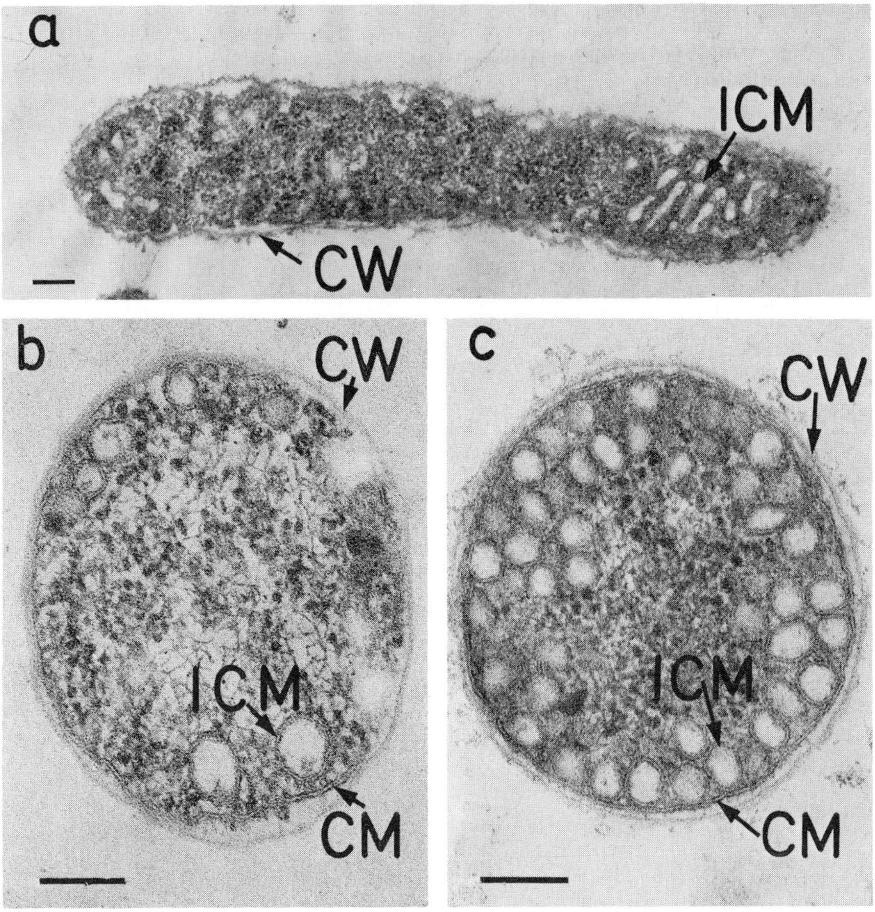

FIG. 1. Sections of *Rhodopseudomonas capsulata,* strain St. Louis (wild type). (a) The cells were grown aerobically in the dark at high oxygen tension, approximately 0.12 μg BChl *a* per milligram cell protein. The tangent longitudinal section shows a few tubular ICM in the polar region of the cell. (b) The cells were grown anaerobically at high light intensity (2000 $W \cdot m^{-2}$) and contained in the mean 172 ICM vesicles per cell, 3.1 μg BChl per milligram cell protein, 24 μg BChl per milligram ICM protein. (c) The cells were grown anaerobically at low light intensities (7 $W \cdot m^{-2}$) and contained in the mean 1090 ICM vesicles per cell, 42 μg BChl per milligram cell protein, and 82 μg BChl per milligram ICM protein. Bar, 100 nm; CW, cell wall; CM, cytoplasmic membrane; ICM, intracytoplasmic membrane. (Data from Golecki *et al.*, 1980; micrographs by J. Golecki.)

calized in chlorosomes, which are attached to the cytoplasmic membrane. The water-soluble light-harvesting BChl *a* complex is presumably localized between membrane and chlorosomes (Pierson and Castenholz,

1978; Staehelin *et al.*, 1978, 1980). The coordinated synthesis of these constituents of the photosynthetic apparatus and its assembly remains to be elucidated.

These few examples demonstrate that among phototrophic bacteria various modes of membrane differentiation have been developed. In *Rs. tenue* all electron transport systems are localized in the cytoplasmic membrane. In other facultative phototrophic bacteria (*Rs. rubrum, Rp. sphaeroides*) the respiratory and the photosynthetic systems are compartmentalized. Although cytoplasmic and intracytoplasmic membrane form a continuous membrane system, they differ with respect to morphology, function, composition, and differentiation patterns (Oelze and Drews, 1972, 1981; Drews, 1978; Kaplan, 1978; Drews and Oelze, 1981).

D. Formation of Precursors and Assembly of the Photosynthetic Apparatus

1. THE INSERTION OF PROTEINS AND PIGMENTS

The biosynthesis of BChl, carotenoids and those polypeptides, which form specific complexes with the pigments, are coordinated. The formation of these constituents follows different routes. BChl is synthesized via the Mg branch of the tetrapyrrol pathway (Jones, 1978; Lascelles, 1978; Rebeiz and Lascelles, 1982). The first enzymes of this pathway (until protoporphyrin IX) are soluble; the enzymes of the Mg branch seem to be membrane-bound. This was demonstrated for the Mg chelatase and methyltransferase (Gorchein, 1972, 1973). It is unknown where the enzymes for carotenoid synthesis are localized. The third constituent of the pigment complexes, the polypeptides, are synthesized on ribosomes and are inserted into the membrane by a co- or post-translational mode (Wickner, 1979). At present, the sequence of steps from translation of these polypeptides to their assembly with pigments in the membrane is unknown.

In the laboratory of G. Drews, the biosynthesis and insertion of polypeptides into the membrane were followed by pulse and pulse-chase experiments and immunoprecipitation of the H polypeptide of reaction center (M. Okamura, G. Feher, and N. Nelson; see Govindjee, 1982) and the two small polypeptides of antenna complex B800–B850 of *Rp. capsulata.* By this technique, no polypeptides of the pigment complexes have been detected in the supernatant of cell extracts after short pulses with ^{14}C-labeled amino acids. All immunoprecipitable material was found exclusively in membrane fractions. The newly synthesized polypeptides were inserted into the light membrane fraction much faster and with a higher proportion than into the heavy membrane fraction. The individual pigment-associated proteins did not exhibit precursor-product relationship between the two membrane fractions (Dierstein *et al.*, 1981). However, radioactivity of a pulse-labeled 11,000 D "precursor" polypeptide has been chased into a 10,000 D polypeptide in the same light membrane fraction. These preliminary results support the idea that at least this polypeptide of the light-harvesting complex B800–B850 is proteolytically modified after or during insertion into the membrane (R. Dierstein, unpublished).

Simultaneously, with the polypeptides of the reaction center and the light-harvesting complex B870, a 45,000 D polypeptide was incorporated into the membrane. But a precursor-product relationship between this heavy polypeptide and the light-harvesting polypeptides could not be demonstrated. Polypeptides of similar size have been detected in a BChl–precursor–protein complex, which has been excreted into the medium (Drews, 1974).

The polypeptides that assemble with BChl and carotenoid in the membrane are hydrophobic (Steiner *et al.*, 1974; Shiozawa *et al.*, 1980). Also, they form oligomeric pigment–protein aggregates (J. A. Shiozawa and G. Drews, unpublished). The B800–B850 complex of *Rp. capsulata*, for example, has a molecular mass of approximately 170,000 D (J. Shiozawa and G. Drews, unpublished). Size and number of integral membrane particles and the concentration of functional subunits in the membrane correlate with each other (Wakim *et al.*, 1978; Golecki *et al.*, 1979). This observation supports the idea that the polypeptide-associated pigments assemble to larger subunits, which can be visualized by freeze-fracture electron microscopy. B800–B850 complexes, isolated from membranes of *Rp. capsulata* and reincorporated into liposomes, have the same size particles as the native membrane (J. Shiozawa, G. Drews, and W. Welte, unpublished).

The whole process of the assembly of pigments and polypeptides seems to be a membrane-associated process which depends on processing and folding of polypeptides and interactions between lipids and proteins and proteins and pigments.

The precursors enter the membrane at specific sites (Dierstein *et al.*, 1981). Whether the intracytoplasmic membranes arise at these specific sites by invagination from the cytoplasmic membrane has still to be determined. As soon as intracytoplasmic membranes have been formed, they are the only, or the privileged structure, for incorporation of newly synthesized precursors of photosynthetic units. The distribution of "growing points" in the membrane system seems to be influenced by the process of growth. It seems possible that the sites, where the photosynthetic units are assembled, are labile structures that become separated by cell fractionation, forming a light fraction (Niederman *et al.*, 1979; Garcia *et al.*, 1981).

2. THE ROLE OF LIPIDS IN THE ASSEMBLY PROCESS

Investigations of the assembly of the photosynthetic apparatus in the membrane have to consider the phospholipids of the membrane because a part of these lipids interacts strongly with the integral membrane proteins; the large functional subunits in the membrane are not freely mobile in the plane of the membrane (Kaplan and Arntzen, 1982). It has been shown by Kaplan and co-workers that in populations of synchronized *Rp. sphaeroides* cells, the accumulation of the photosynthetic apparatus in the membrane is continuous throughout the cell cycle. In contrast, phospholipids are incorporated discontinuously into the membrane system. Consequently, the protein to phospholipid ratio of the isolated membrane fraction undergoes dramatic changes such as from 2:1 to 5:1 (see Kaplan and Arntzen, 1982). Under the condition of a high protein to phospholipid ratio, most of the phospholipids are in the "boundary" form. The rigidity of membranes and the formation of large domains of proteinous subunits might be of importance for membrane fractionation by cell breakage and density gradient centrifugation. Preliminary results from Kaplan's laboratory suggest that during cell division phospholipids move from the cell envelope (cytoplasmic membrane plus cell wall) into the intracytoplasmic membrane. It was described earlier that precursors of the photosynthetic apparatus are incorporated into specific membrane fractions. These results suggest that the assembly of the "photosynthetic membrane" is a multistep process: (1) entry of polypeptides in the membrane; (2) processing, folding of polypeptides, interactions with phospholipids, and association with pigments; (3) formation of oligomeric pigment complexes; (4) lateral movement of pigment complexes, phospholipids and other redox components of the photo-synthetic apparatus; and (5) continuation of rearrangement of subunits and incorporation of new constituents.

E. *The Influence of External Factors on Membrane Differentiation*

1. OXYGEN PARTIAL PRESSURE

It is characteristic of all phototrophic bacteria that the formation of the photosynthetic apparatus depends on a low oxygen tension, but is not dependent on light; it occurs in the dark if the organisms can grow chemotrophically (Lascelles, 1959; Cohen-Bazire and Kunisawa, 1960; Drews and Giesbrecht, 1963). The rates of biosynthesis of BChl, carotenoids, and proteins of the photosynthetic apparatus correlate with the height of oxygen partial pressure. Below strain-specific threshold values of oxygen tension (approximately 1.0–1.5 kPa), the rates of synthesis increase strongly (Biedermann *et al.*, 1967; Dierstein and Drews, 1974). Above the threshold value, the rate of BChl synthesis approaches zero. *Rhodospirillum rubrum* does not synthesize measurable amounts of BChl at 6.6 kPa [pO_2], but *Rp. capsulata* forms small amounts of reaction centers at this oxygen tension. The growth rate is not influenced by the oxygen tension within a broad range of partial pressure (0.7–13 kPa) in the dark (Biedermann *et al.*, 1967).

Changes of oxygen partial pressure around threshold values influence not only the development of the photosynthetic apparatus, but also induce differentiation of the total membrane system.

Upon decrease of oxygen tension, the formation of photosynthetic units is strongly increased, and the formation of respiratory enzymes, especially of the terminal oxidase, is slightly decreased (on the basis of membrane protein). In most species of facultative phototrophic bacteria, the amount of intracytoplasmic membranes per cell is increased. After adaptation to the low oxygen tension, the cells have the capacity for both photophosphorylation and oxidative phosphorylation (Lampe and Drews, 1972; Garcia and Drews, 1980; Garcia *et al.*, 1981). The functional units of the two electron transport systems are in most species distributed on different membranes, i.e., incorporated at specific sites (Dierstein *et al.*, 1981). The size of the photosynthetic unit is dependent on the oxygen tension; it is small at high oxygen tension and vice versa (Schumacher and Drews, 1978; Garcia *et al.*, 1981). The potential activity of photophosphorylation and respiration on the basis of membrane protein decreases upon lowering of the oxygen partial pressure (Throm *et al.*, 1970; Lampe and Drews, 1972; King and Drews, 1975).

During adaptation of phototrophically grown cells to strict aerobiosis, the biosynthesis of the photosynthetic units is immediately inhibited, and the incorporation of respiratory units into the membrane systems are enhanced. In growing cells of *Rs. rubrum,* adapting to aerobiosis, the amount of intracytoplasmic membranes per cell is diminished. Evidently, the cytoplasmic membrane is synthesized faster than the intracytoplasmic membranes.

The specific activities in the respiratory chain of *Rp. palustris,* particularly the terminal cytochrome oxidase, increase considerably if the oxygen tension is increased (King and Drews, 1975; Firsow and Drews,

1977). Although, during adaptation to aerobiosis, the synthesis of new photosynthetic units is strongly reduced, the remaining photosynthetic units are still potentially active. The activities of photophosphorylation in isolated membrane fractions increase on BChl basis, indicating an intimate interrelationship between parts of the photochemical and the respiratory electron transport (Keister and Minton, 1969; J. Oelze and B. Georg, unpublished). The size of the photosynthetic unit in cells of Rhodopseudomonas species decreases during adaptation to aerobiosis, indicating that small amounts of reaction center and light-harvesting B875 are synthesized (Aargaard and Sistrom, 1972; Lien *et al.*, 1973; Firsow and Drews, 1977).

The effect of oxygen partial pressure, described in this section, can only be observed in the facultative phototrophic bacteria. Most members of Chromatiaceae and green bacteria are strictly dependent on anaerobic light conditions. Oxygen does not kill these bacteria but inhibits BChl formation, photophosphorylation, and growth.

2. THE EFFECT OF LIGHT IRRADIANCE

Membrane differentiation in phototrophic bacteria was studied primarily with white light. Preliminary studies with monochromatic light suggested that BChl and carotenoids are the only pigments that absorb light quanta and mediate the influence of light on morphogenesis and membrane differentiation (Drews and Giesbrecht, 1963; Drews and Jäger, 1963; Kaplan, 1978). However, observations of J. Oelze (unpublished) indicate the presence of a receptor which is different from the pigments of the photosynthetic apparatus and which influences pigment synthesis. Early observations have shown that the BChl content of cells is inversely proportional to the incident light energy flux (Cohen-Bazire *et al.*, 1957).

The adaptation to different light intensities is established differently by phototrophic bacteria. *Rhodospirillum rubrum*, which has a photosynthetic unit of unvaried size (constant ratio of BChl B875 to reaction center; no antenna complex B800–B850), reduces, under high light intensities, the amount of intracytoplasmic membrane and the number of photosynthetic units per cell. The BChl content decreases below 12 μg of BChl per mg of cell protein (Oelze *et al.*, 1969; Oelze and Drews, 1972). The activities of respiratory enzymes, of photophosphorylation, and of light dependent NAD^+ reduction with succinate as substrate increase in intracytoplasmic membranes on a BChl basis (Irschik and Oelze, 1973, 1976). The relative proportion of polypeptides associated with pigments decreases per membrane protein upon increase of light intensity (Irschik and Oelze, 1973, 1976).

Rhodopseudomonas sp., which have two antenna complexes, reduce the size of the photosynthetic units, the number of photosynthetic units per membrane protein, and the amount of intracytoplasmic membrane per cell upon increase of light irradiance (Drews and Giesbrecht, 1963; Aagaard and Sistrom, 1972; Lien *et al.*, 1973; Firsow and Drews, 1977; Takemoto and Huang Kao, 1977; Schumacher and Drews, 1979; Golecki *et al.*, 1980). In these species, the light-harvesting BChl–carotenoid–protein complex B800–B850 is the variable component of the photosynthetic unit. The ratio of B800–B850 to reaction center can be modified in the range of 10–100. The molar ratio of BChl to carotenoid is 3:1 in the B800–B850 complex, presumably 1:1 in the B875 complex, and 4:1 in the reaction center (Cogdell, 1978; Cogdell and Thornber, 1979). Consequently, the molar ratio of carotenoids to BChl increases from 0.4 to about 0.9 during adaptation to high light intensities (Sistrom, 1978). At high-incident radiation fluxes, the rates of electron transport, of photophosphorylation and of growth were found to be high (Firsow and Drews, 1977; Schumacher and Drews, 1979). The amount of intracytoplasmic membranes reaches a minimum.

On the other hand, the size of the photosynthetic units is increased (high proportion of B800–B850 complex), the intracytoplasmic membrane system is enlarged, and the rates of photophosphorylation, respiratory activity, and growth decrease upon adaptation to low light intensities (Firsow and Drews, 1977; Sistrom, 1978; Schumacher and Drews, 1979; Golecki *et al.*, 1980). Results from the laboratory of G. Drews on adaptation of *Rp. capsulata* to various light intensities are summarized in Figs. 1 and 2 and Table I. Other purple bacteria adapt in a similar mode to variations of light intensity (Mechler and Oelze, 1978).

The dependence of growth rate and pigment content from the quantity and quality of irradiance was studied by Göbel (1978). The results show that purple bacteria cannot adapt efficiently to very low light intensities. For maximum growth rate, a mean irradiance of about 10 nE $s^{-1} \cdot cm^{-2}$ (860 nm) or 20–30 nE $s^{-1} \cdot cm^{-2}$ (522 nm) is necessary. A variation of the light quanta absorption rate in the cells induces cell differentiation, which primarily comprises modifications of the photosynthetic apparatus including the energy conserving system and, secondarily, the growth-rate determining processes. The flux of light quanta, absorbed and delivered by the pigments of the photosynthetic apparatus, triggers many of the morphogenetic events.

Although the photosynthetic apparatus of green bacteria is differently organized, these bacteria also increase the ratio of light-harvesting BChl *c* per reaction center when adapting to low light irradiance. The number of chlorosomes per area of cytoplasmic membrane is strongly dependent on the light intensity (Holt *et al.*, 1966; Pierson and Castenholz, 1974, 1978; Broche-Due *et al.*, 1978). The number of reaction centers per cell is only slightly modified.

Temperature and nutrition (Dierstein and Drews, 1975) are other factors that influence the membrane differentiation. Low temperature seems to have almost the same effect on the differentiation of the photosynthetic apparatus of *Rhodopseudomonas* sp. as high light intensity and high aeration: small photosynthetic units, low amounts of intracytoplasmic membranes per cell, and high rates of photophosphorylation and respiration have been observed in *Rp. sphaeroides* (Kaiser and Oelze, 1980).

The influence of external factors, such as quality and quantity of light flux, nutrition and temperature on the differentiation of the photosynthetic apparatus, and the thylakoid system of cyanobacteria (blue-green algae) has been studied (see review of Miller and Holt, 1977;

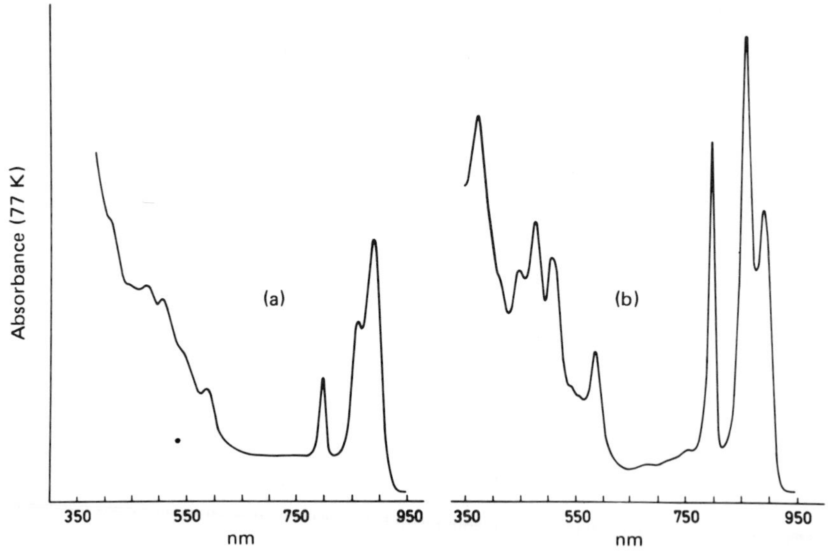

FIG. 2. Absorbance spectra at 77K, of membranes of *Rhodopseudomonas capsulata*, strain 37b4, grown phototrophically at different light intensities [(a) at 2000 $W \cdot m^{-2}$; (b) at 7 $W \cdot m^{-2}$]. In cells grown at high light intensities (a), reaction center and the light-harvesting bacteriochlorophyll–carotenoid–protein complex B875 (890 nm at 77°K) dominate. The size of the photosynthetic unit (moles of total BChl per reaction center) was 46. In cells adapting to low light intensities (b), the variable light-harvesting pigment–protein complex B800–B850 increased considerably. The size of the photosynthetic unit increases to 104. (Data from Schumacher and Drews, 1979.)

TABLE I
The Influence of Light Intensity on the Development of the Photosynthetic Apparatus of *Rhodopseudomonas capsulata*[a]

Property[b]	2000 W·m^{-2}	7 W·m^{-2}	LL/HL
BChl (nmole/mg cell protein)	3.4	46.0	13.5
BChl (nmole/mg ICM protein)	26.4	90.0	3.4
Size of the photosynthetic unit	48	97	2.0
BChl/Car (mole/mole)	1.3	1.9	1.4
Mean number of ICM vesicles per cell	180	1149	6.4
Molecules of BChl per vesicle	7269	13,576	1.8
RC per vesicle	150	139	
Formation of ATP (μmole ATP/min·mg protein)	40	15	

[a]The values are from cells grown for three generations at low light and 10 generations at high light intensity, respectively.
[b]ICM = Intracytoplasmic membrane; LL/HL = values at low light to high light intensities; Car = carotenoid; RC = reaction center; BChl = bacteriochlorophyll. (Data from Golecki *et al.*, 1980; Schumacher and Drews, 1979.)

Goedheer and Kleinenhammans, 1977; Stanier and Cohen-Bazire, 1977; Myers et al., 1978; Sherman, 1978; Ono and Murata, 1979). Of special interest is the influence of light quality and nitrogen nutrition on number and composition of phycobilisomes and the influence of CO_2 on the formation of thylakoids. After CO_2 deprivation, cells of *Synechococcus lividus* lost Chl *a*, c phycocyanin, and thylakoids. Reintroduction of CO_2 into bleached culture resulted in a rapid resynthesis of pigments and thylakoids (Miller and Holt, 1977).

F. Regulation of Differentiation

The mean turnover rates of membrane constituents in growing cells are small compared with the mean rates of incorporation (see Oelze and Drews, 1972; Drews, 1978; Kaplan, 1978; Drews and Oelze, 1981). Thus, the variation in structure, composition, and function during the process of differentiation is due to alterations in the rates, sites, and patterns of incorporation of new constituents into the membrane. It was reported in this chapter that during membrane differentiation in facultative phototrophic bacteria the constituents, which form the pigment–protein complexes, are coordinately synthesized and assembled, i.e., reaction center and light-harvesting complex B875 and, independently, light-harvesting complex B800–B850.

The co-regulation of pigment and protein synthesis is an unsolved problem, and hypotheses about the regulation at the level of transcription and translation have not been confirmed on an experimental basis as yet (Kaplan, 1978; Drews and Oelze, 1981). The observation that mutants, defective in formation of one or more pigment complexes, can be restored by transfer of a small piece of DNA (Drews *et al.*, 1976; Yen and Marrs, 1976; Marrs, 1978) suggests that genes for membrane polypeptides, which assemble with pigments, and genes for enzymes, and which catalyze the last steps of BChl synthesis, are clustered (Marrs, 1981). Moreover, the regulation of BChl on the enzymatic level (Jones, 1978; Lascelles, 1978) seems to affect the modification and assembly of specific proteins in the membrane. A direct effect of precursors of BChl synthesis on the biosynthesis of the proteins cannot be excluded but seems to be unlikely. Further research is necessary to determine if the polypeptides, associated and excreted together with BChl precursors (Lascelles, 1966; Oelze *et al.*, 1970; Drews, 1974), are related to the membrane-bound proteins of the pigment complexes.

G. Concluding Remarks

Many purple bacteria adapt their photosynthetic apparatus to changes in culture conditions by variation of the following parameters: (1) cell size (amount of cytoplasmic membrane per cell in *Rs. tenue*) (2) amount of intracytoplasmic membrane per cell, (3) size of the photosynthetic unit (ratio of total BChl per reaction center), and (4) number of photosynthetic units per membrane protein. The amount of cytochromes and quinones varies in a relatively narrow range. The membrane system of most facultative phototrophic bacteria is compartmentalized, harboring the respiratory chain in the cytoplasmic membrane and the photosynthetic apparatus in the intracytoplasmic membrane. The oxygen partial pressure in the medium is the major external factor regulating differentiation.

The green bacteria can adapt to lower light intensities than the purple bacteria because their light-harvesting pigments are localized in the chlorosomes, which are large antennae attached but not incorporated into the membrane. Thus the cytoplasmic membrane is free to incorporate reaction centers and electron transport constituents.

Organization and composition of the photosynthetic apparatus of cyanobacteria differ completely from the photosynthetic apparatus of purple bacteria. Moreover, the amount of intracytoplasmic membranes (thylakoids) per cell, their composition, and number and composition of phycobilisomes are essentially regulated by quantity and quality of light fluence, CO_2 partial pressure, and temperature.

III. Formation of the Eukaryotic Photosynthetic Membranes

A. Chlorophyll–Protein Complexes and Their Subcellular Origin

Information on the subcellular origin of chloroplast membrane polypeptides is mostly derived from: (a) the analysis of *in vivo* radioactive labeling during growth and greening of algae (Ohad, 1975; Bar-Nun and Ohad, 1977; Chua and Gillham, 1977; Gurevitz *et al.*, 1977; Bingham and Schiff, 1979) in the presence of protein synthesis inhibitors specific for the chloroplast or cytoplast; and (b) synthesis of polypeptides *in vitro* by isolated intact chloroplast (Ellis, 1975, 1977; Zielinski and Price, 1980), or translation of isolated messages (Edelman and Reisfeld, 1978; Grossman *et al.*, 1980, 1981). The conclusions reached, on the basis of these studies, are further supported by analysis of the membrane polypeptide pattern of nuclear or chloroplast mutants deficient in specific polypeptides or functions (Chua and Bennoun, 1975; Chua *et al.*, 1975; Conde *et al.*, 1975; Kretzer *et al.*, 1976).

Isolation of Chl–protein complexes has been obtained by a variety of methods, including fractionation of membranes with digitonin (Wessels, 1977; Argyroudi-Akoyunoglou and Akoyunoglou, 1979), Triton X-100 (Anderson, 1975, Wessels, 1977), deoxycholate (Bar-Nun and Ohad, 1977), and most often, SDS or LDS–polyacrylamide gel electrophoresis (Delepelaire and Chua, 1979; Thornber *et al.*, 1979). The characterization of Chl–protein complexes obtained by the various procedures is based on absorption and fluorescence spectra, dissociation of the complexes, and analysis of their pigment and polypeptide composition, as well as, in certain cases, measurements of photochemical activity (Anderson, 1980). Basically, one can describe the Chl–proteins as being related to either PSII or to PSI complexes.

1. CHLOROPHYLL–PROTEIN COMPLEXES ASSOCIATED WITH PHOTOSYSTEM II

Isolation procedures based on utilization of LDS or SDS polyacrylamide gel electrophoresis yielded the following complexes: CPII containing Chl *a* and *b* in about a 1:1 ratio. Depending on the source of the chloroplast membranes, this complex contains either 2 or 3 polypeptides, all of cytoplasmic origin and having apparent molecular weight (MW) between 22,000 and 28,000 D (Bar-Nun *et al.*, 1977; Chua and Gillham, 1977; Boardman *et al.*, 1978; Burke *et al.*, 1978a, 1979; Thornber *et al.*, 1979). This complex has so far been identified as representing the main light-harvesting (LHC) or antennae system of PSII, exhibiting 686 nm fluorescence at 77°K and apparently participating in the formation of the large (14 nm) particles observed on the EF face of freeze-fractured membranes (Staehelin *et al.*, 1976). The CPII in *Chlamydomonas* has an apparent MW of 28,000 D and as such, it appears to be a monomeric form. Multimeric forms of this complex of higher apparent molecular weight have also been reported (Anderson *et al.*, 1978; Argyroudi-Akoyunoglou and Akoyunoglou, 1979; Delepelaire and Chua, 1979; Machold *et al.*, 1979). Two additional complexes, CPIII and CPIV, containing Chl *a* only and polypeptides associated with the formation of the PSII reaction center, have also been described (Delepelaire and Chua, 1979). These complexes have been regarded as actually representing the PSII reaction center to which the 696 nm fluorescence emission spectrum at 77°K is usually ascribed. Results obtained in the laboratory of I. Ohad (Gershoni and Ohad, 1979) show that, under certain conditions, membranes can be formed that exhibit the presence of the CPII complex and polypeptides associated with the CPIII and CPIV complexes; these membranes are able to perform electron transfer from H_2O to NADP. However, when the CPIII and CPIV Chl-containing complexes are not detectable, the efficiency of electron transfer at low light intensity is drastically impaired. These results indicate that the CPIII and CPIV complexes might play the role of additional antennae systems interconnecting the PSII reaction centers with the LHC complex. Chlorophyll–protein complexes apparently similar to the CPIII and CPIV obtained by different techniques have also been described (Anderson *et al.*, 1978; Argyroudi-Akoyunoglou and Akoyunoglou, 1979; Machold *et al.*, 1979).

2. CHLOROPHYLL–PROTEIN COMPLEXES ASSOCIATED WITH PHOTOSYSTEM I

Again, electrophoresis on polyacrylamide gels in the presence of SDS demonstrates the presence of a Chl *a*–protein complex CPI, apparently containing at least two polypeptides (Herrmann, 1978), the major one with a MW of about 66,000 in *Chlamydomonas reinhardtii*. The apparent MW of the complex in *Chlamydomonas* is about 88,000 (Bar-Nun *et al.*, 1977). Higher molecular weights have been reported for this complex in higher plants (Boardman *et al.*, 1978). Membranes lacking this polypeptide do not exhibit a PSI reaction center nor the CPI complex (Chua *et al.*, 1975). Thus, it was proposed that CPI represents the PSI reaction center. However, membranes can be formed in *Chlamydomonas y-1* mutant in which the CPI complex is absent but the 66,000 polypeptide and Chl *a* are present (Bar-Nun *et al.*, 1977). These membranes are able to perform electron transfer from reduced dichlorophenol-indo-phenol to methyl viologen, as well as from H_2O to NADP. Nevertheless, the efficiency of light utilization in this case is very low. These results again indicate that CPI might represent a PSI reaction center–antennae complex. The long-wave fluorescence emission peak (714 nm) at 77°K in *Chlamydomonas* appears to originate in this complex. Multimeric forms of this complex have also been reported (Anderson *et al.*, 1978; Argyroudi-Akoyunoglou and Akoyunoglou, 1979; Delepelaire and Chua, 1979; Machold *et al.*, 1979).

Treatment of chloroplast membranes from beans by Triton X-100 in a low ionic strength medium, followed by centrifugation on a sucrose gradient, yields two types of particles: one consisting exclusively of LHC, the other representing a PSI reaction center–antennae complex having a ratio of Chl *a*:P700 of approximately 125 (Mullet *et al.*, 1980). This preparation contains several polypeptides, including the 68,000 D polypeptide, as well as polypeptides in the range of 30,000–40,000 D and 16,000–20,000 D. Further treatment of this particle with increasing concentrations of Triton X-100 results in a gradual peeling of some of the polypeptides of the lower molecular range range and reduction in the ratio of Chl *a*:P700, until a final preparation can be obtained containing 45 Chl *a* to one P700. The initial PSI particle in this case was considered to represent the reaction center and at least two additional antennae, which can be removed, leaving a reaction center–antenna complex core. A somewhat similar preparation was also obtained by utilization of deoxycholate (Bar-Nun and Ohad, 1977).

The organization of the various complexes within the membrane in situ is still a matter of controversy. Several models have been proposed describing the spatial and functional relationships of the antennae system and their connection to the reaction centers of various degrees of complexity. The common denominator of these models is the existence, for each photosystem, of a reaction center more or less tightly associated with an immediate small antenna and a second larger antenna existing as a separate entity. The main antenna or LHC is associated with PSII but might also transfer energy to PSI (spill-over) under certain environmental ionic conditions (Thornber *et al.*, 1976; Boardman *et al.*, 1978; Butler, 1979; Steinback *et al.*, 1979; cf. Wong and Govindjee, 1981).

3. ORGANIZATION, ORIGIN, AND INTEGRATION OF THE CHLOROPHYLL–PROTEIN COMPLEXES

Based on analysis of replicas of freeze-fractured chloroplast membranes, two general models have been proposed describing the organization of the eukaryotic thylakoid. In one model, the LHC and PSII reaction center form a tightly bound complex, appearing as the transmembrane large particles remaining on the EF face of freeze-fractured membranes (Staehelin *et al.,* 1976). According to this model, the PSI reaction center–antennae complex is located within one-half of the bilayer forming the membrane PF face. The second model (Simpson, 1979) proposes that the LHC antennae represent small intramembrane particles located on the PF face of the membrane and can become associated with the PSII reaction center. In membranes of mutants possessing the PSII reaction center, but lacking a normal LHC complex, such as it seems to be in the Chl *b*-less mutant of barley, the large particles found on the EF face might represent the PSII reaction center complex alone. Data obtained with *Chlamydomonas reinhardtii y-1* (Miller and Ohad, 1978), in which membranes can be formed lacking both PSII and PSI reaction centers (Bar-Nun *et al.,* 1977), showed that the few particles present on the EF face are significantly smaller (11.5 nm). Their size substantially increased (13 nm) upon insertion *in vivo* of the missing reaction centers of PSII and PSI. These results were interpreted to be in agreement with the first model. However, the EF fracture face of the membranes lacking reaction centers, but having a normal content of LHC, exhibited only a few 11.5-nm particles. Following insertion of the missing reaction centers, not only the size of the particles increased but apparently also their number. This could be in agreement with the second model, on the basis of which one would explain these facts as demonstrating the association of the LHC previously existing as small particles on the PF face with the newly formed PSII reaction centers, which give rise to the more numerous and larger particles appearing now on the EF face. Absence of the large particles on the EF fracture face of thylakoids from a PSII deficient mutant of *Chlamydomonas,* which possess a normal amount of LHC (Olive *et al.,* 1979), were indeed interpreted according to the model of Simpson (1979).

In both models, the intramembrane particles of both photosystems are considered to be free to laterally diffuse and move at random, meaning that LHC can also transfer energy to the PSI complex (spill over) or separate into PSII–LHC domains and PSI enriched domains, in which case energy transfer from LHC to PSI will be impaired. The PSII–LHC domains could further associate by intermembrane interaction, resulting in grana formation or stacked thylakoids. The regulation of this diffusion, phase separation, and interconnection between antennae and reaction centers is ascribed to electrostatic interactions between various charged membrane components. Charge neutralization or masking by mono or divalent cations (Barber, 1980) or low pH (Gerola *et al.*, 1979; cf. Wong *et al.*, 1980) is considered to be responsible for these effects. An important role in this process seems to be played by polypeptides forming the LHC, which are readily exposed to proteolytic digestions (Regitz and Ohad, 1976; Burke *et al.*, 1979).

Recently, it has been demonstrated that the LHC polypeptides are phosphorylated by a light-activated membrane-bound kinase, apparently affected by the membrane redox potential due to PSII activity (Beliveau and Bellemare, 1979; Bennett, 1979b, c; Alfonzo *et al.*, 1980). At the same time, LHC can be dephosphorylated by a soluble phosphatase, apparently located in the chloroplast matrix. The existence of a membrane-bound phosphatase has also been reported (Bennett, 1979b). The exact role of the LHC phosphorylation is still unclear, although it is reasonable to assume that it might be part of the control mechanism of particle movement and energy distribution among the two photosystems (Bennett *et al.*, 1980). So far, most of the work on thylakoid phosphorylation has been carried out in isolated, intact chloroplasts or photosynthetic membranes. Results obtained by Owens and Ohad (1981) indicate that LHC polypeptides can be phosphorylated *in vivo* in the dark as well as in the light. Furthermore, phosphorylation can occur *in vitro* and *in vivo* and also in membranes lacking active PSII and PSI reaction centers.

The polypeptides forming the Chl–protein or light-harvesting complex are of cytoplasmic translation and are coded by the nuclear genome. It is now well established that these polypeptides can be synthesized by free polysomes as soluble precursors, which are larger than the mature polypeptide by 40–50 amino acids (Apel and Kloppstech, 1978; Chua and Schmidt, 1979; Müller *et al.*, 1980a,b; Schmidt *et al.*, 1980), which form an extra piece probably at the N-terminal of the polypeptide. The precursors are further processed, and the extra piece is removed, while they are transported across the chloroplast envelope, and these mature polypeptides become integrated within pre-existing thylakoids. The transport and processing of cytoplasmically synthesized polypeptides across the chloroplast envelope are not restricted to the LHC polypeptides. Thus, the small subunit of the enzyme, ribulosebisphosphate carboxylase, is also synthesized as a soluble precursor in the cytoplasm. The precursor translated *in vitro*, using isolated polyadenylated mRNA from peas or spinach, is processed and transported by each pea or spinach chloroplast (Chua and Schmidt, 1978, 1979; Highfield and Ellis, 1978; Schmidt *et al.*, 1979). However, the precursors translated from *Chlamydomonas* polyadenylated mRNA are not processed or transported by the higher plants chloroplasts (Schmidt *et al.*, 1979). Transport of polypeptide precursors across the chloroplast envelope requires energy, is sensitive to uncouplers (but not to diuron), and can be driven by added ATP (Grossman *et al.*, 1980a,b).

The processing and transport enzymes also appear to be cytoplasmic translates, since they are present in the plastids of the heat-sensitive ray plants that lack active chloroplast ribosomes (Feierabend and Wildner, 1978; Feierabend, 1979; Feierabend *et al.*, 1980). A similar situation can be inferred to exist in *Chlamydomonas,* in which the LHC polypeptides are found in thylakoids of cells lacking active chloroplast ribosomes (Boynton *et al.*, 1972) or in those grown in the presence of chloramphenicol for several generations (Gershoni and Ohad, 1980). Similarly, it has been demonstrated that in *Euglena* the plastid envelope of a mutant lacking chloroplast DNA and ribosomes is basically indistinguishable from that of the wild type, indicating that all these envelope components are cytoplasmic translates of the nuclear genome (Bingham and Schiff, 1979).

Polypeptides known to participate in the formation of the reaction centers of both PSI and PSII are chloroplast translates. The major polypeptide of about 66,000 D appears to be needed for the formation of PSI reaction center component P700. Both PSI polypeptides are present in the Chl *a*–protein complex CPI. Mutants lacking this polypeptide do not exhibit the CPI complex or the P700 activity. However, situations have been described in *Chlamydomonas* that show the absence of the CPI complex but the presence of the CPI polypeptides. In these cases, PSI activity was present (Bar-Nun and Ohad, 1977; Bar-Nun *et al.*, 1977).

Several polypeptides have been considered as participating in the formation of PSII, including at least two, and possibly three polypeptides in the range of 40,000–55,000 D, and one in the range of 26,000 D (Chua *et al.*, 1975; Kretzer *et al.*, 1976; Bar-Nun *et al.*, 1977). Spector and Winget (1980) reported that a polypeptide of 65,000 D might be involved in the binding of Mn required for the H_2O splitting activity. Additional polypeptides of chloroplast translation in the range of about 16,000–20,000 D as well as 28,000–40,000 D might participate in the formation of various components of PSII and PSI antennae (Bar-Nun and Ohad, 1977; Gershoni and Ohad, 1979; Mullet *et al.*, 1980).

Chloroplast membranes of both higher plants and algae contain several polypeptides in the range of 32,000– 35,000 D, whose role is not yet well defined. A 32,000 D polypeptide of chloroplastic origin appears to turn over rapidly in higher plants such as peas (Ellis, 1975), maize (Grebanier *et al.*, 1979), and *Spirodella* (Edelman and Reisfeld, 1978). Its synthesis is light-dependent and is one of the major products of protein synthesis *in vitro* by isolated chloroplasts. The polypeptides of 32,000–35,000 D, as shown by SDS–PAGE, are sensitive to trypsin digestion. Trypsinization of the *Chlamydomonas* membrane under mild conditions results in the removal of these polypeptides, loss of the water splitting activity of PSII, and loss of the NADP reductase activity of PSI. Both reaction centers of PSII and PSI are unimpaired by this treatment and can transfer electrons to artificial acceptors and donors, although electron transfer between PSII and PSI is blocked (Regitz and Ohad, 1976). Proteolysis of the 32,000–35,000 D polypeptides of higher plants thylakoids results in loss of sensitivity toward herbicides that are known to inhibit electron flow at the reducing site of PSII. Trypsin-treated *Chlamydomonas* thylakoids also become insensitive to diuron (Regitz and Ohad, 1976), whereas electron flow in trypsin-treated thylakoids of developing chloroplasts of *Euglena* remain inhibited by diuron (Gurevitz, 1980). Recent studies using photoaffinity labeling by azido derivatives of atrazine or dinoseb have demonstrated that these herbicides bind specifically to the 32,000–35,000 D polypeptide (Oettmeier *et al.*, 1980; Pfister *et al.*, 1981). For a further discussion, see a review by Vermaas and Govindjee (1981). Detailed information on the transcription and translation origin of various plastid polypeptides can be found in the excellent review of Herrmann *et al.* (1980).

B. Development of Photosynthetic Membranes

The eukaryotic biological systems utilized for the investigation of membrane development can be, grosso modo, divided into two categories: (1) continuous development elicited by onset of illumination of dark-grown plants or algae and continued under constant light (Klein *et al.*, 1972; Ohad, 1975; Höyer-Hanson and Simpson, 1977; Boardman *et al.*, 1978; Senger and Strasberger, 1978; Dubertret, 1981a) and (2) stepwise development elicited by alternating changes in environmental conditions, such as exposure to intermittent or light–dark cycles (Schiff, 1970; Gurevitz *et al.*, 1977; Cahen *et al.*, 1978; Lieberman *et al.*, 1978; Argyroudi-Akoyunoglou and Akoyunoglou, 1979; Dubertret, 1981b), alternate exposure to protein synthesis inhibitors specific for the cytoplasm or chloroplast, and various combinations of these treatments using wild-type and/or light- and temperature-sensitive mutants (Eytan and Ohad, 1972b; Kretzer *et al.*, 1976; Schwartzbach *et al.*, 1976; Bar-Nun *et al.*, 1977; Cahen *et al.*, 1977; Dubertret, 1981b).

In the first system, a gradual transition is obtained from the state of membrane primordia or remnants present in the dark-grown cell to that of the mature normally functioning membrane. In such systems, detailed analysis of the process of development of structure–function relationship is difficult, because of the fact that various developmental phenomena, which might not necessarily be immediately related, occur simultaneously or partially overlap in time. Thus, necessary sequential steps in synthesis and assembly of membrane components, as well as the establishment of related functions, might be obstructed or difficult to single out and resolve in detail from other parallel developmental steps. However, such developmental systems provide the necessary basic information as a frame of reference for the other experimental systems.

On the other hand, in the modulated developmental systems, the process can be arrested at various steps; thus, composition–structure–function relationships can be better or more accurately established. Although the information obtained from such modulated systems suffers from a certain degree of doubt in regard to its physiological significance, it nevertheless contributes valuable information on the minimal composition and degree of structural organization required for establishing certain functions and helps to disclose possible transient developmental steps otherwise too short-lived to be detected in the gradually developing systems.

1. SYNTHESIS AND ASSEMBLY OF MEMBRANE COMPONENTS UNDER CONSTANT ENVIRONMENTAL CONDITIONS (CONTINUOUS DEVELOPMENT)

Most of the studies of chloroplast membrane development have been carried out using experimental systems in which Chl synthesis is light dependent. These include a variety of higher plants, among which the greening of etiolated beans, maize, or barley can be considered as representative examples (Boardman, 1977; Höyer-Hansen and Simpson, 1977; Konis *et al.*, 1978; Grebanier *et al.*, 1979). Among the algae in which Chl synthesis and membrane development is light-dependent, wild-type *Euglena* (Schiff, 1970; Bingham and Schiff, 1979) and mutants of *Chlamydomonas* (Ohad, 1975) and *Scenedesmus* are typical (Senger and Strasberger, 1978). According to the type of organism and growth conditions, various stages of plastid development are found in dark-grown cells. Thus, following seed germination in the dark and during the early stages of growth, as well as in continuously dark-grown *Euglena* cells, only proplastids are formed (Klein and Schiff, 1972; Boardman, 1977; Osafune *et al.*, 1980). *Euglena* proplastids are characterized by the presence of a double envelope and a noncrystalline array of tubules forming a prolamellar-like body from which prothylakoids emerge at its periphery. Intricate structural relationships were reported to exist between the prolamellar body, the prothylakoids, and the groups of membrane whorls present within the prolamellar body and between the proplastids and adjacent mitochondria and microbodies (Osafune *et al.*, 1980). The proplastids of higher plants contain only primordia of the prolamellar body and few prothylakoids (Klein and Schiff, 1972). Upon additional growth in the dark, the proplastids develop into etioplasts characterized by the presence of large prolamellar bodies in which tubular arrays are well organized in a crystalline-like form. The prothylakoids extend into the plastid matrix and significantly increase in number.

Both proplastids and etioplasts contain only small amounts of the matrix proteins characteristic of the mature chloroplast. These develop during illumination parallel with the development of the photosynthetic membranes.

A somewhat different situation is found in algae such as *Chlamydomonas* or *Scenedesmus*. The wild type of these algae synthesize Chl and possess a well-differentiated chloroplast when grown in the light or dark. However, in mutants, which have lost the ability to synthesize Chl in the dark and in which the synthesis of Chl depends upon the photoconversion of protochlorophyll(ide), a partially dedifferentiated plastid is formed in the dark (Ohad *et al.*, 1967a; Ohad, 1975). This organelle contains most of the chloroplast matrix proteins, including the enzymes required for the dark CO_2 fixation and the protein synthesizing machinery, as well as soluble and membrane-bound photosynthetic electron carriers such as ferredoxin and cytochrome *f*. Nevertheless, the total amount of thylakoids is reduced following growth in the dark. Also, the presence of noncrystalline prolamellar body-like structures in the plastids of these dark grown cells have been reported (Friedberg *et al.*, 1971). Their role in the greening process is not yet clear.

Neither proplastids nor etioplasts contain Chl. Instead, various amounts of protochlorophyll(ide) are present. The proplastid stage contains less protochlorophyll(ide) as compared with the etioplast in which protochlorophyll(ide) can be photoconverted. Apparently, the protochlorophyll(ide) of the higher plant proplastid is less readily photoconverted (Klein and Schiff, 1972).

As opposed to the system containing proplastids or etioplasts, the dedifferentiated plastid found in dark-grown algae such as the *y-1* mutant of *Chlamydomonas,* contains various amount of Chl residual from the previous growth in the light (Ohad *et al.*, 1967a,b). It must be noted that, as opposed to *Euglena,* the *Chlamydomonas y-1* mutant cannot grow in the dark indefinitely, and in most of the experiments carried out with this mutant, the cells were grown in the dark for 5–7 generations only. The situation appears to be similar for *Scenedesmus* C_2A mutant (Bishop and Senger, 1972; Senger and Bishop, 1972), as well as for a temperature sensitive Chl-less *Chlorella pyrenoidosa* mutant (Lavintman *et al.,* 1978). The protochlorophyll(ide) content of these cells, e.g., the *y-1 Chlamydomonas* mutant, is low as compared with that of etioplast; however, it can be readily photoconverted into Chl.

Initially, it was assumed that the prolamellar body is a storage body of thylakoid components, including the protochlorophyll(ide) and thylakoid polypeptides. This view was apparently in agreement with the observation that the crystalline structure of the etioplast's prolamellar body gradually became disorganized and the whole structure disappeared, whereas primary thylakoids, and then photosynthetic lamellae, developed during illumination. Moreover, the initial structural changes following protochlorophyll(ide) photoconversion were reversible upon return of the plants to the dark and accompanied by reaccumulation of protochlorophyll(ide) (Henningsen and Boynton, 1969, 1974). However, such a correlation between the disappearance of the prolamellar body and formation of thylakoids could not be demonstrated either in *Euglena* or in *Chlamydomonas.*

The isolated tubular material of the prolamellar body consists mostly of saponins (Lütz and Klein, 1979; Lütz, 1980) and their polypeptide pattern, following SDS–PAGE, is different from that of thylakoids (Lütz, 1978). In addition, Lütz and Klein (1979) demonstrated that most of the protochlorophyll(ide) present in etioplast is associated with the primary thylakoids and not with the prolamellar body tubular system. Thus, one could consider that the primary thylakoids are the structural precursors of the photosynthetic membranes.

Konis *et al.* (1978) showed that the maize etioplast already contains the components required for the formation of active PSI reaction centers, which develop during the initial phase of greening, even if Chl synthesis is partially inhibited by levulinic acid. At the same time, development of PSII activity could not be detected in the etioplast before at least 1–2 hours of continuous illumination, and its further development correlated well with that of Chl synthesis. The primary thylakoids also contain the chloroplast ATPase involved in energy conversion (Wellburn *et al.,* 1977). However, light induced pH gradients could be detected in maize etioplast only after illumination for 1–2 hours (Forger and Bogorad, 1973). The state of organization of the Chl–protein complexes during the very early stages of illumination is not as yet well known. In membranes isolated from the dark-grown *Chlamydomonas y-1* mutant, Chl–protein complexes cannot be detected by the SDS–PAGE technique (Bar-Nun *et al.,* 1977). The low-temperature spectrum of these cells shows the presence of only one peak at 680 nm (Burke *et al.,* 1978b; Gershoni and Ohad, 1979), which might be ascribed to the presence of an unorganized light-harvesting complex. This view is based on the fact that the polypeptides of this complex are still present in low amounts in these membranes (Bar-Nun *et al.,* 1977), which also contain chlorophyll *a* and *b* in a ratio of about 2:1 (Burke *et al.,* 1978b). Residual amounts of polypeptides, which form the reaction centers of PSII and PSI, and relatively large amounts of cytochrome *f* are also present (Ohad *et al.,* 1967a). However, these components are not interconnected, and electron transfer from H_2O to NADP or oxygen evolution by whole cells is very low (Cahen *et al.,* 1976). Measurements of the photosynthetic unit size in such membranes gives values of ≥ 2500–5000 Chl molecules per reaction center, but such data should not be interpreted in the usual manner, since the Chls are not connected to the reaction center proper. This is indicated by measurements of variable Chl *a* fluorescence and quantum yield, which show a very low ratio of variable to intrinsic fluorescence (F_v/F_o) and low photosynthetic efficiency. Accordingly, the fluorescence quantum yield is three to five times higher than in a normal membrane (Cahen *et al.,* 1976).

During the early stages of the greening process (first 1–2 hours), Chl synthesis proceeds at a gradually increasing rate (lag period), until it reaches a constant rate after 3–4 hours of illumination (Ohad *et al.,* 1967b). However, drastic changes occur in the thylakoids already during the first hour, even before substantial amounts of Chl have accumulated. The preexisting centers of PSII become connected to the rest of the electron transfer chain, and oxygen evolution in whole cells is re-established. The variable fluorescence parameter F_v/F_o increases, the apparent photosynthetic unit size decreases significantly, and the quantum yield of PSII increases (Cahen *et al.,* 1976). These changes correlate with the disappearance of the 680-nm peak (Burke *et al.,* 1978b). Upon further illumination and Chl accumulation, the 695-nm fluorescence emission peak appears, followed by the development of the 714-nm peak characteristic of these algae, which gradually rises and becomes predominant (Burke *et al.,* 1978b). The quantum yield of both PSII and PSI rises, and their specific activity on a Chl basis—after an initial rise—reaches a plateau and gradually becomes reduced again. These gradual changes indicate, first, a formation of active reaction centers and then of the antennae systems for both PSII and PSI and their interconnection with the reaction centers (Eytan *et al.*, 1974; Ohad, 1975). Concomitantly with the disappearance of the 680 nm peak and the appearance and increase in the fluorescence emission peak at 686 nm, 695 nm, and 714 nm, and the respective room-temperature absorption peaks, Chl– protein complexes CPII, CPI (Bar-Nun *et al.*, 1977), and CPII– CPIV can also be detected. The photosynthetic electron transfer rate at low light intensity increases and becomes similar to that of light-grown cells (Gershoni and Ohad, 1979). These gradual changes also correspond in time with pairing of thylakoids, stacking, and progressive formation of grana (Ohad *et al.*, 1967b).

During the next 2–4 hours of greening, when Chl synthesis continues at constant and maximal rate, newly formed reaction centers and antennae become integrated within the growing membranes, apparently in constant proportions, and thus the value of all the previously mentioned measured parameters remains constant (Eytan et al., 1974; Cahen *et al.*, 1976).

Basically, one could therefore consider the continuous greening process in the *Chlamydomonas y-1* mutant as consisting of two major steps: the first step in which unorganized or disconnected, pre-existing components of the electron-transfer chain reaction centers and antennae become interconnected and functional, and the second step in which membranes grow by addition and functional integration of newly formed components. However, this clear distinction is somewhat artificial, since these two steps overlap in time, and the transition point between them cannot be established accurately.

The major point to be made regarding the continuous greening is that the process, unlike originally assumed, is indeed a stepwise process in which the formations of reaction centers and their activation might occur independently of the formation of the corresponding antennae. Similar results were obtained independently in other experimental systems, including *Euglena* (Cahen *et al.*, 1978), *Scenedesmus* (Bishop and Senger, 1972), and *Chlorella* (Dubertret and Joliot, 1974), as well as in higher plants (Boardman, 1977).

2. SYNTHESIS AND ASSEMBLY OF MEMBRANE COMPONENTS UNDER VARIABLE ENVIRONMENTAL CONDITIONS (STEPWISE DEVELOPMENT)

The usefulness of experimental systems in which the greening process can be modulated or divided into distinct steps, resides in the possibility of obtaining answers to several basic questions such as: Are there any obligatory sequences in the synthesis and assembly of various membrane components? What are the polypeptides pigments and lipids required for the formation of each of the major photosynthetic membrane complexes, the reaction centers and respective antennae of PSII and PSI, and what are the roles of the participating polypeptides? Which are the polypeptides required for the functional assembly and interconnection of the preceding complexes? If a certain sequence of events is mandatory for the formation of the photosynthetic membrane, what are the controlling mechanisms or steps regulating and imposing this process? What are the interrelations between the nuclear and chloroplastic genomes and associated transcription–translation systems in the formation of the eukaryotic photosynthetic membranes.

Ideally an experimental system able to answer such questions will be one in which one could induce and sustain independently the synthesis and integration of cytoplastic and chloroplast translates and in which one could dissociate the process of Chl synthesis from that of synthesis of polypeptides and lipids. Obviously, such an experimental system is not available. However, integration of data obtained in a variety of biological systems fulfilling only some of these ideal expectations can provide tentative answers to at least some of these questions.

Among all the membrane components so far identified, none seems to be absolutely required for the formation of the membrane structure besides Chl and possibly the cytoplastic translates which bind Chl to form the LHC. Mutants deficient in almost any one of the constituents of the photosynthetic electron carriers, reaction centers, and antennae have been described in both algae and higher plants, including the energy transducing ATPase, cytochromes, plastoquinone, NADP reductase, PSI and PSII reaction centers and antennae as well as Chl *b* (Levine, 1969; Chua and Bennoun, 1975; Chua *et al.*, 1975; Bennoun and Chua, 1976; Machold and Höyer-Hansen, 1976; Maroc and Garnier, 1979). Although in the past, it has been shown that synthesis and integration of the cytoplastic translates and Chl is a prerequisite for the formation of the photosynthetic membranes in *Chlamydomonas y-1* mutant (Ohad, 1975), this rule seems to be less stringent in other biological systems. Chloroplast membranes containing only residual amounts of one of the two polypeptides forming the LHC as well as Chl *b*, have been described in barley (Machold and Höyer-Hansen, 1976), and both polypeptides are found only in small amounts in membranes formed in intermittent or alternate light–dark cycles in beans (Akoyunoglou and Argyroudi-Akoyunoglou, 1978) and *Euglena* (Gurevitz *et al.*, 1977). However, inhibition of the synthesis of all cytoplasmic translates and Chl completely blocks development of photosynthetic membranes in the majority of the experimental systems thus far studied.

Of particular interest are temperature-sensitive mutants deficient in only a few membrane polypeptides when grown under nonpermissive conditions. This situation is exemplified by the T_4 *Chlamydomonas* mutant, which lacks a membrane polypeptide in the molecular weight range of 50,000 D as detected by SDS–PAGE using gradient gels (Chua and Bennoun, 1975) and does not possess an active PSII reaction center. The electrophoretic pattern of membranes obtained from 37°C grown cells, and resolved by SDS–PAGE using 7% gels (Kretzer *et al.*, 1976), disclosed that two polypeptides of molecular weight 44,000 D and 47,000 D are missing. Both polypeptides could be specifically synthesized and integrated *in vivo* in the already present membranes at 25°C, thus reestablishing photosynthetic electron flow. The process did not require de novo cytoplastic translation but did require 70 S translation. Apparently, only one of the missing polypeptides was synthesized in the presence of rifampicin, an inhibitor of chloroplastic RNA polymerase. In this case, the reaction center of PSII was activated but the water-splitting activity was not. Both polypeptides could be synthesized in the dark and in the absence of Chl synthesis, but the functional integration of the polypeptide(s) required for activation of the water-splitting enzyme(s) required illumination. These observations indicate that specific polypeptides can be synthesized in the dark and integrated into pre-existing membranes. The activations of the water-splitting enzyme complex might require illumination (membrane energization) in certain cases, and photosynthetic activity is not required for the formation of the chloroplast membrane during the reactivation process.

These conclusions are in agreement with results obtained in other experimental systems. Chloroplast membranes, which are formed in etiolated leaves exposed to repetitive light flashes of millisecond duration, require short but continuous illumination and membrane energization for the activation of the water-splitting enzyme complex, whereas the reaction center of PSII is apparently already active (Dujardin *et al.*, 1970; Remy, 1973; Inoue *et al.*, 1975). The greening process in continuous light is only slightly inhibited by diuron (Schiff *et al.*, 1967). The membranes formed in the presence of diuron are photosynthetically active if the inhibitor is removed by washing. As will be shown later, synthesis of chloroplast translates in *Chlamydomonas y-1* mutant does not require light or Chl synthesis per se.

Stepwise synthesis and assembly of photosynthetic membranes was so far achieved mainly by the modulation of the greening process using two effectors, either separately or combined. These are light and inhibition of protein synthesis either in the chloroplast or in the cytoplast.

When etiolated plants are exposed to a series of light (2 min) and dark (90 min) cycles, Chl and membranes are formed proportional to the sum of the total amount of illumination time. Under these conditions both reaction centers of PSII and PSI are formed as well as components of the PSII and PSI antenna. These are $LHCP_1$, $LHCP_2$ (probably equivalent to CPIII and CPIV), and CPI and CPIa (probably an oligomer of CPI). However in the light-harvesting Chl *a,b*–protein complex, $LHCP_3$ (equivalent to CPII) is not formed. These can be formed following prolonged exposure to light–dark cycles or continuous illumination (Argyroudi-Akoyunoglou and Akoyunoglou, 1979). Thus in this case, formation of active reaction centers and antennae containing polypeptides of chloroplastic translation, can proceed without accumulation of the cytoplasmic translates.

A somewhat similar situation can be induced in *Euglena*. In this organism, the lag period of the greening process extends for 10–16 hr. The lag period can be significantly reduced if the cells are exposed to a short period of illumination and then further incubated in the dark for several hours before the onset of the greening process (Holowinsky and Schiff, 1970). During the initial period of illumination small amounts of Chl and photosynthetically active membranes are formed. Synthesis of Chl and cytoplasmic translates is blocked if the cells are transferred back to the dark, but chloroplast translates required for the formation of reaction centers and water-splitting activity continue to be formed and become integrated into the membranes. Upon re-exposure of the cells to the light, Chl and cytoplastic translates are synthesized again, and the Chl–protein complexes forming the antennae system are assembled. Additional modulation can be imposed on this system by use of protein-synthesis inhibitors. If protein synthesis by 70 S ribosomes is inhibited during the dark incubation, the rise in the activity of reaction centers of PSII is prevented, and the water-splitting activity already developed is lost, whereas PSI activity continues to rise. Photosystem II activity can be "repaired" either in the dark or light if protein synthesis in the chloroplast is resumed. However, the antennae system of both PSII and PSI are formed only if cytoplastic translates and additional Chl are synthesized (Gurevitz *et al.*, 1977). Measurements of reaction centers on a Chl basis, quantum yield and flash yield, as well as partial reaction of electron transfer at the various developmental stages of this system, have demonstrated that the ratio of reaction centers to antennae and interconnection of the reaction centers of PSII with the electron donor and acceptor sites can be modulated extensively (Cahen *et al.*, 1978). By appropriate use of radioactive tracers, electrophoretic separation of polypeptides and Chl–protein complexes, and measurements of fluorescence emission and excitation spectra, it is possible to take advantage of this experimental system in order to identify the antennae and reaction center's polypeptides and their subcellular origin as cytoplastic and chloroplastic, respectively. The results so obtained were in agreement with data obtained independently by similar experiments carried out with greening *Euglena* cells in continuous light (Bingham and Schiff, 1979).

Stepwise development of photosynthetic membranes also can be induced in *Chlamydomonas y-1* mutant. When the dark-grown cells are exposed to the light in presence of chloramphenicol (CAP), only cytoplastic translates and Chl are synthesized and form the light-harvesting Chl *a, b*–protein complex (Ohad, 1975). Since chloroplast translation is inhibited, new reaction centers are not formed, and the photosynthetic activity does not rise above the initial level present at the onset of illumination. Analysis of the Chl–protein complexes by SDS–PAGE reveals the presence of CPII complex only (Bar-Nun *et al.*, 1977). However, the low-temperature fluorescence emission spectrum shows the expected 686 nm peak and a second major peak at 708 nm (Burke *et al.*, 1978b), which might be ascribed to the presence of a PSI antenna component. The high fluorescence yield of these Chl complexes might be due to the absence of quenching by the appropriate reaction centers and eventually additional antennae components (Gershoni and Ohad, 1979) such as antennae of PSII (CPII, CPIII) and PSI (CPI). The absent polypeptides required for the function of the reaction centers and the Chl required for the formation of the antennae can be synthesized and integrated in the pre-existing membranes simultaneously, if both chloroplastic and cytoplastic translation are resumed and the process (repair) is carried out in the light. Alternatively, only reaction centers can be formed and integrated within an operative electron carrier chain if chloroplast translation is resumed, but cytoplastic translation and Chl synthesis are prevented (repair in the dark or in the light in the presence of cyclohexamide). Electron flow from H_2O to NADP and oxygen evolution by whole cells are re-established in both cases. However, energy transfer between the newly formed reaction centers and the pre-existing Chl *a,b*–protein complex, and the presumed PSI antenna component seem to occur only if additional antennae are formed that might establish a link between the reaction center and the light-harvesting complex (CPII). The establishing of these connections is demonstrated by a rise in the electron flow at low light intensity for both photosystems, a reduction in the fluorescence quantum yield at room temperature and at 77°K of both the 686 nm and 708 nm peaks, and appearance of the low steady state fluorescence emission (*Fs*) at room temperature in whole cells ($Fs <$

Fmax) (Gershoni and Ohad, 1979). These changes correlate with appearance of SDS-PAGE detectable Chl protein complexes CPIII, CPIV, and CPI (Gershoni *et al.*, 1981). A schematic representation of the stepwise or continuous development of Chl–protein complexes forming reaction centers and antennae and their interconnections is shown in Fig. 3 and fully explained in the legend.

3. CHANGES IN THE MEMBRANE STRUCTURE DURING DEVELOPMENT

The various developmental stages described earlier, which have been so far identified by the content of their membrane polypeptides, Chl–protein complexes and activities, correspond also to well-defined stages in the structural development of membrane organization. As mentioned earlier, absence or alteration of PSII reaction centers correlates with a reduction in the size (Miller and Ohad, 1978) and number (Olive *et al.*, 1979) of the large particles found on the EF face of freeze-fractured membranes. Following repair of photosynthetic activity of membranes formed in *Chlamydomonas y-1* mutant in presence of chloramphenicol, an increase in the size and number of the EF particles is obtained (Miller and Ohad, 1978). However, the number of small particles present on the PF fracture face of membranes deficient in PSII reaction centers seems to be unchanged or corresponds to that of normal membranes (Olive *et al.*, 1979). On the other hand, membranes deficient in Chl *b* in which the Chl *a,b*–protein complex is altered do exhibit the large EF particles, but show a reduction in the amount of the small particles on the PF fracture face (Simpson, 1979). The interpretation of these results is still controversial (see Kaplan and Arntzen, 1982).

The reversible cation-induced lateral phase separation which occurs in the membranes during grana stacking and results in the formation of separate PSII and PSI enriched domains, seems to be possible or correlate with developmental stages in which the antennae of both photosystems are well developed.

The ability to form stacks depends on the presence of the light-harvesting complex, as demonstrated by electron microscope analysis of thin sections and replicas of freeze-fractured membranes of various algae and higher plant mutants deficient in this complex, as well as by fractionation with digitonin of membranes obtained at various stages of the stepwise greening process (Akoyunoglou and Argyroudi-Akoyunoglou, 1978; Argyroudi-Akoyunoglou and Akoyunoglou, 1979).

As mentioned before, the polypeptides of the LHC are phosphorylated, and thus membrane phosphorylation might be involved in this stacking process (Bennett *et al.*, 1980). The state of membrane phosphorylation at various developmental stages is not yet known. *In vitro*

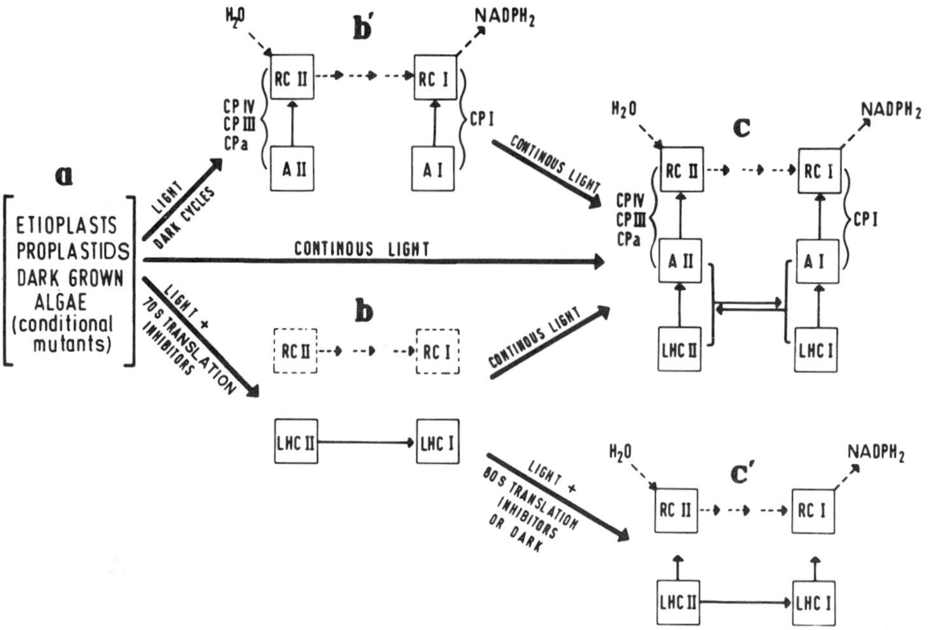

FIG. 3. Schematic representation of possible sequences of events leading to the formation of active photosynthetic units. A long legend is provided to serve as a summary of the process being discussed. (a) Represents the initial stages of development present in various systems. Etioplasts are formed in plants that have been germinating and growing in the dark for several days (6–11 days). Proplastids are found in early stages of leaf growth or in algae such as dark-grown Euglena. The state of plastid development of dark-grown conditional mutants of algae, such as the *y-1* mutant of *Chlamydomonas* (Ohad, 1975) or C_2 of *Scenedesmus* (Senger and Strasberger, 1978) is actually a partially dedifferentiated plastid. According to the type of organism and stage of development or dedifferentiation, the organelle might lack or contain various amounts of electron carriers, reaction centers, and antennae, as well as the machinery required for protein and Chl synthesis and dark CO_2 fixation. The proplastid stage is the least developed, whereas the dedifferentiated plastid stage is very similar to a normal chloroplast, although it contains only reduced amounts of disorganized reaction centers and antennae (Cahen *et al.*, 1976, 1977) which display high fluorescence yield (680 nm at 77°K). (b) Exposure of organisms, which contain organelle in stage (a), to light + CAP. These conditions, which block protein synthesis by the plastid 70 S ribosomes, allow synthesis of Chl *a* and *b* and cytoplasmic translates, which are imported into the plastid and form the Chl *a,b*–protein complex (fluorescence emission 686 nm at 77 K), as well as an additional Chl *a*-containing complex (fluorescence emission 705–708 nm at 77 K) (Gershoni and Ohad, 1979; Burke *et al.*, 1978b). However, no additional reaction centers are formed, and those initially present are not connected to a functional electron carrier chain. No variable fluorescence is detectable, but high fluorescence is emitted (F_0) by the antennae formed under these conditions (the antennae are not connected to reaction centers). The Chl–protein complexes CPIII, CPIV (or their possible equivalent, CPa or $LHCP_2$–$LHCP_1$), and CPI are not detected. However, the light-harvesting Chl *a,b*–protein complex CPII (LHC or $LHCP_3$) is present as in normally developed chloroplast (Bar-Nun *et al.*, 1977). (b′) Exposure of proplastids or etioplasts to alternate light (2 min) and dark (90 min) cycles results in the synthesis of small amounts of Chl (mostly Chl *a*), which become organized into functional reaction centers and the respective closely attached antennae

phosphorylation of isolated membrane proteins by the membrane-bound kinase using [^{32}P] ATP shows that this activity is very low but is present in membranes obtained from dark-grown *Chlamydomonas y-1* mutant cells (Owens and Ohad, 1981, 1982). Polypeptides of the LHC are also phosphorylated *in vitro* in photosynthetically deficient membranes obtained from these cells after greening in the presence of chlo-

---

(Argyroudi-Akoyunoglou and Akoyunoglou, 1979). However, accumulation of the cytoplasmic translates—required for the formation of the Chl *a,b*–protein complex—does not occur; the Chl *a*–protein complexes associated with PSII, CPa or $LHCP_1$–$LHCP_2$ (possibly, equivalent of CPIII–CPIV) as well as CPI associated with PSI are predominant. The ratio of antennae Chl to reaction centers is relatively small (small photosynthetic units); the ratio of variable fluorescence (F_v) to intrinsic fluorescence (F_o) is the highest observed so far (Akoyunoglou and Argyroudi-Akoyunoglou, 1978). It follows that light saturation of electron flow is obtained at relatively high light intensity. (c′) This stage can be obtained if 70 S translation is resumed in plastids presently found in stage (b) whereas Chl synthesis is inhibited (either by incubating the organism in the dark or by adding 80 S translation inhibitors). Under these conditions, reaction centers of both photosystems are formed and connected with a functional electron carrier chain. However, the interconnecting or closely associated antennae of each photosystem are not formed and the corresponding Chl *a*–protein complexes (CPIII, CPIV, and CPI) are not detectable. As a result, the connections between the previously formed antennae of PSII and PSI are not efficiently connected with the newly formed reaction centers. The size of the photosynthetic units is apparently large, the quantum yield remains low, and the electron flow requires high light intensity. Variable fluorescence is detectable, although the ratio F_v/F_o remains low. The high fluorescence yield of stage (b′) is only partially reduced, and the 705–708-nm fluorescence emission peak at 77°K remains prevalent (Gershoni *et al.*, 1981). (c) This stage represents the final development of functional thylakoids under natural conditions of exposure to continuous light via a series of discrete steps, which might include stages (b) or (b′). Plastids of stages (b) or (b′) can also reach state (c), if the organism is exposed to continuous illumination in absence of translation inhibitors. Thus, Chl *a* and *b* as well as cytoplasmic translates are synthesized and accumulated. Both reaction centers are made and connected with a functional electron carrier chain; the newly formed light-harvesting antennae of both photo-systems are connected with the respective reaction centers via the present interconnecting antennae (b′ → c), or the previously formed light-harvesting antennae are now connected with the newly formed interconnecting antennae (b → c). Energy transfer from all chlorophylls to reaction centers is established, the ratio of F_v/F_o and the size of photosynthetic units are similar to that in normally developed plants, and electron flow becomes saturated at low light intensity. All Chl–protein complexes can now be detected as well as the corresponding low-temperature fluorescence emission peaks.

This scheme is based on and adapted from data obtained with algae and higher plants (Anderson *et al.*, 1978; Argyroudi-Akoyunoglou and Akoyunoglou, 1979; Garnier *et al.*, 1979; Gershoni and Ohad, 1979; Dubertret, 1981a,b; Picoud *et al.*, 1981). It represents a generalized picture and does not show details specific to certain organisms or experimental conditions. Heavy lines represent developmental sequences; arrows indicate components of the electron carrier chain (dotted lines) or path of energy transfer (continuous lines). (RC) Reaction centers; (LHC) light-harvesting antennae; (A) interconnecting antennae; (CP) Chl–protein complexes.

ramphenicol, as well as after repair of photosynthetic activity following incubation of these cells in the dark or in the light in presence of cyclohexamide. However, in each case, phosphorylation pattern appears to be different (Owens and Ohad, 1981). Thus, one could assume that the state of stacking in membranes deficient in photosynthetic activity might be related to lack of membrane phosphorylation, since so far it was demonstrated that this process is stimulated by light and seems to require light-induced membrane energization and/or changes in the redox state of plastoquinone (Beliveau and Bellemare, 1979; Bennet, 1979b). However, this might not necessarily be the case *in vivo*. The polypeptides of the light-harvesting complex are phosphorylated *in vivo* if the *Chlamydomonas* cells are incubated in presence of ^{32}P in the light or in the dark. As mentioned earlier, if ATP is supplied, it can occur in photosynthetically inactive membranes obtained from chloramphenicol-treated cells. These observations certainly do not invalidate the role of membrane phosphorylation in the phase separation and stacking process, but they merely point out the possibility that during the development process additional conditions might be required, besides the presence of the LHC and phosphorylation of its polypeptides, in order to obtain the formation of the grana stacks.

The question arises whether phosphorylation of the LHC polypeptides might play a role in their processing, from the cytoplast, and integration into pre-existing membranes. The fact that these polypeptide products of *in vitro* translation can be processed and integrated into CPII complex *in vitro* (Schmidt *et al.*, 1980) seems to indicate that this is not the case. However, direct evidence against or for this possibility is not yet available.

General conclusions, which can be drawn from the analysis of data so far described, are as follows: The formation of the photosynthetic membranes occurs by synthesis and stepwise or sequential integration of newly formed components into pre-existing membranes. There seems to be no general rule governing the various steps sequence. However, growth and normal development of the photosynthetic membrane in the eukaryotic chloroplast requires accumulation of the cytoplastic translates and formation of the Chl–protein complex which might contain various amounts of Chl *b*. Thus in systems that depend on illumination for Chl synthesis, the whole membrane biogenesis process becomes light-dependent. The synthesis and integration of the chloroplast translates required for the formation of the reaction centers, as well as that of electron carrier chain and the coupling factor ATPase, are not directly light-dependent nor do they appear to require Chl synthesis. Of particular interest is the fact that the reaction center's polypeptides synthesized in the dark do become bound to Chl pre-existing in the membranes.

Although chloroplast membrane structures can be formed, within certain limits, irrespective of its exact composition, obviously, the function at each step of development is a direct result of the membrane composition and the ability of its constituents to become organized into appropriate superstructures.

The preceding few "rules," which appear to apply in general, pose several basic questions, which so far remain unasnwered. If the membrane composition and the structure can and do change so widely, what is the mechanism by which this ever-changing entity is recognized as a chloroplast membrane by the nascent polypeptides that are directly incorporated, while being synthesized apparently by membrane-bound 70 S ribosomes or by the polypeptide precursors of cytoplastic translation during or after their processing? It is to be expected that the same rules apply as for the synthesis of other cellular membranes or transport of polypeptides across biological membranes (Chua and Schmidt, 1979). That is, a certain specific and constant feature of the membrane should be recognized by the processed or yet unprocessed polypeptide. So far we do not know what that chloroplast membrane specific feature might be.

The fact that Chl-binding polypeptides (e.g., reaction centers) are synthesized independent of Chl, and then become associated with it implies that either some amount of Chl present in the membrane is free and can be subsequently complexed with the newly formed polypeptides, or if all the Chl is bound, as it is generally assumed, then Chl molecules might dissociate from the existing complexes and redistribute within newly formed ones. So far only limited information is available on the turnover or exchange of Chl bound to Chl–protein complexes, which indicates that this process is very slow and not easily detectable (Bar-Nun *et al.*, 1977; Valane, 1978).

The fact that photosynthetic membranes grow by addition of newly synthesized components is well documented and generally accepted. It is also quite evident that diffusion of proteins and lipoprotein complexes can bring about redistribution and aggregation of specific components and result in the formation of heterogeneous or homogeneous domains. This phenomenon is considered to actually be the basis for the regulation of energy distribution between the two photosystems (e.g., see Murata *et al.*, 1975; Staehelin, 1976; Armond *et al.*, 1977; Burke *et al.*, 1978a; Barber, 1980; Mullet and Arntzen, 1980; Mullet *et al.*, 1980; Sculley *et al.*, 1980; Wolman and Diner, 1980; Wong *et al.*, 1981). Thus, one can assume that any point on the membrane surface can serve as a growing point, and the membrane will expand in all directions in the membrane plane. However, this might not necessarily be the case. As mentioned before, growth of the photosynthetic membrane in bacteria seems to be localized at specific sites. Attempts to find out whether chloroplast membranes in *Chlamydomonas y-1* grow at specific sites or randomly have been made and have shown that newly formed components become randomized within a relatively short time, However, the time required for sampling and the technique used could allow randomization of components even if insertion occurred at specific sites (Eytan and Ohad, 1972a). Analysis of distribution of newly formed lipids and protein along membrane profiles *in situ,* by use of autoradiography at electron microscopic level, indicated that the grana regions were less labeled than the stroma lamellae (Goldberg and Ohad, 1970a,b). This could be in agreement with the fact that in electron micrographs, polyribosomes are more frequently seen close or "bound" to the stroma lamellae or the matrix exposed surface of grana membranes and are not found within the grana partitions. In these experiments, the pulse labeling time was relatively short before the cells were fixed, and thus one could expect that redistribution of the newly inserted components might be limited. Thus, it is not unreasonable to assume that the stroma lamellae might preferentially be the site of membrane growth.

The redistribution of membrane components between stroma and grana lamellae deserves further consideration. In view of the existing models of chloroplast membrane structures, one would expect that lateral diffusion of large membrane constituents (such as reaction centers or antennae Chl–protein complexes) between stroma and grana lamellae might be slow and quite limited. However, one should consider the possibility that grana and stroma structures as such are temporary entities. If dissociation of grana will occur *in vivo,* then the entire thylakoid system will become "stroma." Components of PSI or PSII found initially in the grana region could diffuse and mix with similar components found in the stroma region and then redistribute and form new relatively homogenous domains in another part of the membrane continuum following reformation of new grana and stroma regions.

The other possibility is that in situ components present in a grana region will always remain in the same granum, or if grana dissociation occurs under physiological conditions, the same components would always reassociate in the same way and form the same granum as before. The present day's concepts of membrane structure and fluidity render this second possibility quite unlikely. So far we have no detailed information on the kinetics of grana formation and disassociation in situ. It is quite possible that the lateral diffusion of membrane complexes might be the rate-limiting factor in the process of randomization of the newly formed membrane components across the thylakoid membrane continuum.

4. NUCLEOCYTOPLASTIC INTERRELATIONS AND CONTROL OF MEMBRANE BIOGENESIS

The material so far presented deals with biogenesis of chloroplast membranes in biological systems in which light is the major effector of the developmental process. The complexity of the molecular mechanisms by which light regulates the differentiation of the proplastid into an etioplast and/or plastid and finally into a mature chloroplast is not yet entirely understood.

It is quite certain that light acts upon more than one receptor system and at more than one stage in the preceding sequence of events. Analysis of the action spectra of the various stages of the greening process reveals three major receptor complexes: one absorbing in the blue region of the spectrum; one in the red—the protochlorophyll(ide), and one in the far-red region—the phytochrome (Masoner and Kasemir, 1975; Klein *et al.*, 1977; Senger and Mell, 1977; Brinkman and Senger, 1978; Mohr and Oelze-Karow, 1978). The blue light receptors (Senger and Mell, 1977; Brinkman and Senger, 1978) and the phytochrome systems (Masoner and Kasemir, 1975; Mohr and Oelze-Karow, 1978) have a very complex role and act on the differentiation of the organelle at the early stages of development, which affects the formation of chloroplastic matrix enzymes and the machinery of protein synthesis, as well as the cytoplastic enzymes required for the formation of Chl precursors. The exact molecular mechanism by which the effect of these light-absorbing complexes is exerted is not yet well known. However, a large body of information is available describing a variety of phenomena affected by these systems and their possible interrelations (Mohr and Oelze-Karow, 1978). One should note that these light receptors, and the processes affected by them, are not specific for the chloroplast development and relate to a variety of cellular, systemic, physiological, and developmental responses of the whole plant (Mohr, 1972). On the other hand, the red light receptor systems, the photoconversion of protochlorophyll(ide) to Chl, is specifically involved in the process of membrane development. It is beyond the scope and space of this chapter to deal with the first two receptor systems. The reader is referred to publications treating these subjects (Mohr, 1972; Smith, 1975; Senger, 1979). The mechanism of protochlorophyll(ide) photoconversion and the process of Chl biosynthesis is described by Rebeiz and Lascelles (1982).

The interrelation between the light-dependent synthesis of Chl and that of the cytoplastic translates required for its binding and formation of the Chl *a,b*–protein complex was investigated in both plants and algae. Initially, it was proposed that in the dark-grown *Chlamydomonas y-1*, protochlorophyll–holochrome accumulates and, in combination with chloroplast, translates (whose synthesis is light-independent) act together as a specific repressor of the gene transcription process required for the synthesis of the cytoplastic translates of the chloroplast membrane (Hoober and Stegeman, 1973). The specific inhibition of cytoplastic translation by cycloheximide results in complete inhibition of Chl synthesis. This could be explained by the fact that the enzymes involved in the Chl biosynthesis pathways are found to be cytoplastic translates (Harel, 1978). Since cycloheximide inhibits Chl synthesis within a short time from its addition at all stages of the greening process, it is reasonable to assume that at least one of these enzymes must be short-lived and thus requires continuous translation. This view was further supported by the fact that inhibition of RNA synthesis in the cytoplasm had a similar effect to that of cycloheximide. An appropriate candidate for this key role in the regulation of Chl– protein synthesis could be the enzyme synthesizing δ-amino levulinic acid (δ-ALA), using glycine and succinyl-CoA as substrates, which was shown to be unstable in several systems and had a half-life compatible with the preceding findings (Marver *et al.*, 1966). However, it was found that in higher plant cells, as well as in several algae, synthesis of δ-ALA is carried out by a different enzymatic system using a five-carbon compound as a substrate (Harel, 1978; Kannangara and Gough, 1978). The synthesis of one of the enzymes participating in this process, the glutamate-1-semialdehyde aminotransferase, seems to be light induced. The lifetime of this enzyme once induced seems to be definitely longer than that required by the previously proposed schemes (Kannangara and Gough, 1978). Moreover, it appears now that a chloroplast translate is involved at least in one step, not yet identified, in the reaction chain leading to the formation of Chl. This seems to be the case in both higher plants (Feierabend, 1977, 1979) and algae. Dark-grown *Chlamydomonas y-1* cells can green at least partially in the presence of chloramphenicol, but if chloroplast translation is inhibited for several generations during growth in the dark, the resulting chloramphenicol-degreened (CD) cells cannot green in the light unless 70 S ribosomal activity is resumed (Gershoni and Ohad, 1980). On the other hand, CD cells can green and form significant amounts of membranes containing de novo synthesized Chl, which is integrated into functional Chl–protein complexes in complete absence of cytoplasmic translation. This would indicate that both the cytoplastic translates required for Chl binding and an active δ-ALA synthesizing enzyme were already present in the CD cells. It has been shown that the synthesis of mRNA, coding for the synthesis of the cyotplastic translates involved in the formation of the Chl *a,b*–protein complex, is light induced (Apel and Kloppstech, 1978; Müller *et al.*, 1980a) and persists in light-induced cells for at least several hours after cessation of illumination (Bennett, 1979a). The translation products of this mRNA that is the polypeptide of the LHC found in the membranes became labeled in the dark if radioactively labeled precursors of protein synthesis were added to the cells after an initial illumination. Labeling continues in the dark for a short time, but the translates do not accumulate in the membrane in large quantities (Ohad, 1975; Gershoni *et al.*, 1981).

Finally we should take into account the fact that Chl synthesis is inhibited by a negative feedback mechanism in which a form of protochlorophyll(ide) possibly bound to the holochrome seems to be involved (Nadler and Granick, 1970; Harel, 1978). The inhibition is reversed upon photoconversion of the protochlorophyll(ide) and release of the holochrome.

In view of these and previous data and assuming that the chloroplast translate(s) required for Chl synthesis might be involved in the formation of the regulatory form of protochlorophyll(ide), one could explain the greening of CD cells in the presence of cycloheximide in the following way. During growth in the presence of chloramphenicol, CD cells are depleted of 70 S translates. In this case, the regulatory form of protochlorophyll(ide) does not accumulate, and thus transcription and translation of the nuclear genes involved in the light-regulated formation of membrane protein occurs. The enzyme(s) required for the synthesis of δ-ALA forms and remains active, and a certain but limited amount of apoprotein of the Chl *a,b*–protein complex also accumulates. Upon exposure of CD cells to the light and resumption of 70 S ribosomal activity, protochlorophyll(ide) holochrome is formed and photoconverted. The active δ-ALA synthase already present allows formation of additional precursors and, therefore, continuous synthesis of Chl. Inhibition of protein synthesis in the cytoplast will prevent additional synthesis of the apoproteins of the LHC, but not the binding of the already present polypeptides with the newly formed Chl. Under this condition, the chloroplast-synthesized polypeptides required for the formation of reaction centers will also be synthesized and the photosynthetic electron transfer chain activated (Gershoni and Ohad, 1980).

Measurements of the action spectrum of the greening process in which Chl and the membrane polypeptides of cytoplastic origin are synthesized at constant and maximal rates in *Chlamydomonas y-1* showed a correlation between protochlorophyll(ide) photoconversion and synthesis and accumulation of these polypeptides (Ohad and Drews, 1975; Hoober and Stegeman, 1976). The fact that the mRNA required for the synthesis of Chl *a,b*-binding polypeptides can be found in dark-treated cells, while the translation products fail to accumulate in large quantities, can be explained if we assume that these are unstable and degraded if not complexed to Chl. This might not be the case in organisms in which protochlorophyll is converted into Chl enzymatically in the absence of light (Boynton *et al.*, 1972; Shepherd *et al.*, 1979). This enzymatic process is different from that of the photoconversion and might not require the participation of the chloroplast translates (Griffiths and Mapleston, 1978).

The mechanism by which red light controls the chloroplast membrane development via synthesis and accumulation in eukaryotes may be restricted primarily to the synthesis of Chl itself and only in a secondary way to that of the synthesis and accumulation of the Chl-binding proteins of cytoplastic translation and the chloroplast translates involved in the formation of reaction centers. Again a difference exists between the accumulation of these and that of chloroplast translated cytochromes (Zielinski and Price, 1980) which, as mentioned before, are found in the *Chlamydomonas y-1* dark-grown cells in rather large quantities (Ohad, 1975). The reason for the difference between these two types of membrane polypeptides is not known. Possibly, the chloroplast translates forming reaction centers are also complexed to Chl, and they might obey the same rules as the Chl-binding polypeptides of cytoplastic origin. Thus, it is possible that transcription and translation, once induced, are not necessarily continuously light-regulated, and the appropriate messengers might be present in the dark for some time. However, translation products do not accumulate unless complexed with Chl and properly integrated into the membranes. In the absence of sufficiently available Chl and membrane structure, these polypeptides might turn over rapidly. The fact that almost complete inhibition of chloroplast RNA synthesis by rifampicin has only a limited effect on the formation of active photosynthetic membranes (Jennings and Ohad, 1972) during the greening of *Chlamydomonas y-1* mutant might support the previously mentioned view. In addition, data demonstrating that the transcription process required for the formation of the Chl *a,b*—protein complex can be induced in the *y-1* mutant of *Chlamydomonas* in the dark by temperature shock treatment and that the transcripts persist for several hours (Hoober and Stegeman, 1976) supplies further support. Turnover of a chloroplast translate of 44,000 D required for the formation of water-splitting activity in a temperature-sensitive mutant of *C. reinhardtii* lacking a PSII polypeptide (47,000 D) when grown at 37°C, might account for the results reported by Chua and Bennoun (1975) and Kretzer *et al.* (1976).

Most of the earlier discussion is devoted to an apparently particular case, that of the *y-1* mutant of *Chlamydomonas*. This system was selected not only because of the large body of information available on the greening process, but also because this organism offers a unique situation, namely, by a single gene mutation affecting the enzyme(s) converting protochlorophyll(ide) to Chl in the dark, it has acquired most of the apparent properties of the light-regulated chloroplast development present in higher plants and other algae.

C. Concluding Remarks

On the basis of the few quoted facts and many speculative thoughts presented, one may reach the following conclusion, which should, however, be considered only as a working hypothesis. A close interrelationship exists between the accumulation of newly synthesized polypeptides of both cytoplastic and chloroplastic origin involved in the formation of Chl–protein complexes and the synthesis and accumulation of Chl. Light seems to affect the process of membrane biogenesis primarily by photoconverting protochlorophyll(ide) and thereby reducing the relative concentration of Chl precursors, which may act as inhibitors of various steps of the biogenetic process, as well as permitting the active synthesis of Chl itself.

REFERENCES

Aagaard J., and Sistrom, W. R. (1972). *Photochem. Photobiol.* **15,** 209–225.

Akoyunoglou, G., and Argyroudi-Akoyunoglou, J. H., eds. (1978). "Chloroplast Development." Elsevier/North-Holland Biomedical Press, Amsterdam.

Alfonzo, R., Nelson, N., and Racker, E. (1980). *Plant Physiol.* **65,** 730–734.

Anderson, J. M. (1975). *Biochim. Biophys. Acta* **416,** 191–235.

Anderson, J. M. (1980). *Biochim. Biophys. Acta* **591,** 113–126.

Anderson, J. M., Waldron, J. C., and Thorne, S. W. (1978). *FEBS Lett.* **92,** 227–233.

Apel, K., and Kloppstech, K. (1978). *Eur. J. Biochem.* **85,** 581–588.

Argyroudi-Akoyunoglou, J. H., and Akoyunoglou, G. (1979). *FEBS Lett.* **104,** 78–84.

Armond, T. A., Staehelin, L. A., and Arntzen, C. J. (1977). *J. Cell Biol.* **73,** 400–418.

Baccarini-Melandri, A., and Melandri, B. (1978). *In* "The Photosynthetic Bacteria" (R. K. Clayton and W. R. Sistrom, eds.), pp. 615–628. Plenum, New York.

Barber, J. (1980). *FEBS Lett.* **118,** 1–10.

Bar-Nun, S., and Ohad, I. (1977). *Plant Physiol.* **59,** 161–166.

Bar-Nun, S., Schantz, R., and Ohad, I. (1977). *Biochim. Biophys. Acta* **459,** 451–467.

Beliveau, R., and Bellemare, F. (1979). *Biochem. Biophys. Res. Commun.* **91,** 1377–1382.

Bennett, J. (1979a). *FEBS Adv. Course, Genome Organ. Expression Higher Plants,* No. 62.

Bennett, J. (1979b). *Eur. J. Biochem.* **99,** 133–137.

Bennett, J. (1979c). *FEBS Lett.* **103,** 343–344.

Bennett, J., Steinback, K. E., and Arntzen, J. C. (1980). *Proc. Natl. Acad. Sci. U.S.A.* **77,** 5253–5257.
Bennoun, P., and Chua, N. H. (1976). *In* "Genetics and Biogenesis of Chloroplast and Mitochondria" (T. Bücher, W. Neupert, W. Sebald, and S. Werner, eds.), pp. 33–39. Elsevier/North-Holland Biomedical Press, Amsterdam.
Biedermann, M., Drews, G., Marx, R., and Schröder, J. (1967). *Arch. Mikrobiol.* **56,** 133–147.
Bingham, S., and Schiff, J. A. (1979). *Biochim. Biophys. Acta* **547,** 512–530.
Bishop, N. I., and Senger, H. (1972). *Plant Cell Physiol.* **13,** 937–953.
Boardman, N. K. (1977). *In* "Photosynthesis I" (A. Trebst and M. Avron, eds.), pp. 583–597. Springer-Verlag, Berlin and New York.
Boardman, N. K, Anderson, J. M., and Goodchild, D. J. (1978). *Curr. Top. Bioenerg.* **8,** 35–109.
Bogorad, L. (1975). *In* "Membrane Biogenesis" (A. Tzagaloff, ed.), pp. 201–204. Plenum, New York.
Boynton, J. E., Gillham, N. W., and Chabot, J. F. (1972). *J. Cell Sci.* **10,** 267–305.
Brinkman, G., and Senger, H. (1978). *In* "Chloroplast Development" (G. Akoyunoglou and J. H. Argyroudi-Akoyunoglou, eds.), pp. 201–206. Elsevier/North-Holland Biomedical Press, Amsterdam.
Broch-Due, M., Ormerod, J. G., and Fjerdingen, B. S. (1978). *Arch. Microbiol.* **116,** 269–274.
Broglie, R. M., and Niederman, R. A. (1979). *J. Bacteriol.* **138,** 788–798.
Bull, M. J., and Lascelles, J. (1963). *Biochem. J.* **87,** 15–28.
Burke, J. J., Dito, C. L., and Arntzen, C. J. (1978a). *Arch. Biochem. Biophys.* **187,** 252–263.
Burke, J. J., Steinback, K. E., Ohad, I., and Arntzen, J. C. (1978b). *In* "Chloroplast Development" (G. Akoyunoglou and J. H. Argyroudi-Akoyunoglou, eds.), pp. 413–418. Elsevier/North-Holland Biomedical Press, Amsterdam.
Burke, J. J., Steinback, K. E., and Arntzen, C. J. (1979). *Plant Physiol.* **63,** 237–243.
Butler, W. L. (1979). *Ciba Found. Symp.* [N.S.] **61,** 237–256.
Cahen, D., Malkin, S., Shochat, S., and Ohad, I. (1976). *Plant Physiol.* **58,** 257–267.
Cahen, D., Malkin, S., and Ohad, I. (1977). *Plant Physiol.* **60,** 845–849.
Cahen, D., Malkin, S., Gurevitz, M. and Ohad, I. (1978). *Plant Physiol.* **62,** 1–5.
Carr, N. G., and Whitton, B. A., eds. (1981). "The Biology of Cyanobacteria." Blackwell, Oxford.
Cellarius, R. A., and Peters, G. A. (1968). *Photochem. Photobiol.* **7,** 325–330.
Chua, N. H., and Bennoun, P. (1975). *Proc. Natl. Acad. Sci. U.S.A.* **72,** 2175–2179.
Chua, N. H., and Gillham, N. W. (1977). *J. Cell Biol.* **74,** 441–452.
Chua, N. H., and Schmidt, G. W. (1978). *Proc. Natl. Acad. Sci. U.S.A.* **75,** 6110–6114.
Chua, N. H., and Schmidt, G. W. (1979). *J. Cell Biol.* **81,** 461–483.
Chua, N. H., Matlin, K., and Bennoun, P. (1975). *J. Cell Biol.* **67,** 361–377.
Cogdell, R. J. (1978). *Philos. Trans. R. Soc. London, Ser. B* **284,** 569–579.
Cogdell, R. J., and Thornber, J. P. (1979). *Ciba Found. Symp.* [N.S.] **61,** 61–79.
Cohen-Bazire, G., and Kunisawa, R. (1960). *Proc. Natl. Acad. Sci. U.S.A.* **46,** 1543–1553.
Cohen-Bazire, G., Sistrom, W. R., and Stanier, R. Y. (1957). *J. Cell Comp. Physiol.* **49,** 25–35.
Conde, M. F., Boynton, J. E., and Gillham, N. W. (1975). *Mol. Genet.* **140,** 183–220.
Delepelaire, P., and Chua, N. H. (1979). *Proc. Natl. Acad. Sci. U.S.A.* **76,** 111–115.
Dierstein, R. (1978). *Hoppe-Seyler's Z. Physiol. Chem.* **359,** 1470.
Dierstein, R., and Drews, G. (1974). *Arch. Microbiol.* **99,** 117–128.
Dierstein, R., and Drews, G. (1975). *Arch. Microbiol.* **106,** 227–235.

Dierstein, R., Schumacher, A., and Drews, G. (1981). *Arch. Microbiol.* **128,** 376–383.
Drews, G. (1966). *Z. Naturforsch. B: Anorg. Chem., Org. Chem., Biochem., Biophys., Biol.* **21B,** 1224–1229.
Drews, G. (1974). *Arch. Microbiol.* **100,** 397–407.
Drews, G. (1978). *Curr. Top. Bioenerg.* **8,** 161–207.
Drews, G., and Giesbrecht, P. (1963). *Zentralbl. Bakteriol., Parasitenkd., Infektionskr. Hyg., Abt. I: Orig.* **190,** 508–536.
Drews, G., and Jäger, K. (1963). *Nature (London)* **199,** 1112–1113.
Drews, G., and Oelze, J. (1981). *Adv. Microb. Physiol.* **22,** 1–91.
Drews, G., Dierstein, R., and Schumacher, A. (1976). *FEBS Lett.* **68,** 132–136.
Dubertret, G. (1981a). *Plant Physiol.* **67,** 47–53.
Dubertret, G. (1981b). *Plant Physiol.* **67,** 54–58.
Dubertret, G., and Joliot, P. (1974). *Biochim. Biophys. Acta* **357,** 399–341.
Dujardin, E., De Kouchkovsky, Y., and Sironval, E. (1970). *Photosynthetica* **4,** 223–227.
Edelman, M., and Reisfeld, A. (1978). *In* "Chloroplast Development" (A. Akoyunoglou and J. H. Argyroudi-Akoyunoglou, eds.), pp. 641–652. Elsevier/North-Holland Biomedical Press, Amsterdam.
Ellis, J. R. (1975). *In* "Membrane Biogenesis" (A. Tzagaloff, ed.), pp. 247–276. Plenum, New York.
Ellis, R. J. (1977). *Biochim. Biophys. Acta* **463,** 185–215.
Eytan, G., and Ohad, I. (1972a). *J. Cell Biol.* **247,** 112–121.
Eytan, G., and Ohad, I. (1972b). *J. Biol. Chem.* **247,** 122–129.
Eytan, G., Jennings, R. C., Forti, G., and Ohad, I. (1974). *J. Biol. Chem.* **249,** 738–744.
Feierabend, J. (1977). *Planta* **135,** 83–88.
Feierabend, J. (1979). *Ber. Dsch. Bot. Ges.* **92,** 553–574.
Feierabend, J., and Wildner, G. (1978). *Arch. Biochem. Biophys.* **186,** 283–291.
Feierabend, J., Meschede, D., and Vogel, K. D. (1980). *Z. Pflanzenphysiol.* **98,** 61–78.
Firsow, N. N., and Drews, G. (1977). *Arch. Microbiol.* **115,** 299–306.
Forger, J. M. V., and Bogorad, L. (1973). *Plant Physiol.* **52,** 491–497.
Friedberg, I., Goldberg, I., and Ohad, I. (1971). *J. Cell Biol.* **50,** 268–275.
Garcia, A. G., and Drews, G. (1980). *Arch. Microbiol.* **127,** 157–161.
Garcia, A. G., Drews, G., and Reidl, H. H. (1981). *J. Bacteriol.* **145,** 1121–1128.
Garnier, G., Guyon, D., and Picoud, A. (1979). *Plant Cell Physiol.* **20,** 1013–1027.
Gerola, P. D., Jennings, R. C., Forti, G., and Garlaschi, F. M. (1979). *Plant Sci. Lett.* **16,** 249,254.
Gershoni, J. M., and Ohad, I. (1979). *In* "Membrane Bioenergetics" (C. P. Lee, G. Schatz, and L. Ernstner, eds.), pp. 165–176. Addison-Wesley, Reading, Massachusetts.
Gershoni, J. M., and Ohad, I. (1980). *Cell Biol.* **86,** 124–131.
Gershoni, J. M., Shochat, S., Ohad, I., and Malkin, S. (1982). *Plant Physiol.* (in press).
Göbel, F. (1978). *In* "The Photosynthetic Bacteria" (R. K. Clayton and W. R. Sistrom, eds.), pp. 907–925. Plenum, New York.
Goedheer, J. C., and Kleinenhammans, J. W. (1977). *Acta Bot. Neerl.* **26,** 273–284.
Goldberg, I., and Ohad, I. (1970a). *J. Cell Biol.* **44,** 563–571.
Goldberg, I., and Ohad, I. (1970b). *J. Cell Biol.* **44,** 572–591.
Golecki, J. R., and Oelze, J. (1975). *J. Gen. Microbiol.* **88,** 253–258.
Golecki, J. R., Drews, G., and Bühler, R. (1979). *Cytobiologie* **18,** 381–389.
Golecki, J. R., Schumacher, A., and Drews, G. (1980). *Eur. J. Cell Biol.* **23,** 149–156.
Gorchein, A. (1972). *Biochem. J.* **127,** 97–106.
Gorchein, A. (1973). *Biochem. J.* **134,** 833–845.

Govindjee, ed. (1982). "Photosynthesis: Energy Conversion by Plants and Bacteria, Vol. I." Academic Press, New York.
Grebanier, A. E., Steinback, K. E., and Bogorad, L. (1979). *Plant Physiol.* **63,** 436–439.
Griffiths, W. T., and Mapleston, R. E. (1978). *In* "Chloroplast Development" (G. Akoyunoglou and J. H. Argyroudi-Akoyunoglou, eds.), pp. 99–104. Elsevier/North-Holland Biomedical Press, Amsterdam.
Grossman, A., Bartlet, S., and Chua, N. H. (1980). *Nature (London)* **285,** 625–628.
Grossman, A., Bartlet, S., Weiss, M., and Chua, N. H. (1981). *Proc. Int. Cong. Photosynth. 5th, 1980* Abstracts, p. 233.
Gurevitz, M. (1980). Ph.D. Thesis, Hebrew University
Gurevitz, M., Kratz, H., and Ohad, I. (1977). *Biochim. Biophys. Acta* **461,** 475–488.
Hall, D. O., Coombs, J., and Goodwin, T. W., eds. (1977). "Photosynthesis," Symp. 8, pp. 497–549. Biochem. Soc., London.
Harel, E. (1978). *Prog. Phytochem.* **5,** 127–180.
Henningsen, K. W., and Boynton, J. E. (1969). *J. Cell Sci.* **5,** 757–793.
Henningsen, K. W., and Boynton, J. E. (1974). *J. Cell Sci.* **15,** 31–35.
Herrmann, F. H. (1978). *In* "Chloroplast Development" (G. Akoyunoglou and J. H. Argyroudi-Akoyunoglou, eds.), pp. 221–227. Elsevier/North-Holland Biomedical Press, Amsterdam.
Herrmann, F. H., Börner, T., and Hageman, R. (1980). *Results Probl. Cell Differ.* **10,** 147–177.
Highfield, P. E., and Ellis, R. J. (1978). *Nature (London)* **271,** 420–424.
Higuchi, M., Goto, K., Fujimoto, M., Nakimo, O., and Kikuchi, G. (1965). *Biochim. Biophys. Acta* **95,** 94–110.
Holowinsky, A. W., and Schiff, J. A. (1970). *Plant Physiol.* **45,** 339–347.
Holt, S. C., Conti, S. F., and Fuller, R. C. (1966). *J. Bacteriol.* **91,** 349–355.
Hoober, J. K., and Stegeman, W. J. (1973). *J. Cell Biol.* **56,** 1–12.
Hoober, J. K., and Stegeman, W. J. (1976). *J. Cell Biol.* **70,** 326–337.
Höyer-Hansen, G., and Simpson, D. J. (1977). *Carlsberg Res. Commun.* **42,** 379–389.
Hunter, C. N., van Grondelle, R., Holmes, N. G., Jones, O. T. G., and Niederman, R. A. (1979a). *Photochem. Photobiol.* **30,** 313–316.
Hunter, C. N., Holmes, N. G., Jones, O. T. G., and Niederman, R. A. (1979b). *Biochim. Biophys. Acta* **548,** 253–266.
Inoue, Y., Kobayashi, Y., Sakomoto, E., and Shibata, K. (1975). *Plant Cell Physiol.* **16,** 327–336.
Irschik, H., and Oelze, J. (1973). *Biochim. Biophys. Acta* **330,** 30–89.
Irschik, H., and Oelze, J. (1976). *Arch. Microbiol.* **109,** 307–313.
Jennings, R. C., and Ohad, I. (1972). *Arch. Biochem. Biophys.* **153,** 79–87.
Jones, O. T. G. (1978). *In* "The Photosynthetic Bacteria" (R. K. Clayton and W. R. Sistrom, eds.), pp. 751–778. Plenum, New York.
Kaiser, J., and Oelze, J. (1980). *Arch. Mikrobiol.* **126,** 187–194 and 195–200.
Kannangara, C. G., and Gough, S. P. (1978). *Carlsberg Res. Commun.* **43,** 185–194.
Kaplan, S. (1978). *In* "The Photosynthetic Bacteria" (R. K. Clayton and W. R. Sistrom, eds.), pp. 809–840. Plenum, New York.
Kaplan, S., and Arntzen, C. J. (1982). *In* "Photosynthesis: Energy Conversion by Plants and Bacteria," (Govindjee, ed.), Vol. I, pp. 65–151. Academic Press, New York.
Keister, D. L. (1978). *In* "The Photosynthetic Bacteria" (R. K. Clayton and W. R. Sistrom, eds.), pp. 849–856. Plenum, New York.
Keister, D. L., and Minton, N. J. (1969). *Prog. Photosynth. Res.* **3,** 1299–1305.

King, M.-T., and Drews, G. (1975). *Arch. Microbiol.* **102,** 219–231.
Kirk, J. T. O., and Tilney-Basset, R. A. E. (1978). "The Plastids," pp. 773–789. Elsevier/North-Holland Biomedical Press, Amsterdam.
Klein, S., and Schiff, J. A. (1972). *Plant Physiol.* **49,** 619–626.
Klein, S., Schiff, J. A., and Holowinsky, A. W. (1972). *Dev. Biol.* **28,** 253–273.
Klein, S., Katz, E., and Neeman, E. (1977). *Plant Physiol.* **60,** 335–338.
Konis, Y., Klein, S., and Ohad, I. (1978). *Photochem. Photobiol.* **27,** 177–182.
Kretzer, F., Ohad, I., and Bennoun, P. (1976). *In* "Genetics and Biogenesis of Chloroplast and Mitochondria" (T. Bücher, W. Neupert, W. Sebold, and S. Werner, eds.), pp. 25–32. Elsevier/North-Holland Biomedical Press, Amsterdam.
Lampe, H.-H., and Drews, G. (1972). *Arch. Mikrobiol.* **84,** 1–19.
Lampe, H. H., Oelze J., and Drews, G. (1972). *Arch. Mikrobiol.* **83,** 78–94.
Lascelles, J. (1959). *Biochem. J.* **72,** 508–518.
Lascelles, J. (1966). *Biochem. J.* **100,** 175–183.
Lascelles, J. (1978). *In* "The Photosynthetic Bacteria" (R. K. Clayton and W. R. Sistrom, eds.), pp. 795–808. Plenum, New York.
Lavintman, M., Galling, G., and Ohad, I. (1978). *In* "Chloroplast Development" (G. Akoyunoglou and J. H. Argyroudi-Akoyunoglou, eds.), pp. 875–884. Elsevier/North-Holland Biomedical Press, Amsterdam.
Levine, R. P. (1969). *Annu. Rev. Plant Physiol.* **20,** 523–540.
Lewin, R. A. (1977). *Phycologia* **16,** 217.
Lieberman, J. R., Bose, S., and Arntzen, J. C. (1978). *Biochim. Biophys. Acta* **502,** 417–423.
Lien, S., Gest, H., and San Pietro, A. (1973). *Bioenergetica* **4,** 423–434.
Lütz, C. (1978). *In* "Chloroplast Development" (G. Akoyunoglou and J. H. Argyroudi-Akoyunoglou, eds.), pp. 481–488. Elsevier/North-Holland Biomedical Press, Amsterdam.
Lütz, C. (1980). *Z. Naturforsching,* **35,** 519–521.
Lütz, C., and Klein, S. (1979). *Z. Pflanzenphysiol.* **95,** 227–237.
Machold, O., and Höyer-Hansen, G. (1976). *Carlsberg Res. Commun.* **41,** 359–366.
Machold, O., Simpson, D. J., and Möler, B. L. (1979). *Carlsberg Res. Commun.* **44,** 235–254.
Maroc, J., and Garnier, J. (1979). *Plant Cell Physiol.* **20,** 1029–1040.
Marrs, B. L. (1978). *In* "The Photosynthetic Bacteria" (R. K. Clayton and W. R. Sistrom, eds.), pp. 873–883. Plenum, New York.
Marrs, B. L. (1981). *J. Bacteriol.* **146,** 1003–1012.
Marver, H. S., Collins, A., Tschudy, D. P., and Rechcigl M., Jr. (1966). *J. Biol. Chem.* **241,** 4323–4329.
Masoner, M., and Kasemir, H. (1975). *Planta* **126,** 111–117.
Maudinas, B., Oelze, J., Villcutreix, J., and Reisinger, O. (1973). *Arch. Mikrobiol.* **93,** 219–228.
Mechler, B., and Oelze, J. (1978). *Arch. Microbiol.* **118,** 91–114.
Miller, K., and Ohad, I. (1978). *Cell Biol. Int. Rep.* **2,** 534–550.
Miller, L. S., and Holt, S. C. (1977). *Arch. Microbiol.* **115,** 185–198.
Mohr, H. (1972). "Lectures on Photomorphogenesis." Springer-Verlag, Berlin and New York.
Mohr, H., and Oelze-Karow, H. (1978). *In* "Chloroplast Development" (G. Akoyunoglou and J. H. Argyroudi-Akoyunoglou, eds.), pp. 769–779. Elsevier/North-Holland Biomedical Press, Amsterdam.
Müller, M., Viro, M., Balke, C., and Kloppstech, K. (1980a). *Planta* **148,** 444–447.
Müller, M., Viro, M., Balke, C., and Kloppstech, K. (1980b). *Planta* **148,** 448–452.

Mullet, J., and Arntzen, C. J. (1980). *Biochim. Biophys. Acta* **589,** 100–117.
Mullet, J., Burke, J. J., and Arntzen, C. J. (1980). *Plant Physiol.* **65,** 814–822.
Murata, N., Troughton, J. H., and Forch, D. C. (1975). *Plant Physiol.* **56,** 508–517.
Myers, J., Graham, J.-R., and Wang, R. T. (1978). *J. Phycol.* **14,** 513–518.
Nadler, K., and Granick, S. (1970). *Plant Physiol.* **46,** 240–246.
Niederman, R. A., Mallon, D. E., and Langan, J. J. (1976). *Biochim. Biophys. Acta* **440,** 429–447.
Niederman, R. A., Mallon, D. E., and Parks, L. C. (1979). *Biochim. Biophys. Acta* **555,** 210–220.
Nieth, K.-F., and Drews, G. (1975). *Arch. Microbiol.* **104,** 77–82.
Oelze, J. and Drews, G. (1969). *Biochim. Biophys. Acta* **173,** 448–455.
Oelze, J., and Drews, G. (1972). *Biochim. Biophys. Acta* **265,** 209–239.
Oelze, J., and Drews, G. (1981). *In* "Membranous Structures in Bacterial Cells" (B. K. Ghosh, ed.), pp. 131–195. CRC Press, Inc., Palm Beach, Florida.
Oelze, J., and Pahlke, W. (1976). *Arch. Microbiol.* **108,** 218–285.
Oelze, J., Biedermann, M., Freund-Mölbert, E., and Drews, G. (1969). *Arch. Mikrobiol.* **66,** 154–165.
Oelze, J., Schröder, J., and Drews, G. (1970). *J. Bacteriol.* **101,** 669–674.
Oettmeier, W., Masson, K., and Johanningmeier, U. (1980), *FEBS Lett.* **118,** 267–270.
Ohad, I., (1975). *In* "Membrane Biogenesis" (A. Tzagaloff, ed.), pp. 279–350. Plenum, New York.
Ohad, I., and Drews, G. (1975). *Proc. Int. Cong. Photosynth., 3rd, 1974,* pp. 1907–1912.
Ohad, I., Siekevitz, P., and Palade, G. E. (1967a). *J. Cell Biol.* **35,** 521–552.
Ohad, I., Siekevitz, P., and Palade, G. E. (1967b). *J. Cell Biol.* **35,** 553–584.
Olive, J., Wolman, F. A., Bennoun, P., and Recouvreur, M. (1979). *Mol. Biol. Rep.* **5,** 139–143.
Ono, T.-A., and Murata, N. (1979). *Biochim. Biophys. Acta* **545,** 69–76.
Osafune, T., Klein, S., and Schiff, J. A. (1980). *J. Ultrastructure Res.* **73,** 77–90.
Owens, G. C., and Ohad, I. (1981). *Proc. Int. Symp. Photosynth. 5th, 1980,* **3,** 623–630.
Owens, G. C., and Ohad, I. (1982). *J. Cell Biol.* **93,** 712–718.
Parks, L. C., and Niederman, R. A. (1978). *Biochim. Biophys. Acta* **511,** 70–82.
Pfennig, N. (1978). *In* "The Photosynthetic Bacteria" (R. K. Clayton and W. R. Sistrom, eds.), pp. 3–18. Plenum, New York.
Pfennig, N., and Trüper, H. G. (1974). *In* "Bergey's Manual of Determinative Bacteriology" (R. E. Buchanan and N. E. Gibbons, eds.), pp. 24–64. Williams & Wilkins, Baltimore, Maryland.
Pfister, K., Steinback, K. E., Gardner, G., and Arntzen, C. J. (1981). *Proc. Natl. Acad. Sci. U.S.A.* **78,** 981–985.
Picoud, A., Dubertret, G., Guyon, D., and Hervo, G. (1981). *Proc. Int. Congr. Photosynth., 5th, 1981,* pp. 405–415.
Pierson, B. K., and Castenholz, R. W. (1974). *Arch. Microbiol.* **100,** 283–305.
Pierson, B. K., and Castenholz, R. W. (1978). *In* "The Photosynthetic Bacteria" (R. K. Clayton and W. R. Sistrom, eds.), pp. 179–198. Plenum, New York.
Pradel, J., Lavergne, J., and Moya, I. (1978). *Biochim. Biophys. Acta* **502,** 169–182.
Rebeiz, C., and Lascelles, J. (1982). *In* "Photosynthesis: Energy Conversion by Plants and Bacteria" (Govindjee, ed.), Vol. I, pp. 699–780. Academic Press, New York.
Regitz, G., and Ohad, I. (1976). *J. Biol. Chem.* **251,** 247–252.
Remsen, C. C. (1978). *In* "The Photosynthetic Bacteria" (R. K. Clayton and W. R. Sistrom, eds.), pp. 31–60. Plenum, New York.

Remy, R. (1973). *Photochem. Photobiol.* **18,** 409–416.
Saunders, V. A. (1978). *Microbiol. Rev.* **42,** 357–384.
Schiff, J. A. (1970). *Symp. Soc. Exp. Biol.* **24,** 277–301.
Schiff, J. A., Zeldin, M. H., and Rubman, J. (1967). *Plant Physiol.* **42,** 1716–1725.
Schmidt, G. W., Devilleu-Thiery, A., Dessruiseaux, H., Blobel, G., and Chua, N. H. (1979). *J. Cell Biol.* **83,** 615–622.
Schmidt, G. W., Bartlet, S., Grossman, A. R., Cashmore, A. R., and Chua, N. H. (1980). *In* "Genome Organization and Expression in Plants" (C. Leaver, ed.), pp. 337–351. Plenum, New York.
Schumacher, A., and Drews, G. (1978). *Biochim. Biophys. Acta* **501,** 183–194.
Schumacher, A., and Drews, G. (1979). *Biochim. Biophys. Acta* **547,** 417–428.
Schwartzbach, S. D., Schiff, J. A., and Klein, S. (1976). *Planta* **131,** 1–9.
Sculley, M. J., Duniec, J. T., Thorne, S. W., Chow, W. S., and Boardman, W. K. (1980). *Arch. Biochem. Biophys.* **201,** 339–346.
Senger, H., ed. (1979). "The Blue Light Effect." Springer-Verlag, Berlin and New York.
Senger, H., and Bishop, N. I. (1972). *Plant Cell Physiol.* **13,** 633–649.
Senger, H., and Mell, V. (1977). *Methods Cell Physiol.* **5,** 201–208.
Senger, H., and Strasberger, G. (1978). *In* "Chloroplast Development" (G. Akoyunoglou and J. H. Argyroudi-Akoyunoglou, eds.), pp. 367–378. Elsevier/North-Holland Biomedical Press, Amsterdam.
Shepherd, H. S., Boynton, J., and Gillham, N. W. (1979). *Proc Natl. Acad. Sci. U.S.A.* **76,** 1353–1357.
Sherman, L. A. (1978). *J. Phycol.* **14,** 427–433.
Shiozawa, J. A., Cuendet, P. A., Drews, G., and Zuber, H. (1980). *Eur. J. Biochem.* **111,** 455–460.
Simpson, D. J. (1979). *Carlsberg Res. Commun.* **44,** 305–336.
Sistrom, W. R. (1978). *In* "The Photosynthetic Bacteria" (R. K. Clayton and W. R. Sistrom, eds.), pp. 899–906. Plenum, New York.
Smith, H. (1975). "Phytochrome and Photomorphogenesis." McGraw-Hill, New York.
Spector, M., and Winget, D. G. (1980). *Proc. Natl. Acad. Sci. U.S.A.* **77,** 957–959.
Staehelin, L. A. (1976). *J. Cell Biol.* **71,** 136–158.
Staehelin, L. A., Armond, P. A., and Miller, K. (1976). *Brookhaven Symp. Biol.* **28,** 278–315.
Staehelin, L. A., Golecki, J. R., Fuller, R. C., and Drews, G. (1978). *Arch. Microbiol.* **119,** 269–277.
Staehelin, L. A., Golecki, J. R., and Drews, G. (1980). *Biochim. Biophys. Acta* **589,** 30–45.
Stanier, R. Y., and Cohen-Bazire, G. (1977). *Annu. Rev. Microbiol.* **31,** 225–274.
Steinback, K. E., Burke, J. J., and Arntzen, C. J. (1979). *Arch. Biochem. Biophys.* **195,** 546–557.
Steiner, L. A., Okamura, M. Y., Lopes, A. D., Moskowitz, E., and Feher, G. (1974). *Biochemistry* **13,** 1403–1410.
Takemoto, J. (1974). *Arch. Biochem. Biophys.* **163,** 515–520
Takemoto, J., and Huang Kao, M. Y. (1977). *J. Bacteriol.* **129,** 1102–1109.
Thornber, J. P., and Alberte, R. S. (1978). *In* "Photosynthesis" (A. Trebst and M. Avron, eds.), pp. 574–582. Springer-Verlag, Berlin and New York.
Thornber, J. P., Alberte, R. S., Hunter, F. A., Shiozawa, J. A., and Khan, K. S., (1976). *Brookhaven Symp. Biol.* **28,** 132–148.
Thornber, J. P., Markwell, J. P., and Reinmore, S. (1979). *Photochem. Photobiol.* **29,** 1205–1216.
Throm, E., Oelze, J., and Drews, G. (1970). *Arch. Mikrobiol.* **72,** 361–370.

Trüper, H. G., and Pfennig, N. (1978). In "The Photosynthetic Bacteria" (R. K. Clayton and W. R. Sistrom, eds.), pp. 19–27. Plenum, New York.
Valane, N. (1978). *In* "Chloroplast Development" (G. Akoyunoglou and J. H. Argyroudi-Akoyunoglou, eds.), pp. 241–244. Elsevier/North-Holland Biomedical Press, Amsterdam.
Vermaas, W. F. J., and Govindjee (1981). *Photochem. Photobiol.* **34,** 775–793.
Wakim, B., and Oelze, J., (1980). *FEMS Microbiol. Lett.* **7,** 221–223.
Wakim, B., Golecki, J. R., and Oelze, J. (1978). *FEMS Microbiol. Lett.* **4,** 199–201.
Wellburn, A. R., Quaily, P. H., and Gunning, B. E. S. (1977). *Planta* **134,** 45–52.
Wessels, J. S. C. (1977). *In* "Photosynthesis I" (A. Trebst and M. Avron, eds.), pp. 563–572. Springer-Verlag, Berlin and New York.
Wickner, W. (1979). *Annu. Rev. Biochem.* **48,** 23–45.
Wolman, F. A., and Diner, B. A. (1980). *Arch. Biochem. Biophys.* **201,** 646–659.
Wong, D., and Govindjee (1981). *Photochem. Photobiol.* **33,** 103–108.
Wong, D., Govindjee, and Merkelo, H. (1980). *Biochim. Biophys. Acta* **592,** 546–558.
Wong, D., Merkelo, H., and Govindjee (1981). *Photochem. Photobiol.* **33,** 97–101.
Yen, H. C., and Marrs, B. (1976). *J. Bacteriol.* **126,** 619–629.
Zielinski, R. E., and Price, C. A. (1980). *J. Cell Biol.* **85,** 435–445.

6

Carbon Dioxide Fixation Pathways in Plants and Bacteria

JAMES A. BASSHAM
BOB B. BUCHANAN*

*J. A. Bassham is responsible for aspects of CO_2 fixation pertaining to plant photosynthesis; B. B. Buchanan is responsible for the section on CO_2 fixation in photosynthetic bacteria.

Photosynthesis: Development, Carbon Metabolism, and Plant Productivity, Vol. II

ISBN: 0-12-294302-3

ABBREVIATIONS

AMP	Adenosine monophosphate
ADP	Adenosine diphosphate
ATP	Adenosine triphosphate
C_3	Refers to those (plants) having three-carbon compounds (such as phosphoglyceric acid) as intermediates
C_4	Refers to those (plants) having four-carbon compounds (such as oxaloacetic acid) as intermediates
CAM	Crassulacean acid metabolism
CoA	Coenzyme A
DHAP	Dihydroxyacetone phosphate
DPGA	Diphosphoglyceric acid
E4P	Erythrose 4-phosphate
FBP	Fructose bisphosphate
GAl3P	Glyceraldehyde 3-phosphate
G6P	Glucose 6-phosphate
$NADP^+$	Nicotinamide adenine dinucleotide phosphate
NADPH	Nicotinamide adenine dinucleotide phosphate (reduced)
OPP	Oxidative pentose phosphate (cycle)
P_i	Inorganic phosphate
PP_i	Inorganic pyrophosphate
PAR	Photosynthetically active radiation
PEP	Phosphoenolpyruvic acid
PGA	Phosphoglyceric acid
RPP	Reductive pentose phosphate (cycle)
RuBP	Ribulose bisphosphate
SBP	Sedoheptulose bisphosphate
TCA	Tricarboxylic acid (cycle)
Xu5P	Xylulose 5-phosphate

ABSTRACT

All oxygenic photosynthetic organisms from cyanobacteria to higher plants employ the reductive pentose phosphate cycle (C_3 cycle, Calvin cycle) for the reduction of carbon dioxide to sugar phosphates. In addition, some higher plants carry out preliminary CO_2 fixing cycles in which CO_2 is first fixed into four-carbon dicarboxylic acids, which are later decarboxylated at the site of the reductive pentose phosphate cycle. One such cycle is the C_4 cycle, which occurs in several forms and is found in C_4 plants such as maize and sugar cane, as well as in some dicotyledons. Another C_4 cycle occurs in Crassulaceae (CAM), in which CO_2 is incorporated into C_4 acids at night and is released and refixed by the reductive pentose by day.

The reductive pentose phosphate cycle and the C_4 cycle are described and discussed in terms of reactions, discovery, energetics, regulation, and function. The limitation on photosynthetic rate in C_3 plants due to glycolate formation and photorespiration is considered, and the role of C_4 metabolism in minimizing the detrimental effects of photorespiration is described. Some possible ways of increasing plant productivity through increased rates of photosynthesis are discussed.

Photosynthetic bacteria (other than the cyanobacteria, which are sometimes called blue–green algae) are nonoxygenic but in some cases carry out fixation and reduction of carbon dioxide. Assimilation of carbon dioxide often occurs via the reductive pentose phosphate cycle, but additional reductive carboxylation reactions requiring reduced ferredoxin have been found in a number of organisms. Evidence is reviewed that some photosynthetic bacteria such as the green sulfur bacteria lack key enzymes of the reductive pentose phosphate cycle but instead carry out a reductive carboxylic acid cycle.

I. Introduction

It is widely assumed that life on earth originated in an environment rich in organic compounds formed in the presence of a reducing atmosphere. Various kinds of energy such as ultraviolet and visible radiation, electrical discharges in the atmosphere, vulcanic heat, and energy released by inorganic chemical reactions may have provided the activation for the formation of various organic chemical compounds from the primitive atmosphere. Naturally occurring catalysts, such as certain types of clays and other minerals, may have played roles in the formation of more complex organic and biochemical molecules. Once primitive life arose, an abundance of reduced organic molecules would have been available for supplying the carbon skeletons of molecules essential to life.

It is reasonable to suppose that photosynthetic organisms were among the early life forms. Light energy, absorption, and conversion supply energy for biological conversion of endogenous organic material to biochemicals necessary for life. Other primitive organisms no doubt utilized sources of natural chemical energy occurring in the environment much as certain microorganisms do to this day.

Over 3½ billion years ago, photosynthetic, oxygenic organisms developed. These organisms, probably similar to single cell photosynthetic cyanobacteria that we know today, were able to carry out the complete photosynthetic conversion of water and CO_2 to oxygen and reduced organic compounds. The appearance of large amounts of molecular oxygen in the biosphere must have led rapidly to the development of aerobic organisms and the virtual depletion of the stores of primordial reduced carbon. From that time in the distant past to the present day, life on earth has been dependent to a large extent on the kind of oxygenic, CO_2-fixing photosynthesis exhibited by cyanobacteria, green algae, and all higher plants.

All oxygenic organisms from the simplest prokaryotic cyanobacteria or prochloron species to the most complicated land plant have a common pathway for the reduction of CO_2 to sugar phosphates. This pathway is the reductive pentose phosphate (RPP) cycle, also called the Calvin* cycle or the C_3 pathway of photosynthetic carbon fixation.

There are other supplementary pathways for first fixing CO_2 and then releasing it in the cells where the RPP cycle occurs. One of these pathways, the C_4 path, has evolved in plants suited to environments with high light intensity and warm climates. The occurrence of this pathway, which involves special leaf anatomy and a division of biochemical labor between two kinds of cells, enables the plants endowed with it to avoid wasteful process of photorespiration and consequently to maintain higher rates of net photosynthesis in air under conditions of high light intensity and elevated growing temperatures.

A second supplementary pathway exists in species of Crassulaceae and is called the crassulacean acid metabolism (CAM) (see Chapter 8, this volume). Plants with CAM are often but not exclusively found in arid and semiarid regions. These plants are able to accumulate CO_2 taken in through their open stomata at night by forming C_4 organic acids. The leaves can then close their stomata during the day to conserve water while using absorbed light energy to convert CO_2 released from the carboxyl carbon of the C_4 acids to sugar phosphates via the RPP cycle.

CO_2 fixation is by no means limited to the oxygenic photosynthetic organisms, but is found in many bacteria, both photosynthetic and nonphotosynthetic. Some such microorganisms employ the RPP cycle in a form exactly the same as that of plants, whereas others utilize distinctively different pathways. An interesting different pathway is the reductive tricarboxylic acid (TCA) cycle pathway. In that pathway, reduced ferredoxin, the primary soluble electron acceptor of the light reactions, is used to bring about the reductive carboxylation of succinyl CoA to make α-ketoglutarate in the reversed TCA. Acetyl-CoA produced as a product of the reversed TCA cycle is reductively carboxylated in the presence of another molecule of reduced ferredoxin to give pyruvate, which is eventually converted to carbohydrate via gluconeogenesis.

It seems likely that both the RPP cycle and the reverse TCA cycle may have evolved long before the advent of oxygenic photosynthesis. These cycles would have provided routes whereby primitive microorganisms with adequate supplies of energy but inadequate supplies of reduced carbon skeletons could take up CO_2 from their environment and convert them to useful biochemical products.

In the following sections, we will consider the details of the RPP cycle and the C_4 pathway as well as the carbon-fixing pathways of bacteria.

*Calvin–Benson–Bassham cycle.

Crassulacean acid metabolism will be mentioned only briefly since it is covered in Chapter 8 of this volume. Alternative pathways of CO_2 fixation in bacteria will be described.

II. The Reductive Pentose Phosphate Cycle

A. Occurrence and Location

The reductive pentose phosphate cycle (RPP cycle) (Bassham *et al.*, 1954; Bassham and Calvin, 1957) is apparently ubiquitous in all photoautotrophic green plants (Norris *et al.*, 1955) and also occurs in a variety of photosynthetic bacteria (Stoppani *et al.*, 1955). The RPP cycle is the only pathway in green plants whereby CO_2 can be converted to sugar phosphates. Even in unicellular algae, where rapid growth and division requires substantial intracellular amino acid synthesis (and therefore C_4 acid formation via carboxylation of phosphoenolpyruvate), 70–85% of the newly incorporated CO_2 can be accounted for in intermediates and products of the RPP cycle (Bassham and Kirk, 1960).

In C_3 plants, the entire process of photosynthesis is complete within the chloroplasts (Arnon *et al.*, 1954). Photosynthesis includes capture and conversion of light energy, the oxidation of water to molecular oxygen, and the uptake and reduction of carbon dioxide to starch and to triose phosphates.

The primary reactions of photosynthesis occur in the lamellae, or thylakoid membranes, which contain the pigments, electron carriers, and other constituents involved in light absorption and conversion to chemical energy (see Govindjee, Vol. I, 1982). The chemical energy derived from the primary reactions is used to drive electron transport and photophosphorylation (Arnon, 1958). The overall result of electron transport in membranes is the oxidation of water, giving O_2, and the reduction of the non-heme iron protein, ferredoxin, to its reduced form (Whatley *et al.*, 1963). Photophosphorylation converts ADP and P_i to ATP (Arnon *et al.*, 1954). The soluble co-factors, reduced ferredoxin and ATP, are the source of reducing equivalents and energy for the conversion of carbon dioxide to sugar phosphates in the chloroplasts. Reduced ferredoxin also is required for nitrite and sulfate reduction.

The enzymes catalyzing steps of the RPP cycle are water soluble and are located in the stroma region of the chloroplasts (Allen *et al.*, 1957). Only 3 steps in the cycle out of a total of 13 require co-factors (Bassham *et al.*, 1954; Bassham and Calvin, 1957) (ATP and NADPH), which must be regenerated by the light reactions in the thylakoids. NADPH is formed by reduction of $NADP^+$ coupled to the oxidation of two equivalents of ferredoxin (Fd):

$$2\ Fd^{3+} + NADP^+ + H^+ \rightarrow 2\ Fd^{2+} + NADPH$$

There are no photochemical steps in the RPP cycle. There are, however, indirect effects of the light reactions on steps in the RPP cycle other than those requiring ATP or NADPH. These indirect effects on the catalytic activities of enzymes of the cycle are apparently mediated via changes in concentrations of certain ions, such as Mg^{2+} and H^+, and levels of reduced co-factors, such as Fd and NADPH. They are among the regulatory mechanisms discussed later (Section II, F).

B. The Cyclic Pathway

In the initial step of the cycle, RuBP is carboxylated and hydrolytically split to give two molecules of 3-phosphoglycerate (PGA) (Fig. 1). This C_3 acid then is phosphorylated with ATP to give diphosphoglyceric acid (DPGA) and then reduced to glyceraldehyde-3 phosphate (GA13P) in a reaction using NADPH. This triose phosphate then undergoes a series of isomerizations, condensations, and rearrangements resulting in the conversion of five molecules of triose phosphate to three molecules of pentose phosphate, eventually ribulose 5-phosphate (Ru5P). Phosphorylation with ATP of Ru5P regenerates the carboxylation substrate RuBP, thus completing the cycle.

In a complete cycle, each reaction occurs at least once. Three molecules of RuBP are carboxylated to give six molecules of PGA. Phosphorylation and reduction produces six molecules of GA13P. Only five of these GA13P molecules (15 carbon atoms) are required to regenerate the three RuBP molecules. The sixth GA13P molecule, equivalent in carbon to the three CO_2 molecules fixed, can either be converted to glucose phosphate for starch synthesis, or they can be exported from the chloroplast to the cytoplasm (Heber *et al.*, 1967a; Bassham *et al.*, 1968a; Stocking and Larson, 1969; Werden and Heldt, 1972; Bamberger *et al.*, 1975). Other biosynthetic uses in the chloroplasts are also possible, for example, conversion of GA13P to glycerol phosphate and eventually fats.

Triose phosphates exported from the chloroplasts are mostly converted to sucrose, which is translocated to other parts of the plant (Geiger and Giaquinta, Chapter 8 in this volume). If carbon skeletons are needed for synthesis of cellular components or for energy metabolism, triose phosphate is oxidized to 3-phosphoglycerate in the cytoplasm, yielding ATP and NADH.

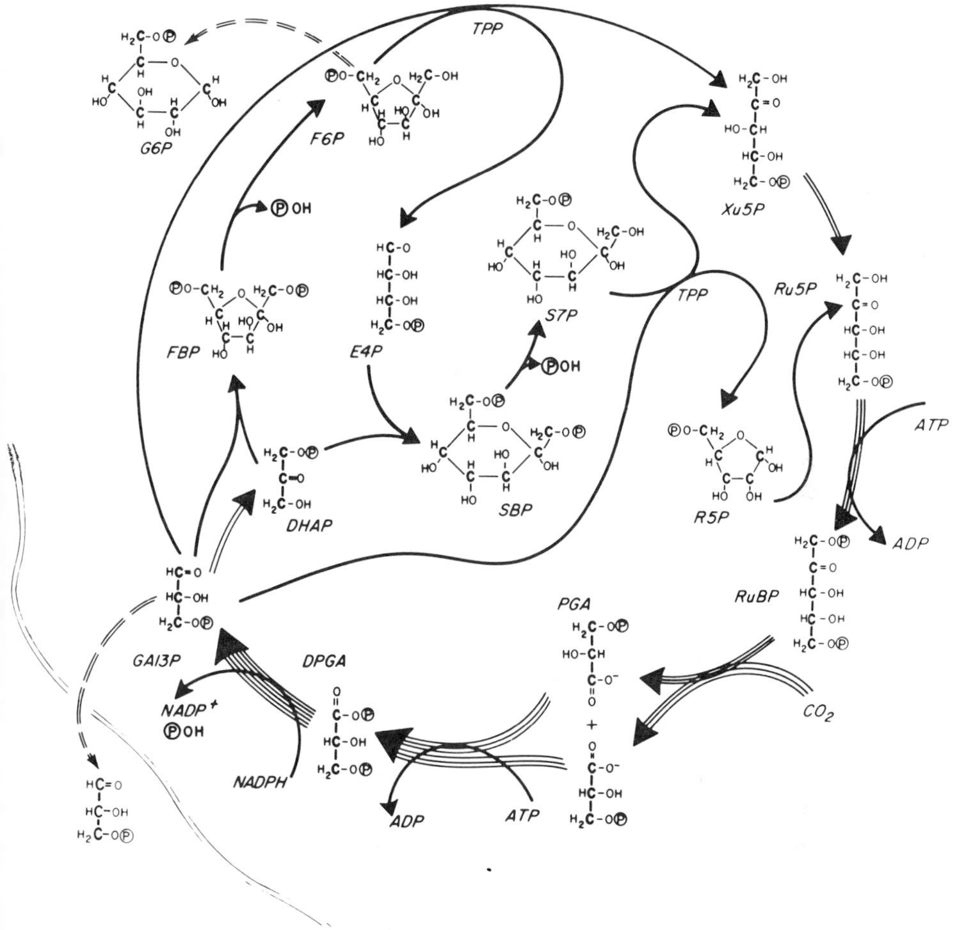

FIG. 1. The reductive pentose phosphate cycle (RPP). The solid lines indicate reactions of the RPP cycle. The number of lines per arrow indicate the number of times each reaction occurs for one complete turn of the cycle in which three molecules of CO_2 are converted to one molecule of GA13P. Each reaction of the cycle occurs at least once. The double dashed lines indicate the principal reactions removing intermediate compounds of the cycle for biosynthesis. Abbreviations: RuBP, ribulose 1,5-bisphosphate; PGA, 3-phosphoglycerate; DPGA, 1,3-diphosphoglycerate; FBP, fructose 1,6-bisphosphate; F6P, fructose 6-phosphate; SBP, sedoheptulose 1,7-bisphosphate; S7P, sedoheptulose 7-phosphate; Xu5P, xylulose 5-phosphate; R5P, ribose 5-phosphate; Ru5P, ribulose 5-phosphate; TPP, thamine pyrophosphate.

C. Stoichiometry and Energetics

For each mole of CO_2 fixed by the cycle, 1 mole of ATP is required for the conversion of Ru5P to RuBP, the substrate for carboxylation. Two more moles of ATP are required in the subsequent activation of the 2 moles of PGA formed, giving DPGA. The reduction of 2 moles of DPGA requires 2 moles of NADPH. For a complete turn of the RPP cycle, 3 moles of CO_2 are taken up, requiring the use of 9 moles of ATP and 6 moles of NADPH to make 1 mole of GA13P. The cycle stoichiometry and energetics can be expressed as the sum of two equations representing the utilization of the co-factors from the light and another equation for the conversion of CO_2, water, and P_i to GA13P.

$$6(\text{NADPH} + 1/2\ O_2 + H^+ \rightarrow \text{NADP}^+ + H_2O) \qquad \Delta G' = -325.5 \text{ kcal}$$
$$9(\text{ATP}^{4-} + H_2O \rightarrow \text{ADP}^{3-} + P_i^{2-} + H^+) \qquad \Delta G' = -\ 68.8 \text{ kcal}$$
$$\text{Total} \qquad \Delta G' = -384.3 \text{ kcal}$$
$$3CO_2 + 3\ H_2O + P_i^{2-} \rightarrow \text{GA13P}^{2-} + 3O_2 \qquad \Delta G' = +350.4 \text{ kcal}$$

The difference between the energy stored in reducing CO_2 to triose phosphate and the energy expended in converting the co-factors in $\Delta G' = -\ 33.9$ kcal. This is the driving energy for one turn of the cycle. The energy efficiency is 350/384 = 91%. This calculation depends on the usual use of physiological free energy changes employed in biochemical energetics. It is somewhat misleading because actual physiological concentrations are always much smaller than required to give activities of 1.0; in fact, they are generally in the range of 10^{-5}–10^{-2} *M*.

When metabolite concentrations, measured and estimated in photosynthesizing *Chlorella pyrenoidosa,* were used to estimate the physiological free energy change (ΔG^s) (Bassham and Krause, 1969), the energy input for 6 moles of NADPH and 9 moles of ATP was calculated as $^-427.0$ kcal, and the energy stored in making GA13P from CO_2, water, and P_i, with O_2 evolution was found to be +353.6 kcal. The chemical free energy converted to heat to drive the cycle was thus 73.4 kcal, and the efficiency of the cycle was 83%.

The high efficiency of the basic reaction of the RPP cycle contrasts with the relatively low overall efficiency of photosynthesis, and even more with the efficiency of plant growth. The production of the 9 ATP and 6 NADPH molecules, along with the oxidation of water to O_2 stores 384.3 kcal, but requires a minimum of 24 einsteins (moles of photons) of photosynthetically active radiation (PAR), with an energy of 1200 kcal. Thus, the energy efficiency of these light reactions is about 32% for PAR. Since PAR is only about 43% of the solar spectrum, the photosynthetic solar energy efficiency of the light reactions is 13.8%. When this is multiplied by the carbon cycle efficiency of 83%, the overall photosynthetic efficiency becomes about 11.4% even before any photorespiratory loss.

Even this efficiency assumes complete absorption of PAR, but actual absorption is seldom more than 80%, further reducing the maximum photosynthetic efficiency to 9%. For plant growth, a further loss of about one-third for respiration and biosynthetic reactions is required, giving an upper limit of about 6% for plant growth energy efficiency in sunlight in the absence of photorespiration loss.

Actual efficiencies for plant growth are much lower yet. The highest efficiency for solar energy conversion to total biomass during the periods of most rapid growth are in the range of 2–3% for C_4 plants and 1–2% for C_3 plants.

D. Discovery of the Cycle

1. IDENTIFICATION OF INTERMEDIATE COMPOUNDS

Little was learned about the biochemistry of photosynthetic CO_2 fixation until the advent of radioisotopic carbon, especially long-lived ^{14}C, discovered in 1940 (see Ruben and Kamen, 1940a,b). From 1946 to 1953 Calvin and co-workers made use of $^{14}CO_2$ to label photosynthetic metabolites, which they then separated by paper chromatography (Benson *et al.*, 1950), a method developed earlier for analysis of amino acids (Martin and Singe, 1941). Radioactively labeled products of photosynthesis on the paper chromatograms were located by exposure of medical X-ray film.

The first compound heavily labeled during photosynthesis with $^{14}CO_2$ proved to be PGA (Benson and Calvin, 1948). At the shortest times, the ^{14}C was mostly in the carboxyl group, indicating formation by carboxylation of an unknown C_2 moiety. Within less than 1 min, the two additional carbon atoms became labeled at an equal rate, suggesting regeneration of the C_2 moiety by a fast running cyclic process, not involving storage products (Schou *et al.*, 1950).

We now know that the plants used in those studies, unicellular algae such as *Scenedesmus obliquus* and higher plants such as barley or soybean, are C_3 plants. Nevertheless, some C_4 acids such as malate were among the labeled early products of $^{14}CO_2$ photosynthesis. Labeling of these compounds could be suppressed with inhibitors without inhibiting CO_2 uptake or labeling of PGA and a number of sugar phosphates, which were prominent among the early products (Bassham *et al.*, 1950). The first of these to be identified were DHAP, FBP, and G6P (Benson *et al.*, 1950).

2. DEGRADATION OF LABELED METABOLITES

Chemical degradation of the hexose phosphates showed the ^{14}C first appearing in the C_3 and C_4 positions, after which the label appeared equally in C_1, C_2, and in smaller amounts in C_5 and C_6 (Bassham *et al.*, 1954). These findings strongly suggest that the PGA, once formed, was converted to hexose monophosphates by the portion of the gluconeogenesis pathway from PGA to glucose. This also meant that the co-factors required from the light reactions should be reduced pyridine nucleotides and ATP.

Other labeled sugar phosphates were identified as SBP, S7P, and RuBP (Benson, 1951; Benson *et al.*, 1951; Calvin *et al.*, 1951). Smaller amounts of Xu5P, Ru5P, and R5P were also found. Degradation of these compounds after short periods of photosynthesis with $^{14}CO_2$ revealed a pattern of labeling requiring the conversion of triose and hexose phosphates to pentose phosphates, via sedoheptulose phosphates by means of a rearrangement of carbon-chain length, as formulated in the final version of the cycle (Bassham *et al.*, 1954; Bassham and Calvin, 1957).

3. STEADY STATE AND TRANSIENT STUDIES

After about 5 min of photosynthesis of *Scenedesmus* with $^{14}CO_2$, the ^{14}C content of intermediate compounds of the cycle no longer increases, indicating that the compounds are fully labeled ("saturated"). After this time, the ^{14}C content may be taken as a measure of the actual concentrations of the compounds in the actively turning over pools in the cells. In one set of experiments, after saturation with ^{14}C, the light was turned off, and more samples were taken in quick succession. The concentration of PGA rose rapidly, indicating that co-factors generated by the light reaction are required for the subsequent conversion of PGA to sugar phosphates as expected (Calvin and Massini, 1952). The rise in level of PGA for 20 sec or longer also indicates that the carboxylation reaction itself proceeds in the dark for a short time, suggesting that co-factors from the light are not directly required for the carboxylation reaction.

When the light was turned off, the most significant decrease was in the level of ^{14}C-labeled RuBP. It was concluded that the step involved in the regeneration of RuBP requires a light-produced co-factor, namely ATP, which is required for the conversion of Ru5P to RuBP with phosphoribulokinse.

Wilson and Calvin (1955) first established steady state photosynthesis with *Scenedesmus* photosynthesizing with $^{14}CO_2$ and then lowered the CO_2 level to nearly zero. In this case, the carboxylation product, PGA, rapidly decreased in concentration, as expected, while the concentration of RuBP was the first among the sugar phosphates to rise. This provided direct *in vivo* evidence that RuBP is the carboxylation substrate in the RPP cycle. Since RuBP is a five-carbon compound, it was concluded that subsequent to the addition of CO_2 there must be a split to two three-carbon molecules. At least one of these products would have to be PGA, since PGA was shown to be the first product of CO_2 fixation. From a consideration of oxidation states of RuBP and CO_2, both three-carbon products must in fact be PGA, if there is no external oxidant or reductant supplied to the reaction.

4. CHARACTERIZATION OF ENZYMES

During approximately the period when the RPP cycle was mapped through the use of labeled carbon, an oxidative pentose phosphate cycle (OPP cycle) was discovered by more classical biochemical methods in which the various enzymes required were isolated and characterized. Several of the reactions postulated for the OPP cycle appeared to be the reverse of reactions of the RPP cycle, and soon many of the required enzyme activities were isolated from green plants. For example, the transketolase enzyme, essential for both cycles, was purified from spinach by Horecker and co-workers in 1953. Also, pentose phosphate isomerase was found in alfalfa (Axelrod and Bandurski, 1953). The finding of these and other enzyme activities of the OPP cycle and glycolysis provided much of the necessary supporting biochemical evidence for the RPP cycle.

There are, however, three enzyme activities unique to RPP Cycle. Of these perhaps the most important in establishing the cycle is the RuBP carboxylase. The enzymatic carboxylation of RuBP *in vitro* was first reported by Quayle and co-workers (1954) who demonstrated the formation of PGA, labeled with ^{14}C in the carboxyl group only, when RuBP and $^{14}CO_2$ were added to a cell-free extract obtained from *Chlorella*. The enzyme was purified and characterized by Weissbach and co-workers (Weissbach and Horecker, 1955; Weissbach *et al.,* 1956) soon afterwards.

A second key enzyme unique to the RPP cycle is phosphoribulokinase, purified from spinach by Hurwitz and co-workers in 1956. The third unique enzyme is sedoheptulose 1, 7-bisphosphatase (SBPase). For a long time it was thought that this enzyme might be identical to fructose

1, 6-bisphosphatase, but recent work by Buchanan *et al.* (1976) suggested that it is derived from FBPase when FBPase is dissociated.

E. Reactions in the Reductive Pentose Phosphate Cycle

1. THE CARBOXYLATION REACTION

In the initial step of the RPP cycle, the enzyme ribulose bisphosphate carboxylase (RuBPCase) (EC 4.1.1.39) catalyzes the addition of CO_2 to C_2 of RuBP (Mulhoffer and Rose, 1965). An unstable enzyme-bound six-carbon intermediate molecule results. Then this intermediate is hydrolytically split with a concurrent transfer of a pair of electrons from C_3 of the RuBP to C_2. The enzyme discriminates to some extent against $^{13}CO_2$ as compared with $^{12}CO_2$, whereas the carboxylation of phosphoenolpyruvate, the first carboxylation step in C_4 and CAM plants, shows little isotopic discrimination. Consequently, negative ^{13}C values of C_3 plants reflect this isotopic discrimination and provide one distinguishing feature by which the class of plants can be characterized (Troughton, 1971).

The forward reaction has a negative Gibbs free-energy change of nearly 10 kcal (Bassham and Krause, 1969). This conversion of chemical energy to heat provides a substantial part of the driving force for the cycle and also facilitates the functioning of this enzyme in important regulatory roles, as will be discussed later. It also means that this reaction is essentially irreversible.

In addition to the carboxylase activity of the enzyme, it can act as an oxygenase. Molecular O_2 is bound by the enzyme and reacts with RuBP at the C_2 position. The resulting products (Lorimer *et al.*, 1973) are 3-phosphoglycerate and 2-phosphoglycolate. The latter compound is converted to glycolate in the chloroplast. The oxygenase reaction appears to be a major source of glycolate for photorespiration (see Chapter 7, this volume). The oxygenase reaction is eliminated or greatly reduced if either O_2 is lowered to 2% (compared to 20% in air) or CO_2 pressure is raised to 0.1% or higher.

2. REDUCTION OF PHOSPHOGLYCERIC ACID TO GLYCERALDEHYDE-3-PHOSPHATE

The reduction of PGA to GA13P occurs in two steps. First, the PGA is converted to the acyl phosphate in a reaction using ATP and mediated by PGA kinase (EC 2.7.2.3). With equal concentrations or activities of reactants this reaction is highly unfavorable ($\Delta G' = +5$ kcal), and it can only proceed as it does in photosynthesis by virtue of the relatively high physiological concentrations of PGA and such low concentrations of phosphoryl 3-phosphoglycerate (DPGA) that the latter compound is not normally detected in experiments *in vivo* or with whole or reconstituted chloroplasts using $^{14}CO_2$ and ^{32}P-labeled phosphate. The DPGA is reduced with NADPH and triosephosphate dehydrogenase (NADPH specific) (EC 1.2.1.13), yielding glyceraldehyde-3-phosphate (GA13P). This reaction is also somewhat unfavorable energetically, but is helped by the fact that there are three products and only two reactants. The ratio NADPH:NADP$^+$ (free plus bound) is probably not more than 3:1 (Lendzian and Bassham, 1976) and the concentration of GA13P may be not less than that of DPGA, but the production of P_i, when its concentration is about 1 m*M*, contributes −4.1 kcal to the reaction (Bassham and Krause, 1969). The overall reaction, whereby PGA is phosphorylated and reduced to GA13P with ATP and NADPH, proceeds in the light under highly reversible conditions.

3. CONVERSION OF TRIOSE PHOSPHATES TO PENTOSE PHOSPHATES

A series of isomerizations and rearrangements are required for the conversion of five triose phosphate molecules to three pentose phosphate molecules. None of these reactions utilize light-generated co-factors (ATP and NADPH), and most steps are highly reversible. Two steps which liberate P_i are rate-limiting and have substantial negative ΔGs values. Both are sites of regulation.

Two molecules of GA13P (per turn of the cycle) are converted with triosephosphate isomerase (EC 5.3.1.1) to DHAP. In the presence of aldolase (EC 4.1.2.13), the GA13P and DHAP condense to give fructose 1,6-bisphosphate (FBP) in a reversible step. FBP is then converted to fructose 6-phosphate (F6P) with fructose bisphosphatase (EC 3.1.1.11). This step has an estimated physiological $\Delta G = -7.0$ kcal (Bassham and Krause, 1969).

The conversion of triose phosphates and hexose phosphates to pentose phosphates is initiated with transketolase (EC 2.2.1.1), which transfers C_1 and C_2 (bound on the enzyme as the thiamine pyrophosphate adduct) from F6P to GA13P, leaving the four-carbon sugar phosphate, erythrose 4-phosphate (E4P), and forming xylulose 5-phosphate (Xu5P). This reaction is reversible.

The four-carbon aldose phosphate (E4P) can then condense with DHAP in a second reaction mediated by aldolase (EC 4.1.2.13) to give sedoheptulose 1,7-bisphosphate (SBP). Like RuBP, this compound is unique to the RPP cycle and is not an intermediate compound in the OPP cycle (also called the phosphogluconate pathway or the hexose monophosphate shunt). SBP is converted to sedoheptulose 7-phosphate and P_i with sedoheptulose 1,7-bisphosphatase. The reaction has a negative ΔG^s under physiological conditions of about −7 kcal (Bassham and Krause, 1969) and is a regulated and rate-limiting step.

A second transketolase-mediated step follows in which C_1 and C_2 of S7P are transferred to GA13P to give two pentose phosphates: Xu5P and ribose 5-phosphate (R5P). This completes the conversion of five GA13P molecules to three pentose phosphate molecules. The two molecules of Xu5P are each converted with L-ribulosephosphate 4-epimerase (EC 5.1.3.4), to ribulose 5-phosphate (Ru5P), whereas R5P is converted to Ru5P with D-ribose 5-phosphate ketol-isomerase (EC 5.3.1.6).

4. FORMATION OF RIBULOSE BISPHOSPHATE

The final step in the RPP cycle is the conversion of Ru5P to RuBP with ATP and phosphoribulokinase (EC 2.7.1.19). This reaction has an estimated ΔG^s of about −4 kcal, so that it is intermediate between those reactions that are clearly reversible (ΔG^s = 0 to −2 kcal) and those that are almost completely irreversible (ΔG^s = −6 to −10 or more kcal).

F. *Metabolic Regulation of the Reductive Pentose Phosphate Cycle*

1. *IN VIVO* STEADY STATE STUDIES

The methods of kinetic analysis of measuring levels of labeled metabolites have also proved useful in the identification of sites of metabolic regulation. The steady state levels of radioactive intermediate compounds can be used to estimate concentrations of metabolites in the chloroplasts (Bassham and Krause, 1969). These concentrations are then used as approximations of activities in order to calculate the physiological free-energy changes (ΔG^s) for a specified plant and set of physiological conditions. This information provides a direct measure of the reversibility of the reactions as they are occurring *in vivo*. It can be easily shown that the relation between ΔG^s and the reversibility of the reaction is given by

$$\Delta G^s = -RT \ln(f/b)$$

where f is the forward reaction rate and b the back reaction rate.

In order for such measurements to be meaningful, accurate procedures for the maintenance of steady state conditions and continuous measurement of CO_2, specific radioactivity, rapid sampling and killing, and quantitative analysis of radioactivity in each compound as a function of the amount of tissue samples were developed (Bassham and Kirk, 1960). Initially, the steady state kinetic method was used to demonstrate amino acid formation directly from photosynthate without the inter-

TABLE I
Free-Energy Changes of the RPP Cycle[a]

Reaction	$\Delta G'$	ΔG^s
$CO_2 + RuBP^{4-} + H_2O \rightarrow 2\ PGA^{3-} + 2\ H^+$	−8.4	−9.8
$H^+ + PGA^{3-} + ATP^{4-} + NADPH \rightarrow ADP^{3-} + GAl3P^{2-} + NADP^+ + P_i^{2-}$	+4.3	−1.6
$GAl3P^{2-} \rightarrow DHAP^{2-}$	−1.8	−0.2
$GAl3P^{2-} + DHAP^{2-} \rightarrow FBP^{4-}$	−5.2	−0.4
$FBP^{4-} + H_2O \rightarrow F6P^{2-} + P_i^{2-}$	−3.4	−6.5
$F6P^{2-} + GAl3P^{2-} \rightarrow E4P^{2-} + Xu5P^{2-}$	+1.5	−0.9
$E4P^{2-} + DHAP^{2-} \rightarrow SBP^{4-}$	−5.6	−0.2
$SBP^{4-} + H_2O \rightarrow S7P^{2-} + P_i^{2-}$	−3.4	−7.1
$S7P^{2-} + GAl3P^{2-} \rightarrow R5P^{2-} + Xu5P^{2-}$	+0.1	−1.4
$R5P^{2-} \rightarrow Ru5P^{2-}$	+0.5	−0.1
$Xu5P^{2-} \rightarrow Ru5P^{2-}$	+0.2	−0.1
$Ru5P^{2-} + ATP^{4-} \rightarrow RuBP^{4-} + ADP^{3-} + H^+$	−5.2	−3.8
$F6P^{-2} \rightarrow G6P^{2-}$	−0.5	−0.3

[a]The standard physiological Gibbs free-energy changes ($\Delta G'$) were calculated for unit activities, except $[H^+] = 10^{-7}$. The physiological free-energy changes at steady state (ΔG^s) are for a 1% w/v suspension of *Chlorella pyrenoidosa* photosynthesizing with 0.04% $^{14}CO_2$ in air and with other conditions as described by Bassham and Krause (1969). The stroma concentrations were assumed to be four times the average cellular concentrations, and are used as approximations for activities.

mediacy of sucrose or starch in photosynthesizing *Chlorella pyrenoidosa* (Smith *et al.*, 1961).

Some years later, the method was used to determine the ΔG^s values for reactions of the RPP cycle, as summarized in Table I (Bassham and Krause, 1969). The reactions shown to be rate-limiting in the light (during photosynthesis) were those mediated by RuBPCase, FBPase, SBPase, and phosphoribulokinase.

2. REGULATION BETWEEN LIGHT AND DARK: *IN VIVO* STUDIES

Metabolism in the chloroplasts might be expected to change dramatically between light, when photosynthesis is occurring, and dark, when breakdown of starch might be expected. Chloroplasts lack the enzymes of the TCA cycle, but operation of either of the OPP cycle or of glycolysis could convert starch to PGA in the dark.

When the light is turned off, following a period of steady state photosynthesis by algae long enough to obtain ^{14}C saturation of photosynthetic intermediates, rapid sampling in the dark and analysis of labeled metabolites revealed interesting transient changes. Further changes were observed when the light was again turned on, and additional information was obtained by employing not only $^{14}CO_2$, but also ^{32}P-labeled inorganic phosphate (Pedersen *et al.*, 1966; Bassham and Kirk, 1968; Bassham, 1971). Among the transient changes observed were the rapid appearance of 6-phosphogluconate in the dark and an equally rapid disappearance in the light. 6-Phosphogluconate is unique to the OPP cycle.

With whole cells, the operation of the OPP cycle, indicated by the appearance of 6-phosphogluconate, could be occurring in the chloroplasts, the cytoplasm, or both. In fact, Heber *et al.* (1967b; also see 1967a) found that the unique enzymes of the OPP cycle, glucose 6-phosphate dehydrogenase and 6-phosphogluconate dehydrogenase, are present in both the cytoplasm and the chloroplasts of spinach and *Elodea,* with the larger amounts located in the cytoplasm. Studies with isolated chloroplasts photosynthesizing with $^{14}CO_2$ and then inhibited by vitamin K_5 addition demonstrated 6-phosphogluconate formation (Krause and Bassham, 1969).

The timing of disappearance of labeled starch and sucrose in *Chlorella* in the dark following a long period of photosynthesis with $^{14}CO_2$ revealed the sources of respiratory carbon in the chloroplasts and in the cytoplasm (Kanazawa *et al.*, 1972). When the light was turned off, the level of labeled starch immediately began to decline, and it continued to decline for the duration of the experiment at a substantial but *constant* rate. Immediately after the light was turned off, the level of 6-phosphogluconate rose, and other changes in the sugar phosphate levels were indicative of operation of the OPP cycle. Since starch in *Chlorella* is in the chloroplasts, this OPP cycle activity is presumed to occur inside the chloroplasts.

When the light was turned off, labeled sucrose remained constant for many minutes but later began a steady decline precisely when 1 m*M* NH_4^+ was added. Coincident with this change was a second rise in 6-phosphogluconate. These and other changes showed that the utilization of sucrose, located in the cytoplasm, was unaffected by darkness but dependent on intracellular NH_4^+, with utilization probably occurring via the OPP cycle in the cytoplasm. Such data strongly support the likelihood that the OPP cycle in the chloroplasts can operate in the dark in mobilization of chloroplast starch for chloroplast biosynthesis and export of carbon for mitochondrial respiration.

The finding of phosphofructokinase activity in chloroplasts (Kelly and Latzko, 1975) suggests that metabolism of sugar phosphates formed during starch breakdown could also proceed via glycolysis. Failure to find detectable labeled 6-phosphogluconate or $^{14}CO_2$ in darkened intact spinach chloroplasts following photosynthetic labeling of starch (Peavey *et al.*, 1977) has led to the suggestion that glycolysis but not the OPP cycle is important in chloroplast-dark metabolism (Preiss and Levy, 1979). It may be, however, that *isolated* spinach chloroplasts in the dark have no need for the extra reduced NADPH produced by the OPP cycle. Experiments with reconstituted spinach chloroplasts in the dark showed that under conditions that activate glucose 6-phosphate dehydrogenase, 6-phosphogluconate, CO_2, and pentose monophosphates are formed from ^{14}C-labeled glucose but that the pentose phosphates formed were not converted to triose phosphates and PGA. There was no lack of transaldolase and other required activities, but the concentration ratios of hexose monophosphates to pentose monophosphates favored conversion of hexose phosphates to pentose phosphates rather than the reverse (Kaiser and Bassham, 1979c).

In rapidly growing and dividing green cells, such as those of *Chlorella,* biosynthesis of chloroplast fatty acids in the dark may be substantial, and demand for NADPH could require OPP cycle operation. In mature spinach leaf cell chloroplasts, little synthesis of new chloroplasts occurs, and hence little lipid synthesis is required. Chloroplasts of such cells in the dark seem likely to employ mainly glycolysis for the conversion of starch via sugar phosphates to triose phosphates and PGA.

The existence of OPP cycle and of glycolytic activity in the dark in chloroplasts raises the necessity for metabolic regulation leading to inactivation in the light of the key steps in these pathways. These appear to be reactions mediated by glucose 6-phosphate dehydrogenase (for the OPP cycle) and phosphofructokinase (for glycolysis). More importantly, several reactions of the RPP cycle should be stopped in the dark to avoid futile cycles and hence energy wastage. These are the reactions mediated by FBPase, SBPase, phosphoribulokinase, and RuBP carboxylase. There is both kinetic and enzymatic evidence for the dark inactivation of each of these. In addition, there is enzymatic evidence for the inactivation of triose phosphate dehydrogenase and of phosphoglycerate kinase in the dark (see Section II, F,3).

Kinetic *in vivo* evidence for dark inactivation of FBPase and SBPase came from the examination of light–dark and dark–light transient changes in levels of metabolites in *Chlorella,* previously labeled with ^{14}C and ^{32}P during photosynthesis (Pedersen *et al.*, 1966; Bassham and Kirk, 1968). When the light was turned on again after 10-min darkness, there was a very rapid build up in the levels of FBP and SBP (as well as DHAP) for about 30 sec, with the levels reaching higher than steady state light levels. Then there was an equally rapid drop in these levels for another 30 sec, followed by damped oscillations leading to a steady state light level equal to that achieved in the previous light period.

The interpretation of these interesting kinetics is that when light was turned on, producing ATP and NADPH, there was a rapid conversion of PGA to triose phosphates, which were rapidly converted to FBP and SBP. The "overshoot" in the levels of FBP and SBP is attributed to the bisphosphatases having become inactive in the dark period, and requiring about 30 sec in the light to become reactivated. During this period, the level of F6P and S7P also dropped—further indicating that the bisphosphatases were inactive. After 30 sec, when the bisphosphatases became fully active, levels of these bisphosphates fell sharply, as these compounds were converted to sugar monophosphates and eventually to RuBP.

Light–dark kinetic tracer experiments also indicated dark inactivation of RuBP carboxylase. In *Chlorella pyrenoidosa*, photosynthesizing with $^{14}CO_2$ in a total CO_2 pressure of 0.03% (air level) under steady state conditions, the level of RuBP is relatively high: over 0.4 m*M* in the cells and probably more than 2 m*M* in the stroma region of the chloroplasts (Bassham and Krause, 1969). When the light is turned off, RuBP level declines rapidly and then after 2 min levels off at about 0.1 m*M*, from which it declines only very slowly (Pedersen *et al.*, 1966). Since the K_m RuBP for the fully activated enzyme is about 0.035 m*M* (Chu and Bassham, 1975), and the $\Delta G'$ for the carboxylation reaction is -8.4 kcal (Bassham and Krause, 1969), this failure for the reaction to continue after 2 min of darkness means that the enzyme activity has greatly declined.

The light–dark inactivation of the RuBP carboxylase is also evident with isolated spinach chloroplasts (Bassham and Kirk, 1968) where, following a period of photosynthesis with $^{14}CO_2$, the level of the RuBP in the dark declined to about one-half the light value and then remained constant. When the light was again turned on, the level of RuBP rose very rapidly for 30 sec and then declined to the light level. This behavior is analogous to that of the changes in FBP and SBP levels described earlier and attributed to dark inactivation of bisphosphatase activity, followed by light reactivation requiring 30 sec.

When the drop in RuBP level in the isolated spinach chloroplasts was prevented by the addition of ATP to the suspending medium just after the light was turned off, very little uptake of $^{14}CO_2$ occurred as long as the light was off (even though there was as much RuBP present in the chloroplasts as in the light). When the light was turned on again, high rates of $^{14}CO_2$ uptake resumed (Jensen and Bassham, 1968). Although the rate of entry of ATP into whole chloroplasts may be low compared to the requirements of photosynthesis (Heber and Santarius, 1970; Stokes and Walker, 1971; Heldt *et al.*, 1972), this low rate is apparently sufficient to maintain the level of RuBP once the RuBP carboxylase is inactivated.

In vivo light–dark transient studies are not so revealing with respect to regulation of the conversion of Ru5P to RuBP since the level of the substrate ATP declines rapidly in the dark. *In vivo* evidence for phosphoribulokinase inactivation in the dark came from studies in which vitamin K_5 was added to photosynthesizing *Chlorella pyrenoidosa* (Krause and Bassham, 1969). The result was that electrons were diverted from the reduction of ferredoxin to the reduction of the oxidized form of vitamin K_5, but there was little effect on the level of ATP, which remained high. Upon the addition of vitamin K_5 to the algae, there was an immediate increase in 6-phosphogluconate and in pentose monophosphates, but a rapid drop in the level of RuBP. Thus, it appears that the OPP cycle was activated but that the conversion of Ru5P to RuBP ceased. This is in agreement with the known properties of the isolated enzyme (phosphoribulokinase), since its activity depends on a high redox level in the chloroplast (see following discussion of FBPase and SBPase activation).

The regulation of the enzymes involved in the reduction of PGA to GA13P was recognized from studies of the properties of the isolated enzymes, not from kinetic studies with whole cells. *In vivo,* the direction of metabolism between PGA and GA13P already is controlled by the levels of ATP and ADP, and of NADPH and $NADP^+$. In the light, with high levels of ATP and of NADPH, the reactions proceed in the direction of reduction of PGA and GA13P, but with very little steady state free-energy change. That is, the two steps are nearly reversible (Bassham and Krause, 1969). In the dark, however, the reverse oxidative reaction may proceed with a substantial negative free-energy change, requiring that enzyme activities be diminished.

Prevention of the OPP cycle in the chloroplasts during photosynthesis requires that the glucose 6-phosphate dehydrogenase be inactivated. The sudden appearance of 6-phosphogluconate in the dark and its disappearance in the light have already been mentioned. Not surprisingly, glucose 6-phosphate dehydrogenase is inactivated with increasing ratios of $NADPH:NADP^+$ (Lendzian and Bassham, 1975; Wildner, 1975), and with changes in this ratio equal to those actually observed in chloroplasts between light and dark, there is a large change in the activity of this enzyme. The activity is further affected by RuBP and by pH in the directions expected to inactivate in the light (Lendzian and Bassham, 1975).

3. LIGHT–DARK REGULATION: ENZYMOLOGY

FBPase and SBPase, like several other enzymes of the RPP cycle and the OPP cycle in chloroplasts, seem to be regulated by more than one factor that changes in concentration between light and dark. In common with some other regulated chloroplast enzymes, FBPase and SBPase respond to changes in pH and Mg^{2+}. Increased Mg^{2+} lowers the pH optima of these enzymes (Preiss *et al.*, 1967; Garnier and Latzko, 1972). Since both Mg^{2+} (Lin and Nobel, 1971; Barber *et al.*, 1974; Hind *et al.*, 1974; Krause, 1974; Barber, 1976) and pH (Heldt *et al.*, 1973) increase in light in chloroplasts, the combined change has a substantial effect on enzyme activity.

A second major regulation of FBPase and SBPase depends on another important change between light and dark: the ratio of levels of reduced to oxidized co-factors. Two systems have been proposed for such mediation by redox levels. A protein factor found to be responsible for ferredoxin-dependent activation of FBPase and SBPase (Buchanan *et al.*, 1967b, 1971, 1976; Schurmann and Buchanan, 1975) has been resolved into two components. One of these is thioredoxin and the other an enzyme ferredoxin-thioredoxin reductase (Buchanan and Wolosiuk, 1976; Schurmann *et al.*, 1976; Wolosiuk and Buchanan, 1977; Holmgren, 1977). Thioredoxin is involved in many kinds of cells with ribonucleotide reductase, and is a general disulfide reductant. Two other regulated enzymes of the RPP cycle, phosphoribulokinase, and glyceraldehyde phosphate (triose phosphate) dehydrogenase are also activated by reduced thioredoxin. It has been known for several years that dithiothreitol activates some of these enzymes (Anderson, 1974). Apparently, this disulfhydryl reagent can substitute for the naturally occurring thioredoxin.

A second "redox" system has been proposed in which a membrane-bound vicinal–dithiol-containing factor or light effect mediator (LEM) would accept electrons from the photosynthetic electron transport system (both before and after ferredoxin) and would activate several regulated enzymes in the chloroplast (Anderson and Avron, 1976; Anderson and Duggan, 1976). The activated enzymes are generally the same ones as those activated by the thioredoxin system.

The activity of PGA kinase is apparently regulated by energy charge and specifically is decreased by ADP (K_i = 53 μm) (Pacold and Anderson, 1973, 1975; Lavergne *et al.*, 1974).

The properties of glyceraldehyde phosphate dehydrogenase have been extensively studied (see Latzko and Kelly, 1979). There has been much interest in its synthesis in response to light and possible interconversion of different forms of the enzyme between light and dark. Activation of the enzymes by preincubation with NADP, NADPH, or ATP has been reported (Muller, 1970; Pupillo and Giuliani-Piccari, 1973; Wolosiuk and Buchanan, 1976). As already mentioned, glyceraldehyde phosphate dehydrogenase apparently is also regulated by the ferredoxin–thioredoxin system and by the LEM which accepts electrons from the photoelectron transport system (Anderson, 1979).

Like triosephosphate dehydrogenase, SBPase and FBPase, phosphoribulokinase is activated by dithiol reagents (Latzko and Gibbs, 1969; Anderson, 1973) probably as a simulation of physiological regulation by the ferredoxin–thioredoxin system and/or the LEM system (Anderson *et al.*, 1979).

Like the activities of FBPase and SBPase, RuBP carboxylase activity depends in part on pH, Mg^{2+}, and reduced co-factors, but the mechanisms are different. The control of the carboxylase activity is complicated by the necessity for the plants to avoid, at least to some extent, the wasteful conversion of RuBP by the oxygenase activity of this enzyme. Isolated RuBP carboxylase is activated by preincubation with CO_2 or bicarbonate and high levels of Mg^{2+} (e.g., 10 m*M*) *before* the enzyme is exposed to RuBP (Pon *et al.*, 1963; Chu and Bassham, 1973, 1974, 1975). Preincubation with physiological levels of RuBP in the absence of either bicarbonate or Mg^{2+} results in conversion of the enzyme to an inactive form with high K_m values for CO_2, and the enzyme does not recover its activity for many minutes upon subsequent exposure to physiological levels of bicarbonate and Mg^{2+} (Chu and Bassham, 1973, 1974). Full activation of the isolated purified enzyme requires that the preincubation with CO_2 and Mg^{2+} also be carried out in the presence of either 0.5 m*M* NADPH or 0.05 m*M* 6-phosphogluconate, that is, at physiological levels of each (Chu and Bassham, 1974, 1975). The fact that various reports of activation of RuBP by chloroplast metabolities have sometimes been at odds has led to some skepticism about the physiological significance of such activation (Akazawa, 1979). In the case of activation by NADPH or by 6-phosphogluconate, however, there is an increase of two- or threefold in activity over that seen with preincubation with Mg^{2+} and HCO_3^- only, and such activation has been seen in several laboratories. Moreover, both carboxylase and oxygenase activities are stimulated (Chollet and Anderson, 1976).

With respect to light–dark regulation, it seems clear that the changes in reduced co-factors, Mg^{2+} levels, and pH in the chloroplasts, which affect FBPase and SBPase activities, also result in changes in RuBP carboxylase activity. The pH optimum of the isolated enzyme shifts toward the pH actually found in chloroplasts in the light (about 8.0) with increased Mg^{2+}, and the value of K_m for CO_2 is lower at pH 8.0 than at pH 7.2 (Bassham *et al.*, 1968b; Sugiyama *et al.*, 1968; Lorimer *et al.*, 1976).

The activation of RuBP carboxylase by NADPH seems to be yet another part of the light–dark regulation. Carbon dioxide fixation by soluble chloroplast enzymes (stroma enzymes) is increased by the presence of 1 m*M* NADPH (Bassham *et al.*, 1978). Fixation by a "reconstituted chloroplast" system (soluble enzymes plus thylakoid membranes) with added ferredoxin and 0.5 m*M* RuBP is strongly dependent on illumination (Lendzian, 1978).

The activation by 6-phosphogluconate is, at first, surprising since this compound appears in the dark. Kinetic studies show that the 6-phosphogluconate is still present during the first 2 min of light after a dark period (Bassham and Kirk, 1968), and it may be that a useful activation occurs during the first light exposure, while the level of NADPH is still being built up (Chu and Bassham, 1973). In the dark, 6-phosphogluconate would not activate the carboxylase since the optimal conditions of pH and Mg^{2+} levels would not be met.

Although it appeared for many years that K_m CO_2 for RuBP carboxylase is too low to support the RPP cycle, a number of laboratories have shown in the past several years that the K_m CO_2 is sufficiently low (Bahr and Jensen, 1974; Badger *et al.*, 1975).

4. REGULATION DURING PHOTOSYNTHESIS

Besides the substantial regulation required for transition from photosynthesis to respiration, finer tuning of rate-limiting steps is required to keep in balance the concentrations of intermediate compounds as the physiological needs and rates of photosynthesis of the cells change (Kanazawa *et al.*, 1970). The possibilities for factors in the cytoplasm (e.g., P_i) to influence the relative amounts of triose phosphates exported from the chloroplasts (see Section II, G,2) mean that the rates of formation and conversion of GA13P via the cycle must be adjustable. The steady state free-energy data (Table I) show that the most rate-limiting steps in the light are the formation and the carboxylation of RuBP and the conversions of FBP and SBP to F6P and S7P, respectively (Bassham and Krause, 1969). Thus, the rate of the carboxylation reaction relative to the bisphosphatase reactions determines the rates of formation and utilization of triose phosphates. The reduction of PGA to triose phosphates and the conversion of triose phosphates to FBP and SBP are highly reversible in the light and therefore modulation of triosephosphate dehydrogenase *activity* plays no role in controlling triose phosphate concentrations during active, unimpaired photosynthesis. A drop in PGA concentration, for whatever reason, would result in a decreased rate of PGA reduction (by mass action) and could lead to an increase in NADPH:$NADP^+$ ratio, since NADPH would not be used as rapidly. This increased NADPH might in turn stimulate RuBP carboxylase, thus restoring the level of PGA.

Probably, there is much more to be learned about the way in which triose phosphate concentration is regulated, but one possible mechanism can be found in the sigmoidal dependence of FBPase activity on FBP concentration (Preiss et al., 1967). If the GA13P level were to drop too much due to triose phosphate export, then the levels of DHAP and of FBP would also drop, the FBPase activity would decline and triose phosphate concentration would build up. Undoubtedly, other mechanisms are required as well for maintaining the levels of triose phosphates and also for the levels of hexose, heptose, and pentose monophosphates.

G. Utilization of Carbon Cycle Metabolites

1. CHLOROPLAST REACTIONS

Within the chloroplasts, intermediates of the RPP cycle are starting points for a variety of biosynthetic reactions. For example, DHAP can be reduced to provide glycerol phosphate for phospholipid synthesis. Glucose 6-phosphate, made from F6P, can be converted to galactose monophosphate also for phospholipid synthesis. E4P may be used along with phosphoenolpyruvate in shikimic acid and thence aromatic amino acid synthesis. R5P is available for ribonucleotide and deoxyribonucleotide synthesis. It appears that leaf chloroplasts make only small amounts of alpha keto acids from CO_2, since isolated spinach chloroplasts make very little labeled amino acids or fatty acids (Everson and Gibbs, 1967; Stumpf and Boardman, 1970). Some amino acid synthesis does occur in isolated chloroplasts, however (Kirk and Leech, 1972; Murphy and Leech, 1978). Moreover, tracer studies with isolated spinach leaf cells provide indirect evidence for the chloroplasts as primary sites of synthesis of several amino acids (Larsen et al., 1981), for example, the aromatic amino acids.

In "mature" chloroplasts in fully developed leaf cells, the principal fate of most reduced carbon not exported as triose phosphates or glycolate is in conversion to chloroplast starch (Ghosh and Preiss, 1965). In the pathway to starch, F6P is converted to glucose 6-phosphate (G6P) with glucose phosphate isomerase (EC 5.3.1.9). G6P is converted to G1P with phosphoglucomutase (EC 2.7.5.1). The next step is the reaction of G1P with ATP, mediated by ADP glucose pyrophosphorylase and yielding ADP glucose and inorganic pyrophosphate (PP_i).

$$\text{F6P} \rightleftarrows \text{G6P} \rightleftarrows \text{G1P}$$
$$\text{G1P} + \text{ATP} \rightarrow \text{ADP glucose} + \text{PP}_i$$

The reaction catalyzed by ADP glucose pyrophosphorylase is an important regulatory point (Ghosh and Preiss, 1966). The enzyme is activated by PGA and is strongly inhibited by high levels of inorganic phosphate (P_i). The concentrations of both metabolites can change and are believed responsible for starch regulation. Also, along with the rapid drop in ATP in the dark, decreased activity of the enzyme resulting from increased levels of P_i could account for the immediate drop in the level of ADP glucose when the light is turned off (Bassham, 1972).

Metabolite levels were measured in intact isolated chloroplasts in the light and again in the dark (Kaiser and Bassham, 1979a). These levels were then each maintained in two solutions of the soluble components from lysed intact chloroplasts (Kaiser and Bassham, 1979b). The "light" solution was adjusted to have 1 m*M* ATP and 4 m*M* PGA plus 50 m*M* dithiothreitol (to simulate the reducing conditions in the light), whereas the "dark" solution had 0.5 m*M* ATP and 1.4 m*M* PGA. Each solution contained 4 m*M* P_i. The rate of formation of ADP glucose from G6P was 130 times faster in the light solution than in the dark solution. Thus, although the P_i level is doubtless important in controlling starch synthesis in chloroplasts, other regulatory mechanisms are available as well. Once formed, ADP glucose can transfer glucose to lengthen the amylose chain of a starch molecule.

$$(C_6H_{10}O_5)_{n+1} + \text{ADPglucose} \rightarrow (C_6H_{10}O_5)_n + \text{ADP}$$

2. TRIOSE PHOSPHATE EXPORT

The triose phosphates, GA13P and DHAP, were found to be the intermediate compounds of the chloroplasts that appeared to the largest extent in the medium of isolated spinach chloroplasts carrying out high rates of complete photosynthetic reduction of CO_2 (Bassham *et al.*, 1968a). This and other studies (Stocking and Larson, 1969; Heber and Santarius, 1970; Werden and Heldt, 1972) suggested that these compounds are the photosynthetic products exported to the cytoplasm. There is a specific phosphate translocator whereby the transport out of the chloroplast of DHAP, GA13P, or PGA is balanced by the movement into the chloroplast of P_i (Heldt and Rapley, 1970; Werden and Heldt, 1972; Fliege *et al.*, 1978). PGA is not exported by isolated photosynthesizing chloroplasts nearly as rapidly as GA13P or DHAP (Bassham *et al.*, 1968a), and this discrimination against PGA efflux is seen only in illuminated chloroplasts (Heldt *et al.*, 1978). Fliege *et al.* (1978) suggest that the PGA efflux may be inhibited by the proton gradient present when chloroplasts are in the light and stroma pH is about 1 unit higher than the external space. The triose phosphates, lacking a carboxyl group, would not be so affected, since the phosphate groups are exchanged.

This mechanism would allow retention of PGA in the chloroplasts during photosynthesis, but would allow export of PGA in the dark during respiratory metabolism. Depending on the relative needs of chloroplast and cytoplasm for ATP and NAD(P)H in the dark, the proportion of triose phosphate to PGA exported could be controlled. Regulatory properties of phosphoglycerate kinase and triosephosphate dehydrogenase may be involved in limiting the oxidation of triose phosphates to PGA inside the chloroplasts in the dark.

3. GLYCOLATE FORMATION AND EXPORT

Under most conditions of photosynthesis in chloroplasts, some glycolate is formed (Schou *et al.*, 1950; Benson *et al.*, 1951). Under conditions that favor photorespiration, a large part of the carbon fixed by the RPP cycle can be converted to glycolate (Zelitch, 1974). These conditions generally include high light intensity, low CO_2 pressure, more than a few percent O_2, and temperatures above normal (Wilson and Calvin, 1955; Bassham and Kirk, 1962). Unicellular algae such as *Chlorella pyrenoidosa,* which do not evolve photorespiratory CO_2, can form large amounts of glycolate under these conditions, and much of the glycolate is excreted into the medium. With higher plants, as much as 40% (Jensen and Bahr, 1976) to 60% (Zelitch, 1957) of photosynthetically incorporated CO_2 may be converted to glycolate. Even C_4 plants, which exhibit little or no photorespiration, can form some glycolate in amounts that increase with the conditions just listed. In the case of these plants, CO_2 formed by photorespiratory-type reactions is mostly recaptured in the leaves.

The two most widely proposed pathways by which glycolate can be formed in chloroplasts involve oxidation of sugar phosphates that are intermediate compounds of the RPP cycle. There is evidence that both paths may operate in *Chlorella pyrenoidosa* (Bassham and Kirk, 1973).

Wilson and Calvin (1955) observed greatly accelerated formation of glycolate in photosynthesizing algae when the level of CO_2 in the gas bubbling through the algae was suddenly depleted. Since this was accompanied by an increase in levels of sugar monophosphates, Wilson and Calvin proposed that in the presence of transketolase the glycolyl moiety normally transferred from a ketose monophosphate to an aldose phosphate acceptor is instead oxidized to give glycolate. The dihydroxyethyl–thiamine pyrophosphate intermediate in the transketolase reaction can be converted to glycolate with ferricyanide (Bradbeer and Racker, 1961). Shain and Gibbs (1971) described a reconstituted preparation containing fragments of spinach chloroplasts, transketolase, and co-factors that are capable of rapid conversion of F6P or dihydroxyethylthiamine pyrophosphate to glycolate in the light.

Since phosphoglycolate is also seen upon the increase of O_2 concentration and decrease of CO_2 with photosynthesizing *Chlorella pyrenoidosa,* Bassham and Kirk (1962) suggested that phosphoglycolate could be formed *in vivo* by the oxidation of RuBP.

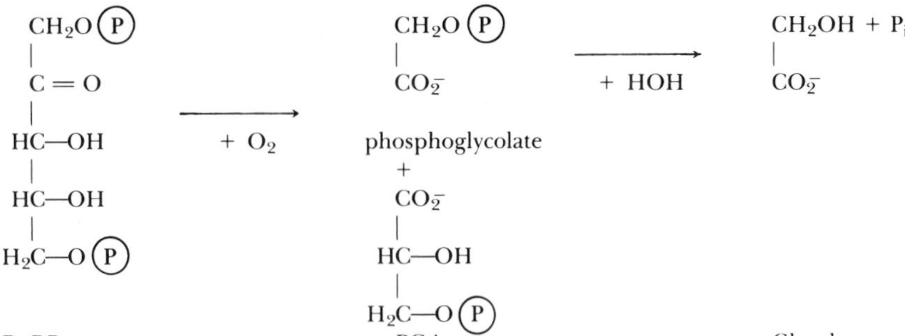

RuBP PGA Glycolate

Bowes *et al.* (1971, 1975) found that RuBP carboxylase also has an oxygenase activity that catalyzes the reaction producing phosphoglycolate. In the presence of $^{18}O_2$, this reaction produces glycolate with ^{18}O in the carboxyl group (Lorimer *et al.*, 1973). Chloroplasts contain a specific phosphoglycolate phosphatase (Richardson and Tolbert, 1961), which would convert phosphoglycolate to glycolate.

It seems possible that the relative importance of the two pathways for glycolate synthesis varies with physiological conditions (see Beck, 1979). A detailed understanding of the amount of glycolate produced under physiological conditions via the different pathways is of considerable importance, since substantial efforts are being made to find ways to reduce photorespiration in C_3 plants (those plants with the RPP cycle but no C_4 cycle). One of the best ways to reduce photorespiration in such plants would seem to be to diminish glycolate formation from sugar phosphate intermediates of the RPP cycle. Since the reactions for oxidation of sugar phosphates to glycolate or to phosphoglycolate are essentially irreversible, any diminution of photorespiratory loss would appear to require inhibition or minimization of the initial oxidative step.

Photorespiratory pathways, whereby glycolate is converted to CO_2 and to carbon compounds, such as PGA that can be reassimilated into the RPP cycle, are reviewed in Chapter 7 of this volume.

III. The Four-Carbon (C_4) Intermediate Pathways*

A. *Occurrence, Location, and Function*

The C_4 pathways (there are several variations) occur in a variety of higher plants (Hatch *et al.*, 1967; Slack and Hatch, 1967; Johnson and Hatch, 1968). First discovered in sugar cane (Kortschak *et al.*, 1957, 1965; Hatch and Slack, 1966) and initially thought to be limited to tropi-

*Also known as the Hatch and Slack pathway.

cal grasses and a species of *Cyperus*, the C_4 cycle was later found in species of the dicotyledons, *Ameranthus (Amaranthaceae)*, and *Atriplex (Chenopodiaceae)*.

Although widely distributed among the higher plant kingdom, species with C_4 metabolism are characterized by their ability to photosynthesize at high rates under conditions of bright light and warm temperatures, by the absence of appreciable photorespiration that other plants exhibit under such conditions (Forrester *et al.*, 1966a,b), and by a radial leaf anatomy (Kranz anatomy) in which the bundle sheath cells around the vascular tissue are surrounded by mesophyll cells. This special anatomy is involved in a biochemical division of labor between the two cell types, with a part of the cycle occurring in mesophyll cells and another part in the bundle sheath cells (Downes and Hesketh, 1968; Downton and Tregunna, 1968; Johnson and Hatch, 1968; Laetsch, 1968).

Often, there are striking ultrastructural differences between the chloroplasts of the bundle sheath cells, which may be agranal, and the mesophyll cells, which may have regular grana consisting of stacked thylakoids (Ray and Black, 1979). Such a differentiation is not required for C_4 photosynthesis in all cases, however (Black and Mollenhauer, 1971).

Most of the steps in the C_4 pathway occur in the chloroplasts of the two cell types, but the very important carboxylation reaction appears to be located in the cytoplasm of the mesophyl cells (Gibbs *et al.*, 1970; Baldry *et al.*, 1971). Some species decarboxylate malic acid in the mitochondria of the bundle sheath cells (Kagawa and Hatch, 1975).

The function of the C_4 pathway is to capture CO_2 in the outer tissues of the leaf and release CO_2 (and reducing power as NADPH) in the inner tissues, that is, the bundle sheath cells. Since chloroplasts of the latter cells are the exclusive sites of the complete RPP cycle and especially of RuBP carboxylase, raising the CO_2 pressure within these chloroplasts allows CO_2 to compete effectively with O_2 for the CO_2:O_2 binding sites on the enzyme. This in turn virtually eliminates phosphoglycolate production, a principal source of glycolate for photorespiration in leaves. CO_2 fixation by isolated bundle sheath cells is inhibited by O_2, and such cells do exhibit RuBP oxygenase activity and produce glycolate (Chollet, 1976).

Bundle sheath cells also possess enzymes of the photorespiratory pathway (Liu and Black, 1972; Burris and Black, 1976; see Chapter 7, this volume). Glycolate formed by the alternative pathway, oxidation of the C_2 moiety of sugar monophosphates during the transketolase-mediated reaction, could serve as substrate for photorespiration in the bundle sheath cells. Any CO_2 produced by photorespiration in C_4 plants would have to find its way out past the mesophyll cells where it would be recaptured by the C_4 carboxylation reaction.

B. The C_4 Cycles

In the C_4 cycles (Fig. 2), the initial step is the carboxylation of phosphoenolpyruvic acid (PEP), in the cytoplasm of the mesophyll cells. From this point on, there are alternative pathways, but let us consider first the earliest correct version of C_4 metabolism. The product of this carboxylation, oxaloacetic acid, can then be converted as shown in Fig. 2

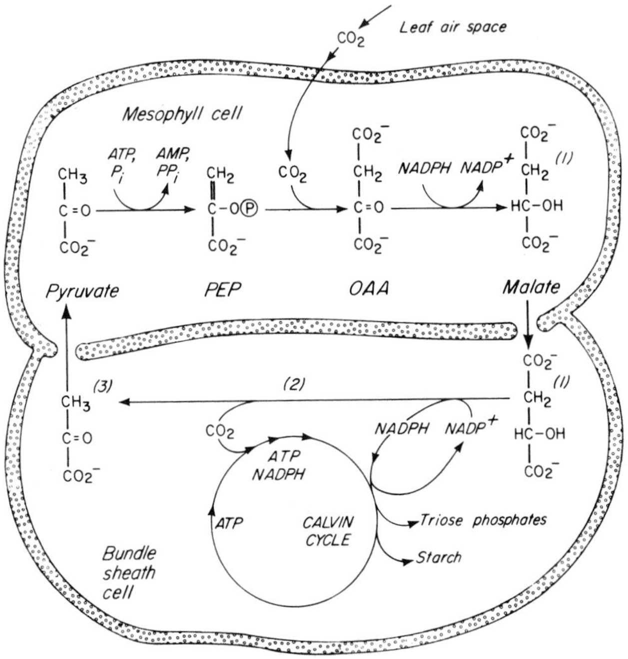

FIG. 2. The C_4 cycle of CO_2 fixation in photosynthesis—malate/pyruvate type. The pathway shown is that occurring in Type-1 C_4 plants such as *Zea mays*. In Type-2 plants, such as *Atriplex spongiosa,* and in Type-3 plants, such as *Paniccun maximum,* malate (1) is replaced as the C_4 acid transported by aspartate, which is formed by transamination of oxaloacetate (OAA). After transport to the bundle sheath cells, aspartate is converted to OAA by a second transamination. The OAA is reduced to malate, and the malate is decarboxylated by a NADP$^+$-specific malic enzyme, giving NADPH, CO_2, and pyruvate; (2) In Type-3 plants, OAA is directly decarboxylated with ATP and the PEP carboxykinase, yielding PEP, CO_2, and ADP; (3) PEP then replaces pyruvate as metabolite for movement of C_3 acid from bundle sheath cells back to mesophyll cells. In Types 2 and 3, where aspartate is the metabolite translocated, nitrogen balance must be maintained between cells, perhaps by alanine transport. Abbreviations: PEP, phosphoenol pyruvate; OAA, oxaloacetate.

to malic acid by reduction with NADPH in the mesophyll chloroplasts (Black, 1973). Malic acid is then translocated to bundle sheath cells, where it is oxidatively decarboxylated with $NADP^+$, producing pyruvate, CO_2, and NADPH. The pyruvate is translocated back to the mesophyl cells, while the CO_2 and NADPH are used for the RPP cycle in the chloroplasts of the bundle sheath cells. In the mesophyll cells, pyruvate is converted with ATP and P_i to phosphoenolpyruvate, with the formation of AMP and inorganic pyrophosphate, PP_i. The PP_i is hydrolized to 2 P_i, while AMP reacts with another ATP to make 2 ADP molecules. The energetic cost of the C_4 cycle is therefore the conversion of 2 ATP molecules to 2 ADP and 2 P_i molecules per CO_2 transported.

$$2\ ATP^{4-} + 2\ HOH \rightarrow 2\ ADP^{3-} + 2\ P_i^{2-} + 2\ H^+$$
$$\Delta G' = -15.29\ \text{kcal}, \qquad \Delta G^s = -26\ \text{kcal}$$

Possibly, translocation of metabolites back and forth between mesophyll and bundle sheath cells may cost additional energy, but the amount, if any, is not known. Clearly these costs have to be borne in addition to the $\Delta G' = -384.3$ kcal of CO_2 reduction in C_3 plants via the RPP cycle. This small additional investment in energy is more than made up in C_4 plants by the elimination of photorespiration under conditions where photorespiration would otherwise occur. The sources of the required additional ATP appear to be noncyclic, pseudocyclic, or cyclic photophosphorylation, with the relative amounts of each depending on species and conditions (Edwards *et al.*, 1976). The C_4 cycle just described is found in plants such as *Zea mays* (Maize) and *Digitaria sanguinalis*.

Besides the C_4 pathway just discussed, there are at least two other forms. The variations in C_4 path biochemistry have been summarized by Ray and Black (1979) as shown in Table II.

Plants representative of type 2 are *Atriplex spongiosa* and *Portulaca oleracea*. Type 3 is represented by *Panicum maximum* and *Sporobulus poiretii* (Ray and Black, 1979). Aside from the operation of the C_4 cycle, PGA moves from the bundle sheath cells to the mesophyll cells in all three types of plants to support synthesis of hexoses and starch as needed. Most of the reduced carbon formed in the bundle sheath cells of mature leaves is probably converted to sucrose, which is loaded into the vascular tissue for export.

C. Discovery

Early work with C_3 species had shown that C_4 acids are often significant among the early product of $^{14}CO_2$ uptake during photosynthesis, and for a time it was speculated that C_4 acids might be intermediates in the basic path of carbon fixation in such plants (Calvin and Benson,

TABLE II
Variations in the Biochemistry of C_4 Photosynthesis Found in Specific C_4 Plants[a]

Type	Bundle sheath cell decarboxylase	Co-factors produced (+) or used (−) in decarboxylation	Major substrate moving from: Mesophyll to bundle sheath cells	Major substrate moving from: Bundle sheath to mesophyll cells
1	$NADP^+$ malic enzyme	+ 1 NADPH	Malate	Pyruvate
2	NAD^+ malic enzyme	+ 1 NADH	Aspartate	Alanine/pyruvate
3	PEP carboxylase	− 1 ATP	Aspartate	Phosphoenol pyruvate[b]

[a]After Ray and Black, 1979.
[b]Nitrogen balance also must be maintained via an aminotransferase-type shuttle, probably involving alanine. PEP also may be converted to pyruvate with formation of ATP in bundle sheath cells.

1948). Such possibilities were soon eliminated for C_3 plants by inhibitor studies (Bassham *et al.*, 1950) demonstrating the continuing cycle for CO_2 reduction with malate formation blocked. Kortschak *et al.* (1957, 1965) found that in sugar cane the first ^{14}C-labeled products were mostly the C_4 acids, malate and aspartate, and not PGA. Similar findings for maize (Tarchevskii and Karpilov, 1963) were reported. Hatch and Slack (1966) confirmed the primacy of C_4 acids in sugar cane and proposed a cycle for CO_2 fixation, which had several features of the C_4 cycle as it is now known. These features include carboxylation of a C_3 acid to give C_4 acids and subsequent transfer of the newly incorporated carbon to a sugar phosphate to give PGA, a deduction based in part on degradation of sugar phosphate and C_4 acids. At first it was proposed that the labeled carboxyl group was transferred to a sugar phosphate acceptor, possibly RuBP via a transcarboxylation reaction, because there seemed to be a low activity of RuBP carboxylase activity in C_4 plants as compared with C_3 plants (Slack and Hatch, 1967). Then, a key development was the discovery of pyruvate P_i dikinase activity, which catalyzes the reaction of ATP and P_i with pyruvate to give phosphoenolpyruvate, AMP and PP_i (Hatch and Slack, 1968). This reaction is not found in C_3 plants, and it provides the energy to drive the C_4 cycle (though there is an additional energy input in type 3 C_4 plants).

The range of plant species possessing the C_4 cycle was rapidly extended. Slack (1969) and Slack *et al.* (1969) soon found that the enzyme activities of mesophyll cells differ from those of bundle sheath cells, with PEP carboxylase mostly in mesophyll cells and RuBP carboxylase in bundle sheath cells. Later techniques for isolating mesophyll cells and bundle sheath cells (Edwards *et al.*, 1970; Edwards and Black, 1971) permitted quantitative determination of the division of enzyme activities between the two cell types. The compartmentation of the enzymes between the two cell types has been supported by a great amount of data since that time (Kanai and Black, 1972; Chen *et al.*, 1973; Kanai and Edwards, 1973; Chen *et al.*, 1974; Huber and Edwards, 1975a,b; Burris and Black, 1976; Hatch and Kagawa, 1976). There now seems to be little doubt, despite earlier controversy, that the C_4 cycle biochemistry is indeed divided between cell types as summarized in Table II.

D. C_4 Reactions, Enzymes, and Regulation

1. THE CARBOXYLATION REACTION

The initial carboxylation is catalyzed by PEP carboxylase (EC 4.1.1.31), which mediates the addition of HCO_3^- to PEP:

$$PEP^{3-} + HCO_3^- \rightarrow \text{oxaloacetate}^{2-} + P_i^{2-}$$

The reaction is essentially irreversible, since $\Delta G'$ is about −8.5 kcal (Paul *et al.*, 1978). The enzyme is now considered to be in the mesophyll cell cytoplasm. The enzyme from C_4 plants has a higher K_m for PEP and Mg^{2+} and a higher V_{max} than the enzyme from C_3 plants (Ting and Osmond, 1973a,b). The low isotopic discrimination against $^{13}CO_2$ seen with C_4 plants is exhibited by the isolated enzyme in contrast to the large discrimination against $^{13}CO_2$ found with RuBP carboxylase (Whelan *et al.*, 1973).

There are a number of reports of regulation of the enzyme by metabolites: activation by sugar phosphates (Coombs and Baldry, 1972; Coombs *et al.*, 1973; Ting and Osmond, 1973c; Goatly and Smith, 1974), activation by glycine (Nishikido and Takanashi, 1973; Bhagvat and Sane, 1976; Uedan and Sugiyama, 1976), and inhibition by organic acids (Ting, 1968; Lowe and Slack, 1971; Huber and Edwards, 1975c).

Feedback inhibition by C_4 acids could of course be a response to overproduction of acids by carboxylation in comparison to their rate of transport to the bundle sheath cells. Perhaps the most consistent effect is activation by G6P, though the physiological significance is not established. Possibly G6P buildup in the mesophyll cells is a consequence of PEP carboxylase activity not keeping up with the flow of C_3 compounds to the mesophyll cells from the bundle sheath cells, either as pyruvate, PEP, or PGA. The C_3 compounds accumulated in the mesophyll cells would then be reduced to triose phosphates, which in turn would be converted to G6P. All the necessary enzymes are present in mesophyll cells for this conversion. Increased G6P could further activate the carboxylase, thus bringing metabolite flow between cells into balance.

2. REDUCTION

Oxaloacetic acid is reduced by NADP malic dehydrogenase (EC 1.1.1.82) present in mesophyll chloroplasts (Johnson and Hatch, 1970). In common with some regulated enzymes of the RPP cycle, this enzyme requires dithiol reagents such as DTT (dithiothreitol) in the extraction media for full activity. Not surprisingly, it has recently been found that the enzyme can be activated by the thioredoxin–ferredoxin–thioredoxin reductase system, which also activates those enzymes of the RPP cycle stimulated by thiol reagents (Buchanan *et al.*, 1978; Wolosiuk *et al.*, 1978). This provides a direct route for light activation of the enzyme via reduced ferredoxin.

3. TRANSAMINATION

As an alternative pathway to the type-1 plant malate formation and translocation, types 2 and 3 form aspartate from oxaloacetate with aspartate aminotransferase (EC 2.6.1.1):

$$\text{L-aspartate} + \text{ketoglutarate} \rightleftharpoons \text{oxaloacetate} + \text{L-glutamate}$$

Since this reaction is reversible, it is unlikely to be regulated. The supply of nitrogen to the two cell types could provide a mechanism for regulation, however.

4. DECARBOXYLATION BY NADP–MALIC ENZYME

This enzyme, malate dehydrogenase (EC 1.1.1.40), oxidatively decarboxylates L-malate in bundle sheath cell chloroplasts:

$$\text{Malate}^{2-} + \text{NADP}^{+} \rightleftharpoons \text{pyruvate}^{1-} + CO_2 + \text{NADPH}$$

The $\Delta G'$ for this reaction is close to zero, so that the reaction would be highly reversible were it not for the fact that there are two reactants and three products. Considering CO_2 as the third product and assuming that [pyruate] [NADPH]/[malate][$NADP^+$] is not far from unity the actual free-energy change would be perhaps -5 kcal at air levels of CO_2 (Paul *et al.*, 1978). A 10-fold increase in CO_2, which would almost completely block the oxygenase reaction of RuBP carboxylase: oxygenase would still leave a ΔG^s of -3 kcal or so, more than enough to drive the reaction in the forward direction. Although the enzyme from a number of species is inhibited by P_i, triose phosphate, and AMP (Coombs, 1979), it is not clear what regulatory role, if any, it plays.

5. DECARBOXYLATION BY NAD–MALIC ENZYME

In species such as *Amaranthus* and *Atriplex*, this enzyme (EC 1.1.1.39) is responsible for malate decarboxylation. Curiously, it requires Mn^{2+} rather than Mg^{2+} for maximum activity (Hatch and Kagawa, 1974). Like the NADP-specific enzyme, it has a $\Delta G'$ near zero but is driven in the forward direction by having a third product (CO_2) at low activity compared to unity.

6. DECARBOXYLATION BY PHOSPHOENOL PYRUVATE CARBOXYKINASE

This enzyme (EC 4.1.1.49) catalyzes the reversible reaction:

$$\text{Oxaloacetate}^{2-} + \text{ATP}^{4-} \rightarrow \text{PEP}^{3-} + CO_2 + \text{ADP}^{3-}, \ \Delta G' = \text{O}$$

The enzyme from the C_4 plant *Panicum maximum* (Ray and Black, 1976; Hatch and Mau, 1977) has been characterized. As in the case of the previous two decarboxylations, the ΔG^s is -3 to -5 kcal depending on CO_2 concentration maintained in the bundle sheath chloroplasts.

Although this decarboxylation costs an ATP, the conversion of PEP to pyruvate prior to translocation back to mesophyll cells could regenerate ATP. If PEP is translocated intact, then the energy cost becomes only 1 ATP in the bundle sheath cells and no ATPs would be required in the mesophyll cells. This seems suspiciously too efficient.

7. PYRUVATE SUPPLY TO MESOPHYLL CELLS

Where pyruvate is the direct product of decarboxylation in the bundle sheath, it can be translocated back to the mesophyll cells or it can be converted first to alanine with alanine aminotransferase. In plants in which aspartate rather than malate is translocated from mesophyll to bundle sheath, there is a necessity for $-NH_2$ groups to be moved back to mesophyll to maintain nitrogen balance. For this reason also (see Section III,D,6), the translocation of PEP would seem to require additional translocation mechanisms as compared with those in plants which translocate alanine or pyruvate.

8. CONVERSION OF PYRUVATE TO PHOSPHOENOL PYRUVATE

The conversion of pyruvate to PEP in the mesophyll cells (unless PEP is translocated instead of pyruvate; see Section III,D,7 earlier) is mediated by pyruvate P_i dikinase (EC 2.7.9.1) (Andrews and Hatch, 1969; Hatch and Slack, 1969).

$$\text{Pyruvate}^{-} + \text{ATP}^{4-} + \text{P}_i^{2-} \rightarrow \text{PEP}^{3-} + \text{AMP}^{2-} + \text{PP}_i^{2-}$$

The enzyme is first phosphorylated with ATP and P_i, producing AMP, PP_i, and enzyme–phosphate; the latter reacts with pyruvate to produce PEP. The overall reaction is reported to be reversible, but a calculation of the expected $\Delta G'$ suggests that the reaction should be thermodynamically unfavorable in the absence of PP_i hydrolysis. The hydrolysis of PEP to give pyruvate and P_i has a $\Delta G'$ of -13.2 kcal, some 5.6 kcal more negative than that of ATP to ADP and P_i. If we assume that conversion of ADP and P_i to AMP and PP_i would have a $\Delta G'$ of -1 kcal, the reaction mediated by pyruvate P_i dikinase would have a $\Delta G' = +4.5$ kcal. A value for the ratio of [ATP][P_i][pyruvate]/[AMP][PP_i][PEP] of 1000 would bring the reaction to about equilibrium.

E. Physiological Regulation

The regulation of C_4 photosynthetic metabolism presumably must satisfy all or most of the conditions that were required for C_3 metabolism: light–dark regulation and adjustment of rate-limiting steps to accommodate changing physiological needs of the plant during photosynthesis, such as sucrose synthesis and export versus starch formation in the leaves. In addition, C_4 metabolism must be regulated to maintain the concentrations and rates of flow between mesophyll and bundle sheath cells of the various translocated compounds: malate or aspartate, pyruvate or alanine (or perhaps PEP), and the balance of reduced nitrogen (NH_3 or $-NH_2$ group) and of phosphates. In a previous section (Section III,C), some suggestions have been made about ways in which regulation of enzyme activities could maintain such a balance, but physiological proof is lacking, and many questions remain.

C_4 plants respond to their environment in ways different from C_3 plants. C_4 plant photosynthesis is not saturated with respect to increasing light intensity until nearly full sunlight is reached and is not susceptible to limitation by photorespiration at air levels of O_2 and CO_2. Increasing temperature favors photorespiratory reactions and presumably requires increasing CO_2 shuttle in C_4 plants to compensate. No doubt such characteristics impose further regulatory requirements on C_4 plants.

IV. Crassulacean Acid Metabolism

The third major route of initial carbon dioxide fixation in green plants is Crassulacean acid metabolism (CAM). Since CAM is discussed later (see Chapter 8, this volume) it is only necessary at this point to relate CAM to C_4 and C_3 pathways.

Essentially CAM employs a similar biochemical strategy to C_4 plants: CO_2 is initially fixed by carboxylation of PEP to produce C_4 acids; these C_4 acids are later decarboxylated, and the resulting CO_2 is reincorporated via the RPP cycle.

The Crassulaceae differ from C_4 plants in that C_4 plants separate initial CO_2 fixation from the RPP cycle *spatially* (mesophyll cells and bundle sheath cells), whereas CAM plants separate CO_2 fixation from the RPP cycle *temporaly* (night and day). There are, of course, great differences in leaf anatomy and also in some aspects of biochemistry. Like the C_4 plants, CAM plants utilize the RPP cycle for net reduction of CO_2 to sugar phosphates. CAM plants do not avoid photorespiration, but since their stomata are closed during the day (or part of it), the photorespiratory CO_2 cannot escape, except by diffusion loss through the thick cuticle. Like C_4 plants, CAM plants exhibit several variations within their carbon-fixing pathway.

V. Possibilities for Plant Yield Improvement

The plant world exhibits a remarkable environmental adaptability, and this is admirably illustrated by the pathways for photosynthetic CO_2 fixation and their regulation. Since photosynthesis is the foundation for all plant growth and production, it is an obvious choice for research directed to increased productivity.

High rates of photosynthesis would seem to be a necessary, though clearly not a sufficient, condition for high productivity. A great many other factors influence productivity: partitioning of photosynthate, translocation, development of seeds or roots, plant structure and form, disease resistance, efficiency of water use and mineral uptake, and so on through a long list. Many of the factors interact in some way with CO_2 fixation.

Since the discovery of C_4 metabolism and the fact that it practically abolishes photorespiration under good growing conditions in the C_4 species, there has been considerable interest in imparting such metabolism or some aspect of it to the many agriculturally important crops that do not have C_4 metabolism. Considering the special Kranz leaf anatomy required, as well as the complexity of the C_4 pathway itself, this would seem to be a difficult if not impossible undertaking in many cases. There are, however, some 18 plant genera in which there are both C_3 and C_4 species (Burris and Black, 1976), and investigations of crosses between closely related C_4 and C_3 species have been performed (Nobs *et al.*, 1971; Nobs, 1976; Ray and Black, 1979). Also, naturally occurring intermediates C_3/C_4 species have been studied (Kennedy and Laetsch, 1974; Brown and Brown, 1975; Brown, 1976).

Imparting some C_4 characteristics to C_3 plants does not in many cases improve their fixation rates under conditions favoring photorespiration (Ray and Black, 1979). Nevertheless, there are C_3 plants in which response of photosynthesis to high light intensities is between that of C_3 and C_4 plants: An example is *Helianthus annuus* (sunflower) (Hesketh, 1963). Further study of plants with leaf anatomy or photosynthetic and photorespiratory response to light, CO_2, and temperature uncharacteristic of their presumed plant type might allow discovery or even improvement of new types of C_3/C_4 intermediate plants.

A minimum requirement for minimizing photorespiratory CO_2 loss could be the presence of high PEP carboxylase activity to reincorporate CO_2 within the cell, C_4 decarboxylation activity in the chloroplasts, and perhaps some form of energy input, such as that provided by the P_i pyruvate dikinase. An active transport of HCO_3^- from cytoplasm to chloroplast stroma could accomplish the same purpose in a much simpler way. Such a system may in fact exist in some unicellular algae. For example in *Chlamydamonas reinhardtii*, there is an active accumulation of HCO_3^-/CO_2 in the cell (Badger *et al.*, 1980). However, it is not yet clear that this transport is across the chloroplast outer membrane.

Another way to avoid or decrease photorespiration would be for the oxygenase activity of RuBP carboxylase and other reactions generating glycolate by oxidation of sugar phosphates to be decreased or eliminated. It may be of course that evolution has already gone as far as possible in this regard, but research is continuing in pursuit of this goal.

A simple, though not in most cases presently economic, way to overcome photorespiratory loss is to increase the CO_2 pressure by a factor of 3 or so to 0.1–0.2% CO_2 in air. This has been advocated as one of several important advantages for future controlled environment agriculture (CEA). Other advantages would be water conservation, year round growth outside the tropics, disease control, and greatly increased N_2 fixation in legumes (Bassham, 1971). Since the inflatable plastic covers would have to be replaced periodically, a part of the cellulosic agricultural waste product would be converted to plastic via the path cellulose → glucose → ethanol → ethylene → polyethylene.

Among the many kinds of research to be done before this is feasible would be research to develop plants that would grow substantially faster with CO_2 enrichment. A problem that has to be faced with many plants improved by either CO_2 enrichment or photorespiration would be subsequent limitation on translocation of sugars and filling of storage organs (Evans, 1973). Whatever the means used to increase photosynthetic CO_2 fixation rates in plants, it is clear that such success will be only a first step toward increased productivity.

VI. Carbon Dioxide Fixation in Photosynthetic Bacteria: Reductive Pentose Phosphate Cycle and Associated Reactions

Photosynthetic bacteria appear to assimilate CO_2 via two basically different pathways: (1) the RPP cycle (C_3 cycle or Calvin cycle) and its associated reactions as in green plants; and (2) ferredoxin-linked reactions of the reductive carboxylic acid cycle that are unique to prokaryotic cells. Each of these routes is discussed later in relation to bacterial photosynthesis. General aspects of the reductive pentose phosphate cycle are more fully discussed earlier.

The RPP cycle appears to function as the main path of CO_2 assimilation in two of the three major groups of photosynthetic bacteria, i.e., the purple sulfur bacteria (represented by *Chromatium vinosum*) and the purple nonsulfur bacteria (represented by *Rhodospirillum rubrum*) (Buchanan, 1972, 1973; McFadden, 1973; Fuller, 1978) but, as described later, not in the third major group of these organisms, i.e., the green sulfur bacteria (represented by *Chlorobium thiosulfatophilum*). The RPP cycle appears to be the cyclic pathway of CO_2 assimilation in the cyanobacteria (blue–green algae) (McFadden, 1973; Stanier and Cohen-Bazire, 1977; Ohmann, 1979).

In photosynthetic bacteria in which the RPP cycle is predominant, an important carboxylase, i.e., one functional in addition to RuBP carboxylase, is PEP carboxylase. In these organisms, PEP carboxylase appears to function in the formation of C_4 acids, analogous to its role in C_4 plants (see Section II,B). These bacteria also possess ferredoxin-linked pyruvate synthase and, in some cases, other ferredoxin-linked carboxylases described later.

VII. Ferredoxin-Linked Carbon Dioxide Assimilation in Photosynthetic Bacteria

A role for ferredoxin in the assimilation of CO_2 emerged from studies in 1964 (Bachofen *et al.*, 1964). Since that time, six different carboxylation reactions have been described. In the following discussion, the ferredoxin-linked carboxylases are collectively referred to as "synthases" and are identified for individual reactions by the name of the α-keto acid product formed. An alternate name, one that indicates reversibility of the reactions, is "oxidoreductase." In the following equations, reversibility of the reactions is not indicated.

A. *Synthesis of Pyruvate*

$$\text{Acetyl-CoA} + CO_2 + \text{ferredoxin}_{\text{reduced}} \rightarrow \text{pyruvate} + \text{CoA-SH} + \text{ferredoxin}_{\text{oxidized}}$$

Pyruvate synthase (pyruvate:ferredoxin oxidoreductase) is widely distributed in anaerobic organisms. It has been found in each of the three main types of photosynthetic bacteria, in various types of fermentative bacteria, in cyanobacteria, and in methanogenic bacteria (Buchanan, 1972, 1973; Zeikus *et al.*, 1977; Fuchs and Stupperich, 1978; Fuchs *et al.*, 1978). In most of these organisms, pyruvate synthase appears to be important in the assimilation of exogenous acetate and CO_2. There is evidence that CO_2, rather than bicarbonate, is the active species fixed by pyruvate synthase (Thauer *et al.*, 1975). Based on inhibition studies with glyoxylate, pyruvate synthase appears to occupy the key position in CO_2 assimilation in the green sulfur bacteria (Quandt *et al.*, 1978).

Pyruvate synthase has been purified from two photosynthetic bacteria (*Chlorobium thiosulfatophilum* and *Chromatium vinosum*) and, following specific treatment to release the co-factor from the enzyme, it was shown to require thiamine pyrophosphate (Buchanan *et al.*, 1965). A similar enzyme has been highly purified from the fermentative bacterium *Clostridium acidi-urici* and shown to contain an iron–sulfur chromophore in addition to thiamine pyrophosphate (Uyeda and Rabinowitz, 1971a,b). This chromophore appears to couple directly to ferredoxin in both the synthesis and breakdown of pyruvate. It is noteworthy that a corresponding enzyme purified from mixed cultures of rumen microorganisms showed only synthase activity (Bush and Sauer, 1976; Sauer *et al.*, 1976). A second enzyme, also partially purified, was required for pyruvate breakdown. These findings raise the possibility that at least some organisms may use one enzyme for pyruvate synthesis and another enzyme for pyruvate breakdown.

B. Synthesis of α-Ketoglutarate

$$\text{Succinyl-CoA} + CO_2 + \text{ferredoxin}_{\text{reduced}} \xrightarrow[\text{synthase}]{\alpha\text{-ketoglutarate}} \alpha\text{-ketoglutarate} + \text{CoA-SH} + \text{ferredoxin}_{\text{oxidized}}$$

α-Ketoglutarate synthase (α-ketoglutarate: ferredoxin oxidoreductase) occurs in *C. thiosulfatophilum, Rs. rubrum,* and in a mixed culture of *Prostheocochloris aesturii* (Buchanan and Evans, 1965; Buchanan, 1972, 1973). α-Ketoglutarate synthase has also been demonstrated in fermentative bacteria, namely those of the rumen (Allison and Robinson, 1970; Milligan, 1970; Allison *et al.*, 1979). However, to date there is no evidence for this enzyme in photosynthetic purple sulfur bacteria, such as *Chromatium vinosum.*

Apart from its role in the reductive carboxylic acid cycle described later, α-ketoglutarate synthase appears to function in the assimilation of exogenous succinate and CO_2. The principal products formed from succinate and CO_2 by this reaction are amino acids, especially glutamate, which is derived directly from α-ketoglutarate by transamination (Buchanan, 1972, 1973).

α-Ketoglutarate synthase has been purified from *C. thiosulfatophilum* (Gehring and Arnon, 1972). The enzyme at its highest state of purity was free of pyruvate synthase activity and catalyzed the breakdown as well as the synthesis of α-ketoglutarate. Like pyruvate synthase, α-ketoglutarate synthase shows a requirement for thiamine pyrophosphate.

C. *Synthesis of α-Ketobutyrate*

$$\text{Propionyl-CoA} + CO_2 + \text{ferredoxin}_{\text{reduced}} \xrightarrow[\text{synthase}]{\alpha\text{-ketobutyrate}} \alpha\text{-ketobutyrate} + \text{CoA-SH} + \text{ferredoxin}_{\text{oxidized}}$$

α-Ketobutyrate synthase (α-ketobutyrate:ferredoxin oxidoreductase) occurs in both photosynthetic and fermentative bacteria (Buchanan, 1969). In these organisms, α-ketobutyrate synthase appears to function in a novel pathway for the biosynthesis of isoleucine and α-aminobutyrate. Several lines of evidence indicate that α-ketobutyrate synthase is a separate enzyme and is not associated with pyruvate synthase.

D. *Synthesis of Phenylpyruvate*

$$\text{Phenylacetyl-CoA} + CO_2 + \text{ferredoxin}_{\text{reduced}} \xrightarrow[\text{synthase}]{\text{phenylpyruvate}} \text{phenylpyruvate} + \text{CoA-SH} + \text{ferredoxin}_{\text{oxidized}}$$

Phosphoenol pyruvate synthase (Gehring and Arnon, 1971), which has been described in green photosynthetic sulfur bacteria, appears to function in the synthesis of aromatic amino acids via a pathway that is independent of the shikimate pathway established for aerobic cells (Buchanan, 1972). Phenylpyruvate synthase has not been purified but evidence evidence suggests that this activity is due to a specific enzyme.

E. *Synthesis of α-Ketoisovalerate*

$$\text{Isobutyryl-CoA} + CO_2 + \text{ferredoxin}_{\text{reduced}} \xrightarrow[\text{synthase}]{\alpha\text{-ketoisovalerate}} \alpha\text{-ketoisovalerate} + \text{CoA-SH} + \text{ferredoxin}_{\text{oxidized}}$$

α-Ketoisovalerate synthase (Allison and Peel, 1971) was found in cell-free extracts from two different fermentative bacteria. The α-ketoisovalerate formed in this reaction, which is dependent on thiamine pyrophosphate, is converted to valine by transamination. As with the other ferredoxin-linked carboxylation reactions that lead to amino acids, the α-ketoisovalerate synthase mechanism for valine biosynthesis does not involve steps of the pathway previously established for aerobic cells (Buchanan, 1972). The presence of α-ketoisovalerate synthase in photosynthetic cells has not been reported.

F. Synthesis of Formate

$$CO_2 + \text{ferredoxin}_{\text{reduced}} \xrightarrow[\text{reductase}]{\text{carbon dioxide}} \text{formate} + \text{ferredoxin}_{\text{oxidized}}$$

Carbon dioxide reductase (reduced ferredoxin:CO_2 oxidoreductase) was discovered in cell-free extracts of the fermentative bacterium *C. pasteurianum* (Jungermann *et al.*, 1970) and has so far not been reported to occur in photosynthetic bacteria. Like pyruvate synthase, the active species fixed by CO_2 reductase is CO_2 rather than bicarbonate (Thauer *et al.*, 1975). Growth and inhibitor studies suggest that, despite the reversibility of the reaction, CO_2 reductase functions in the synthesis of formate rather than in its degradation (Thauer *et al.*, 1974), and that molybdenum is an essential component of the enzyme (Thauer *et al.*, 1973). CO_2 reductase is the only known case in which reduced ferredoxin specifically promotes the fixation of CO_2 via a reaction that does not involve an acyl co-enzyme derivative.

G. The Reductive Carboxylic Acid Cycle

The reductive carboxylic acid cycle was proposed in 1966 as a cyclic pathway for the assimilation of CO_2 by the photosynthetic bacteria *C. thiosulfatophilum* (Evans *et al.*, 1966) and *Rs. rubrum* (Buchanan *et al.*, 1967a). On the basis of the influence of different colors of light on photosynthetic products, it has been suggested that the reductive carboxylic acid cycle functions also in higher plants (Punnett, 1976; Punnett and Kelly, 1976). The confirmation of this proposal awaits a demonstration in leaves of the enzymes associated with the cycle.

The reductive carboxylic acid cycle is in effect a reversal of the oxidative citric acid cycle of Krebs and in one turn yields one molecule of acetyl co-enzyme A from two molecules of CO_2 (Fig. 3). Reduced ferredoxin is needed to form (via α-ketoglutarate synthase) α-ketoglutarate, a key intermediate of the cycle. The formation of pyruvate from acetyl co-enzyme A and CO_2 by pyruvate synthase is also driven by reduced ferredoxin. The pyruvate formed in this manner is used for a variety of biosynthetic reactions, including the synthesis of amino acids and carbohydrates (Buchanan *et al.*, 1972; Sirevåg, 1974). In both cases, *C. thiosulfatophilum* would use the enzyme pyruvate, P_i dikinase for the synthesis of phosphoenolpyruvate prior to the formation of sugars by a reversal of glycolysis (Buchanan, 1974) or of amino acids by carboxylation–transamination reactions (Buchanan *et al.*, 1972). Pyruvate, P_i dikinase is also found in *Chromatium vinosum* and *Rs. rubrum* (Buchanan, 1974).

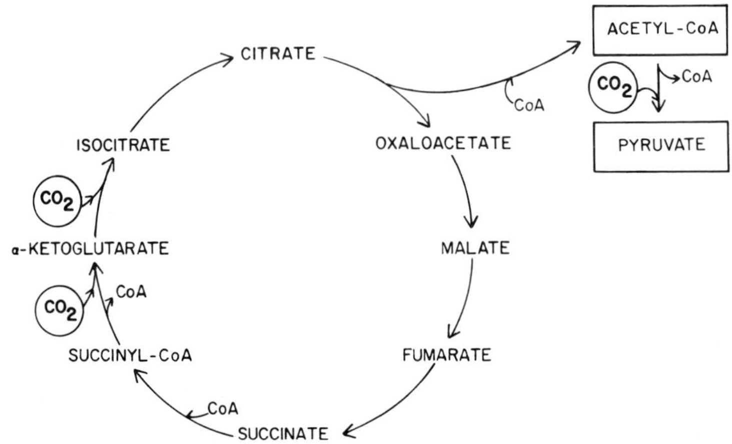

FIG. 3. CO_2 **fixation pathway in photosynthetic bacteria. CoA stands for co-enzyme A.**

Aside from demonstration of the formation of the intermediates of the reductive carboxylic acid cycle in $^{14}CO_2$ short-exposure experiments with whole cells, evidence was presented for the occurrence of the enzymes of, and associated with, the cycle in cell-free extracts of *C. thiosulfatophilum* (Evans *et al.*, 1966) and *Rs. rubrum* (Buchanan *et al.*, 1967a) (acetyl-CoA synthetase; pyruvate synthase; pyruvate,P_i dikinase; phosphoenolpyruvate carboxylase; malate dehydrogenase; fumarate hydratase; succinate dehydrogenase; succinyl-CoA synthetase; α-ketoglutarate synthase; isocitrate dehydrogenase; aconitate hydratase; and citrate lyase). The doubt that was earlier expressed concerning the citrate cleavage point in the cycle (Beuscher and Gottschalk, 1972) has been recently eliminated by Russian investigators who reported evidence for an ATP-linked citrate lyase in Chlorobium (Ivanovsky *et al.*, 1980). It is noteworthy in this connection that a nonphotosynthetic methanogenic bacterium was recently reported to contain all of the enzymes of the reductive carboxylic acid cycle except citrate lyase (Fuchs and Stupperich, 1978). In the methanogens, it is believed that acetate is formed by direct reduction of CO_2 (Fuchs and Stupperich, 1978) and feeds into an incomplete reductive carboxylic acid cycle (i.e., ending with α-ketoglutarate formation) for biosynthesis (Fuchs *et al.*, 1978).

VIII. Path of Carbon Dioxide Assimilation in Photosynthetic Green Bacteria

In light of the widely held view that the RPP cycle is present universally in photosynthetic cells, a comment on the existence of this pathway in the photosynthetic green bacteria seems appropriate. Although there is not complete agreement on the issue (McFadden, 1973; Fuller, 1978), most experiments suggest that the green bacteria lack this carbon reduction mechanism that is otherwise considered to be present in all autotrophic cells. Evidence for this conclusion rests on the absence in these organisms of the two enzymes peculiar to the RPP cycle (i.e., RuBP carboxylase and phosphoribulokinase) (Buchanan *et al.*, 1972; Sirevåg, 1974; Bondar *et al.*, 1976; Buchanan and Sirevåg, 1976; Quandt *et al.*, 1977; Takabe and Akazawa, 1977) and on $^{12}C/^{13}C$ isotope discrimination studies, which indicated that CO_2 assimilation via the cycle is minimal at best (Bondar *et al.*, 1976; Quandt *et al.*, 1977; Sirevåg *et al.*, 1977). Thus, in line with other evidence (Quandt *et al.*, 1978), it now appears that the ferredoxin-linked carboxylation reactions, in particular pyruvate synthase, constitute the major routes of CO_2 assimilation in this group of photosynthetic organisms.

IX. Concluding Remarks

A large number of plants and bacteria assimilate CO_2 via the RPP cycle (the Calvin cycle). Many plants initially fix CO_2 via PEP carboxylation and then refix it via the Calvin cycle either at a different time (CAM) or at a different location (C_4 plants). Green sulfur bacteria seem to lack key enzymes of the Calvin cycle. These and others utilize reductive carboxylation reactions (reversed Krebs cycle) for CO_2 assimilation.

It is hoped that the readers of this chapter will formulate new ideas on how to manipulate these systems to improve the overall rate of CO_2 assimilation—that may ultimately lead to improved productivity.

REFERENCES

Akazawa, T. (1979). *Encycl. Plant Physiol., New Ser.* **6,** 208–229.

Allen, M. B., Whatley, F. R., Rosenberg, J. R., Capindale, J. B., and Arnon, D. I. (1957). *In* "Research in Photosynthesis" (H. Gaffron, A. H. Brown, C. C. French, R. Livingston, E. I. Rabinowitch, B. C. Strehler, and N. E. Tolbert, eds.), pp. 288–295. Wiley (Interscience), New York.

Allison, M. J., and Peel, J. L. (1971). *Biochem. J.* **121,** 431–437.

Allison, M. J., and Robinson, I. M. (1970). *J. Bacteriol.* **104,** 50–56.

Allison, M. J., Robinson, I. M., and Baetz, A. L. (1979). *J. Bacteriol.* **140,** 980–986.

Anderson, L. E. (1973). *Plant Sci. Lett.* **1,** 331–334.

Anderson, L. E. (1974). *Biochem. Biophys. Res. Commun.* **59,** 907–913.

Anderson, L. E. (1979). *Encycl. Plant Physiol., New Ser.* **6,** 271–280.

Anderson, L. E., and Avron, M. (1976). *Plant Physiol.* **57,** 209–213.

Anderson, L. E., and Duggan, J. X. (1976). *Plant Physiol.* **58,** 135–139.

Anderson, L. E., Hansen, M. J., and Anderson, J. B. (1979) *Plant Physiol.* **63,** Suppl., 2.
Andrews, T. J., and Hatch, M. D. (1969). *Biochem. J.* **114,** 117–125.
Arnon, D. I. (1958). *Brookhaven Symp. Biol.* **11,** 181–235.
Arnon, D. I., Allen, M. B., and Whatley, F. R. (1954). *Nature (London)* **174,** 394–398.
Axelrod, B., and Bandurski, R. S. (1953). *J. Biol. Chem.* **204,** 939–948.
Bachofen, R., Buchanan, B. B., and Arnon, D. I. (1964). *Proc. Natl. Acad. Sci. U.S.A.* **51,** 690–694.
Badger, M. R., Andrews, T. J., and Osmond, C. B. (1975). *Proc. Int. Congr. Photosynth., 3rd, 1974,* Vol. II, pp. 1421–1429.
Badger, M. R., Kaplan, A., and Berry, J. A. (1980). *Plant Physiol.* **63,** 407–413.
Bahr, J. T., and Jensen, R. G. (1974). *Plant Physiol.* **53,** 39–44.
Baldry, C. W., Bucke, C., and Coombs, J. (1971). *Planta* **97,** 310–319.
Bamberger, E. S., Ehrlich, B. A., and Gibbs, M. (1975). *Plant Physiol.* **55,** 1023–1030.
Barber, J. (1976). *Trends Biochem. Sci.* **1,** 33–36.
Barber, J., Mills, J., and Nicholson, J. (1974), *FEBS Lett.* **49,** 106–110.
Bassham, J. A. (1971). *Science* **172,** 526–534.
Bassham, J. A. (1972). *Photosynth., Two Centuries Its Discovery Joseph Priestley, Proc. Int. Congr. Photosynth. Res., 2nd, 1971,* pp. 1723–1735.
Bassham, J. A., and Calvin, M. (1957). "The Path of Carbon in Photosynthesis." Prentice-Hall, Englewood Cliffs, New Jersey.
Bassham, J. A., and Kirk, M. R. (1960). *Biochim. Biophys. Acta* **43,** 447–464.
Bassham, J. A., and Kirk, M. R. (1962). *Biochem. Biophys. Res. Commun.* **9,** 376–380.
Bassham, J. A., and Kirk, M. R. (1968). *In* "Comparative Biochemistry and Biophysics of Photosynthesis" (K. Shibata, A. Takamiya, A. T. Jagendorf, and R. C. Fuller, eds.), pp. 365–378. Univ. of Tokyo Press, Tokyo.
Bassham, J. A., and Kirk, M. R. (1973). *Plant Physiol.* **52,** 407–411.
Bassham, J. A., and Krause, G. H. (1969). *Biochim. Biophys. Acta* **189,** 207–221.
Bassham, J. A., Benson, A. A., and Calvin, M. (1950). *J. Biol. Chem.* **185,** 781–787.
Bassham, J. A., Benson, A. A., Kay, L. D., Harris, A. Z., Wilson, A. T., and Calvin, M. (1954). *J. Am. Chem. Soc.* **76,** 1760–1770.
Bassham, J. A., Kirk, M., and Jensen, R. G. (1968a). *Biochim. Biophys. Acta* **153,** 211–218.
Bassham, J. A., Sharp, P., and Morris, I. (1968b). *Biochim. Biophys. Acta* **153,** 901–902.
Bassham, J. A., Krohne, S., and Lendzian, K. J. (1978) *In* "Photosynthetic Carbon Assimilation" (H. W. Siegelman and G. Hind, eds.) pp. 77–93. Plenum, New York.
Beck, E. (1979). *Encycl. Plant Physiol., New Ser.* **6,** 327–335.
Benson, A. A. (1951). *J. Am. Chem. Soc.* **73,** 2971.
Benson, A. A., and Calvin, M. (1948). *Cold Spring Harbor Symp. Quant. Biol.* **13,** 6–10.
Benson, A. A., Bassham, J. A., Calvin, M., Goodale, T. C., Hass, V. A., and Stepka, W. (1950). *J. Am. Chem. Soc.* **72,** 1710–1718.
Benson, A. A., Bassham, J. A., and Calvin, M. (1951). *J. Am. Chem. Soc.* **73,** 2970.
Beuscher, N., and Gottschalk, G. (1972). *Z. Naturforsch., B: Anorg. Chem., Org. Chem., Biochem., Biophys., Biol.* **27B,** 967–973.
Bhagvat, A. S., and Sane, P. V. (1976). *Indian J. Exp. Biol.* **14,** 155–158.
Black, C. C. (1973). *Annu. Rev. Plant Physiol.* **24,** 253–286.
Black, C. C., and Mollenhauer, H. H. (1971). *Plant Physiol.* **47,** 15–23.
Bondar, V. A., Gogotova, G. I., and Ziakum, A. M. (1976). *Dokl. Biol. Sci. (Engl. Transl.)* **228,** 223–228.
Bowes, G., Ogren, W. L., and Hageman, R. H. (1971). *Biochem. Biophys. Res. Commun.* **45,** 716–722.
Bowes, G., Ogren, W. L., and Hageman, R. H. (1975). *Plant Physiol.* **56,** 630–633.

Bradbeer, J. W., and Racker, E. (1961). *Fed. Proc., Fed. Am. Soc. Exp. Biol.* **20,** 88.
Brown, R. H. (1976). *In* "CO_2 Metabolism and Plant Productivity" (R. H. Burris and C. C. Black, eds.), pp. 311–325. University Park Press, Baltimore, Maryland.
Brown, R. H., and Brown, W. V. (1975). *Crop. Sci.* **15,** 681–685.
Buchanan, B. B. (1969). *J. Biol. Chem.* **244,** 4218–4223.
Buchanan, B. B. (1972). *In* "The Enzymes" (P. D. Boyer, ed.), 3rd ed., Vol. 6, pp. 193–216. Academic Press, New York.
Buchanan, B. B. (1973). *In* "Iron-Sulfur Proteins" (W. Lovenberg, ed.), Vol. 1, pp. 129–150. Academic Press, New York.
Buchanan, B. B. (1974). *J. Bacteriol.* **119,** 1066–1068.
Buchanan, B. B., and Evans, M. C. W. (1965). *Proc. Natl. Acad. Sci. U.S.A.* **54,** 1212–1218.
Buchanan, B. B., and Sirevag, R. (1976). *Arch. Microbiol.* **109,** 15–19.
Buchanan, B. B., and Wolosiuk, R. A. (1976). *Nature (London)* **264,** 669–670.
Buchanan, B. B., Evans, M. C. W., and Arnon, D. I. (1965). *In* "Non-Heme Iron Proteins: Role in Energy Conversion" (A. San Pietro, ed.), pp. 175–188. Antioch Press, Yellow Springs, Ohio.
Buchanan, B. B., Evans, M. C. W., and Arnon, D. I. (1967a). *Arch. Mikrobiol.* **59,** 23–49.
Buchanan, B. B., Kalberer, P. P., and Arnon, D. I. (1967b). *Biochem. Biophys. Res. Commun.* **29,** 74–79.
Buchanan, B. B., Schurmann, P., and Kalberer, P. P. (1971). *J. Biol. Chem.* **246,** 5952–5959.
Buchanan, B. B., Schurmann, P., and Shanmugam, K. T. (1972). *Biochim. Biophys. Acta* **283,** 136–145.
Buchanan, B. B., Schurmann, P., and Wolosiuk, R. A. (1976). *Biochem. Biophys. Res. Commun.* **69,** 970–978.
Buchanan, B. B., Wolosiuk, R. A., Crawford, N. E., and Yee, B. C. (1978). *Plant Physiol.* **61,** 385.
Burris, R. H., and Black, C. C., eds. (1976). "CO_2 Metabolism and Plant Productivity." University Park Press, Baltimore, Maryland.
Bush, R. S., and Sauer, F. D. (1976). *Biochem. J.* **157,** 325–331.
Calvin, M., and Benson, A. A. (1948). *Science* **107,** 476–480.
Calvin, M., and Massini, P. (1952). *Experientia* **8,** 445–457.
Calvin, M., Bassham, J. A., Benson, A. A., Lynch, V., Ouellet, C., Schou, L., Stepka, W., and Tolbert, N. W. (1951). *Symp. Soc. Exp. Biol.* **5,** 284–305.
Chen, T. M., Campbell, W. H., Dittrich, P., and Black, C. C. (1973). *Biochem. Biophys. Res. Commun.* **51,** 461–467.
Chen, T. M., Dittrich, P., Campbell, W., and Black, C. C. (1974). *Arch. Biochem. Biophys.* **163,** 246–262.
Chollet, R. (1976). *In* "CO_2 Metabolism and Plant Productivity" (R. H. Burris and C. C. Black, eds.), pp. 327–341. University Park Press, Baltimore, Maryland.
Chollet, R., and Anderson, L. E. (1976). *Arch. Biochem. Biophys.* **176,** 344–351.
Chu, D. K., and Bassham, J. A. (1973). *Plant Physiol.* **52,** 373–379.
Chu, D. K., and Bassham, J. A. (1974). *Plant Physiol.* **54,** 556–559.
Chu, D. K., and Bassham, J. A. (1975). *Plant Physiol.* **55,** 720–726.
Coombs, J. (1979). *Encycl. Plant Physiol., New Ser.* **6,** 251–261.
Coombs, J., and Baldry, C. W. (1972). *Nature (London)* **238,** 268–270.
Coombs, J., Baldry, C. W., and Bucke, C. (1973). *Planta* **110,** 95–107.
Downes, R. W., and Hesketh, J. D. (1968). *Planta* **78,** 79–84.
Downton, W. J. S., and Tregunna, E. B. (1968). *Can. J. Bot.* **46,** 207–215.
Edwards, G. E., and Black, C. C. (1971). *In* "Photosynthesis and Photorespiration" (M. D.

Hatch, C. B. Osmond, and R. O. Slayter, eds.), pp. 153–168. Wiley (Interscience), New York.

Edwards, G. E., Lee, S. S., Chen, T. M., and Black, C. C. (1970). *Biochem. Biophys. Res. Commun.* **39,** 389–395.

Edwards, G. E., Huber, S. C., Ku, S. B., Gutierrez, M., Rathnam, C. K., and Mayne, B. C. (1976). *In* "CO_2 Metabolism and Plant Productivity" (R. H. Burris, and C. C. Black, eds.), pp. 83–112. University Park Press, Baltimore, Maryland.

Evans, L. T. (1973). *In* "Crop Physiology: Some Case Histories" (L. T. Evans, ed.), pp. 327–356. Cambridge Univ. Press, London and New York.

Evans, M. C. W., Buchanan, B. B., and Arnon, D. I. (1966). *Proc. Natl. Acad. Sci. U.S.A.* **55,** 928–934.

Everson, R. G., and Gibbs, M. (1967). *Plant Physiol.* **43,** 1153–1154.

Fliege, R., Flugge, U. I., Werden, K., and Heldt, H. W. (1978). *Biochim. Biophys. Acta* **502,** 232–247.

Forrester, M. L., Krotkov, G., and Nelson, C. D. (1966a). *Plant Physiol.* **41,** 422–427.

Forrester, M. L., Krotkov, G., and Nelson, C. D. (1966b). *Plant Physiol.* **41,** 428–431.

Fuchs, G., and Stupperich, E. (1978). *Arch. Microbiol.* **118,** 121–126.

Fuchs, G., Stupperich, E., and Thauer, R. K. (1978). *Arch. Microbiol.* **117,** 61–66.

Fuller, R. C. (1978). *In* "The Photosynthetic Bacteria" (R. K. Clayton and W. R. Sistrom, eds.), pp. 691–705. Plenum, New York.

Garnier, R. V., and Latzko, E. (1972). *Photosynth., Two Centuries Its Discovery Joseph Priestley, Proc. Int. Congr. Photosynth. Res. 2nd, 1971,* pp. 1839–1845.

Gehring, U., and Arnon, D. I. (1971). *J. Biol. Chem.* **246,** 4518–4522.

Gehring, U., and Arnon, D. I. (1972). *J. Biol. Chem.* **247,** 6963–6969.

Ghosh, H. P., and Preiss, J. (1965). *J. Biol. Chem.* **240,** 960–961.

Ghosh, H. P., and Preiss, J. (1966). *J. Biol. Chem.* **241,** 4491–4504.

Gibbs, M., Latzko, E., O'Neil, D., and Hew, C. S. (1970). *Biochem. Biophys. Res. Commun.* **40,** 1356–1361.

Goatley, M. B., and Smith, H. (1974). *Planta* **117,** 67–73.

Govindjee, ed. (1982). "Photosynthesis: Energy Conversion by Plants and Bacteria" Vol. I. Academic Press, New York.

Hatch, M. D., and Kagawa, T. (1974). *Aust. J. Plant Physiol.* **1,** 357–369.

Hatch, M. D., and Kagawa, T. (1976). *Arch. Biochem. Biophys.* **175,** 39–53.

Hatch, M. D., and Mau, S. L. (1977). *Aust. J. Plant Physiol.* **4,** 207–216.

Hatch, M. D., and Slack, C. R. (1966). *Biochem. J.* **101,** 103–111.

Hatch, M. D., and Slack, C. R. (1968). *Biochem. J.* **106,** 141–146.

Hatch, M. D., and Slack, C. R. (1969). *Biochem. J.* **112,** 549–558.

Hatch, M. D., Slack, C. R., and Johnson, J. S. (1967). *Biochem. J.* **102,** 417–422.

Heber, U., Santarius, K., Hudson, M., and Hallier, U. (1967a). *Z. Naturforsch., B: Anorg. Chem., Org. Chem., Biochem., Biophys., Biol.* **22B,** 1189–1199.

Heber, U., Hallier, U. W., and Hudson, M. A. (1967b). *Z. Naturforsch,* **22,** 1200–1215.

Heber, U. W., and Santarius, K. A. (1970). *Z. Naturforsch., B: Anorg. Chem., Org. Chem., Biochem., Biophys., Biol.* **25B,** 718–728.

Heldt, H. W., and Rapley, L. (1970). *FEBS Lett.* **10,** 143–148.

Heldt, H. W., Sauer, F., and Rapley, F. (1972). *Photosynth., Two Centuries Its Discovery Joseph Priestley, Proc. Int. Congr. Photosynth. Res., 2nd, 1971,* Vol. 2, pp. 1345–1355.

Heldt, H. W., Werdan, K., Milovancev, M., and Geller, G. (1973). *Biochim. Biophys. Acta* **314,** 224–241.

Heldt, H. W., Flugge, U. I., and Fliege, R. (1978). *In* "Mechanism of Proton and Calcium Pumps" (M. Avron, G. F. Axxone, J. Metcalfe, E. Quagliariello, and N. Siliprandi, eds.), pp. 105–114. Elsevier/North-Holland, Amsterdam.

Hesketh, J. D. (1963). *Crop. Sci.* **3,** 493–496.
Hind, G., Nakatani, H. Y., and Izawa, S. (1974). *Proc. Natl. Acad. Sci. U.S.A.* **71,** 1484–1488.
Holmgren, A. (1977). *J. Biol. Chem.* **252,** 4600–4606.
Horecker, B. L., Smyrniotis, P. Z., and Klenow, H. (1953). *J. Biol. Chem.* **205,** 661–682.
Huber, S. G., and Edwards, G. E. (1975a). *Plant Physiol.* **55,** 835–844.
Huber, S. G., and Edwards, G. E. (1975b). *Plant Physiol.* **56,** 324–331.
Huber, S. G., and Edwards, G. E. (1975c). *Can. J. Bot.* **53,** 1925–1933.
Hurwitz, J., Weissbach, A., Horecker, B. L., and Smyrniotis, P. Z. (1956). *J. Biol. Chem.* **218,** 769–783.
Ivanovsky, R. N., Sintov, N. V., and Kondratieva, E. N. (1980). *Arch. Microbiol.* **128,** 239–241.
Jensen, R. G., and Bahr, J. T. (1976). *In* "CO_2 Metabolism and Plant Productivity" (R. H. Burris and C. C. Black, eds.), pp. 3–18. University Park Press, Baltimore, Maryland.
Jensen, R. G., and Bassham, J. A. (1968). *Biochim. Biophys. Acta* **153,** 227–234.
Johnson, H. S., and Hatch, M. D. (1968). *Photochemistry* **7,** 375–380.
Johnson, H. S., and Hatch, M. D. (1970). *Biochem. J.* **119,** 273–280.
Jungermann, K., Kirshniawy, E., and Thauer, R. K. (1970). *Biochem. Biophys. Res. Commun.* **41,** 682–689.
Kagawa, T., and Hatch, M. D. (1975). *Arch. Biochem. Biophys.* **167,** 687–696.
Kaiser, W. M., and Bassham, J. A. (1979a). *Plant Physiol.* **63,** 105–108.
Kaiser, W. M., and Bassham, J. A. (1979b). *Plant Physiol.* **63,** 109–113.
Kaiser, W. M., and Bassham, J. A. (1979c). *Planta* **144,** 193–200.
Kanai, R., and Black, C. C. (1972). *In* "Net Carbon Dioxide Assimilation in Higher Plants" (C. C. Black, ed.), pp. 75–93. Cotton Inc., Raleigh, North Carolina.
Kanai, R., and Edwards, G. E. (1973). *Plant Physiol.* **51,** 1133–1137.
Kanazawa, T., Kanazawa, K., Kirk, M. R., and Bassham, J. A. (1970). *Plant Cell Physiol.* **11,** 149–160.
Kanazawa, T., Kanazawa, K., Kirk, M. R., and Bassham, J. A. (1972). *Biochim. Biophys. Acta* **256,** 656–669.
Kelly, G. J., and Latzko, E. (1975). *Nature (London)* **256,** 429–430.
Kennedy, R. A., and Laetsch, W. M. (1974). *Science* **184,** 1087–1089.
Kirk, P. R., and Leech, R. M. (1972). *Plant Physiol.* **50,** 228–239.
Kortschak, H. P., Hartt, C. E., and Burr, G. D. (1957). *Proc. Hawaii Acad. Sci.,* p. 21.
Kortschak, H. P., Hartt, C. E., and Burr, G. D. (1965). *Plant Physiol.* **40,** 209–213.
Krause, G. H. (1974). *Biochim. Biophys. Acta* **333,** 301–313.
Krause, G. H., and Bassham, J. A. (1969). *Biochim. Biophys. Acta* **172,** 553–565.
Laetsch, W. M. (1968). *Am. J. Bot.* **55,** 875–883.
Larsen, P. O., Cornwell, K. L., Gee, S. L., and Bassham, J. A. (1981). *Plant Physiol.* **68,** 292–299.
Latzko, E., and Gibbs, M. (1969). *Prog. Photosynth. Res., Proc. Int. Congr. [1st], 1968,* Vol. 3, pp. 1624–1630.
Latzko, E., and Kelly, G. J. (1979). *Encyl. Plant Physiol., New Ser.* **6,** 239–249.
Lavergne, D., Bismuth, E., and Champigny, M. L. (1974). *Plant Sci. Lett.* **3,** 391–397.
Lendzian, K. J. (1978). *Planta* **143,** 291–296.
Lendzian, K., and Bassham, J. A. (1975). *Biochim. Biophys. Acta* **396,** 260–275.
Lendzian, K., and Bassham, J. A. (1976). *Biochim. Biophys. Acta* **430,** 478–489.
Lin, D. C., and Nobel, P. S. (1971). *Arch. Biochem. Biophys.* **145,** 622–632.
Liu, A., and Black, C. C. (1972). *Arch. Biochem. Biophys.* **149,** 269–280.
Lorimer, G. H., Andrews, T. J., and Tolbert, N. E. (1973). *Biochemistry* **12,** 18–23.
Lorimer, G. H., Badger, M. R., and Andrews, T. J. (1976). *Biochemistry* **15,** 529–536.
Lowe, J., and Slack, C. R. (1971). *Biochim. Biophys. Acta* **235,** 207–209.

McFadden, B. (1973). *Bacteriol. Rev.* **37,** 289–319.
Martin, A. J. P., and Singe, R. L. M. (1941). *Biochem. J.* **35,** 1358–1368.
Milligan, L. P. (1970). *Can. J. Biochem.* **48,** 463–468.
Mulhoffer, G., and Rose, I. A. (1965). *J. Biol. Chem.* **240,** 1341.
Muller, B. (1970). *Biochem. Biophys. Res. Commun.* **53,** 126–133.
Murphy, D. J., and Leech, R. M. (1978). *FEBS Lett.* **88,** 192–196.
Nishikido, T., and Takanashi, H. (1973). *Biochem. Biophys. Res. Commun.* **53,** 126–133.
Nobs, M. A. (1976). *Year Book—Carnegie Inst. Washington* **75,** 421–423.
Nobs, M. A., Bjorkman, D., and Pearcy, R. W. (1971). *Year Book—Carnegie Inst. Washington* **69,** 625–629.
Norris, L., Norris, R. E., and Calvin, M. (1955). *J. Exp. Bot.* **6,** 64–74.
Ohmann, E. (1979). *Plant Physiol., New Ser.* **6,** 691–705.
Pacold, I., and Anderson, L. E. (1973). *Biochem. Biophys. Res. Commun.* **51,** 139–143.
Pacold, I., and Anderson, L. E. (1975). *Plant Physiol.* **55,** 168–171.
Paul, J. S., Cornwell, K. L., and Bassham, J. A. (1978). *Planta* **142,** 49–54.
Peavey, D. G., Steup, M., and Gibbs, M. (1977). *Plant Physiol.* **60,** 305–308.
Pedersen, T. A., Kirk, M., and Bassham, J. A. (1966). *Biochim. Biophys. Acta* **112,** 189–203.
Pon, N. G., Rabin, B. R., and Calvin, M. (1963). *Biochem. Z.* **383,** 7–9.
Preiss, J., and Levy, C. (1979). *Encycl. Plant Physiol., New Ser.* **6,** 239–249.
Preiss, J., Biggs, M., and Greenberg, E. (1967). *J. Biol. Chem.* **242,** 2292–2294.
Punnett, T. (1976). *Fed. Proc. Fed. Am. Soc. Exp. Biol.* **35,** 1597.
Punnett, T., and Kelly, J. H. (1976). *Plant Physiol.* **54,** 59.
Pupillo, P., and Giuliani-Piccari, G. (1973). *Arch. Biochem. Biophys.* **154,** 324–331.
Quandt, L., Gottschalk, G., Ziegler, H., and Stichler, W. (1977). *FEMS Microbiol. Lett.* **1,** 125–128.
Quandt, L., Pfennig, N., and Gottschalk, G. (1978). *FEMS Microbiol. Lett.* **3,** 277–230.
Quayle, J. R., Fuller, R. C., Benson, A. A., and Calvin, M. (1954). *J. Am. Chem. Soc.* **76,** 3610–3611.
Ray, T. B., and Black, C. C. (1976). *J. Biol. Chem.* **251,** 5824–5826.
Ray, T. B., and Black, C. C. (1979). *Encycl. Plant Physiol., New Ser.,* **6,** 77–98.
Richardson, K. E., and Tolbert, N. E. (1961). *J. Biol. Chem.* **236,** 1285–1290.
Ruben, S., and Kamen, M. D. (1940a). *Phys. Rev.* **57,** 549.
Ruben, S., and Kamen, M. D. (1940b). *J. Am. Chem. Soc.* **62,** 3451–3455.
Sauer, F. D., Bush, R. S., and Stevenson, L. L. (1976). *Biochim. Biophys. Acta* **445,** 518–520.
Schou, L., Benson, A. A., Bassham, J. A., and Calvin, M. (1950). *Physiol. Plant.* **3,** 487–495.
Schurmann, P., and Buchanan, B. B. (1975). *Biochim. Biophys. Acta* **376,** 189–192.
Schurmann, P., Wolosiuk, R. A., Breazeale, V. D., and Buchanan, B. B. (1976). *Nature (London)* 263, 257–258.
Shain, Y., and Gibbs, M. (1971). *Plant Physiol.* **48,** 325–330.
Sirevåg, R. (1974). *Arch. Microbiol.* **98,** 3–18.
Sirevåg, R., Buchanan, B. B., Berry, J. A., and Troughton, J. H. (1977). *Arch. Microbiol.* **112,** 35–38.
Slack, C. R. (1969). *Phytochemistry* **8,** 1387–1391.
Slack, C. R., and Hatch, M. D. (1967). *Biochem. J.* **103,** 660–665.
Slack, C. R., Hatch, M. D., and Goodchild, D. F. (1969). *Biochem. J.* **114,** 489–498.
Smith, D. C., Bassham, J. A., and Kirk, M. R. (1961). *Biochim. Biophys. Acta* **48,** 299–313.
Stanier, R. Y., and Cohen-Bazire, G. (1977). *Annu. Rev. Microbiol.* **31,** 225–274.
Stocking, C. R., and Larson, S. (1969). *Biochem. Biophys. Res. Commun.* **3,** 278–282.
Stokes, D. M., and Walker, D. A. (1971). *In* "Photosynthesis and Photorespiration" (M. D. Hatch, C. B. Osmond, and R. O. Slayter, eds.), pp. 226–231. Wiley (Interscience), New York.

Stoppani, A. O. M., Fuller, R. C., and Calvin, M. (1955). *J. Bacteriol.* **69,** 491–501.
Stumpf, P. K., and Boardman, N. K. (1970). *J. Biol. Chem.* **245,** 2579–2587.
Sugiyama, T., Nakayama, N., and Akazawa, T. (1968). *Biochem. Biophys. Res. Commun.* **30,** 118–123.
Takabe, T., and Akazawa, T. (1977). *Plant Cell Physiol.* **18,** 753–765.
Tarchevskii, I. A., and Karpilov, Y. S. (1963). *Fiziol. Rast.* **10,** 229–231.
Thauer, R. K., Fuchs, G., Schnitker, U., and Jungermann, K. (1973). *FEBS Lett.* **38,** 45–48.
Thauer, R. K., Fuchs, G., and Jungermann, K. (1974). *J. Bacteriol.* **118,** 758–760.
Thauer, R. K., Kaufer, B., and Fuchs, G. (1975). *Eur. J. Biochem.* **55,** 111–117.
Ting, T. P. (1968). *Plant Physiol.* **43,** 1919–1924.
Ting, T. P., and Osmond, C. B. (1973a). *Plant Physiol.* **51,** 439–447.
Ting, T. P., and Osmond, C. B. (1973b). *Plant Physiol.* **51,** 448–453.
Ting, T. P., and Osmond, C. B. (1973c). *Plant Sci. Lett.* **1,** 123–128.
Troughton, J. H. (1971). *Planta* **100,** 87–92.
Uedan, K., and Sugiyama, T. (1976). *Plant Physiol.* **57,** 906–910.
Uyeda, K., and Rabinowitz, J. C. (1971a). *J. Biol. Chem.* **246,** 3111–3119.
Uyeda, K., and Rabinowitz, J. C. (1971b). *J. Biol. Chem.* **246,** 3120–3125.
Walker, D. A. (1976). *In* "The Intact Chloroplast" (J. Barber, ed.), pp. 236–278. Elsevier/North-Holland Biomedical Press, Amsterdam.
Weissbach, A., and Horecker, B. L. (1955). *Fed. Proc. Fed. Am. Soc. Exp. Biol.* **14,** 302–303.
Weissbach, A., Horecker, B. L., and Hurwitz, J. (1956). *J. Biol. Chem.* **218,** 795–810.
Werden, K., and Heldt, H. W. (1972). *Photosynth., Two Centuries Its Discovery Joseph Priestley, Proc. Int. Congr. Photosynth. Res., 2nd, 1971,* Vol. 2, pp. 1337–1344.
Whatley, F. R., Tagawa, K., and Arnon, D. I. (1963). *Proc. Natl. Acad. Sci. U.S.A.* **49,** 266–270.
Whelan, T., Sackett, W. M., and Benedict, C. R. (1973). *Plant Physiol.* **51,** 1051–1054.
Wildner, G. F. (1975). *Z. Naturforsch. C. Biosc.* **30C,** 756–760.
Wilson, A. T., and Calvin, M. (1955). *J. Am. Chem. Soc.* **77,** 5948–5957.
Wolosiuk, R. A., and Buchanan, B. B. (1976). *J. Biol. Chem.* **251,** 6456–6461.
Wolosiuk, R. A., and Buchanan, B. B. (1977). *Nature (London)* **266,** 565–567.
Wolosiuk, R. A., Nishizawa, A. N., and Buchanan, B. B. (1978). *Plant Physiol.* **61,** 975.
Zeikus, J. G., Fuchs, G., Kenealy, W., and Thauer, R. K. (1977). *J. Bacteriol.* **132,** 604–613.
Zelitch, I. (1957). *J. Biol. Chem.* **224,** 251–260.
Zelitch, I. (1974). *Arch. Biochem. Biophys.* **163,** 367–377.

7

Photorespiration

WILLIAM L. OGREN
RAYMOND CHOLLET

ABBREVIATIONS

α-HPMS	α-Hydroxypyridinemethane sulfonate
BHB	Butyl ester of HBA
CAM	Crassulacean acid metabolism
C_3 (C_4) plants	Plants having initial products of photosynthesis containing three-carbons (or four-carbons)
Γ	CO_2 compensation point
HBA	2-Hydroxy-3-butynoate
INH	Isonicotinic acid hydrazide
MBA	Methyl ester of HBA
OAA	Oxaloacetate
PEP	Phosphoenolpyruvate
3-PGA	3-Phosphoglycerate
PSII	Photosystem II
RuBP	Ribulose bisphosphate

ABSTRACT

A major objective in photorespiration research is to reduce or eliminate this process, so that the carbohydrate that is lost can be stored in the economic plant parts rather than being oxidized to CO_2. The data available at this time do not provide a definitive answer

Photosynthesis: Development, Carbon Metabolism, and Plant Productivity, Vol. II

ISBN: 0-12-294302-3

on whether this can be achieved, but progress toward an answer has been made. It has been demonstrated that the ratio of ribulose bisphosphate (RuBP) carboxylase/oxygenase activities has been altered *in vivo*—perhaps through evolution—and powerful new screening systems to select for reduced photorespiration have been devised. The identification of a CO_2-concentrating mechanism in algae and cyanobacteria provides some optimism that, even if RuBP oxygenase activity cannot be significantly reduced, it may be possible to incorporate such a mechanism into C_3 leaves. Unlike C_4 photosynthesis, the algal and cyanobacterial CO_2-concentrating mechanisms do not appear to require a specialized anatomy or compartmentation, and no enzymes not already present in C_3 leaves have yet been associated with the active bicarbonate uptake system. The concept that photosynthetic efficiency and crop productivity can be increased by reducing photorespiration continues to be an eminently attractive one.

I. Introduction

The process of photorespiration, as defined in this chapter, comprises the oxygenation of RuBP and the integrated metabolic pathways taken by the P-glycolate carbon produced in this oxygenation. Photorespiration is an integral, inseparable aspect of photosynthesis in those higher plants which fix CO_2 by the C_3 pathway. The process is less important in C_4 plants, CAM plants, algae, and cyanobacteria because these organisms possess CO_2-concentrating mechanisms, and high CO_2 concentrations inhibit photorespiration by competitively inhibiting RuBP oxygenation.

In terms of agricultural productivity, photorespiration is important because it occurs in most crop plants and, when experimentally abolished, photosynthetic CO_2 uptake and dry matter production at atmospheric CO_2 concentrations increases by about 45%. Considerable experimentation has been directed toward understanding this process and on procedures to reduce its adverse effect on net photosynthesis. Recent research indicates that photorespiration has been reduced in nature by alteration of the kinetic properties of RuBP carboxylase/oxygenase and by the creation of CO_2-concentrating mechanisms. The multiplicity of naturally occurring mechanisms provides optimism that similar solutions can be derived and applied to increase the productivity of C_3 plants.

II. Photorespiration in C_3 Plants

A. Biochemistry

A scheme of photorespiratory carbon and nitrogen metabolism, and the approximate stoichiometric relationship of these processes to photosynthetic CO_2 fixation in air (320 μl/liter CO_2, 21% O_2) at 25°C is given in Fig. 1. Much of the evidence on which this scheme is based, and several variations on the scheme, have been reviewed in numerous articles (Chollet and Ogren, 1975; Zelitch, 1975, 1979; Schnarrenberger and Fock, 1976; Andrews and Lorimer, 1978; Beck, 1979; Lea and Miflin, 1979; Tolbert, 1980). This section will emphasize three aspects of the process, which have been the most controversial and for which definitive information has recently become available. These three aspects are (1) the source of glycolate; (2) the site(s) of photorespiratory CO_2 release; and (3) the pathway of photorespiratory NH_3 reassimilation.

Photorespiratory carbon and nitrogen metabolism (Fig. 1) requires the integration of biochemical pathways in three separate leaf cell organelles (Tolbert, 1971). Glycolate is synthesized in the chloroplast and is transported to the peroxisome. In the peroxisome, glycolate is oxidized to glyoxylate, followed by transamination to glycine. Glycine then enters the mitochondrion, where it is converted to serine and photorespiratory CO_2. Serine returns to the peroxisome where it is deaminated and reduced to glycerate. Then, glycerate enters the chloroplast and is phosphorylated to reenter the C_3 photosynthesis cycle (see Chapter 6, this volume, for details).

Nitrogen enters the cycle through amination of glyoxylate by glutamate in the peroxisome. One-half the nitrogen is released as NH_3 during glycine oxidation in the mitochondrion, and the other half is returned to the peroxisome in serine and donates the amino group to glyoxylate (Tolbert, 1971). The ammonia released during glycine decarboxylation is incorporated into glutamate by the sequential action of glutamine synthetase, in the cytosol or chloroplast, and glutamate synthase in the chloroplast (Keys *et al.*, 1978; Wallsgrove *et al.*, 1980). Glutamate completes the cycle on returning to the peroxisome. Biochemical pathways of the carbon and nitrogen cycles are well defined at present, but almost nothing is known about the mechanisms regulating and directing transport between the organelles involved in the process.

1. GLYCOLATE BIOSYNTHESIS

Glycolic acid was the first photorespiratory intermediate to be identified (Zelitch, 1959; Moss, 1968), and the synthesis of this compound has been the subject of intensive investigation. Early theories held that glycolate was produced by oxidation of transketolase intermediates of the Calvin cycle, perhaps by oxidized lipoic acid (Bassham, 1963b; Bassham *et al.*, 1963), by H_2O_2 (Coombs and Whittingham, 1966; Shain and Gibbs, 1971), by the oxidant generated by PSII (Shain and Gibbs, 1971) or, in a more recent suggestion, by O_2^- (Takabe *et al.*, 1980). Alter-

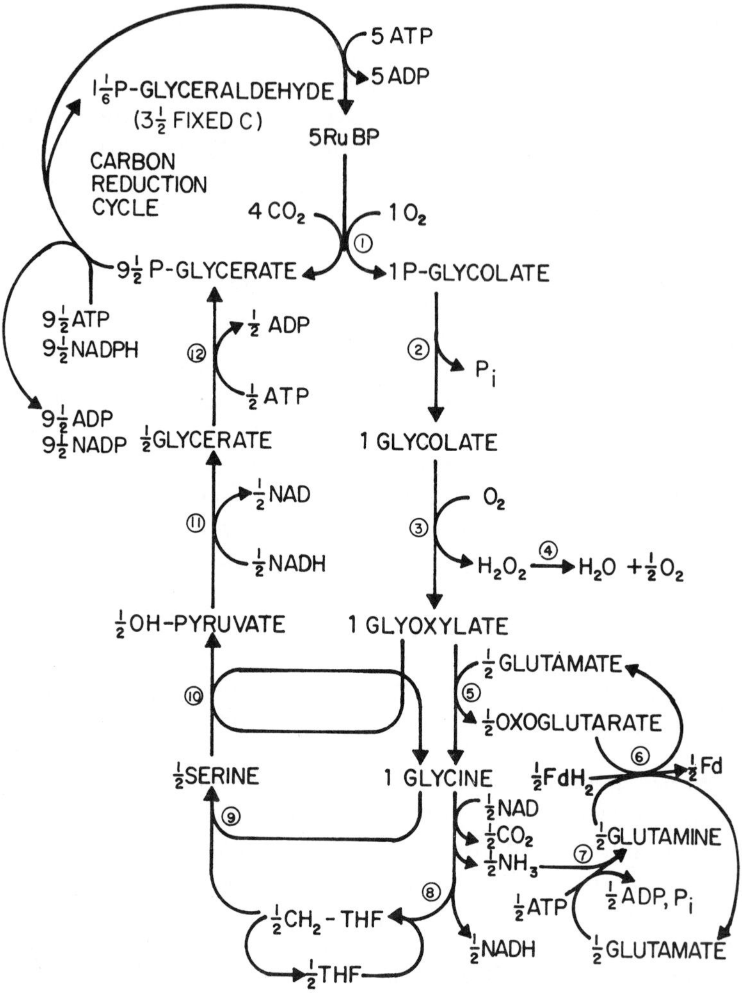

FIG. 1. Stoichiometry of the photorespiratory carbon oxidation pathway at air levels of CO_2 and O_2. The enzymes, indicated by the encircled numerals, and the organeller location of each enzyme (c, chloroplast; p, peroxisome; m, mitochondrion) are given below. 1, RuBP oxygenase (c); 2, P-glycolate phosphatase (c); 3, glycolate oxidase (p); 4, catalase (p); 5, glutamate-oxoglutarate aminotransferase (p); 6, glutamate synthase (c); 7, glutamine synthetase (c, cytosol); 8, glycine decarboxylase (m); 9, serine transhydroxymethylase (m); 10, serine-glyoxylate aminotransferase (p); 11, hydroxypyruvate reductase (p); 12, glycerate kinase (c). Abbreviations: FdH_2 (Fd), reduced (oxidized) ferredoxin; THF, tetrahydrofolate. (For details of the carbon reduction cycle (Calvin cycle), see Fig. 1 in Chapter 6, this volume.)

natively, an undefined reaction sequence producing glycolate by the reductive condensation of two CO_2 molecules has been suggested (Zelitch, 1965, 1971). Acceptance of these theories was hampered by the absence of enzymes, which carry out the proposed reactions, and by the presence of a chloroplast enzyme specific for the substrate P-glycolate (Richardson and Tolbert, 1961). The occurrence of the phosphatase indicated a strong likelihood that P-glycolate was the metabolic precursor of glycolate.

A source of P-glycolate was identified with the discovery (Bowes *et al.*, 1971) that RuBP carboxylase also catalyzed an oxygenase reaction in the presence of O_2, yielding P-glycolate as a product. Additionally, CO_2 and O_2 were found to be mutually competitive for RuBP (Ogren and Bowes, 1971; Bowes and Ogren, 1972; Badger and Andrews, 1974; Laing *et al.*, 1974), accounting for the antagonistic effects of CO_2 and O_2 on photosynthesis first observed by Otto Warburg in 1920. The sequential action of RuBP oxygenase and P-glycolate phosphatase has been generally acknowledged as a route of photorespiratory glycolate synthesis (Chollet and Ogren, 1975; Beck, 1979; Zelitch, 1979; Tolbert, 1980), but there is some difference of opinion as to whether it is the primary pathway (Chollet and Ogren, 1975; Andrews and Lorimer, 1978; Tolbert, 1980) or one of multiple significant pathways (Bassham and Kirk, 1973; Beck, 1979; Zelitch, 1979).

The most compelling evidence on the source of glycolate was obtained from studies on a mutant strain of the C_3 plant *Arabidopsis thaliana* (L.) Heynh., which is deficient in P-glycolate phosphatase activity (Somerville andOgren, 1979). Although no accumulation of this compound was observed in wild-type plants, this mutant accumulated large amounts of P-glycolate at air levels of CO_2 and O_2. In the presence of HBA (2-hydroxy-3-butynoate), a specific and irreversible inhibitor of glycolate oxidase (Jewess *et al.*, 1975; Servaites and Ogren, 1977), glycolate accumulated in wild-type plants in substantial quantity but little accumulated in mutant plants. Also, there was no photorespiratory CO_2 evolved by the mutant. These observations indicated that there was no significant source of photorespiratory glycolate other than P-glycolate. These experiments do not exclude the possibility of a source of P-glycolate other than RuBP oxygenase, but no other reactions that synthesize P-glycolate have been demonstrated.

In addition to the reduction of net photosynthesis by photorespiratory CO_2 evolution, O_2 acts as a direct, competitive inhibitor of photosynthesis with respect to CO_2 (Laing *et al.*, 1974; Chollet and Ogren, 1975; Servaites and Ogren, 1978). The rate of photosynthesis is defined by the amount of CO_2 incorporated, so each O_2 molecule that is consumed by the RuBP oxygenase reaction prevents the fixation of one CO_2 molecule (Laing *et al.*, 1974; Ogren, 1975, 1978). One molecule of O_2 incorporated by RuBP oxygenase prevents the fixation of one CO_2 molecule and leads to the photorespiratory evolution of one-half of a CO_2 molecule. Thus, the direct inhibition of photosynthesis by O_2 accounts for two-thirds of the total O_2 inhibition of photosynthetic CO_2 fixation (Laing *et al.*, 1974; Chollet and Ogren, 1975; Ogren, 1975).

Although the findings that O_2 competitively inhibited RuBP carboxylase with respect to CO_2 (Ogren and Bowes, 1971) and substituted for CO_2 in the reaction to yield P-glycolate (Bowes *et al.*, 1971) are relatively new, the interaction of O_2 and other oxidants with RuBP carboxylase had been considered previously. Tamiya and Huzisige (1949) proposed that O_2 reversibly inhibited the carboxylating enzyme, but they later discarded this theory for one in which O_2 competed with CO_2 for a photochemically produced reductant R* (Miyachi *et al.*, 1955; Tamiya *et al.*, 1957; Hirokawa *et al.*, 1958). The possibility of O_2 interaction with RuBP carboxylase was one of several possibilities considered by Turner and Brittain (1962) to explain O_2 inhibition of photosynthesis. Bassham and Kirk (1962) speculated that P-glycolate might be produced from RuBP during carboxylation by an unidentified oxidant. Bassham subsequently asserted that the oxidant was a disulfide compound (Bassham, 1963a; Bassham *et al.*, 1963), perhaps lipoic acid (Bassham, 1963b). The postulated role of O_2 in this mechanism, now known to be incorrect, was to reoxidize the dithiol compound produced during glycolate synthesis (Bassham, 1963a).

2. SITE(S) OF CO_2 RELEASE

In the glycolate oxidase reaction, glyoxylate and H_2O_2 are produced. These two products will react spontaneously to form CO_2 and formic acid, and this reaction was the first suggested source of photorespiratory CO_2 (Zelitch, 1966). Glycolate oxidase was subsequently localized in the peroxisome which also contains catalase. The catalase activity in the peroxisome is sufficient to decompose all the H_2O_2 produced during glycolate oxidation (Kisaki and Tolbert, 1969), and it is unlikely that any photorespiratory CO_2 is produced in this manner.

Glycine and serine are products of glycolate metabolism, and because

*This should not be confused with the intermediate R (or B) in the electron transport pathway between photosystem II and I (PSII and PSI) (Govindjee, 1982).

CO_2 is released in the conversion of glycine to serine, this mitochondrial reaction is an obvious potential source of photorespiratory CO_2 (Tolbert, 1971). The reaction requires NAD and tetrahydrofolic acid as co-factors, and the NADH produced can be utilized to synthesize ATP (Bird *et al.*, 1972; Douce *et al.*, 1977). The photorespired CO_2 arises from the carboxyl group of glycine, and the methylene group is transferred to a second glycine molecule by the action of serine transhydroxymethylase (Tolbert, 1971; Fig. 1).

Inhibitor studies of the glycolate pathway have provided evidence suggesting that a site of CO_2 release exists in addition to the glycine–serine conversion. The addition of INH (isonicotinic acid hydrazide), an inhibitor of glycine oxidation, to isolated soybean leaf cells in the presence of $^{14}CO_2$ caused the accumulation of ^{14}C in glycine and inhibited serine synthesis. However, twice as much label accumulated in glycolate when the glycolate oxidase inhibitor BHB (butyl ester of HBA) was present (Servaites and Ogren, 1977). These observations suggested that only one-half the glycolate is metabolized to glycine. From the rates of CO_2 release from glycolate and glycine added exogenously to soybean leaf cells in the presence of MHA (methyl ester of HBA), glycidate (inhibitor of glutamate–glyoxylate aminotransferase) or INH, Oliver (1979) concluded that one-half the photorespiratory CO_2 came from glycine oxidation and the other one-half from the oxidation of glyoxylate.

Experiments with *Arabidopsis* mutants that were deficient in serine transhydroxymethylase strongly suggest that glycine oxidation is the only physiological source of photorespiratory CO_2 (Somerville and Ogren, 1981). In these mutants, glycine was an end product of photosynthesis and accumulated in large quantity. Photorespiratory CO_2 release was initially suppressed in these mutants, but in continuous illumination the rate of CO_2 evolution gradually increased to high rates. Addition of exogenous NH_3 completely stopped photorespiration. The interpretation of these findings was that, in the mutant plant, amino nitrogen accumulated in glycine because glycine was not oxidized. After several minutes in the light, there was no longer sufficient NH_3 or primary amino groups to aminate the glyoxylate being produced in photorespiration. Glyoxylate did not accumulate, so it is presumably unstable *in vivo* and was oxidized to CO_2. The addition of exogenous NH_3 provided a source of reduced nitrogen for amination of oxoglutarate to glutamate, glyoxylate was transaminated to glycine, so CO_2 evolution ceased. The total absence of any effect of exogenous NH_3 on photorespiration rate in wild-type plants indicated that glyoxylate oxidation did not occur in normal plants. This mechanism can also account for the apparently anomalous results obtained with isolated soybean leaf cells (Servaites and Ogren, 1977; Oliver, 1979). In the presence of the glycine oxidation inhibitor, an identical accumulation of NH_3 in glycine will occur and glyoxylate, whether derived from RuBP via P-glycolate and glycolate *in vivo* or from exogenously added glycolate, cannot be aminated, and, thus it will be decarboxylated. The site and mechanism of glyoxylate decarboxylation in the absence of amino donors is not known. Inasmuch as glyoxylate was never observed to accumulate in significant quantity in photosynthetic tissue, it is evidently unstable *in vivo*.

3. AMMONIA REASSIMILATION

In glycine decarboxylation, equimolar amounts of CO_2 and NH_3 are released (Keys *et al.*, 1978; Woo *et al.*, 1978). The CO_2 can either be refixed by photosynthesis or released to the atmosphere, but NH_3 is toxic and must be reassimilated. Potential leaf-enzyme systems, which could assimilate NH_3, are glutamate dehydrogenase and the glutamine synthetase–glutamate synthase coupled system. NADH–glutamate dehydrogenase is located in the mitochondrion (Lea and Thurman, 1972), the site of NH_3 release, and would thus appear to be the enzyme of choice. However, the $K_m(NH_3)$ values for leaf glutamate dehydrogenase are greater than 10 m*M*, and it has been argued that this high value is inconsistent with the low NH_3 concentrations, which must be maintained in the leaf. From this information, and an observed NH_3 release from glycine in leaf tissue treated with an inhibitor of glutamate synthase, Keys *et al.* (1978) and Wallsgrove *et al.* (1980) concluded that photorespiratory NH_3 was reassimilated by the glutamine synthetase– glutamate synthase system discussed earlier.

The requisite role of this enzyme system in photorespiration was confirmed by the isolation of several *Arabidopsis* mutants deficient in glutamate synthase (Somerville and Ogren, 1980a). The phenotype of these mutants, growth only under atmospheric conditions that do not support photorespiration (1% CO_2), was identical to the phenotype of several other photorespiration mutants deficient in P-glycolate phosphatase (Somerville and Ogren, 1979), serine–glyoxylate aminotransferase (Somerville and Ogren, 1980b), and serine transhydroxymethylase (Somerville and Ogren, 1981). Under high (1%) CO_2 or low (2%) O_2, little or no photorespiratory glycine was synthesized so little or no NH_3 was released in glycine oxidation. In air, however, NH_3 was liberated in glycine oxidation and accumulated to toxic levels (Somerville and Ogren, 1980a). The activity of glutamate dehydrogenase in the mutants was equal to wild-type activity. Thus, it is evident that glutamate synthase fulfills an essential role in photorespiratory NH_3 reassimilation. Because plant growth is indistinguishable from the wild type under 1% CO_2, it was further concluded that leaf glutamate synthase is not necessary for any function other than reassimilation of photorespiratory NH_3. Leaf glutamate synthase does not appear to be required for the assimilation of the much smaller amounts of NH_3 required for the biosynthesis of nitrogen-containing components of the plant.

B. Regulation

Since reducing photorespiration may be agronomically beneficial, much effort has gone into attempts to alter the process chemically or genetically. Two general approaches have been suggested, inhibiting the photorespiratory pathway prior to the loss of CO_2 (Zelitch, 1966, 1979) or altering the kinetic parameters of RuBP carboxylase so that the ratio of carboxylation to oxygenation is increased (Chollet and Ogren, 1975; Ogren, 1975, 1978). It has been clearly demonstrated that the first approach will not be successful.

1. INHIBITION

Chemical inhibitors have been described and characterized for three enzymes of the photorespiratory pathway, glycolate oxidase (Zelitch, 1966; Jewess *et al.*, 1975; Kumarasinghe *et al.*, 1977; Servaites and Ogren, 1977; Servaites *et al.*, 1978; Oliver, 1979; Rathnam and Chollet, 1980a), glycine decarboxylase (Pritchard *et al.*, 1962; Kumarasinghe *et al.*, 1977; Servaites and Ogren, 1977; Oliver, 1979), and glutamine synthetase (Keys *et al.*, 1978). Inhibitors of glycolate oxidase and glycine decarboxylase did not increase CO_2 fixation (Kumarasinghe *et al.*, 1977; Servaites and Ogren, 1977; Servaites *et al.*, 1978; Doravari and Canvin, 1980; Rathnam and Chollet, 1980a) as suggested (Zelitch, 1966). Rather, they increased the O_2 inhibition of photosynthesis. This inhibition could be overcome by preventing photorespiration by assaying under high CO_2 or low O_2 concentrations (Servaites and Ogren, 1977; Rathnam and Chollet, 1980a). Servaites and Ogren (1977) concluded that photorespiration could not be reduced after the initial reaction in the process, P-glycolate synthesis, had occurred and that photorespiration can be reduced only by reducing the rate of P-glycolate synthesis. Additional evidence that photorespiration cannot be controlled by inhibiting enzymes of the pathway comes from the isolation of mutant strains of *Arabidopsis* deficient in several photorespiratory enzymes (Sections II, A, 1–3). In all cases, these mutants function normally under conditions where photorespiration cannot occur and show inhibition of photosynthetic CO_2 fixation under conditions where photorespiration does occur.

The mechanism of inhibition of photosynthesis by the chemicals is not known. The restoration of full photosynthetic activity in the presence of saturating CO_2 or low (1%) O_2 concentrations precludes direct chemical inhibition of photosynthesis. It was suggested (Servaites and Ogren, 1977) that accumulation of carbon in the substrates of the inhibited reactions prevented recycling of carbon back to the photosynthesis cycle, thereby reducing the level of intermediate metabolites. The observation that leaf protoplast photosynthesis was restored by ribose 5-phosphate, a photosynthesis intermediate, supports this conclusion (Rathnam and Chollet, 1980a). This mechanism may also explain the inhibition of photosynthesis observed in certain of the *Arabidopsis* photorespiration mutants, specifically those deficient in glycine decarboxylase (Somerville and Ogren, 1981) and serine–glyoxylate aminotransferase (Somerville and Ogren, 1980b). P-glycolate is an effective inhibitor of triose phosphate isomerase and may account for inhibition in the P-glycolate phosphatase deficient mutant (Somerville and Ogren, 1979). In the case of the glutamate synthase mutant, NH_3 may accumulate to toxic levels (Somerville and Ogren, 1980a).

2. RuBP CARBOXYLASE/OXYGENASE

The relative rates of photosynthesis and photorespiration are determined by the kinetic properties of RuBP carboxylase/oxygenase (Laing *et al.*, 1974; Ogren, 1975, 1978). This relationship is defined by the equation:

$$v_c/v_o = V_c K_o C / V_o K_c O$$

where v_c and v_o are the velocities of the carboxylase and oxygenase reactions, respectively, V_c and V_o are the maximal velocities of the two reactions, K_c is the $K_m(CO_2)$, K_o is the $K_m(O_2)$, C is the CO_2 concentration, and O is the O_2 concentration. Thus any regulation of the photorespiration rate requires that one or more of the components of the equation be changed. As mentioned elsewhere in this chapter, C_4 plants, algae, and cyanobacteria reduce photorespiration by increasing CO_2 at the site of RuBP carboxylase. The ratio of the reactions, and therefore the relative rates of photosynthesis and photorespiration, are also affected by temperature (Laing *et al.*, 1974; Badger and Collatz, 1977) and by substituting Mn^{2+} for Mg^{2+} as the enzyme metal co-factor (Christeller and Laing, 1979; Wildner and Henkel, 1979; Jordan and Ogren, 1981a).

Several reports of chemical alteration of the ratio of the two activities have appeared, but, with the exception of Mn^{2+}, they could not be confirmed when care was taken to assay the two reactions under identical conditions (Chollet and Anderson, 1976; Brown *et al.*, 1980; Jordan and Ogren, 1981a). RuBP oxygenase activity is difficult to quantitate, and this problem has led to much confusion in the literature. Enzyme assays are typically conducted at saturating substrate and optimal cofactor concentrations. These conditions cannot be met in RuBP oxygenase assays because the enzyme requires CO_2 for activation (Pon *et al.*, 1963; Andrews *et al.*, 1975; Laing *et al.*, 1975; Lorimer *et al.*, 1976), yet CO_2 competitively inhibits oxygenase activity with respect to O_2 (Badger and Andrews, 1974; Laing *et al.*, 1974). To attempt to overcome this difficulty in typical assays, the enzyme is activated by preincubation in saturating CO_2, and a small aliquot is added to the reaction mixture to dilute the CO_2 (Lorimer *et al.*, 1976). This permits the reaction to be initiated with activated enzyme, but activity is rapidly lost because the half-time of deactivation under these conditions is 1.0–1.5 min (Ogren and Hunt, 1978). Furthermore, the $K_m(O_2)$ is 40–60% O_2, so O_2 concentration is not saturating unless the assay is conducted at several atmospheres O_2 pressure. Since it is difficult to assay under this condition, it has been done only once (Laing *et al.*, 1975). Thus, oxygenase assays are routinely conducted at a suboptimal substrate concentration in the presence of a competitive inhibitor, and at an unknown, changing, level of enzyme activation. It is the uncertainty about the enzyme activation state in oxygenase assays that has led to the several claims of differential regulation of the two activities catalyzed by RuBP carboxylase/oxygenase.

Although it is not possible to assay oxygenase under optimal conditions, it is possible to accurately determine the ratio of the two activities if they are measured simultaneously in the same reaction vessel. Such assays have been developed (Laing, 1974; Kent and Young, 1980; Jordan and Ogren, 1981a) and successfully used (Laing *et al.*, 1974, 1975; Chollet and Anderson, 1976; Christeller and Laing, 1979; Jordan and Ogren, 1981a,b). These assays are somewhat complex, so they have not been routinely employed. But for definitive statements on the putative differential alteration of the two enzyme activities by chemicals or genetics, it is essential to use such an assay.

3. GENETIC VARIATION

If photorespiration can be reduced, it would be desirable to do so genetically. Several reports indicating differences in photorespiration relative to photosynthesis in higher plants have been published, but none have yet received convincing confirmation. Zelitch and Day (1968) reported that a slow-growing tobacco mutant possessed high rates of photorespiration, but photorespiration was measured by an unacceptable method (Chollet and Ogren, 1975; Chollet, 1978). Okabe *et al.* (1977), in more definitive studies with a related tobacco mutant, found no differences in the relative rates of photosynthesis and photorespiration. Genetic variation in the RuBP carboxylase/oxygenase ratio was reported for *Nicotiana* species (Kung and Marsho, 1976; Rhodes *et al.*, 1980) and a *Chlamydomonas* mutant (Nelson and Surzycki, 1976), but the activities were assayed under different conditions and are therefore equivocal.

Garrett (1978) reported that ploidy level altered the $K_m(CO_2)$ of RuBP carboxylase in ryegrass, whereas the $K_i(O_2)$ and $V_{max}(CO_2)$ remained constant. He found that the $K_m(CO_2)$ was about two times as high in diploid cultivars as in tetraploid lines. From the equation $v_c/v_o = V_cK_oC/V_oK_cO$ (Laing *et al.*, 1974), this indicated that the rate of carboxylation relative to oxygenation was increased in the tetrapolid lines, which therefore photorespired at a reduced rate. This observation with the purified protein was not repeated elsewhere (Rejda *et al.*, 1981; McNeil *et al.*, 1981), although it was supported by findings that O_2 inhibition of photosynthetic $^{14}CO_2$ fixation by leaf slices and isolated protoplasts was less for tetraploid than for diploid material and that a greater percentage of the carbon fixed in diploid protoplasts was incorporated into glycolate in the presence of a glycolate oxidase inhibitor (Rathnam and Chollet, 1980a).

The specificity of RuBP carboxylase/oxygenase with respect to CO_2 was found to differ greatly between enzymes isolated from diverse sources (Jordan and Ogren, 1981b). Enzymes from photosynthetic bacteria were found to have the lowest specificity, and enzymes from higher plants showed the highest specificity for CO_2. The specificities of enzymes from cyanobacteria and green algae were found to be intermediate, with the algal enzymes having a greater specificity for CO_2 than the cyanobacterial enzymes. Thus the CO_2 specificity for RuBP carboxylase/oxygenase enzymes increased in the order photosynthetic bacteria < cyanobacteria < green algae < higher plants, suggesting that the enzyme has evolved over time to adapt to the geological shift from an atmosphere containing high CO_2 and low O_2 concentrations to one consisting of low CO_2 and high O_2 concentrations.

Attempts to create genetic alteration of the carboxylase through mutagenesis have been made without success to date. The magnitude of the O_2-sensitive CO_2 compensation point (Γ) indicates the relative rate of photorespiration to photosynthesis in healthy leaves (Laing *et al.*, 1974). An early screening system was based on the premise that plants with reduced photorespiration would have a reduced Γ and, therefore, would survive at a CO_2 concentration less than Γ, conditions under which normal C_3 plants die (Widholm and Ogren, 1969). More recently, mutants of *Arabidopsis thaliana* with defects in the photorespiratory pathway have been isolated (Somerville and Ogren, 1979, 1980a,b, 1981). Because these mutations are lethal under photorespiratory but not non-photorespiratory conditions, it is possible to apply strong selection pressure for second-site revertants, which survive because they no longer oxygenate RuBP (Somerville and Ogren, 1980b). In another approach, it may be possible to locate alterations in RuBP carboxylase or oxygenase activity by selecting for revertants of *Alcaligenes eutrophus* (Andersen, 1979) or *Chlamydomonas reinhardtii* (Spreitzer and Mets, 1980) mutants with defective RuBP carboxylase/oxygenase. The success of these attempts to create reduced oxygenase activity requires that the high photorespiration rates observed in C_3 plants are neither necessary (Osmond and Björkman, 1972; Heber and Krause, 1980) nor inevitable (Andrews and Lorimer, 1978).

C. Function

As indicated, attempts to reduce or eliminate photorespiration are based on the premise that photorespiration is not an essential process. This presumption may not be correct, even though dramatic increases in dry matter accumulation are observed when C_3 plants are grown under elevated CO_2 or reduced O_2 concentrations (Björkman *et al.*, 1968; Parkinson *et al.*, 1974; Quebedeaux and Chollet, 1977). Photorespiration may be necessary to protect the chloroplast against photooxidative damage under conditions of low CO_2 supply (Schnarrenberger and Fock, 1976; Cornic, 1978; Powles and Osmond, 1978; Powles *et al.*, 1979; Heber and Krause, 1980). Such damage is suggested to occur because there may be no physiological electron acceptor at very low CO_2, a condition that might occur if the stomates completely closed (Osmond and Björkman, 1972). This theory is supported by the observation that illuminating leaves at high light intensity for several hours at low O_2 concentration and CO_2 concentrations less than Γ causes photoinhibition of subsequent photosynthesis. However, the chloroplast may be able to dissipate energy in the absence of CO_2 by more direct and less wasteful mechanisms. For example, algae are capable of reducing O_2 in the absence of carboxylase activity, perhaps by a Mehler-type reaction (Radmer and Kok, 1976), and such a mechanism may also occur in leaves (Canvin *et al.*, 1980). Pretreatments at O_2 concentrations less than 21% O_2, generally required for photoinhibition, may prevent these other potential protective mechanisms from functioning. A causal relationship between photoinhibition and the absence of photorespiration, if one exists, remains to be resolved.

Andrews and Lorimer (1978) suggested that RuBP oxygenation is an inevitable consequence of the RuBP carboxylation mechanism. It has been shown, however, that the two reactions are differentially altered by Mn^{2+} substitution for Mg^{2+} (Christeller and Laing, 1979; Wildner and Henkel, 1979; Jordan and Ogren, 1981a) by temperature (Laing *et al.*, 1974; Badger and Collatz, 1977) and in enzymes from different groups within the plant kingdom (Jordan and Ogren, 1981b). Thus, inevitability is not an absolute in this case and the stoichiometry of the two reactions is not immutable. Previous consideration of the physical characteristics of the CO_2 and O_2 molecular structures led to the suggestion that differences in size, shape, and charge distribution between the two gaseous substrates may permit the enzyme to be beneficially altered (Ogren, 1978). At the present time, any conclusion on whether or not the enzyme can be altered to catalyze a greater rate of carboxylation relative to oxygenation than is found in higher plants is largely a matter of personal outlook, with the definitive experiment(s) yet to be done.

III. Photorespiration in C_4 Plants

Nonsucculent higher plants can be divided into two major groups, C_3 and C_4 species, based on the initial products of photosynthetic CO_2 fixation (see Chapter 6, this volume). In C_3 plants such as wheat and soybean, CO_2 is initially fixed into the three-carbon compound 3-PGA by RuBP carboxylase. In C_4 species, atmospheric CO_2 is initially fixed into the four-carbon dicarboxylic acids (C_4 acids) OAA, malate, and aspartate by the combined action of PEP carboxylase, NADP-malate dehydrogenase, and aspartate aminotransferase, thus providing the biochemical basis for the designation of C_3 and C_4 plants. The C_4 pathway of photosynthesis has been found in at least 18 major plant families of both monocots and dicots (Ehleringer, 1979). All the families that contain C_4 species also contain C_3 plants and, in addition, at least 19 genera contain both C_3 and C_4 species, including *Panicum, Mollugo, Atriplex,* and *Flaveria.* Two of these latter genera (*Panicum, Mollugo*) have been reported to also contain naturally occurring intermediate C_3–C_4 species (see Section IV). No algae, bryophytes, lower vascular plants, or gymnosperms have been found to possess the C_4 pathway. This taxonomic distribution suggests that C_4 photosynthesis is a derived condition, having evolved polyphyletically among several of the more advanced orders of C_3 monocots and dicots (Brown and Smith, 1972, Björkman, 1976).

Even with a given genus (e.g., *Atriplex*), the C_4 pathway appears to have arisen independently more than once (Björkman, 1976).

A. Anatomical Features

Probably the most easily determined feature of all plants with C_4 photosynthesis is a "Kranz" or wreathlike leaf anatomy (Laetsch, 1974). The anatomy of a typical C_3 dicot is characterized by a distinct dorsiventrality of the internal leaf structure as a result of well-defined palisade and spongy mesophyll cell layers, both of which contain numerous chloroplasts. In marked contrast is the wreathlike leaf anatomy of a C_4 grass such as maize in which the vascular tissue is surrounded by a concentric layer of large bundle sheath cells containing numerous mitochondria, peroxisomes, and starch-filled chloroplasts; this layer is in turn surrounded by one or more rings of mesophyll cells, which also contain numerous chloroplasts (Black and Mollenhauer, 1971). In general, the mesophyll and bundle sheath cells are situated such that each mesophyll cell is separated, at most, by one cell from the nearest bundle sheath cell (Hattersley and Watson, 1975). In addition, numerous plasmodesmata traverse the mesophyll-bundle sheath interfacial cell wall, providing a direct means for symplastic transport between the two photosynthetic cell types (Osmond and Smith, 1976). This specialized type of leaf anatomy is found in all naturally occurring C_4 plants, regardless of their taxonomic and phylogenetic relationships, although the detailed structure may show considerable variation among different C_4 species. For example, two radial layers of functional mesophyll cells have been noted in the tribe Fimbristylideae of the Cyperaceae (Carolin *et al.*, 1977). The outer layer of mesophyll cells is considerably larger than the inner layer; the mesophyll cells are in turn separated from an inner Kranz sheath by a nonchlorophyllous layer of cells. Also noted were Kranz anatomy dicots with palisade and spongy mesophyll layers (Rathnam *et al.*, 1976; Carolin *et al.*, 1978). However, in all cases of documented C_4 photosynthesis two distinct photosynthetic cell types are present.

B. Biochemistry

The biochemical mechanism for net CO_2 fixation in leaves of C_4 plants involves two complete, interdependent pathways of photosynthetic carbon metabolism, the C_4 cycle and the conventional C_3 cycle. The C_4 pathway, operating as an appendage of the C_3 cycle, serves as a mechanism for initially fixing atmospheric CO_2 and ultimately concentrating it at the site of RuBP carboxylase. Figure 2 represents a much

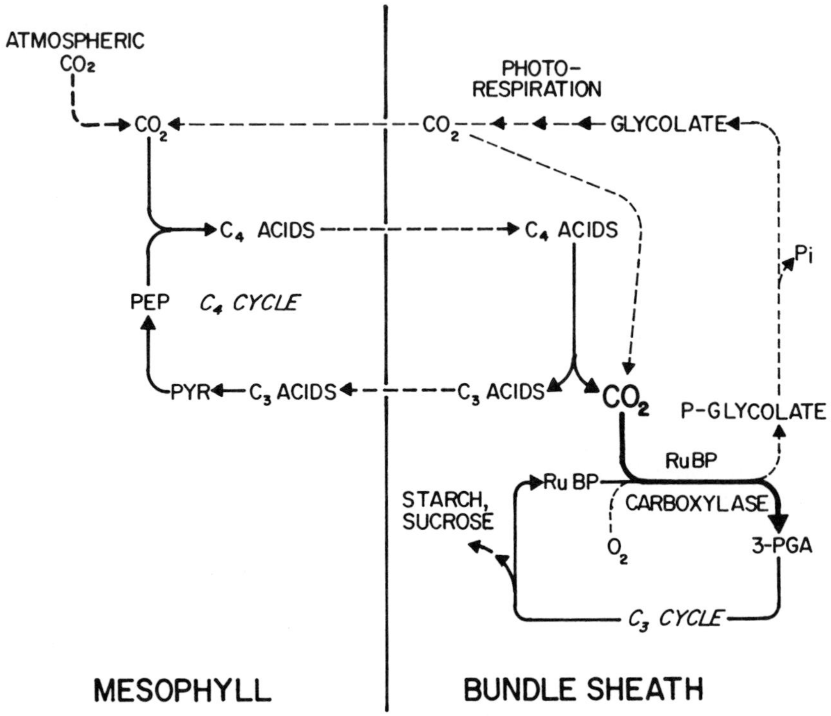

FIG. 2. A simplified scheme of the sequence and intercellular compartmentation of the basic reactions of C_4 photosynthesis and the suggested mechanism for C_4 control of photorespiration. Abbreviation: PYR, pyruvate. Adapted from Chollet (1976), by permission of University Park Press. (For another version of this cycle, see Fig. 2 in Chapter 6, this volume.)

simplified outline of the sequence and intercellular compartmentation of the basic reactions of C_4 photosynthesis. For further details and an account of the experimental evidence supporting this scheme, see Chapter 6 in this Volume and elsewhere (Hatch and Osmond, 1976; Hatch, 1977; Edwards and Huber, 1979; Ray and Black, 1979; Rathnam and Chollet, 1980b). In summary, atmospheric CO_2 is initially fixed in the mesophyll by PEP carboxylase, with the resulting OAA being predominantly reduced to malate or aminated to aspartate via NADP–malate dehydrogenase and aspartate aminotransferase, respectively. These latter C_4 acids occur in large pools compared with OAA and are transferred to the bundle sheath cells probably by simple diffusion through plasmodesmata and the cell symplasm (Hatch and Osmond, 1976; Osmond and Smith, 1976). In the bundle sheath cells, the C_4 acids are enzymically decarboxylated at the C_4 carboxyl position by the combined action of NADP– and NAD–malic enzymes, PEP carboxykinase and NAD–malic enzyme, or NAD–malic enzyme (Rathnam and Chollet, 1980b), and the resulting CO_2 is refixed and metabolized via RuBP carboxylase and the C_3 cycle, which are exclusively localized in this cell type (Hattersley *et al.*, 1977). The precise metabolic steps involved in C_4 acid decarboxylation in the bundle sheath vary between different C_4 species, leading to the classification of C_4 plants into three subgroups based on the major C_4 acid decarboxylase present—NADP–malic enzyme type, PEP carboxykinase type and NAD–malic enzyme type (Gutierrez *et al.*, 1974; Hatch *et al.*, 1975). The C_3 acid remaining after C_4 acid decarboxylation in the bundle sheath (ultimately pyruvate or alanine) is transported back to the mesophyll cell where it serves as a precursor of the initial carboxylation substrate PEP. This regenerative phase of the C_4 cycle is catalyzed by the mesophyll chloroplast enzyme pyruvate, P_i dikinase, the most diagnostic marker enzyme for C_4 photosynthesis in nonsucculent higher plants.

Regardless of the specifics of the decarboxylation mechanism in the bundle sheath, it is evident that the C_4 pathway of photosynthesis is characterized by having two sequential, but spatially compartmented, carboxylations. The first, utilizing atmospheric CO_2, is catalyzed by PEP carboxylase in the mesophyll cytosol. The second, utilizing (probably exclusively) CO_2 generated internally by the various C_4 acid decarboxylation reactions, is catalyzed by RuBP carboxylase in the bundle sheath chloroplasts. The PEP carboxylase reaction in the mesophyll functions as an efficient trap for fixation of CO_2 from the atmosphere. This fixed carbon is transferred in the form of C_4 acids to the bundle sheath where it is enzymically released as CO_2. Thus the C_4 cycle serves as a biochemical CO_2 pump, increasing the CO_2 concentration at the site of RuBP carboxylase in the bundle sheath above that in free equilibrium with the leaf. From intact leaf CO_2-exchange studies (Ehleringer and Björkman, 1977) and measurements of the size of the intermediate $^{14}CO_2$ pool developed during steady state C_4 photosynthesis in $^{14}CO_2$ (Hatch, 1971), it has been estimated that the concentration of free CO_2 developed in the bundle sheath cells under normal atmospheric concentrations of CO_2 and O_2 is at least seven times that in the atmosphere surrounding the leaf ($\sim 8\ \mu M$). This elevated level of CO_2 ($> 60\ \mu M$) is in marked contrast to the situation in C_3 species in which the CO_2 concentration at the site of RuBP carboxylase ($\sim 6\ \mu M$) is in more direct equilibrium with the external atmosphere. Thus the combination of anatomical and biochemical features of C_4 species offers advantages at the cellular level which, in turn, explains the distinctive physiological characteristics of C_4 plants discussed in Sections III,C and D. Due to the high carboxylation potential of PEP carboxylase, rapid rates of CO_2 fixation from the external atmosphere can be maintained in spite of the commonly prevailing situation in C_4 leaves of low stomatal conductances to CO_2 diffusion and thus low steady state levels of CO_2 in the mesophyll ($\sim 1\ \mu M$) (Hatch and Osmond, 1976; Rathnam and Chollet, 1980b). In addition, the subsequent steps of the C_4 cycle provide sufficient CO_2 in the bundle sheath to maintain rates of net CO_2 assimilation by the C_3 cycle commensurate with the initial rates of CO_2 fixation.

C. Control of Photorespiration

In addition to the preceding anatomical and biochemical features, C_4 species can be distinguished from C_3 plants on the basis of several photosynthetic CO_2 exchange responses (Chollet and Ogren, 1975; Canvin, 1979). Most notable is the observation that levels of O_2 of up to 60% have little or no inhibitory effect on the light-limited or light-saturated rate of photosynthesis by C_4 plants (Forrester *et al.*, 1966; Bull, 1969; Osmond and Björkman, 1972; D'Aoust and Canvin, 1973; Brown and Brown, 1975; Brown, 1980; Morgan *et al.*, 1980). In addition, the leaves of C_4 species do not evolve CO_2 in the light and also have an O_2-insensitive Γ of less than 10 μl/liter; therefore, they appear to lack photorespiration and the associated O_2 inhibition of photosynthesis characteristic of C_3 species (Canvin, 1979). Similarly, the results from long-term growth studies indicate that dry matter accumulation in C_4 plants is essentially unaffected by either CO_2 fertilization or O_2 levels of up to 40% (Quebedeaux and Chollet, 1977; Imai and Murata, 1979). However, studies with RuBP carboxylase/oxygenase extracted from C_4 leaf tissue indicate that the kinetic properties of the CO_2/Mg^{2+}-activated enzyme do not differ significantly from its C_3 counterpart (Bahr and Jensen, 1974; Matsumoto *et al.*, 1977; Jordan and Ogren, 1981b; but see Yeoh *et al.*, 1980), which is consistent with the observation that photosynthetic CO_2 fixation by isolated C_4 bundle sheath strands is competitively inhibited by O_2 with respect to CO_2 (Chollet, 1976; Rathnam, 1978). Additionally, it has been demonstrated that C_4 leaf tissue can metabolize exogenous glycolate or glycine to CO_2 via an O_2-sensitive route (Kisaki *et al.*, 1972; Rathnam, 1979), an observation consistent with the presence of peroxisomes, mitochondria, and photorespiratory enzyme activity in the tissue. More detailed analyses at the cellular–subcellular level indicate that, although isolated C_4 mesophyll protoplasts and mitochondria do not oxidize glycolate or glycine to CO_2 (Neuburger and Douce, 1977; Woo and Osmond, 1977; Rathnam,

1979), isolated C_4 bundle sheath strands photosynthesize glycolate and readily metabolize exogenous glycolate and glycine to CO_2 in an O_2-sensitive manner (Chollet, 1974, 1976; Rathnam, 1978, 1979). Glycine oxidation in isolated C_4 bundle sheath mitochondria, as in C_3 species (Section II, A,2), is linked to the electron transport chain and is coupled to three phosphorylation sites (Woo and Osmond, 1977; Rathnam, 1979). Although the frequency of peroxisomes and the activities of photorespiratory enzymes (all of which are primarily localized in the bundle sheath; Frederick and Newcomb, 1971; Osmond and Harris, 1971; Ku and Edwards, 1975) are generally lower than those observed in C_3 plants, there is little doubt that C_4 species have the potential for photosynthesizing and metabolizing P-glycolate to CO_2 in the bundle sheath via the photorespiratory cycle.

The apparent lack of photorespiratory CO_2 release and absence of an inhibitory effect of 21% O_2 on net photosynthesis in C_4 plants have been attributed to a variety of biochemical mechanisms related to the C_4 pathway. Based on previous metabolite feeding experiments with C_3 leaf disks (Oliver and Zelitch, 1977) and more recent studies with isolated maize bundle sheath strands (Oliver, 1978), Oliver has suggested that glycolate synthesis and thus photorespiration in C_4 plants may be decreased by the presence of inhibitory endogenous concentrations of aspartate and glutamate in the bundle sheath. However, since the extent of involvement of these amino acids in bundle sheath photosynthetic carbon metabolism varies depending on the specific C_4 subgroup (Hatch and Osmond, 1976; Rathnam and Chollet, 1980b), whereas photorespiration is undetectable in all three subgroups, we consider this explanation to be unlikely. The absence of detectable photorespiration in intact leaves of C_4 plants has most frequently been attributed to an efficient CO_2 refixation mechanism coupled to the Kranz-type leaf anatomy. That is, photorespiratory CO_2 evolved in the bundle sheath in photorespiration is refixed by PEP carboxylase in the surrounding mesophyll before it escapes from the leaf (Black, 1973; Kestler *et al.*, 1975). However, there are several lines of experimental evidence that argue strongly against refixation as the primary mechanism by which C_4 plants reduce photorespiration. First, the internal recycling of photorespiratory CO_2 would be associated with additional energy inputs for the regeneration of PEP, the initial carboxylation substrate. Consequently, when photosynthesis in a C_4 plant is light-limited, generation and refixation of photorespiratory CO_2 in response to increasing O_2 concentration would be expected to reduce the energy available for CO_2 uptake from the ambient atmosphere and thereby inhibit the rate of net photosynthesis. The failure to detect an inhibitory effect of atmospheric levels of O_2 on the light-limited rate of photosynthesis in C_4 plants (Bull, 1969; Osmond and Björkman, 1972; Brown, 1980) suggests that refixation of photorespiratory CO_2, if it occurs, is negligible in relation to total CO_2 uptake. Similarly, Ehleringer and Björkman (1977) reported that in contrast to C_3 photosynthesis, the quantum yield of C_4 photosynthesis (i.e., CO_2 fixed per absorbed quantum) is insensitive to changes in temperature and the concentrations of O_2 and CO_2, environmental factors known to perturb photorespiration in C_3 plants. These comparative measurements of the quantum yields of C_3 and C_4 photosynthesis demonstrate that photorespiration and concurrent refixation of the released CO_2 do not occur in C_4 plants. Consistent with these observations are the findings from studies dealing with the kinetics of photosynthetic $^{14}CO_2$ uptake under steady state conditions, combined with analysis of ^{14}C flow through key photosynthetic and photorespiratory intermediates (Mahon *et al.*, 1974; Blackwood and Miflin, 1976; Lawlor anf Fock, 1978; Canvin, 1979; Morot-Gaudry *et al.*, 1980). The results from these intact leaf experiments indicate that in 21% O_2 the flux of carbon through glycine and serine in the photorespiratory cycle is markedly reduced in C_4 plants compared to that in C_3 species. This observation suggests that internal photorespiratory CO_2 released during C_4 photosynthesis is small compared to that released in C_3 photosynthesis since the glycine to serine conversion is considered to be the major source of CO_2 evolution during photorespiration (Section II,A,2). Additional whole-leaf radiotracer evidence, which argues against substantial internal photorespiratory CO_2 recycling, is derived from the intramolecular labeling pattern of serine. The observation that 50–80% of the radiocarbon in serine following short-term photosynthesis in $^{14}CO_2$–air is located in the carboxyl position (Johnson and Hatch, 1969; Osmond, 1972; Morot-Gaudry *et al.*, 1980) indicates that a substantial portion of the serine pool in C_4 plants is derived from 3-PGA via the glycerate pathway rather than from glycolate via the photorespiratory cycle.

An alternative explanation for the apparent lack of photorespiration and O_2 inhibition of C_4 photosynthesis involves the C_4 cycle-mediated CO_2 pump that concentrates CO_2 in the bundle sheath at the site of RuBP carboxylase/oxygenase (Section III,B) (Bowes and Ogren, 1972; Chollet and Ogren, 1975; Chollet, 1976). As discussed previously, O_2 inhibits RuBP carboxylase activity and can substitute for CO_2 in the reaction to yield 3-PGA and P-glycolate. Furthermore, the O_2 effects on RuBP carboxylase are competitively inhibited by CO_2. At atmospheric levels of O_2, photosynthetic CO_2 fixation by the C_4 cycle in the mesophyll cell layer proceeds unhindered (Huber and Edwards, 1975; Chollet, 1976; Rathnam, 1978), leading to an increased CO_2 concentration in the bundle sheath. Since the intracellular O_2 concentration during active photosynthesis presumably remains unchanged (Steiger *et al.*, 1977), the elevated CO_2/O_2 ratio in the bundle sheath would allow CO_2 to compete more effectively with O_2 for RuBP carboxylase during bundle sheath photosynthesis, thus reducing O_2 inhibition of net CO_2 uptake in leaves of C_4 plants. Similarly, an increased level of CO_2 in the bundle sheath would reduce the RuBP oxygenase-catalyzed formation of P-glycolate from RuBP, thereby decreasing the amount of glycolate available for photorespiratory oxidation to CO_2. If any photorespiratory activity does occur, the CO_2 released by this process would be refixed by PEP carboxylase and/or RuBP carboxylase before it could exit from the leaf (Fig. 2).

Supportive evidence for this proposal was provided by a variety of studies, including experiments with C_4 leaf slices (Rathnam, 1978; Rathnam and Chollet, 1979a). Photosynthesis by control leaf slices having a functional C_4 cycle was, as expected, not inhibited by 21% O_2 at low CO_2. However, $^{14}CO_2$ fixation by C_4 leaf slices pretreated with enzyme inhibitors to prevent the operation of the C_4 cycle (e.g., inhibitors of PEP carboxylase or the C_4 acid decarboxylases) is markedly inhibited by 21% O_2 at low CO_2 concentrations. The appearance of ^{14}C in photorespiratory intermediates under these conditions documents the potential for photorespiration in C_4 leaf tissue. O_2 inhibition and the associated labeling of photorespiratory intermediates are, however, overcome by increased CO_2. Since the only difference between the control and inhibitor-treated leaf slices is presumably the presence of a functional C_4 cycle in the former, it appears that the C_4 pathway regulates photorespiration in leaves of C_4 plants by increasing the CO_2 concentration in the bundle sheath. In related studies using the photorespiratory inhibitors α-HPMS (α-hydroxypyridinemethane sulfonate) and BHB, Zelitch (1973) and Servaites *et al.* (1978) reported a reduced rate of glycolate formation relative to photosynthesis in C_4 plants compared to C_3 species. However, these results must be interpreted with caution since the inhibitors added to the leaf tissue to block the subsequent metabolism of glycolate also markedly inhibit photosynthesis, especially in C_4 plants (Osmond and Avadhani, 1970; Servaites *et al.*, 1978).

Evidence, which tends to refute a reduced rate of glycolate formation during C_4 photosynthesis, has been presented by Dimon *et al.* (1977) from the results of an *in vivo* $^{18}O_2$-incorporation study with maize leaf tissue. In 30% O_2 and CO_2-free conditions the half-time for ^{18}O-labeling of the carboxyl groups of glycine and serine was 1–2 min and 3–4 min, respectively. These observations would be consistent with a rapid synthesis of these amino acids from glycolate produced by the RuBP oxygenase reaction and thus would indicate rapid photorespiration during C_4 photosynthesis. Confirmation of these data at physiological levels of CO_2 and O_2 is essential as the incorporation of ^{18}O into glycine and serine and rapid flux through the photorespiratory cycle are inconsistent with the flux experiments described earlier using $^{14}CO_2$ (Canvin, 1979). Along these lines, Canvin *et al.* (1980) reported a light-stimulated uptake of $^{18}O_2$ by intact leaves of a C_4 plant *Amaranthus edulis* at physiological levels of CO_2 and O_2. However, the rates are lower and the uptake much less sensitive to CO_2 than with C_3 plants, suggesting that perhaps a Mehler-type reaction is primarily responsible for O_2 uptake in C_4 species.

From the data discussed earlier, it is evident that leaves of C_4 plants have the potential for glycolate synthesis and metabolism to CO_2 in the bundle sheath by photorespiration, and for refixing the photorespired CO_2 by PEP carboxylase in the surrounding mesophyll. Nevertheless, we conclude that the lack of photorespiration and the associated O_2 inhibition of photosynthesis in C_4 plants is mainly due to the increased CO_2 concentration at the site of RuBP carboxylase/oxygenase resulting from the CO_2-concentrating mechanism of the C_4 cycle (Fig. 2). The refixation of photorespired CO_2 by PEP carboxylase is probably a contributing factor, but is not considered a major component *in vivo*.

D. Related CO_2-Exchange Features

Although C_4 photosynthesis is relatively insensitive to changes in oxygen concentration from 2–60% O_2, CO_2 fixation is inhibited at levels of O_2 approaching 100% (Forrester *et al.*, 1966; Poskuta, 1969; Lewanty *et al.*, 1971; Gale and Tako, 1976; Ku and Edwards, 1980). However, this inhibitory effect of O_2 on C_4 photosynthesis appears distinct from that in C_3 plants in that it is neither readily reversible (Forrester *et al.*, 1966; Gale and Tako, 1976; Ku and Edwards, 1980) nor relieved by increased levels of CO_2 (Poskuta, 1969). Several possible sites of interaction of elevated O_2 with photosynthesis in C_4 plants have been suggested, including a decrease in stomatal conductance to gaseous diffusion (Gale and Tako, 1976; but see Gauhl and Björkman, 1969) and an inhibition of PEP regeneration (Chollet, 1976; but see Huber and Edwards, 1975) or a C_4 acid decarboxylation and refixation of the released CO_2 by RuBP carboxylase (Lewanty *et al.*, 1971; Ku and Edwards, 1980). Oxygen has also been shown to exert an effect on the distribution of ^{14}C between the C_4 acids during photosynthesis in $^{14}CO_2$. In representatives of all three C_4 subgroups, low O_2 concentration favors aspartate labeling, whereas malate labeling predominates at high levels of O_2 (Hickman and Keys, 1972; Foster and Black, 1977; Höhler and Schaub, 1979). Furthermore,

Creach (1979) reported that increasing O_2 also alters the intramolecular labeling pattern of malate and aspartate during light-enhanced dark $^{14}CO_2$ uptake by maize leaf segments. The nature of these diverse O_2 effects on the perturbation of C_4 acid labeling is unknown.

A near-zero CO_2 compensation point in air and the absence of a detectable release of CO_2 in the light are usually regarded as diagnostic gas exchange features of C_4 plants (Chollet and Ogren, 1975; Canvin, 1979), but several developmental studies suggest that these characteristics may be altered during leaf ontogeny. Kennedy and co-workers (Kennedy, 1976; Williams and Kennedy, 1977) reported that senescent C_4 leaf tissue has relative rates of photorespiration approaching those in C_3 plants based on Γ and the light–dark $^{14}CO_2$ efflux assay of photorespiration. However, the validity of the photorespiratory values obtained by this latter technique is questionable unless the results are confirmed with independent measures of photorespiratory CO_2 exchange such as an O_2-sensitive CO_2 compensation point, the percentage of O_2 inhibition of photosynthesis, or the rate of total CO_2 evolution (μmol CO_2 evolved) in the light (Chollet and Ogren, 1975; Chollet, 1978). Along these lines, Williams and Kennedy (1977) reported that senescence in maize does not result in a typical C_3 oxygen sensitivity of net photosynthesis or Γ which would be predicted for an actively photorespiring leaf. Therefore, it appears that these and similar reports (Crespo *et al.*, 1979), which claim increased photorespiration with increasing leaf age in C_4 plants, are experimentally deficient. It is likely that the high CO_2-compensation points observed with such tissue at 21% O_2 (from 22–43 μl/liter) are due to a combination of dark respiration and a senescent photosynthetic apparatus rather than photorespiration per se.

In addition to the CO_2 exchange characteristics described earlier and in Section III, C, C_4 plants can be distinguished from C_3 species on the basis of their respective photosynthetic gas exchange responses to temperature and CO_2 at atmospheric levels of O_2. At high light intensity and low temperatures the rate of photosynthesis is essentially the same in C_3 and C_4 species, but as temperature is increased C_4 plants become increasingly superior (Ludlow and Wilson, 1971; Björkman, 1973; Bird *et al.*, 1977; Ray and Black, 1979). At 30°–35°C, the rate of photosynthesis in C_4 species is nearly twice that in C_3 plants. It is likely that the ability of C_4 plants to fix CO_2 rapidly at high temperatures is related to suppression of photorespiration by the CO_2-concentrating mechanism of the C_4 cycle since the limitations imposed by photorespiration and the associated O_2 inhibition of photosynthesis become progressively more pronounced as temperature increases (see Section II). Suppression of RuBP oxygenation by high levels of CO_2 in the bundle sheath clearly enables C_4 plants to exploit the higher temperature range where photorespiration would otherwise seriously reduce the efficiency of net CO_2 fixation catalyzed by RuBP carboxylase. Comparative measurements of the temperature dependence of the quantum yields of C_3 and C_4 photosynthesis highlight this phenomenon (Ehleringer and Björkman, 1977). At temperatures below 30°C, the observed quantum yield of C_3 photosynthesis at 21% O_2 is superior to that of the C_4 pathway, presumably due to the higher intrinsic energy requirement of C_4 photosynthesis (Hatch and Osmond, 1976; Hatch, 1977). However, above 30°C the positions are reversed, presumably due to the increasing energy demands of photorespiration during C_3 photosynthesis. It is likely that the lower quantum yield in C_4 plants at low and moderate temperatures is an important factor in limiting the geographical distribution of C_4 species in nature (Ehleringer, 1978; Tieszen *et al.*, 1979).

As to the effect of CO_2 concentration on net photosynthesis, C_4 species are considerably more efficient in utilizing low levels of CO_2 than are C_3 plants. Also, the CO_2 concentration required for saturation of photosynthesis is much lower in C_4 plants, although the rate of net photosynthesis at CO_2 saturation is essentially the same in C_3 and C_4 species (Björkman, 1973; De Jong, 1978; Long and Woolhouse, 1978; Brown, 1980). Most of the differences in photosynthetic CO_2 exchange characteristics between C_3 and C_4 plants discussed earlier become less evident when the O_2 level is reduced to 1–2% (Björkman, 1973). So, the differences observed at atmospheric levels of O_2 may be largely attributed to the O_2 effects on C_3 photosynthesis at the level of RuBP carboxylase/oxygenase. This concept is consistent with that discussed earlier (Section III,B and C): The C_4 pathway of photosynthesis serves as a metabolic CO_2-concentrating mechanism, suppressing the O_2 effects on RuBP carboxylase/oxygenase in the bundle sheath and increasing the rate of net CO_2 fixation catalyzed by this bifunctional enzyme.

IV. Photorespiration in C_3–C_4 Intermediate Plants

From the foregoing discussion, it is evident that the leaves of species with C_4 photosynthesis possess a full syndrome of anatomical, physiological, and biochemical characteristics that are clearly distinguished from those present in C_3 plants. All available evidence indicates that C_4 plants have evolved from C_3 species and that this has occurred independently many times during evolution (Björkman, 1976). One would thus predict that there might exist naturally occurring intermediate species that provide a link between these two distinct photosynthetic groups.

Until recently, though, no intermediate species had been detected and C_4 plants appeared to have arisen without leaving any trace of their evolutionary development. Attempts have been made to create hybrids between C_3 and C_4 plants using conventional breeding techniques in the genera *Panicum, Euphorbia, Zygophyllum,* and *Atriplex.* To date, such hybridizations have met with success only with *Atriplex* species (Björkman, 1976). The successful hybridization between C_3 and C_4 species of *Atriplex* indicates that the genetic diversity between C_3 and C_4 plants is not necessarily great. Indeed, analyses of segregating populations of the hybrids suggest that only a small number of genes determine each of the major components of the C_4 syndrome, e.g., Kranz-type leaf anatomy. From these studies, several F_2 and F_3 generation hybrids have been shown to possess several anatomical and biochemical characteristics intermediate between the C_4 and C_3 parents, but these features were either not properly compartmented or causally coordinated to lead to a reduction in photorespiration and the associated O_2 inhibition of photosynthesis (Björkman, 1976). This requirement for a complete coordination of the anatomical and biochemical properties of the leaf will make it exceedingly difficult, if not impossible, to introduce the C_4 pathway per se into C_3 plants via genetic manipulation.

Attempts to find naturally occurring intermediate species have met with some success. Three species of the *Laxa* group of the genus *Panicum* (*P. milioides, P. decipiens,* and *P. schenckii*) have been positively identified as being intermediate between C_3 and C_4 plants with respect to leaf anatomy and photorespiration estimated by the O_2 sensitivity of net photosynthesis and Γ (Brown and Brown, 1975; Kanai and Kashiwagi, 1975; Brown, 1976, 1980; Keck and Ogren, 1976; Quebedeaux and Chollet, 1977; Morgan and Brown, 1979; Morgan *et al.*, 1980). Of these, only *P. milioides* has been extensively characterized with respect to photosynthetic carbon metabolism. Based on comparative studies with intact leaves, thin leaf slices and isolated mesophyll and bundle sheath cell types, Rathnam and Chollet (1978, 1979a,b) proposed that CO_2 fixation in *P. milioides* is mediated via two photosynthetic pathways, a limited, but functional, C_4 pathway and the conventional C_3 cycle. The mechanism of CO_2 fixation by the limited C_4 cycle is basically similar to that in NAD–malic enzyme-type C_4 plants (Hatch and Osmond, 1976; Rathnam and Chollet, 1980b) with respect to the sequence of reactions leading to the synthesis and subsequent decarboxylation of malate and aspartate, the refixation of the released CO_2, and the regeneration of PEP, the initial CO_2 acceptor. However, it has yet to be determined whether the limited C_4-like PEP carboxylation–regeneration system is present in *all* mesophyll cells or only in the radially arranged cells immediately adjacent to the chloroplast-containing bundle sheath (Rathnam and Chollet, 1979b). Although RuBP carboxylase and a functional C_3 cycle are present in both cell types (Hattersley *et al.*, 1977; Rathnam and Chollet, 1978, 1979b), the likely role of the limited C_4 cycle is to concentrate CO_2 at the site of bundle sheath, but not mesophyll, RuBP carboxylase. This interpretation is consistent with several lines of experimental evidence, which indicate that NAD–malic enzyme, the only C_4 acid decarboxylase detected in this and the two other intermediate *Panicum* species (Rathnam and Chollet, 1979a; C. K. M. Rathnam and R. Chollet, unpublished results), is exclusively localized in the bundle sheath (Rathnam and Chollet, 1978, 1979b). It thus appears that reduced photorespiration in this C_3–C_4 intermediate species is due to a limited degree of NAD–malic enzyme-type C_4 photosynthesis permitting an increase in CO_2 concentration at the site of bundle sheath, but not mesophyll, RuBP carboxylase/oxygenase. The reduced rate of glycolate formation relative to photosynthesis reported for *P. milioides* (Servaites *et al.*, 1978) is consistent with this proposal. Similarly, preliminary experiments with thin leaf slices of *P. decipiens* and *P. schenckii* suggest that a limited C_4-like CO_2 pump (mediated by a PEP carboxylase/NAD–malic enzyme reaction sequence similar to that in *P. milioides*) is also responsible for the reduced photorespiration and O_2 sensitivity of net photosynthesis in these related intermediate species (C. K. M. Rathnam and R. Chollet, unpublished results).

Earlier attempts by other laboratories to demonstrate a limited potential for C_4 photosynthesis in *P. milioides* met with little success. Leaf extracts of this C_3–C_4 intermediate species were reported to have, at most, only slightly higher activities of the three C_4 acid decarboxylases, including NAD–malic enzyme, in comparison to C_3 plants (Ku *et al.*, 1976; Morgan *et al.*, 1980), although the activity of pyruvate, P_i dikinase was not determined. In addition, the initial studies with mesophyll protoplasts and bundle sheath strands isolated by standard enzymic procedures developed for C_4 leaf tissue did not show any clear compartmentation of PEP carboxylase, RuBP carboxylase, or C_4 acid decarboxylase activity between the two cell types (Ku *et al.*, 1976). These findings, together with the results from $^{14}CO_2$-pulse and pulse-chase labeling experiments with intact leaves (Kanai and Kashiwagi, 1975; Kestler *et al.*, 1975), prompted the conclusion that there was no C_4 photosynthesis in *P. milioides* despite the presence of C_4-like leaf anatomy. However, reevaluation of these earlier studies suggests some possible deficiencies in experimental protocol. For example, with respect to the initial intercellular compartmentation studies (Ku *et al.*, 1976), it has been demonstrated that the conventional "Onozuka" cellulase–pectinase enzyme digestion system and sequential filtration protocol that have been successfully employed for isolating pure leaf cell types from C_4 plants yield cross-contaminated mesophyll and bundle sheath preparations from *P. milioides* (Rathnam and Chollet, 1979b; J. C. Servaites, personal communication). Furthermore, the procedure used previously to assay for NAD–malic enzyme activity (Morgan *et al.*, 1980) would probably not have activated this Mn^{2+}-dependent enzyme. Similarly, it is likely that the earlier $^{14}CO_2$-pulse and pulse-chase photosynthesis experiments (Kanai and Kashiwagi, 1975; Kestler *et al.*, 1975) failed to detect significant labeling of malate and aspartate in *P. milioides* due to the high levels of total CO_2 employed in these studies (425–710 μl/liter) (Rathnam and Chollet, 1979b).

Although *P. milioides* is the only C_3–C_4 intermediate species that has been characterized in some detail with respect to its photosynthetic carbon metabolism, several other reports have appeared describing additional higher plants with one or more intermediate characteristics. Kennedy and Laetsch (1974) reported that *Mollugo verticillata* is a C_3–C_4 intermediate species based on the initial products of photosynthesis in $^{14}CO_2$, leaf anatomy, and the relative rate of photorespiration determined by the light–dark $^{14}CO_2$ efflux assay. Four ecotypes of *M. verticillata* have now been described with differences in the percentage of label in C_4 acids, Γ, and the percentage of O_2 inhibition of photosynthesis (Sayre and Kennedy, 1977). Although all four ecotypes, namely, Kansas, Iowa, Mexico, and Massachusetts, exhibit reduced O_2 inhibition of photosynthesis (11–17% inhibition by 21% O_2), only the Kansas ecotype has an intermediate CO_2 compensation point of 25 μl/liter, the other three being similar to C_3 species. Since reduced O_2 sensitivity of photosynthesis would be expected to be associated with a lower Γ (as in the three intermediate C_3–C_4 *Panicum* species), the lack of such a correlation in the Iowa, Mexico, and Massachusetts ecotypes is puzzling. Moreover, the same researchers have reported differences in C_4 cycle enzyme activity (including PEP carboxylase, NAD–malic enzyme and aspartate and alanine aminotransferases) between the four ecotypes (Sayre *et al.*, 1979) and suggested that the observed variation in enzyme activity is well correlated with the previously reported ecotypic differences in $^{14}CO_2$-labeling patterns and photosynthetic/photorespiratory CO_2 exchange (Sayre and Kennedy, 1977). However, the activity of PEP carboxylase in the four populations is similar to that in C_3 plants and either varied little between the ecotypes (on a chlorophyll basis) or was 55% greater in the Kansas population compared to the other three (on a fresh weight basis). In contrast, the previous $^{14}CO_2$-labeling studies indicated that the percentage of the total ^{14}C fixed entering the C_4

acid pool during 3-sec pulse photosynthesis was two to three times greater in the Kansas *and* Iowa ecotypes compared to the Mexico and Massachusetts populations. From a related study, Raghavendra *et al.* (1978) have reported the simultaneous occurrence of C_3 and C_4 photosynthesis in *M. nudicaulis* with the older leaves being C_4, the young leaves C_3, and the medium-aged leaves intermediate between C_3 and C_4. The transition from the C_3 cycle in young leaves to the C_4 pathway in old leaves was manifest in at least five features—leaf anatomy, initial photosynthetic products, rates of photosynthesis, rates of photorespiration (based on the light–dark ^{14}C-assay), and the degree of starch-staining in the bundle sheath. However, preliminary studies by C. K. M. Rathnam and R. Chollet (unpublished results) have failed to confirm this reported variation of leaf anatomy in *M. nudicaulis* in that all the leaves exhibit typical C_4 Kranz anatomy, irrespective of their position along the stem. In addition, the validity of the relative photorespiratory values obtained with the light–dark ^{14}C assay employed in these *Mollugo* studies is questionable unless the results are confirmed with independent measures of photorespiratory CO_2 exchange (Chollet, 1978). Similarly, incorporation of label into C_4 acids during short-term photosynthesis in $^{14}CO_2$ is not a valid indication of C_4 photosynthesis per se unless evidence is also provided for the transfer of radiocarbon from the C_4 acids to C_3 cycle intermediates during a chase in $^{12}CO_2$–air (Ray and Black, 1979; Rathnam and Chollet, 1980b).

Several reports have appeared describing reduced photorespiration in *Moricandia arvensis,* a crucifer. Crookston (1972) and Krenzer *et al.* (1975) observed that *M. arvensis* grown from seeds obtained from the Royal Botanic Gardens at Kew exhibited intermediacy with respect to the relative rate of photorespiration (estimated by Γ and the rate of CO_2 efflux into CO_2–free air in the light) compared to representative C_3 and C_4 plants, including several C_3 species of the Cruciferae family. This plant material, though, is qualitatively C_3-like with respect to the initial products of photosynthesis and stomatal and mesophyll conductances to gaseous diffusion (Crookston, 1972). From related studies, it was reported (Apel *et al.*, 1978; Apel and Ohle, 1979) that *M. arvensis* grown from seeds obtained from the Botanical Garden at the University of Leipzig exhibited C_3–C_4 intermediacy with respect to the CO_2 compensation point and the O_2 sensitivity of Γ. At 21% O_2, Γ was about 24 μl/liter compared to 50 μl/liter in a related C_3 species, *M. foetida.* Furthermore, in *M. arvensis* a plot of Γ versus O_2 concentration was curvilinear, with a break at about 15% O_2. At less than 15% O_2, Γ responded little to increasing O_2, and at O_2 concentrations between 15–50% O_2, Γ was proportional to O_2, but the slope was considerably less than that observed with *M. foetida*. These results are strikingly similar to related CO_2 exchange studies by several groups with the intermediate C_3–C_4 *Panicum* species (Keck and Ogren, 1976; Quebedeaux and Chollet, 1977; Morgan *et al.*, 1980) and indicate that photosynthesis in *M. arvensis* is less sensitive to O_2 than is photosynthesis in typical C_3 plants and that photorespiration, relative to photosynthesis, is reduced. In addition to these physiological similarities between the Leipzig cultivar and the intermediate *Panicum* species, Apel and Ohle (1979) and Holaday *et al.* (1981) have also observed a sheath of chloroplast-containing cells around the leaf vascular bundles of *M. arvensis*. This C_4-like leaf anatomy is highlighted further at the ultrastructural level by the presence of numerous granal, starch-containing chloroplasts and prominent mitochondria in the "bundle sheath" (Holaday *et al.*, 1981). At the biochemical level, studies from P. Apel's laboratory (Bauwe and Apel, 1979) indicate that there is essentially no difference between *M. arvensis* and *M. foetida* with respect to the kinetic properties of partially purified, CO_2/Mg^{2+}-activated RuBP carboxylase, the leaf content of RuBP carboxylase protein, and the $^{13}C/^{12}C$ carbon isotope fractionation ratio. However, the PEP carboxylase activity in crude leaf extracts prepared from *M. arvensis* is two to three times that in representative C_3 species (Bauwe and Apel, 1979; Holaday *et al.*, 1981). In summary, reasonable evidence has been provided by several laboratories to indicate that both cultivars of *M. arvensis* are intermediate between C_3 and C_4 plants with respect to photorespiration, although the mechanism(s) by which photorespiration has been reduced remains to be elucidated.

V. Photorespiration in Other Plants and Bacteria

A. Algae

Green algae photosynthesize via the C_3 pathway (Bassham and Calvin, 1957; Hogetsu and Miyachi, 1979), yet when grown at air levels of CO_2, O_2 sensitivity of photosynthesis is much less than that observed in terrestrial C_3 plants (Lloyd *et al.*, 1977; Shelp and Canvin, 1980b). Additionally, the CO_2 compensation point in these species is close to zero (Lloyd *et al.*, 1977; Birmingham and Colman, 1979; Tsuzuki and Miyachi, 1979). When grown at elevated CO_2 concentrations (greater than 1% CO_2), O_2 inhibition of photosynthesis is similar to that observed in C_3 plants, and the algae synthesize large amounts of glycolate (Bowes and Berry, 1972). Therefore, algae are able to substantially reduce photorespiratory activity when cultured at a low CO_2 concentration.

Kinetic analysis of photosynthetic CO_2 uptake indicated that the $K_m(CO_2)$ for photosynthesis was much lower in cells grown at low CO_2 than in those grown at high CO_2 (Bowes and Berry, 1972; Berry *et al.*, 1976; Findenegg, 1976; Raven and Glidewell, 1978); thus, the low CO_2-grown cells have a greater affinity for CO_2. The $K_m(CO_2)$ for RuBP carboxylase was similar for both high and low CO_2-grown cells, so the difference cannot be explained by differences in this enzyme. Since the K_m (CO_2) for carboxylase is greater than the $K_m(CO_2)$ for photosynthesis in low CO_2-grown cells (Berry *et al.*, 1976; Shelp and Canvin, 1980a), the difference cannot be due to greater diffusion resistance in the high CO_2-grown cells but must be due to an increased CO_2 concentration in the low CO_2-grown cells. Analysis of the internal CO_2 pool indicates that low CO_2-grown cells are able to concentrate CO_2 inside the cells by an order of magnitude or more (Badger *et al.*, 1978, 1980). The increased CO_2 concentration stimulates RuBP carboxylase and inhibits RuBP oxygenase activity, so photosynthesis is increased and photorespiration is reduced in low CO_2-grown cells. The CO_2 pump is not present in high CO_2-grown cells (Badger *et al.*, 1978, 1980).

The mechanism of the CO_2-concentrating mechanism is not known, but HCO_3^-/OH^- exchange systems have been described for algae (Raven, 1970; Findenegg, 1979; Miyachi and Shiraiwa, 1979). The pumping mechanism is susceptible to inhibitors and uncouplers of photosynthetic electron transport so these two processes may be linked (Badger *et al.*, 1980). An additional, unresolved biochemical phenomenon associated with the ability to concentrate CO_2 is the activity of cellular carbonic anhydrase. This enzyme, which catalyzes the interconversion between CO_2 and bicarbonate, is present in much higher activity in low CO_2-grown cells (Findenegg, 1976; Ingle and Colman, 1976; Hogetsu and Miyachi, 1979). Activity increases as high CO_2-grown cells adapt to low CO_2 concentrations (Findenegg, 1976) and vice versa (Hogetsu and Miyachi, 1979), and these changes parallel photosynthetic activity. In the presence of carbonic anhydrase inhibitors, the affinity of low CO_2-grown cells for CO_2 is reduced to that observed in high CO_2-grown cells (Hogetsu and Miyachi, 1979; Badger *et al.*, 1980). These observations suggest that carbonic anhydrase is involved in concentrating CO_2, although the mechanism is not evident. It may participate in HCO_3^- uptake, which has been suggested as being necessary to maintain the high photosynthesis rates of low CO_2-grown algae at low CO_2 concentrations (Findenegg, 1976) or to speed the conversion of the HCO_3^- taken up to CO_2 (Hogetsu and Miyachi, 1979), the substrate of carboxylation (Cooper *et al.*, 1969).

B. Bacteria and Cyanobacteria

The mechanism of glycolate biosynthesis in a variety of cyanobacteria and photo- and chemolithotrophic bacteria appears identical to that in algae and higher plants. All microbial RuBP carboxylases examined to date also catalyze the oxygenase reaction (Akazawa, 1979) and the activity of this enzyme and P-glycolate phosphatase are sufficient to account for the *in vivo* rates of glycolate formation (Codd and Turnbull, 1975; Codd *et al.*, 1976, 1980; Lorimer *et al.*, 1978; Takabe *et al.*, 1979). Consistent with these observations are the numerous reports of O_2-dependent glycolate production (i.e., excretion) during CO_2 fixation by bacteria and cyanobacteria, which is decreased by increased levels of CO_2 (Han and Eley, 1973; Asami and Akazawa, 1974; Codd and Turnbull, 1975; Codd *et al.*, 1976; Ingle and Colman, 1976; Cohen *et al.*, 1979; King and Andersen, 1980). In more critical *in vivo* experiments using $^{18}O_2$, it has been demonstrated that the major pathway of glycolate synthesis by *Chromatium vinosum* and *Rhodospirillum rubrum* involves reaction(s) that bring about the incorporation of one atom of molecular oxygen into the carboxyl group of glycolate (Lorimer *et al.*, 1978; Takabe *et al.*, 1979), an observation indicative of the operation of the RuBP oxygenase reaction *in vivo*.

Following the synthesis of glycolate via the oxygenase reaction, the metabolic fate of this photorespiratory substrate markedly differs from that in higher plants. In a variety of cyanobacteria and chemo- and photolithotrophic bacteria glycolate is largely excreted into the surrounding medium. This excretion is generally enhanced in the presence of α-HPMS (Codd and Turnbull, 1975; Asami and Akazawa, 1976; Codd *et al.*, 1976; Ingle and Colman, 1976), suggesting that these microorganisms are also capable of further metabolizing this photorespiratory substrate via a glycolate-oxidizing enzyme, presumably glycolate dehydrogenase. Consistent with this notion are glycolate excretion by a glycolate dehydrogenase mutant of *Alcaligenes eutrophus* (King and Andersen, 1980) and the detailed investigations by T. Akazawa's group of glycolate metabolism in *Chromatium*. During photosynthetic $^{14}CO_2$ fixation by this purple sulfur bacterium in the presence of O_2, both glycolate and glycine become labeled and are largely excreted, whereas little ^{14}C is detected in serine (Asami and Akazawa, 1975, 1976). In addition, the metabolism of exogenous glycolate or glycine to CO_2 and glycine, or CO_2, respectively, is O_2 independent and involves little or no concomitant serine formation (Asami *et al.*, 1977; Sado *et al.*, 1980), suggesting that the route of glycolate and glycine metabolism to CO_2 in *Chromatium* differs from that in higher plants (Section II,A,2 and Fig. 1). It appears that once glycolate is photosynthesized via the oxygenase reaction, this product is both excreted and metabolized internally via an undefined O_2-insensitive pathway to CO_2 and glycine, the latter of which is also excreted (Asami *et al.*, 1977). Although O_2 is a competitive inhibitor of *Chromatium* photosynthesis with respect to CO_2 (Takabe and Akazawa, 1977), further investigation is needed to determine if, indeed, there is a concomitant light-induced efflux of CO_2 during photosynthesis *in vivo*.

In the cyanobacteria, glycolate oxidation activity is associated mainly with the thylakoid membranes and appears to be coupled indirectly to O_2 via the terminal electron carriers of the respiratory electron transport system (Grodzinski and Colman, 1976, Codd and Sallal, 1978). Whereas the metabolism of exogenous glycolate to CO_2 is inhibited by α-HPMS, neither glycolate excretion nor metabolism is affected by INH (Codd and Stewart, 1973; Han and Eley, 1973; Ingle and Colman, 1976), suggesting that the glycine to serine condensation reaction is of minor importance. Indeed, Codd and Stewart (1973) have proposed that in illuminated *Anabaena cylindrica* glycolate is metabolized mainly via glyoxylate → tartronic semialdehyde → glycerate → 3-PGA in which the major source of CO_2 production is from the condensative decarboxylation of 2 moles of glyoxylate to tartronic semialdehyde catalyzed by tartronate-semialdehyde synthase. Although the presence of this alternate pathway of glycolate metabolism has yet to be investigated in other cyanobacteria, it is interesting to note that it has been detected in the marine diatom *Cylindrotheca fusiformis* (Paul and Volcani, 1976).

Despite the potential to photosynthesize and excrete glycolate or metabolize it, in part, to CO_2, photorespiration per se (estimated by Γ, the inhibition of photosynthesis by 21% O_2 and light-induced CO_2 release) has not been detected in a variety of cyanobacteria (Cheng and Colman, 1974; Miyachi and Okabe, 1976; Lloyd *et al.*, 1977; Birmingham and Colman, 1979; Ray *et al.*, 1979; Coleman and Colman, 1980a). Several workers have proposed, based mainly on enzyme activity and inhibitor experiments and the initial products of $^{14}CO_2$ fixation, that the cyanobacteria possess the C_4 pathway of photosynthesis (Döhler, 1974b; Colman *et al.*, 1976; Colman and Coleman, 1978). However, this notion is inconsistent with the detailed *in vivo* ^{14}C-labeling studies by several groups, which indicate that these microorganisms fix CO_2 via the C_3 cycle (Pelroy and Bassham, 1972; Miyachi and Okabe, 1976; Pelroy *et al.*, 1976; Ray *et al.*, 1979; Coleman and Colman, 1980b). It is now evident that several cyanobacteria and algae (Section V,A) when adapted to a low CO_2 environment have a mechanism for concentrating inorganic carbon from the surrounding medium (Badger *et al.*, 1978; Kaplan *et al.*, 1980;

Miller and Colman, 1980a,b). This CO_2-concentrating mechanism apparently involves an active HCO_3^- uptake system because it is light-dependent and sensitive to diuron and uncouplers, and its induction is often accompanied by a coordinate induction of carbonic anhydrase activity (Döhler, 1974a; Ingle and Colman, 1976; but see Kaplan *et al.*, 1980). Thus, like C_4 plants, these microorganisms appear to control photorespiration by a metabolic CO_2-concentrating mechanism, which effectively suppresses the deleterious effects of O_2 on C_3 photosynthesis at the level of RuBP carboxylase/oxygenase. Although this system for concentrating inorganic carbon superficially appears less complicated than the higher plant C_4 acid mechanism (Fig. 2), which involves changes in leaf anatomy, photosynthetic enzymes, and the inter- and intracellular compartmentation of component reactions for achieving the same result, this generalization must await more detailed characterization of the active HCO_3^- influx system.

C. Crassulacean Acid Metabolism Plants and Submerged Aquatic Macrophytes

Photorespiration and O_2 inhibition are probably not significant factors in the photosynthetic productivity of crassulacean acid metabolism (CAM) plants, although they undoubtedly occur. In CAM photosynthesis (see Chapter 8, this volume) most of the net CO_2 uptake occurs during the dark. CO_2 is incorporated into malic acid via PEP carboxylase and NAD–malate dehydrogenase, and stored in the vacuole. The stomates close in the light, the malic acid is enzymically decarboxylated, and the CO_2 released during decarboxylation is refixed by RuBP carboxylase and incorporated into products of the C_3 cycle (Osmond, 1978; Klüge, 1979). Oxygen does not affect the rate of CO_2 uptake in the dark (Osmond and Björkman, 1975) because PEP carboxylase is insensitive to O_2 (Bowes and Ogren, 1972; Chollet, 1976).

At the onset of the light phase, stomates close and the internal CO_2 concentration rises dramatically as the malic acid is decarboxylated (Cockburn *et al.*, 1979; Spalding *et al.*, 1979) at a rate greatly exceeding the rate of RuBP carboxylation. The O_2 concentration inside the plants also increases, but the ratio of the CO_2/O_2 concentrations is greater than in air so RuBP oxygenase activity, and thus photorespiration, is probably reduced to very low levels (Spalding *et al.*, 1979). The concentration of stored malate decreases during the period of illumination, reducing the rate of malate decarboxylation and the internal CO_2 concentration. As the CO_2 concentration decreases, the stomates open (Cockburn *et al.*, 1979; Spalding *et al.*, 1979) and the internal ratio of CO_2/O_2 concentrations approaches that in the air, so RuBP oxygenase activity and photorespiration occur. The occurrence of photorespiration is indicated by an O_2-sensitive Γ (Osmond and Björkman, 1975). When the stomates are open in the light, photosynthesis and photorespiration in CAM plants are equivalent to that which occurs in C_3 plants.

The submerged aquatic macrophyte *Hydrilla verticillata* possesses some characteristics of CAM photosynthesis, particularly high rates of dark CO_2 fixation, when collected in the summer or grown at a warm (27°C) temperature (Holaday and Bowes, 1980). Plants from these environments also show diurnal fluctuations in titratable acid, a low Γ, and high rates of C_4 acid synthesis from CO_2. There is also a high activity of enzymes associated with CAM and C_4 photosynthesis, including PEP carboxylase, pyruvate, P_i dikinase, and NAD–malic enzyme. When this species is collected in winter or grown at a cool (11°C) temperature, it possesses a high Γ and a decreased PEP/RuBP carboxylase ratio in enzyme extracts. It was concluded that the mode of *H. verticillata* photosynthesis may adapt to the changing environmental conditions encountered during the year. On summer days, when the temperature and light intensity are high, the CO_2 concentration in the native environment approaches zero (Van *et al.*, 1976), so it would be advantageous to store, in malate, CO_2 produced at night by respiration and then incorporate this CO_2 into carbohydrate during the day (Holaday and Bowes, 1980).

The present chapter has dealt with the biochemistry, regulation, and function of photorespiration in C_3 and C_4 plants. We have also discussed photorespiration in C_3–C_4 intermediate plants, algae, cyanobacteria, photosynthetic bacteria, CAM plants, and submerged aquatic macrophytes. Despite the apparent lack of success, thus far, the concept of improving crop productivity by eliminating or reducing photorespiration continues to be a promising idea.

REFERENCES

Akazawa, T. (1979). *Encycl. Plant Physiol., New Ser.* **6,** 208–229.
Andersen, K. (1979). *Biochim. Biophys. Acta* **585,** 1–11.
Andrews, T. J., and Lorimer, G. H. (1978). *FEBS Lett.* **90,** 1–9.
Andrews, T. J., Badger, M. R., and Lorimer, G. H. (1975). *Arch. Biochem. Biophys.* **171,** 93–103.
Apel, P., and Ohle, H. (1979). *Biochem. Physiol. Pflanz.* **174,** 68–75.
Apel, P., Tichá, I., and Peisker, M. (1978). *Biochem. Physiol. Pflanz.* **172,** 547–552.
Asami, S., and Akazawa, T. (1974). *Plant Cell Physiol.* **15,** 571–576.
Asami, S., and Akazawa, T. (1975). *Plant Cell Physiol.* **16,** 631–642.
Asami, S., and Akazawa, T. (1976). *Plant Cell Physiol.* **17,** 1119–1129.
Asami, S., Takabe, T., and Akazawa, T. (1977). *Plant Cell Physiol.* **18,** 149–159.
Badger, M. R., and Andrews, T. J. (1974). *Biochem. Biophys. Res. Commun.* **60,** 204–210.

Badger, M. R., and Collatz, G. J. (1977). *Year Book—Carnegie Inst. Washington* **76,** 355–361.
Badger, M. R., Kaplan, A., and Berry, J. A. (1978). *Yearbook—Carnegie Inst. Washington* **77,** 251–261.
Badger, M. R., Kaplan, A., and Berry, J. A. (1980). *Plant Physiol.* **66,** 407–413.
Bahr, J. T., and Jensen, R. G. (1974). *Biochem. Biophys. Res. Commun.* **57,** 1180–1185.
Bassham, J. A. (1963a). *Adv. Enzymol.* **25,** 39–117.
Bassham, J. A. (1963b). *In* "Photosynthetic Mechanisms of Green Plants" (B. Kok and A. T. Jagendorf, eds.), pp. 635–647. Natl. Acad. Sci.—Natl. Res. Counc., Washington, D.C.
Bassham, J. A., and Calvin, M. (1957). "The Path of Carbon in Photosynthesis." Prentice-Hall, Englewood Cliffs, New Jersey.
Bassham, J. A., and Kirk, M. (1962). *Biochem. Biophys. Res. Commun.* **9,** 376–380.
Bassham, J. A., and Kirk, M. (1973). *Plant Physiol.* **52,** 407–411.
Bassham, J. A., Egeter, H., Edmonston, F., and Kirk, M. (1963). *Biochem. Biophys. Res. Commun.* **13,** 144–149.
Bauwe, H., and Apel, P. (1979). *Biochem. Physiol. Pflanz.* **174,** 251–254.
Beck, E. (1979). *Encycl. Plant Physiol., New Ser.* **6,** 327–337.
Berry, J. A., Boynton, J., Kaplan, A., and Badger, M. R. (1976). *Yearbook—Carnegie Inst. Washington* **75,** 423–432.
Bird, I. F., Cornelius, M. J., Keys, A. J., and Whittingham, C. P. (1972). *Biochem. J.* **128,** 191–192.
Bird, I. F., Cornelius, M. J., and Keys, A. J. (1977). *J. Exp. Bot.* **28,** 519–524.
Birmingham, B. C., and Colman, B. (1979). *Plant Physiol.* **64,** 892–895.
Björkman, O. (1973). *Photophysiology* **8,** 1–63.
Björkman, O. (1976). *In* "CO_2 Metabolism and Plant Productivity" (R. H. Burris and C. C. Black, eds.), pp. 287–309. University Park Press, Baltimore, Maryland.
Björkman, O., Hiesey, W. M., Nobs, M., Nicholson, F., and Hart, R. W. (1968). *Yearbook—Carnegie Inst. Washington* **66,** 228–232.
Black, C. C., Jr. (1973). *Annu. Rev. Plant Physiol.* **24,** 253–286.
Black, C. C., Jr., and Mollenhauer, H. H. (1971). *Plant Physiol.* **47,** 15–23.
Blackwood, G. C., and Miflin, B. J. (1976). *J. Exp. Bot.* **27,** 735–747.
Bowes, G., and Berry, J. A. (1972). *Yearbook—Carnegie Inst. Washington* **71,** 148–158.
Bowes, G., and Ogren, W. L. (1972). *J. Biol. Chem.* **247,** 2171–2176.
Bowes, G., Ogren, W. L., and Hageman, R. H. (1971). *Biochem. Biophys. Res. Commun.* **45,** 716–722.
Brown, H. M., Rejda, J. M., and Chollet, R. (1980). *Biochim. Biophys. Acta* **614,** 545–552.
Brown, R. H. (1976). *In* "CO_2 Metabolism and Plant Productivity" (R. H. Burris and C. C. Black, eds.), pp. 311–325. University Park Press, Baltimore, Maryland.
Brown, R. H. (1980). *Plant Physiol.* **65,** 346–349.
Brown, R. H., and Brown, W. V. (1975). *Crop Sci.* **15,** 681–685.
Brown, W. V., and Smith, B. N. (1972). *Nature (London)* **239,** 345–346.
Bull, T. A. (1969). *Crop Sci.* **9,** 726–729.
Canvin, D. T. (1979). *Encycl. Plant Physiol., New Ser.* **6,** 368–396.
Canvin, D. T., Berry, J. A., Badger, M. R., Fock, H., and Osmond, C. B. (1980). *Plant Physiol.* **66,** 302–307.
Carolin, R. C., Jacobs, S. W. L., and Vesk, M. (1977). *Bot. Gaz. (Chicago)* **138,** 413–419.
Carolin, R. C., Jacobs, S. W. L., and Vesk, M. (1978). *Aust. J. Bot.* **26,** 683–698.
Cheng, K. H., and Colman, B. (1974). *Planta* **115,** 207–212.
Chollet, R. (1974). *Arch. Biochem. Biophys.* **163,** 521–529.
Chollet, R. (1976). *In* "CO_2 Metabolism and Plant Productivity" (R. H. Burris and C. C. Black, eds.), pp. 327–341. University Park Press, Baltimore, Maryland.

Chollet, R. (1978). *Plant Physiol.* **61,** 929–932.
Chollet, R., and Anderson, L. L. (1976). *Arch. Biochem. Biophys.* **176,** 344–351.
Chollet, R., and Ogren, W. L. (1975). *Bot. Rev.* **41,** 137–179.
Christeller, J. T., and Laing, W. A. (1979). *Biochem. J.* **183,** 747–750.
Cockburn, W., Ting, I. P., and Sternberg, L. O. (1979). *Plant Physiol.* **63,** 1029–1032.
Codd, G. A., and Sallal, A.-K. J. (1978). *Planta* **139,** 177–181.
Codd, G. A., and Stewart, W. D. P. (1973). *Arch. Mikrobiol.* **94,** 11–28.
Codd, G. A., and Turnbull, F. (1975). *Arch. Microbiol.* **104,** 155–158.
Codd, G. A., Bowien, B., and Schlegel, H. G. (1976). *Arch. Microbiol.* **110,** 167–171.
Codd, G. A., Okabe, K., and Stewart, W. D. P. (1980). *Arch. Microbiol.* **124,** 149–154.
Cohen, Y., de Jonge, I., and Kuenen, J. G. (1979). *Arch. Microbiol.* **122,** 189–194.
Coleman, J. R., and Colman, B. (1980a). *Plant Physiol.* **65,** 980–983.
Coleman, J. R., and Colman, B. (1980b). *Planta* **149,** 318–320.
Colman, B., and Coleman, J. R. (1978). *Plant Sci. Lett.* **12,** 101–105.
Colman, B., Cheng, K. H., and Ingle, R. K. (1976). *Plant Sci. Lett.* **6,** 123–127.
Coombs, J., and Whittingham, C. P. (1966). *Proc. R. Soc. London, Ser. B.* **164,** 511–520.
Cooper, T. G., Filmer, D., Wishnick, M., and Lane, M. D. (1969). *J. Biol. Chem.* **244,** 1081–1083.
Cornic, G. (1978). *Can. J. Bot.* **56,** 2128–2137.
Creach, E. (1979). *Plant Physiol.* **64,** 435–438.
Crespo, H. M., Frean, M., Cresswell, C. F., and Tew, J. (1979). *Planta* **147,** 257–263.
Crookston, R. K. (1972). Ph.D. Thesis, University of Minnesota, Minneapolis.
D'Aoust, A. L., and Canvin, D. T. (1973). *Can. J. Bot.* **51,** 457–464.
De Jong, T. M. (1978). *Oecologia* **36,** 59–68.
Dimon, B., Gerster, R., and Tournier, P. (1977). *C. R. Hebd. Seances Acad. Sci. Ser. D.* **284,** 297–299.
Döhler, G. (1974a). *Planta* **117,** 97–99.
Döhler, G. (1974b). *Planta* **118,** 259–269.
Doravari, S., and Canvin, D. T. (1980). *Plant Physiol.* **66,** 628–631.
Douce, R., Moore, A. L., and Neuburger, M. (1977). *Plant Physiol.* **60,** 625–628.
Edwards, G. E., and Huber, S. C. (1979). *Encycl. Plant Physiol., New Ser.* **6,** 102–112.
Ehleringer, J. R. (1978). *Oecologia* **31,** 255–267.
Ehleringer, J. R. (1979). *HortScience* **14,** 217–222.
Ehleringer, J., and Björkman, O. (1977). *Plant Physiol.* **59,** 86–90.
Findinegg, G. R., (1976). *Z. Pflanzenphysiol.* **79,** 428–437.
Findinegg, G. R. (1979). *Plant Sci. Lett.* **17,** 101–108.
Forrester, M. L., Krotkov, G., and Nelson, C. D. (1966). *Plant Physiol.* **41,** 428–431.
Foster, A., and Black, C. C. (1977). *Plant Cell Physiol., Spec. Issue,* 325–340.
Frederick, S. E., and Newcomb, E. H. (1971). *Planta* **96,** 152–174.
Gale, J., and Tako, T. (1976). *Photosynthetica* **10,** 89–92.
Garrett, M. K. (1978). *Nature (London)* **274,** 913–915.
Gauhl, E., and Björkman, O. (1969). *Planta* **88,** 187–191.
Govindjee (ed.). (1982). "Photosynthesis: Energy Conversion by Plants and Bacteria" Vol. I. Academic Press, New York.
Grodzinski, B., and Colman, B. (1976). *Plant Physiol.* **58,** 199–202.
Gutierrez, M., Gracen, V. E., and Edwards, G. E. (1974). *Planta* **119,** 279–300.
Han, T.-W., and Eley, J. H. (1973). *Plant Cell Physiol.* **14,** 285–291.
Hatch, M. D. (1971). *Biochem. J.* **125,** 425–432.
Hatch, M. D. (1977). *Trends Biochem. Sci.* **2,** 199–202.

Hatch, M. D., and Osmond, C. B. (1976). *Encycl. Plant Physiol., New Ser.* **3,** 143–184.
Hatch, M. D., Kagawa, T., and Craig, S. (1975). *Aust. J. Plant Physiol.* **2,** 111–128.
Hattersley, P. W., and Watson, L. (1975). *Phytomorphology* **25,** 325–333.
Hattersley, P. W., Watson, L., and Osmond, C. B. (1977). *Aust. J. Plant Physiol.* **4,** 523–539.
Heber, U., and Krause, G. H. (1980). *Trends Biochem. Sci.* **5,** 32–34.
Hickman, S. A., and Keys, A. J. (1972). *Proc. Int. Congr. Photosynth., 2nd, 1971,* pp. 2225–2231.
Hirokawa, T., Miyachi, S., and Tamiya, H. (1958). *J. Biochem. (Tokyo)* **45,** 1005–1010.
Hogetsu, D., and Miyachi, S. (1979). *Plant Cell Physiol.* **20,** 747–756.
Höhler, T., and Schaub, H. (1979). *Biochem. Physiol. Pflanz.* **174,** 58–67.
Holaday, S., and Bowes, G. (1980). *Plant Physiol.* **65,** 331–335.
Holaday, A. S., Shieh, Y.-J., Lee, K. W., and Chollet, R. (1981). *Biochim. Biophys. Acta* **637,** 334–341.
Huber, S. and Edwards, G. (1975). *Biochem. Biophys. Res. Commun.* **67,** 28–34.
Imai, K., and Murata, Y. (1979). *Jpn. J. Crop Sci.* **49,** 58–65.
Ingle, R. K., and Colman, B. (1976). *Planta* **128,** 217–223.
Jewess, P. J., Kerr, M. W., and Whitaker, D. P. (1975). *FEBS Lett.* **53,** 292–296.
Johnson, H. S., and Hatch, M. D. (1969). *Biochem. J.* **114,** 127–134.
Jordan, D. B., and Ogren, W. L. (1981a). *Plant Physiol.* **67,** 237–245.
Jordan, D. B., and Ogren, W. L. (1981b). *Nature (London)* **291,** 513–515.
Kanai, R., and Kashiwagi, M. (1975). *Plant Cell Physiol.* **16,** 669–679.
Kaplan, A., Badger, M. R., and Berry, J. A. (1980). *Planta* **149,** 219–226.
Keck, R. W., and Ogren, W. L. (1976). *Plant Physiol.* **58,** 552–555.
Kennedy, R. A. (1976). *Planta* **128,** 149–154.
Kennedy, R. A., and Laetsch, W. M. (1974). *Science* **184,** 1087–1089.
Kent, S. S., and Young, J. D. (1980). *Plant Physiol.* **65,** 465–468.
Kestler, D. P., Mayne, B. C., Ray, T. B., Goldstein, L. D., Brown, R. H., and Black, C. C. (1975). *Biochem. Biophys. Res. Commun.* **66,** 1439–1446.
Keys, A. J., Bird, I. F., Cornelius, M. J., Lea, P. J., Wallsgrove, R. M., and Miflin, B. J. (1978). *Nature (London)* **275,** 741–743.
King, W. R., and Andersen, K. (1980). *Arch. Microbiol.* **128,** 84–90.
Kisaki, T., and Tolbert, N. E. (1969). *Plant Physiol.* **44,** 242–250.
Kisaki, T., Yano, N., and Hirabayashi, S. (1972). *Plant Cell Physiol.* **13,** 581–584.
Klüge, M. (1979). *Encycl. Plant Physiol., New Ser.* **6,** 113–125.
Krenzer, E. G., Moss, D. N., and Crookston, R. K. (1975). *Plant Physiol.* **56,** 194–206.
Ku, S. B., and Edwards, G. E. (1975). *Z. Pflanzenphysiol.* **77,** 16–32.
Ku, S. B., and Edwards, G. E. (1980). *Planta* **147,** 277–282.
Ku, S. B., Edwards, G. E., and Kanai, R. (1976). *Plant Cell Physiol.* **16,** 669–679.
Kumarasinghe, K. S., Keys, A. J., and Whittingham, C. P. (1977). *J. Exp. Bot.* **28,** 1163–1168.
Kung, S. D., and Marsho, T. V. (1976). *Nature (London)* **259,** 325–326.
Laetsch, W. M. (1974). *Annu. Rev. Plant Physiol.* **25,** 27–52.
Laing, W. A. (1974). Ph.D. Thesis, University of Illinois, Urbana-Champaign.
Laing, W. A., Ogren, W. L., and Hageman, R. H. (1974). *Plant Physiol.* **54,** 678–685.
Laing, W. A., Ogren, W. L., and Hageman, R. H. (1975). *Biochemistry* **14,** 2269–2275.
Lawlor, D. W., and Fock, H. (1978). *J. Exp. Bot.* **29,** 579–596.
Lea, P. J., and Miflin, B. J. (1979). *Encycl. Plant Physiol., New Ser.* **6,** 445–456.
Lea, P. J., and Thurman, D. A. (1972). *J. Exp. Bot.* **23,** 440–449.
Lewanty, Z., Maleszewski, S., and Poskuta, J. (1971). *Z. Pflanzenphysiol.* **65,** 469–472.

Lloyd, N. D. H., Canvin, D. T., and Culver, D. A. (1977). *Plant Physiol.* **59,** 936–940.
Long, S. P., and Woolhouse, H. W. (1978). *J. Exp. Bot.* **29,** 567–577.
Lorimer, G. H., Badger, M. R., and Andrews, T. J. (1976). *Biochemistry* **15,** 529–536.
Lorimer, G. H., Osmond, C. B., Akazawa, T., and Asami, S. (1978). *Arch. Biochem. Biophys.* **185,** 49–56.
Ludlow, M. M., and Wilson, G. L. (1971). *Aust. J. Biol. Sci.* **24,** 1065–1075.
McNeil, P. H., Foyer, C. H., Walker, D. A., Bird, I. F., Cornelius, M. J., and Keyes, A. J. (1981). *Plant Physiol.* **67,** 530–534.
Mahon, J. D., Fock, H., Hohler, T., and Canvin, D. T. (1974). *Planta* **120,** 113–123.
Matsumoto, K., Nishimura, M., and Akazawa, T. (1977). *Plant Cell Physiol.* **18,** 1281–1290.
Miller, A. G., and Colman, B. (1980a). *J. Bacteriol.* **143,** 1253–1259.
Miller, A. G., and Colman, B. (1980b). *Plant Physiol.* **65,** 397–402.
Miyachi, S., and Okabe, K.-I. (1976). *Plant Cell Physiol.* **17,** 973–986.
Miyachi, S., and Shiraiwa, Y. (1979). *Plant Cell Physiol.* **20,** 341–348.
Miyachi, S., Izawa, S., and Tamiya, H. (1955). *J. Biochem. (Tokyo)* **42,** 221–244.
Morgan, J. A., and Brown, R. H. (1979). *Plant Physiol.* **64,** 257–262.
Morgan, J. A., Brown, R. H., and Reger, B. J. (1980). *Plant Physiol.* **65,** 156–159.
Morot-Gaudry, J. F., Farineau, J. P., and Huet, J. C. (1980). *Plant Physiol.* **66,** 1079–1084.
Moss, D. N. (1968). *Crop Sci.* **8,** 71–76.
Nelson, P. E., and Surzycki, S. J. (1976). *Eur. J. Biochem.* **61,** 475–480.
Neuburger, M., and Douce, R. (1977). *C. R. Acad. Sci., Ser. D* **285,** 881–884.
Ogren, W. L. (1975). *In* "Environmental and Biological Control of Photosynthesis" (R. Marcelle, ed.), pp. 45–52. Junk, The Hague.
Ogren, W. L. (1978). *Proc. Int. Congr. Photosynth., 4th, 1977,* pp. 721–733.
Ogren, W. L., and Bowes, G. (1971). *Nature (London), New Biol.* **230,** 159–160.
Ogren, W. L., and Hunt, L. D. (1978). *In* "Photosynthetic Carbon Assimilation" (H. W. Siegelman and G. Hind, eds.), pp. 127–138. Plenum, New York.
Okabe, K., Schmid, G. H., and Straub, J. (1977). *Plant Physiol.* **60,** 150–156.
Oliver, D. J. (1978). *Plant Physiol.* **62,** 690–692.
Oliver, D. J. (1979). *Plant Physiol.* **64,** 1048–1052.
Oliver, D. J., and Zelitch, I. (1977). *Plant Physiol.* **59,** 668–694.
Osmond, C. B. (1972). *Proc. Int. Congr. Photosynth., 2nd, 1971,* pp. 2233–2239.
Osmond, C. B. (1978). *Annu. Rev. Plant Physiol.* **29,** 379–414.
Osmond, C. B., and Avadhani, P. N. (1970). *Plant Physiol.* **45,** 228–230.
Osmond, C. B., and Björkman, O. (1972). *Yearbook—Carnegie Inst. Washington* **71,** 141–148.
Osmond, C. B., and Björkman, O. (1975). *Aust. J. Plant Physiol.* **2,** 155–162.
Osmond, C. B., and Harris, B. (1971). *Biochim. Biophys. Acta* **234,** 270–282.
Osmond, C. B., and Smith, F. A. (1976). *In* "Intercellular Communication in Plants: Studies on Plasmodesmata" (B. E. S. Gunning and A. W. Robards, eds.), pp. 229–241. Springer-Verlag, Berlin and New York.
Parkinson, K. J., Penman, H. L., and Tregunna, E. B. (1974). *J. Exp. Bot.* **25,** 132–144.
Paul, J. S., and Volcani, B. E. (1976). *Arch. Microbiol.* **110,** 247–252.
Pelroy, R. A., and Bassham, J. A. (1972). *Arch. Mikrobiol.* **86,** 25–38.
Pelroy, R. A., Levine, G. A., and Bassham, J. A. (1976). *J. Bacteriol.* **128,** 633–643.
Pon, N. G., Rabin, B. R., and Calvin, M. (1963). *Biochem. Z.* **338,** 7–19.
Poskuta, J. (1969). *Physiol. Plant.* **22,** 76–85.
Powles, S. B., and Osmond, C. B. (1978). *Aus. J. Plant Physiol.* **5,** 619–629.
Powles, S. B., Osmond, C. B., and Thorne, S. W. (1979). *Plant Physiol.* **64,** 982–988.
Pritchard, G., Griffin, W., and Whittingham, C. P. (1962). *J. Exp. Bot.* **13,** 176–184.
Quebedeaux, B., and Chollet, R. (1977). *Plant Physiol.* **59,** 42–44.

Radmer, R. J., and Kok, B. (1976). *Plant Physiol.* **58,** 336–340.
Raghavendra, A. S., Rajendrudu, G., and Das, V. S. R. (1978). *Nature (London)* **273,** 143–144.
Rathnam, C. K. M. (1978). *Z. Pflanzenphysiol.* **87,** 65–84.
Rathnam, C. K. M. (1979). *Planta* **145,** 13–23.
Rathnam, C. K. M., and Chollet, R. (1978). *Biochem. Biophys. Res. Commun.* **85,** 801–808.
Rathnam, C. K. M., and Chollet, R. (1979a). *Arch. Biochem. Biophys.* **193,** 346–354.
Rathnam, C. K. M., and Chollet, R. (1979b). *Biochim. Biophys. Acta* **548,** 500–519.
Rathnam, C. K. M., and Chollet, R. (1980a). *Plant Physiol.* **65,** 489–494.
Rathnam, C. K. M., and Chollet, R. (1980b). *Prog. Phytochem.* **6,** 1–48.
Rathnam, C. K. M., Raghavendra, A. S., and Das, V. S. R. (1976). *Z. Pflanzenphysiol.* **77,** 283–291.
Raven, J. A. (1970). *Biol. Rev. Cambridge Philos. Soc.* **45,** 167–221.
Raven, J. A., and Glidewell, S. M. (1978). *Plant, Cell Environ.* **1,** 185–197.
Ray, T. B., and Black, C. C. (1979). *Encycl. Plant Physiol., New Ser.* **6,** 77–101.
Ray, T. B., Mayne, B. C., Toia, R. E., and Peters, G. A. (1979). *Plant Physiol.* **64,** 791–795.
Rejda, J. M., Johal, S., and Chollet, R. (1981). *Arch. Biochem. Biophys.* **210,** 617–624.
Rhodes, P. R., Kung, S. D., and Marsho, T. V. (1980). *Plant Physiol.* **65,** 69–73.
Richardson, K. E., and Tolbert, N. E. (1961). *J. Biol. Chem.* **236,** 1285–1290.
Sado, M., Asami, S., Nishimura, M., and Akawaza, T. (1980). *Plant Cell Physiol.* **21,** 1077–1084.
Sayre, R. T., and Kennedy, R. A. (1977). *Planta* **134,** 257–262.
Sayre, R. T., Kennedy, R. A., and Pringnitz, D. J. (1979). *Plant Physiol.* **64,** 293–299.
Schnarrenberger, C., and Fock, H. (1976). *Encycl. Plant Physiol., New Ser.* **3,** 185–234.
Servaites, J. C., and Ogren, W. L. (1977). *Plant Physiol.* **60,** 461–466.
Servaites, J. C., and Ogren, W. L. (1978). *Plant Physiol.* **61,** 62–67.
Servaites, J. C., Schrader, L. E., and Edwards, G. E. (1978). *Plant Cell Physiol.* **19,** 1399–1405.
Shain, Y., and Gibbs, M. (1971). *Plant Physiol.* **48,** 325–330.
Shelp, B. J., and Canvin, D. T. (1980a). *Plant Physiol.* **65,** 774–779.
Shelp, B. J., and Canvin, D. T. (1980b). *Plant Physiol.* **65,** 780–784.
Somerville, C. R., and Ogren, W. L. (1979). *Nature (London)* **280,** 833–836.
Somerville, C. R., and Ogren, W. L. (1980a). *Nature (London)* **286,** 257–259.
Somerville, C. R., and Ogren, W. L. (1980b). *Proc. Natl. Acad. Sci. U.S.A.* **77,** 2684–2687.
Somerville, C. R., and Ogren, W. L. (1981). *Plant Physiol.* **67,** 666–671.
Spalding, M. H., Stumpf, D. K., Ku, M. S. B., Burris, R. H., and Edwards, G. E. (1979). *Aust. J. Plant Physiol.* **6,** 557–567.
Spreitzer, R. J., and Mets, L. (1980). *Nature (London)* **285,** 114–115.
Steiger, H. M., Beck, E., and Beck, R. (1977). *Plant Physiol.* **60,** 903–906.
Takabe, T., and Akazawa, T. (1977). *Plant Cell Physiol.* **18,** 753–765.
Takabe, T., Osmond, C. B., Summons, R. E., and Akazawa, T. (1979). *Plant Cell Physiol.* **20,** 233–241.
Takabe, T., Asami, S., and Akazawa, T. (1980). *Biochemistry* **19,** 3985–3989.
Tamiya, H., and Huzisige, H. (1949). *Acta Phytochim.* **1,** 83–103.
Tamiya, H., Miyachi, S., and Hirokawa, T. (1957). *In* "Research in Photosynthesis" (H. Gaffron, A. H. Brown, C. C. French, R. Livingston, E. I. Rabionwitch, B. C. Strehler, and N. E. Tolbert, eds.), pp. 213–223. Wiley (Interscience), New York.
Tieszen, L. L., Senyimba, M. M., Imbamba, S. K., and Troughton, J. H. (1979). *Oecologia* **37,** 337–350.
Tolbert, N. E. (1971). *Annu. Rev. Plant Physiol.* **22,** 45–74.

Tolbert, N. E. (1980). *In* "The Biochemistry of Plants" (D. D. Davies, ed.), Vol. 2, pp. 487–523. Academic Press, New York.
Tsuzuki, M., and Miyachi, S. (1979). *FEBS Lett.* **103,** 221–223.
Turner, J. S., and Brittain, E. G. (1962). *Biol. Rev. Cambridge Philos. Soc.* **37,** 130–170.
Van, T. K., Haller, W. T., and Bowes, G. (1976). *Plant Physiol.* **58,** 761–768.
Wallsgrove, R. M., Keys, A. J., Bird, I. F., Cornelius, M. J., Lea, P. J., and Miflin, B. J. (1980). *J. Exp. Bot.* **31,** 1005–1017.
Warburg, O. (1920). *Biochem. Z.* **103,** 188–217.
Widholm, J. M., and Ogren, W. L. (1969). *Proc. Natl. Acad. Sci. U.S.A.* **63,** 668–675.
Wildner, G. F., and Henkel, J. (1979). *Planta* **146,** 223–228.
Williams, L. E., and Kennedy, R. A. (1977). *Z. Pflanzenphysiol.* **81,** 314–322.
Woo, K. C., and Osmond, C. B. (1977). *Plant Cell Physiol., Spec. Issue,* pp. 315–323.
Woo, K. C., Berry, J. A., and Turner, G. L. (1978). *Yearbook—Carnegie Inst. Washington* **77,** 240–245.
Yeoh, H.-H., Badger, M. R., and Watson, L. (1980). *Plant Physiol.* **66,** 1110–1112.
Zelitch, I. (1959). *J. Biol. Chem.* **234,** 3077–3081.
Zelitch, I. (1965). *J. Biol. Chem.* **240,** 1869–1876.
Zelitch, I. (1966). *Plant Physiol.* **41,** 1623–1631.
Zelitch, I. (1971). "Photosynthesis, Photorespiration, and Plant Productivity." Academic Press, New York.
Zelitch, I. (1973). *Plant Physiol.* **51,** 299–305.
Zelitch, I. (1975). *Annu. Rev. Biochem.* **44,** 123–145.
Zelitch, I. (1979). *Encycl. Plant Physiol., New Ser.* **6,** 353–367.
Zelitch, I., and Day, P. R. (1968). *Plant Physiol.* **43,** 1838–1844.

8

Crassulacean Acid Metabolism (CAM)

MANFRED KLUGE

ABBREVIATIONS

AMP	Adenosine monophosphate
ADP	Adenosine diphosphate
ATP	Adenosine triphosphate
CAM	Crassulacean acid metabolism
DPGA	Diphosphoglyceric acid
F 6-P	Fructose 6-phosphate
FDP	Fructose diphosphate
G-1(or6)-P	Glucose 1 (or 6-)-phosphate
NAD^+	Nicotinamide adenine dinucleotide
NADH	Nicotinamide adenine dinucleotide (reduced)
$NADP^+$	Nicotinamide adenine dinucleotide phosphate
NADPH	Nicotinamide adenine dinucleotide phosphate (reduced)
OAA	Oxaloacetic acid (oxaloacetate)
P	Phosphate
P_i	Inorganic phosphate
PP_i	Pyrophosphate
P(I–IV)	Phase (I–IV)
PEP	Phosphoenol pyruvate
PGA	Phosphoglyceric acid (phosphoglycerate)
RuBP; RuDP	Ribulose bisphosphate; the older terminology is Ribulose diphosphate

Photosynthesis: Development, Carbon Metabolism, and Plant Productivity, Vol. II

ISBN: 0-12-294302-3

ABSTRACT

This chapter describes crassulacean acid metabolism (CAM) as an ecologically relevant modification of the usual photosynthetic carbon assimilation pathway. After the description of some basic criteria of CAM, the metabolic sequences and its compartmentation and the control of CAM are considered. The gas-exchange linked with CAM and its implications for the ecology of CAM are additional aspects discussed.

I. Introduction

The crassulacean acid metabolism (CAM) was discovered in 1815 by Heyne (see Wolf, 1960; Kluge and Ting, 1978). Intensive investigation of the CAM phenomenon during the past two decades has led to the concept that CAM represents a modification of the photosynthetic pathway by which certain terrestrial plants harvest carbon dioxide from the atmosphere. Furthermore, as is also true for the C_4 photosynthesis, CAM provides a dramatic example of "strategies" that enable plants to consolidate their water and carbon balance in arid environments; this allows these plants to conquer ecological niches where water is deficient.

CAM has recently attracted world-wide interest mainly because of its ecological importance and because of the interesting aspects of its comparative biochemistry and physiology. It is therefore not amazing that numerous reviews (see, e.g., Ranson and Thomas, 1960; Wolf, 1960; Osmond, 1978, Osmond and Holtum, 1981) and even a monograph (Kluge and Ting, 1978) dealing with CAM have been published. Considering this activity, Osmond (1978) stated that "CAM is now in serious danger of being reviewed to an extent that is difficult to justify in terms of its significance in the biosphere [p. 379]." However, even if CAM plays quantitatively a minor role in the total primary productivity of the biosphere, any presentation that aims at an integrated approach to photosynthesis would be incomplete without a consideration of CAM; its inclusion here is not only justified, but necessary.

This chapter shall deal with the basic aspects of CAM and with recent findings in this field. The reader should also consult the reviews and books cited (see also Chapters 6 and 7, this volume).

II. Basic Phenomena of CAM

Crassulacean acid metabolism represents, as indicated earlier, a modification of the normal photosynthetic pathway. It was first discovered in Crassulacean species; however, the view is now well supported that CAM

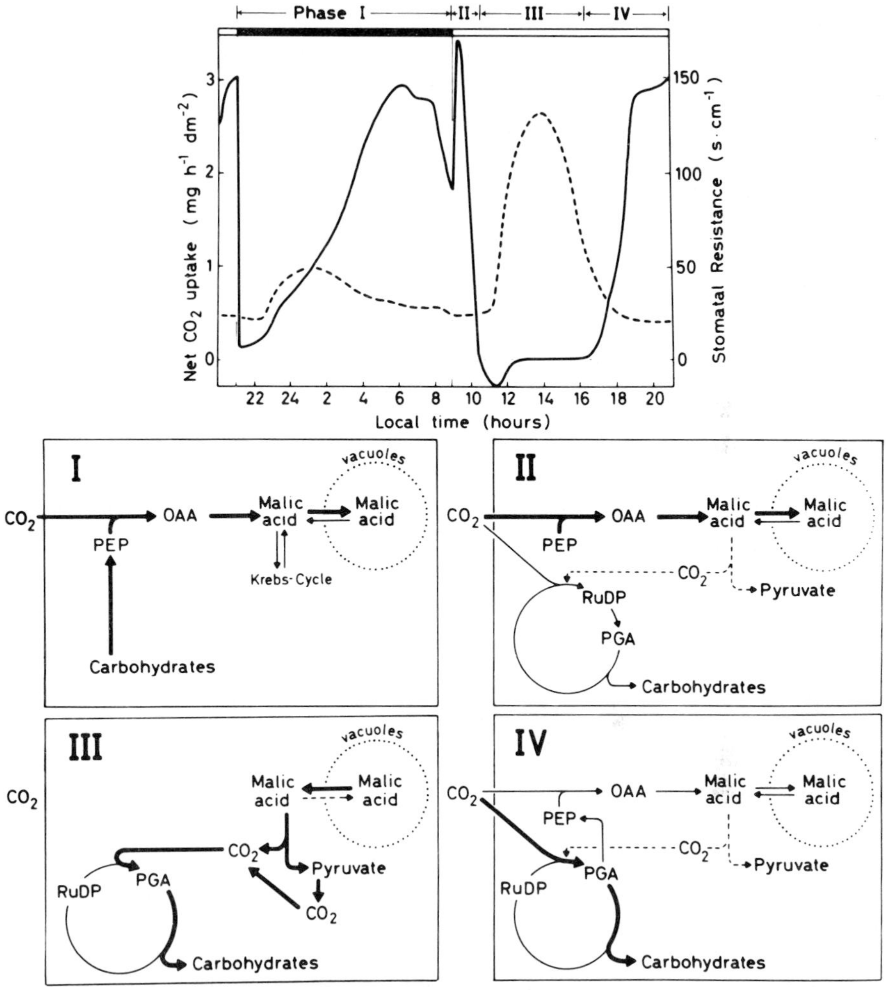

FIG. 1. Upper part: diurnal pattern of CO_2 exchange (———) and stomatal resistance (– – –) in a typical CAM plant (*Kalanchoe daigremontiana*) under 12:12 hr dark:light regime. The gas exchange curve is divided in typical phases (I–IV). Lower part: carbon flow pathways during the above four phases of the CO_2 exchange curve. (Compare with Fig. 2 for C_4 pathways in Chapter 6, this volume.) For definitions of symbols, see list of abbreviations.

is the result of polyphyletic evolution, and hence it is distributed among various families of higher plants having succulent leaves or stems (see Kluge and Ting, 1978).

Superficially, the CAM type of photosynthesis behaves paradoxically (Fig. 1). In contrast to other plants, CAM plants fix external CO_2 mainly at night; they open their stomata at night and keep them closed during many hours of the day. Finally, they exhibit a distinct diurnal oscillation of their malic acid content. This malic acid rhythm is characterized by the nocturnal increase of the malic acid concentrations up to 200 $\mu Eq \cdot g^{-1}$ fresh weight, and the disappearance of that acid during the day. This diurnal rhythm is coupled with an inverse rhythm of the starch level in the CAM performing cells.

Two groups of CAM plants may be distinguished, i.e., those which perform CAM permanently (obligate CAM) and those in which CAM can be induced (facultative CAM) by environmental signals such as photoperiod and water stress or by endogenous ontogenetic factors. The capacity and pattern of both the obligate and facultation CAM can be modulated by environmental factors (Kluge and Ting, 1978; Osmond, 1978).

III. The Metabolic Sequence of CAM

In this section, the main pathways of carbon flow in CAM shall be considered (Fig. 2). A more detailed discussion of individual reactions and enzymes involved has been provided by Kluge and Ting (1978), Dittrich (1979a), and Kluge (1979a).

A. The Metabolic Pathways of the Dark Period

1. THE PRIMARY PROCESS: DARK CO_2 FIXATION

The initial step of the diurnal CAM cycle consists in the fixation of CO_2 by β-carboxylation of phosphoenol pyruvate (PEP) according to Eq. 1.

$$\text{PEP} + CO_2 + H_2O \xrightarrow{\text{PEP carboxylase}} \text{oxaloacetate} + P_i \tag{1}$$

This reaction is mediated by PEP carboxylase, an enzyme that can be considered to be the key enzyme of CAM. The bulk of the oxaloacetate (OAA) resulting from the preceding reaction is further reduced to malate by NADH dependent malate dehydrogenase (Eq. 2).

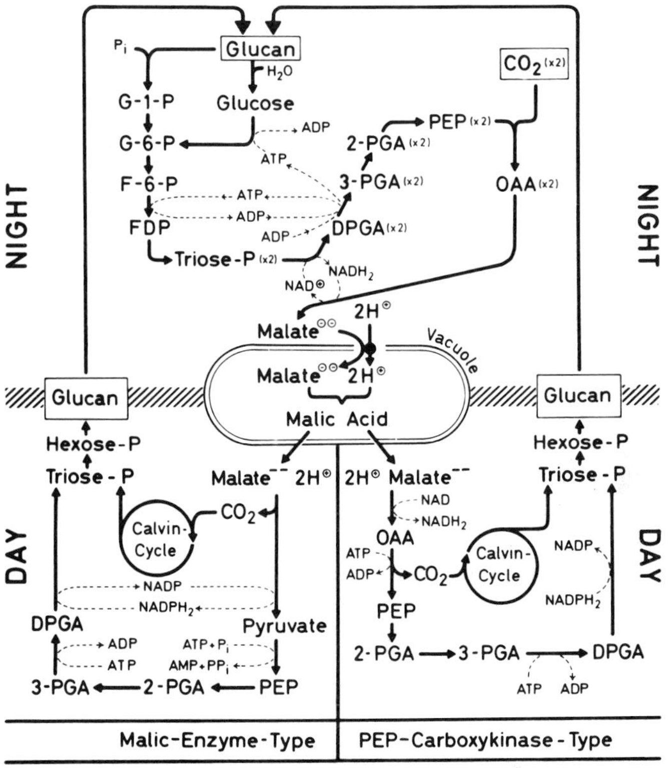

FIG. 2. Scheme of total carbon flow in the two metabolic variants of CAM ("malic–enzyme type" and "PEP–carboxykinase type"). During the night, the metabolic pathways are identical in both types of CAM. (For details on Calvin cycle, see Fig. 1 in Chapter 6, this volume.) See list of abbreviations for definitions of symbols.

$$\mathrm{OAA} + \mathrm{NADH} + \mathrm{H}^+ \xrightleftharpoons{\text{malate dehydrogenase}} \mathrm{malate} + \mathrm{NAD}^+ \tag{2}$$

In tracer experiments, normally 80–95% of the radiocarbon harvested by the plant via $^{14}CO_2$ dark fixation can finally be found in malate. Therefore, this substance can be considered as the end product of the nocturnal CO_2 fixation. It is transported together with two protons into the vacuoles of the cells (see Sections III,C,2) where it accumulates as malic acid during the night. In addition to malate, aspartate is also a stable product of dark CO_2 fixation. However as indicated by tracer experiments, the flow of carbon through aspartate is, compared with malate, low during the nocturnal CAM. It is supposed that aspartate is generated directly from OAA, the product of the β-carboxylation (Eq. 3).

$$\text{Glutarate} + \text{OAA} \underset{}{\overset{\text{aspartate aminotransferase}}{\rightleftharpoons}} \alpha\text{-ketoglutarate} + \text{aspartate} \qquad (3)$$

2. THE ORIGIN OF THE SUBSTRATES CONSUMED IN DARK CO_2 FIXATION

Under normal conditions, most of the CO_2 consumed in dark CO_2 fixation is derived from the ambient air and is taken up via the stomata. This explains why the nocturnal malate synthesis is severely inhibited in CO_2-free air. However, endogenous CO_2 produced by respiration is also fixed; under stress conditions, which force the stomata to close, CAM may even persist exclusively at the expense of respiratory CO_2 (Szarek *et al.*, 1973). Apart from earlier hypotheses (Bradbeer *et al.*, 1958; see also Ranson and Thomas, 1960; Wolf, 1960), it is now almost certain that PEP, the other substrate of β-carboxylation, is generated via glycolysis from reserve glucan, i.e., mainly from starch (Fig. 2). This can be concluded from the fact that during the night glucan disappears to the same extent as malate is synthesized (Ranson and Thomas, 1960; Wolf, 1960; Sutton, 1975a, b; Osmond, 1978). The genesis of PEP from stored carbohydrate is also indicated by ^{14}C pulse-chase experiments (Kluge *et al.*, 1975) and finally by the finding that the nocturnal consumption of glucan becomes drastically inhibited if the demand for the CO_2-acceptor PEP is kept low either by keeping the CAM performing plant during the night in CO_2-free air or by otherwise inhibiting dark CO_2 fixation (Kluge, 1969a; Sutton, 1975a, b).

The glycolytic breakdown of glucan to PEP has been investigated in detail by Sutton (1975a, b) and Pierre and Queiroz (1979). The enzymatic machinery for glycolysis is present in CAM-performing cells. The mobilization of starch as the initial step seems to be mediated both by α-amylase and phosphorylase (Sutton, 1975a,b; Vieweg and de Fekete, 1977; Schilling and Dittrich, 1979) (see Fig. 2). The investigation of the composition of stable carbon isotopes in metabolites of the CAM pathway led recently to the conclusion that during the night two strictly separated glycolytic pathways operate simultaneously in CAM-performing cells (Deleens and Garnier-Dardart, 1977; Deleens *et al.*, 1979); one of these two pathways is assumed to produce the CO_2-acceptor PEP to operate at the expense of glucan, the other one is part of the respiratory pathway and consumes soluble sugars.

3. SECONDARY PRODUCTS OF DARK CO_2 FIXATION

Tracer experiments have revealed that up to about 10% of the carbon fixed via PEP carboxylase is transferred into Krebs cycle intermediates including citric, isocitric, succinic, and fumaric acids (Kluge and Ting, 1978). These products may be derived not only directly from oxaloacetate produced by the β-carboxylalion reaction, but also from the malate stored in the vacuole, which exchanges slowly with the cytoplasmic malate pool (M. Kluge, unpublished). This process is increased by temperature (B. Heininger and H. Ziegler, personal communication). Thus, at high nocturnal temperatures more malate is converted by respiration than synthesized and stored in the vacuole. As a consequence, high temperatures at night can prevent CAM completely (Kaplan *et al.*, 1976, 1979; von Willert, 1979).

B. The Metabolic Pathways of the Light Period

1. THE CONSUMPTION OF MALIC ACID BY DECARBOXYLATION

For further metabolic consumption the malic acid stored during the previous night has to be released from the vacuole. This process will be discussed later in more detail. Once released from the vacuole, the four-carbon skeleton of malate undergoes decarboxylation yielding free CO_2 and a three-carbon residue. With respect to the pathway by which malate is broken down, two types of CAM plants (see Fig. 2) may be distinguished (Osmond, 1976, 1978; Dittrich, 1979a, b; Kluge, 1979a). In the majority of CAM plants, the consumption is initiated by oxidative decarboxylation according to Eq. 4.

$$\underset{}{\text{Malate}} + \underset{(\text{NAD}^+)}{\text{NADP}^+} \xrightarrow[]{\substack{\text{malic}\\ \text{enzyme}}} CO_2 + \text{pyruvate} + \underset{(\text{NADH}+\text{H}^+)}{\text{NADPH}+\text{H}^+} \tag{4}$$

This reaction is catalyzed by NADP$^+$ or NAD$^+$ linked malic enzyme (Garnier-Dardart, 1965; Dittrich, 1975, 1979a,b; Osmond, 1978). In the CAM performing Liliaceae, Bromeliaceae, Asclepiadaceae, and some CAM plants from other families, malate is converted first to oxaloacetate (reverse of Eq. 2), which undergoes decarboxylation mediated by PEP–carboxykinase (Eq. 5).

$$\text{OAA} + \text{ATP} \underset{}{\overset{\text{PEP carboxykinase}}{\rightleftharpoons}} CO_2 + \text{PEP} + \text{ADP} \tag{5}$$

2. FURTHER PATHWAY OF CARBON DERIVED FROM MALATE

There is no doubt that the CO_2 produced in malate decarboxylation is refixed by the C_3 pathway of photosynthesis in the chloroplasts (Fig. 2). At temperatures below 25°C and high light intensities, the endogenous CO_2 may be refixed completely. However, at higher temperatures, the rate of decarboxylation may exceed photosynthetic refixation of CO_2; thus, even net loss of CO_2 during the day can be observed in certain CAM plants (Lange *et al.*, 1975; Kluge, 1976a,b; Nose *et al.*, 1977; André *et al.*, 1979; Schuber and Kluge, 1979).

The further metabolism of the C_3 residue remaining from the decarboxylation reaction is still under discussion. In the malic enzyme-type CAM plants, pyruvate remains as product of decarboxylation, whereas in the PEP carboxykinase-type CAM plants, PEP is produced. The view that the remaining C_3 fragment can be used in toto for synthesis of carbohydrates is increasingly well supported. It has been proposed that this could proceed via the reversed pathway of glycolysis (Fig. 2; see Haidri, 1955; Holtum, 1979; Osmond and Holtum, 1981). In the case of the malic enzyme-type CAM plants, such gluconeogenesis requires as the initial step the conversion of pyruvate to PEP. This step could be mediated by pyruvate $-P_i-$dikinase (Eq. 6) which has been shown to be active in malic enzyme-type CAM plants (Kluge and Osmond, 1971; Sugiyama and Laetsch, 1975; Holtum, 1979; Spalding *et al.*, 1979a).

$$\text{pyruvate} + \text{ATP} + P_i \xrightarrow{\text{pyruvate}-P_i-\text{dikinase}} \text{PEP} + \text{AMP} + PP_i \quad (6)$$

In those CAM plants where PEP is the direct product of decarboxylation, pyruvate–P_i–dikinase activity is lacking (Holtum, 1979). Apart from the preceding arguments which favor the view of direct conversion of the three-carbon fragments into carbohydrates, there are also findings which suggest that at least a part of the C_3 fragment pool is further broken down to CO_2 by complete oxidation, which finally becomes substrate of the Calvin cycle (Champigny, 1960; Szarek and Ting, 1974b; André *et al.*, 1979).

3. CONSUMPTION OF EXOGENOUS CO_2 IN THE LIGHT

Although CAM plants may harvest CO_2 from the atmosphere under certain conditions exclusively at night, net CO_2 uptake during the light period can, normally, also be observed. However, as it can be seen from Fig. 1 and as it will be discussed later in more detail, this daytime CO_2 uptake occurs mainly at the beginning of the light period [phase II of the curve shown in Fig. 1] and towards its end (PIV of the curve). There is increasing evidence that in each of these phases the external CO_2 is metabolized differently (see Fig. 2) (Osmond, 1976, 1978; Kluge, 1979a,b; Meyer, 1979). During the "morning burst" (PII) of CO_2 uptake, malate from the vacuole does not seem to be available for light-dependent consumption (Kluge, 1968a,b), and malate synthesis and accumulation may even be continued during PII (M. Kluge, unpublished). Tracer experiments performed by Osmond and Allaway (1974) and Meyer (1979) suggest that this malate synthesis may proceed at the expense of external CO_2 fixed by PEP carboxylase (Figs. 1 and 2). Some of the exogenous carbon might enter directly into the Calvin cycle (Figs. 1 and 2), and it has been shown by Meyer (1979) that the proportion of CO_2 fixed directly by the C_3 pathways increases with the duration of the morning burst.

Phase III of the gas exchange curve (Fig. 1) is characterized by the decarboxylation of malate and refixation of malate-derived CO_2. Cockburn *et al.* (1979), Spalding *et al.* (1979b), and Kluge *et al.* (1981a) have shown that the production of CO_2 from malate increases the internal CO_2 concentration up to 4% (vol/vol). It is, therefore, not amazing that the endogenous CO_2 is preferentially used and exogenous CO_2 is more or less excluded from the sites of photosynthesis during PIII.

Phase IV of the gas exchange curve (Fig. 1) represents the only situation during the diurnal CAM cycle where a sort of steady state net CO_2 uptake may be maintained (see Osmond and Allaway, 1974; Osmond and Björkmann, 1975). Tracer experiments suggest that the bulk of the external CO_2 flows directly via the Calvin cycle into carbohydrates (Osmond and Allaway, 1974). However, together with Calvin cycle products the malate pool also gains external carbon substantially; at light intensities below 10,000 lux malate may represent even the main labeled product if $^{14}CO_2$ is fed during PIV to CAM-performing leaves, and thus the malate level may increase (Kluge, 1969a,b). It can be therefore concluded that, in contrast to PIII, the PEP-carboxylase pathway is capable of operating in PIV, i.e., during the late phase of the light period. Results of Björkman and Osmond (1974), Kluge *et al.* (1975), and Osmond and Björkman (1975) suggest that during PIV at least parts of the PEP needed for the PEP–carboxylase mediated CO_2 fixation might be derived from a C_3 pool generated by the Calvin cycle.

4. GENERAL ASPECTS OF CARBON FLOW IN CAM

As true for C_4 photosynthesis (see Bassham and Buchanan, this volume), CAM does not represent an alternative to the Calvin cycle, i.e., the C_3 pathway of photosynthesis. Rather, the Calvin cycle is also an essential constitutent of CAM. It is only by this pathway that CO_2 is reduced to carbohydrates at the expense of solar energy. All the metabolic steps apart from the Calvin cycle contributing to CAM represent nothing more than an auxiliary mechanism, which is capable of facilitating photosynthetic carbon assimilation under arid environmental conditions (see Section V).

The malic acid, which accumulates during the night, represents a storage for CO_2, which is filled during the night by the uptake and the fixation of external CO_2. This storage pool is emptied afterwards during the day by decarboxylation of the malic acid and the final assimilation of the malate-derived CO_2 in photosynthesis. These interrelationships are clearly evidenced by pulse-chase tracer experiments (Fig. 3).

CAM allows not only reversible storage of CO_2 in the dark until the light period, but in CAM, light energy is trapped by photosynthesis and stored in photosynthetic products (mainly starch) for final use in the dark period. Starch derived from photosynthesis provides the source for the energy-rich CO_2-acceptor PEP. The high energy content of PEP allows the nocturnal CO_2 fixation to proceed in an exergonic reaction. Hence, the reversible storage of CO_2 and light energy allows a separation in time of the PEP carboxylase mediated primary CO_2 fixation, from the final CO_2 assimilation by the C_3 pathway of photosynthesis. It must be kept in mind that it is this temporal separation of the two CO_2-fixing steps that provides the basis of the ecological advantage of CAM (see section V).

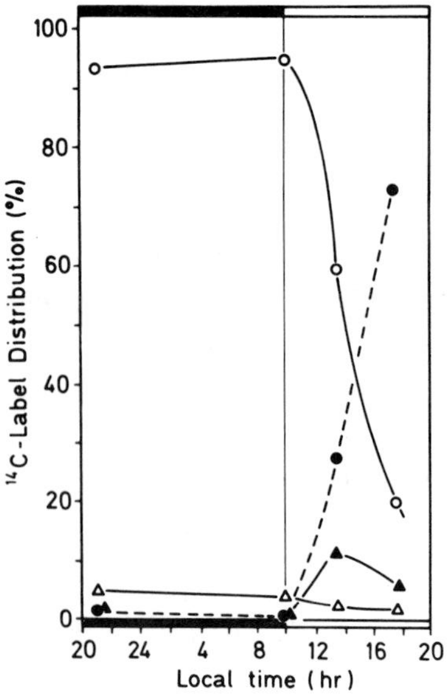

FIG. 3. **Flow of carbon in CAM shown by a ^{14}C pulse-chase experiment. The CAM plant (*Kalanchoë tubiflora*) was labeled by allowing the plant to fix $^{14}CO_2$ for 30 min at the beginning of the night. Samples were taken and analyzed for label distribution among metabolites at indicated times. ○, malate; ●, glucan; ▲, soluble sugars; △, amino acids. Experimental conditions: 15°C at night, 25°C during the light period; light intensity: 25,000 lux.**

5. PHOTORESPIRATION IN CAM PLANTS

It is now well established that the CAM plants are capable of photorespiration (Osmond, 1976, 1978; see Chapter 7, this volume). As true in other plants, RuBP carboxylase of CAM plants has oxygenase activity (Björkmann and Osmond, 1974; Badger *et al.*, 1975; Osmond and Björkmann, 1975); the enzymes of the glycolyte pathway have also been shown to be active in CAM plants (Osmond, 1976; Herbert *et al.*, 1978); and the glyoxysomes of CAM cells have been studied by electron microscopy (Kapil *et al.*, 1975). Since during the phase of malate consumption (PIII of Fig. 1) the internal CO_2 concentration may rise up to 4% (Cockburn *et al.*, 1979; M. Kluge, unpublished) and the internal O_2/CO_2 ratio is low during this phase (Spalding *et al.*, 1979b), it is conceivable that oxygenation of RuBP and thus photorespiration is reduced and the initial reaction of the Calvin cycle is shifted more in the direction of carboxylation during PIII (Osmond, 1976; Spalding *et al.*, 1979b).

C. The Compartmentation of CAM

It is a characteristic of CAM that its individual steps (i.e., malate synthesis, storage of malic acid, and malate consumption by photosynthesis) are located together in one and the same cell. However, the complexity of CAM can be only understood if we assume cellular compartmentation of the metabolic processes involved. The details of CAM compartmentation are still under investigation and remain controversial.

1. GENERATION OF PHOSPHOENOL PYRUVATE AND NOCTURNAL CO_2 FIXATION

In contrast to the photosynthetic refixation of malate-derived CO_2, which must occur in the chloroplast, the location of the primary CO_2 fixation (Eq. 1), i.e. of the PEP carboxylase, is much less clear. Schnarrenberger *et al.* (1980) interpreted results on separation and characterization of cell organelles from CAM plants as showing PEP carboxylase to be exclusively located in the chloroplasts. However, the interpretation of Schnarrenberger *et al.* is contradictory to that of the others which favor the view that the PEP carboxylase in CAM performing cells is not associated with chloroplasts, mitochondria, or peroxisomes, but with the cytoplasm (Kluge and Ting, 1978; Spalding *et al.*, 1979a; K. Winter, personal communication).

Based on their elegant compartmentalization analysis of the key enzymes, Spalding *et al.* (1979a) proposed the metabolic sequence leading from glucan to 3-PGA (see Fig. 2) to be located in the chloroplasts, from where 3-PGA is exported into the cytoplasm. The further conversion of 3-PGA to PEP, the β-carboxylation of PEP, and finally the reduction of OAA to malate is assumed, by these authors, to proceed in the cytoplasm. The preceding hypothesis, which assumes that the glycolytic conversion of glucan to PEP might take place at least partially in the chloroplast, is in agreement with the postulate of two independent and spatially separated glycolytic pathways operating simultaneously during the nocturnal phase of CAM (Deleens and Garnier-Dardart, 1977; Deleens *et al.,* 1979).

2. MALIC ACID TRANSPORT ACROSS THE TONOPLAST AND ITS STORAGE IN THE VACUOLE

It is widely accepted that malate, once synthesized by dark CO_2 fixation, is transported into the vacuole where it is kept temporarily out of further metabolism, and it is in the vacuole where it accumulates during the night. However, it is difficult to understand in this picture why ^{14}C-label, which is transferred by dark $^{14}CO_2$ fixation primarily into the carbon atom number 4 of the malate molecule (Sutton and Osmond, 1972), becomes randomized afterwards so that about one-third of the label occurs in carbon number 1, whereas two-thirds remain in carbon number 4. This label randomization is most likely due to the activity of fumarase (Sutton and Osmond, 1972; Dittrich, 1976; Winter and Kandler, 1976). As true in other plants a fumarase, being a constitutent of the Krebs cycle, should be located in the mitochondria. Hence, randomization due to fumarase would suggest quick exchange of the freshly synthesized malate between cytoplasm and mitochondria, before it is finally transported into the vacuole. However, even the malic acid stored in the vacuole may exchange, at a low rate, with the cytoplasm (Dittrich, 1976; Osmond, 1978; M. Kluge, unpublished), which should result in further label randomization.

The arguments that the vacuoles are the sites where malate is stored in the form of malic acid come from indirect (Kluge and Heininger, 1973) or direct compartmentalization analysis (Buser and Matile, 1977; Kenyon *et al.,* 1978) and, even more convincingly, from biochemical considerations (cf. Kluge and Ting, 1978; Osmond, 1978; Lüttge and Ball, 1979). The intracellular transport and storage of malic acid must be part of the regulatory circuit that controls the diurnal CAM cycle (Wolf, 1960; Kluge and Heininger, 1973; Lüttge *et al.,* 1975; Kluge, 1976a,b; Lüttge and Ball, 1977; Kluge and Ting, 1978; Osmond, 1978).

Loading of the vacuole with malic acid is obviously an active process, but its mobilization from the vacuole proceeds passively (Denius and Homann, 1972; Lüttge and Ball, 1974, 1979; Lüttge *et al.,* 1975). Lüttge

(1980) and Lüttge and Ball (1979) showed that the primary process of loading the vacuole with malic acid consists of an active transport of protons across the tonoplast linked with a passive cotransport of the malate anion (Fig. 2). These authors also assumed that the passive transport of malate anions might be mediated by a yet unknown translocator and hence represent an example of "catalyzed diffusion."

The mechanism of the increased release of malic acid from the vacuole, during the light period, is less well understood than the loading process. There are arguments in favor of light signals being the trigger for the increased malic acid export from the vacuole (Nalborczyk *et al.*, 1975). It has been deduced from experiments where malate efflux from leaf slices (suspended in media of varied water potential) was measured that the turgor, increased by malate uptake and concomitant flow of water into the vacuole, may control export of malic acid (Lüttge *et al.*, 1975). However, the relevance of this turgor mechanism to the *in vivo* operation remains to be evaluated (Lüttge *et al.*, 1977; see also Osmond, 1978).

The diurnally changing level of malic acid in the vacuoles implies a considerable fluctuation of the osmotically determined water potential in the compartment bordering the cytoplasm. The question arises if the cytoplasm can compensate these fluctuations of the vacuolar water potential. Such a compensation appears to be necessary in order to avoid dehydration of the cytoplasm when the malate accumulation drops the vacuolar water potential (cf. Lüttge, 1981). Considering that in CAM plant cells the cytoplasm occupies less than 1% of the vacuolar volume (J.A.C. Smith, personal communication), it is obvious that the accumulation of only small amounts of osmotically effective substances in the cytoplasm would be sufficient to approach the cytoplasmic water potential to that of the vacuole filled by malic acid. It is conceivable that the glycolytic breakdown of starch during the night could, at least, partially account for the compensation of the vacuolar solute potential in the cytoplasm. Also, the finding of Morel *et al.* (1980) that the level of quarternary polyamines increases in correlation with the malate accumulation in the vacuole could be interpreted in terms of cytoplasmic osmoregulation during the diurnal CAM cycle, provided that these polyamines are located in the cytoplasm and the molecules are small enough to be osmotically effective.

3. METABOLIC STEPS OF THE LIGHT PERIOD

There is no doubt that the refixation of CO_2 derived from malate proceeds in the chloroplasts where also the end product of photosynthesis, glucan, is stored. The location of the malate decarboxylation steps is much more difficult to find. According to Dittrich (1979a,b), the PEP carboxykinase (Eq. 5) is not in the chloroplast, but most likely in the cytosol and mitochondria. We do not yet know where the product of the PEP carboxykinase reaction, PEP, is further converted (see Fig. 2). The situation in the case of the malic enzyme-type of CAM is likewise poorly understood. According to Spalding *et al.* (1979a), Dittrich (1979a,b), and Arron *et al.* (1979), the $NADP^+$ malic enzyme exists in the cytosol, and the NAD^+ malic enzyme mainly in the mitochondria. It is therefore assumed (Arron *et al.*, 1979) that malate, as far as not decarboxylated in the cytosol, might partially be taken up into the mitochondria and decarboxylated there. Because CAM plant mitochondria have only a poor capacity to oxidize pyruvate, the product of malate decarboxylation, Arron *et al.* (1979) postulated that pyruvate is transported into the cytoplasm for further reactions (see Fig. 2) instead of being completely oxidized to CO_2 in mitochondria. On the other hand, von Willert and Schwöbel (1978) demonstrated an extraordinarily high capability of CAM plant mitochondria to oxidize malate completely to CO_2.

If malate is only decarboxylated to pyruvate (Eq. 4) instead of being oxidized to CO_2, the residual pyruvate has to be converted into PEP (Eq. 6). This process should proceed in the chloroplasts because, by analogy to the situation in the C_4 plants, pyruvate–P_i–dikinase, the enzyme responsible for this step, may be located in the chloroplast (Hatch and Osmond, 1976).

D. Control of CAM

The problem of CAM control has become one of the most intensively investigated fields. We are still far away from a generally accepted concept (see the reviews of Queiroz, 1974, 1975, 1979; Kluge, 1976b; Kluge and Ting, 1978; Osmond, 1978; Dittrich, 1979a,b). It is generally accepted that the CAM enzymes (see Dittrich, 1979a) are of great importance in the mechanisms controling CAM. However, the whole complex of CAM control cannot be expected to be interpretable solely in terms of pure enzymology. Rather, the cellular compartmentation of enzymes and metabolites together with intracellular transport processes can be predicted to be significantly involved. As the knowledge in this field is still rather poor, all present explanations of CAM regulation are based necessarily on tentative models.

The problem of CAM regulation has to be considered under two aspects, i.e., regulation during a long duration of time (for instance, throughout the seasons) and regulation of CAM throughout its diurnal cycle.

1. INDUCTION AND SEASONAL MODULATION OF CAM

The features of CAM such as gas exchange pattern and diurnal acid rhythm may undergo considerable alterations during the seasons, along the duration of stress, or during varying photoperiods (see the reviews cited earlier). In many cases, CAM may be induced, i.e., plants may shift from C_3 photosynthesis to CAM by water and salt stress (Winter and von Willert, 1972; Winter and Lüttge, 1976, 1979; von Willert, 1979), or by photoperiodic treatment (Gregory *et al.*, 1954; Queiroz, 1974, 1975, 1979). The analysis of the control mechanisms behind these phenomena is difficult because ontogenetic processes may interfere with the induction phenomena (Winter *et al.*, 1978; von Willert, 1979). Nevertheless, it appears that the long term regulation of CAM is mediated to a great extent via variations of enzyme capacity* (i.e., the maximal extractable enzyme activity) of PEP carboxylase (Queiroz and Morel, 1974a,b). During the induction of CAM either by stress or photoperiod, the capacity of this enzyme increases drastically (see the reviews by Queiroz, 1974, von Willert, 1979; Winter and Lüttge, 1979). In addition, during the increase of PEP carboxylase, a new protein having PEP-carboxylase activity is synthesized, which can be distinguished by its different electrophoretic and kinetic properties (von Willert *et al.*, 1976, 1977a; Brulfert *et al.*, 1979). The induction of CAM by photoperiod in *Kalanchoe blossfeldiana* is under control of the phytochrome system (Brulfert *et al.*, 1973, 1975).

In obligate CAM plants , the activity of CAM fluctuates with the seasons (Bartholomew, 1973; Szarek and Troughton, 1976; Schuber and Kluge, 1979), and the period of high CAM activity is correlated with a high level of PEP carboxylase (M. Schuber and D. Müller, in M. Kluge's laboratory, unpublished). Under natural conditions, seasonal fluctuations of CAM may also largely be determined by the hydroperiod (see Section V).

2. CONTROL OF THE DIURNAL CAM CYCLE

There is an overwhelming number of arguments in favor of the view that even in the diurnal control of CAM the regulation of PEP carboxylase plays a central role. During the night, the activity of this enzyme should be high because of the efficient initiation of the CAM cycle by high rates of dark CO_2 fixation. During the day, however, PEP-carboxylase activity should be low in order to avoid competition with the Calvin

*The capacity of an enzyme is defined as the maximum achievable *in vitro* enzyme activity (Queiroz, 1979). This term has to be distinguished from the *in vivo* activity of the enzyme, which is not known.

cycle when the malate derived CO_2 has to be fixed via this cycle. Queiroz and his co-workers postulated endogenous rhythms (i.e., rhythms controlled by a biological clock) in the capacity of all enzymes responsible for such a regulation (see the reviews by Brulfert *et al.*, 1973, 1975; Queiroz, 1974, 1975, 1979; Queiroz and Morel, 1974a,b; Pierre and Queiroz, 1979). Other models favor the view that the allosteric properties of PEP carboxylase, together with the transport processes of malate, are the main factors in the regulatory process (see Kluge, 1976a,b; Kluge and Ting, 1978; Osmond, 1978). Both endogenous rhythmicity and allosteric regulation might cooperate in controlling the diurnal CAM cycle (Queiroz and Morel, 1974a,b).

a. Control during the Night. Both the carboxylation step (Eq. 1) and the generation of the CO_2 acceptor PEP are regulated. The PEP carboxylase from CAM plants have been purified and characterized (Jones *et al.*, 1978); it appears to be an allosteric enzyme, which can be activated and inactivated by various effectors. Malate is a powerful inhibitor of this enzyme with a K_i in the range of 0.1–0.4 m*M* (Buchanan-Bollig *et al.*, 1980). Now the question arises as to how, during the night, the feedback inhibition of PEP carboxylase by malate is prevented in order to allow the continuation of malic acid synthesis. During the night, the vacuole represents a strong sink for malate (Kluge, 1977b), thus malate is quickly removed from the cytoplasm where PEP carboxylase is assumed to be located. It has further been shown that the sensitivity of this enzyme to malate is lower during the night than in the day (von Willert *et al.*, 1977b, 1979; Kluge *et al.*, 1980, 1981b; Winter, 1980, 1981). The activity of this enzyme may further change as it is activated by glucose 6-P (Ting and Osmond, 1973) and inorganic phosphate (von Willert, 1975). The level of both metabolites have been shown to be particularly high during the night in CAM-performing cells (von Willert, 1975; Cockburn and McAulay, 1977). Finally, it has been found (Kluge *et al.*, 1981b; Winter, 1981) that affinity of the PEP carboxylase to PEP undergoes drastic diurnal changes; it is higher during the dark than during the light period. The enhanced affinity of PEP carboxylase to its substrate PEP would provide a further mechanism to maintain high rates of dark CO_2 fixation, i.e., malate synthesis.

The phosphofructokinase-catalyzed reaction is the crucial step where glycolysis is regulated. This enzyme is normally severely inhibited by PEP. However, in CAM plants the sensitivity of phosphofructokinase to inhibition by PEP is two orders of magnitude lower if compared with other plants (Sutton, 1975a,b), i.e., effective glycolytic production of PEP may be maintained during nocturnal CAM even with PEP concentrations being high. Schilling and Dittrich (1979) have shown that the phosphorolytic cleavage of starch, i.e., the initial step of PEP production from glucan, is also under metabolic control.

b. Control during the Day. All hypotheses dealing with the CAM-regulation mechanism of the light period should explain how competition for the malate-derived CO_2 between the PEP-carboxylase pathway and the Calvin cycle is avoided. Here, again, the regulation of PEP carboxylase seems to be centrally involved. Since after the dark period malate is released rigorously from the vacuole, it is expected that malate concentration will increase in the cytoplasm to such an extent that it will become inhibitory to the enzyme. It has been shown that the sensitivity of this enzyme against malate increases during the day (von Willert *et al.*, 1979; Winter, 1980, 1981; Kluge *et al.*, 1981b). Greenway *et al.* (1978) have suggested that this effect might be due to two forms of the enzyme having different properties, one form being active during the night, and the other during the day. However, von Willert and von Willert (1979) and, von Willert *et al.* (1979) suggest that light deactivates this enzyme in certain Mesembryanthemaceae. Since there is a considerable lag phase in the effect, it could only be an indirect effect. The same holds true for the inhibitory effects of high temperatures on this enzyme *in vivo,* as shown by B. Heininger and H. Ziegler (unpublished): the temperature effect might reflect primarily an increased export of malic acid from the vacuole with concomitant inhibition of the enzyme by the exported malate.

The activity of PEP carboxylase during the day may also be diminished by a short supply of PEP because C_3 skeletons needed for PEP synthesis might be preferentially used in photosynthesis (Osmond and Allaway, 1974). The decrease in the affinity of PEP carboxylase to PEP, which occurs in CAM plants during the day (Kluge *et al.*, 1981b; Winter, 1981), could be a part of this latter mechanism.

Regulation of CAM during PIV (see Fig. 1) seems to be particularly complicated. Because malate has now been consumed, no further feedback inhibition of PEP carboxylase by malate can be expected. Indeed, labeling patterns obtained after $^{14}CO_2$ fixation in the light reveal that carbon can now substantially be transferred also into malate. This indicates a successful competition for CO_2 of the PEP carboxylase with the Calvin cycle enzymes (Kluge, 1969a,b).

IV. Gas Exchange Linked with CAM

A. Pattern of CO_2 Exchange and Its Modulation

The classical pattern of CO_2 exchange of CAM plants is shown in Fig. 1. It is characterized by net CO_2 uptake during the night (PI) and a depression of CO_2 uptake during the day (PIII). In the light period, net CO_2 uptake may occur at the beginning of the day (PII) and at its end (PIV). This basic pattern of CAM-linked gas exchange is easily modified by external or developmental factors (Neales, 1975; Kluge and Ting, 1978). For instance, the CO_2 uptake during the day can be completely suspended, and thus external CO_2 be fixed exclusively at night, or, CAM plants may lack for a limited time nocturnal CO_2 fixation and fix CO_2 during the day, and, thus behave like a C_3 plant.

In a well-irrigated CAM plant, the CO_2 exchange is primarily determined by the diurnal course of CAM proceeding in the mesophyll cells rather than by the behavior of the stomata (V,B). External factors, which influence the metabolic machinery of CAM, must therefore influence also the gas exchange pattern. This postulate does not, however, exclude that under certain conditions, for instance water stress, the stomatal movements may interfere and modify the diurnal pattern of CO_2 exchange. On the basis of the available data (Kluge and Ting, 1978; Meyer, 1979), the following generalization can be deduced: All external factors or experimental manipulations that stimulate malate consumption and photosynthesis during the day will enhance CO_2 fixation during the following night, and any stimulation of nocturnal CO_2 fixation, and thus malic acid accumulation, will extend the depression of CO_2 uptake (i.e., PIII) during the following day.

1. NOCTURNAL CO_2 EXCHANGE

The amount of the nocturnal CO_2 exchange is modulated mainly by temperature, by water status, by leaf age, and by the light intensity of the preceding day. Night temperatures higher than 20°C tend to reduce the nocturnal CO_2 uptake and concomitant malic acid synthesis. Kaplan *et al.* (1976, 1977) have shown that this effect is due to the temperature-dependent increase of respiration, which competes with CO_2 fixation and is not caused by direct thermal inhibition of PEP carboxylase-mediated β-carboxylation. Brinckmann and von Willert (1979) and von Willert *et al.* (1979) observed that this effect might be relevant to the ecology of certain CAM plants in the Namib desert (see Section V).

High light intensities during the day, particularly when combined with high temperatures, increase the dark CO_2 fixation during the following night. In such situations, the vacuoles, the storage organelle for malic acid, are effectively emptied, and a high level of glucan is produced from where PEP can be generated during the dark period. This is followed by increased CO_2 uptake.

In certain CAM plants, the photoperiod is of great significance. The best example is that of *Kalanchoe blossfeldiana* var. Tom Thumb. Here the nocturnal CO_2 fixation increases drastically when these plants are cultivated under short days (see the pioneering work of Gregory *et al.*, 1954, and the reviews by Queiroz, 1974, 1979).

As discussed earlier, the nocturnal net CO_2 uptake can be induced by water and salt stress (see Winter and Lüttge, 1976, 1979) in Mesembryanthemaceae, in *Sedum acre* (Kluge, 1977a, 1978), and in *Portulacaria afra* (Ting and Hanscom, 1977). In obligate CAM species of *Tillandsia,* dark CO_2 fixation may be enhanced by water stress (Kluge *et al.*, 1973; Medina *et al.*, 1977). However, in the majority of CAM plants, water stress tends to reduce the nocturnal CO_2 fixation (Kluge and Fischer, 1967; Meinzer and Fundel, 1973; Meyer, 1979). Under severe drought, dark CO_2 fixation may disappear completely because stomata are kept closed, and then CAM depends completely on the refixation of respiratory CO_2 (Szarek and Ting, 1974a, b).

Leaf age has a considerable effect on dark CO_2 fixation. Generally, the youngest leaves, which are not yet fully expanded, show lower rates of dark CO_2 fixation than mature leaves. With progressing senescence, the capability of the leaves to fix CO_2 in darkness decreases again (see, e.g., Meyer, 1979). It is conceivable that water stress accelerates maturing of the leaves and thus of the occurrence of CAM including nocturnal CO_2 fixation. Hence, the preceding induction of CAM by water stress may finally reflect ontogenetic processes (Winter *et al.*, 1978; von Willert, 1979).

2. CO_2 EXCHANGE DURING THE DAY

It is well documented that the typical depression of CO_2 uptake during the day is caused by the preferential usage of malate-derived CO_2 in photosynthesis (Kluge, 1969a, b, Meyer, 1979) as outlined in detail by Kluge and Ting (1978). The duration of the depression of daytime CO_2 fixation (i.e., PIII, see Fig. 1) depends therefore on the amount of malate accumulated during the preceding night, i.e., on external stimuli that influence the nocturnal malate accumulation. It also depends upon factors that influence the velocity of malate consumption. For instance, light intensities and high temperature tend to accelerate the removal of malate and thus shorten the duration of PIII, whereas low light intensity and low temperature cause the opposite.

At high temperatures, the rate of CO_2 production by malate decarboxylation may become so high that refixation of the endogenous CO_2 by photosynthesis becomes saturated and net CO_2 output may occur during PIII (Lange *et al.*, 1975; André *et al.*, 1979; Meyer, 1979; Schuber and Kluge, 1979).

Water stress tends to reduce CO_2 uptake during the day. As dark CO_2

exchange is less susceptible to drought, CAM plants can harvest external CO_2 exclusively during the night if water becomes deficient (Kluge and Fischer, 1967; Kluge, 1976a).

The initial burst of daytime CO_2 uptake (PII in Fig. 1) is also easily modulated by external factors. It disappears under water stress and after prolonged dark periods (Kluge and Fischer, 1967; Marcelle, 1975; Meyer, 1979). The duration of PII is extended after increase in night temperature that inhibits malate accumulation, and its duration is shortened with increase in day temperatures (Meyer, 1979).

B. The Behavior of Stomata and Transpiration in CAM

1. GENERAL PHENOMENOLOGY

One of the most interesting and ecologically relevant features of gas exchange control in CAM plants is that the stomata may open at night and close during the day. A typical pattern of the diurnal course in the stomatal resistance is given in Fig. 1. However, as true for the CO_2 exchange pattern, any given CAM plant may show variations in this basic pattern of stomatal movement, mainly in response to environmental factors. For example, the stomata may open only toward the end of the night and may remain more or less open during the entire light period, or the stomata may be open throughout the night and close during the entire day (Neales, 1975; Kluge and Ting, 1978).

2. CONTROL OF STOMATA MOVEMENTS

It has been demonstrated by Kluge and Fischer (1967) that the peculiar CAM pattern of stomatal movements must be determined by the CO_2 exchange of the mesophyll cells, instead of the CO_2 exchange pattern being primarily determined by the stomata (see also Cockburn *et al.*, 1979; Spalding *et al.*, 1979b; Kluge *et al.*, 1981a). According to Kluge and Fischer (1967), the CAM processes proceeding in the mesophyll cells and the movement of the stomata are assumed to be coupled by the CO_2 concentrations in the substomatal intercellular spaces. It is well known that stomata open at low intercellular CO_2 concentrations, and close at high CO_2 concentrations (Raschke, 1975). Hence, during the night the stomata can be expected to open because the powerful dark CO_2 fixation keeps the CO_2 concentrations low. On the other hand, during the day decarboxylation of malic acid creates high intercellular CO_2 concentrations which should force the stomata to close as long as CO_2 is produced from malic acid. This hypothesis has been experimentally supported by Kluge (1968b) who showed that the duration of the depression of CO_2 uptake during the day (i.e., the duration of stomatal closure) depends on the duration of malate decarboxylation. It has been further supported by direct measurements of internal CO_2 concentrations in CAM-performing tissues (Cockburn *et al.*, 1979; Spalding *et al.*, 1979b; Kluge *et al.*, 1981b).

The preceding causal relationship between carbon metabolism governing the CO_2 exchange by the mesophyll cells and the stomatal movements imply that CAM is capable of regulating rather directly the transpiration, i.e., the water balance of the plant (Allaway *et al.*, 1974). The ecological relevance of the close relationship between carbon metabolism and water balance will be discussed later.

There is evidence that the preceding model of primary control of stomatal movements by CAM-determined internal CO_2 concentration is only valid for conditions where the plant water potential is optimal. Under water stress, plant water potential becomes the dominant factor controlling the stomatal movements. Even a rather small drop in the plant water potential is sufficient to cause a tendency of the stomata to close (Osmond, 1978; Osmond *et al.*, 1979b). The daytime opening of stomata is more sensitive to water stress than dark CO_2 fixation; hence, CO_2 may be fixed by CAM plants exclusively at night (Kluge and Fischer, 1967; Neales *et al.*, 1968; Neales, 1975; Osmond, 1976, 1978). This situation can be maintained for longer durations of time. Under more severe water stress, CAM plants also close the stomata at night, and then CAM depends completely upon the recycling of respiratory CO_2 (Szarek *et al.*, 1973). It is conceivable that the sensitivity of the response of the stomata to a small drop in the leaf water potential is mediated by the extremely high water conductance, which is typical for CAM plant cells (Steudle *et al.*, 1980).

In well-watered CAM plants at night, the stomata may respond also to changes in the H_2O vapor pressure difference between leaf and ambient air. High air humidity (i.e., a small vapor pressure difference) tends to open the stomata (Schulze *et al.*, 1974; von Willert *et al.*, 1977b, 1979; Lange and Medina, 1979; Osmond *et al.*, 1979b) in the night. Such positive effects of air humidity on the opening of stomata during the day are much less clear (Osmond *et al.*, 1979a, b). It should also be noted that in not all CAM plants do stomata respond to the changes in water vapor concentrations of the air. For instance, the stomata of *Agave deserti* were found to behave inertly against any drop in the air humidity (Nobel and Hartsock, 1979).

Temperature is another factor modulating the pattern of stomatal movements in CAM plants. Normally, an increase of the leaf temperature above 15°C tends to close the stomata both at night and day (Ting *et al.*, 1967; Neales, 1973; Nobel and Hartsock, 1979; see also Kluge and Ting, 1978). However, it is difficult to distinguish between direct temperature effects on stomata from indirect effects mediated by temperature dependent alterations of water status and mesophyll carbon metabolism.

3. TRANSPIRATION, WATER USE EFFICIENCY, AND WATER RELATIONS IN CAM PLANTS

Since both the CO_2 exchange and the transpiration are under the control of stomata, higher plants, in particular when growing in arid habitats, face a physiological dilemma: Water is necessarily lost to the environment by transpiration if the stomata are opened to allow CO_2 uptake for assimilation. CAM may reasonably be interpreted as a mechanism capable of minimizing the loss of water, because it is obvious that the uptake of external CO_2 through the opening of stomata at night, when the evaporative demand of the atmosphere is clearly lower, should substantially reduce the transpiratory water loss (Allaway *et al.*, 1974). During the day, however, CAM plants may close the stomata thus reducing transpiration. Nevertheless, even behind the closed stomata, photosynthesis does not stop because the nocturnally stored malic acid serves as an endogenous source of CO_2. Hence, CAM allows harvesting of carbon from the atmosphere with only a low cost of water.

The previously mentioned view is supported by many observations showing clearly that CAM plants have a much higher "water use efficiency" (i.e., a low transpiration ratio; see Table I) for carbon assimilation than other higher plants. This is particularly true if the environmental conditions prevent CO_2 uptake during the day (PII and PIV of the gas exchange curve, see Fig. 1), and external CO_2 is taken up exclusively during the night. CO_2 uptake during the day can proceed only at the expense of an increased transpiratory water loss, i.e., the water use efficiency of daytime CO_2 uptake approaches that of C_3 plants (Osmond, 1978; see Table I). In this context, it is worth noting that CAM plants, when under water stress, restrict first the daytime CO_2 uptake (i.e., day time opening of stomata) and only then nocturnal CO_2 uptake (Kluge and Fischer, 1967; Neales *et al.*, 1968; Bartholomew, 1973; Kluge, 1976a).

In CAM plants, the water-saving CAM syndrome is supported by a variety of anatomical and morphological adaptations, such as high cuticular resistances (Bartholomew and Kadzimin, 1975; Nobel, 1976, 1977; Kluge and Ting, 1978), low number of stomata per surface unit (Ting *et al.*, 1972), peculiarities of the stomatal anatomy (Kluge and Ting, 1978), reduction of the transpiring surfaces (Ting *et al.*, 1972; Kluge and Ting, 1978), and morphological properties, which are capable of minimizing heat absorption, such as stem ribs in cacti (Nobel,

TABLE I
CO_2 and H_2O Exchange Parameter in CAM, C_4, and C_3 Plants[a]

	CAM				
	Soil water potential (bar)				
Property	−1	−5	−50	C_4	C_3
Water use efficiency, expressed as transpiration ratio (g H_2O released/g CO_2 assimilated)	50–600	18–50	Zero	250–350	450–950
Growth rates ($g \cdot m^{-2} \cdot day^{-1}$)	5–20	0.5–1.5	Zero	400–500	50–200
Δ malic acid (% maximum)	100	50–80	25	—	—
Exchange of external CO_2/H_2O					
Phase I	+	+	±		
Phase II	+	+	−		
Phase IV	+	−	−		

[a]For the CAM plants, the values show the response to decreasing soil water potentials. The PI, PII, and PIV refer to the gas exchange curve shown in Fig. 1. After Osmond (1978) and Black (1973).

1980). Finally, under water stress, CAM plants may even reduce its area of active roots thus isolating the plant body from the soil having very low water potentials (Kausch, 1965; Szarek and Ting, 1974a, 1975).

As a consequence of all of the preceding biochemical and morphological adaptations, the water potential in the physiologically active tissues can be maintained high even if the water potential of the environment, i.e., soil and atmosphere, drops down to the very low values. Szarek and Ting (1974a, b) found that in *Opuntia* the tissue water potential never fell below −15 bars even if the soil water potentials remained for months as low as −80 bars (Kluge and Ting, 1978, p. 154).

Another factor that helps to maintain a high water potential in the CAM-performing tissue is the high capability of CAM plants to transfer water between cells and tissues, i.e., the high water conductance (Steudle *et al.,* 1980). Whenever water is lost, it can be replaced from other parts of the plant, which then die (S. R. Szarek, cited in Kluge and Ting, 1978; Nobel, 1977). The strategy behind this property could be the attempt to keep at least the apical meristems and parts of the assimilatory tissue alive when the water availability from the soil becomes zero for long durations of drought. There is evidence that the high effectiveness of water transfer in CAM plant tissues is based upon an extremely high hydraulic conductivity of the plasmalema and high elastic modulus of CAM plant cells (Steudle *et al.,* 1980).

V. Ecology of CAM

The geographical distribution of CAM plants (cf. Kluge and Ting, 1978) supports the interpretation that CAM represents a variant of photosynthesis that permits harvesting of carbon with low cost of water; this allows these plants to conquer ecological niches where water availability may become limited. Typically, CAM plants inhabit arid areas of the world or exist in seemingly mesic habitats in locally dry environments, for instance within rocks in shallow soils, or as epiphytes (Schuber and Kluge, 1979). In the Namib Desert and in the Richtersveld (Southwest Africa), more than 80% of the plant species perform CAM (D. von Willert, personal communication). Also in the deserts of the New World, for example in California and Arizona (Ting and Jennings, 1976), CAM plants are widely abundant. CAM plants may be found also in the dry bush of Madagascar (Winter, 1979).

Functioning of CAM as water-saving mechanism anticipates that the plants must have more or less regularly the opportunity to take up water, which is then stored in the succulent organs. In fact, the typical habitat of CAM plants are semideserts where seasonal rainfalls (either summer or winter) alternate with periods of drought. True deserts, which more or less lack regular precipitations, therefore exclude CAM plants. All the field studies on CAM (see Kluge and Ting, 1978; Osmond, 1978; von Willert, 1979) are consistent with the results of laboratory studies outlined earlier. Also, *in situ* CAM is a highly flexible system that is largely affected by environmental factors, in particular by soil water potential and by temperature. Some of these responses emphasize that CAM is capable of adaptation. The best example is the response of CO_2 uptake to decreasing soil water potential. As long as high soil water potential allows maintenance of high plant water potential, CO_2 uptake during the day contributes considerably to the total carbon gain of the CAM plant, thus increasing its productivity (Hanscom and Ting, 1977, 1978). However, under water stress, daytime CO_2 uptake is reduced, and finally external CO_2 is harvested exclusively by the water-saving dark CO_2 fixation. This adaptive tendency can be observed both along climatic gradients (Osmond, 1976) or during seasonal changes in water availability (Bartholomew, 1973; Szarek and Ting, 1974a,b; Nobel, 1976, 1977; Winter *et al.*, 1978; Osmond *et al.*, 1979a,b).

The high water-storage capacity of the succulent CAM plant organs permits nocturnal opening of the stomata, hence continuation of CO_2 uptake for weeks after the soil water potential has dropped below the plant water potential (Nobel, 1976). Under severe water deficits, CAM plants may finally hermetically seal the stomata throughout the diurnal cycle and, hence, may suspend any gas exchange with the environment (Szarek and Ting, 1974a,b). In this situation, CAM functions at the expense of nocturnally refixed respiratory CO_2. In particular, this latter example demonstrates conclusively that CAM represents ultimately a survival mechanism.

The question arises about the adaptive significance of CAM induction occurring in facultative CAM plants. The extended investigations by Winter and Troughton (1978), Winter *et al.* (1978), and von Willert (1979) on *Mesembryanthemacean* species in situ showed that the young plants develop during the rainy season and perform C_3 photosynthesis. During this time, they have high growth rates and productivity. With the onset of the dry season, the plants change from having C_3 photosynthesis to CAM. It is reasonable to postulate that the seasonal shift from the C_3 option of photosynthesis to CAM is accelerated by the decreasing water availability. However, as indicated earlier, it is very likely that ontogenetic factors also play a role along with the water stress-mediated CAM induction (von Willert, 1979; Winter and Lüttge, 1979). It has not yet been observed in situ that facultative CAM plants once induced to perform CAM can completely reverse to the C_3 pattern of photosynthesis (Winter, 1973).

CAM induction seems to play its major role in a seasonal adaptation of photosynthesis rather than as a quick response of the plant to alterations of the microclimate. It is worth noting that seasonally determined changes from C_3 to CAM-type photosynthesis may also be achieved by seasonal dimorphism of the plant rather than by shifts in the metabolic machinery (Lange and Zuber, 1977).

The preceding outline does not exclude that CAM plants may also be capable of fast responses to changes in the microclimate. When remaining under long duration of drought, CAM succulents such as *Agava americana* or *Opuntia basilaris* may seal their stomata completely and, hence, suspend virtually any CO_2 exchange and transpiration (Szarek *et al.*, 1973; Nobel, 1977). On the other hand, by developing shallow functional roots within a few hours, the plants can make use of even a single rainfall to recover the plant water potential by water uptake (Szarek *et al.*, 1973). This allows the reopening of the stomata thus reestablishing the nocturnal CO_2 uptake and high amplitude of the diurnal acid rhythms; this has to be contrasted with the conditions before reirrigation when CAM had to operate exclusively at the expense of recycled respira-

The sensitivity of stomata against changes in the humidity of the ambient air (see Section III,B,2) or, more precisely, against the water vapor pressure difference between plant organ and ambient air, provides a further interesting possibility for short-term control of water use in situ. If the vapor pressure difference is small, the stomatal resistance is low, and when the difference is large, the resistance is high. This suggest that not only the humidity of the ambient air itself, but also changes in the temperature of the plant organ (i.e., due to reradiation or changes in convection), may contribute to the optimization of water use via control of stomatal movements (Osmond *et al.*, 1979b). In *Agave deserti* growing in situ, no response to open stomata on water vapor gradients between leaf and air could be observed (Nobel and Hartsock, 1979).

Temperature is among the major factors modulating CAM in situ. However, the pattern of how CAM plants growing in nature respond to temperature is very varied. Temperature differences between night and day are not an essential prerequisite for CAM performance in situ (Bartholomew, 1973). Generally, high night temperature tends to reduce nocturnal CO_2 uptake and malate accumulation. This effect might be due to the increased stomatal resistance (Neales, 1973; Nobel, 1977; Nobel and Hartsock, 1979) caused by increased temperature. However, as mentioned in IV,A,1, in some CAM plant species high temperature stimulates respiration to such an extent that malic acid is converted via the Krebs cycle rather than being stored (Kaplan *et al.*, 1976, 1977). A dramatic example of this behavior was observed by von Willert (1979) and Brinckmann and von Willert (1979) in Mesembryanthemaceae growing in its natural stands in Richtersveld (Southwest Africa). When hot desert winds raise the night temperature above 30°C, no malic acid rhythm occurs. Rather, the plants lower their malate contents down to hardly measurable levels. In other plants, CAM is less susceptible to inhibition by high night temperatures. For example, nocturnal malic acid accumulation does not change in plants growing in their natural stands along an altitudinally determined temperature gradient in the Californian desert during the hot season (Gulmon and Bloom, 1979).

Some CAM plants while growing in situ show net loss of CO_2 during several hours of the day. This loss was shown to be temperature dependent (Lange *et al.*, 1975; Schuber and Kluge, 1979) and can approach such levels that the total carbon balance of the plant becomes negative, in particular, if the night temperatures are also sufficiently high to inhibit dark CO_2 fixation (Lange *et al.*, 1975). Thus, extremely temperature-sensitive CAM plants are restricted to shady habitats (gaps and crevices between rocks), where heating of the leaves by the full sun radiation is avoided; this is true for *Caralluma negevensis* (Lange *et al.*, 1975), *Dudleya arizonica,* and *Dudleya saxosa* (I. Ting, personal communication).

VI. Productivity of CAM Plants

From the preceding considerations it should have become clear that CAM enables the plants having it to behave in situ as typical drought-resisters (Levitt, 1972). This drought resistance is due to the ability of these plants to collect water during times when it is easily available, to store it in specialized tissues, and finally to minimize the loss of the stored water by the peculiar carbon metabolism called CAM. Hence, as already mentioned CAM is typically a survival mechanism. CAM plants have lower productivity than found in other higher plants (Table I), but they survive in unusual conditions.

The reduced productivity of the CAM plants can reasonably be interpreted in terms of the depression of the daytime CO_2 fixation. As long as there is enough water, the daytime CO_2 fixation (PII and PIV, Fig. 1) contributes substantially to the total productivity of the CAM plant (see Hanscom and Ting, 1977, 1978). Under such optimal conditions, this productivity may approach the lower limit of C_3 plants. For example, the pineapple, *Ananas comosus,* may produce about 44 tons ha^{-1} year^{-1}. However, it must be mentioned that the productivity of pineapple is exceptionally high among CAM plants. Even irrigated *Opuntia* has growth rates of only about 25% of that of *Ananas; Ferrocactus* and *Agave* have under desert rainfall only 8% of pineapple growth rates (see Osmond, 1978).

Considering the inhibition of daytime CO_2 fixation by water stress (see Section III,B,2), it is logical that the productivity of CAM plants decreases with decreasing soil water potential (see Table I). In wet seasons, which allow substantial daytime CO_2 fixation, many CAM plants show high productivity (Bloom and Troughton, 1979). The facultative CAM plants, such as *Mesembryanthemum crystallinum* (Winter *et al.*, 1978), and those with seasonal dimorphism (see the example of *Frerea indica* investigated by Lange and Zuber, 1977) during the wet seasons perform C_3 photosynthesis allowing high productivity. The dry season is then overbridged by performing mainly the water-saving dark CO_2 fixation, but with less productivity, which finally may approach zero (if under water stress, CO_2 uptake is totally suspended). Hence, CAM plants under permanently arid conditions, which allow only dark CO_2 fixation, grow very slowly. For example, a saguaro cactus, *Carnegia gigantea,* having a final height of about 12 m grows about 7 cm in height per year (Hastings and Alcon, 1961); Nobel (1977) calculated from the productivity values that a moderately large barrel cactus about 90 cm tall (and having 35 cm diameter) would be about 54 years old.

The relatively low productivity of CAM plants may be one of the reasons why their economic exploitation (see Kluge and Ting, 1978) has been rather limited. By far, *Ananas* is the most important agriculturally used CAM plant. Also some *Agave* and *Yucca* species are cultivated, mainly for the production of fibers for rope and cord. The flat-stemmed cacti *Opuntia ficus-indica* and *Nopalea* sp. are used as cattle fodder, and the edible fruits of these plants are consumed by man.

We predict that because of the ability of CAM plants to be productive under arid conditions, their economic exploitation will be extended after their potential for production of food, fiber, raw cellulose, and drugs has been more systematically explored.

REFERENCES

Allaway, W. G., Austin, B., and Slatyer, R. O. (1974). *Aust. J. Plant Physiol.* **1,** 397–405.

André, M., Thomas, D. A., von Willert, D. J., and Gerbaud, A. (1979). *Planta* **147,** 141–145.

Arron, G., Spalding, M., and Edwards, G. (1979). *Plant Physiol.* **64,** 182–186.

Badger, M. R., Andrews, T. J., and Osmond, C. B. (1975). *Proc. Int. Congr. Photosynth., 3rd, 1974,* pp. 1421–1429.

Bartholomew, B. (1973). *Photosynthetica* **7,** 114–120.
Bartholomew, B., and Kadzimin, S. B. (1975). *In* "Ecophysiology of Tropical Crops" (P. T. Alvin, ed.), pp. 152–164. CEPLAC, Brazil.
Björkman, O., and Osmond, C. B. (1974). *Year Book—Carnegie Inst. Washington,* **73,** 852–858.
Black, C. C. (1973) *Ann. Rev. Plant Physiol.* **24,** 253–286.
Bloom, A. J., and Troughton, J. H. (1979). *Oecologia* **38,** 35–43.
Bradbeer, J. W., Ranson, S. L., and Stiller, M. (1958). *Plant Physiol.* **33,** 66–70.
Brinckmann, E., and von Willert, D. (1979). *Naturwissenschaften* **66,** 526–527.
Brulfert, J., Guerrier, D., and Queiroz, O. (1973). *Plant Physiol.* **51,** 220–222.
Brulfert, J., Guerrier, D., and Queiroz, O. (1975). *Planta* **125,** 33–44.
Brulfert, J., Arrabaca, M. C., Guerrier, D., and Queiroz, O. (1979). *Planta* **146,** 129–133.
Buchanan-Bollig, I. C., Kluge, M., and Lüttge, U. (1980). *Z. Pflanzenphysiol.* **97,** 457–470.
Buser, C., and Matile, P. W. (1977). *Z. Pflanzenphysiol.* **82,** 462–466.
Champigny, M. L. (1960). Ph.D. Thesis, University of Paris.
Cockburn, W., and McAulay, M. A. (1977). *Plant Physiol.* **59,** 455–458.
Cockburn, W., Ting, J. P., and Sternberg, C. O. (1979). *Plant Physiol.* **63,** 1029–1032.
Deleens, E., and Garnier-Dardart, J. (1977). *Planta* **135,** 241–248.
Deleens, E., Garnier-Dardart, J., and Queiroz, O. (1979). *Planta* **146,** 441–449.
Denius, H. R., and Homann, P. (1972). *Plant Physiol.* **49,** 873–880.
Dittrich, P. (1975). *Plant Physiol.* **57,** 310–314.
Dittrich, P. (1976). *Plant Physiol.* **58,** 288–291.
Dittrich, P. (1979a). *Encycl. Plant Physiol., New Ser.* **6,** 263–270.
Dittrich, P. (1979b). *Ber. Dtsch. Bot. Ges.* **92,** 109–116.
Garnier-Dardart, J. (1965). *Physiol. Veg.* **3,** 215–227.
Greenway, H., Winter, K., and Lüttge, U. (1978). *J. Exp. Bot.* **29,** 547–559.
Gregory, F. G., Spear, J., and Thimann, K. V. (1954). *Plant Physiol.* **29,** 220–228.
Gulmon, S. L., and Bloom, A. J. (1979). *Oecologia* **38,** 217–222.
Haidri, D. (1955). *Plant Physiol.* **30,** Suppl. IV.
Hanscom, Z., and Ting, I. P. (1977). *Bot. Gaz. (Chicago)* **138,** 159–167.
Hanscom, Z., and Ting, I. P. (1978). *Oecologia* **33,** 1–15.
Hastings, J. R., and Alcon, S. M. (1961). *J. Ariz. Acad. Sci.* **2,** 32–39.
Hatch, M. D., and Osmond, C. B. (1976). *Encycl. Plant Physiol., New Ser.* **3,** 144–184.
Herbert, M., Burkhard, C., and Schnarrenberger, C. (1978). *Planta* **143,** 279–284.
Holtum, J. A. R. (1979). Ph.D. Thesis, Australian National University, Canberra.
Jones, R., Wilkins, M. B., Coggins, J. R., Fewson, C. H., and Malcolm, A. D. B. (1978). *Biochem. J.* **175,** 391–406.
Kapil, R. N., Pugh, T. D., and Newcomb, E. H. (1975). *Planta* **124,** 231–244.
Kaplan, A., Gale, J., and Poljakoff-Mayber, A. (1976). *J. Exp. Bot.* **27,** 220–230.
Kaplan, A., Gale, J., and Poljakoff-Mayber, A. (1977). *Aust. J. Plant Physiol.* **4,** 745–752.
Kausch, W. (1965). *Planta* **66,** 229–238.
Kenyon, W., Kringstad, R., and Black, C. (1978). *FEBS Lett.* **94,** 281–283.
Kluge, M. (1968a). *Planta* **80,** 255–263.
Kluge, M. (1968b). *Planta* **80,** 359–377.
Kluge, M. (1969a). *Planta* **86,** 142–150.
Kluge, M. (1969b). *Planta* **88,** 113–129.
Kluge, M. (1976a). *Ecol. Stud.* **19,** 313–323.
Kluge, M. (1976b). *In* "CO_2 Metabolism and Plant Productivity" (R. H. Burris and C. C. Black, eds.), pp. 205–216. University Park Press, Baltimore, Maryland.
Kluge, M. (1977a). *Oecologia* **29,** 72–83.
Kluge, M. (1977b). *Symp. Soc. Exp. Biol.* **31,** 155–175.

Kluge, M. (1978). *Proc. Int. Congr. Photosynth., 4th, 1977,* pp. 335–345.
Kluge, M. (1979a). *Encycl. Plant Physiol., New Ser.* **6,** 113–124.
Kluge, M. (1979b). *Ber. Dtsch. Bot. Ges.* **92,** 95–107.
Kluge, M., and Fischer, K. (1967). *Planta* **77,** 212–223.
Kluge, M., and Heininger, B. (1973). *Planta* **113,** 333–343.
Kluge, M., and Osmond, C. B. (1971). *Naturwissenschaften* **58,** 414–415.
Kluge, M., and Ting, I. (1978). "Crassulacean Acid Metabolism: Analysis of an Ecological Adaptation" *Ecol. Stud.* **30,** Springer-Verlag: Berlin, and New York.
Kluge, M., Lange, O. L., von Eichmann, M., and Schmid, R. (1973). *Planta* **112,** 357–372.
Kluge, M., Bley, L., and Schmid, R. (1975). *In* "Environmental and Biological Control of Photosynthesis" (R. Marcelle, ed.), pp. 281–288. Junk Publ., The Hague.
Kluge, M., Böcher, M., and Jungnickel, G. (1980). *Z. Pflanzenphysiol.* **97,** 197–205.
Kluge, M., Böhlke, C., and Queiroz, O. (1981a). *Planta* **152,** 87–92.
Kluge, M., Brulfert, J., and Queiroz, O. (1981b). *Plant, Cell Environ.* **4,** 251–256.
Lange, O. L., and Medina, E., (1979). *Oecologia* **40,** 357–364.
Lange, O. L., and Zuber, M. (1977). *Oecologia* **31,** 67–72.
Lange, O. L., Schulze, E. D., Kappen, L., Evenari, M., and Buschbom, U. (1975). *Photosynthetica* **9,** 318–326.
Levitt, J. (1972). "Responses of Plants to Environmental Stresses," pp. 1–16. Academic Press, New York.
Lüttge, U. (1980). *In* "Plant Membrane Transport: Current Conceptual Issues" (R. M. Spanswick, W. J. Lucas, and J. Dainty, eds.), pp. 49–60. Elsevier/North Holland Publishing Co., Amsterdam.
Lüttge, U., and Ball, E. (1974). *Z. Pflanzenphysiol.* **73,** 326–338.
Lüttge, U., and Ball, E. (1977). *Z. Pflanzenphysiol.* **83,** 43–54.
Lüttge, U., and Ball, E. (1979). *J. Membr. Biol.* **47,** 401–422.
Lüttge, U., Kluge, M., and Ball, E. (1975). *Plant Physiol.* **56,** 613–616.
Lüttge, U., Ball, E., and Greenway, H. (1977). *Plant Physiol.* **60,** 521–523.
Marcelle, R. (1975). *In* "Environmental and Biological Control of Photosynthesis" (R. Marcelle, ed.), pp. 349–356. Junk, The Hague.
Medina, E., Delgado, M., Troughton, J. H., and Medina J. D. (1977). *Flora (Jena)* **166,** 137–152.
Meinzer, F. C., and Rundel, P. W. (1973). *Photosynthetica* **7,** 358–364.
Meyer, C. P. (1979). Ph.D. Thesis, University of Melbourne, Australia.
Morel, C., Villezmueva, V. R., and Queiroz, O. (1980). *Planta* **149,** 440–444.
Nalborczyk, E., La Croix, L. J., and Hill, R. D. (1975). *Can. J. Bot.* **53,** 1132–1138.
Neales, T. F. (1973). *Aust. J. Biol. Sci.* **26,** 705–714.
Neales, T. F. (1975). *In* "Environmental and Biological Control of Photosynthesis" (R. Marcelle, ed.), pp. 299–310. Junk, The Hague.
Neales, T. F., Hartney, V. J., and Patterson, A. A. (1968). *Nature (London)* **219,** 469–472.
Nobel, P. S. (1976). *Plant Physiol.* **58,** 576–582.
Nobel, P. S. (1977). *Oecologia* **27,** 117–133.
Nobel, P. S. (1980). *Oecologia* **45,** 160–166.
Nobel, P. S., and Hartsock, T. L. (1979). *Plant Physiol.* **63,** 63–66.
Nose, A., Shiroma, M., Miyazato, K., and Murayama, S. (1977). *Jpn. J. Crop Sci.* **46,** 579–588.
Osmond, C. B. (1976). *In* "CO_2 Metabolism and Plant Productivity" (R. H. Burris and C. C. Black, eds.), pp. 217–233. University Park Press, Baltimore, Maryland.
Osmond, C. B. (1978). *Annu. Rev. Plant Physiol.* **29,** 379–414.
Osmond, C. B., and Allaway, W. G. (1974). *Aust. J. Plant Physiol.* **1,** 503–511.

Osmond, C. B., and Björkman, O. (1975). *Aust. J. Plant Physiol.* **2,** 155–162.
Osmond, C. B. and Holtum, J. A. M. (1981). *In* "The Biochemistry of Plants" (P. K. Stampf and E. E. Conn, eds.), **8,** pp. 283–328, Academic Press, New York.
Osmond, C. B., Nott, D. L., and Firth, P. M. (1979a). *Oecologia* **40,** 331–350.
Osmond, C. B., Ludlow, M. M., Dawis, R., Cowan, I. R., Powles, S. B., and Winter, K. (1979b). *Oecologia* **41,** 65–76.
Pierre, J., and Queiroz, O. (1979). *Planta* **144,** 143–151.
Queiroz, O. (1974). *Annu. Rev. Plant Physiol.* **24,** 115–134.
Queiroz, O. (1975). *In* "Environmental and Biological Control of Photosynthesis" (R. Marcelle, ed.), pp. 357–368. Junk, The Hague.
Queiroz, O. (1979). *Encycl. Plant Physiol., New Ser.* **6,** 126–137.
Queiroz, O., and Morel, C. (1974a). *Plant Physiol.* **53,** 596–602.
Queiroz, O., and Morel C. (1974b). *J. Interdiscipl. Cycle Res.* **5,** 217–222.
Ranson, S. L., and Thomas, M. (1960). *Annu. Rev. Plant Physiol.* **11,** 81–110.
Raschke, K. (1975). *Annu. Rev. Plant Physiol.* **26,** 309–340.
Schilling, N., and Dittrich, P. (1979). *Planta* **147,** 210–215.
Schnarrenberger, C., Gross, D., Burkhard, O., and Herbert, M. (1980) *Planta* **147,** 477–485.
Schuber, M., and Kluge, M. (1979). *Flora (Jena)* **168,** 209–216.
Schulze, E. D., Lange, O. L., Evenari, M., Kappen, L., and Buschbom, U. (1974). *Oecologia* **17,** 159–170.
Spalding, M. H., Schmitt, M. R., Ku, S. B., and Edwards, G. E. (1979a). *Plant Physiol.* **63,** 738–743.
Spalding, M. H., Strempf, D. K., Ku, M. S. B., Burris, R. H., and Edwards, G. E. (1979b). *Aust. J. Plant Physiol.* **6,** 557–569.
Steudle, E., Smith, J. A. C., and Lüttge, U. (1980). *Plant Physiol.* **66,** 1155–1163.
Sugiyama, T., and Laetsch, W. M. (1975). *Plant Physiol.* **56,** 605–607.
Sutton, B. G. (1975a). *Aust. J. Plant Physiol.* **2,** 377–387.
Sutton, B. G. (1975b). *Aust. J. Plant Physiol.* **2,** 389–402.
Sutton, B. G., and Osmond, C. B. (1972). *Plant Physiol.* **50,** 360–365.
Szarek, S. R., and Ting, I. (1974a). *Plant Physiol.* **54,** 76–81.
Szarek, S. R., and Ting, I. (1974b). *Plant Physiol.* **54,** 829–834.
Szarek, S. R., and Ting, I. (1975). *Am. J. Bot.* **62,** 602–609.
Szarek, S. R., and Troughton, J. H. (1976). *Plant Physiol.* **58,** 367–370.
Szarek, S. R., Johnson, H. B., and Ting, I. (1973). *Plant Physiol.* **52,** 539–541.
Ting, I., and Hanscom, Z. (1977). *Plant Physiol.* **59,** 511–514.
Ting, I., and Jennings, W. (1976). "Deep Canyon, A Desert Wilderness for Science." Palm Desert: Boyd Deep Canyon Desert Research Center. University of California.
Ting, I., and Osmond, C. B. (1973). *Plant Sci. Lett.* **1,** 123–128.
Ting, I., Dean, M. C., and Dugger, W. M. (1967). *Nature (London)* **213,** 526–527.
Ting, I., Johnson, H. B., and Szarek, S. R. (1972). *In* "Net Carbon Dioxide Assimilation in Higher Plants" (C. C. Black, ed.), pp. 26–53. Cotton Inc., Raleigh, North Carolina.
Vieweg, G. H., and de Fekete, M. A. R. (1977). *Z. Pflanzenphysiol.* **81,** 74–79.
von Willert, D. J. (1975). *Planta* **122,** 273–280.
von Willert, D. J. (1979). *Ber. Dtsch. Bot. Ges.* **92,** 133–144.
von Willert, D. J., and Schwöbel, H. (1978). *In* "Plant Mitochondria" (G. Ducet and L. Lance, eds.), pp. 403–410. Elsevier/North-Holland Biomedical Press, Amsterdam.
von Willert, D. J., and von Willert, K. (1979). *Z. Pflanzenphysiol.* **95,** 43–50.
von Willert, D. J., Treichel, S., Kirst, G. O., and Curdts, E. (1976). *Phytochemistry* **15,** 1435–1436.

von Willert, D. J., Curdts, E., and von Willert, K. (1977a). *Biochem. Physiol. Pflanz.* **171,** 101–107.
von Willert, D. J., Thomas, D. A., Lobin, W., and Curdts, E. (1977b). *Oecologia* **29,** 67–76.
von Willert, D. J., Brinckmann, E., Scheitler, B., Thomas, D. A., and Treichel, S. (1979). *Planta* **147,** 31–36.
Winter, J., and Kandler, O. (1976). *Z. Pflanzenphysiol.* **78,** 103–112.
Winter, K. (1973). *Ber. Dtsch. Bot. Ges.* **86,** 467–476.
Winter, K. (1979). *Oecologia* **40,** 104–112.
Winter, K. (1980). *Plant Physiol.* **65,** 792–796.
Winter, K. (1981). *Abstr. Commun., Int. Congr. Photosynth., 5th, 1980* 636.
Winter, K., and Lüttge, U. (1976). *Ecol. Stud.* **19,** 323–332.
Winter, K., and Lüttge, U. (1979). *Ber. Dtsch. Bot. Ges.* **92,** 117–132.
Winter, K., and Troughton, J. (1978). *Z. Pflanzenphysiol.* **88,** 153–162.
Winter, K., and von Willert, D. J. (1972). *Z. Pflanzenphysiol.* **67,** 166–170.
Winter, K., Lüttge, U., Winter, E., and Troughton, J. (1978). *Oecologia* **34,** 225–237.
Wolf, J. (1960). *In* "Handbuch der Pflanzen-physiologie" (W. Ruhland, ed.), Vol. 12, pp. 809–889. Springer-Verlag, Berlin and New York.

9

Environmental Regulation of Photosynthesis

JOSEPH A. BERRY
W. JOHN S. DOWNTON

ABBREVIATIONS

Pathways:

CAM	Crassulacean acid metabolism
C_3 plants	Plants that use RuBP C'ase as the initial enzyme for CO_2 fixation
C_4 plants	Plants which use PEP C'ase as the initial enzyme for CO_2 fixation
TCA cycle	Tricarboxylic acid cycle

Photosynthesis: Development, Carbon Metabolism, and Plant Productivity, Vol. II

ISBN: 0-12-294302-3

Chemicals:	
ABA	Abscisic acid
ATP	Adenosine triphosphate
DCIP	Dichlorophenolindophenol
DCMU	3-(3′,4′-dichlorophenyl)-1,1-dimethylurea
FBP	Fructose 1,6-bisphosphate
DPC	Diphenylcarbazide
NADPH	Nicotinamide adenine dinucleotide phosphate (reduced)
PAN	Peroxyacyl nitrates
PEP	Phosphoenolpyruvate
3-PGA	3-Phosphoglycerate
RuBP	Ribulose 1,5-bisphosphate
Enzymes:	
RuBP C'ase	Ribulosebisphosphate carboxylase/oxygenase
PEP C'ase	Phosphoenolpyruvate carboxylase
FBP P'ase	Fructose-1,6-bisphosphate phosphatase
Symbols:	
A	Net CO_2 uptake (assimilation), (μmole m^{-2} sec^{-1})
A^{mes}/A	Ratio of mesophyll cell surface area to leaf area
C_o	CO_2 concentration, ambient (μbar)
C_i	CO_2 concentration, intercellular (μbar)
ΔC	CO_2 concentration gradient (C_o-C_i) (μbar)
E	Transpiration of water (mmole m^{-2} sec^{-1})
g	Leaf conductance to water vapor (mole m^{-2} sec^{-1})
g'	Leaf conductance to CO_2; $g' = g/1.6$ (mole m^{-2} sec^{-1})
I	Photon fluence rate (light intensity) (μmole m^{-2} sec^{-1})
Γ	The CO_2 compensation point (μbar)
PS	Photosystem
R_t	Leaf resistance to CO_2 uptake (m^2 sec $mole^{-1}$)
R_s	Stomatal resistance to CO_2 uptake (m^2 sec $mole^{-1}$)
R_m	Mesophyll "resistance" to CO_2 uptake (m^2 sec $mole^{-1}$)
W_o	Water vapor concentration, ambient (mbar)
W_i	Water vapor concentration, intercellular (mbar)
ΔW	Leaf-air humidity gradient (W_i-W_o) (mbar)
$\delta^{13}C$	^{13}C:^{12}C ratio of a sample relative to a standard (see Lehrman, 1975) (o/oo)
ϕ·	Ratio of oxygenation to carboxylation of RuBP in photosynthetic carbon metabolism (see Laing *et al.*, 1974) (mole/mole)
Φ_i	Yield for net CO_2 fixation (based upon incident quanta, 400–700 mm) (mole/mole)
Φ_a	Quantum yield for net CO_2 fixation (based upon absorbed quanta) (mole/mole)
ψ_w	Water potential *MPa* or bar; *MPa* = 10 bars)

ABSTRACT

The photosynthetic productivity of plants, especially those growing under natural conditions, is strongly influenced by factors of the environment. For example, photosynthetic tissues may experience wide variations in temperature which affect the rate and integrity of many component reactions of photosynthesis. The availability of essential resources for photosynthesis (light, water, CO_2, and nutrients) also varies with time and habitat. In addition, environmental stresses such as those imposed by drought, salinity, nutrient deficiency, pollutants, or excessively high or low temperatures have direct effects upon photosynthetic capacity. In this chapter, the complex interactions between plants and their environments are approached primarily from the viewpoint of identifying the mechanisms that underlie the plant responses. Another important concern, the description and analysis of the physical environment of plants, is not considered here.

Drawing on the accumulated knowledge about the photosynthetic process summarized in these volumes, we first consider how net CO_2 uptake, as it is observed with whole leaves, is determined by the capacities of component steps or reactions that comprise the process of photosynthesis. Studies that consider the relative importance of the stomatal conductance to CO_2 and the capacity for CO_2 uptake of the leaf mesophyll cells in determining the rate of CO_2 uptake by intact leaves are reviewed. We also consider how the properties of the mesophyll cells are determined by subcellular characteristics, such as the activity of carboxylating enzymes and the capacities of the chloroplast membranes for the primary steps in photosynthetic energy conversion. Some of the responses of these component steps of photosynthesis to environmental conditions are reviewed. We then turn to studies which characterize the responses of intact leaves or whole plants to some to the major environmental factors. In reviewing these responses, we have emphasized those experiments that attempt to probe the mechanistic basis of the responses in an attempt to separate out effects associated with stomata, soluble enzymes, or photosynthetic membranes.

Adaptive mechanisms that may enable some plants to better cope with environmental limitations are also considered. Comparative studies of the photosynthetic responses of plants adapted to or acclimated to different environments have revealed many interesting variations on the basic process of photosynthesis. These variations provide useful insight into the inner workings of the process; they help to explain the successes of plants in occupying the diversity of natural habitats, and they represent a biological resource for possibly extending agricultural production into less favorable climates.

I. Introduction

Plant productivity is to a large extent determined by the rate and efficiency with which plants are able to conduct photosynthesis, given the environmental conditions and the resources available at the site where they happen to be growing. In this chapter, we will review the functional relationship between photosynthesis and the major environmental resources (water, CO_2, and light) and environmental stresses (temperature, drought, salinity, nutrient deficiency, and environmental pollutants). We will also consider functional specialization, which may serve to adapt or acclimate specific plants to specific environmental conditions.

II. Analysis of Environmental Responses

Photosynthesis as a physiological process can be considered as an approximately stepwise series of subprocesses that begin with the absorption of light energy and ultimately leads to fixation of CO_2, which diffuses from the atmosphere into the internal air spaces of the leaf.

Important subprocesses include: (a) diffusive transport; (b) biochemical reactions, which occur in the stroma of the chloroplast; (c) electron and ion transport reactions of the chloroplast membranes; and (d) photochemical reactions of the leaf pigments. Parts (a) and (b) are covered in this volume, whereas (c) and (d) are discussed by several authors in Volume I, also edited by Govindjee (1982). All of these can be studied in simple systems *in vitro*. However, in the intact leaf, each subprocess is in part conditioned by other factors of the leaf. That is, other subprocesses provide reactants to, or use products of, the subprocess in question. To understand photosynthesis, we must bring together knowledge of the basic mechanisms of the component processes and of the leaf factors that influence these processes.

Several approaches to these problems have been taken. The most direct is to work toward a kinetic understanding of the mechanisms that contribute to the responses observed at a whole leaf level. This is a sequential procedure, and obviously a given process would be difficult to relate to CO_2 uptake unless the intervening processes are fairly well understood. This approach is most suited for studies of processes closely coupled to CO_2 uptake, namely, gaseous diffusion, CO_2 fixation (see Bassham and Buchanan, Chapter 6), and photorespiration (Ogren and Chollet, Chapter 7, this volume). Other processes, which are more completely embedded in the sequence, are less accessible to such direct kinetic analysis.

Another approach, which is less dependent upon complete sequential development, is based upon comparative studies (see Björkman, 1973). In such studies, correlation is sought between a difference in some capacity of intact leaves for photosynthesis and a difference observed at a mechanistic level. These studies might utilize leaves of plants from contrasting natural environments or leaves that have or have not been exposed to some stress. The extent to which a given change or difference in some photosynthetic property of the intact leaf correlates with some change in a biochemical or photochemical characteristic of that leaf is taken as an indication that the characteristic may be a determinant of the rate. This type of experiment has been extremely valuable in analyzing the mechanistic basis for certain photosynthetic responses. However, it is important that such studies be complemented with quantitative studies of the relevant kinetic relationships.

A. Stomatal and Nonstomatal Limitations

Net CO_2 uptake of a leaf reflects (1) the intrinsic photosynthetic capacity of the cells of the leaf mesophyll, and (2) the availability of CO_2

that has diffused through the leaf boundary layer and stomata to these cells. An indication of the CO_2 requirement for photosynthesis may be obtained from studies of net CO_2 uptake by leaves, (A), as the CO_2 concentration of the air surrounding the leaf (C_o), is varied under otherwise optimal conditions for photosynthesis (Fig. 1a). These curves, which compare the responses of typical C_3 and C_4 species of higher plants apply only for the particular stomatal limitation that prevailed during the measurements. Obviously, if stomata were to open or close for some reason, a different relationship between C_o and A would be obtained, since the photosynthetic cells of the leaf are responding to the CO_2 concentration in the intercellular air spaces (C_i).

The gradient in CO_2 concentration, which develops during photosynthesis, cannot be directly measured. However, this gradient can be calculated from simultaneous measurements of transpiration (E) and photosynthesis (A) by assuming that the physical pathway for water-vapor diffusion out of the leaf through the stomata and boundary layer is identical to that for the diffusion of CO_2 into the leaf. Required in addition to E and A are: (a) the water vapor content of the air surrounding the leaf (W_o); (b) the leaf temperature, which specifies the water vapor concentration of the intercellular air spaces (W_i) (assuming 100% relative humidity in the intercellular air spaces of the leaf), and (c) the ratio of the diffusion coefficients, D_{H_2O}/D_{CO_2}, for H_2O and CO_2 determined empirically to be 1.6 (Jarvis, 1971).

According to Fick's Law, H_2O and CO_2 transport should be proportional to the concentration gradient, ΔW or ΔC, and the leaf conductance (g or g'), for H_2O or CO_2, respectively.

$$E = \Delta W \cdot g \qquad (1)$$

$$A = \Delta C \cdot g' \qquad (2)$$

These expressions can be combined and solved for ΔC since $g = 1.6g'$.

$$\Delta C = \Delta W \cdot 1.6\ A/E \qquad (3)$$

Measurements of CO_2 uptake as a function of ambient CO_2 concentration (Fig. 1a) if made with simultaneous measurements of water vapor exchange may then be expressed as a function of intercellular CO_2 concentration (Fig. 1b) since

$$C_i = C_o - \Delta CO_2 \qquad (4)$$

These curves provide an indication of the *true* response of the photosynthetic cells of the leaf to CO_2. The relationship of A to C_i should be independent of the aperture of the stomata.

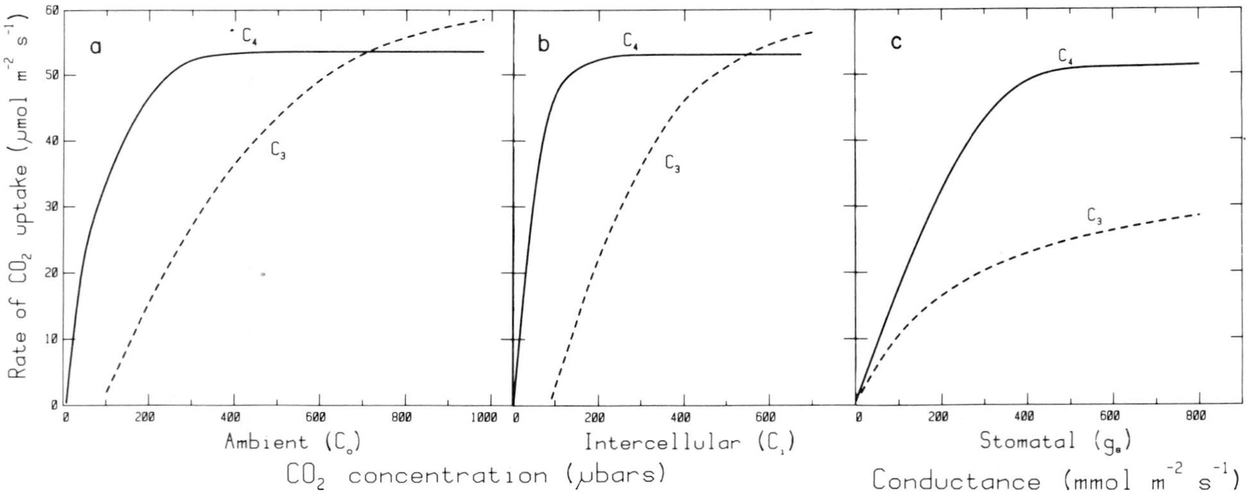

FIG. 1. Net CO_2 uptake by C_3 and C_4 plants in response to CO_2 concentration. (a), The response to ambient CO_2 concentration; (b), the response expressed according to the calculated intercellular CO_2 concentration; and (c), the response to stomatal conductance at a constant ambient CO_2 concentration (320 μbar) calculated from the curves in (b). These responses are from measurements of net CO_2 uptake of *Tidestromia oblongifolia* (C_4) and *Larrea divaricata* (C_3) at 40°C and 2000 μmole photons $m^{-2}sec^{-1}$ (O. Björkman, W. J. S. Downton, and C. S. Pike, unpublished).

Although the stomatal conductance or its reciprocal, the stomatal resistance (R_s), are not required per se to evaluate the concentration gradient for CO_2, these parameters are often used to compare the relative aperture of the stomata of different leaves or of the same leaf under different conditions. The conductance has units of centimeters per second if E and A are defined in terms of mass per unit time and the concentration gradients are defined in terms of mass per unit volume (Jarvis, 1971). Recently, Cowan (1977b) has advocated the use of units based upon mole fraction rather than upon density. When E and A are expressed as moles $m^{-2}sec^{-1}$ and ΔW and ΔC are expressed in mole fraction (e.g., mole/mole, partial pressure or volume fraction), conductance has units of mole $m^{-2}sec^{-1}$. The latter convention leads to simpler mathematical expressions and will be used here. A conductance of 1 cm sec^{-1} is approximately equivalent to 400 mmole $m^{-2}sec^{-1}$ at 25°C.

The true responses of the cells of the leaf to the intercellular CO_2 concentration (Fig. 1b) may be used to estimate what would happen to CO_2 uptake of a leaf if the stomatal conductance were varied while C_o and all other factors were held constant (Fig. 1c). This type of response curve is obviously relevant to understanding the control of photosynthesis by stomatal aperture under natural conditions. By studying the relation between CO_2 uptake and intercellular CO_2 concentration under a variety of conditions (such as different light intensities; see Fig. 2), the interaction between stomatal and other limiting factors may be estimated. The extent to which a given stomatal conductance may be limiting photosynthesis, obviously, is very dependent upon the photosynthetic characteristics of the leaf cells and the conditions under which the measurements are conducted. These other parameters may be experimentally evaluated as earlier; however, this is often not practical. A useful index of the extent to which stomatal diffusion is important under a given condition can be estimated from the relative drop in CO_2 concentration, $\Delta C/C_o$, that occurs across the stomata. This is conveniently expressed as $1 - C_i/C_o$.

Historically, the approach to expressing the extent of rate limitations by stomatal and nonstomatal factors was based upon an electrical-analog model (Gaastra, 1959). Empirically, the rate of CO_2 uptake may be treated as if it were simply controlled by diffusion of CO_2 from the air to a site within the leaf maintained at the CO_2 compensation point (Γ) via a pathway with a resistance (R_t).

$$A = (C_o - \Gamma)/R_t \tag{5}$$

That portion of R_t that is due to the resistance of the stomata and associated boundary layer, R_s, can be evaluated from simultaneous mea-

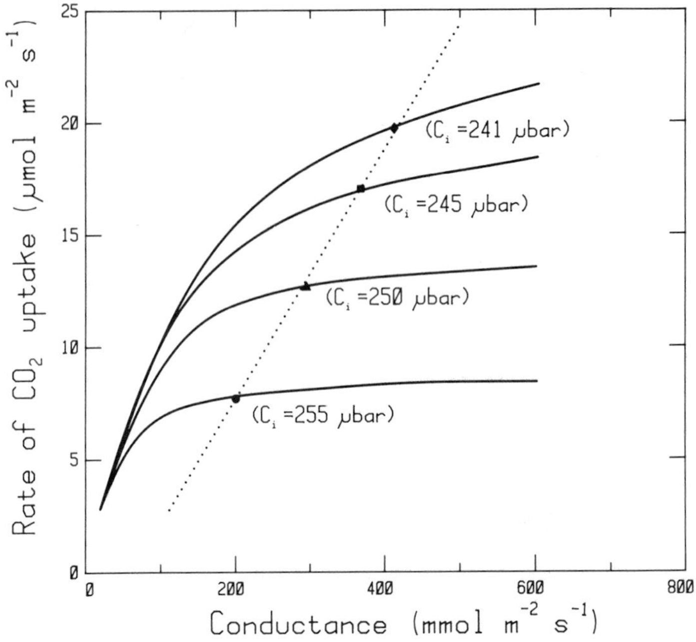

FIG. 2. Net CO_2 uptake as a function of stomatal conductance for the same leaf at differing light intensities (♦, 2000 μmole photons $m^{-2}sec^{-1}$; ■, 960 μmole photons $m^{-2}sec^{-1}$; ▲, 550 μmole photons $m^{-2}sec^{-1}$; ●, 250 μmole photons $m^{-2}sec^{-1}$). The actual conductance and net CO_2 uptake measured with this leaf at steady state at each of these light intensities is indicated by a data point on the corresponding curve. The intercellular CO_2 concentration at each point is shown. Ambient CO_2 concentration was held constant at 320 μbar, the humidity gradient at 20 mbar, and temperature at 30°C. Experiments were with *Eucalyptus pauciflora*, a C_3 plant. (Redrawn from Wong *et al.*, 1978.)

surements of transpiration ($R_s = 1/g'$; see Eq. 2). The total resistance to CO_2 uptake can then be separated into a portion associated with the stomata (R_s) and a portion associated with the mesophyll cells (R_m). It should be noted, however, that the term R_m is most likely a reflection of the enzymatic fixation of CO_2. Intracellular diffusion of CO_2 probably plays only a minor role in determining the mesophyll "resistance" (see Section II,C,2). These resistances may be considered in series.

$$A = (C_o - \Gamma)/(R_s + R_m) \tag{6}$$

Some studies report values for R_s and R_m and attribute responses of photosynthesis to changes either in R_s or in R_m. The relative importance of stomatal aperture in limiting photosynthesis may be obtained by taking the ratio of R_s to the total resistance $R_s/(R_s + R_m)$. The larger the ratio, the larger the relative restriction of CO_2 uptake by stomatal diffusion.

The ratio of resistances can be restated in terms of the CO_2 concentration gradients associated with the respective portions of the pathway.

$$R_s/(R_s + R_m) = (C_o - C_i)/(C_o - \Gamma) \tag{7}$$

The CO_2 compensation point (Γ) is nearly 0 for C_4 plants, hence the preceding ratio becomes $(C_o - C_i)/C_o$ or $1 - C_i/C_o$. With C_3 plants, the ratio is approximately equal to $1 - C_i/C_o$ when C_o is several fold larger than Γ. The ratio of the intercellular to the ambient CO_2 concentrations (C_i/C_o) is easily obtained from leaf gas-exchange measurements, and as shown here, this ratio provides an index of stomatal and nonstomatal limitations of photosynthesis that is equivalent to the Gaastra analysis.

B. The Dual Role of Stomata

As stated by Raschke (1976), "Land plants are in a dilemma throughout their lives: Assimilation of CO_2 from the atmosphere requires intensive gas exchange; the prevention of excessive water loss demands that gas exchange be kept low [p. 551]." Plants must strike some compromise between these two opposing objectives. This compromise determines the relative limitation of photosynthesis by stomatal conductance. As the relative stomatal limitation is increased, that is, C_i/C_o decreases, the efficiency of water use, that is, A/E, increases, but total CO_2 uptake decreases. The exact compromise that occurs in nature between restricting water loss through stomata versus maintaining a high carbon gain must depend upon a number of factors including the availability of water to the plants. Cowan and Farquhar (1977) provide a mathematical treatment that may be used to predict the optimal short-term responses of stomatal conductance to environmental variations, given a long-term water-use pattern. Although the details of this analysis are complex, the result is quite clear and simple. Their analysis predicts that the stomatal apparatus should respond to changes in environmental factors (that directly affect photosynthesis) in such a way as to keep the relative stomatal limitation about constant, that is, C_i/C_o should be constant. Experimental support for this hypothesis is now appearing; Wong *et al.* (1978) determined the responses of a leaf at different light intensities as a function of conductance, from the measured responses to intercellular

CO_2 concentration (Fig. 2). The actual conductances that the leaf assumed at these light intensities are indicated as points on these response curves. The corresponding intercellular CO_2 concentration at each point (assuming a constant ambient CO_2 concentration of 320 μbar) is indicated in brackets. Apparently, the mechanism that controlled the stomatal response kept the intercellular CO_2 concentration about constant as the light intensity was changed. Wong *et al.* (1979) showed that the mechanism, which accomplishes this regulation, operates such that the ratio of C_i/C_o tends to be constant as C_o is varied (contrary to earlier assumptions that C_i would be regulated). The mechanisms that allow this control are not yet resolved (see Wong *et al.*, 1978, 1979). These workers also demonstrated that this form of regulation applies for longer term changes in the photosynthetic capacity of leaves, which may occur as a result of different nutrition levels, different light intensities throughout growth, or with changing leaf age. Figure 3 shows the intercellular CO_2 concentration plotted against the corresponding rate of net CO_2 uptake at a constant condition for a C_4 plant, *Zea mays,* and for a C_3 plant, *Gossypium hirsutum.* Each point respresents a separate individual that developed at different levels of nitrogen nutrition. The stomata of these plants have apparently adjusted in each case to give about the same relative stomatal limitation of photosynthesis, that is, C_i/C_o or $R_m/(R_m + R_s)$ is about constant. This ratio is generally 0.6–0.8 for C_3 species and 0.2–0.4 for C_4 species.

It should be noted that in the preceding analysis changes in factors that affect photosynthesis have been considered whereas environmental variations that would effect transpiration, but not photosynthesis, have been held constant. Cowan and Farquhar (1977) in their analysis predict that under conditions where factors controlling photosynthesis are held constant and factors which affect water relations are manipulated, the relative stomatal limitation and the ratio C_i/C_o should change. Abscisic acid (ABA), a hormone that often accumulates in response to water stress (Dubbe *et al.,* 1978), caused stomatal limitation of photosynthesis to increase when fed to the petiole of a detached leaf (Fig. 4). Also, increasing the leaf-to-air humidity gradient that drives transpiration (Eq. 1) caused an increase in the relative stomatal limitation of photosynthesis and a decrease in the ratio C_i/C_o, as predicted. Since stomatal limitation increased, photosynthesis decreased (Fig. 4).

These studies provide tentative support for the theoretical analysis of Cowan and Farquhar (1977), to be considered further (Section VI,A,3). They also illustrate that the responses of stomata to environmental fac-

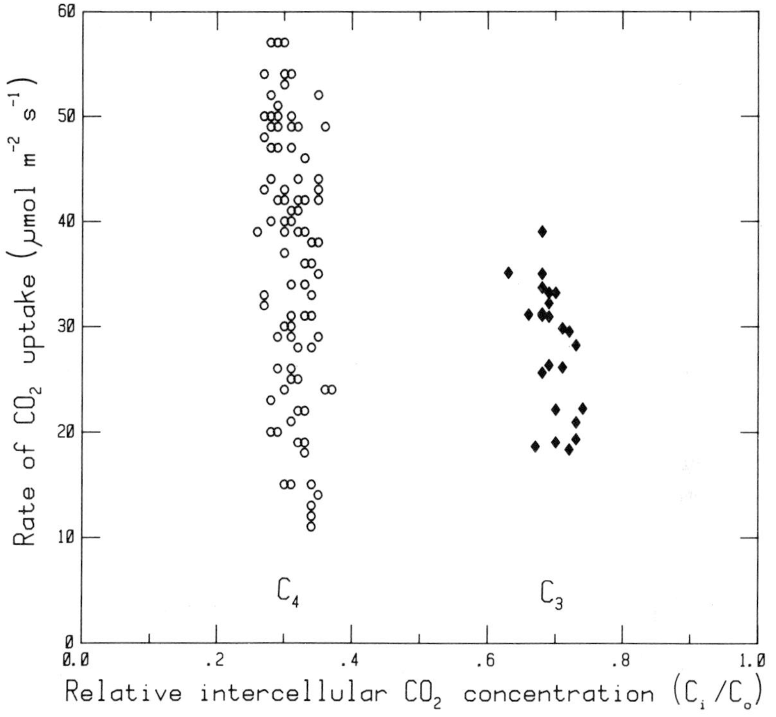

FIG. 3. Net CO_2 uptake versus the intercellular CO_2 concentration (expressed as a ratio to the ambient CO_2 concentration) in a C_4 plant *Zea mays* (○), and a C_3 plant *Gossypium hirsutum* (◆). Each point represents a separate determination with an individual plant, which developed at differing levels of nitrogen nutrition. Although assimilation of CO_2 varies, intercellular CO_2 concentration is fairly constant. Measurements were made at an ambient CO_2 of 320 μbar, leaf air humidity gradient of 20 mbar, leaf temperature of 30°C, and 2000 μmole photons $m^{-2}sec^{-1}$. (Redrawn from Wong, 1980b.)

tors may be considered as being interrelated with the response of photosynthesis to environmental factors. However, it should also be recognized that there are some responses of stomata that are clearly not linked to the responses of photosynthesis, e.g., stomatal closure upon loss of leaf turgor during water stress (Hsiao, 1973), stomatal opening at extremely high temperature (Drake *et al.*, 1970), and failure of stomata to close at very low temperature (Wilson, 1976). The mechanisms of stomatal control in higher plants are not considered here (for reviews, see Raschke, 1975, 1976; Hall *et al.*, 1976; Cowan, 1977a,b; Sheriff, 1979).

C. Limitations by Subcellular Characteristics

1. THE LAW OF LIMITING FACTORS

According to the concept of limiting factors, attributed to Blackman (1905), the rate of cellular photosynthesis under any given conditions would be expected to be limited by the step that proceeds at the slowest rate under those conditions. Response curves, which relate the photo-

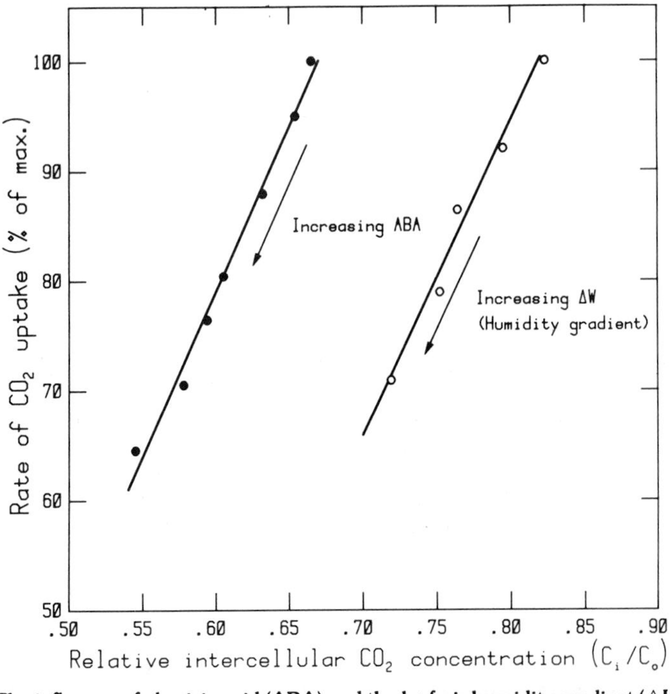

FIG. 4. The influence of abscisic acid (ABA) and the leaf-air humidity gradient (△*W*) on the intercellular CO_2 concentration and net CO_2 assimilation. (The response to ABA is taken from Wong *et al.*, 1979.) $10^{-6}M$ ABA was provided to the petiole of a detached leaf of *Gossypium hirsutum*. The points represent differing periods of accumulation of ABA. Conditions as in Fig. 2. (The response to humidity gradient is taken from C. Field and J. A. Berry, unpublished.) A leaf of *Lepechinia calycina* was held at constant temperature, 20°C; light intensity, 1000 μmole photons $m^{-2}sec^{-1}$; CO_2 concentration, 356 μbar. The humidity gradient was increased from 3–18 mbar in steps allowing the leaf to come to steady state at each step. The difference in the relative intercellular CO_2 concentration of the controls of the two response curves is in part related to a difference in the air-leaf humidity gradient at that point (20 mbar for the ABA curve and 3 mbar for the humidity response curve).

synthetic capacity to a single environmental variable under otherwise nonlimiting conditions, are an important starting point for the analysis of environmental control of photosynthesis. Typical responses to CO_2 concentration (Fig. 1), temperature (Fig. 11, p. 295), and light intensity (Fig. 9, p. 291) are shown. Although the concept of limiting factors is very useful as a first approximation, it is important to recognize that interactions between environmental factors, which are not anticipated by this theory, are commonly observed in responses of photosynthesis. These interactions are not confined to regions of transition between limiting factors or to the optimum where interaction might be expected (Verduin, 1952). For example, several environmental factors (temperature, CO_2, and O_2 concentrations) and exposure to some stresses (drought, heat, or cold) may affect the photosynthetic rate obtained at strictly rate-limiting light intensities. This is an apparent contradiction of the concept of limiting factors since light is already strongly limiting. This type of interaction can be understood by recognizing that the rate may be determined not only by the *capacity* of the slowest step but also by the *efficiency* of that step. Thus, the rate of photosynthesis under rate-limiting light intensities is a function of both the rate at which light is absorbed and the efficiency with which the absorbed light is utilized for CO_2 fixation. The factors mentioned earlier affect the quantum yield for net CO_2 fixation.

The relations between subcellular characteristics and the photosynthetic responses of leaf cells to the major environmental factors are explored in the following sections. For the most part, the responses at a cellular level are inferred from leaf gas-exchange studies; however, direct studies utilizing isolated leaf cells or chloroplasts prepared from cells are also useful. Photosynthetic rates may be obtained by measuring O_2 exchange or $^{14}CO_2$ fixation by an aqueous suspension of cells or chloroplasts; the concentrations of CO_2 and O_2 in such suspensions can be controlled, and the suspensions can be subsampled for simultaneous measurement of photosynthetic and biochemical parameters (see Portis *et al.*, 1977; Collatz and Badger, 1978; Servaites and Ogren, 1978). These procedures offer many advantages for correlating whole leaf properties and biochemical characteristics. However, the rates of photosynthesis of these preparations are usually lower on an equivalent basis than in the intact tissue. Also, many plants contain noxious substances or have leaf structures that damage the leaf cells or chloroplasts during isolation. As these problems are solved, cellular and subcellular preparations will be increasingly useful in studies of environmental responses of photosynthesis.

2. INTRACELLULAR DIFFUSION

CO_2 must be transported from the intercellular air spaces to the site of its fixation in the chloroplast stroma, and some gradient of CO_2 concentration between these sites is requisite for this transport. Since this gradient is in the liquid phase or at the gas–liquid interface, it cannot be measured by the techniques used for measurement of gaseous-phase diffusion. Hall (1971) calculated that the resistance of the intracellular portion of the diffusive pathway (0.25–0.5 m^2 sec $mole^{-1}$ for *Atriplex patula*) is about one-tenth of the combined resistance of boundary layer and stomata when the stomata are fully open. Farquhar *et al.* (1982) concluded from the observed carbon isotope fractionation during photosynthetic CO_2 fixation that the CO_2 concentration gradient between the site of CO_2 fixation in the chloroplast and the intercellular air spaces must be small in comparison to that developed across the stomata and leaf boundary layer. On the other hand, Nobel and coworkers (1975; Longstreth *et al.*, 1980) demonstrated a correlation between the photosynthetic capacity of leaves and surface area of the cells within the leaf. This correlation suggests that the capacity of some rate-limiting step increases in direct proportion to the surface area of mesophyll cells. Although this step could be diffusion, the correlation by itself does not require this, and the change in surface area may be related to other less obvious constraints upon leaf structure (see Björkman, 1981). Nevertheless, the internal structure of leaves is such that a very large surface area is exposed—typically 10–30 times the projected leaf area (El-Sharkawy and Hesketh, 1965; Longstreth and Nobel, 1980)—and this feature may be an adaptation to permit efficient transport from the air to the site of CO_2 fixation.

Intercellular diffusion probably plays a more important role in the photosynthetic mechanisms of C_4 plants (Berry and Farquhar, 1978) and some algae (Badger *et al.*, 1980; Kaplan *et al.*, 1980). In these organisms, metabolic mechanisms occur, which appear to maintain CO_2 concentration at the site of the RuBP carboxylase reaction many fold *above* that of the ambient concentration. There are two requirements for this putative increase in CO_2 concentration to occur: (a) There must be a metabolically driven CO_2 transport mechanism; and (b) there must be a restriction to "back-diffusion" of CO_2 away from the site of accumulation and RuBP carboxylase fixation. Based on a theoretical treatment, Berry and Farquhar (1978) conclude that the resistance to "back-diffusion" of CO_2 from the site of decarboxylation in the bundle sheath of C_4 plants to the intercellular air spaces should, for maximum benefit of C_4 metabolism, be about the same magnitude as the resistance of an unstirred layer of water 50 μm thick. Hatch and Osmond (1976) discuss transport pathways between the mesophyll and bundle sheath cells. Peisker (1979) derived a means to estimate this resistance to CO_2 transport (from the CO_2 compensation point), and his estimate is similar to that proposed on theoretical grounds.

3. BIOCHEMICAL REACTIONS OF PHOTOSYNTHESIS

a. C_3 Plants. Net uptake of CO_2 by illuminated cells of C_3 plants (ignoring dark respiration) is determined by the rates of carboxylation or oxygenation of ribulose 1,5-bisphosphate (RuBP) and by the subsequent metabolism of the products of these reactions by the photosynthetic carbon reduction cycle (PCR) or the photorespiratory carbon oxidation pathway (PCO) as diagrammed in Fig. 5. It is beyond the scope of this chapter to consider the biochemical details of carbon metabolism (see Bassham and Buchanan, Chapter 6, and Ogren and Chollet, Chapter 7, this volume). Our concern here is to consider some reactions of this sequence in relation to the rate of net CO_2 assimilation. Major points where limitations are most likely to occur are at the level of energy input from the electron transport reactions and at the reactions of RuBP carboxylase/oxygenase (RuBP C'ase). There is little evidence to suggest that major restrictions upon rate occur in the pathways of photorespiration after the initial step of oxygenation; however a regulatory and perhaps a rate-limiting role for the phosphatase enzymes, which operate on fructose and sedoheptulose bisphosphates, may also be important.

The rate of RuBP C'ase reactions *in vivo* is a function of (a) the quantity of enzyme present in the chloroplast; (b) the concentration of the substrates (CO_2, O_2, and RuBP) available to the enzyme; (c) the kinetic constants which apply under the conditions of the chloroplast stroma; and (d) the state of activation of the enzyme. These parameters are not fully elucidated, but considerable progress has been made in studies of this enzyme and many features of the regulation of this reaction can now be explained. For example, Farquhar *et al.* (1980a), using experimentally determined kinetic constants for RuBP C'ase of spinach and by assuming 220 μliter/liter of CO_2 concentration at the site of the carboxylation reaction, calculated that about 3 g m^{-2} RuBP carboxylase would be required to support a photosynthetic rate of 20–25 μmole $m^{-2}sec^{-1}$, a rate typically observed at rate saturating light intensity and normal air with healthy leaves of a C_3 plant such as spinach. Leaves of healthy spinach plants typically contain 2.5–3.5 g m^{-2} RuBP carboxylase (J. G. Collatz, unpublished). This comparison is tentative since constants used for the calculation may be slightly different than those that apply *in vivo* and the intracellular diffusive resistance is not known; however, the amount of carboxylase typically present in spinach is simi-

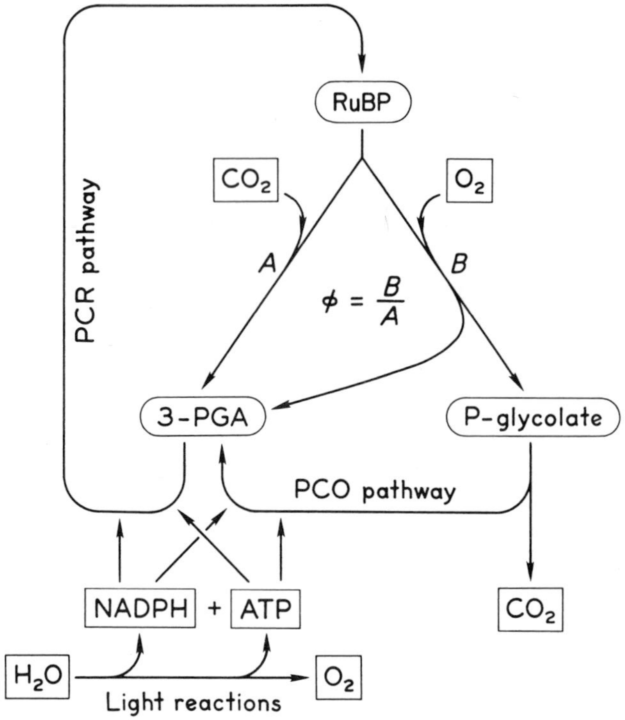

FIG. 5. Schematic diagram indicating the key relationships of the photosynthetic carbon reduction (PCR) and the photorespiratory carbon oxidation (PCO) pathways. The balance between the RuBP carboxylase (A) and RuBP oxygenase (B) reactions controls the relative rates of the two pathways (ϕ). NADPH and ATP produced by the photosynthetic membranes are used to drive each pathway. For each carboxylation, 3 ATP and 2 NADPH are required to regenerate RuBP; for each oxygenation 3.5 ATP and 2 NADPH are required to regenerate RuBP, and ½ CO_2 is produced by the PCO pathway for each oxygenation. (Modified from Berry and Björkman, 1980.)

lar to that required for photosynthesis. Furthermore, comparative studies with many C_3 plants have found a good correlation between the substrate-saturated catalytic activity (V_{max}) of RuBP carboxylase and photosynthetic capacity. Björkman (1981) compiled data from several studies, yielding an overall correlation coefficient of 0.96 between these parameters. This single enzyme is a very large fraction of the soluble protein of leaves (Jensen and Bahr, 1977; Seemann *et al.*, 1981). The efficiency with which the potential activity of this protein is utilized would seem to be a very important component of the overall protein economy of the leaf.

It would be a mistake, however, to single out the level of RuBP carboxylase as the only factor regulating the maximum capacity of the cycles illustrated in Fig. 5. For example, Björkman *et al.* (1978) provide evidence that fructose 1,6-bisphosphate phosphatase (FBP P'ase) is a rate-limiting enzyme in C_3 plants exposed to low temperature. Björkman and Badger (1979) also showed that a twofold change in maximum photosynthetic capacity at 20°C of fully expanded leaves of *Nerium oleander* occurs upon transfer from hot to cool growth conditions (or *vice versa*), and this is paralleled by a change in FBP P'ase. Changes in the activity of several other enzymes of carbon metabolism including RuBP carboxylase were also followed but these did not appear to be large enough to explain the change in rate. The relation of RuBP carboxylase and FBP P'ase activities as rate-limiting factors is not yet understood. It may be that these enzymes will each be found limiting but over different ranges of temperature or CO_2 concentration. More information concerning the differential responses of the activity of these enzymes to temperature under the conditions that prevail *in vivo* are required to evaluate this possibility.

It is known that several key chloroplast enzymes are inactive in the dark and are activated by reactions that occur in the chloroplast upon illumination. Activation may involve light/dark changes in sulfhydryl/disulfide bonds of the enzyme proteins (Wolosiuk and Buchanan, 1977), changes in levels of ions, pH, and (possibly) other mechanisms (for a review, see Anderson, 1980). Although these processes play an important role in light/dark and dark/light transitions, the role of activation processes in regulating the steady state rate of photosynthetic carbon metabolism has not yet been resolved.

Under many circumstances, the rate of energy supply to the process of carbon metabolism must restrict the rate of reactions. It is most likely that supply of the substrate, RuBP, limits the rate of the enzyme reaction under such "energy limited" conditions. Farquhar (1979a) developed kinetic expressions which incorporate the known dependence of RuBP C'ase activity upon RuBP concentration and the kinetics which apply in the chloroplast stroma where the concentration of active sites available to bind RuBP exceeds (by about two orders of magnitude) the K_m concentration of RuBP. Much of the RuBP present in chloroplast would, thus, be bound to the enzyme, and the concentration of RuBP present in a leaf could be considered rate-saturating only if this concentration exceeded that of the available binding sites (~ 4 m*M*). Collatz (1978, 1980), Sicher and Jensen (1979), and Hitz and Stewart (1980) measured the influence of environmental conditions on the steady state pool of RuBP during photosynthesis. Collatz (1980), for example, showed that the decrease in photosynthesis, which occurred at a temperature above the temperature optimum, was associated with a decline in the steady state pool of RuBP present in leaf cells during photosynthesis to a concentration *below* that of the enzyme-active sites. On this basis, it was suggested that the rate of photosynthesis under these conditions was limited by the rate of regeneration of RuBP.

The influence of changes in temperature on the reactions of carbon metabolism is complex. The rate of enzymatic reactions are generally assumed to be strongly temperature-dependent yielding a Q_{10} somewhere near 2 over temperatures at which the enzyme is stable. This generalization is true *only* if the substrate concentrations are rate-saturating. This may be illustrated by considering the rate of an enzymatic reaction at strictly rate-limiting substrate concentrations (well below the K_m concentration). Under these conditions, the rate of reaction (V) is proportional to the substrate concentration [S] and the maximum activity (V_{max}) and is inversely proportional to the Michaelis–Menten constant for substrate.

$$V \cong [S] \cdot V_{max}/K_m \tag{8}$$

If [S] is held constant and temperature is changed, the rate will change according to the ratio of the changes in the V_{max} and K_m terms. Badger and Collatz (1977) determined kinetic constants and their temperature dependence for the reactions catalyzed by RuBP C'ase. The V_{max} increases with temperature yielding a Q_{10} of about 2.2; however the $K_m(CO_2)$ also increases with a Q_{10} of about 2.2. From Eq. 8, the rate of CO_2 uptake under strictly rate-limiting CO_2 concentrations should be independent of temperature (i.e., Q_{10} = 1.0). In this regard, it is of interest that Ku and Edwards (1977) reported that the initial slope of the CO_2 response of wheat leaves is independent of temperature. The temperature dependence of the RuBP C'ase reaction should increase with the CO_2 concentration from a Q_{10} of 1 at very low substrate concentration to a Q_{10} of 2.2 at rate-saturating CO_2 concentration. Since the CO_2 concentration in the chloroplast is a function of the ambient CO_2 concentration and since the effect of temperature on the kinetic constants of the enzyme have been determined (Laing *et al.*, 1974; Badger and Collatz, 1977), it is possible to predict the temperature response of this enzyme under conditions that prevail *in vivo* (Berry and Farquhar, 1978; Farquhar *et al.*, 1980a). Berry and Björkman (1980) presented simulations of the temperature dependence of net CO_2 uptake (assuming that it is limited by RuBP C'ase) at different CO_2 concentrations. The temperature responses of other key enzymes of photosynthetic carbon metabolism are virtually unknown.

Enzymes of carbon metabolism are, in general, more stable than other constituents to the extremes of high or low temperature. Soluble enzymes may be denatured by high temperature treatment; however, key enzymes of carbon metabolism such as RuBP C'ase are stable to temperatures that result in an irreversible loss of photosynthetic capacity (Björkman *et al.*, 1976, 1978; Björkman and Badger, 1977; Berry and Björkman, 1980). Although the loss of photosynthetic capacity at high temperature seems to be related to damage of the chloroplast membranes, a portion of the soluble protein of the leaf is lost and light activated enzymes such as phosphoribulokinase become deactivated with heat treatment. No function has been assigned to the former, and the latter appears to be an indirect result of heat damage to the electron transport reactions required to maintain the enzymes in an active form (Björkman and Badger, 1977). Huner and MacDowall (1979a) characterized a form of RuBP C'ase of cold-hardened rye seedlings that is more stable at near freezing temperature than the enzyme from unhardened seedlings.

The *efficiency* as well as the *rate* of reactions of photosynthetic carbon metabolism of C_3 plants may be influenced by factors of the environment. The branching of photosynthetic carbon metabolism to either the carboxylation or oxygenation of RuBP (Fig. 5) affects the net uptake of CO_2 per RuBP consumed, and the energy input (as ATP or NADPH) required for each net CO_2 taken up. These effects may be quantitatively related to the branching ratio, ϕ (there are ϕ oxygenations per carboxylation), by expressions presented by Farquhar *et al.* (1980a). Since one CO_2 is released in photorespiration for every two oxygenations, net CO_2 uptake, $A = (1 - \phi/2) \times$ (the rate of carboxylation of RuBP). Since 3 ATPs and 3.5 ATPs are required to regenerate RuBP from the products of the carboxylation or oxygenation reactions respectively (Berry and Farquhar, 1978), the ratio of $A/\mathrm{ATP} = (1 - \phi/2)/(3 + 3.5\phi)$. An analogous expression for the efficiency of NADPH use in CO_2 fixation can be written, $A/\mathrm{NADPH} = (1 - \phi/2)/(2 + 2\phi)$ since 2 NADPH are required to regenerate RuBP from either carboxylation or oxygenation (Berry and Farquhar, 1978). These indices of efficiency decrease as ϕ increases.

Factors, which determine the ratio ϕ, are the kinetic constants of RuBP C'ase, the ratio of the concentrations of CO_2 and O_2, and the temperature (Laing *et al.*, 1974; Ogren and Chollet, Chapter 7, this volume).

Evidence for the branching of photosynthetic carbon metabolism as proposed in Fig. 5, has been obtained from studies of O_2 and CO_2 exchange of leaves using mass spectrometry (Mulchi *et al.*, 1971; Canvin *et al.*, 1980). Ordinarily the production of CO_2 and the uptake of CO_2, which occur in photorespiration, would be obscured by the simultaneous exchange of these gases in photosynthesis. If, however, the air surrounding the leaf is labeled with $^{18}O_2$, then processes that consume oxygen will use $^{18}O_2$, whereas photosynthetic oxidation of water would yield $^{16}O_2$. Figure 6 illustrates the rates of net CO_2 uptake, net O_2 production, and gross uptake and production of O_2 by a leaf of a C_3 plant as determined by mass spectrometry. Changes in the rate of O_2 uptake with CO_2 concentration probably mostly reflect the control of RuBP oxygenase by the ratio of the CO_2 and O_2 concentrations. These responses should be compared with the dependence of the quantum yield for CO_2 fixation by C_3 species upon the CO_2 concentration (see Fig. 8). It has been suggested earlier that increases in ϕ as CO_2 concentration is decreased lead to a decrease in the yield of net CO_2 uptake per unit of energy supplied by the light reactions, thereby decreasing the quantum yield for net CO_2 uptake. The decrease in the quantum yield of C_3 species with increasing temperature is probably also related to an increase of ϕ with increased temperature.

Because of these large effects of photorespiration on both net CO_2 uptake and the energy requirement of photosynthesis, it is of great interest to measure its rate. The most reliable method, in our opinion, is mass spectrometry. Ludwig and Canvin (1971) developed a procedure for measuring photorespiratory CO_2 production that should in principle yield comparable results, but separate studies of the control of photorespiration by CO_2 concentration yielded contradictory interpretations (Bravdo and Canvin, 1979; Canvin *et al.*, 1980). Many other studies relied upon less direct measures of photorespiration; these include (a) enhancement of net CO_2 uptake by low O_2 concentration; (b) measurement of CO_2 release into CO_2-free air; (c) measurement of the CO_2 compensation point; and (d) use of various metabolic tracers. It should be recognized that these indirect approaches depend upon assumptions that are not easily tested. For example, in order to interpret measurements of the O_2 enhancement in terms of photorespiration, it is necessary to assume that lowering the O_2 concentration has no effect upon photosynthesis except that of inhibiting photorespiration. Direct measurements of O_2 exchange (Canvin *et al.*, 1980) indicate that this assumption is not valid. An emerging area of interest is in defining the role of photorespiration in the response of photosynthesis to stress. The available technical approaches for measuring the specific effects of environ-

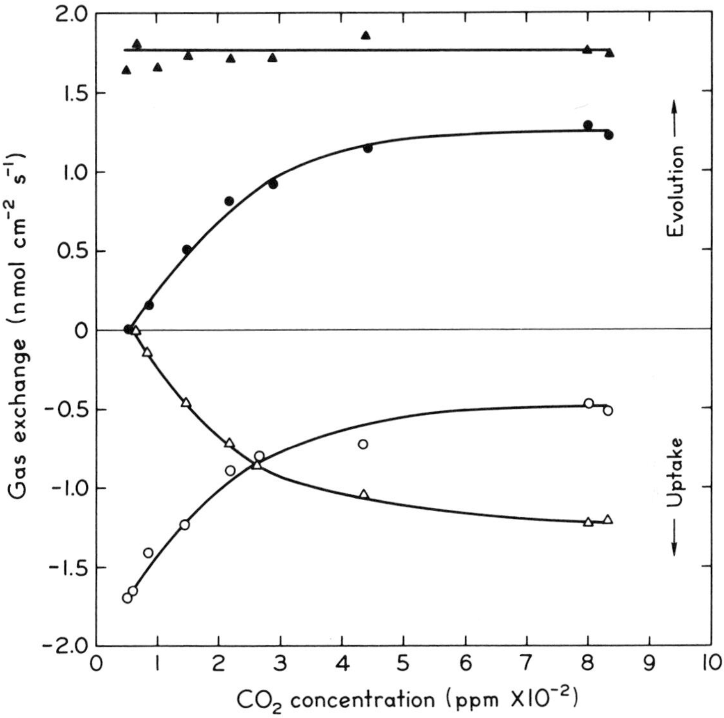

FIG. 6. Simultaneous measurement of oxygen uptake (○); oxygen production (▲); net oxygen production (●); and net CO_2 uptake (△) by a detached leaf of *Hirshfeldia incana*, a C_3 plant, as a function of CO_2 concentration. Measurements were conducted at 25°C and a light intensity of 200 μmole photons $m^{-2}sec^{-1}$. (From Berry and Badger, 1979.)

mental factors on photorespiration are a limitation in assessing these possibilities.

b. C_4 Plants. The photosynthetic metabolism of C_4 plants is essentially similar to that of C_3 plants except that C_4 plants possess additional biochemical steps which transport CO_2 to the site of the RuBP carboxylase reaction within the cells of the bundle sheath (Fig. 7; see Bassham and Buchanan, Chapter 6, this volume). The CO_2-concentrating function of the C_4 system most likely alters the local conditions under which the RuBP carboxylase reaction occurs, such that the oxygenase activity is largely suppressed and the rate of the carboxylation reaction under normal environmental conditions, is probably almost rate-saturated with respect to CO_2 concentration.

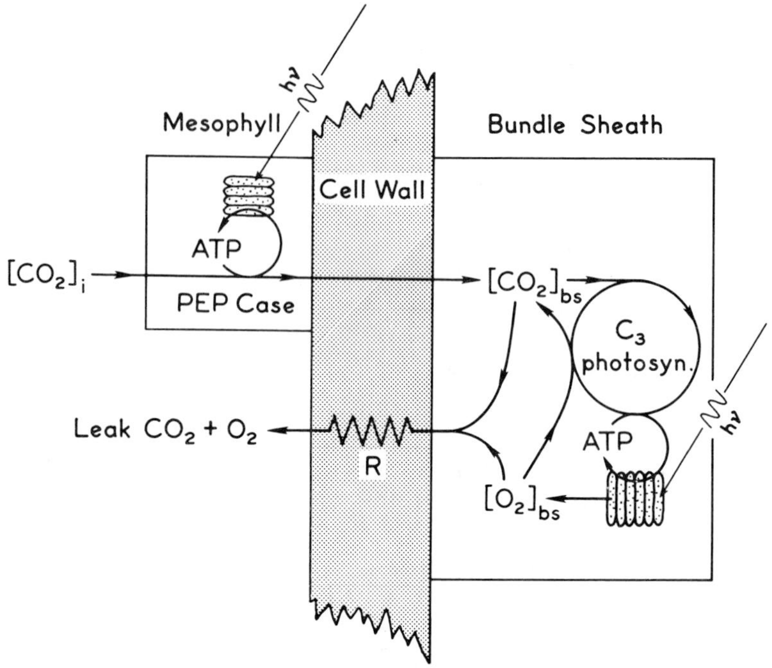

FIG. 7. **Schematic diagram illustrating the CO_2-concentrating function of C_4 photosynthesis. An elevated concentration of CO_2 can be maintained in the bundle sheath compartment if the rate of CO_2 transport via the C_4 pathway exceeds the rate of net CO_2 fixation in the bundle sheath cells. The excess CO_2 diffuses across the cell wall back to the intercellular air spaces. A resistance *R* is associated with this pathway.**

Despite the fact that RuBP carboxylase is *not* the initial carboxylating enzyme, Pearcy (1977) and Björkman and Badger (1977) showed a strong correlation between RuBP C'ase activity and photosynthetic capacity of C_4 plants when measured at suboptimal temperature. From the scheme in Fig. 7, it can be deduced that in order for the CO_2 concentration to build up in the bundle sheath cells, the rate of CO_2 transport by the C_4 cycle should exceed the rate at which CO_2 can be fixed in these cells. The activity of PEP carboxylase, the initial carboxylating enzyme in C_4 plants, is generally 5–10 times higher than that of RuBP carboxylase (Hatch and Osmond, 1976).

The C_4 cycle requires the equivalent of 2 ATP per CO_2 transported, hence a substantial waste of energy could occur if the C_4 cycle brought CO_2 to the site of RuBP carboxylation at a rate much greater than the rate at which the second carboxylation occurred (the excess would diffuse back to the mesophyll cells; Section II,C,2). Thus we expect that the rates of the two cycles must be regulated in such a way as to maintain an appropriate balance between these cycles as light, intercellular CO_2 concentration, and temperature change in a dynamic environment. Hatch and Osmond (1976) suggest that the rate of the C_4 cycle exceeds that of the C_3 cycle by about 10%.

Because of the additional steps in the C_4 pathway, more energy is required to run it than would be required to run C_3 photosynthesis under ideal conditions. A benefit of extra energy used in the C_4 process is that (by virtue of the CO_2-concentrating function) photorespiration and its concomitant energy consumption is reduced to a level that can be ignored. The balance between the energy cost and CO_2 concentration benefit is difficult to determine on theoretical grounds; however, measurements of the quantum yield for CO_2 fixation (Fig. 8) of C_3 and C_4 plants, at normal atmospheric concentration of O_2 and CO_2, suggest that the energy saved in suppressing photorespiration is about the same (at 30°C) as that required for C_4 metabolism (Ehleringer and Björkman, 1977; Section III,B).

As a result of the CO_2-concentrating function, the RuBP C'ase of C_4 plants is most likely exposed to nearly rate-saturating concentrations of CO_2. In contrast to C_3 plants, the temperature dependence of this reaction should be steeper and the catalytic efficiency of a given quantity of this enzyme should be greater in C_4 plants—especially at high temperature. Enzymes of C_4 carbon metabolism are about as stable to high temperature treatment as are those of C_3 metabolism (Björkman and Badger, 1977). Pyruvate, P_i dikinase, an enzyme unique to C_4 metabolism, is unstable *in vitro* at low temperature (Hatch, 1979), and differences in the stability of this enzyme from different species correlates with their ability to grow at low temperature (Sugiyama *et al.*, 1979). Graham *et al.* (1979) reported that PEP carboxylase is denatured at low temperature.

4. MEMBRANE REACTIONS OF PHOTOSYNTHESIS

The rate of photosynthetic metabolism must often be limited by the energy supply in the form of ATP and NADPH, which are produced by the chloroplast membranes. The production of ATP and NADPH is linked to electron and ion transport reactions driven by the energy from absorbed light (see Govindjee, 1982). Although the maximum rates of electron transport determined *in vitro* are usually comparable to that required to support the observed rates of CO_2 uptake observed *in vivo*, these assays generally employ a variety of artificial electron acceptors or donors, and the rates are more variable than the corresponding assays for the enzymes of carbon metabolism. At present there are no reliable

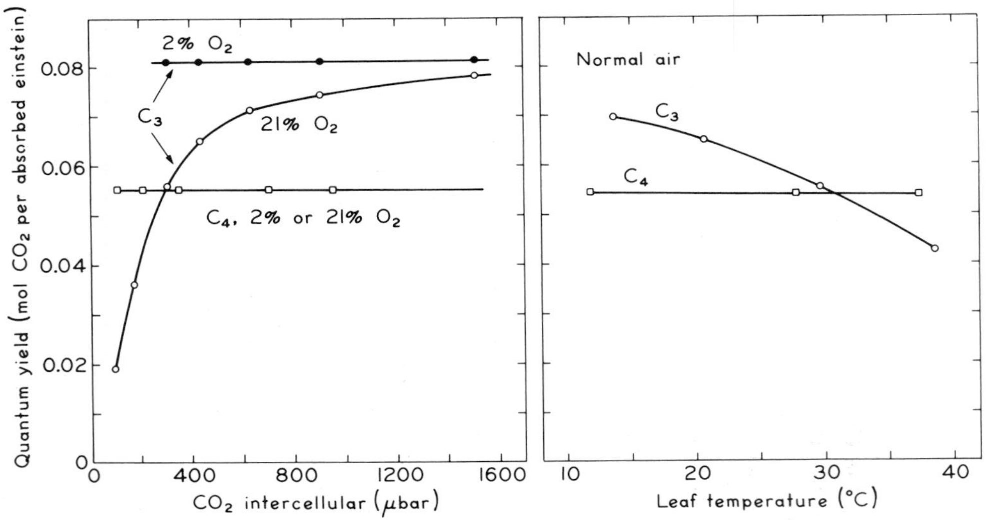

FIG. 8. The quantum yield for net CO_2 uptake of C_3 and C_4 plants as a function of the CO_2 and O_2 concentration (left panel) and temperature at normal CO_2 and O_2 concentrations (right panel). The quantum yield is expressed on the basis of absorbed photons in the wave band 400–700 mm (Φ_a). (Data from Ehleringer and Björkman, 1977; figure from Björkman, 1981.)

means of estimating the electron transport capacity of chloroplasts *in vivo*, although measurement of $^{16}O_2$ production (see Fig. 6) might be used for this purpose. Consequently, considerable caution is required in attempting to relate the capacities of isolated chloroplast membranes *in vitro* to their corresponding capacities *in vivo*.

The maximum capacity for electron transport is probably dependent on the amount of various constituents of the chloroplast membranes. The rate-limiting steps are generally considered to be between the two photosystems (see Govindjee, 1982). Membrane proteins may participate both as catalysts and as substrates, and in addition the properties of the lipid matrix of the membrane may also influence these reactions. Comparative studies of the electron transport capacity of chloroplasts from plants which developed under different light regimes show correlation between the light-saturated electron transport capacity and the amounts of *b*- and *f*-type cytochromes, plastoquinone, and coupling factor associated with the membranes (for a review, see Björkman, 1981).

The kinetics of the response of electron transport reactions to light intensity has received little attention since the review by Kok (1965). The response of electron transport at higher light intensities appears to follow a rectangular hyperbola with half-saturation occurring at different light intensities for chloroplasts from different sources (Farquhar and von Caemmerer, 1981). The initial slope of the light response curve (the quantum yield for electron transport) is probably fairly constant among species (Björkman, 1981).

Membrane reactions are highly temperature dependent. At low temperatures, the rate of electron transport may respond with a Q_{10} of 2 or higher. However, as temperature is increased, the temperature dependence becomes less steep (usually in steps, e.g., Raison, 1974; Murata and Fork, 1976; Nolan and Smillie, 1977), and at temperatures in excess of about 30°C, the rate of electron transport may begin to be inhibited by increased temperature. These responses are highly dependent upon the species and the nature of the electron transport assay used. No adequate treatment of the kinetics of the temperature response of electron transport reactions is available. Farquhar *et al.* (1980) used an equation that assumes temperature-dependent, reversible denaturation of a rate-limiting enzyme to simulate the temperature dependence of whole-chain electron transport. Experiments of Armond *et al.* (1978a) suggest that reversible inhibition of Photosystem II (PSII) reaction centers may occur at high and low temperatures.

Membrane-associated reactions, such as PSII and photophosphorylation, suffer irreversible inhibition by excessively high temperatures, and damage to these functions has been implicated in the irreversible inhibition of photosynthesis by exposure to high leaf temperature. The basis of this sensitivity to high temperature is not understood. However comparative studies have implicated differences in lipid properties with species dependent and growth temperature dependent differences in thermal stability (Section IV,C).

Several studies have addressed the question of whether carbon metabolism or membrane reactions are limiting the rate of CO_2 uptake under optimal conditions. Lilley and Walker (1975) provided evidence that the maximum capacity of intact isolated chloroplasts at rate-saturating CO_2 concentration is considerably lower than the maximum catalytic capacity of the RuBP C'ase (obtained by osmotic shock) of those chloroplasts when the enzyme is provided with a nonlimiting concentration of RuBP. However, at CO_2 concentrations that the chloroplast would normally be exposed to during photosynthesis *in vivo,* there was about equivalent capacity for CO_2 uptake with intact chloroplasts or with exogenous RuBP. Studies of rate limitation by the RuBP pool size *in vivo* led to a similar conclusion (Collatz, 1978, 1980). Terry (1980) used specific effects of iron deficiency stress upon the capacity for electron transport to probe the effect of changes in electron transport capacity on net CO_2 uptake by sugarbeet leaves. These studies indicate that differences in the electron-transport capacity affect leaf performance at light intensities approaching rate saturation. Von Caemmerer and Farquhar (1981) suggested that a point of equivalence between light reaction capacity and carbon metabolism capacity corresponds to a change in the slope of the CO_2 response curve observed by them at intercellular CO_2 concentrations near 250 μbar. The preceding studies support the assumption that the capacity of these two processes are about equivalent under normal physiological conditions. Also, comparative studies show a correlation between the temperature optimum for whole leaf photosynthesis and for electron transport by isolated chloroplasts *in vitro* (Armond *et al.,* 1978a,b; Björkman *et al.,* 1980a).

Integrative models that relate whole leaf responses to the kinetics and capacities of individual steps of the photosynthetic process and its organization are being developed (see Farquhar and von Caemmerer, 1982). These may in the future provide a more quantitative basis for understanding and analyzing the environmental control of photosynthesis.

III. Control of Photosynthesis by Light Intensity

A. Absorption of Light

Leaves typically absorb 0.8–0.85 of the incident light in the useful waveband (400–700 nm). Much of that not absorbed is reflected. Epidermal characteristics such as leaf hairs or salt glands may have a large influence on the reflection of light by leaves. Pubescent leaves of *Encelia farinosa* (with normal content of chlorophyll per unit area) absorb as little as 0.30 of the incident light, which is only 40% of that absorbed by a glabrous leaf (Ehleringer and Björkman, 1978).

Leaves typically contain 400–600 mg chlorophyll m^{-2}. Quantitative aspects of the absorption of light by chloroplasts and especially intact leaves differ considerably from those of a chlorophyll solution. These differences result because chlorophyll molecules occur in optically dense packets (the chlorophyll protein complexes) and because the effective path length of light through a leaf is increased by multiple scattering (see Butler, 1964). Variations in chlorophyll concentration usually have the largest effects upon absorption in the green and far-red regions of the spectrum, where chlorophyll absorbs less strongly. It is these wavelengths that are most abundant in naturally shaded environments where the light is filtered through intervening layers of vegetation (Tasker and Smith, 1977), and plants native to deeply shaded environments often have higher than normal concentrations of chlorophyll (Björkman *et al.*, 1972a).

Differences between leaves in their absorptance to photosynthetically active light must be taken into account when considering the light requirement for photosynthesis. All else being equal, a leaf having a higher absorptance should have a higher apparent yield based upon incident light (Φ_i) than a leaf of lower absorptance. For example, a difference in Φ_i between pubescent and glabrous leaves of *Encelia farinosa* (Ehleringer, 1977) or between leaves with differing amounts of chlorophyll (Terry, 1980) could be entirely explained by differences in absorptance, since the quantum yield based upon absorbed light (Φ_a) was constant in both instances.

B. The Quantum Yield

On theoretical grounds, at least four quanta are required for reduction of one NADPH. The quantum requirements for ATP synthesis depends upon the stoichiometry of the reactions linking photophosphorylation to electron transport and upon the role of cyclic electron transport *in vivo*. These are still not fully resolved. If we assume that the ATP requirement can be met without invoking cyclic electron transport, then the minimum quantum requirement should be 8 quanta (2 NADPH) per CO_2 fixed. Measurements of O_2 production by algae approach the maximum value of Φ_a = ⅛ = 0.125 (Kok, 1948; Emerson, 1958; Govindjee *et al.*, 1968). With higher plants, maximum Φ_a = 0.07 to 0.10 for net CO_2 fixation by healthy leaves of C_3 plants at low O_2 concentration have been obtained (Mohanty and Boyer, 1976; Ehleringer and Björkman, 1977; Terry, 1980). The slightly lower apparent quantum yield of the higher plants may in part be attributed to (a) absorption of some light by nonphotosynthetic pigments; (b) continued photorespiration under the conditions used to measure quantum yields; (c) light-dependent changes in the use of ATP or NADPH for other cellular functions such as nitrate or sulfate reduction; and (d) the possibility that some portion of the electron transport occurs via a cyclic path. Regardless of the explanation, the apparent quantum requirement for C_3 species of higher plants is remarkably constant for various plants when measured at low O_2 concentration (see Björkman, 1981).

In the presence of normal atmospheric O_2 (21%) the quantum yield of C_3 species is lower, typically $\Phi_a \simeq 0.05$ at 30°C and 330 μbar CO_2. Φ_a varies with the temperature and CO_2 concentration as these affect ϕ, the ratio of RuBP oxygenase to carboxylase reactions (Section II,C,3). C_4 plants have an apparent quantum yield that is equivalent to that of a C_3 plant in normal air and at 30°C. The quantum yields of C_3 are higher than C_4 plants at <30°C leaf temperature and are lower at >30°C leaf temperature (Fig. 8). These differences may have important implications for the habitat preference of C_3 and C_4 species (Berry and Raison, 1981).

C. Sun–Shade Adaptation

Plants are capable of growing in habitats that are so completely shaded by other plants that the light available for photosynthesis is less than 1% of that available in an exposed habitat. Any single leaf would not be able to function efficiently over this large dynamic range of light intensities. Typical light-response curves for leaves of plants adapted to contrasting light regimes are shown in Fig. 9. These responses are replotted in terms of the efficiency of light utilization on a log scale in Fig. 10a. The efficiency index used here is the actual rate of net CO_2 uptake relative to the theoretical maximum at that light intensity (dotted line, Fig. 9). The efficiency is 0 at the light compensation point (the light intensity at which net CO_2 exchange is zero). The efficiency increases with intensity to a maximum and then declines as rate saturation occurs. The light intensity of maximum efficiency is somewhat lower than the maximum intensity which the respective plants received during growth (arrows, Fig. 10). This analysis indicates that leaves from these sun and shade species achieve approximately the same efficiencies of net solar energy conversion over the ranges of light intensities, which would normally prevail in their respective native habitats. Presumably, it is this net

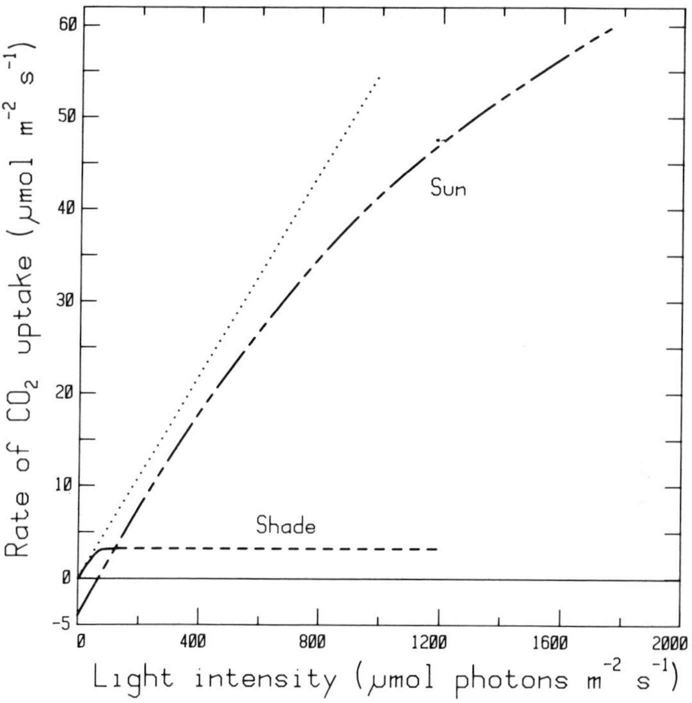

FIG. 9. Net CO_2 assimilation as a function of incident light intensity for a shade plant (*Alocasia macrorrhiza*) measured in its native habitat, a rain forest in Queensland, Australia (Björkman *et al.*, 1972a), and a sun plant, (*Camissonia claviformis*) measured in its native habitat, Death Valley, California (Mooney *et al.*, 1976). The dotted line is the maximum theoretical rate of gross CO_2 uptake (ignoring respiration); $A = 0.055\ I$.

efficiency, rather than the maximum possible rate of net CO_2 uptake, which would impart a competitive advantage in a given habitat. Any given leaf can maintain a relatively high efficiency over about one log unit (an order of magnitude) of light intensities. Taken together these plants span about 3 orders of magnitude in light intensity.

The physiological specializations, which result in this pattern of optimization, appear to be mutually exclusive; modifications, which result in a high rate of photosynthesis at high light intensity, seem to be linked to a high light compensation point, and modifications, which enable a shade plant to have a very low light compensation point, seem to be linked to a low maximum capacity for photosynthesis at moderate light intensities. The reason for this correlation is not clear. However, the leaves of sun plants contain higher concentrations of soluble protein, RuBP carboxylase, and membrane components (Björkman, 1973, 1981),

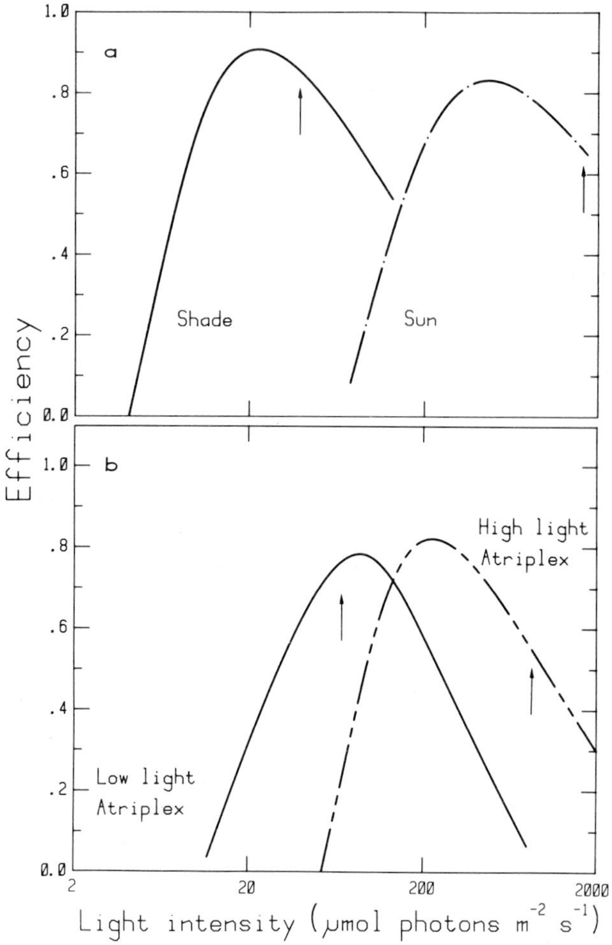

FIG. 10. **The relative efficiency of solar energy conversion of leaves of sun and shade plants (a) or high light and low light grown *Atriplex triangularis* (b). The efficiency is the ratio of the maximum achievable rate of gross CO_2 uptake (Fig. 9) to the actual measured rate of net CO_2 uptake. Arrows indicate the maximum light intensities for growth. (Data for a from Fig. 9; data for b from Björkman *et al.*, 1972b.)**

which are necessary to support the higher photosynthetic capacity. It has been suggested (Björkman, 1973) that a portion of the dark respiration of leaves may be associated with "maintenance" of these protein components. Studies of the quantum yield for photosynthesis do not indicate any substantial differences between sun and shade plants in the efficiency of light utilization (Björkman *et al.*, 1972a,b). The pigment systems of shade plants are apparently adapted to provide better absorption of the green and far-red light of shaded environments (Section III,A). Furthermore, the number of chlorophylls associated with each PSII reaction center, "the photosynthetic unit" size, may be larger in shade plants than in sun plants (see Fork and Govindjee, 1980), and the ratio of PSII to PSI reaction centers may be higher (~ 3:1) in shade plants than in sun plants (~ 2:1). It has been suggested that these differences which should lead to preferential excitation of PSII may serve to compensate for the decreased abundance of PSII light in shaded environments (Melis and Harvey, 1981). Some researchers have emphasized changes in the internally exposed surface area of mesophyll cells per unit leaf area (A^{mes}/A) with growth at different light intensities as the basis for differences in photosynthetic characteristics of sun and shade leaves (Nobel *et al.*, 1975; Nobel, 1977). This difference may affect the maximum photosynthetic capacity (see Section II,C,2), but intracellular diffusion would be unlikely to affect the net efficiency of light use at very low intensities. This is also an important aspect of shade adaptation. A review by Björkman (1981) provides a more detailed discussion of sun–shade adaptation.

The phenomenon of optimization for different light intensities is not restricted to different genotypes, which are native to contrasting habitats. The leaves of any given genotype are capable of adjustments to the light intensity at which the leaf develops. These environmentally induced modifications are similar to those that separate true shade plants from sun plants. The range of modification is, however, narrower. The efficiency index of *Atriplex triangularis* leaves grown at different light intensities are summarized in Fig. 10b. These responses have obvious relevance to productivity in canopies of even the same plant. Leaves at the bottom of a canopy are shaded by the leaves above. Productivity of the entire plant would be increased if leaves throughout the canopy could operate optimally at the light intensity that they each receive. The environmental cue of this acclimation appears to be the total daily quantum dose (Chabot *et al.*, 1979). The possibility that species may have different capacities to acclimate to gradients of light intensity has received little attention thus far.

D. Photoinhibition

Leaves of higher plants may be damaged by exposure to abnormally high light intensities or even by normal light intensities if reactions that normally serve as sinks for energy trapped by the photoacts are inhibited. Photoinhibition is frequently observed when a leaf of a shade plant is exposed for an extended period to light intensities approaching full sunlight. Damage from exposure to high light intensities as well as restricted spans of efficient light utilization place limits on the range of light intensities that a leaf can tolerate (Björkman, 1968). Leaves exposed to photoinhibiting conditions exhibit a lower quantum yield for CO_2 uptake and a lower maximum rate of net CO_2 uptake at light saturation. Damage at the chloroplast level affects PSII electron transport and photophosphorylation (Jones and Kok, 1966a,b; Powles and Critchley, 1980).

Presumably, leaves are subject to photoinhibition when the rate of electron-transport reactions is not adequate to use the energy from a highly reactive intermediate formed at a reaction center when light is absorbed. Shade plants which have a lower maximum capacity of electron transport reactions than sun plants are, thus, more easily photoinhibited. Factors that may influence the capacity for electron transport may also affect photoinhibition. Lack of suitable electron acceptors or the effect of another stress such as low temperature (Section IV,B) or low water potential may block normal electron transport and result in an increased sensitivity to photoinhibition.

Stomatal closure during illumination, which may occur during drought stress, and its possible role in photoinhibition is discussed in Section V.

IV. Control of Photosynthesis by Temperature

Typical responses of net CO_2 uptake to temperature are shown in Figs. 11 and 12. Temperature effects upon photosynthesis are generally reversible if the temperature range does not exceed the range in which key leaf components are stable. The range of stability varies with genotype and prehistory of the leaf but extends from 10°–35°C for nearly all plants. Exposure to temperatures that exceed the limits of tolerance of the physiological mechanisms of the leaf results in irreversible loss of photosynthetic capacity.

A. *Reversible Responses*

Increasing temperature stimulates dark respiration thereby increasing the portion of gross CO_2 uptake lost through respiration. This increase in dark respiration may be very important to carbon balance at

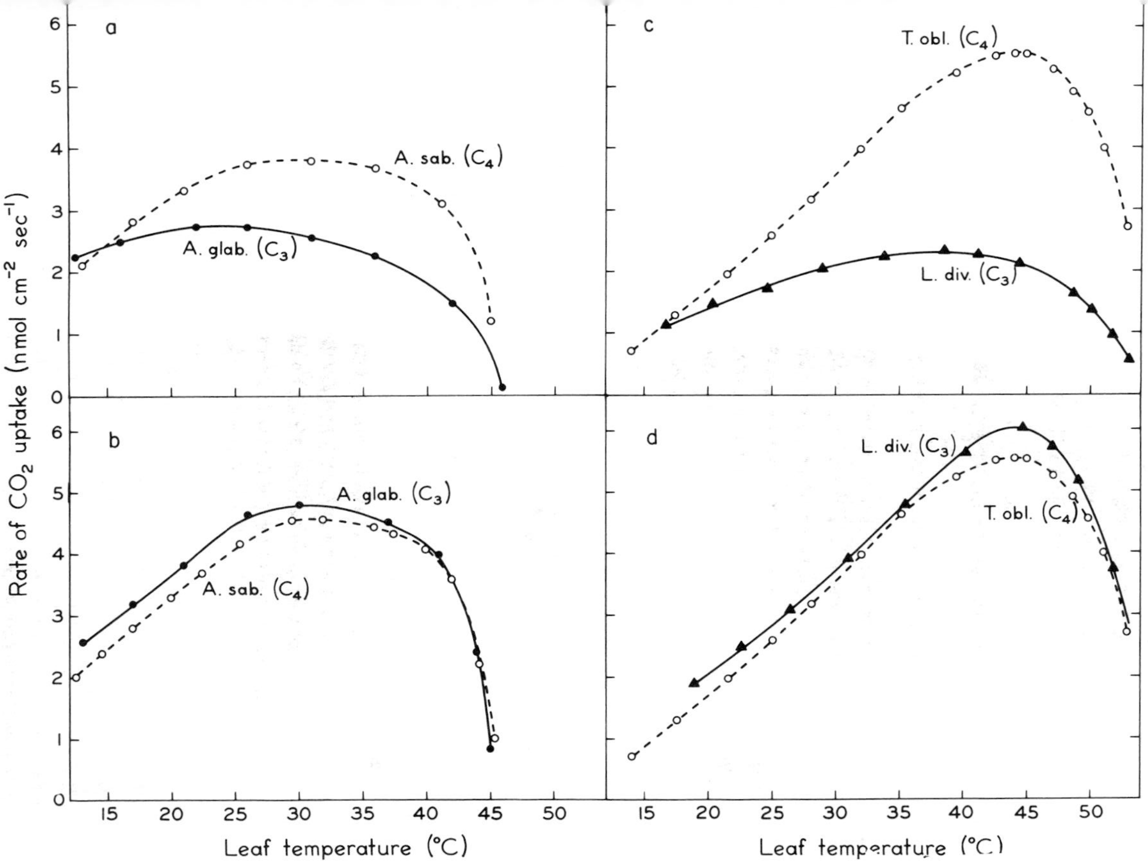

FIG. 11. Temperature dependence of net CO_2 assimilation at 330 μbar CO_2 (top) and 1000 μbar CO_2 (bottom) of pairs of C_3 and C_4 species native to and grown at cool temperatures (left) or high temperatures (right). *Atriplex sabulosa* (C_4) and *Atriplex glabriuscula* (C_3) were grown at 16°C. *Tidestromia oblongifolia* (C_4) and *Larrea divaricata* (C_3) were grown at 45°C. (Used by permission from Osmond *et al.*, 1980.)

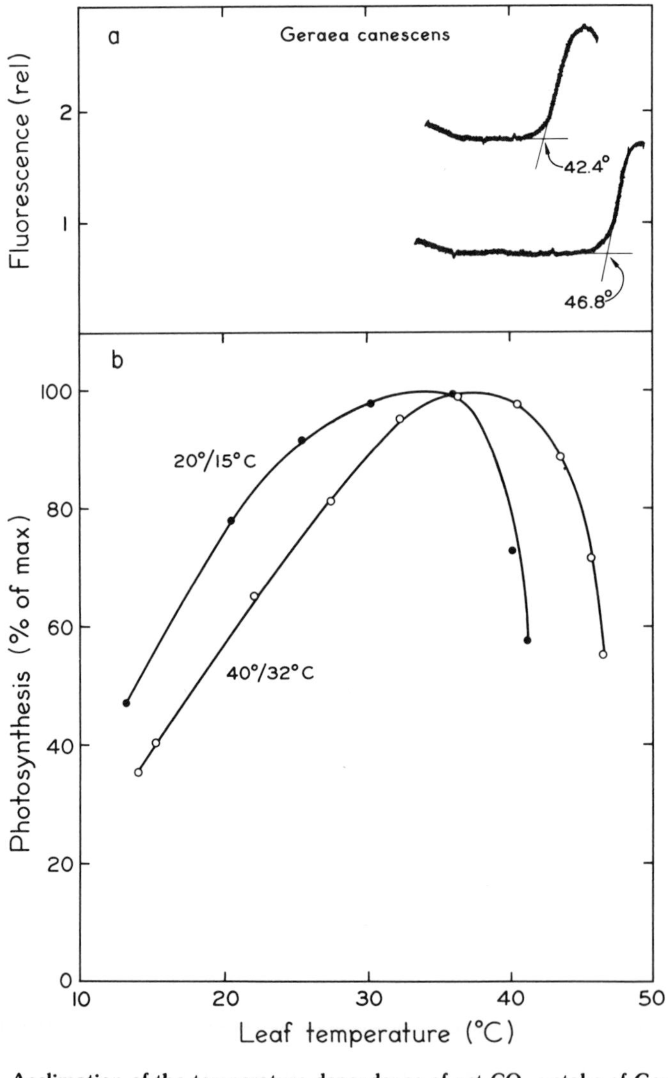

FIG. 12. Acclimation of the temperature dependence of net CO_2 uptake of *Geraea canescens* (b) grown either at 20 or 40°C maximum day temperature. Measurements were made at high (1000 μbar) CO_2. Time dependent inhibition of photosynthesis by high temperature was observed at the highest temperature on each curve. Chlorophyll *a* fluorescence versus temperature curves for similar leaves of these plants are shown (a). The sharp increase in fluorescence at high temperature is thought to occur when chlorophyll containing proteins become detached from their photochemical reaction centers (Schreiber and Armond, 1978). This occurs at about the same temperature as irreversible inhibition of net CO_2 uptake, and it is used as an index for the thermal stability of photosynthetic membranes. (J. R. Seemann, J. Berry, and W. J. S. Downton, unpublished data.)

low light intensities; however, respiration of leaves is rather small in comparison to the light-saturated rate of photosynthesis. Therefore, temperature dependent changes in the rate of (dark) respiration (contrary to earlier assumptions) do not appear to have a large effect upon the temperature response of net photosynthesis at light saturation (e.g., Pearcy, 1977). Pearcy's studies and those of Mooney *et al.* (1978) also demonstrate that the observed decline in photosynthesis at high temperature is not due to increases with temperature in the stomatal limitation of photosynthesis. In these and many other studies, intercellular CO_2 concentration actually increases as the temperature is increased above optimum. Temperature effect is thus primarily on the intrinsic reactions of photosynthesis.

Since temperature affects all of the biochemical reactions which contribute to the photosynthetic process, the overall response of net CO_2 uptake to temperature is complex. It seems reasonable to assume that the capacity of the various steps of photosynthesis are approximately balanced at some temperature, perhaps the normal operating temperature of that particular genotype. As temperature departs from this point the capacity of some reactions may decrease in capacity and become rate-limiting. Obvious candidates are the capacity of the photosynthetic membranes to generate ATP and NADPH and the capacity of rate-limiting reactions in the pathway of CO_2 fixation. Another factor of importance is the change with temperature in photorespiration and the energy requirement for net CO_2 fixation by the C_3 pathway (Section II,C,3,a). As temperature increases more energy is required per net CO_2 fixed. This is most clearly seen in studies of the quantum yield (Fig. 8), but the same considerations should also apply under conditions where the maximum capacity of the photosynthetic membranes to supply ATP or NADPH rate-limits. The temperature dependence of some components may change with environmental conditions; for example, at elevated CO_2 concentration the activity of a given quantity of RuBP C'ase responds more sharply to temperature and photorespiration is largely suppressed. C_3 plants at high CO_2 and C_4 plants are, thus, likely to have fundamentally different rate-limiting steps than C_3 plants at normal CO_2 and O_2 concentrations. No clear resolution of the relative importance of various steps to the overall temperature response is yet available. Comparative studies of plants native to or acclimated to contrasting thermal regimes provide some insight into this question (for a review see Berry and Björkman, 1981).

Temperature response of C_3 and C_4 species native to warm or cool habitats and measured at normal 330 μbar or high 1000 μbar CO_2 concentration (see Fig. 11) permit some resolution of the relative importance of carbon metabolism and other factors in adaptation of plants to contrasting thermal regimes. When measured at normal CO_2, the C_4 species have superior photosynthetic capacities to the corresponding C_3 species from the warm or cool habitats. However, this difference is eliminated when the measurements are made at high CO_2 concentration that should negate the advantage of the CO_2-concentrating mechanism of the C_4 plants. This mechanism apparently enables C_4 plants such as *T. oblongifolia* to attain very high rates of photosynthesis at high temperature. However, the C_4 plant *A. sabulosa* attains high rates of photosynthesis at low temperature, and this plant is unable to tolerate high temperatures. In these respects it resembles *A. glabriuscula,* a C_3 plant from the same environment. C_4 photosynthesis is an advantage at high temperature, but it is not necessarily a disadvantage at low temperature. This comparison shows that the cool- and warm-adapted species have many similarities despite their differences in the pathway for CO_2 fixation. The C_3 and C_4 species adapted to warm habitats have a higher temperature optimum and are not strongly inhibited by high temperature until somewhat higher temperature than the cool-adapted species. Since the temperature optimum is most likely related to an inhibition of a reaction by high temperature (see Berry and Raison, 1981), these responses suggest a greater stability of some essential component of the photosynthetic process in the warm-adapted species. Comparative studies suggest that the stability of chloroplast membrane reactions are involved (Section II,C,4). Further, Collatz (1980) demonstrated that the steady state concentration of RuBP during photosynthesis in isolated leaf cells declines at temperatures above the temperature optimum, suggesting that carbon metabolism becomes increasingly limited by the energy supply (Section II,C,3,a).

At low temperature, the cool-adapted species maintain a higher rate of photosynthetic CO_2 uptake than the corresponding warm-adapted species. This difference by itself suggests that cool-adapted species as compared to warm-adapted species may have higher contents of soluble enzymes or membrane components that catalyze essential steps. Comparative studies suggest that the activity of RuBP C'ase is rate-limiting in C_4 species at suboptimal temperatures (Section II,C,3,b). Pearcy (1977) reported that the temperature dependence of photosynthesis of the C_4 species *Atriplex lentiformis* is parallel to that of RuBP C'ase at rate-saturating CO_2 concentrations and that changes in RuBP C'ase with growth temperatures are associated with proportional changes in net CO_2 uptake. The temperature response of C_3 species at normal CO_2 and O_2 concentrations is typically less steep than that of C_4 plants at low temperatures. This may reflect the lower temperature-dependence of the RuBP C'ase reaction rate under CO_2-limiting conditions (Section II,C,3,a). Other enzymes of carbon metabolism, especially FBP P'ase appear to be rate-limiting when C_3 species grown at high temperature are measured at low temperature (Björkman *et al.*, 1978; Björkman and Badger, 1979).

Since the temperature response of these various primary interactions of temperature at a subcellular level can be described from studies conducted *in vitro* (Section II,C,3), it is possible to summarize these effects in a model. Farquhar *et al.* (1980) used the measured responses of whole-chain electron transport and a model of photosynthetic carbon metabolism and photorespiration (based upon the measured effects of temperature on the kinetic constants of RuBP C'ase) to simulate the temperature response of a C_3 leaf. This model, which assumes a linear sequence limited either by the potential rate of carbon metabolism or by that of whole chain electron transport (depending upon the condition), provides fairly accurate simulations of the temperature response of C_3 plants, and the rather complex interactions that occur with temperature, CO_2 concentration and light intensity. Much more quantitative work is required, however, to develop the biochemical basis of the model.

B. Irreversible Responses

Structural entities used in the photosynthetic process, such as enzymes, pigment–protein complexes, reaction centers, and membranes, are quite fragile. Specific damage to photosynthetic capacity results when leaves are exposed to temperatures outside of the normal range of their habitat temperatures. Breakdown of other cellular processes may occur, but usually not until the temperatures are extreme (Berry *et al.*, 1975; Björkman, 1975). Just beyond the threshold for irreversible damage, changes in photosynthetic capacity have been found to correlate with specific changes in chloroplast membrane and enzyme activities. The properties of certain components of the chloroplast, thus, appear to determine the overall tolerance limits of the photosynthetic process.

1. HIGH TEMPERATURE

Photosynthesis is rapidly and irreversibly inhibited by exposure to temperatures above a rather sharp threshold. This threshold may be reached at a temperature where photosynthetic rate still attains about one-half the maximum (optimum) value (Bauer, 1978; Seemann *et al.*,

1979). A good indicator of thermal damage is a time-dependent decline of the photosynthetic rate at a stable but high temperature (Pearcy, 1977). Leaves exposed to damaging temperatures and then returned to normal physiological conditions do not immediately recover their original photosynthetic capacity. Depending upon the length and severity of the treatment, the leaf may recover over a period of several hours to days. Harsher treatments can result in death (Bauer and Senser, 1979). There is a loss of PSII electron-transport capacity in chloroplasts isolated from heat-treated leaves. The threshold for this damage is proportional to and occurs at the same temperature as the loss of photosynthetic capacity in intact leaves (Björkman *et al.*, 1976, 1980a). Chloroplasts are more labile to heat when heated *in vitro* than when heated *in vivo*. This is especially so if they have lost their outer membrane (Krause and Santarius, 1975). While more laborious, the procedure of applying the heat treatment to the intact leaf and then isolating chloroplasts for study is more relevant to understanding the basis of heat tolerance *in vivo* than are studies that heat chloroplasts or enzymes *in vitro*.

A number of studies, which attempt to identify the specific nature of the heat damage to the chloroplast membranes, have been reviewed (Berry and Björkman, 1980). These studies lead to the conclusion that heat damage affects the organization of membrane proteins (chlorophyll–proteins, reaction center proteins, and the oxygen-evolving proteins), which together function as PSII. A specific structural association of several separate polypeptides imbedded in, and attached to, the thylakoid membrane is probably required for activity. Analysis of thylakoid membranes by freeze fracture electron microscopy showed that heat treatment caused a progressive change in the size distribution of membrane particles, consistent with the hypothesis that proteinaceous units within the membrane are being altered by heat treatment (Armond *et al.*, 1979). Raison and Berry (1979) and Raison *et al.* (1980) demonstrated that the fluidity of chloroplast membrane lipids influences the temperature tolerance of PSII of the corresponding membranes *in vivo*. Santarius (1973) and Armond and Hess (1979) showed that changes in the suspending medium (such as substitution of D_2O for H_2O) that increase the thermal stability of soluble proteins also increase the thermal stability of the lipid–protein complexes of the chloroplast membrane. These results indicate that thermal denaturation probably involves changes in the interaction of peptide chains with water (possibly unfolding), whereas the correlation between denaturation temperatures and lipid fluidity indicates that membrane lipids influence the thermodynamics of membrane protein denaturation.

Associated with the apparent denaturation of the PSII complexes are changes in the fluorescence of chlorophyll within the intact leaf. As the linkage between the pigment protein complexes and the photochemical reactions of PSII are disrupted by heat, there is a sharp increase in the fluorescence of chlorophyll. Figure 12 compares the fluorescence-versus-temperature curves and photosynthesis curves for leaves of two plants of *Geraea canescens* that had been grown under different conditions. Leaves are heated slowly (~ 1°C/min), and fluorescence and temperature are displayed on an *x-y* recorder. The correspondence between the break in the fluorescence-versus-temperature curve and the sharp decline in photosynthesis rate at high temperature has been shown to hold for several other species (Seemann *et al.*, 1979). This technique is a valuable and rapid method to screen for differences in adaptation or acclimation by plants to heat stress (Pearcy *et al.*, 1977; Schreiber and Berry, 1977; Seemann *et al.*, 1980).

2. LOW TEMPERATURE

Prolonged exposure of plants to low temperature may also result in inhibition of the photosynthetic capacity. There is no indication that injury associated with the freezing of tissue water is specific to photosynthesis or that frost tolerance of plants is related to photosynthetic capacity. This discussion will be restricted to specific effects of low temperature on photosynthetic capacity. Specific damage to photosynthesis may, with some sensitive species, occur at temperatures well above 0°C or, in the case of frost-tolerant species, be observed only at subfreezing temperatures. It is important to note that damage to the photosynthetic processes of leaves at low temperature (unlike that at high temperature) usually requires prolonged exposure. Exposure for only a few minutes is often not damaging. Also, low temperature in combination with light is more damaging than is low temperature alone. Öquist *et al.* (1980) showed that photosynthetic capacity of frost-hardened seedlings of *Pinus sylvestris* declined slowly over several days of exposure to simulated winter conditions. Analysis of chloroplast reactions indicated specific damage to PSII reactions parallel to the loss of photosynthetic capacity. Furthermore, winter damage is more severe at high than at low light intensities. Under natural conditions, damage to the photosynthetic membranes appears to accumulate (perhaps as a result of some direct effect of light) over the winter and ultimately leads to bleaching of chlorophyll and general breakdown of the photosynthetic membranes. RuBP carboxylase activity of the leaves declined to about 50% of the control during midwinter but was much more stable than membrane activity. Needles exposed to these conditions regain photosynthetic capacity during the spring. Very similar events occur when plants of tropical origin are exposed to cold, but nonfreezing, temperatures, and these responses have been shown to interact strongly with light intensity (Taylor and Rowley, 1971; Van Hasselt and Van Berlo, 1980; Powles *et al.*, 1980b). Low temperature also results in disruption of normal development of the photosynthetic capacity of new leaves. Chlorophyll in some species is not incorporated into the membrane at low temperatures (McWilliam and Naylor, 1967). Slack *et al.* (1974) suggested that inability of sorghum leaves to form normal chloroplast membrane components at low temperature may be related to malfunctions of the chloroplast ribosomes at chilling temperatures. Smillie (1976) attributed abnormal development of barley seedlings at low temperature to the differential effect of temperature on the rates of reactions that lead to biosynthesis of the chloroplast membranes (for a background on biosynthesis of membranes, see Ohad and Drews, Chapter 5, this volume).

It has long been suspected that solidification of membrane lipids might be related to cold damage. The hypothesis that lipid-phase changes may play a key role in determining the sensitivity of plants to low temperatures (Lyons, 1973; Lyons *et al.*, 1979) is simple and attractive. With the development of physical techniques, it has become possible to detect this phenomenon. With the complex mixtures of lipids that make up higher plant membranes, the phase change from fluid to solid occurs over a temperature range of several degrees. Within this range, the two phases separate laterally to form a bilayer of mixed phase (Linden *et al.*, 1973). Evidence now indicates that membrane lipids of higher plants are usually in the fluid phase at normal physiological temperatures and that phase separation may begin to occur near the lower boundary for normal physiological functioning (Raison *et al.*, 1980). Membranes become much more permeable to ions when in mixed phase, protein–lipid interactions are changed, and the diffusion of lipid-soluble substances is restricted (Raison, 1980). All of these would be expected to have substantial effects upon the physiological functions of a membrane. Changes in the activation energy of membrane-associated photosynthetic reactions (Shneyour *et al.*, 1973; Raison, 1974; Murata *et al.*, 1975; Jursinic and Govindjee, 1977) have been observed to occur at the phase separation temperature. Loss of membrane-bound Mn^{2+} essential for PSII activity occurs (Margulis, 1972; Kaniuga *et al.*, 1978), and the association of chloroplast ribosomes with membranes is altered (Millerd *et al.*, 1969) during exposure to temperatures below the point of phase separation.

In addition to its effect on lipid structure, low temperature affects the interactions that maintain native protein conformation; some enzymes that participate in photosynthetic carbon metabolism have been shown to be cold labile (Section II,C,3,a). Differences in the stability of soluble proteins to low-temperature denaturation (Graham *et al.*, 1979; Huner and MacDowall, 1979a; Sugiyama *et al.*, 1979) or in the kinetics of enzyme reactions at low temperature (Huner and MacDowall, 1979b) may play a role in addition to any effect on lipid structure at low temperature. Studies comparing the low-temperature limits to physiological responses of whole leaves and the stability of components (membrane and soluble proteins) of those leaves are required to assess the significance of the various biochemical effects of low temperature.

C. Adaptation to Contrasting Thermal Regimes

Plants native to (or grown in) thermally contrasting habitats generally have photosynthetic responses that reflect adaptation to the respective growth environments. These changes encompass differences in the photosynthetic capacity over specific temperature ranges together with adjustments of the limits of the leaf to tolerate either high or low temperature extremes. Like the physiological responses to high or low irradiance these adjustments may be environmentally induced (Fig. 12) or genotypically fixed (Fig. 11). The adjustments to temperature extremes usually result in poorer performance at a contrasting temperature. From the preceding section, it may be inferred that adaptation to high temperature is in part related to increases in the temperature tolerance of the chloroplast membranes, whereas adaptation to low temperature, in part, involves increases in the activity of rate-limiting enzymes. There is no reason to expect *a priori*, that a plant could not simultaneously increase both the quantity of enzymes and the thermal stability of the chloroplast membranes. However, these parameters generally change in opposite directions during environmentally induced temperature acclimation (for a review, see Berry and Björkman, 1980).

Lipids play a role in determining the sensitivity of chloroplast membranes to thermal denaturation. Changing the growth temperature of fully expanded *Nerium oleander* leaves from low (20°C) to high (45°C) or vice versa caused rapid upward or downward adjustments in the apparent thermal stability of chloroplast membranes. These changes in stability correlated with changes in lipid fluidity and fatty acid composition of the acyl lipids of chloroplast membrane (Raison *et al.*, 1982). The same correlation has been found in other species, which because of genetic differences or differences in growth temperature had different thermal stabilities. In *Spinacea oleracea*, however, Santarius and Muller (1979) found no correlation between an increase in thermal stability of about 3°C and changes in fatty acid composition. They suggested that additional factors may cause the acclimation. Hellmuth (1971) suggested that there was a linkage between osmotic adjustment of several Australian shrubs and increasing tolerance to high temperatures. Seemann *et al.* (1980) also reported a strong correlation between the osmotic potential of leaf water and the apparent thermal stability in situ of the chloroplast membranes of desert winter annuals. It is not yet known if the lipids of these plants also change during acclimation.

Associated with the change in lipid properties at high temperatures are corresponding changes in the phase separation temperatures of membrane phospholipids. It is unlikely that the phase separation is related per se to the thermal stability of chloroplast membranes as the phase separation occurs at a much lower temperature. However, these two parameters may reflect changes in chloroplast membrane composition. Since the phase separation temperature is more easily determined and summarized, it may be useful to consider the correlation between the thermal tolerance and the phase separation temperature. Smillie and Nott (1979) examined the thermal stability (using chlorophyll fluorescence from intact leaves as a probe) of representative plants from alpine, temperate, and tropical Australia. Raison *et al.* (1979) examined the phase separation temperature (using spin-labelled probes) of membrane lipids from temperate and tropical plants of Australia. Tropical species tended to have higher phase separation temperatures and higher thermal stabilities than did the temperate species. Comparable studies of lipids (Pike and Berry, 1980) and thermal stability (Downton *et al.*, 1980b) of selected groups of warm- or cool-season annual species indicated a correlation between habitat temperatures, thermal stability, and lipid phase separation temperature. Desert evergreen species, which experience large changes in habitat temperature from summer to winter, have large and correlated changes in lipid phase separation temperature and thermal stability (Downton *et al.*, 1980b; Pike and Berry, 1980). These studies are summarized in Table I.

The preceding differences in lipid properties can be interpreted in terms of adaptation to accommodate the high or the low temperatures likely to be encountered in a natural habitat. Most likely, both limits are affected to some extent by lipid properties, and the complex lipid mixture of plant membranes may reflect selective pressures to extend both tolerance limits in opposite directions.

There are significant differences among plants in their ability to adapt to changing growth temperatures. It may be that these differences are related to differences in the ability of plants to adjust the fatty acid composition of the chloroplast membranes. No definitive test of this postulate is yet available, but there is a general tendency for the propor-

TABLE I
The Phase Separation Temperature of Membrane Phospholipids and the Thermal Stability of Chloroplast Membranes *in Vivo* of Warm and Cool-Season Plants[a]

Species	Separation temperature (°C)	Thermal stability (°C)
Cool-season annuals		
Monocots		
Avena fatua	−9	39.8
Avena sativa	−11	41.4
Bromus rigidus	−10	41.4
Hordeum vulgare	−6	41.5
Mean	−9 ± 2	41.0 ± .8
Dicots		
Cryptantha angustifolia	2	41.6
Lepidium lasiocarpum	−1	42
Perityle emoryi	3	39.6
Mean	1.3 ± 2	41.0 ± 1.3
Warm-season annuals		
Monocots		
Chloris virgata	4	47.5
Digitaria sanguinalis	8	48
Panicum texanum	7	47.8
Zea mays	9	46.5
Mean	7 ± 2.2	47.5 ± .7
Dicots		
Boerhaavea coccinea	12	44.6
Mollugo verticillata	17	44.8
Pectis papposa	13	45.8
Portulaca oleracea	11	45.1
Mean	13.2 ± 2.6	45.1 ± 0.5
Evergreen perennials		
Cool season		
Atriplex hymenelytra	−15	48.2
Larrea divaricata	−8	48.1
Nerium oleander	−4	43
Warm season		
Atriplex hymenelytra	0	50.6
Larrea divaricata	9	52.4
Nerium oleander	7	53

[a]The annual species were grown at a common growth temperature, (28/21°C day/night); *A. hymenelytra* and *L. divaricata* were sampled from Death Valley, California, in spring and summer; *N. oleander* was grown at 45/32 or 20/15°C day/night. Data taken from Pike and Berry (1980) and W. J. S. Downton, J. R. Seemann, and J. A. Berry, unpublished.

tion of unsaturated fatty acids to increase with hardening at low temperature (Willemot, 1979) and to decrease with growth at higher temperatures (Pearcy, 1978; Raison *et al.*, 1982). The ability of a particular genotype to adapt to changing temperature is correlated with the variance of temperature in its habitat. Evergreen species from temperate regions may change the thermal stability of photosynthetic membranes of their leaves by as much as 10°C in response to stresses imposed by growth at high or low temperature. Plants from more stable thermal environments seem to have a smaller capacity to change thermal stability in response to growth temperature (Downton *et al.*, 1980b).

The C_4 pathway of photosynthesis has often been referred to as an adaptation to high temperature. The superior photosynthetic capacity of C_4 species at rate-saturating and rate-limiting light intensities when temperatures exceed about 30°C is the basis of this assertion. Furthermore, the improved water economy of C_4 versus C_3 plants (Section VII,A,3) would also favor C_4 species at high temperature. In view of these differences, it is surprising that C_4 plants are not a more dominant component of the flora of many regions (Teeri and Stowe, 1976; Stowe and Teeri, 1978; Doliner and Jolliffe, 1979). It might be suggested that C_4 plants lack the capacity to acclimate to low temperature. Osmond *et al.* (1980), Pearcy (1977), and Caldwell *et al.* (1977) nevertheless document that at least some C_4 species are capable of acclimation to function at low temperature, and at low temperature may have photosynthetic rates at least comparable to C_3 species (see Fig. 11). Berry and Björkman (1980) suggest that C_4 plants may have evolved in warm tropical or subtropical regions, where the physiological features of C_4 photosynthesis would be of maximum advantage, and that C_4 photosynthesis might therefore co-occur with other adaptations to warm and stable thermal regions. The success of C_4 plants in other climates may depend upon additional factors which are not immediately related to the pathway used for CO_2 fixation.

V. Water Stress and Photosynthesis

Plant water deficits may arise during the course of a day if transpiration exceeds the rate of water movement to the leaf, or seasonally if soil moisture reserves become depleted. Bulk leaf water potential exerts a major influence on stomatal conductance, and in the absence of compensating mechanisms such as osmotic adjustment (Section VI), loss of turgor leads to stomatal closure.

A. Consequences of Stomatal Closure

Transpiration and uptake of CO_2 by leaves are greatly reduced when stomata close. While stomatal closure serves to prevent or delay further dessication of the leaf during an interruption of its water supply, the leaf temperature may increase (see Section VI), and it is necessary for the leaf to continue the normal electron transport reactions which occur in the chloroplast or possibly suffer photoinhibition (Section III,D). With closed stomates, leaf cells no longer have access to an external supply of CO_2, and thus, electron transport can no longer be linked to net uptake of CO_2. The energy-consuming aspects of the photorespiratory cycle have been incorporated into a hypothesis that this metabolic pathway serves to dissipate excess excitation energy under CO_2-limiting conditions, thereby protecting the leaf from photoinhibition (Osmond and Björkman, 1972). According to this hypothesis, if complete stomatal closure occurs, the internal CO_2 concentration would fall to the CO_2 compensation point. At this point, net CO_2 exchange is 0, and photorespiratory production of CO_2 equals CO_2 uptake ($\phi = 2$, Fig. 5). The metabolism of C_3 plants is such that NADPH and ATP continue to be consumed in the absence of net assimilation of CO_2 (Osmond and Björkman, 1972; Lorimer *et al.*, 1978), and electron transport reactions are able to continue at a substantial rate (see Canvin *et al.*, 1980). Experimental support for this hypothesis shows that, if C_3 leaves are exposed to prolonged illumination in the absence of CO_2 and low O_2, damage to the photosynthetic apparatus ensues, as manifested by a loss of quantum yield (Powles and Osmond, 1978). The availability of CO_2 for fixation and atmospheric levels of O_2 for photorespiration are sufficient to protect a C_3 leaf at normal irradiance. C_4 plants also become photoinhibited in CO_2-free conditions, but oxygen tension is not critical (Powles *et al.*, 1980a). A flow of carbon from mesophyll to bundle sheath seems necessary to avoid photoinhibition in these plants, and the lack of an oxygen effect is consistent with a lower capacity for photorespiratory O_2 uptake by C_4 plants (Canvin *et al.*, 1980). For CAM plants, which may recycle endogenously produced CO_2 for a large part of the year (Szarek *et al.*, 1973), an additional opportunity for energy dissipation would exist through daytime reassimilation of respiratory CO_2, which is trapped as malic acid during the night (see also Kluge, Chapter 8, this volume). CO_2 would recycle at the CO_2 compensation point once the malate pool became depleted (Jones and Mansfield, 1972).

Despite these recycling devices, which may play a role in protecting C_3, C_4, and CAM plants from photoinhibition when deprived of a CO_2

supply, a loss of quantum yield (Mohanty and Boyer, 1976) and pigments (Alberte *et al.*, 1977) have been recorded for water-stressed plants. The primary effect is most likely a direct effect of water stress itself (as discussed in Section V,B) rather than the result of stomatal closure. However, it is likely that nonstomatal inhibition of electron transport (and carbon cycling) at low water potential (Lawlor and Fock, 1975; Lawlor, 1976a) would (in the presence of strong light) lead to photoinhibition (Section III,D). The decline in light-harvesting chlorophyll *a*/*b* protein in mesophyll cells of maize subjected to 8 days of water stress (Alberte *et al.*, 1977) may have resulted from water-stress-induced photoinhibition.

B. Nonstomatal Inhibition

Although stomatal closure and restriction of the CO_2 supply is a well-documented response to water stress, several reports have shown that water deficit can directly alter the efficiency of component processes in photosynthesis (Potter and Boyer, 1973; Keck and Boyer, 1974; Fellows and Boyer, 1976; Mohanty and Boyer, 1976; Younis *et al.*, 1979; Björkman *et al.*, 1980b; Govindjee *et al.*, 1981).

1. INHIBITION OF MEMBRANE REACTIONS

Most of the reports on changes in photochemistry accompanying water stress are from experiments with sunflower by J. S. Boyer and co-workers. Leaves experiencing rapid water stress suffer inhibition of the quantum yield, but they recover upon rewatering (Fig. 13). Chloroplasts isolated from these leaves and assayed for PSII electron transport show parallel changes in quantum yield (Mohanty and Boyer, 1976; Fig. 13). The effect of low-water potential *in vivo* can be differentiated from a direct effect of low-water potential as observed *in vitro* upon exposure of isolated plastids to lower water potential. While the latter treatment of plastids inhibits DCIP reduction, ferricyanide reduction, and CO_2 fixation, transfer to a medium of higher water potential reverses the effect (Fry, 1972; Plaut and Bravdo, 1973; Potter and Boyer, 1973). Chloroplasts inhibited by water stress *in vivo* are not restored by increasing the water potential of the chloroplast suspending medium, only rehydration of the water-stressed leaf before isolation of the chloroplasts may result in restored photochemical activity. In previously unstressed sunflower, both electron transport and photophosphorylation become inhibited around −11 bar (−1.1 MPa). At −1.7 MPa, both cyclic and noncylic phosphorylation decline to 0, whereas a stable, residual electron transport capacity remains. The loss of photophosphorylation is associ-

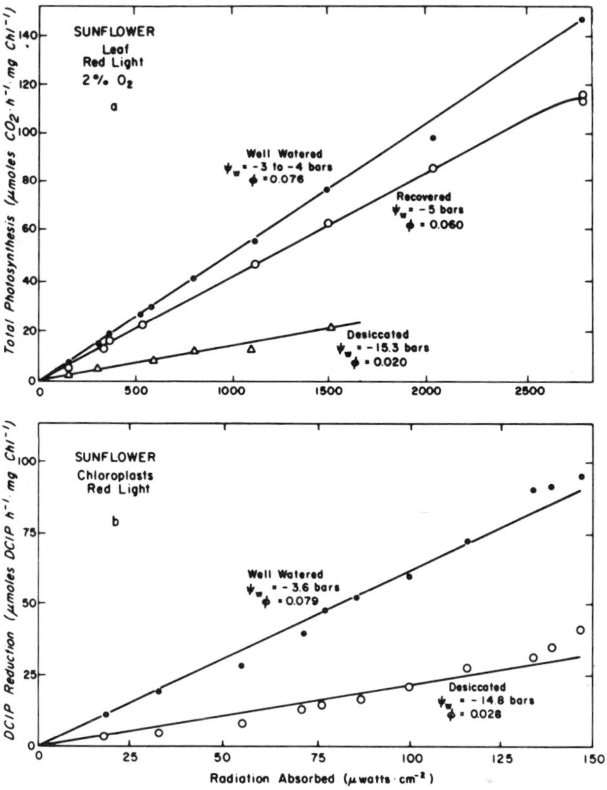

FIG. 13. Photosynthesis at limiting intensities of absorbed radiation as influenced by water potential (Ψ_w). (a) Total CO_2 assimilation by an attached sunflower leaf having different Ψ_w. Low Ψ_w was imposed by withholding water from the soil for 3 days. Water was then resupplied and measurements were repeated 15 hours later. (b) Photoreduction of DCIP in red light by chloroplasts from two halves of a sunflower leaf each having different Ψ_w. Quantum yields (Φ_a) for the intact leaf and for the isolated chloroplasts were determined from the slopes of the response curves and are expressed on an equivalent basis. (Used by permission from Mohanty and Boyer, 1976.)

ated with uncoupling of electron transport (Keck and Boyer, 1974) and changed conformation of coupling factor (Younis *et al.*, 1979). The loss of quantum yield in water-stressed leaves and its recovery upon rehydration resembles damage to photosynthetic membranes by photoinhibition (Section III,D) and heat damage (Section IV). The mechanistic basis of these effects on the membrane are not fully understood.

Fellows and Boyer (1976) detected a thinning of thylakoid membranes in water-stressed leaves, which supports the notion that conformational changes are responsible for loss of quantum yield. Their ultrastructural study is an important one since osmotic shock and rehydration artifacts were guarded against by including osmotic support in the fixation medium to match the water potential of the tissue. Unlike earlier ultrastructural investigations, which overlooked this precaution, this study showed no loss in structural integrity of thylakoid membranes.

2. EFFECTS ON PHOTOSYNTHETIC ENZYMES

A number of photosynthetic enzymes, which are known to be light-activated, show thermal inhibition at temperatures similar to those causing irreversible decreases in PSII activity and CO_2 uptake (Björkman *et al.*, 1976; Björkman and Badger, 1977). Therefore, enzymes such as pyruvate, P_i dikinase (Hatch and Slack, 1968), NADP-malate dehydrogenase (Johnson and Hatch, 1970), and NADP-glyceraldehyde phosphate dehydrogenase (Ziegler *et al.*, 1969; Steiger *et al.*, 1971), which are sensitive to light, might be expected to lose activity if photochemical efficiency becomes impaired by water stress. This hypothesis warrants testing in view of Stewart and Lee's (1972) observation that NADP-glyceraldehyde phosphate dehydrogenase activity was the most sensitive of 10 enzymes assayed in mosses to low water potential. The more drought-resistant races were able to maintain the sulfhydryl groups of this enzyme in a more active state, perhaps because these species retained greater capacity for photosynthetic electron transport under water stress conditions. RuBP carboxylase does not seem to be much affected by low water potential in the short term at least (Huffaker *et al.*, 1970; Plaut, 1971; Stewart and Lee, 1972; O'Toole *et al.*, 1977; Björkman *et al.*, 1980b). Longer term reductions probably reflect reduced synthesis and adjustments to a lower level of photosynthetic activity (Björkman, 1968; Björkman *et al.*, 1972b; Jones, 1973). Other photosynthetic enzymes such as ribose-phosphate isomerase in barley and PEP carboxylase in sorghum are not very responsive to water deficit (Huffaker *et al.*, 1970; Shearman *et al.*, 1972).

3. EFFECTS ON PHOTORESPIRATION

Photorespiration is an integral part of the photosynthetic cycle (see Fig. 5), but complexities in measuring photorespiration have hindered analysis of effects of environmental influences upon it. Using the carbon isotope technique of Ludwig and Canvin (1971), Lawlor and Fock (1975) found true photosynthesis, net photosynthesis, and photorespiration to decline in short-term water-stress experiments. However, photorespiration *increased as a proportion* of net photosynthesis. The $^{14}CO_2$ evolved in photorespiration was of lower specific activity with increasing stress. TCA-cycle respiration (insensitive to O_2 concentration above 1–2%) increased, which accounted for virtually all of the CO_2 produced in the light as water potential in sunflower fell to −1.8 MPa. Consequently, the oxygen-sensitive CO_2 evolution from the photorespiratory pathway became totally inhibited, similar to the situation when photosynthesis is poisoned by DCMU (Downton and Tregunna, 1968). A stimulation of respiration as inferred from an increased CO_2 compensation concentration has often been observed in water-stressed C_3 and C_4 plants (e.g., Meidner, 1967; Shearman *et al.,* 1972; Lawlor, 1976a). Consistent with this gas exchange data, a lower total but greater relative proportion of ^{14}C label accumulated in glycine and serine, and relatively less label was accumulated by organic acids, 3-PGA, sugar phosphates, and sugars (Lawlor, 1976b; Lawlor and Fock, 1977a) during water stress. Measurements of total pool sizes showed greater amounts of glycine and serine, but decreased soluble carbohydrates as water potential decreased (Lawlor and Fock, 1977b).

In these short-term studies, the major influence of water stress seems to stem from stomatal closure, which restricts CO_2 availability and balance between carboxylation and oxygenation by RuBP carboxylase. Hence the flux through glycolate, glycine, and serine increases, and photorespiratory CO_2 evolution is a greater proportion of net CO_2 fixation. Nonetheless, photorespiration decreases as stress increases. This cannot be ascribed simply to increased internal recycling of photorespiratory CO_2 associated with low stomatal conductance because reducing the oxygen concentration causes little enhancement of photosynthesis in the most severely stressed plants (Lawlor and Fock, 1975; Lawlor, 1976a).

The generalization of these findings to carbon balance in crops encountering water deficits must remain tentative until longer term responses have been explored in naturally water-stressed plants. However, these studies indicate that severe and direct inhibition of photosynthesis is likely to occur as water stress develops.

VI. Drought Resistance

The preceding section dealt with ways in which water deficit may directly affect partial processes contributing to overall carbon assimilation. Plants have evolved the capacity to accommodate drought to varying extents, and though this commonly takes the form of morphological and physiological adaptations, which allow plants to minimize water stress, some xerophytes and "resurrection" plants actually tolerate partial or complete desiccation.

A. Drought Avoidance

1. MORPHOLOGICAL ADAPTATIONS

Evaporation of water lost through transpiration helps to dissipate the heat of solar radiation absorbed by leaves. A leaf, which closes its stomata in order to conserve water during a period of drought, may thus experience a considerable increase in temperature. At the very least, this would increase the gradient of water vapor concentration, which drives transpiration [ΔW, (Eq. 1)]. The increase in temperature could, if coupled with high ambient temperatures, cause leaf temperatures to rise to lethal limits. Several adaptive morphological characteristics that affect the leaf energy balance enable plants to tolerate drought. Leaves of drought resistant plants are typically small and exchange heat efficiently with the air, which tends to minimize overheating. Some plants also restrict the absorption of light.

Reflective leaf surfaces and appropriate leaf orientation may be used to reduce the absorption of solar radiation during periods of drought. *Atriplex hymenelytra,* an evergreen shrub, undergoes a substantial reduction in absorptance during hot periods as a consequence of a reflective coating of salt crystals, which form on the leaf when salt bladders collapse. In addition, the leaves of this species tend to be oriented at a steep angle to the sun and thus absorb less light than would a leaf perpendicular to the sun (Mooney *et al.,* 1977). Leaves of *Encelia farinosa* enhance reflection of light by forming a dense layer of leaf hairs. Reflectance of this species is closely correlated with drought (Ehleringer *et al.,* 1976).

These mechanisms also result in less light being available for photosynthesis of the leaf. In the case of *A. hymenelytra,* only low levels of light are required to saturate photosynthesis. With *E. farinosa,* on the other hand, the increased leaf reflectance definitely results in lower photosynthesis. The apparent trade-off for absorbing less light is that the leaf temperature may be lower and nearer to the optimum for photosynthesis during hot periods, less water may be lost, and the leaf may avoid potentially lethal high temperatures (Ehleringer, 1980).

Several workers have emphasized the importance of leaf movements, which tend to keep the leaves of some species parallel to the sun's rays during drought stress (Shackel and Hall, 1979; Ehleringer and Forseth, 1980). Other characteristics that may be associated with water conservation include waxy coatings on leaf surfaces (Chatterton *et al.,* 1975), deposits of wax in stomatal antechambers (Jeffree *et al.,* 1971), and leaf rolling (O'Toole and Cruz, 1980).

For CAM plants, where gas exchange is predominantly a nocturnal activity (Osmond, 1978) and no transpirational cooling occurs during the day, structures such as spines and ribs may facilitate heat loss (Lewis and Nobel, 1977). In other desert perennials, leaves may be shed during drought periods to conserve water. Photosynthetic stems can make an important contribution to the carbon economy of plants such as the palo verde, *Cercidium microphyllum.* As much as 40–70% of net yearly carbon gain of this C_3 species may be derived from bark photosynthesis under natural conditions (Adams and Strain, 1969; Szarek and Woodhouse, 1978). The water-use efficiency for growth of *palo verde* was about two-fold greater than for other desert trees and shrubs (McGinnies and Arnold, 1939). This improved water economy has not been explained, but it may be related to the capacity for refixation of CO_2 from dark respiration within the stem (Schaedle, 1975).

2. STOMATAL ADJUSTMENTS

The daily pattern of stomatal opening (and water loss) may change dramatically as a species accommodates to drought stress (Fig. 14). When provided with abundant water, the drought tolerant shrub *Nerium oleander* has a high Ψ_w (curve 1), and stomatal conductance follows the light intensity, but when these plants are subjected to water stress (curves 2 and 3), the stomata open for only a brief interval in the morning and close again when the leaf water potential falls. As a result of stomatal closure, total transpiration during the day for treatments 2 and 3 was reduced to 4% and 1%, respectively, of that of treatment 1. Although leaf water potential of the drought-stressed plants was low, it did not fall to levels which caused severe damage, and these plants could resume photosynthetic activity rapidly upon rewatering (Björkman *et al.*, 1980b). The ability of these plants to restrict water loss and prevent desiccation of leaf tissue apparently permits this species to avoid or at least delay the development of severe water stress during periods of drought. All plants probably have some capacity to avoid water stress by closing stomata; however, xerophytic plants such as *Nerium oleander* are much more effective than are mesophytes.

The mechanisms that contribute to stomatal closure before the advent of severe water stress are complex. Stomata of a wide range of species respond directly to changes in ambient humidity (Schulze *et al.*, 1972; Hall and Kaufmann, 1975; Aston, 1976; Fig. 15). The strength of the response varies considerably from species to species. Increases in the leaf–air humidity gradient as the relative humidity of the air falls during midday could lead to a strong midday depression of photosynthesis and transpiration with the more responsive of these species.

Osmotic adjustments play a role in maintaining leaf turgor as water stress develops, and as a result the threshold water potential for stomatal

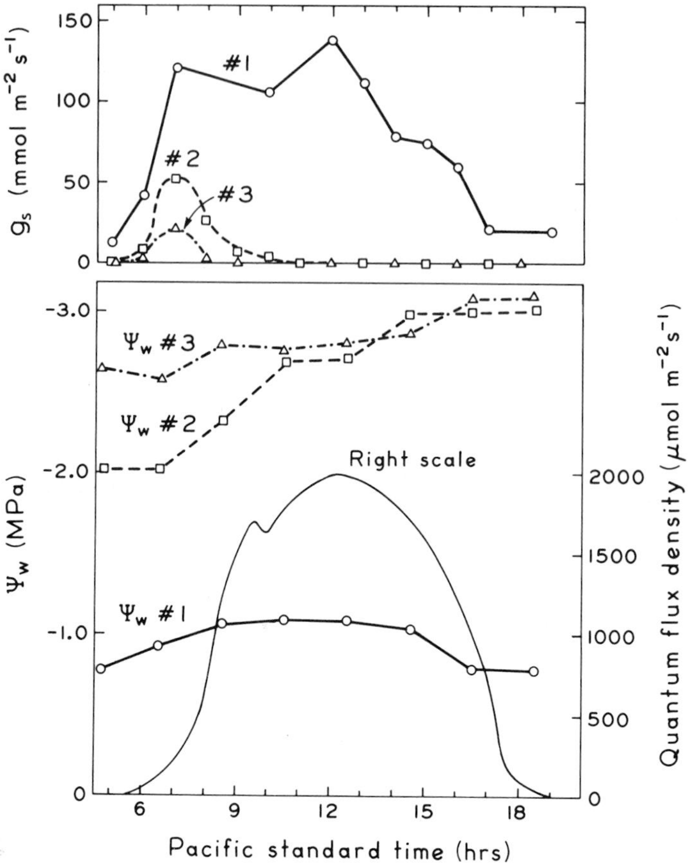

FIG. 14. The daily course of solar radiation and water potential (Ψ_w) (bottom panel) and of stomatal conductance (g) (top panel) for plants of *Nerium oleander* grown at different levels of water supply. (From Björkman *et al.*, 1980b.)

closure is shifted (Hsiao *et al.*, 1976; Jones and Turner, 1978). Leaf conductances of field-grown plants are commonly less responsive to water deficits than controlled environment material or plants grown in small containers (Begg and Turner, 1976; Ludlow, 1976). When unhardened plants encounter water stress for the first time, stomatal closure occurs over a narrow range of water potentials. When water stress develops gradually over time or when plants are subjected to repeated cycles of stress, osmotic adjustments occur, and stomatal closure may be shifted to more negative ψ_w values (Brown *et al.*, 1976; Jones and Rawson, 1979; Ackerson, 1980).

Previous drought conditions also increase stomatal sensitivity to CO_2,

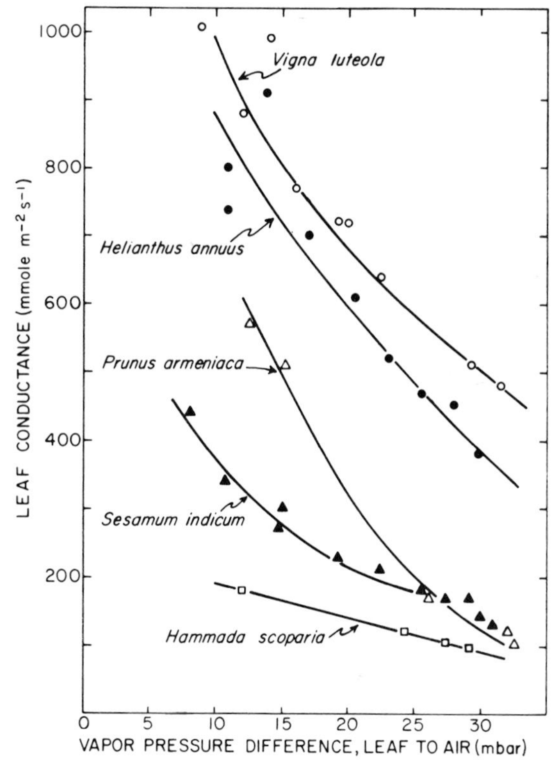

FIG. 15. Calculated leaf conductances from steady state, gas-exchange responses at a leaf temperature of 30°C and moderate to high irradiances with well-watered plants. (Used by permission from Hall *et al.*, 1976.)

and abscisic acid (ABA) has been implicated here (Raschke, 1975; Dubbe *et al.*, 1978; Björkman *et al.*, 1980b). Increasing levels of ABA in the leaf accompanying stress where osmotic adjustment fails to keep pace with turgor loss can result in stomatal closure (Ackerson, 1980). Stomatal closure not only blocks further water loss, which may be crucial to the survival of the plant, but also provides an opportunity for osmotic adjustment, if the species has this capability. This closure, however, also restricts photosynthetic CO_2 uptake (Figs. 1c, 2, and 4), and if the plant is to restrict water loss, it must also forego photosynthetic productivity.

3. WATER-USE EFFICIENCY

Nearly all of the water used by plants is lost via transpiration as an inevitable consequence of CO_2 uptake. If a plant is to absorb CO_2 from the atmosphere, it must at the same time permit water vapor to escape to the air. The efficiency of this exchange is critical to the water economy of plants with a limited water supply. This topic has been the subject of a great deal of analysis, both practical and theoretical, which is presented in reviews by Fischer and Turner (1978), Cowan (1977a,b), and Cowan and Farquhar (1977).

The relevant index of water-use efficiency is the ratio of total carbon gained:total water lost over an extended time interval (days or weeks) in the natural environment. Environmental factors such as light intensity and temperature, which affect the instantaneous capacity of the photosynthetic reactions of the leaf, vary continuously according to random and diurnal patterns. In addition the factors governing transpiration (the humidity of the air, wind speed, stomatal resistance, and leaf temperature) also vary. Thus, the efficiency obtained is a very complex function that is affected both by plant responses and by environmental conditions.

An appreciation of the factors that lead to relative differences in the water-use efficiency can be obtained using a simplified expression for the instantaneous water-use efficiency of a leaf—recognizing, however, that the leaf and environment parameters are not constant in nature. Using the resistance analog model for expressing photosynthesis [Eq. (6)] and transpiration [Eq (1)], we may express the ratio of net CO_2 uptake (A) to transpiration (E) as

$$A/E = [R_s/(R_s + R_m)] \times (C_o - \Gamma)/1.6\ \Delta W \qquad (9)$$

where R_s and R_m are the stomatal and mesophyll "resistances" to transport of CO_2, C_o is the CO_2 concentration of the air, Γ is the CO_2 compensation point, ΔW is the water vapor pressure gradient between the leaf and the air, and 1.6 is the constant relating the diffusivity of water vapor to that of CO_2. The term $[R_s/(R_s + R_m)]$ is the extent of stomatal limitation of photosynthesis and is approximately equivalent to $1 - C_i/C_o$ as discussed in Section II,A. According to Eq. (9) the water-use efficiency can approach the ratio of $C_o/1.6\ \Delta W$ as the ratio C_i/C_o tends to zero. At $C_o = 320$ μbar and $\Delta W = 20$ mbar, (the conditions used in the experiments of Fig. 3) this maximum would be 0.01 mole CO_2/mole H_2O. Of course the water-use efficiency would increase if the C_o were increased or if the ΔW were decreased, but these factors are beyond the control of the plant. It is the extent of stomatal limitation which the plant may control. As noted (Section II,A) plants appear to have developed mechanisms that can maintain this balance relatively constant as environmental or leaf factors change, and as a result the ratio C_i/C_o is relatively constant. Using 0.7 and 0.3 as representative values for C_i/C_o of cotton and maize (Fig. 3), and assuming Γ is zero for maize and 50 μbar for cotton, we may calculate the corresponding water use efficiency. This is 0.007 for maize, a C_4 plant, and 0.003 for cotton, a C_3 plant.

Measurements of the integrated water-use efficiency for growth (dry matter accumulation) of plants (reviewed by Fischer and Turner, 1978) always show a sharp difference between C_3 and C_4 species as predicted from the photosynthetic characteristics. This efficiency varies with environmental conditions, being greater in cool humid climates than in warm arid climates. However, the relative agreement between that predicted on the basis of the instantaneous approximation [Eq. (9)] and that observed over long-term experiments is good. This agreement between a short-term index and the integral must indicate that the physiological mechanisms, which govern the stomatal responses, effectively translate short-term capabilities into long-term accomplishments. Elucidating the details of these control mechanisms is one of the most fascinating problems of environmental physiology.

Cowan and Farquhar (1977) have analyzed the problem of optimizing stomatal responses, assuming that the aim of the leaf is to gain the maximum quantity of CO_2 for a given quantity of water to be expended over the course of a day. Using mathematical arguments, they show that the plant should respond to the environmental changes of a day so as to keep the marginal cost of water for carbon gain ($\partial E/\partial A$) a constant (see Farquhar and Sharkey, 1982). This theory is consistent with a number of observations of stomatal responses. For example, responses that keep C_i/C_o constant as factors affecting photosynthesis change, and responses which cause C_i/C_o to decrease as the ΔW increases (Figs. 2, 3 and 4) tend to keep $\partial E/\partial A$ constant. Also the theory predicts that, under some conditions, the stomata of plants should tend to close during midday when leaf temperature and ΔW are highest, and this occurs with some species (see Fischer and Turner, 1978). It would be premature to conclude that stomatal responses in nature can be explained by this theory, but it is one of the most interesting and stimulating examples of the application of mathematics to biology.

It is also important to consider the basis of the two- to threefold difference between C_3 and C_4 species in their water-use efficiency. As shown, Fig. 1, the initial slope of the response of photosynthesis to intercellular CO_2 concentration is steeper and the intercept (Γ) is lower for C_4 plants than for C_3 plants. We may estimate from Eq. (9) that the difference in Γ has only a small effect (13%) on the water-use efficiency. The major difference lies in the fact that C_4 plants can close their stomata so that the ratio C_i/C_o is lower without substantially restricting their capacity for CO_2 uptake. This is a direct result of the steep slope of the CO_2 response curve. Recall from Eq. (8) that the initial slope of the rate of an enzymatic reaction versus substrate concentration is proportional to the V_{max} activity of the enzyme present and inversely proportional to the K_m of the enzyme for the substrate. The enzymes which fix CO_2 in C_3 and C_4 plants differ. Hatch and Osmond (1976) concluded that the steeper slope of C_4 plants was because they contain 5–10 times higher V_{max} activity per unit leaf area of PEP C'ase than the corresponding activity of RuBP C'ase of C_3 plants. We might ask why C_3 plants do not form more RuBP C'ase protein. C_3 plants already devote about 20% of their total protein to this single enzyme function (Seemann *et al.*, 1981). The amount of total leaf protein devoted to the PEP C'ase function in C_4 plants is probably less than 10% (Uedan and Sugiyama, 1976). The reason that C_4 plants can have such a high activity while investing so little protein lies in the higher activity of PEP C'ase per unit protein (25 μmol $mg^{-1}min^{-1}$; Uedan and Sugiyama, 1976) versus RuBP C'ase (2 μmol $mg^{-1}min^{-1}$; Badger and Andrews, 1975). It would require a massive increase in leaf protein, given the low efficiency of RuBP C'ase, for C_3 plants to increase their capacity for CO_2 fixation at limiting CO_2 concentration to levels achieved by C_4 plants.

B. Drought Tolerance

Not all plants osmoregulate when confronted with water deficit (e.g., Turner *et al.*, 1978) and a number of sclerophyllous xerophytes seemingly endure desiccation to the extent that turgor decreases linearly with falling water potential, reaching 0 at around −3.5–−5.0 MPa and may become negative (Tunstall and Connor, 1975; Adams *et al.*, 1978). Stomata of leaves of jojoba, a desert sclerophyll, stay open far below that of most mesophytes, showing marked closure only at −4.0 MPa. Similar responses have been noted for brigalow (*Acacia harpophylla*), a dominant forest tree in the subtropical arid regions of eastern Australia. Gas exchange and maximal opening of stomata in brigalow remains virtually independent of bulk tissue water potential down to −5.5 MPa. Photosynthesis ceases around −7.5 MPa irrespective of whether plants are laboratory or field grown (Ludlow, 1976). It would appear that these plants may experience negative turgor (however, see Tyree, 1976).

Desiccation-tolerant plants (Gaff, 1980) can revive from the air-dry state. These include aerophilic algae, lichens, reviviscent mosses, epiphytic ferns, and a relatively few angiosperms (Gaff and Hallam, 1974; Bewley, 1979). Despite an ability of the protoplasm to survive air-drying, for angiospermous "resurrection" plants the rate of dehydration is critical to the development of desiccation tolerance. Rapid drying is fatal. For plants, such as *Borya nitida,* the fully hydrated tissue is not immediately desiccation-tolerant; it must undergo a time-dependent, tolerance-endowing process in the vicinity of −5.6 MPa water potential (Gaff and Churchill, 1976). The plant's xeromorphic features, which slow down the rate of dehydration and allow the development of full drought tolerance, are, thus, also adaptations for drought avoidance. Some of the angiosperms retain their photosynthetic pigments and organelles in the dried state, whereas others, such as *Borya,* lose chlorophyll and become yellow, retaining chloroplasts and mitochondria with only indistinct bounding membranes (Gaff *et al.,* 1976). During rehydration, the degraded plastids appear to repair bounding membranes and produce new thylakoids. A description of gas-exchange behavior during leaf drying and rehydration for these vascular plants is not yet available.

VII. Responses to Limiting Nutrients and Salinity

A. Nutrient Deficiencies (Nitrogen, Potassium, Phosphorous, and Iron)

The review of Nátr (1972) highlighted many of the problems associated with the study of the effect of nutrient deficiencies on photosynthesis and emphasized a need for more detailed investigations where particular deficiencies can be associated with the failure of specific components in the photosynthetic system. Some progress has been made in this direction. In the following section, experiments are discussed where deficiencies of nitrogen, phosphorus, potassium and iron have resulted in reduced whole leaf photosynthesis, which can be attributed to effects on particular components of the photosynthetic process.

Nitrogen deficiency in *Atriplex patula* was shown by Medina (1970) to reduce both the CO_2-saturated and CO_2-dependent (at a constant intercellular concentration of 200 μliter/liter of CO_2) rates of photosynthesis. Correspondingly, photosynthetic rates of leaves differing in nitrogen status were closely correlated with extractable RuBP C'ase activity. These data are consistent with those from nitrogen-deficient maize, cotton, and bean where reduced photosynthesis was a consequence of increased mesophyll "resistance" to CO_2 fixation rather than to stomatal limitations (Ryle and Hesketh, 1969).

Brown (1978) presented evidence for greater efficiency of nitrogen utilization by C_4 plants compared to C_3 plants based on dry matter production and on CO_2 fixation per unit of nitrogen invested in leaf material. C_4 species are known to allocate a smaller percentage of total soluble protein to RuBP C'ase and to synthesize less of the enzyme compared to C_3 plants (Björkman *et al.,* 1976; Ku *et al.,* 1979). Berry and

Farquhar (1978) discussed the basis for the differential efficiency of RuBP carboxylase in C_3 and C_4 plants and the influence of temperature. To obtain the benefit of the CO_2-concentrating mechanism, C_4 plants must form additional enzymes. The amount of nitrogen invested into protein of C_4 cycle enzymes is probably less than the change in RuBP C'ase protein, but this remains to be determined. Since the photosynthetic capacity of C_3 plants such as *Atriplex patula* can be limited by nitrogen availability under natural conditions through restriction of RuBP C'ase activity (Medina, 1970), the greater nitrogen efficiency of C_4 plants may give them a competitive advantage on sites low in nitrogen and high in temperature.

Potassium-deficient alfalfa and sugarbeet leaves experienced increases in mesophyll "resistance" to CO_2 uptake prior to increases in stomatal resistance (Terry and Ulrich, 1973b; Peoples and Koch, 1979), despite the well-established central role of potassium in guard-cell function. In alfalfa, the reduction in photosynthesis was ascribed to reduced carboxylase activity rather than to limitations in electron transport (Peoples and Koch, 1979). Unfortunately, the different data bases used to express rates of photosynthesis (area), enzyme activity (protein), and light reaction capacity (chlorophyll) in this study do not permit resolution of the matter.

Phosphorus-deficiency in sugarbeet was also associated with increased mesophyll impedance to CO_2 uptake before stomatal effects became pronounced (Terry and Ulrich, 1973a). It has not been determined whether a short fall in ATP synthesis due to restrictive P_i levels limits photosynthetic capacity in these leaves.

The data of Longstreth and Nobel (1980) on nitrogen-, potassium-, and phosphorus-deficient cotton leaves confirmed the importance of nonstomatal factors in reducing photosynthetic rates. These investigators subdivided mesophyll conductance (reciprocal of mesophyll "resistance") into a geometrical conductance term, which takes account of alterations in leaf anatomy arising from growth conditions, and into a cellular CO_2 conductance term which embodies photochemical, biochemical, and diffusion components in CO_2 fixation. Geometrical considerations were not found to influence assimilation in the nutrient-deficient cotton leaves, implying a reduced capacity for photochemistry or carbon metabolism.

Photosynthesis per unit area in iron-deficient sugarbeet leaves declined as a function of chlorophyll content (Terry, 1980). Although RuBP carboxylase/oxygenase activity showed some decrease with development of iron deficiency, noncyclic photophosphorylation was more severely affected, suggesting that photosynthesis was limited (at light saturation) by electron transport. Iron deficiency did not greatly affect the size of the photosynthetic unit; rather, it led to a substantial reduction in the number of units per leaf area (Spiller and Terry, 1980). As a result the rate of photosynthesis per unit chlorophyll did not decrease in iron deficient plants.

B. Salinity Effects

The impact of salinity on world crop production can be gauged from estimates that for irrigated agriculture alone, one-third of the 160×10^6 ha under cultivation are salt affected. Man's crop plants are predominantly sensitive to high concentrations of electrolytes in the soil, and salinity problems are especially manifest in semiarid zones where soils may already be saline or become so from irrigation. As demands on water resources for irrigated agriculture increase, there is also pressure to utilize waste waters from such sources as agricultural drains and sewage treatment works. These often contain considerable salt loads and their reuse may need to be restricted to more tolerant crop plants—or perhaps extended to biomass for energy production.

Halophytes are species that are adapted to, and thrive in, saline habitats. Most crop species are sensitive to salinity and are referred to as *glycophytes.* In some cases, halophytes accumulate high concentrations of salt without apparent harm. This life-style seems to be linked to strict subcellular compartmentation of ions, since the metabolism of halophytes has proven to be no more resistant to salinity than that of salt-sensitive species (Flowers *et al.,* 1977). Though glycophytes may also accumulate electrolytes, their sensitivity to salinity seems to reside in poor synchronization between ion uptake and subsequent compartmentation within cells. Consequently, the more salt tolerant cultivars of nonhalophytes tend to be salt excluders (Greenway and Munns, 1980). Most halophytes are also to some extent salt excluders (Osmond *et al.,* 1980).

Most plants are capable of osmotic adjustment, thereby maintaining a water potential gradient from the saline soil solution; thus salt stress is no longer considered purely in terms of "physiological drought" (Gale, 1975). Nonetheless, not all species or plant parts fully adjust osmotically, and adverse water relations may develop, especially under fluctuating levels of salinity. It is a common observation that plant growth is reduced under saline conditions, even though turgor may be maintained by osmotic adjustment. The data of Hoffman *et al.* (1980) pointed to a failure of osmotic adjustment to keep pace with volume changes required for cell enlargement in developing leaves. The prevention of normal cell enlargement, arising from suboptimal turgor pressure, restricts leaf area increase. Consequently, growth may be retarded even if photosynthetic rates per unit area remain unaffected by salinity. Salinity may also alter rates of respiration and this can have a large effect on net daily CO_2 fixation (Gale, 1975).

1. GAS EXCHANGE

a. Halophytes. Halophytes such as the C_4 species *Spartina anglica* and *Spartina alternifolia* show remarkable plasticity to extremes of salinity, and high rates of photosynthesis can be sustained both in nonsaline and seawater environments (Mallott *et al.*, 1975; Longstreth and Strain, 1977). Nonetheless, the presence of salt influences photosynthetic behavior. For *S. anglica*, CO_2 uptake in salt-depleted plants decreased sharply above 33°C, whereas that in high salt plants was maintained at optimum rates over a broad temperature range (30–44°C). Growth of *S. alternifolia* under low light, highly saline conditions led to a substantial reduction in photosynthesis, which did not occur when plants were grown at high light (Longstreth and Strain, 1977). When cultured in a saline medium, growth of the C_3 halophyte, *Salicornia rubra,* is greatly improved. This is associated with a reduced light compensation point and enhanced CO_2 uptake (Tiku, 1976). Reductions in mesophyll "resistance" parallel increases in stomatal resistance in salt-treated *Atriplex halimus* (C_4) plants, such that CO_2 uptake remains unaltered by salinity down to −0.9 MPa (Gale and Poljakoff-Mayber, 1970). *Atriplex patula,* a C_3 saltbush, maintains stable rates of photosynthesis down to −1.2 MPa and then declines at −1.6 MPa due to small increases in both stomatal and mesophyll "resistances." Although leaf thickness increased at the highest salinities, there was no change in the ratio of mesophyll cell surface area to leaf surface area (A^{mes}/A) (Longstreth and Nobel, 1979).

b. Glycophytes. Unlike halophytes, where photosynthetic capability may improve when growth occurs in a saline environment, glycophytes respond adversely to salinity. In many cases, the primary effect is increased stomatal resistance to CO_2 diffusion (Gale *et al.*, 1967; Longstreth and Nobel, 1979; Walker *et al.*, 1979). This could result from such factors as the failure of guard cells to osmotically adjust, ionic interference with stomatal function or increased endogenous concentrations of abscisic acid.

Salinity effects on gas exchange are often more complicated than this and nonstomatal components are involved. For example, Boyer (1965) noted at 25% reduction in photosynthesis of cotton grown at −0.85 MPa NaCl, even though stomatal resistance was unaffected by salinity. Gale *et al.* (1967) detected stomatal closure in salt-stressed cotton, but photosynthesis was also inhibited at low light intensities and at elevated CO_2 concentrations, indicating salt damage to biochemical or photochemical processes. Longstreth and Nobel (1979) also noted increased mesophyll "resistance" in salt-affected cotton even though leaf thickness and A^{mes}/A increased. Thus, the change in internal cell surface area, which could have reduced mesophyll "resistance," was overridden by salt-induced decreases in some of the metabolic components of photosynthesis.

About two-thirds of the reduction in photosynthesis of beans treated with up to −0.35 MPa KNO_3 was ascribed to nonstomatal causes. Salinity exerted no effect on photochemical efficency, but mesophyll "resistance," determined on an intercellular CO_2 basis, rose, pointing to increased biochemical limitations to CO_2 fixation (Jensen, 1975, 1977). Likewise in grapevine, increased mesophyll "resistance" accounted for most of the inhibition of light-saturated photosynthesis (Downton, 1977). This was accompanied by accumulation of radiocarbon in intermediates of the photorespiratory (glycolate) pathway (see Section VII,B,2) and could be related to the chloride status of leaves. Although a stimulation of photorespiration might be taken to indicate a low intercellular CO_2 concentration, it can be calculated from diffusive resistance equations that the vine leaves with increased chloride content (and reduced photosynthetic rates) actually experienced higher intercellular levels of CO_2 than leaves containing lower concentrations of chloride. This situation contrasts with water-stressed plant material (Section VI) where similar alterations in carbon metabolism have been related to reduced internal CO_2 concentration resulting from partial stomatal closure (Lawlor and Fock, 1975, 1977a,b; Lawlor, 1976b). The changes in stomatal resistance, which usually accompany alterations in mesophyll "resistance" in salt-affected plants, may represent an adjustment of stomatal aperture to the capacity of the leaves to fix CO_2, thus balancing transpiration and carbon gain (Wong *et al.*, 1979).

2. CARBON METABOLISM

A change in the pathway of CO_2 fixation in response to salinity stress has been documented for some members of the *Aizoaceae*. Mature leaves from *Mesembryanthemum cyrstallinum* plants receiving saline treatment for a number of days shift from C_3 photosynthesis to CAM. The development of CAM is characterized by a diurnal fluctuation in malate production, the ability to fix CO_2 at night through nocturnal opening of stomata, and increased activity of PEP carboxylase (von Willert *et al.*, 1976; Winter and Lüttge, 1976). Young leaves on salt-treated plants do not develop CAM features until they have reached a certain stage of development and mature leaves of low salt plants exhibit only weak development of CAM. The induction of CAM appears to be triggered by the development of water deficits within the plant, since exposure of roots to high salt levels, reduced temperature, or anoxia elicits the response (Winter and Lüttge, 1976; Osmond, 1978). Bloom (1979), however, presented evidence that indicates that salt may play an integral role in the reactions of CAM of this species. The impact of this metabolic flexibility on the productive potential of this species in its native environment has been studied by Bloom and Troughton (1979) and Winter *et al.* (1978).

With the exception of these halophytic succulents, there is no reliable evidence to suggest that the relative ratios of PEP and RuBP carboxylation reactions in C_3 plants shift with salinity. Claims that some C_4 grasses fix CO_2 by the C_3 pathway when they are sodium deficient and shift to C_4 photosynthesis upon the application of NaCl (Shomer-Ilan and Waisel, 1973, 1976) have not been confirmed (Osmond and Greenway, 1972; Downton and Törökfalvy, 1975; Kennedy, 1977; Boag and Brownell, 1979). On the basis of $\delta^{13}C$ value, CO_2 compensation point, and percentage of label in C_4 dicarboxylic acids during short-term photosynthesis, Boag and Brownell (1979) confirmed C_4 photosynthesis in *Kochia childsii* and *Chloris barbata* grown under extremely sodium-deficient conditions. Sodium is an essential micronutrient element for C_4 plants and deficiency symptoms can be alleviated by 0.02 m*M* NaCl (Brownell and Crossland, 1972; P. F. Brownell, personal communication). Substantial changes in the distribution of ^{14}C among photosynthetic products in sodium-deficient plants may be largely due to changes in pool size of C_4 acids (C. B. Osmond, P. F. Brownell, and C. J. Crossland, unpublished), and this type of response may have occurred in the salt-depleted *Aeluropus litoralis* (Shomer-Ilan and Waisel, 1973, 1976). Parallel CO_2 fixation by PEP carboxylase and RuBP carboxylase may occur in developing leaves of C_4 plants, where the functions of the mesophyll and bundle sheath cells are not yet fully integrated (Perchorowicz and Gibbs, 1980). Sankhla and Huber (1974) have observed some minor changes with salinity on the distribution of ^{14}C between the organic acid and amino acid fractions in germinating *Pennisetum typhoides*. For fully differentiated material, salt and water stress cause no major changes in the distribution of ^{14}C among metabolites during short-term exposure of *Zea mays* and *Portulaea oleracea* to $^{14}CO_2$ (Kennedy, 1977). Likewise, Osmond and Greenway (1972) failed to find an effect of salinity in the proportion of ^{14}C incorporated into C_4 acids in *Atriplex* and *Zea mays*, which were grown for several weeks on saline media. In grapevine C_3, the proportion of label in malate and aspartate decreases with increased chloride concentration during short-term exposure to $^{14}CO_2$ (Downton,

1977); RuBP carboxylase activity does not change (Walker *et al.*, 1981). The extremely low PEP carboxylase activity found in leaves of sodium deficient *Aeluropus litoralis* (Shomer-Ilan and Waisel, 1973) was probably a consequence of the extraction procedure used. Hatch and Oliver (1978) reported that this enzyme is unstable, and high levels of PEP carboxylase have been detected (Downton and Törökfalvy, 1975) in this species.

The most notable effect of salinity stress on photosynthetic carbon metabolism of glycophytes is that labeling of RuBP and the photorespiratory intermediates glycolate, glycine, and serine are stimulated at the expense of sucrose synthesis (Downton, 1977). As noted earlier, this was apparently not a result of decreased intercellular CO_2 concentration. Since photorespiration is initiated by the oxygenase function of RuBP carboxylase/oxygenase, its enhancement relative to carbon fixation in salt-affected material suggests altered internal concentrations of CO_2 and O_2 or differential effects of accumulated ions on carboxylase and oxygenase activity. At present, there is no information on *in vitro* effects of salinity on the kinetic parameters of carboxylation and oxygenation. Obviously salinity can alter CO_2/O_2 ratios if stomata close. However, the accumulation of salts in the cell wall (Oerti, 1968; Greenway and Munns, 1980) or in other cell compartments (Larkum and Hill, 1970) could also have far-reaching effects on cellular CO_2/O_2 levels. Not only are the solubilities of CO_2 and O_2 in water affected somewhat differently by dissolved salts, but more significantly the first apparent dissociation constant of carbonic acid (pK) is shifted downward, from about 6.4 in pure water to about 6.0 in seawater (Kester, 1975, Skirrow, 1975). The consequence of such a shift in pK would be a drop in the proportion of CO_2 in equilibrium with bicarbonate at physiological pH. This may affect the catalytic activities of RuBP carboxylase such that oxygenation is stimulated relative to carboxylation (Downton, 1977; Passera and Albuzio, 1978) even in the absence of stomatal changes. It would be profitable to continue to explore salinity effects on RuBP carboxylase and the glycolate pathway since gas exchange and metabolic studies show photorespiration to be sensitive to salt stress.

3. MEMBRANE REACTIONS

Cations are known to play an important role in the regulation of primary photochemical processes (see, e.g., Wong *et al.*, 1981), yet estimates of ionic concentrations within chloroplasts are few and variable (Barber, 1976). Thus the impact on photosynthesis of increasing ion concentrations within chloroplasts is uncertain. *In vitro* studies on spinach showed monovalent salts above 100 m*M* to inhibit ferricyanide reduction in uncoupled thylakoids, but the coupled reaction was salt-tolerant and little affected up to 600 m*M* (Baker, 1978). Salinity also reduced the effective concentration of plastoquinone and slowed the rate of PSII primary photochemistry. It is now feasible to examine light-harvesting processes *in vivo* by means of fluorescence methodology (Dominy and Baker, 1980), and this should be applied to salt-stressed leaves.

VIII. Responses to Environmental Pollutants

A. Gaseous Pollutants

The addition of toxic gases to the atmosphere by human activities and volcanos has had a significant impact upon plants and by all indications will be an increasing consideration in the future. The impact of these pollutants upon vegetation is a function of many complex meterological and plant factors. A complete treatment of these subjects is beyond the scope of this review, and the focus will be primarily upon the mechanisms whereby these pollutants may affect photosynthesis. Changes in the photosynthetic activity of leaves provide a rapid and sensitive assay for damage to leaves by pollutants. Also, photosynthetic reactions appear to be a primary site of damage caused by these agents.

Much of the work to date has been concerned with the concentration of pollutants required to cause damage to plant tissue. Several studies (Bressan *et al.*, 1978; Winner and Mooney, 1980a–d) emphasized the importance of quantitative measurement of the interaction of pollutants with leaves. Winner and Mooney (1980a) distinguished between the *absorption* of substances by the mesophyll cells of the leaf via the stomata and *adsorption* to exterior surface of the leaf. Although there is evidence that some pollutants may react directly with the leaf cuticle (Godzik and Sassen, 1978; Black and Black, 1979), the physiological effects of these substances is most likely to be *via* absorption into the mesophyll cells. Substances such as SO_2 are extremely soluble in water, and nearly complete absorption must occur upon contact with the wet surfaces of the cells of the leaf interior. By assuming that the intercellular concentration of SO_2 is kept nearly zero by absorption, Winner and Mooney (1980a) calculated the rate at which SO_2 enters the leaf interior from measurements of the ambient concentration of SO_2 and the stomatal conductance (determined by water vapor exchange). The relative proportion of uptake by adsorption and absorption differs considerably among species (Winner and Mooney, 1980a, b). Because the effective concentrations of pollutants is quite low (< 2 ppm) and because these substances are highly reactive with water (and many materials typically used in construction of gas-exchange systems, e.g., acrylic plastics, aluminum, nickel plated brass, etc.), measurement of uptake of pollutants by leaves demands careful technical approaches. Winner and Mooney (1980a) describe a gas-exchange system designed for measurement of SO_2 exchange simultaneous with measurement of photosynthesis and transpiration.

1. SULFUR DIOXIDE

Sulfur dioxide (SO_2) emitted during the combustion of fossil fuels is injurious to crop plants, and photosynthesis may be reduced without the appearance of visible damage to tissue. The study of SO_2 effects on plant growth and photosynthesis has been frought with controversy and numerous negative responses have been recorded in fumigation experiments (Cowling and Koziol, 1978; Bell, 1980). Effects by SO_2 are apparently very sensitive to prevailing environmental conditions (Black and Unsworth, 1979a), and toxic action is enhanced by slow growth (Davies, 1980) and reduced boundary layer conditions in fumigation chambers (Ashenden and Mansfield, 1977). Variations in susceptibility of plants from various ecosystems have been characterized (for references, see Bressen *et al.*, 1978; Winner and Mooney, 1980a).

As originally proposed (Thomas and Hill, 1935), the preceding differences in sensitivity between species and between conditions of exposure are probably related to differences in the rate of *absorption* of SO_2 during exposure. For example, differences in the sensitivity of two shrubs to external SO_2 concentration (Fig. 16a) diminish when the sensitivity is compared on the basis of absorbed flux of SO_2 during the fumigation period (Fig. 16b) (Winner and Mooney, 1980b).

An important factor governing the uptake is the stomatal conductance at the time of exposure and the response of stomata to the exposure. There are numerous accounts of SO_2 increasing stomatal conductance (Majernik and Mansfield, 1970; Unsworth *et al.*, 1972; Black and Unsworth, 1979b; Muller *et al.*, 1979; Barton *et al.*, 1980). Winner and Mooney (1980c) showed that while stomatal conductance may double in response to 0.5 ppm SO_2 with one species of *Atriplex*, the stomata of another species of *Atriplex* are hardly affected. Obviously, increases in stomatal conductance in response to SO_2 would increase the potential for physiological damage by SO_2.

Studies of the sensitivity of native Hawaiian vegetation to volcanic SO_2 (Winner and Mooney, 1980d) highlight the importance of stomatal responses. Only mature leaves of *Metrosideros collina* were able to survive the high (>100 ppm) concentrations of SO_2, which resulted from an eruption at Pauaki Crater. Winner and Mooney (1980d) found that leaves of only this species had low stomatal conductances during the

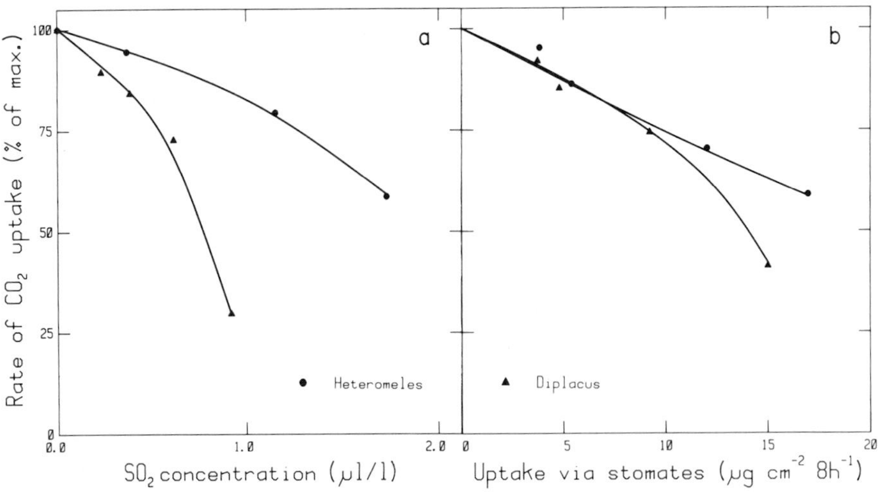

FIG. 16. **SO_2-induced changes in net CO_2 assimilation at 20°C, 1000 μmole photons $m^{-2}sec^{-1}$ during an 8-hour fumigation at the concentrations indicated (a). The changes are related to the quantities of SO_2 absorbed at each treatment during the 8-hour fumigation (b). Species used were *Diplacus aurantiacus* and *Heteromeles arbutifolia*. (Redrawn by permission from Winner and Mooney, 1980b.)**

period of active SO_2 venting. One week after the eruption the conductance of *M. collina,* leaves in the volcanic area were similar to those of the same or other species in unaffected areas. The survival of this species in proximity to continuous volcanic activity is apparently related to mechanisms that result in stomatal closure (and consequently restrict SO_2 absorption) during volcanic activity.

Increase in SO_2 resistance among *Petunia* cultivars has been associated with an abundance of trichomes (Elkiey and Ormrod, 1979), and there is correlative evidence that some plant species evolve trichomes in polluted areas (Sharma, 1975; Sharma and Butler, 1975). These structures may adsorb SO_2 from the boundary layer of the leaf thus reducing the local concentration of SO_2 available for absorption by the mesophyll cells of the leaf.

Biochemical differences may also influence the susceptibility of leaves to SO_2. A C_3 species of *Atriplex* was more sensitive to SO_2 than a C_4 species (Winner and Mooney, 1980c). This difference may be related to differences in the sensitivity of RuBP and PEP carboxylase to this pollutant (Ziegler, 1972, 1973, 1975).

Concentrations of SO_2, which inhibit photosynthesis, result in a depression of the quantum yield for CO_2 uptake (Taniyama *et al.*, 1972;

Shimazaki and Sugahara, 1979; Winner and Mooney, 1980c). With *Atriplex triangularis* (Winner and Mooney, 1980c), it is clear that the effect on quantum yield is not due to a decrease in the intercellular CO_2 concentration, because the stomata actually open in response to the fumigation. Spinach chloroplasts isolated from leaves receiving 1–2 ppm SO_2 were suppressed in electron flow driven by PSII, but they showed normal light-induced pH changes, phenazine methosulfate-catalyzed cyclic photophosphorylation, and PSI activity (Shimazaki and Sugahara, 1979). At the ultrastructural level, Wellburn *et al.* (1972) noted a swelling of thylakoids in *Vicia faba* leaves exposed to either SO_2 or NO_2. As with other stresses, susceptibility to photoinhibitory or photooxidative events as a result of the primary effect on electron transport may exacerbate the initial damage caused by SO_2.

At very low concentrations, SO_2 treatment may stimulate dark respiration (Taniyama *et al.*, 1972; Koziol and Jordon, 1978; Black and Unsworth, 1979b). For example, the light response curves for *Vicia faba* leaves fumigated with low concentrations of SO_2 (up to 0.175 ppm = 500 $\mu g\ m^{-3}$) show the same initial slope (apparent quantum yield) as control leaves, but rates are displaced by the extent to which dark respiration is increased in the treated leaves (Black and Unsworth, 1979b). At low irradiances, the inhibition of photosynthesis is independent of SO_2 concentrations between 100 and 500 $\mu g\ m^{-3}$. Low levels of SO_2 might thus stimulate respiratory losses by crop canopies.

2. NITROGEN DIOXIDE

Nitrogen dioxide (NO_2) is released together with SO_2 during the combustion of fossil fuels, having its origin in the heat-induced combination of atmospheric N_2 and O_2 yielding nitric oxide (NO), which is rapidly oxidized in light to NO_2. In combination with SO_2, NO_2 further depresses rates of photosynthesis (Bull and Mansfield, 1974) in pea plants. This combination closes stomates in bean leaves, even though SO_2 and NO_2 individually increase stomatal conductance (Ashenden, 1979). Very little is known about the singular effects of NO_2 on photosynthesis. The most comprehensive gas-exchange study at present is that of Srivastava *et al.* (1975a, b) for bean plants. Photosynthesis and dark respiration were both inhibited by 1–7 ppm NO_2, but transpiration was little affected. Increased concentrations of NO_2 reduced CO_2 production into CO_2-free air, but the percentage inhibition of apparent photosynthesis was not altered by lowering oxygen tension, indicating proportional effects of NO_2 upon photosynthesis and photorespiration.

Atmospheric concentrations of CO_2 usually increase with the level of pollutants such as SO_2 and NO_2. CO_2 enrichment around leaves can partly counteract the inhibitory effects of NO_2 (Srivastava *et al.*, 1975b) as well as the combination of NO_2 and SO_2 (Hou *et al.*, 1977). A reduction in stomatal aperture by elevated CO_2 thereby restricting pollutant uptake (Majernik and Mansfield, 1972; Srivastava *et al.*, 1975b) might occur, or the stimulation of photosynthesis by the increased CO_2 levels might lead to greater resistance of the tissue to pollutants. Capron and Mansfield (1976) demonstrated that the amounts of NO and NO_2 generated by hydrocarbon burners used to provide CO_2 enrichment for glasshouses are sufficient to substantially reduce photosynthesis in tomato plants. NO, which can accumulate at low light when the photochemical conversion to NO_2 is slow, appears to be very toxic to plants (A. R. Wellburn, personal communication).

3. PHOTOCHEMICAL OXIDANTS

Effects of peroxyacyl nitrates (PAN) on photosynthesis have received virtually no attention since the topic was reviewed some 10 years ago (Dugger and Ting, 1970); the reader is referred to this article.

The other major oxidizing pollutant of photochemical origin, ozone (O_3), was featured in a few photosynthesis studies. It is clear that stomatal responses to O_3 are of great importance (Hill and Littlefield, 1969). A number of species with cultivars differing in O_3 sensitivity apparently achieve resistance through reduced stomatal frequency and partial closure of stomata when exposed to ozone (Butler and Tibbitts, 1979). The failure of the sensitive cultivars to adjust stomatal aperture permits greater access of the pollutant to the mesophyll cells, which results in photosynthetic decline. Other stresses such as salinity, which cause a lower stomatal conductance, may thereby protect against ozone injury (Hoffman *et al.*, 1975).

Chlorophyll fluorescence induction kinetics have been used to probe the *in vivo* effects of O_3 on the light reactions (Schreiber *et al.*, 1978). The first apparent effect of O_3 on bean chloroplasts is a reduction in water-splitting activity, thereby restricting the donation of electrons to PSII. A decline in electron transport between the photosystems follow. Ozone has also been observed to reduce RuBP carboxylase/oxygenase activity in rice leaves 12–24 hours after exposure to 0.12 ppm O_3 for 2–3 hours. Activity recovered to some extent after 48 hours in younger leaves, but not in older leaves (Nakamura and Saka, 1978). In loblolly and white pine, significant alterations in the products of $^{14}CO_2$ fixation accompanied O_3 treatment and soluble sugars decreased as a proportion of total carbon fixed, whereas sugar phosphates and amino acids, particularly alanine, increased (Wilkinson and Barnes, 1973).

4. CARBON DIOXIDE

Carbon dioxide concentration in the atmosphere is increasing at a rate of about 1–1.5 μbar per annum, largely from the combustion of fossil fuel. Although CO_2 in this context might be regarded as a pollutant by those concerned about the climatic implications of an increased "greenhouse effect," elevated levels of CO_2 are likely to be beneficial to agriculture. In fact, horticulturalists have for some years practiced CO_2 enrichment of enclosed crops to increase yield. Although enhancement of photosynthesis by increased CO_2 at light saturation is understandable in terms of limiting factors and suppression of oxygenase activity relative to carboxylase activity (Jolliffe and Tregunna, 1968; Laing *et al.*, 1974; Ku *et al.*, 1977), it has remained a paradox that CO_2 enrichment should also be of benefit during the winter when photosynthesis is often severely light limited (Heath and Meidner, 1967; Gifford, 1977). The work of Ehleringer and Björkman (1977) provided an explanation: The quantum yield (Φ_a) in C_3 plants is dependent upon the intercellular CO_2 concentration (see Fig. 8) rising from about 0.05 mole CO_2 fixed per absorbed einstein at 325 μbar CO_2 to about 0.075 at 1000 μbar CO_2. At 200 μbar CO_2, an internal concentration commonly measured in C_3 plants photosynthesizing in air, Φ_a may be less than 0.04. The consequence for productivity is that when light is limiting, for example, in glasshouses during winter or in most canopies in the field, an increase in external CO_2 concentration will improve the efficiency of CO_2 uptake per unit of absorbed light by raising intercellular CO_2. CO_2 enrichment also has the effect of lowering the light compensation point (Heath and Meidner, 1967), which extends the duration of net carbon gain and possibly the longevity of leaves lowermost in the canopy. Gifford (1979) attributed certain long-term trends of yield increase in wheat to the global rise of atmospheric CO_2. Carbon dioxide enrichment apparently offers no advantage to light-limited photosynthesis in C_4 species since photorespiration is minimized by the CO_2-concentrating mechanism, which functions to raise CO_2 levels around RuBP C'ase in the bundle sheath cells (Hatch, 1977). With this as hindsight, the C_4 pathway can be regarded as a successful evolutionary experiment in CO_2 enrichment.

The preceding discussion has considered ways in which CO_2 enrichment enhances carbon acquisition, i.e., through improved quantum yield and lowered light compensation point under light-limited conditions and by providing more CO_2 at light saturation. However, there are also indications that growth at high CO_2 alters photosynthetic characteristics, which are retained when plants are transferred to normal CO_2 levels (Bishop and Whittingham, 1968; Hofstra and Hesketh, 1975;

Frydrych, 1976; Kriedemann *et al.*, 1976; Hicklenton and Jolliffe, 1978). An interesting example of this was observed in grapevines grown at high temperature to eliminate viruses; CO_2 enrichment resulted in a reduction of photosynthetic activity (measured at 300 μbar CO_2) to approximately one-half of that in heat-treated plants grown in normal air (Kriedemann *et al.*, 1976). As the CO_2 concentration during measurement was raised above approximately 500 μbar, however, the plants grown at high CO_2 became superior in photosynthesis to the unenriched plants. This difference in CO_2 response was interpreted in terms of anatomical changes, since growth at high CO_2 doubled the thickness of the hypostomatous leaves such that atmospheric concentration of CO_2 apparently became insufficient to saturate the more remote photosynthetic sites closer to the upper epidermis. RuBP C'ase levels were also examined in these plants and found to decline with CO_2 enrichment, but the activities recovered were insufficient (due to release of inhibitors during the extraction procedure) to account for rates of photosynthesis at high CO_2. There is, nevertheless, convincing evidence that the level of RuBP C'ase adjusts to altered CO_2 concentrations during plant growth. Leaves of *Nerium oleander* grown at twice atmospheric CO_2 concentration were 25–30% greater in fresh weight and soluble protein per unit area than in untreated leaves (Downton *et al.*, 1980a). RuBP C'ase activity per unit area did not differ between the treatments. A radioimmune assay (Collatz *et al.*, 1979) confirmed a 25% difference in proportion of soluble protein allocated to RuBP C'ase. Though the specific activity of RuBP C'ase decreased in the enriched plant, another photosynthetic enzyme, FBP phosphatase, remained unaffected by the growth conditions. Control plants achieved a greater rate of photosynthesis (per unit area) at atmospheric concentrations of CO_2, but at higher CO_2 tensions there was little difference in rates between the treatments, as would be expected from measured carboxylase activity. On the other hand, when *Atriplex triangularis* (*patula*) was grown at twice atmospheric CO_2, there was little change in leaf fresh weight per unit area, but soluble protein per unit fresh weight was reduced to 77% of control plants (Downton *et al.*, 1980a). Specific activity of RuBP C'ase in the enriched plants declined by only 10%. Consequently CO_2 enrichment led to a substantial reduction of carboxylase activity (65–70% of control plants) expressed on a leaf area, fresh weight or chlorophyll basis. Photosynthesis of leaf cells isolated from the high CO_2 grown plants showed only 70% of the activity of cells from the control plants per unit of chlorophyll (G. W. Harvey, personal communication). Cotton plants grown at high CO_2 also exhibited reductions in assimilation rate and RuBP C'ase activity per unit leaf compared to plants grown in normal air (Wong, 1980a).

These examples illustrate two different ways in which increased atmospheric concentrations of CO_2 may affect photosynthesis (1) by reducing the fraction of soluble protein allocated to RuBP carboxylase/oxygenase; and (2) by reducing enzyme concentration per unit leaf area, chlorophyll, or fresh weight. Both of these trends are evident in the levels of RuBP carboxylase in C_4 plants (Björkman *et al.*, 1976; Ku *et al.*, 1979).

B. Heavy Metals

Cadmium, lead, nickel, zinc, thallium, and other heavy metals are released into the environment from automobile emissions, smelters, and other industrial and manufacturing operations. These substances settle out with dust and contaminate both foliage and soil. In other instances, sewage sludge enriched in heavy metals has been applied to crop plants as a fertilizer. Despite man's concern about heavy metals entering food chains, little effort seems to have been expended investigating the consequences for photosynthesis of heavy metal accumulation in land plants. Especially noticeable is that very few studies have used plants grown in the presence of heavy metals. More frequently, solutions of metals have been fed to detached plant parts or added to isolated organelles.

In a study where maize and sunflower were grown on nutrient solution containing 0–10 ppm of heavy metals, thallium was found to be most toxic to photosynthesis, followed by cadmium and nickel (Carlson *et al.*, 1975). Nickel seemed to depress photosynthesis through stomatal closure, but thallium caused additional direct effects upon photosynthesis. Studies with detached leaves showed the inhibition by T1, Ni, Cd, and Pb to be closely tied to stomatal function (Bazzaz *et al.*, 1974a,b; Lamoreaux and Chaney, 1978).

Growth of tomato plants on excessive concentrations of cadmium resulted in reduced photosynthesis and chlorophyll content per unit area (Baszynski *et al.*, 1980). Chloroplasts isolated from these plants were less active in electron transport and noncyclic photophosphorylation compared to cadmium-free plants. In accordance with earlier observations for isolated chloroplasts exposed to cadmium *in vitro* (Bazzaz and Govindjee, 1974), addition of diphenylcarbazide (DPC) to chloroplasts isolated from the cadmium-grown plants restored PSII activity to control values indicative of a cadmium effect on the water side of PSII. Cadmium treatment had no apparent effect on PSI activity or cyclic photophosphorylation, and transfer of cadmium-grown plants to a manganese-enriched medium led to the restoration of PSII activity (Baszynski *et al.*, 1980). In contrast, Lucero *et al.* (1976) found both cyclic and noncyclic photophosphorylation to be inhibited when cadmium was applied to isolated chloroplasts. Also, Li and Miles (1975) were unable to restore fluorescence yield by adding DPC or manganese to chloroplasts exposed to cadmium *in vitro*. Whereas most of the evidence points to an inhibitory effect of cadmium on PSII, the exact site of action remains unresolved. (Some of the differences in results are due to differences in concentrations of Cd used by various authors.) Disorganization of grana in chloroplasts of cadmium-treated plants has been observed (Baszynski *et al.*, 1980). Structure could be partially restored when extra manganese was added to the cadmium-containing growth medium.

Photosynthetic $^{14}CO_2$ fixation by isolated spinach chloroplasts was inhibited noncompetitively by cadmium and zinc added to the reaction medium (Hampp *et al.*, 1976). The concentration for 50% inhibition (K_i) for zinc (22.5 μM) was much higher than the K_i for cadmium (3.5 μM). Although these metals also restricted DCIP reduction (PSII assay), substantial inhibition occurred only at concentrations higher than 100 μM, with cadmium remaining the more effective inhibitor.

It is clear that while *in vitro* studies provide a measure of the potential damage to photosynthesis by heavy metals, these should be accompanied by whole plant investigations to determine the actual damage to photosynthesis and growth associated with heavy metal accumulation. For example, it is possible that heavy metals might accumulate in vacuoles in intact systems, whereas cytoplasmic compartments might be overwhelmed when heavy metals are fed directly into the transpiration stream of excised leaves.

IX. Conclusions

In this chapter we have considered the control of photosynthesis by major factors of the environment; where possible we have attempted to focus attention on the fundamental mechanisms which underlie these responses and upon adaptive mechanisms which enable some plants to adjust to better cope with the limitations of their environment. This is a very broad topic which brings together information from all aspects of the photosynthesis literature. Most of these topics are treated in more complete detail in other chapters of these volumes and in more specialized reviews which we have cited. In this chapter we have attempted to use the current understanding of photosynthesis to analyze and explain the photosynthetic performance of whole leaves of plants in natural conditions. This approach, if nothing else, provides a good focus for trying to integrate knowledge of the photosynthetic process, and it is a route by which our studies of photosynthetic mechanisms might ultimately have an impact in more applied areas such as crop physiology, plant breeding, or resource management.

Acknowledgments

The assistance of Tim Ball and Jeffrey Seemann with the many details of preparing this manuscript are gratefully acknowledged. John Boyer, A. E. Hall, and W. E. Winner provided original prints of figures used in this chapter. Steve Powles, Graham Farquhar, and Winslow Briggs provided useful criticisms of the manuscript. Olle Björkman, who also provided original figures, has been a guiding influence in the development of this topic.

Financial assistance from the National Science Foundation (DEB-78-10724) and the USDA Competitive Grants Program (Agreement 5901-0410-0128) to Joseph Berry is gratefully acknowledged.

REFERENCES

Ackerson, R. C. (1980). *Plant Physiol.* **65,** 455–459.

Adams, J. A., Bingham, F. T., Kaufmann, M. R., Hoffman, G. J., and Yermanos, D. M. (1978). *Agron. J.* **70,** 381–387.

Adams, M. S., and Strain, B. R. (1969). *Photosynthetica* **3,** 55–62.

Alberte, R. S., Thornber, J. P., and Fiscus, E. L. (1977). Plant Physiol. **59,** 351–353.

Anderson, L. E. (1980). *Encycl. Plant Physiol., New Ser.* **6,** 271–281.

Armond, P. A., and Hess, J. L. (1979). *Year Book—Carnegie Inst. Washington* **78,** 168–171.

Armond, P. A., Badger, M. R., and Björkman, O. (1978a). *In* "Chloroplast Development" (G. Akoyunoglou and J. H. Argyroudi-Akoyonoglou, eds.), pp. 857–862. Elsevier/North-Holland Biomedical Press, Amsterdam.

Armond, P. A., Schreiber, U., and Björkman, O. (1978b). *Plant Physiol.* **61,** 411–415.

Armond, P. A., Björkman, O., and Staehelin, L. A. (1979). *Year Book—Carnegie Inst. Washington* **78,** 292–293.

Ashenden, T. W. (1979). *Environ. Pollut.* **18,** 45–50.

Ashenden, T. W., and Mansfield, T. A. (1977). *J. Exp. Bot.* **28,** 729–735.

Aston, M. J. (1976). *Aust. J. Plant Physiol.* **3,** 489–501.

Badger, M. R., and Andrews, T. J. (1974). *Biochem. Biophys. Res. Commun.* **60,** 204–210.

Badger, M. R., and Collatz, G. J. (1977). *Year Book—Carnegie Inst. Washington* **76,** 355–361.

Badger, M. R., Kaplan, A., and Berry, J. A. (1980). *Plant Physiol.* **66,** 407–413.

Baker, N. R. (1978). *Plant Physiol.* **62,** 889–893.

Barber, J. (1976). *In* "The Intact Chloroplast" (J. Barber, ed.), pp. 89–134. Elsevier-North-Holland Biomedical Press, Amsterdam.

Barton, J. R., McLaughlin, S. B., and McConathy, R. K. (1980). *Environ. Pollut., Ser. A* **21,** 255–265.

Baszynski, T., Wajda, L., Krol, M., Wolinska, D., Krupa, Z., and Tukendorf, A. (1980). *Physiol. Plant.* **48,** 365–370.

Bauer, H. (1978). *Physiol. Plant.* **44,** 400–406.

Bauer, H., and Senser, M. (1979). *Z. Pflanzenphysiol.* **91,** 359–369.

Bazzaz, F. A., Carlson, R. W., and Rolfe, G. L. (1974a). *Environ. Pollut.* **7,** 241–246.

Bazzaz, F. A., Rolfe, G. L., and Carlson, R. W. (1974b). *Physiol. Plant.* **32,** 373–376.

Bazzaz, M. B., and Govindjee (1974). *Environ. Lett.* **6,** 1–12.
Begg, J. E., and Turner, N. C. (1976). *Adv. Agron.* **28,** 161–217.
Bell, J. N. B. (1980). *Nature (London)* **284,** 399–400.
Berry, J. A., and Badger, M. R. (1979). *Year Book—Carnegie Inst. Washington* **78,** 175–178.
Berry, J. A., and Björkman, O. (1980). *Annu. Rev. Plant Physiol.* **31,** 491–543.
Berry, J. A., and Farquhar, G. D. (1978). *Proc. Int. Congr. Photosynth., 4th, 1977,* pp. 119–131.
Berry, J. A., and Raison, J. K. (1981). *Encycl. Plant Physiol., New Ser.* **12A,** 277–338.
Berry, J. A., Fork, D. C., and Garrison, S. (1975). *Year Book—Carnegie Inst. Washington* **74,** 751–759.
Bewley, J. D. (1979). *Annu. Rev. Plant Physiol.* **30,** 195–238.
Bishop, P. M., and Whittingham, C. P. (1968). *Photosynthetica* **2,** 31–38.
Björkman, O. (1968). *Physiol. Plant.* **21,** 84–99.
Björkman, O. (1973). *Photophysiology* **8,** 1–63.
Björkman, O. (1975). *Year Book—Carnegie Inst. Washington* **74,** 748–751.
Björkman, O. (1981). *Encycl. Plant Physiol., New Ser.* **12A,** 57–107.
Björkman, O., and Badger, M. (1977). *Year Book—Carnegie Inst. Washington* **76,** 346–354.
Björkman, O., and Badger, M. (1979). *Year Book—Carnegie Inst. Washington* **78,** 262–275.
Björkman, O., Ludlow, M. M., and Morrow, P. A. (1972a). *Year Book—Carnegie Inst. Washington* **71,** 94–107.
Björkman, O., Boardman, N. K., Anderson, J. M., Thorne, S. W., Goodchild, D. J., and Pyliotis, N. A. (1972b). *Year Book—Carnegie Inst. Washington* **71,** 115–135.
Björkman, O., Boynton, J., and Berry, J. (1976). *Year Book—Carnegie Inst. Washington* **75,** 400–407.
Björkman, O., Badger, M. R., and Armond, P. A. (1978). *Year Book—Carnegie Inst. Washington* **77,** 262–282.
Björkman, O., Badger, M. R., and Armond, P. A. (1980a). *In* "Adaptation of Plants to Water and High Temperature Stress" (N. C. Turner and P. K. Kramer, eds.), pp. 233–250. Wiley (Interscience), New York.
Björkman, O., Downton, W. J. S., and Mooney, H. A. (1980b). *Year Book—Carnegie Inst. Washington* **79,** 150–157.
Black, C. R., and Black, V. J. (1979). *J. Exp. Bot.* **30,** 291–298.
Black, V. J., and Unsworth, M. H. (1979a). *J. Exp. Bot.* **30,** 81–88.
Black, V. J., and Unsworth, M. H. (1979b). *J. Exp. Bot.* **30,** 473–483.
Blackman, F. F. (1905). *Ann. Bot. (London)* **19,** 218–295.
Bloom, A. J. (1979). *Plant Physiol.* **63,** 749–753.
Bloom, A. J., and Troughton, J. H. (1979). *Oecologia* **38,** 35–43.
Boag, T. S., and Brownell, P. F. (1979). *Aust. J. Plant Physiol.* **6,** 431–434.
Boardman, N. K. (1977). *Annu. Rev. Plant Physiol.* **28,** 355–377.
Boyer, J. S. (1965). *Plant Physiol.* **40,** 229–234.
Bravdo, R. A., and Canvin, D. T. (1979). *Plant Physiol.* **63,** 399–401.
Bressan, R. A., Wilson, L. G., and Filner, P. (1978). *Plant Physiol.* **61,** 761–767.
Brown, K. W., Jordan, W. R., and Thomas, J. C. (1976). *Physiol. Plant.* **37,** 1–5.
Brown, R. H. (1978). *Crop Sci.* **18,** 93–98.
Brownell, P. F., and Crossland, C. J. (1972). *Plant Physiol.* **49,** 794–797.
Bull, J. N., and Mansfield, T. A. (1974). *Nature (London)* **250,** 443–444.
Butler, L. K., and Tibbitts, T. W. (1979). *J. Am. Soc. Hortic. Sci.* **104,** 213–216.
Butler, W. (1964). *Annu. Rev. Plant Physiol.* **15,** 451–470.
Caldwell, M. M., Osmond, C. B., and Nott, D. (1977). *Plant Physiol.* **60,** 157–164.

Canvin, D. T., Berry, J. A., Badger, M. R., Fock, H., and Osmond, C. B. (1980). *Plant Physiol.* **66,** 302–307.
Capron, T. W., and Mansfield, T. A. (1976). *J. Exp. Bot.* **27,** 1181–1186.
Carlson, R. W., Bazzaz, F. A., and Rolfe, G. L. (1975). *Environ. Res.* **10,** 113–120.
Chabot, B. F., Jurik, T. W., and Chabot, J. F. (1979). *Am. J. Bot.* **66,** 940–945.
Chatterton, N. J., Hanna, W. W., Powell, J. B., and Lee, D. R. (1975). *Can. J. Plant Sci.* **55,** 641–643.
Collatz, G. J. (1978). *Year Book—Carnegie Inst. Washington* **77,** 248–251.
Collatz, G. J. (1980). Ph.D. Dissertation, Stanford University, Stanford, California.
Collatz, G. J., and Badger, M. R. (1978). *Year Book—Carnegie Inst. Washington* **77,** 245–257.
Collatz, G. J., Badger, M. R., Smith, C., and Berry, J. A. (1979). *Year Book—Carnegie Inst. Washington* **78,** 171–175.
Cowan, I. R. (1977a). *In* "Water, Planets, Plants, and People" (A. K. McIntyre, ed.), pp. 71–107. Aust. Acad. Sci., Canberra.
Cowan, I. R. (1977b). *Adv. Bot. Res.* **4,** 117–228.
Cowan, I. R., and Farquhar, G. D. (1977). *Symp. Soc. Exp. Biol.* **31,** 471–505.
Cowling, D. W., and Koziol, M. J. (1978). *J. Exp. Bot.* **29,** 1029–1036.
Davies, T. (1980). *Nature (London)* **284,** 483–485.
Doliner, L. H., and Jolliffe, P. A. (1979). *Oecologia* **38,** 23–34.
Dominy, P. J., and Baker, N. R. (1980). *J. Exp. Bot.* **31,** 59–74.
Downton, W. J. S. (1977). *Aust. J. Plant Physiol.* **4,** 183–192.
Downton, W. J. S., and Törökfalvy, E. (1975). *Z. Pflanzenphysiol.* **75,** 143–150.
Downton, W. J. S., and Tregunna, E. B. (1968). *Plant Physiol.* **43,** 923–929.
Downton, W. J. S., Björkman, O., and Pike, C. S. (1980a). *In* "Carbon Dioxide and Climate: Australian Research" (G. I. Pearman, ed.), pp. 143–151. Aust. Acad. Sci., Canberra.
Downton, W. J. S., Seemann, J. R., and Berry, J. A. (1980b). *Year Book—Carnegie Inst. Washington* **79,** 143–146.
Drake, B. G., Raschke, K., and Salisbury, F. B. (1970). *Plant Physiol.* **46,** 324–330.
Dubbe, D. R., Farquhar, G. D., and Raschke, K. (1978). *Plant Physiol.* **62,** 413–417.
Dugger, W. M., and Ting, I. P. (1970). *Annu. Rev. Plant Physiol.* **21,** 215–234.
Ehleringer, J. R. (1977). *Year Book—Carnegie Inst. Washington* **76,** 367–369.
Ehleringer, J. R. (1980). *In* "Adaptation of Plants to Water and High Temperature Stress" (N. C. Turner and P. J. Kramer, eds.), pp. 295–308. Wiley (Interscience), New York.
Ehleringer, J. R., and Björkman, O. (1977). *Plant Physiol.* **59,** 86–90.
Ehleringer, J. R., and Björkman, O. (1978). *Oecologia* **36,** 151–162.
Ehleringer, J. R., and Forseth, I. (1980). *Science* **210,** 1094–1098.
Ehleringer, J. R., Björkman, O., and Mooney, H. A. (1976). *Science* **192,** 376–377.
Elkiey, T., and Ormrod, D. P. (1979). *Plant Physiol.* **63,** Suppl., 150.
El-Sharkawy, M., and Hesketh, J. D. (1965). *Crop Sci.* **5,** 517–521.
Emerson, R. (1958). *Annu. Rev. Plant Physiol.* **9,** 1–24.
Farquhar, G. D. (1979a). *Arch. Biochem. Biophys.* **193,** 456–468.
Farquhar, G. D. (1979b). *In* "Photosynthesis and Plant Development" (R. Marcelle, H. Clijsters, and M. van Poucke, eds.), pp. 321–328. Junk, The Hague.
Farquhar, G. D., and Sharkey, T. D. (1982). *Annu. Rev. Plant Physiol.* **33,** 317–345.
Farquhar, G. D., and von Caemmerer, S. (1982). *Encycl. Plant Physiol., New Ser.* **12B,** (in press).
Farquhar, G. D., von Caemmerer, S., and Berry, J. A. (1980). *Planta* **149,** 78–90.
Farquhar, G. D., O'Leary, M. H., and Berry, J. A. (1982). *Aust. J. Plant Physiol.,* **9,** 121–137.
Fellows, R. J., and Boyer, J. S. (1976). *Planta* **132,** 229–239.

Fischer, R. A., and Turner, N. C. (1978). *Annu. Rev. Plant Physiol.* **29,** 277–317.

Flowers, T. J., Troke, P. F., and Yeo, A. R. (1977). *Annu. Rev. Plant Physiol.* **28,** 89–121.

Fork, D. C., and Govindjee (1980). *Naturwissenschaften* **67,** 510–511.

Fry, K. E. (1972). *Crop Sci.* **12,** 698–701.

Frydrych, J. (1976). *Photosynthetica* **10,** 335–338.

Gaastra, P. (1959). *Meded. Landbouwhogesch. Wageningen* **59,** 1–68.

Gaff, D. F. (1980). *In* "Adaptation of Plants to Water and High Temperature Stress" (N. C. Turner and P. K. Kramer, eds.), pp. 207–230. Wiley (Interscience), New York.

Gaff, D. F., and Churchill, D. M. (1976). *Aust. J. Bot.* **24,** 209–224.

Gaff, D. F., and Hallam, N. D. (1974). *Bull.—R. Soc. N.Z.* **12,** 389–393.

Gaff, D. F., Zee, S.-Y., and O'Brien, T. P. (1976). *Aust. J. Bot.* **24,** 225–236.

Gale, J. (1975). *Ecol. Stud.* **15,** 168–185.

Gale, J., and Poljakoff-Mayber, A. (1970). *Aust. J. Biol. Sci.* **23,** 937–945.

Gale, J., Kohl, H. C., and Hagan, R. M. (1967). *Physiol. Plant.* **20,** 408–420.

Gifford, R. M. (1977). *Aust. J. Plant Physiol.* **4,** 99–110.

Gifford, R. M. (1979). *Aust. J. Plant Physiol.* **6,** 367–378.

Godzik, S., and Sassen, M. A. (1978). *Environ. Pollut.* **17,** 13–18.

Govindjee, ed. (1982). "Photosynthesis: Energy Conversion by Plants and Bacteria" Vol. 1, Academic Press, New York.

Govindjee, R., Rabinowitch, E., and Govindjee (1968). *Biochim. Biophys. Acta* **162,** 539–544.

Govindjee, Downton, W. J. S., Fork, D. C., and Armond, P. A. (1981). *Plant Sci. Lett.* **20,** 191–194.

Graham, D., Hockley, D. G., and Patterson, B. D. (1979). *In* "Low Temperature Stress in Crop Plants: The Role of the Membrane" (J. M. Lyons, D. Graham, and J. K. Raison, eds.), pp. 435–462. Academic Press, New York.

Greenway, H., and Munns, R. (1980). *Annu. Rev. Plant Physiol.* **31,** 149–190.

Hall, A. E. (1971). *Year Book—Carnegie Inst. Washington* **70,** 530–540.

Hall, A. E., and Kaufmann, M. R. (1975). *Plant Physiol.* **55,** 455–459.

Hall, A. E., Schulze, E.-D., and Lange, O. (1976). *Ecol. Stud.* **19,** 169–188.

Hampp, R., Beulich, K., and Ziegler, H. (1976). *Z. Pflanzenphysiol.* **77,** 336–344.

Hatch, M. D. (1977). *Trends Biochem. Sci.* **2,** 199–202.

Hatch, M. D. (1979). *Aust. J. Plant Physiol.* **6,** 607–619.

Hatch, M. D., and Oliver, I. R. (1978). *Aust. J. Plant Physiol.* **5,** 571–580.

Hatch, M. D., and Osmond, C. B. (1976). *Encycl. Plant Physiol., New Ser.* **3,** 143–184.

Hatch, M. D., and Slack, C. R. (1968). *Biochem. J.* **106,** 141–146.

Heath, O. V. S., and Meidner, H. (1967). *J. Exp. Bot.* **18,** 746–751.

Hellmuth, E. O. (1971). *J. Ecol.* **59,** 225–259.

Hicklenton, P. R., and Jolliffe, P. A. (1978). *Can. J. Plant Sci.* **58,** 801–817.

Hill, A. C., and Littlefield, N. (1969). *Environ. Sci. Technol.* **3,** 52–56.

Hitz, W. D., and Stewart, C. R. (1980). *Plant Physiol.* **65,** 442–446.

Hoffman, G. H., Maas, E. V., and Rawlins, S. L. (1975). *J. Environ. Qual.* **4,** 326–331.

Hoffman, G. J., Shalhevet, J., and Meiri, A. (1980). *Physiol. Plant.* **48,** 463–469.

Hofstra, G., and Hesketh, J. D. (1975). *In* "Environmental and Biological Control of Photosynthesis" (R. Marcelle, ed.), pp. 71–80. Junk, The Hague.

Hou, L.-Y., Hill, A. C., and Soleimani, A. (1977). *Environ. Pollut.* **12,** 7–16.

Hsiao, T. C. (1973). *Annu. Rev. Plant Physiol.* **24,** 519–570.

Hsiao, T. C., Acevedo, E., Fereres, E., and Henderson, D. W. (1976). *Philos. Trans. R. Soc. London, Ser. B* **273,** 479–500.

Huffaker, R. C., Radin, T., Kleinkopf, G. E., and Cox, F. L. (1970). *Crop Sci.* **10,** 471–474.

Huner, N. P. A., and MacDowall, F. D. H. (1979a). *Can. J. Biochem.* **57,** 155–164.
Huner, N. P. A., and MacDowall, F. D. H. (1979b). *Can. J. Biochem.* **57,** 1036–1041.
Jarvis, P. G. (1971). *In* "Plant Photosynthetic Production. Manual of Methods" (Z. Sestak, J. Catsky, and P. Jarvis, eds.), pp. 566–631. Junk, The Hague.
Jeffree, C. E., Johnson, R. P. C., and Jarvis, P. G. (1971). *Planta* **98,** 1–10.
Jensen, C. R. (1975). *Acta Agric. Scand.* **25,** 3–10.
Jensen, C. R. (1977). *Acta Agric. Scand* **27,** 159–164.
Jensen, R. G., and Bahr, J. T. (1977). *Annu. Rev. Plant Physiol.* **28,** 377–400.
Johnson, H. S., and Hatch, M. D. (1970). *Biochem. J.* **119,** 273–280.
Jolliffe, P. A., and Tregunna, E. G. (1968). *Plant Physiol.* **43,** 902–906.
Jones, H. G. (1973). *New Phytol.* **72,** 1095–1105.
Jones, L. W., and Kok, B. (1966a). *Plant Physiol.* **41,** 1037–1043.
Jones, L. W., and Kok, B. (1966b). *Plant Physiol.* **41,** 1044–1049.
Jones, M. B., and Mansfield, T. A. (1972). *Planta* **103,** 134–146.
Jones, M. M., and Rawson, H. M. (1979). *Physiol. Plant.* **45,** 103–111.
Jones, M. M., and Turner, N. C. (1978). *Plant Physiol.* **61,** 122–126.
Jursinic, P., and Govindjee (1977). *Photochem. Photobiol.* **26,** 617–628.
Kaniuga, Z., Zabek, J., and Sochanowicz, B. (1978). *Planta* **144,** 49–56.
Kaplan, A., Badger, M. R., and Berry, J. A. (1980). *Planta* **149,** 219–226.
Keck, R. W., and Boyer, J. S. (1974). *Plant Physiol.* **53,** 474–479.
Kennedy, R. A. (1977). *Z. Pflanzenphysiol.* **83,** 11–24.
Kester, D. R. (1975). *In* "Chemical Oceanography" (J. P. Riley and G. Skirrow, eds.), 2nd ed., Vol. 1, pp. 1–192. Academic Press, New York.
Kok, B. (1948). *Enzymologia* **13,** 1–56.
Kok, B. (1965). *In* "Plant Biochemistry" (J. Bonner and J. Varner, eds), pp. 904–960. Academic Press, New York.
Koziol, M. J., and Jordon, C. F. (1978). *J. Exp. Bot.* **29,** 1037–1043.
Krause, G. H., and Santarius, K. A. (1975). *Planta* **127,** 285–299.
Kriedemann, P. E., Sward, R. J., and Downton, W. J. S. (1976). *Aust. J. Plant Physiol.* **3,** 605–618.
Ku, S. B., and Edwards, G. E. (1977). *Plant Physiol.* **59,** 991–999.
Ku, S. B., Edwards, G. E., and Tanner, C. B. (1977). *Plant Physiol.* **59,** 868–872.
Ku, S. B., Schmitt, M. R., and Edwards, G. E. (1979). *J. Exp. Bot.* **30,** 89–98.
Laing, W. A., Ogren, W. L., and Hageman, R. H. (1974). *Plant Physiol.* **54,** 678–685.
Lamoreaux, R. J., and Chaney, W. R. (1978). *Physiol. Plant.* **43,** 231–236.
Larkum, A. W. D., and Hill, A. E. (1970). *Biochim. Biophys. Acta* **203,** 133–138.
Lawlor, D. W. (1976a). *Photosynthetica* **10,** 378–387.
Lawlor, D. W. (1976b). *Photosynthetica* **10,** 431–439.
Lawlor, D. W., and Fock, H. (1975). *Planta* **126,** 247–258.
Lawlor, D. W., and Fock, H. (1977a). *J. Exp. Bot.* **28,** 320–328.
Lawlor, D. W., and Fock, H. (1977b). *J. Exp. Bot.* **28,** 329–337.
Lehrman, J. C. (1975). *In* "Environmental and Biological Control of Photosynthesis" (R. Marcelle, ed.), pp. 323–335. Junk, The Hague.
Lewis, D. A., and Nobel, P. S. (1977). *Plant Physiol.* **60,** 609–616.
Li, E. H., and Miles, C. D. (1975). *Plant Sci. Lett.* **5,** 33–40.
Lilley, R. M., and Walker, D. A. (1975). *Plant Physiol.* **54,** 1087–1092.
Linden, C. D., Wright, K. L., McConnell, H. M., and Fox, C. F. (1973). *Proc. Natl. Acad. Sci. U.S.A.* **70,** 2271–2275.
Longstreth, D. J., and Nobel, P. S. (1979). *Plant Physiol.* **63,** 700–703.

Longstreth, D. J., and Nobel, P. S. (1980). *Plant Physiol.* **65,** 541–543.
Longstreth, D. J., and Strain, B. R. (1977). *Oecologia* **31,** 191–199.
Longstreth, D. J., Hartstock, T. L., and Nobel, P. S. (1980). *Physiol. Plant.* **48,** 494–498.
Lorimer, G. H., Woo, K. C., Berry, J. A., and Osmond, C. B. (1978). *Proc. Int. Congr. Photosynth., 4th, 1977,* pp. 311–322.
Lucero, H. A., Andreo, C. S., and Vallejos, R. H. (1976). *Plant Sci. Lett.* **6,** 309–313.
Ludlow, M. M. (1976). *Ecol. Stud.* **19,** 386–388.
Ludwig, L. J., and Canvin, D. T. (1971). *Plant Physiol.* **48,** 712–719.
Lyons, J. M. (1973). *Annu. Rev. Plant Physiol.* **24,** 445–466.
Lyons, J. M., Graham, D., and Raison, J. K., eds. (1979). "Low Temperature Stress in Crop Plants: The Role of the Membrane." Academic Press, New York.
McGinnies, W. G., and Arnold, J. F. (1939). *Ariz., Agric. Exp. Stn., Tech. Bull.* **30,** 167–246.
McWilliam, J. R., and Naylor, A. W. (1967). *Plant Physiol.* **42,** 1711–1715.
Majernik, O., and Mansfield, T. A. (1970). *Nature (London)* **227,** 377–378.
Majernik, O., and Mansfield, T. A. (1972). *Environ. Pollut.* **3,** 1–7.
Mallott, P. G., Davy, A. J., Jefferies, R. L., and Hutton, M. J. (1975). *Oecologia* **20,** 351–358.
Margulis, M. M. (1972). *Biochim. Biophys. Acta* **267,** 96–103.
Medina, E. (1970). *Year Book—Carnegie Inst. Washington* **69,** 655–662.
Meidner, H. (1967). *J. Exp. Bot.* **18,** 177–186.
Melis, A., and Harvey, G. W. (1981). *Biochim. Biophys. Acta* **637,** 138–145.
Millerd, A., Goodchild, D. J., and Spencer, D. (1969). *Plant Physiol.* **44,** 567–583.
Mohanty, P., and Boyer, J. S. (1976). *Plant Physiol.* **57,** 704–709.
Mooney, H. A., Ehleringer, J. R., and Berry, J. A. (1976). *Science* **194,** 322–324.
Mooney, H. A., Ehleringer, J. R., and Björkman, O. (1977). *Oecologia* **29,** 301–310.
Mooney, H. A., Björkman, O., and Collatz, G. J. (1978). *Plant Physiol.* **61,** 406–410.
Mulchi, C. L., Volk, R. J., and Jackson, W. A. (1971). *In* "Photosynthesis and Photorespiration" (M. D. Hatch, C. B. Osmond, and R. O. Slatyer, eds.), pp. 35–50. Wiley (Interscience), New York.
Muller, R. N., Miller, J. R., and Sprugel, D. G. (1979). *J. Appl. Ecol.* **16,** 567–576.
Murata, N., and Fork, D. C. (1976). *Biochim. Biophys. Acta* **461,** 365–378.
Murata, N., Troughton, J. H., and Fork, D. C. (1975). *Plant Physiol.* **56,** 508–517.
Nakamura, H., and Saka, H. (1978). *Jpn. J. Crop Sci.* **47,** 707–714.
Nátr, L. (1972). *Photosynthetica* **6,** 80–99.
Nobel, P. S. (1977). *Physiol. Plant.* **40,** 137–144.
Nobel, P. S., Zaragosa, L. J., and Smith, W. K. (1975). *Plant Physiol.* **55,** 1067–1070.
Nolan, W. G., and Smillie, R. M. (1977). *Plant Physiol.* **59,** 1141–1145.
Oerti, J. J. (1968). *Agrochemica* **12,** 461–469.
Öquist, G., Brunes, L., Hallgren, J.-E., Gezelius, K., Hallen, M., and Malmberg, G. (1980). *Physiol. Plant.* **48,** 526–531.
Osmond, C. B. (1978). *Annu. Rev. Plant Physiol.* **29,** 379–414.
Osmond, C. B., and Björkman, O. (1972). *Year Book—Carnegie Inst. Washington* **71,** 141–148.
Osmond, C. B., and Greenway, H. (1972). *Plant Physiol.* **49,** 260–263.
Osmond, C. B., Björkman, O., and Anderson, D. J. (1980). "Physiological Processes in Plant Ecology, Toward a Synthesis with *Atriplex.*" Springer-Verlag, Berlin and New York.
O'Toole, J. C., and Cruz, R. T. (1980). *Plant Physiol.* **65,** 428–432.
O'Toole, J. C., Ozbun, J. L., and Wallace, D. H. (1977). *Physiol. Plant.* **40,** 111–114.
Passera, C., and Albuzio, A. (1978). *Can. J. Bot.* **56,** 121–126.
Pearcy, R. W. (1977). *Plant Physiol.* **59,** 795–799.

Pearcy, R. W. (1978). *Plant Physiol.* **61,** 484–486.
Pearcy, R. W., Berry, J. A., and Fork, D. C. (1977). *Plant Physiol.* **59,** 873–878.
Peisker, M. (1979). *Photosynthetica* **13,** 198–207.
Peoples, T. R., and Koch, D. W. (1979). *Plant Physiol.* **64,** 878–881.
Perchorowicz, J. T., and Gibbs, M. (1980). *Plant Physiol.* **65,** 802–809.
Pike, C. S., and Berry, J. A. (1980). *Plant Physiol.* **66,** 238–241.
Plaut, Z. (1971). *Plant Physiol.* **48,** 591–595.
Plaut, Z., and Bravdo, B. (1973). *Plant Physiol.* **52,** 28–32.
Portis, A. R., Chon, C. J., Mosbach, A., and Heldt, H. M. (1977). *Biochim. Biophys. Acta* **461,** 313–325.
Potter, J. R., and Boyer, J. S. (1973). *Plant Physiol.* **51,** 993–997.
Powles, S. B., and Critchley, C. (1980). *Plant Physiol.* **65,** 1181–1187.
Powles, S. B., and Osmond, C. B. (1978). *Aust. J. Plant Physiol.* **5,** 619–629.
Powles, S. B., Chapman, K. S. R., and Osmond, C. B. (1980a). *Aust. J. Plant Physiol.* **7,** 737–747.
Powles, S. B., Berry, J. A., and Björkman, O. (1980b). *Year Book—Carnegie Inst. Washington* **79,** 157–160.
Raison, J. K. (1974). *Bull—R. Soc. N.Z.* **12,** 487–497.
Raison, J. K. (1980). *In* "Biochemistry of Plants" (P. K. Stumpf and E. E. Conn, eds.), Vol. 4. pp. 57–83. Academic Press, New York.
Raison, J. K., Chapman, E. A., Wright, L. C., and Jacobs, S. W. L. (1979). *In* "Low Temperature Stress in Crop Plants: The Role of the Membrane" (J. M. Lyons, D. Graham, and J. K. Raison, eds.), pp. 177–186. Academic Press, New York.
Raison, J. K., Berry, J. A., Armond, P. A., and Pike, C. S. (1980). *In* "Adaptations of Plants to Water and High Temperature Stress" (N. C. Turner and P. K. Kramer, eds.), pp. 261–273. Wiley (Interscience), New York.
Raison, J. K., Roberts, J. K. M., and Berry, J. A. (1982). *Biochim. Biophys. Acta* **688,** 218–228.
Raschke, K. (1975). *Annu. Rev. Plant Physiol.* **26,** 309–340.
Raschke, K. (1976). *Philos. Trans. R. Soc. London, Ser. B* **273,** 551–560.
Ryle, G. J. A., and Hesketh, J. D. (1969). *Crop Sci.* **9,** 451–454.
Sankhla, N., and Huber, W. (1974). *Biochem. Physiol. Pflanz.* **166,** 181–187.
Santarius, K. A. (1973). *Planta* **113,** 105–114.
Santarius, K. A., and Muller, M. (1979). *Planta* **146,** 529–538.
Schaedle, M. (1975). *Annu. Rev. Plant Physiol.* **26,** 101–115.
Schreiber, U., and Armond, P. A. (1978). *Biochim. Biophys. Acta* **502,** 138–151.
Schreiber, U., and Berry, J. A. (1977). *Planta* **136,** 233–238.
Schreiber, U., Vidaver, W., Runeckles, V. C., and Rosen, P. (1978). *Plant Physiol.* **61,** 80–84.
Schulze, E.-D., Lange, O. L., Buschbom, U., Kappen, L., and Evenari, M. (1972). *Planta* **108,** 259–270.
Seemann, J. R., Downton, W. J. S., and Berry, J. A. (1979). *Year Book—Carnegie Inst. Washington* **78,** 157–162.
Seemann, J. R., Berry, J. A., and Downton, W. J. S. (1980). *Year Book—Carnegie Inst. Washington* **79,** 141–143.
Seemann, J. R., Tepperman, J. M., and Berry, J. A. (1981). *Year Book—Carnegie Inst. Washington* **80,** 67–72.
Servaites, J. C., and Ogren, W. L. (1978). *Plant Physiol.* **61,** 62–67.
Shackel, K. A., and Hall, A. E. (1979). *Aust. J. Plant Physiol.* **6,** 265–276.
Sharma, G. K. (1975). *Can. J. Bot.* **53,** 2313–2314.

Sharma, G. K., and Butler, J. (1975). *Ann. Bot. (London)* **39,** 1087–1090.
Shearman, L. L., Eastin, J. D., Sullivan, C. Y., and Kinbacher, E. J. (1972). *Crop Sci.* **12,** 406–409.
Sheriff, D. W. (1979). *Plant, Cell Environ.* **2,** 15–22.
Shimazaki, K., and Sugahara, K. (1979). *Plant Cell Physiol.* **20,** 947–955.
Shneyour, A., Raison, J. K., and Smillie, R. (1973). *Biochim. Biophys. Acta* **292,** 152–161.
Shomer-Ilan, A., and Waisel, Y. (1973). *Physiol. Plant* **29,** 190–193.
Shomer-Ilan, A., and Waisel, Y. (1976). *Z. Pflanzenphysiol.* **77,** 272–273.
Sicher, R. C., and Jensen, R. G. (1979). *Plant Physiol.* **64,** 880–883.
Skirrow, G. (1975). *In* "Chemical Oceanography" (J. P. Riley and G. Skirrow, eds.), 2nd ed., Vol. 2, pp. 1–192. Academic Press, New York.
Slack, C. R., Roughan, P. G., and Bassett, H. C. M. (1974). *Planta* **118,** 57–73.
Smillie, R. M. (1976). *In* "Genetics and Biogenesis of Chloroplast and Mitochondria" (T. H. Bücher, W. Neupert, W. Sebald, and S. Werner, eds.), pp. 103–110. Elsevier/North-Holland Biomedical Press, Amsterdam.
Smillie, R. M., and Nott, R. (1979). *Aust. J. Plant. Physiol.* **6,** 135–141.
Spiller, S., and Terry, N. (1980). *Plant Physiol.* **65,** 121–125.
Srivastava, H. S., Jolliffe, P. A., and Runeckles, V. C. (1975a). *Can. J. Bot.* **53,** 466–474.
Srivastava, H. S., Jolliffe, P. A., and Runeckles, V. C. (1975b). *Can. J. Bot.* **53,** 475–482.
Steiger, E., Ziegler, I., and Ziegler, H. (1971). *Planta* **96,** 109–118.
Stewart, G. R., and Lee, J. A. (1972). *New Phytol.* **71,** 461–466.
Stowe, L. G., and Teeri, J. A. (1978). *Am. Nat.* **112,** 609–623.
Sugiyama, T., Schmitt, M. R., Ku, S. B., and Edwards, G. E. (1979). *Plant Cell Physiol.* **20,** 965–971.
Szarek, S. R., and Woodhouse, R. M. (1978). *Oecologia* **37,** 221–229.
Szarek, S. R., Johnson, H. B., and Ting, I. P. (1973). *Plant Physiol.* **52,** 539–541.
Taniyama, T., Arikado, H., Iwata, Y., and Sawanaka, K. (1972). *Proc. Crop. Sci. Soc. Jpn.* **41,** 120–125.
Tasker, R., and Smith, H. (1977). *Photochem. Photobiol.* **26,** 487–491.
Taylor, A. O., and Rowley, Y. A. (1971). *Plant Physiol.* **47,** 713–718.
Teeri, J. A., and Stowe, L. G. (1976). *Oecologia* **23,** 1–12.
Terry, N. (1980). *Plant Physiol.* **65,** 114–120.
Terry, N., and Ulrich, A. (1973a). *Plant Physiol.* **51,** 43–47.
Terry, N., and Ulrich, A. (1973b). *Plant Physiol.* **51,** 783–786.
Thomas, M. D., and Hill, G. R. (1935). *Plant Physiol.* **10,** 291–307.
Tiku, B. L. (1976). *Physiol. Plant.* **37,** 23–28.
Tunstall, B. R., and Connor, D. J. (1975). *Aust. J. Plant Physiol.* **2,** 489–499.
Turner, N. C., Begg, J. E., Rawson, H. M., English, S. D., and Hearn, A. B. (1978). *Aust. J. Plant Physiol.* **5,** 179–194.
Tyree, M. T. (1976). *Can. J. Bot.* **54,** 2738–2746.
Uedan, K., and Sugiyama, T. (1976). *Plant Physiol.* **57,** 906–910.
Unsworth, M. H., Biscoe, P. V., and Pinckney, H. R. (1972). *Nature (London)* **239,** 458–459.
Van Hasselt, P. R., and Van Berlo, H. A. C. (1980). *Physiol. Plant.* **50,** 52–56.
Verduin, J. (1952). *Science* **115,** 23–24.
von Caemmerer, S., and Farquhar, G. D. (1981). *Planta* **153,** 376–387.
von Willert, D. J., Kirst, G. O., Treichel, S., and von Willert, K. (1976). *Plant Sci. Lett.* **7,** 341–346.
Walker, R. R., Kriedemann, P. E., and Maggs, D. H. (1979). *Aust. J. Agric. Res.* **30,** 477–488.
Walker, R. R., Törökfalvy, E., Scott, N. S., and Kriedemann, P. E. (1981). *Aust. J. Plant Physiol.* **8,** 359–374.

Wellburn, A. R., Majernik, O., and Wellburn, F. A. M. (1972). *Environ. Pollut.* **3,** 37–49.
Wilkinson, T. G., and Barnes, R. L. (1973). *Can. J. Bot.* **51,** 1573–1578.
Willemot, C. M. (1979). *In* "Low Temperature Stress in Crop Plants: The Role of the Membrane" (J. M. Lyons, D. Graham, and J. K. Raison, eds.), pp. 441–430. Academic Press, New York.
Wilson, J. M. (1976). *New Phytol.* **76,** 257–270.
Winner, W. E., and Mooney, H. A. (1980a). *Oecologia* **44,** 290–295.
Winner, W. E., and Mooney, H. A. (1980b). *Oecologia* **44,** 296–302.
Winner, W. E., and Mooney, H. A. (1980c). *Oecologia* **46,** 49–54.
Winner, W. E., and Mooney, H. A. (1980d). *Science* **210,** 789–790.
Winter, K., and Lüttge, U. (1976). *Ecol. Stud.* **19,** 323–334.
Winter, K., Lüttge, U., Winter, E., and Troughton, J. H. (1978). *Oecologia* **34,** 225–237.
Wolosink, R. A., and Buchanan, B. B. (1977). *Nature (London)* **266,** 656–567.
Wong, D., Merkelo, H., and Govindjee (1981). *Photochem. Photobiol.* **33,** 97–101.
Wong, S. C. (1980a). *Oecologia* **44,** 68–74.
Wong, S. C. (1980b). Ph.D. Dissertation, Australian National University, Canberra.
Wong, S. C., Cowan, I. R., and Farquhar, G. D. (1978). *Plant Physiol.* **62,** 670–674.
Wong, S. C., Cowan, I. R., and Farquhar, G. D. (1979). *Nature (London)* **282,** 424–426.
Younis, J. M., Boyer, J. S., and Govindjee (1979). *Biochim. Biophys. Acta* **548,** 328–340.
Ziegler, H., Ziegler, I., Schmidt-Clausen, H. J., Muller, B., and Dorr, I. (1969). *In* "Progress in Photosynthesis Research" (H. Metzner, ed.), pp. 1636–1645. I.U.B.S., Tübingen.
Ziegler, I. (1972). *Planta* **103,** 155–163.
Ziegler, I. (1973). *Phytochemistry* **12,** 1027–1030.
Ziegler, I. (1975). *Residue Rev.* **56,** 79–105.

10

Translocation of Photosynthate

DONALD R. GEIGER
ROBERT T. GIAQUINTA

ABBREVIATIONS

ABA	Abscisic acid
IAA	Indoleacetic acid
LPI	Leaf plastochron index
NCE	Net carbon exchange
ψ_p	Phloem turgor
RuBP C'ase	Ribulosebisphosphate carboxylase
SPP	Sucrose phosphate phosphatase
SPS	Sucrose phosphate synthetase
SS	Sucrose synthase
UDP	Uridine diphosphate
	For other abbreviations, see the legend of Fig. 5, p. 377.

ABSTRACT

Translocation of organic nutrients is an integrative process from a number of viewpoints. In terms of crop yield, translocation forms a practical link between photosynthesis

Photosynthesis: Development, Carbon Metabolism, and Plant Productivity, Vol. II

ISBN: 0-12-294302-3

and harvest index. Within the plant, translocation is responsible, in part, for communication between plant parts and the correlation of plant structure and physiology. The role of translocation in linking plant parts and processes should be considered when we study mechanisms for control of translocation and of allocation of the products of photosynthesis.

The material presented in this chapter reflects the need for a holistic and integrative approach. Topics ranging from responses of starch synthesis to photosynthetic duration in intact plants to the molecular aspects of proton cotransport will be considered. It is the challenge to current and future workers in this field to be well informed of the biophysical aspects of membrane function, the biochemistry of assimilate metabolism, hormone studies, plant anatomy, and most fundamentally, an appreciation of the intricacies of control in integrated systems found in higher plants. To seriously work at understanding translocation necessitates dedication to studies which go beyond the safety of narrowly delimited disciplines and a willingness to communicate with investigators in a wide spectrum of fields of study. It is hoped that this chapter has gone a short distance in promoting these attitudes and approaches.

I. Photosynthesis, Translocation, and Crop Yield: Introductory Considerations

A. Three Approaches to Increasing Crop Yield

1. INCREASING CANOPY PHOTOSYNTHESIS

Canopy photosynthesis, net carbon fixation by the assemblage of shoot tissue, is affected by a composite of factors. An increase of seasonal canopy photosynthesis can be brought about by agricultural practices affecting leaf area index, photosynthetic duration, and rapidity of canopy closure. The effect of spacing on seasonal canopy photosynthesis of soybean is shown in Fig. 1. Average yield per plot for two seasons, six plots each, is in the ratio of 100:71:61:32 for 6, 12, 18, and 24 inch spacing, respectively (see Chapter 14, this volume). Because the increased canopy photosynthesis due to closer spacing came largely prior to pod fill, it supports the view that translocation is important in producing increased yield. In addition to improving yield by optimizing productivity for a stand of plants, it is also possible to improve individual plant performance. Two approaches are mentioned in the next two sections.

2. INCREASING PHOTOSYNTHETIC EFFICIENCY

Enhancement of photosynthesis and improvement of its efficiency can improve productivity and increase yield. Other chapters in this volume address topics that relate to increasing the rate and efficiency of photosynthesis. This chapter deals with a third approach, improving yield by modifying translocation, the link between photosynthesis, productivity, and yield.

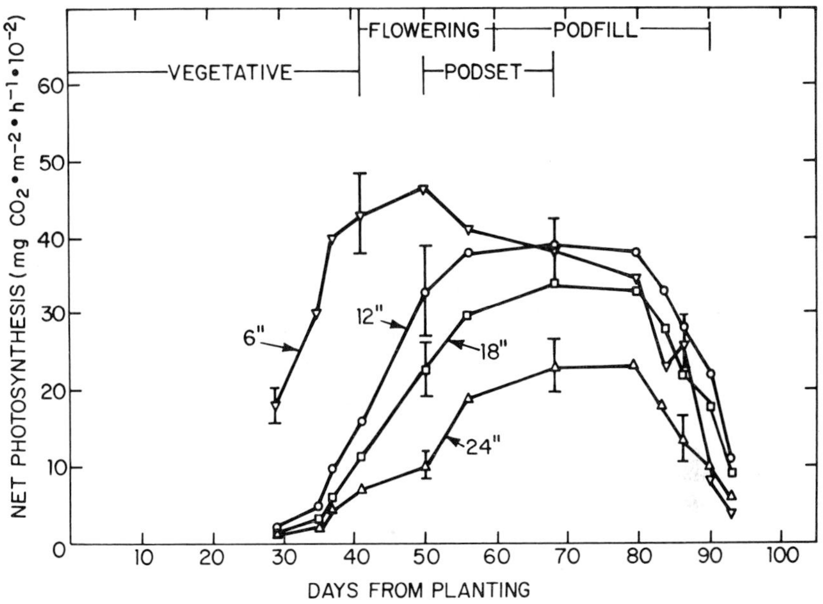

FIG. 1. Net carbon-fixation rate of soybean plants during an entire season. Plants are in 30-in rows at the spacing indicated. (Data from Christy and Porter, Chapter 14, this volume.)

3. IMPROVING EFFECTIVENESS OF DISTRIBUTION WITHIN PLANTS

Control of distribution of fixed carbon into useful plant parts offers another approach to increasing crop yield. To be able to effect a substantial increase in yield by changing the pattern and rate of translocation of the products of photosynthesis requires knowledge of controlling factors. To help attain this knowledge certain considerations, discussed later, appear to be critical to guide our approach.

B. Plants Are Integrated, Adapted Systems

1. SYSTEM GOALS

Successful adaptation of plants has led to acquisition of patterns of response, which generally bring about and maintain certain states that enable the plant to succeed under a range of normal circumstances. These engendered responses can be regarded as system goals that are part of the plant's genetic constitution (Geiger, 1979). Balanced growth and metabolism in a plant require close control over nutrition of the various parts of the plant. This control is brought about by regulation of allocation of the products of photosynthesis. A plant must maintain a positive balance of carbon in a readily accessible form such as starch or sucrose, both in source leaves and throughout the plant as a whole and this under a variety of environmental conditions.

More specifically, sufficient carbon must be exported to provide nutrients for new structures and new synthetic apparatus in sink regions. Likewise nutrients are needed in these places for maintenance. Sufficient newly fixed carbon must be retained in the source region to provide a relatively steady supply of nutrients to sink regions under alternating light/dark cycles of various duration and under periodic events such as shading or defoliation. Not only is this allocation crucial for the survival of the plant but it also substantially affects yield. Modeling of sugarbeet growth by Hunt and Loomis (1979) demonstrates the large differences in yield caused by changes in partitioning resulting from changes in growth and maintenance parameters.

2. ALLOCATION MANAGEMENT

By virtue of adaptation to a wide variety of situations, the genetic constitution of higher plants has accumulated a variety of mechanisms that enable the plant to adjust to a wide spectrum of conditions. As in a well-managed corporation, responses are made that result in adjustments throughout the entire system. In effect, in a successful green plant there are a number of "interests" that cannot be neglected without a cost to the plant as a whole. Allocation of carbon simply in a competitive mode will favor a part of the plant to the ultimate detriment of the whole. A single-minded approach to increasing yield by attacking an "inefficient" plant function viewed from the standpoint of yield may, in the final analysis, prove counterproductive.

To help appreciate the implications of manipulations of plant function, a corporation analogy can provide insights (Fig. 2). In one sense, the various allocations stand in a competitive relationship with each other because what is allocated to one comes at the expense of what is allocated to another. In another sense, the distribution is governed by a higher process, which was selected because of success in integrating the functioning of the green plant. Appreciation of the levels of organization of plant responses and their interactions, as well as the implication of these, help us both to understand plant function and to avoid over-simple approaches to increasing yield. Strategies for improving yield must deal with the various demands made on the allocation of the products of photosynthesis and their importance in overall operation of the plant.

Development and selection of plants for high yield can draw on a

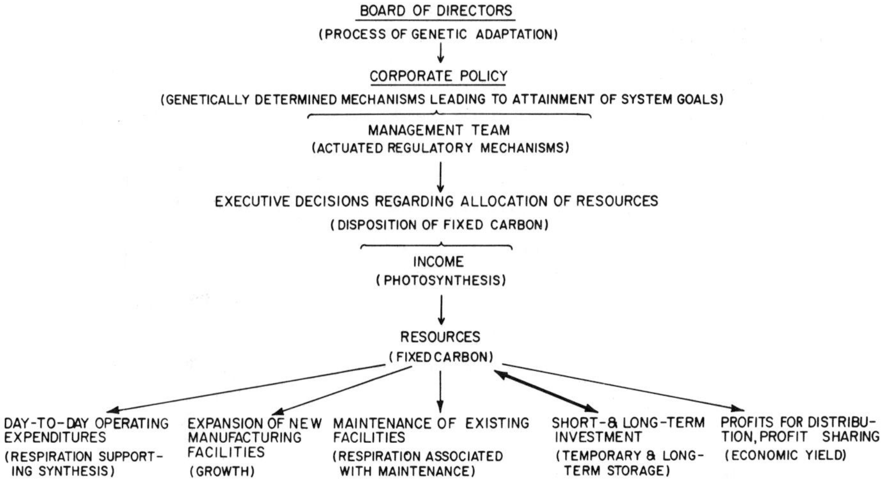

FIG. 2. Corporate analogy for allocation of resources in higher plants.

large range of genetically determined adaptive responses. But plant improvement efforts are faced with the fact that, during the long history of shaping of adaptive responses, success was measured in terms of propagation of the species and not of high economic yield.

Because of the difference in goals, we need to understand the relationship between photosynthetic carbon fixation and control of allocation of its products as a basis for improving the efficiency of distribution for increased yield.

II. Relation between Translocation and Photosynthesis

In examining the relationships between photosynthesis and translocation, a somewhat artificial dichotomy will be made, one that looks separately at the two directions of interaction. First, the extent to which translocation limits photosynthesis will be considered. We will review evidence for and against a type of end-product inhibition of photosynthesis and examine alternative hypotheses. Second, the effect of photosynthesis on translocation will be looked at. We will discuss factors affecting setting of partitioning ratios that control export of products of photosynthesis. Biochemical details of processes related to controlling export will be covered in later sections. In all these sections, control mechanisms and processes that aid in the integration of photosynthesis, translocation, and plant function will be emphasized.

A. Effect of Translocation on Photosynthesis

The rate at which export removes the products of photosynthesis can conceivably affect the rate of photosynthesis by a process analogous to end-product inhibition of biochemical reactions. In an often-cited review of this topic, Neales and Incoll (1968) refer to an early statement of this hypothesis proposed by Boussingault in 1868, ". . . the accumulation of assimilates in an illuminated leaf may be responsible for a reduction in the net photosynthesis rate of that leaf [p. 107]." The frequency with which this idea has been voiced has increased in recent years to the point that it is not practical to list all the references to it. Extensive studies of the various phenomena, which should occur if the hypothesis is true, are rarely done. Geiger (1976a) reviewed studies relating to this hypothesis and put forth several generalizations. In relatively intact systems, changes in photosynthesis resulting from onset of a new developmental stage appear to coincide with changes in sink–source ratio and, by inference, with export. Most manipulations such as partial defoliation do not produce an immediate change in net carbon exchange; in a number of cases photosynthesis undergoes an adjustment over a several day period (Fraser and Bidwell, 1974; Thorne and Koller, 1974). On the whole, a simple feedback inhibition mechanism does not seem to be favored by the literature reviewed. A review of several studies will serve as examples of responses.

In a number of studies, the export from a leaf has been drastically reduced by girdling its petiole. The treatment generally does not produce a decline in the net carbon exchange predicated by some form of end-product inhibition. Carmi and Koller (1977) showed that petiole girdling did not cause a decrease in net carbon exchange but that dry matter accumulation in the source leaf did increase, including a conspicuous increase in starch grains in the lamina of the treated leaf. It appears that leaves of at least some plant species are able to avoid inhibition of photosynthesis, even though starch or sugar accumulates in them. It also seems that net carbon exchange does not respond to the rate of export or accumulation of products of photosynthesis as such.

In most cases, when treatments are given, which produce both a change in export rate and a corresponding change in the net carbon-exchange rate, factors other than feedback end-product inhibition of photosynthesis appear to be at work or at least are not easily ruled out. Carmi and Koller (1978) found that when they removed 70–80% of the roots of young bean plants, *Phaseolus vulgaris*, L, there was a considerable reduction of net carbon exchange by primary leaves over a 6-day period. Their previous study dismissed the likelihood that inhibition was a result of accumulation of products of photosynthesis. These authors also concluded that the effect on photosynthesis was not the result of a decreased mineral supply. Exogenous benzyladenine supplied at 2-day intervals increased chlorophyll content of leaves but failed to compensate for the effect of root removal on photosynthesis. These results and a number of other studies have led to a search for alternative explanations for changes in photosynthesis triggered by experimental treatments likely to change translocation.

A growing number of studies provide evidence for regulation of photosynthesis by some type of communication between sinks and photosynthesizing leaves. Generally this communication does not seem to be simply the export of major transport molecules from the source leaf. In the study involving root removal mentioned earlier, Carmi and Koller (1978) concluded that roots probably supply essential substances for activation of photosynthesis or for maintaining its level. In a subsequent study, Carmi and Koller (1979) found that excision of the shoot above the primary node and removal of auxillary buds caused net photosynthesis, chlorophyll per area, RuBP C'ase per unit protein, protein per fresh weight, leaf thickness, and leaf area to increase in the primary leaves over a 7- to 8-day period. Heat girdling above the primary node did not produce this effect, indicating that removal of translocation sinks, as such, was not responsible. Removal of major transpiration surfaces by defoliation of leaves above the primary node produced the same effect as decapitation above the primary node. Carmi and Koller concluded, as did Wareing *et al.* (1968), that partial defoliation leads to the diversion of increasing amounts of some promotive factor such as cytokinins from the roots to the remaining leaves, thus increasing their photosynthetic capacity over a period of several days.

A related but different mechanism appears to operate in soybean plants, *Glycine max* (L.) Merr., following pod removal. Koller and Thorne (1978) observed that when they excised rapidly growing pods there was an increase in stomatal resistance over a several day period; this phenomenon was also observed by other workers in other species. Several studies provide evidence of hormonal involvement. Setter *et al.* (1980a) observed that pod removal and petiole girdling produced 70% and 90% reductions in leaf CO_2-exchange rate, respectively. Similar changes were observed in stomatal diffusive resistance but mesophyll conductivity and assimilation of ^{14}C were not substantially affected by the treatments. Sucrose and glucose accumulated to a greater extent in the depodded and girdled plants than in the controls (Setter *et al.*, 1980b). Starch, which is present to a much greater extent than sucrose or glucose, did not accumulate at a noticeably faster rate. When photosynthesis was prevented for 24 hours after girdling or depodding, inhibition of carbon exchange and reduction of stomatal conductivity still occurred. The effect does not appear to be simply the result of inhibition of export or of accumulation of major products of photosynthesis. Free abscisic acid levels increased within 3 hours of treatment and rose 10-fold in leaves from depodded plants and 25-fold in girdled leaves (Setter *et al.*, 1980b). Setter and co-workers concluded that increased free abscisic acid in leaves, independent of water stress, is responsible for the inhibition of photosynthesis. Obstruction of translocation of abscisic acid out of the leaves seems to be responsible for the increase in its level (Setter *et al.*, 1980b).

The studies cited earlier offer alternative explanations to end-product inhibition for the results of studies that appear to be supportive of feedback inhibition of photosynthesis as a result of altered translocation. Failure of experimenters to rigorously establish cases of inhibition of photosynthesis by accumulation of products of photosynthesis has considerably weakened the case for the hypothesis proposed by Boussingault. On the other hand, the alternative mechanisms described earlier have firm experimental basis. In addition, they appear to be much more compatible with the correlation between sinks and sources, which is commonly observed in the balanced growth of plants.

B. *Effect of Photosynthesis on Translocation*

It is not easily disputed that the rate of translocation in light depends on the rate of photosynthesis. At photon flux densities somewhat above light compensation point, the rate of export of recent products of photosynthesis is clearly proportional to the rate of net photosynthesis in sugarbeet (Servaites and Geiger, 1974). Recent products of photosynthesis appear to be the major source of exported carbon under these conditions (Fondy and Geiger, 1980). In these studies 25–40% of the carbon, which was fixed, was exported immediately. Ho (1976a) observed a linear dependence of export of carbon on photon flux density; as the light compensation point was approached, export of previously fixed carbon, at a rate independent of the light intensity, became important. Translocation rate is clearly determined by the rate of net carbon fixation, the primary source of material exported during photosynthesis, but a major portion of the carbon, which is fixed, is not immediately exported. It is far less clear how proportion of newly fixed carbon, which is to be exported, is controlled. There are a number of factors and circumstances that affect the proportion of newly fixed carbon exported in the light and the rate of mobilization of reserves during darkness. Some of these factors are themselves dependent on aspects of photosynthesis.

1. NET CARBON-EXCHANGE RATE

Ho (1977) demonstrated that a source leaf at a given developmental stage exports carbon at a given net carbon-exchange (NCE) rate, which is higher in leaves grown previously at a higher NCE rate. Likewise, translocation in the dark is higher for leaves previously grown at higher NCE rates. Presumably these latter plants have a larger reserve of carbohydrates available for export. When plants are shifted to a new level of photon flux density or CO_2, NCE rate was steady and determined primarily by these two factors. On the other hand, the partition of fixed carbon into export changed with time and achieved a new steady value within 2–10 days. These data for tomato plants indicate the key orientation of control mechanisms to produce balanced distribution of carbon within leaves and throughout the plant under differing rates of net carbon fixation.

Ryle and Powell (1976) observed that the pattern of distribution of exported material also shifted when the rate of NCE was changed. Complete adjustment of assimilate distribution to a new light regime in plants of *Lolium temulentum* and of uniculm barley was attained in approximately 7 days. The changed pattern suggests a priority of sinks, with the terminal meristem favored over roots and tillers in this case. In both sets of studies cited it is clear that the influence of NCE rate goes beyond simply supplying material for immediate export.

2. PHOTOSYNTHETIC DURATION AND PHOTOPERIOD

Plants differ in the way in which material that is exported is distributed in the various plant parts. Genetically determined differences between several growth forms of *Beta vulgaris*, L. dramatize the variety in partitioning priorities for spinach, beet, chard, and sugarbeet. Snyder and Carlson (1978) described selections of sugarbeet that differ in tap root fresh weight per unit leaf lamina fresh weight by a factor of approximately 2. These basic patterns can be further altered by growth conditions such as photoperiod and light quality. Milford and Lenton (1976) observed that extending the photoperiod of sugarbeet plants beyond a 12-hour photosynthetic period by 4 hours of low intensity tungsten light increased leaf area by 47%. Assimilate distribution between root and leaves was unchanged, with the extra leaf growth coming at the expense of the crown. Photomorphogenic effects modify distribution of products of photosynthesis as they condition genetic determination. In this case, lengthening photoperiod, but not photosynthetic duration, by 4 hours brought a 25% increase in total plant dry weight.

In addition to photoperiod effects, photosynthetic duration also affects export. Chatterton and Silvius (1979) observed that diurnal starch accumulation rates in soybean leaves were inversely related to the duration of the daily photosynthetic period but not to photoperiod as such. Increased starch accumulation came at the expense of residual dry weight and export of recent products of photosynthesis. This adjustment to the new pattern of partition took place over a 4-day period. The change in partitioning appears to be a programmed response adapted to the energy demand of the diurnal dark period.

The pattern of response does not support the concept that increased starch accumulation is the result of a limitation in ability to synthesize and export sucrose. The decrease in export resulted in an increased shoot–root ratio in plants under long photosynthetic duration. The plant responded to the shortened photosynthetic duration by maintaining uninterrupted export during the diurnal dark period and by minimally reducing photosynthetic area under the reduced total net diurnal photosynthetic carbon fixation. Challa (1976) observed a similar effect of photosynthetic duration on rate of starch accumulation in cucumber (*Cucumis sativus,* L.).

Chatterton and Silvius (1979) concluded from their photon flux density studies that partitioning was affected by duration of the photosynthetic period, and perhaps the duration of the dark period rather than absolute amount of light energy received per day. The mechanism controlling the proportion of recent products of photosynthesis going to starch synthesis is not known; it seems reasonable to look to the adaptive goals achieved for clues to the type of control mechanism operating.

3. SINK–SOURCE RATIO

Many translocation studies have been based on the working hypothesis that an increase in sink–source ratio will lead to increased export of carbon compounds; Thorne and Koller (1974) lend some support to this view. They increased the sink–source ratio by shading all but one source leaf of the soybean plant. Translocation in treated plants was only measured on the 8th day after treatment by which time photosynthesis per unit area had increased by approximately 50%. Following the increase in sink–source ratio, the export of pulse-labeled photosynthetic products indicated that a greater proportion of the newly synthesized carbon was being exported. In addition, newly fixed carbon compounds potentially available for export increased. They showed that net carbon fixation increased gradually over an 8-day period, but they did not examine the proportion of the products of photosynthesis exported during the transition period.

Several studies examined the time course and extent of changes in translocation when the sink–source ratio is increased. Borchers-Zampini *et al.* (1980) increased the sink–source ratio for a primary leaf of bean plants by excising or darkening the opposite primary leaf. Within less than 1 hour, distribution of labeled compounds to a monitored sink leaflet increased without decrease in import by the main leaflet sink. Export from the illuminated source leaf was inferred to increase but NCE rate did not change over the several hour period of the experiment. A follow-up study was performed by Fondy and Geiger (1980) in which export from the primary leaf was measured as well as the accumulation of exported material by leaflet and root sinks. Import into the less directly connected sink leaflet of a young trifoliate leaf increased a short time after shading commenced, at the expense of import into the roots. In some cases export of newly fixed carbon from the illuminated source leaf increased by approximately 20%. On the other hand, decreasing sink–source ratio below the level to which the plant was then accommodated by girdling the translocation path to sinks caused a relatively rapid, large decrease in export from a source leaf (Fondy and Geiger, 1980).

A similar study of effects of increasing sink–source ratio was made on sugarbeet plants. Export of labeled carbon from an illuminated source leaf was monitored, and the effect of darkening the other 7–10 source leaves was recorded. In this case, treatment neither increased export of recently fixed carbon compounds nor did it decrease the net rate of storage of total starch. Distribution of exported carbon among the various sinks changed shortly after the sink–source ratio was increased. It appears that the short-term effect of increasing sink–source ratio is to change the distribution of translocate according to as yet unexplained sink priorities, whereas the export of carbon increases little or not at all. In bean plants, the largest increases were observed when some of the original sinks were previously blocked by girdling (Fondy and Geiger, 1980). In the short-term, it seems that both the amount of recent photosynthetic products exported and the absolute amount of carbon translocated are closely limited. The adaptations appear to buffer plant systems against responding in a major way to changes in sink–source ratio temporarily. It seems likely that frequent transient responses would result in disadvantageous disruption of metabolism.

What seems to be a more suitable response is a gradual adjustment of the various processes associated with translocation of the products of photosynthesis. The gradual changes in NCE, sucrose concentration, and inorganic phosphate content of source leaves, observed when Thorne and Koller (1974) increased sink–source ratio by shading the other source leaves, appear to exemplify this response. Increased NCE rate can be expected both to increase the proportion of recent products of photosynthesis exported (Ho, 1977) and increase the amount of carbon available. In cases where the increase in sink–source ratio is the result of plant development and conversion to reproductive state, changes in NCE appear to anticipate or coincide with the increase rather than lag for several days (Fraser and Bidwell, 1974).

In summary, it appears that the proportion of products of recent photosynthesis, which is exported, is not rapidly nor readily changed by increasing sink–source ratio. Also, mobilization of leaf reserves is not easily increased by these treatments. Adjustments to treatments that increase the sink–source ratio appear to occur over a number of days and to involve a number of aspects of plant function. Decreasing sink–source ratio below the accommodated level does bring about a rapid decrease in export. Responses to such decreases have not been studied over a longer period but long-term adjustments can also be expected. The results described earlier point to the need to examine means by which export is controlled. Likewise, studying the wide variety of physiological responses promises to help us to understand the relation between photosynthesis and export.

III. Potential Means of Controlling Export

Translocation rate and the pattern of distribution of exported materials are determined by processes that occur in source regions as well as by those that occur in sink regions and along the translocation path (Fig. 3). This chapter deals primarily with processes in the source regions, although these also may be strongly affected by events outside the source region. Potentially, the export in source regions can be controlled in three ways: biochemical or compartmental limitation of availability of material for export, limitation of movement of this material to minor veins, and restriction of entry of these substances into minor vein phloem.

A. Availability of Compounds for Export

Availability for export has a biochemical facet—synthesis of those compounds that can be actively loaded into the phloem or can permeate it—and a physical facet—compartmentation in a region from which compounds can readily move to the phloem.

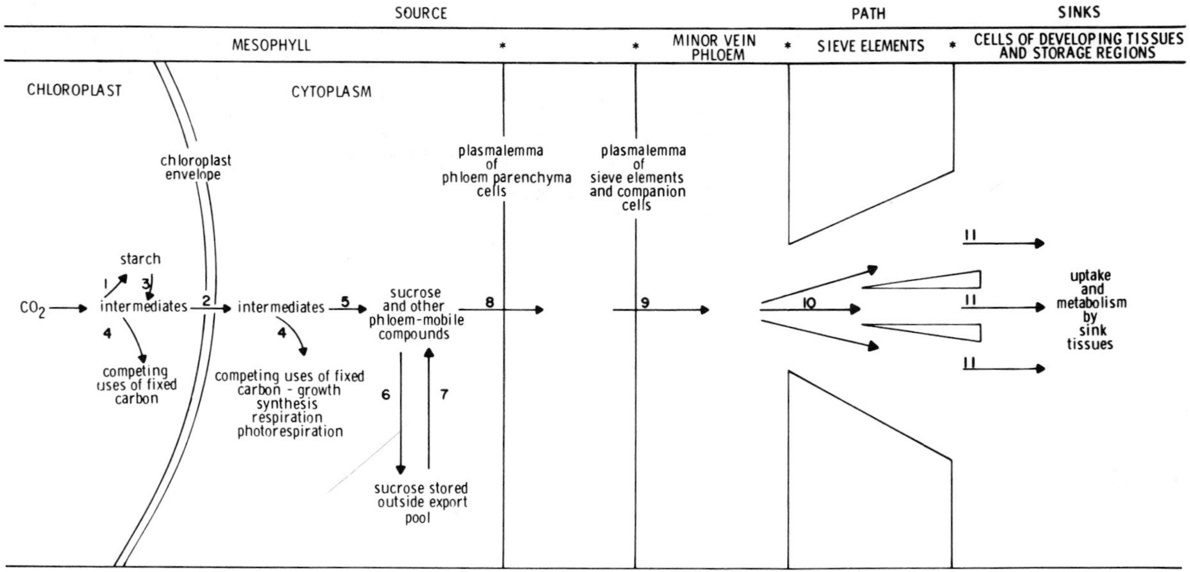

FIG. 3. Possible control points for regulating the partitioning of assimilated carbon and the export of phloem-mobile compounds. (From Geiger, D. R. (1979), *Bot. Gaz.* 140, p. 243, Fig. 1. Copyright © 1979 by The University of Chicago Press.)

1. METABOLIC CONTROL

Response of export to environmental factors appears to involve control of synthesis of molecules capable of entering the phloem readily. These responses appear to be adaptive mechanisms that maintain balanced distribution of carbon among source leaves and the various sinks (Geiger, 1979). Metabolic control of sucrose synthesis appears to involve transporter shuttles for exchange of materials through the chloroplast envelope (Walker and Herold, 1977). Stoichiometric relationships among CO_2 fixation, sucrose synthesis, starch synthesis and mobilization, production and utilization of triose phosphate, and release and use of inorganic phosphate seem to be key components of metabolic control of sucrose availability (Giaquinta, 1980a). These interrelationships will be discussed in detail in later parts of this chapter, which deal with starch and sucrose metabolism.

2. COMPARTMENTATION IN MESOPHYLL

Studies, largely based on histochemical techniques, have provided data on distribution of sucrose in leaves of C_3 plants; these studies locate the bulk of sucrose in the sieve element–companion cell complex, in the cytoplasm outside the chloroplasts, and in the vacuoles of mesophyll cells. Outlaw and Fisher (1975) determined that the palisade and spongy mesophyll of *Vicia faba*, L. leaves do not differ qualitatively in photosynthetic carbon metabolism. Differences in relative photosynthetic rates between these tissues are due largely to a light intensity gradient through the leaf. Fisher and Outlaw (1979) examined sucrose distribution in sections from regions of *Vicia* leaf tissue enriched in vacuoles or in cytoplasm and organelles as a result of centrifugation. Sucrose concentration was found to be three times higher in the extra-chloroplast cytoplasm than in the vacuoles. The amount of sucrose contained in vacuoles was approximately fivefold more abundant than that in the cytoplasm because of the large relative volume of the vacuoles.

Although the sieve element–companion cell complex of the minor veins of the leaf is less than 1% of the leaf volume, a major part of the sucrose in the leaf is present there (Geiger *et al.*, 1973; Geiger, 1975). Fisher *et al.* (1978) used quantitative autoradiography to determine the ready availability of sucrose from source leaf pools. They demonstrated a close correspondence between the ^{14}C content of the companion cells in minor veins of morning glory and soybean leaves and arrival of ^{14}C in sink tissue. These data confirm that sucrose in the minor vein companion cells represents a major part of the transport sucrose pool in the leaves of C_3 plants.

Cytoplasmic sucrose of mesophyll cells appears also to be part of the transport sucrose pool. Fondy and Geiger (1980) used steady state labeling of sugarbeet source leaves in $^{14}CO_2$ followed by transfer to CO_2 to study export of sucrose from various source leaf pools. Total sucrose in the source leaf blade remained unchanged throughout all but the early part of the 14-hour photoperiod. Disappearance of [^{14}C]sucrose from the leaf blade revealed only two discernable pools with 22- and 132- min half-times. The latter pool is likely to be made up of the sucrose in mesophyll vacuoles, whereas the former is likely to be the sucrose in the extra-chloroplast cytoplasm of the mesophyll as well as the sucrose of the sieve element–companion cell complexes of the minor veins. The half-time of the pool, which is turning over more rapidly, corresponds to the kinetics of the transport sucrose pool observed during steady state labeling of sugarbeet leaves with $^{14}CO_2$ (Geiger and Swanson, 1965).

Pool size of the transport sucrose pool can be calculated from values for half-time of the export sucrose pool and export rate. To support the observed rate of 0.4 μg C $cm^{-2}min^{-1}$, a transport sucrose pool of 13 μg C cm^{-2} leaf is required, given a half-time of 22 min (Fondy and Geiger, 1980). A transport sucrose pool of 13 μg C cm^{-2} is 60% of the 22 μg C cm^{-2} observed in source leaf sucrose, which indicates that 40% of the sucrose is in the vacuolar storage sucrose pool. The sucrose concentration of the sugarbeet source leaves remained nearly constant after a buildup at the start of the light period. These observations provide evidence that both the transport and the vacuolar pools are turning over, the latter at one-sixth the rate of the former (Fondy and Geiger, 1980). It seems doubtful that there is a large static pool of storage sucrose in sugarbeet source leaves. It is likely that the vacuolar storage pool changes size with changes in conditions.

Silvius *et al.* (1978) and Giaquinta (1978) observed a pattern of change in allocation of carbon into starch and sucrose in soybean and sugarbeet leaves, respectively, as a function of leaf age. In soybeans, as leaves progressed from leaf plastochron index (LPI) 1 through 7.5, sucrose in the lamina increased from 1–2.5% of leaf dry weight. At the same time, leaf starch content at the end of the 11 hour photoperiod decreased from 14% at LPI 1 to 10% at LPI 7.5. Sucrose progressed from being 1/14 as abundant as starch at the end of the light period in young leaves to being 1/4 as abundant in old leaves (Silvius *et al.*, 1978). Fondy and Geiger (1980) reported values of 750 μg starch and 50 μg sucrose cm^{-2} source leaf at the end of a 14-hour light period, a ratio of 15:1. Dark-period translocation and respiration reduced starch to approximately 100 μg cm^{-2} by the end of the 10-hour night.

Partition of carbon is also affected by photosynthetic rate and duration. Wardlaw and Marshall (1976) observed that a reduction in photon flux density caused a decreased incorporation of ^{14}C into sucrose and an increased accumulation of label in amino acids in leaves of *Lolium temulentum,* a C_3 species and *Sorghum sudanense,* a C_4 species. The proportion of ^{14}C incorporated into starch increased at high photon flux density. Observation of a gradual increase in incorporation of carbon into starch when soybean plants were placed under short photosynthetic duration (Chatterton and Silvius, 1979) was discussed earlier.

The diurnal course of starch and sugar content of leaves under 8- and 14-hour photoperiods was studied by Challa (1976). Graphs for leaf starch under 8- and 14-hour photoperiods in air and 8-hour photoperiod under enriched CO_2 concentration are given in Fig. 4. The higher rate of starch accumulation under short photosynthetic duration results in an end-of-day starch level, which is only slightly less than that for the longer photoperiod. Similarly, sugar levels under the two sets of conditions are not markedly different. Sugar and starch fall to considerably lower levels by the end of the night period under short photoperiods. Increasing the level of CO_2 to 1700 μl liter^{-1} increases the level of starch by the end of the photoperiod but does not alter the level of starch present by the end of the night period. Partitioning of products of current photosynthesis into starch appears to be a well-controlled function of photosynthetic duration or duration of the period with negative carbon balance, or perhaps both. Description of partitioning of carbon into various compartments of the mesophyll and elucidation of mechanisms for controlling allocation are both topics for much needed research.

3. RELATIVE IMPORTANCE OF CURRENT AND STORED PHOTOSYNTHATE

Under some conditions, export of products of current photosynthesis may be supplemented by mobilization of stored material. Ho (1976a) observed that export of carbon from leaves decreased with a decrease in photon flux density and reached a lower limit as net photosynthetic carbon fixation approached 0 and became negative. Export was always maintained, mainly at the expense of starch. In the early part of the dark period, reserve sucrose probably plays a role in maintaining export also (Geiger and Batey, 1967; Fondy and Geiger, 1982). It seems to be generally assumed that starch is the major source of translocation during the dark period. This conclusion is supported by the large regular cycling of starch and low carry-over of starch to the next day (Challa, 1976; Chatterton and Silvius, 1979). Vacuolar sucrose may support export during the initial part of a period of low light intensity.

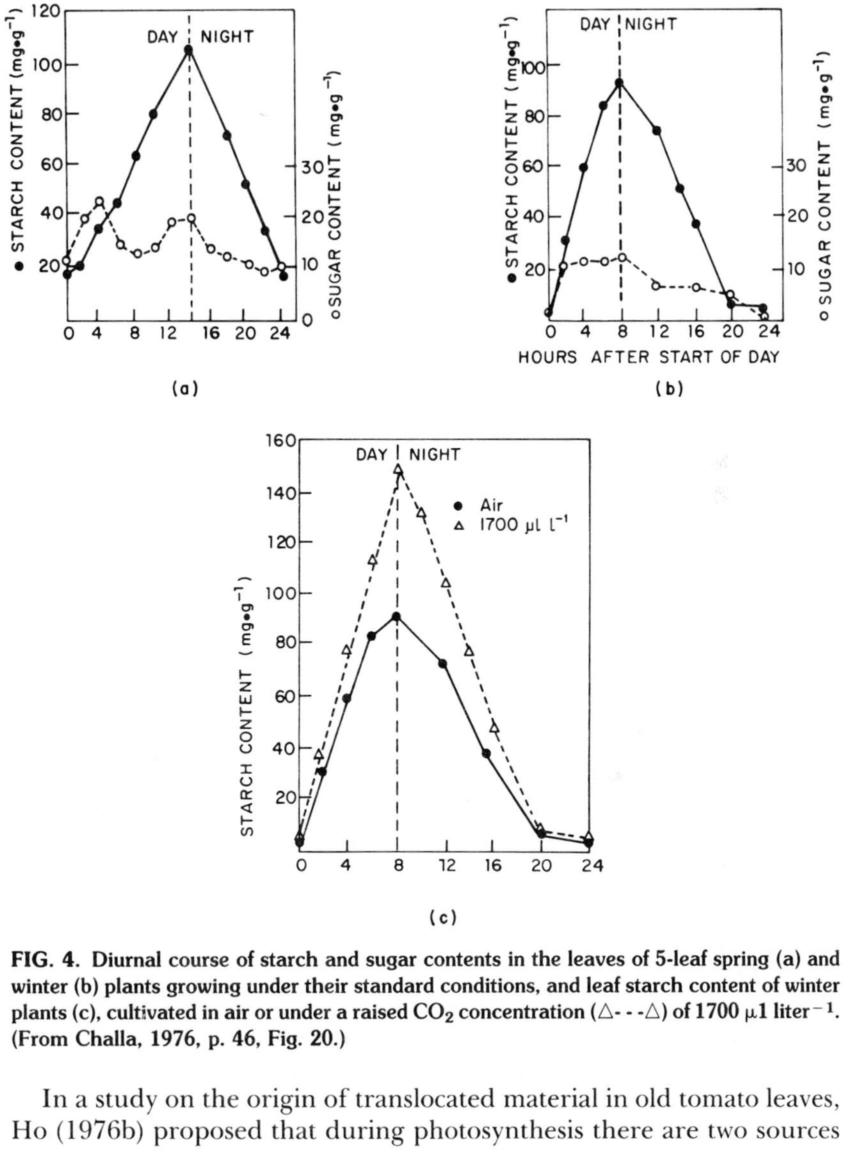

FIG. 4. Diurnal course of starch and sugar contents in the leaves of 5-leaf spring (a) and winter (b) plants growing under their standard conditions, and leaf starch content of winter plants (c), cultivated in air or under a raised CO_2 concentration (△- - -△) of 1700 μl liter^{-1}. (From Challa, 1976, p. 46, Fig. 20.)

In a study on the origin of translocated material in old tomato leaves, Ho (1976b) proposed that during photosynthesis there are two sources of the material that are being exported. One portion is derived from current products of photosynthesis, and the rate of its production is proportional to the current rate of carbon fixation. A second portion is derived from leaf reserves and is proportional to the reserve content of the leaf, particularly starch. By studying tomato plants under a variety of CO_2 and light regimes, Ho (1978) was able to evaluate the proportion of current photosynthate exported (18–52%) and the basal rate of postulated export of material derived from reserves. The latter figures varied over a wide range. The postulated turnover of starch reserves during photosynthesis and their contribution to export during the light period requires further verification.

Pearson (1974) used compartmental analysis to study sources of label exported at various times from *Vicia faba* leaves. He concluded that 50% of the carbon fixed during the light period was exported during the current photoperiod, 14% during the dark period, and 5% during the following photoperiod. Though the author does not suggest it, this latter amount may represent label carried over from [^{14}C]starch mobilized during the dark period. The data of Challa (1976) and Chatterton and Silvius (1979) showed that starch is reduced to a low level by the end of each dark period under uniform conditions and that little of the current photosynthate is carried over to the next day as starch. Sucrose provides only a small contribution to the assimilate pool available for translocation on the following day. Fondy and Geiger (1980) observed that approximately 4 mg C dm^{-2} as starch or 8% of the total carbon fixed was carried over to the next day by sugarbeets under uniform conditions of illumination and photoperiod. During transition to different photoperiod conditions carry-over may change gradually (Ho, 1978; Chatterton and Silvius, 1979). Export during the light period appears to originate mainly from currently produced sucrose. Under conditions of low light, reserve sucrose may be important temporarily in addition to starch. An open question is the extent to which starch, produced during the current light period, turns over and contributes to export. A related question is whether a threshold of photosynthesis rate exists below which starch is mobilized or whether contribution occurs at all rates.

A study of distribution and utilization of recently fixed carbon was carried out by Gordon *et al.* (1980). A summary of the major categories, which are not mutually exclusive, is given in Table I. At the stage of uniculm barley used (two fully expanded leaves), most of the carbon was exported from the observed source leaf. Much of the current day's assimilate was respired, presumably mostly by the growing regions of the young plants, with little being carried over in the source leaf as reserves. The study by Ho (1978) demonstrated that the amount of reserves carried over to subsequent photoperiods depends on the current rate of net carbon fixation and the stage of acclimatization of the plant to changing photoperiod. The data of Challa (1976), Chatterton and Silvius (1979), and Fondy and Geiger (1980) pointed out that plants that are adjusted to a photoperiod regime under steady lighting carry uniform and relatively

TABLE I
Distribution and Utilization of Assimilate during One Diurnal Period (8.5/15.5) by a Single Young Mature Leaf of Uniculm Barley[a]

	Rate		Total	
	$mgC\ dm^{-2}hr^{-1}$	%	mgC	%
Net carbon fixed	8.9	100	75.3	100
Export, light period	5.3	60	45.4	60
Export, dark period	1.7	18	25.9	34
Current assimilate remaining in leaf	—	—	4.0	6
Respiration, light period	1.8	20	15.4	20
Respiration, dark period	1.4	16	20.9	28
Current assimilate remaining in plant	—	—	39.0	52

[a]Gordon *et al.*, 1980, Table 2.

small amounts of starch and sucrose reserves over to the subsequent photoperiod.

B. Movement to Site of Entry into Phloem

Compounds in the mesophyll that are available for export on the basis of biochemical and cytological compartmentation must make their way to the site of entry into the phloem before they can be exported. Consequently, movement to the site of entry constitutes a potential process by which export from a leaf can be controlled. Several facets of the process will be examined.

1. MOVEMENT TO MINOR VEINS

At present, little definitive data exists on this topic. Tyree (1970) cites structural and thermodynamic evidence that supports the possibility of symplastic transport of sugar across the bulk cytoplasm of cells and through cytoplasm-filled plasmodesmata. Current evidence indicates that a portion of the newly synthesized sucrose is present in the mesophyll cytoplasm outside the chloroplasts and should be readily available for transport via the symplast. Movement to the minor veins probably accounts for the several minute delay in loading observed in C_3 plants (Geiger, 1976b). These data describe possible symplastic transport without demonstrating the extent to which it occurs.

Data from studies of distribution in source leaves of C_3 and C_4 plants (Geiger *et al.*, 1973; Evert *et al.*, 1978) suggested uniformity of solute concentration throughout the leaf mesophyll and an abruptly higher concentration in the minor vein phloem. The abrupt step up in concentration in the minor vein phloem at the membranes of the sieve elements and companion cells, the placement of cell types, and plasmadesmata frequency between these cells have been used as evidence in forming working hypotheses for transport of compounds to be translocated to the site of loading (Geiger *et al.*, 1973; Kuo *et al.*, 1974; Evert *et al.*, 1978). On the basis of these data, plus solute efflux data (Geiger *et al.*, 1974) and studies of phloem loading (Giaquinta, 1976; Doman and Geiger, 1979), it seems likely that movement to the minor veins occurs, in part, in the apoplast. Studies with isolated cells seems to support the view that release of products of photosynthesis from the mesophyll plasmalemma is very low and is not specific for transport molecules (Kaiser *et al.*, 1979). Exit of sucrose and other molecules destined for export may well occur from specialized cells located near the minor vein phloem with movement to these sites occurring in the symplast.

Several studies of structural features of leaves as they relate to movement of solutes to the phloem have been made. Kuo *et al.* (1974) studied the distribution of plasmodemata and pit fields in the mestome-sheath cells of wheat leaves and concluded that transfer of sugar across the mestome sheath of the longitudinal veins occurs by plasmodesmata. These data support a symplastic route at least up to the phloem cells where the sugar may enter the apoplast.

Evert *et al.* (1977, 1978) employed structural and plasmolytic methods to examine the path of sucrose to the companion cell–sieve tube complexes in leaves of *Zea mays*, a C_4 plant. They found that movement of products of photosynthesis from the mesophyll to the bundle-sheath cells is restricted largely to the symplastic pathway, whereas transpirational water is restricted largely or entirely to the cell walls (Evert *et al.*, 1977). Cell walls between bundle-sheath cells and vascular parenchyma cells appear to have unsuberized regions, and inner tangential walls of sheath cells contiguous to thin-walled sieve tubes and companion cells are only partially suberized. Plasmodesmata between sheath cells or vascular parenchyma cells and companion cells or thin-walled sieve tubes are sparse further suggesting that sucrose enters the apoplast from bundle sheath cells or vascular parenchyma cells or both in the leaf of this C_4 plant. Lush (1976) reported that the maximum direct path from the site of fixation of CO_2 to the nearest vein is 55 μm in *Panicum maximum*, a C_4 plant and 168 μm in *Lolium temulentum*, a C_3 plant.

From the preceding data, largely dealing with structure, it seems likely that in both C_3 and C_4 plants, products of photosynthesis move to the vicinity of the minor veins in the symplast and then enter the apoplast prior to being loaded into the phloem. Detailed structural and physiological studies are needed to resolve the question for different species of plants.

2. PROCESS OF ENTRY INTO THE FREE SPACE

Techniques for sampling free space have been used to demonstrate selective entry of sucrose into the free space in exporting leaves of *Zea mays* (C_4) and *Beta vulgaris* (C_3). Geiger *et al.* (1974) used an isotope trapping method to reveal the turnover of sucrose in the free space of sugarbeet leaves. Rate of turnover of free space sucrose paralleled changes in export rate. The results of the study confirmed the conclusions of Kursanov and Brovchenko (1970) that free space is filled with sugar from the mesophyll prior to export from the leaf. Heyser *et al.* (1978) observed that sucrose was the only sugar present in xylem exudate obtained from the cut surface of a leaf of *Zea mays* by reduced pressure. These data confirm that sucrose is the normal major constituent of the free space solution at sites of phloem loading. The concentration of sucrose in the xylem exudate increases following photosynthesis and sucrose becomes labeled when $^{14}CO_2$ is administered. The concentration of sucrose in the xylem exudate varied from 1–1.5 m*M* in illuminated attached leaves.

Evaluation of the flux of sucrose into the free space needed to support phloem loading at the observed rates requires a rapid process such as facilitated entry of sucrose into the free space (Doman and Geiger, 1979). This conclusion is even more imperative if efflux is occurring mainly in the vicinity of the companion cell–sieve tube complexes; fluxes are many times those expected for passive permeation. The mechanism of this entry process needs to be examined and clarified.

3. FACTORS AFFECTING ENTRY INTO FREE SPACE

The point of entry of sucrose into the free space prior to phloem loading is a potential point of control in export by the phloem. Brovchenko *et al.* (1975) proposed that the exit of products of photosynthesis into the free space is promoted by photophosphorylation (mainly sucrose) and light dependent oxidation (mainly hexoses). Doman and Geiger (1979) observed that K^+ or Na^+ added to leaf free space at concentrations up to 30 m*M* promoted export of sucrose from sugarbeet leaves. Release of labeled sugar into the solution bathing the leaf paralleled promotion of export. These results indicate that K^+ promotes export by increasing release of sucrose into the apoplast prior to phloem loading. These observations appear to relate to the results of a study of the effect of mild K^+ deficiency on sucrose translocation by Amir and Reinhold (1971). These investigators concluded that K^+ is involved in promotion of entry of sucrose into the translocation pool following its formation by photosynthesis. The K^+ did not promote loading of exogenous sugar, an observation confirmed by Doman and Geiger (1979).

It seems beneficial to plant function that release of sucrose into the free space be adjusted to correspond to the rate of phloem loading. Data from several studies provide the basis for constructing a working hypothesis dealing with a mechanism for regulating release of sucrose into the free space. Phloem loading appears to involve sucrose–proton cotransport and, under at least some circumstances, exchange of K^+ and H^+ (Giaquinta, 1977, 1979; Hutchings, 1978; Malek and Baker, 1978). Van Bel and Van Erven (1979) observed that below a free space pH of 5.5, K^+ antiport is coupled to sucrose–H^+ cotransport into the phloem. If this occurs loading of sucrose would result in an increase in free space K^+ in the vicinity of the minor vein phloem. This increased K^+ may, in turn, cause exit of sucrose from mesophyll or vascular parenchyma cells near the minor veins. In this way, the rates of loading and release of sucrose could be matched through the mediation of K^+. Alternatively, exit of sucrose may be the result of lowering of free-space sucrose concentration as a result of phloem loading. These possibilities need to be investigated further.

Increased potassium supply has been observed to increase translocation of sucrose and cause increased yield (Mengel and Viro, 1974; Mengel and Haeder, 1977). The authors of these studies attribute the effect to the promotive effect of K^+ on cyclic photophosphorylation. An alternative explanation is that K^+ content of leaves may affect release of sucrose and thereby influence phloem loading rate. Increased arrival of K^+ in sugar beet source leaves as a result of increasing K^+ supplied from 2 to 10 mM failed to produce increased NCE or translocation (Conti and Geiger, 1982). Kamanina and Anisimov (1977) concluded that conditions of nitrogen and phosphorus metabolism exert a selective influence on release of sucrose into the leaf free space and affect export from the leaf. The role of plant inorganic nutrition in regulating export and possible associated regulatory mechanisms appear to merit further study.

C. Phloem Loading

The site of entry of sucrose into the phloem certainly constitutes another important control point for assimilate distribution within the plant. How changes in the translocation status of the plant are perceived at the phloem membrane level are very poorly understood, but it is likely that these events figure prominently in the control of assimilate distribution in response to sucrose availability and sink demand. The molecular nature of any control mechanism would seem to be intimately related to the loading mechanism itself. (See earlier reviews by Geiger (1975) and Giaquinta (1980a, c) for a detailed treatment of the structures, cellular pathways, and carrier characteristics related to loading.) We shall explore how control *may* be exerted at the phloem membrane level in light of what is known about the loading process.

1. MECHANISM OF PHLOEM LOADING

It has been suggested that sucrose loading into the phloem is coupled to the cotransport of protons. The driving force for loading is viewed as the electrochemical potential gradient of protons (proton motive force), which is generated by an asymmetric proton efflux into the free space at the loading sites (Giaquinta, 1977, 1979, 1980b; Baker, 1978). The way in which the electrochemical proton gradient interacts with the putative sucrosyl carrier is not known. Presumably a neutral or negatively charged transport carrier sequentially binds sucrose and proton(s). The resulting ternary complex migrates, possibly via a conformational change, across the membrane by dissipating the energy stored in the transmembrane proton gradient (Giaquinta, 1980b, and references therein). Cho and Komor (1980) have shown that the transient influx of protons accompanying sucrose proton symport in *Ricinus* cotyledons is also accompanied by potassium efflux into the external medium. The K^+ efflux, however, was not mechanistically coupled to the sucrose–H^+ transporter (e.g., through a H^+/K^+ exchanging ATPase) but instead resulted from a passive movement for charge compensation in response to the membrane depolarization. Thus, sucrose transport is electrogenic since there is a net movement of positive charge by the sucrose proton transport system. Therefore, the components of the loading mechanism which warrant consideration as control points include the carrier protein itself, ion fluxes associated with loading, and the resulting transmembrane electrical potential.

2. CONTROL OF PHLOEM LOADING

Direct modulation of the preceding components of the loading process can be mediated through changes in the solute concentration as well as turgor within the sieve tubes. The data and our understanding of phloem loading are both too scarce at present to allow anything more than speculation as to what are likely control mechanisms. Nevertheless, this speculative approach can provide some insights and testable hypotheses concerning control of phloem loading at the membrane level.

The internal sucrose concentration of the sieve tubes certainly has the potential to directly or indirectly exert control of the carrier mechanism.

It is well documented that high internal concentrations of ions, amino acids, and sugars can inhibit the further entry of these solutes into bacteria, yeasts, and animal and plant cells (see Giaquinta, 1980a,c). This regulatory phenomenon, which is referred to as transinhibition, presumably occurs because the high internal solute concentrations prevent the dissociation of the hypothetical carrier–substrate complex at the inner membrane surface by mass action effects. This results in less "free" carrier at the external membrane surface available for additional substrate binding because of the slow migration of the carrier–substrate complex. High internal sucrose concentrations inhibit the subsequent accumulation of sucrose in castor bean cotyledons (Komor, 1977) and in sugarbeet source leaves (Giaquinta, 1980c). Conversely, a respiration-induced lowering of the internal sucrose concentration in these tissues was accompanied by an increased rate of sucrose loading, suggesting that loading can respond to changes in internal solute concentrations.

The intracellular level of solutes can also exert allosteric control over the transport carrier. In this regard, Glass and Dunlop (1979) have proposed that high internal concentrations of K^+ in plant roots allosterically control the influx of external K^+ by binding to allosteric sites on the carrier complex. Saturation of these binding sites was thought to cause a conformation change that reduced the affinity of the carrier for external K^+. Similarly, Hodges (1973) has suggested that K^+ influx and K^+-stimulated plasmalemma ATPase show negative cooperativity kinetics toward K^+. Negative cooperativity assumes a multisubunit enzyme in which the binding of one ligand (K^+) to a subunit induces a conformational change in a second subunit which in turn decreases the affinity of that subunit to the second ligand. This results in a modulation of the kinetics of the carrier toward the substrate at varying internal and external substrate concentrations. It remains to be determined whether changes in the internal sucrose or K^+ concentrations of the sieve tubes (or in the free space for that matter) can alter the kinetics or velocity of the carrier mechanism by the aforementioned ways.

Other lines of evidence suggest that control of loading is mediated through the osmotic characteristics, particularly the turgor pressure of the phloem. In this scheme, an increased sink demand would lower the hydrostatic pressure in the sieve tubes in the source region, which in turn would cause a compensatory increase in the loading rate (Milburn, 1974). The hypothesis that phloem loading responds to changes in phloem turgor is strengthened by the studies of Smith and Milburn (1980a, b, c) on phloem sap exudation in *Ricinus*. These authors found that the decrease in phloem sucrose concentration in the exudate of plants placed in continual darkness for 3 days (12% of the control) was accompanied by an increase in sap K^+ levels and a partial maintenance of phloem turgor (60% of control). The maintenance of phloem turgor at low internal sucrose concentrations by a compensatory increase in K^+ loading into the phloem was interpreted as an osmoregulatory response during periods of restricted sucrose availability.

Loading was also found to respond to changes in turgor induced by phloem incisions and water stress. Although successive bark incisions into the stem resulted in large variations in the rate of sap exudation, the solute concentration of the phloem sap remained remarkably constant. The constancy of solute concentration (ψ_s) was principally due to the maintenance in the levels of sucrose and K^+ which together with the associated anions accounted for 75% of the phloem solute potential. This suggests that the rate of phloem loading was able to respond to the incision-induced changes in flux and maintain the phloem osmotic potential accordingly. The incisions, which represent an artificially high sink demand, presumably reduced the phloem pressure potential to near 0 (atmospheric pressure) at the site of the incision. The drop in phloem turgor (ψ_p) is thought to be rapidly propagated along the sieve tube pathway to the loading sites where the decrease in ψ_{total} causes a compensatory increase in loading (Smith and Milburn, 1980b).

These authors also found that exudation continued during severe water stress in *Ricinus,* mainly because of an increased solute concentration of phloem which allowed the phloem to maintain a positive phloem turgor (Smith and Milburn, 1980c). Phloem loading, therefore, is regulated by cell turgor whereby a decrease in ψ_p of the phloem causes a compensatory increase in solute loading. Thus, changes in sink demand in the intact plant may regulate phloem loading through changes in turgor at the loading sites. It is also possible that the increased loading in response to a decrease in turgor may be more related to the change in the rate of water entry into the sieve tubes (Fondy and Geiger, 1980). At this time, it is not possible to distinguish between a direct affect of ψ_p on sucrose loading, water conductivity, or a dependence of loading on cell volume or elastic modulus of the cell wall (Zimmermann and Steudle, 1978; Smith and Milburn, 1980c). In light of the preceding material, the osmotic characteristics of the phloem in relation to the regulation of loading warrant considerable attention.

It is important to address the question of how these purported changes in turgor may be transduced to the actual mechanism of sucrose loading. Changes in the hydrostatic pressure difference in many osmoregulating organisms in itself are probably too small to be a significant driving force for solute transport (Cram, 1976). This suggests that the pressure or turgor change is a signal that has to be transduced and amplified at the membrane level in order to affect a change in solute loading. Increased loading is needed for long-term increases in export. Based on what is known about the phloem-loading mechanism, as well as possible pressure-sensing mechanisms in nonphloem systems, the membrane potential seems to be a likely candidate as the transducer in the sieve tubes. In this regard, plant cell membranes may contain a pressure sensing mechanism which is involved in osmoregulation (Coster *et al.*, 1977).

Increasing turgor pressure has been proposed to cause a marked compression in cell membrane thickness (Coster *et al.*, 1977), and sudden changes in turgor have been shown to cause changes in the membrane potential (Cram, 1976). The turgor-induced membrane deformation with its resulting effect on membrane carriers may be the pressure transducing mechanism involved in solute transport and osmoregulation associated with phloem loading. Interestingly, the changes in membrane compression are also thought to be sensitive to changes in the membrane potential. Because of the relationship between electric field and membrane compression, Coster *et al.* (1977) proposed that control of turgor by solute transport may be sensitive to concentrations of specific ions which do not necessarily contribute significantly to the total osmolarity of the system. They, for example, suggested that slight changes in K^+ concentration may reduce the membrane potential with resulting changes in membrane conformation. It remains to be established whether these events are applicable to phloem loading. These examples offer some insights into possible ways in which a change in the demand for assimilates can exert an influence on export. These effects on membranes may be associated with the loading of assimilates into the phloem as well as with a possible compensating efflux of sucrose into the free space in the region of phloem loading. Regulation of phloem loading at the membrane level promises to be an exciting and important area for future research in translocation. The role of hormones, such as ABA and IAA, in relation to the membrane potential and solute fluxes may have important implications in phloem loading and unloading and also warrant further study.

IV. Processes Affecting Export of Products of Photosynthesis

The previous sections dealt with the ability of source leaves to alter the partitioning of assimilated carbon between export and nonexport carbohydrate pools in response to changes in sink demand, environment, and plant ontogeny. These adaptive responses necessitate that the metabolic pathways governing assimilate metabolism and partitioning be able to respond to changes in the translocation status of the intact plant. Although it is far from certain how this information flow is achieved, some insight into this problem can be gained by examining the biochemical control of the synthesis and utilization of starch and sucrose, the principle carbohydrates related to assimilate partitioning. It is important to stress that even though many aspects of regulation of carbohydrate metabolism have been established at the biochemical level, the relationship between these cellular control mechanisms and the dynamics of assimilate distribution in the intact plant remains to be established.

A. Starch Biosynthesis in Leaves

The predominant pathway of starch biosynthesis in the chloroplast stroma involves the synthesis of the sugar nucleotide ADP glucose, from hexose phosphate derived from the photosynthetic reduction cycle (Preiss and Levi, 1979). This reaction, which is catalyzed by ADP glucose pyrophosphorylase, favors the synthesis of ADP glucose because of the subsequent hydrolysis of pyrophosphate (Reaction 1). Starch synthase then catalyzes the transfer of the glucosyl residue of ADP glucose to the elongating glucose chain on the starch granule via a α-(1 → 1, 4) glucoside linkage (Reaction 2). Branching enzyme or Q enzyme is responsible for the α-(1 → 6) linkage found in amylopectin (Reaction 3).

$$\text{ADP} + \text{glucose 1-P} \underset{}{\overset{\text{ADP glucose pyrophosphorylase}}{\rightleftharpoons}} \text{ADPglucose} + \text{PP}_i \qquad (1)$$

$$\text{ADP glucose} + (\text{glucosyl})_n \xrightarrow{\text{starch synthase}} \text{ADP} + (\text{glucosyl})_{n+1} \qquad (2)$$

$$\alpha\text{-}(1\rightarrow 4)\text{ linear glucose} \xrightarrow{\text{Branching enzyme}} \alpha\text{-}(1\rightarrow 6)\text{-amylopectin} \qquad (3)$$

1. ADP GLUCOSE PYROPHOSPHORYLASE

Regulation of starch biosynthesis is mainly mediated through control of ADP glucose formation. Preiss and co-workers (1967; Preiss and Kosuge, 1970; Preiss and Levi, 1979) have shown that ADP glucose pyrophosphorylase isolated from a variety of plants is subject to allosteric activation by phosphoglycerate (PGA) and other glycolytic intermediates and allosteric inhibition by inorganic phosphate. The activation at *in vivo* PGA concentrations (7–370 μM) increased both the apparent affinity and maximum velocity of the pyrophosphorylase for its substrates, whereas *in vivo* phosphate concentrations inhibited pyrophosphorylase activity and also reversed the enzyme activation by PGA. The allosteric control of this enzyme by the intracellular PGA:P_i ratio is currently viewed as one of the principal control mechanisms for regulating starch synthesis at the biochemical level.

There are several lines of experimental support for the pivotal role of the intracellular triose phosphate:Pi ratio in the regulation of starch synthesis via its allosteric control of ADP glucose pyrophosphorylase (glucose-1-P adenylyltransferase) (Preiss and Levi, 1979). Studies of metabolite concentrations in intact chloroplasts have shown that the stromal phosphate concentration in the dark is near 10 m*M* and decreases by 30–50% during illumination (Santarius and Heber, 1965). The decrease in phosphate concentration is accompanied by an increase in the levels of various glycolytic intermediates, ATP, and reduced pyridine nucleotides (Heldt *et al.*, 1977). Based on studies of the purified ADP glucose pyrophosphorylase, a decrease in phosphate concentration of this magnitude (that is from 10 to 5 m*M* phosphate at a PGA concentration of 5 m*M*) would result in a 23-fold increase in the activity of ADP glucose pyrophosphorylase (Preiss and Levi, 1979). Similarly, Heldt *et al.* (1977) demonstrated that a stromal concentration of 10 m*M* phosphate completely inhibited starch synthesis from CO_2 in isolated chloroplasts with the inhibition being reversed by exogenous PGA. Moreover, these authors reported that the changes in PGA and phosphate levels and ratios in the chloroplast stromal compartment (and subsequent starch synthesis) were similar to the concentrations and ratios of these metabolites required for the *in vitro* activation of ADP glucose phosphorylase.

These studies show that alterations in the PGA:Pi ratio over physiological concentrations can have profound effects on the rate of ADP glucose pyrophosphorylase activity and thus may provide a sensitive control mechanism for starch synthesis *in vivo*. Next, we shall consider the additional role of TP/P_i as a major determinant of metabolite fluxes across the chloroplast envelope.

2. STARCH SYNTHASE

Control of starch synthesis is also possible at the level of starch synthase, the enzyme catalyzing the transfer of glucose residues from ADP glucose to the elongating linear glucan on the starch granule. Starch synthase exists in two forms: one bound to the starch granule, which does not display absolute glucosyl donor specificity, and a soluble form with higher activity and absolute specificity for ADP glucose. Preiss and

Levi (1979) speculate that *in vivo* the soluble synthase is localized at the site of starch granule formation, whereas the bound form represents a fraction of the soluble enzyme that becomes entrapped in amylose during granule formation. Hawker *et al.* (1974) suggested that regulation of starch synthase may be achieved through changes in the ionic composition around the enzyme. They showed that the bound, but not the soluble, form of starch synthase isolated from sugarbeet, bean, and saltbush leaves was stimulated twofold by K^+. Sugarbeet plants grown in the presence of 5 m*M* potassium in the nutrient solution contained nine times as much foliar starch as did plants grown with 0.5 m*M* potassium/4.95 m*M* sodium. In sugarbeet, Na^+ could replace K^+ without any adverse effects on plant growth. The decrease in starch content in the high Na^+ plants was attributed to the K^+ requirement of the starch granule–bound ADP glucose starch synthase. In contrast to the interpretation of Preiss and Levi (1979), it has been proposed that dissociation of the loosely associated soluble synthase from the granules during extraction may have altered the ability of the enzyme to respond to K^+. Similarly, the observations that high nutrient K^+:Na^+ ratios in sugarbeets favor leaf growth at the expense of beet growth, whereas low K^+:Na^+ ratios favor sucrose translocation to the beet (El-Sheikh and Ulrich, 1970) may reflect the retention of dry matter within the leaves because of the K^+ requirement of starch synthase. Although other interpretations are possible, it has been suggested that the cellular ionic environment can exert physiological control by altering electrostatic potentials of many types of proteins, including allosteric enzymes, and membrane carriers (Donzou and Maurel, 1977). Although the regulatory role of starch synthase in starch biosynthesis in leaves should not be overlooked, there is more experimental support for control of starch synthesis being exerted at the ADP glucose pyrophosphorylase level and at the chloroplast envelope.

3. CHLOROPLAST–MESOPHYLL INTERACTION

In addition to the allosteric control of the phosphorylase, the triose P:P_i ratio plays an important role in the partitioning of assimilated carbon between starch synthesis in the chloroplast and sucrose synthesis in the cytoplasm. There is compelling evidence that the initial products of CO_2 fixation, such as dihydroxyacetone phosphate and triose phosphate, are exported across the inner chloroplast envelope to the cytoplasm for sucrose synthesis (Walker, 1976; Heldt *et al.*, 1977). The export of triose phosphate is obligatorily and stoichiometrically coupled to a 1:1 counter exchange of inorganic phosphate via a specific "phosphate translocator" in the inner chloroplast envelope (Fliege *et al.*, 1978).

At low cytoplasmic phosphate concentrations, triose phosphate is retained within the chloroplast because of the lack of exchangeable phosphate for triose phosphate export. The resulting increase in PGA within the chloroplast stroma (high PGA:P_i ratio) would in turn favor starch synthesis by allosterically activating ADP glucose pyrophosphorylase. Alternatively, high cytoplasmic phosphate not only promotes the export of triose phosphate to the cytoplasm for sucrose synthesis because of the presence of exchangeable phosphate for the "translocator," but also decreases starch synthesis within the chloroplast because of the reduced availability of substrate and the allosteric inhibition of ADP glucose pyrophosphorylase by the decrease in stromal PGA:P_i (Heldt *et al.*, 1977). The well-documented inhibition of photosynthesis in isolated chloroplasts by high external phosphate can be readily explained by an accelerated export of triose phosphate from the chloroplasts which prevents the accumulation of carbon cycle intermediates needed to sustain photosynthesis at optimal rates (Walker, 1976). Addition of PGA can restore photosynthesis by reducing the rate of rapid export of triose phosphate. Thus, it is quite clear, at least at the biochemical level, that regulation of the PGA:P_i can alter the partitioning of photosynthetic carbon between starch and sucrose as well as control the activity of the key enzyme in starch synthesis, ADP glucose pyrophosphorylase.

B. Starch Degradation in Leaves

Compared to starch biosynthesis, the pathways and metabolic control of starch remobilization in leaves are very poorly understood. Starch degradation seems to involve three distinct stages: (1) degradation of the insoluble starch granule to soluble maltodextrins; (2) hydrolysis of these dextrins to hexoses and hexose phosphates; and (3) metabolism of hexose and hexose phosphates to products that can be exported from the chloroplast (Preiss and Levi, 1979). Although amylase is responsible for conversion of the insoluble starch to soluble maltodextrins, there are conflicting reports on whether subsequent degradation is amylolytic or phosphorolytic (Stitt *et al.*, 1978). The majority of evidence, however, favors the phosphorolytic route as the predominant pathway, although both pathways seem to be necessary for maximum starch degradation in spinach chloroplasts (Okita *et al.*, 1979).

$$\text{Starch granule} \xrightarrow{\text{amylase}} \text{linear dextrins} \quad (4)$$

$$\text{Dextrin} \xrightarrow{\text{phosphorylase}} \text{glucose 1-P} \longrightarrow \text{triose phosphate} \quad (5)$$

Based on isolated chloroplast studies, the major products of starch degradation are dihydroxyacetone phosphate, 3 PGA, maltose, and glucose (Levi and Gibbs, 1976). Steup *et al.* (1976) showed that starch degradation was dependent on phosphate and that the P_i dependency for degradation differed from that for synthesis. Subsequent work by Heldt *et al.* (1977) also demonstrated that starch degradation was promoted by phosphate and that the major products of the phosphate-dependent remobilization were hexose monophosphate and ultimately triose phosphates, which could be exported from the chloroplast. The rates of conversion for the phosphate-independent conversion of starch to maltose and glucose were much slower than the phosphorolytic pathway.

Although the evidence for the regulation of starch synthesis by triose phosphate:P_i at both the "translocator" and ADP glucose pyrophosphorylase levels appears quite convincing, the direct control of starch degradation in the dark remains to be resolved. Regulation may be mediated by the phosphate requirements of phosphorylase, light-driven pH activation of stromal enzymes, or regulation by intermediates of the carbon metabolism pathways, such as phosphofructokinase or hexose kinase (Okita *et al.*, 1979; Preiss and Levi, 1979). The possibility also exists that direct regulation of degradation does not exist (i.e., it occurs continually) and the level of starch in the leaves is regulated solely by reactions occurring in the light period.

C. Sucrose Synthesis and Degradation in Leaves

The level of sucrose in photosynthesizing leaves is governed by the activities of four enzymes: (1) sucrose phosphate synthetase (SPS); (2) sucrose phosphate phosphatase (SPP); (3) sucrose synthase (SS); and (4) invertase (Pontis, 1978).

$$\text{UDPglucose} + \text{fructose-6-phosphate} \xrightarrow{\substack{\text{sucrose}\\\text{phosphate}\\\text{synthetase}}} \text{sucrose phosphate} + \text{UDP} \quad (5)$$

$$\text{Sucrose-P} \xrightarrow{\text{sucrose phosphate phosphatase}} \text{sucrose} + P_i \quad (6)$$

$$\text{Fructose} + \text{UDPglucose} \underset{}{\overset{\text{sucrose synthase}}{\rightleftarrows}} \text{sucrose} + \text{UDP} \quad (7)$$

$$\text{Sucrose} \xrightarrow{\text{invertase}} \text{glucose} + \text{fructose} \quad (8)$$

Collectively, the metabolic regulation and cellular compartmentation of these reactions can be envisioned as a "sucrosestat" mechanism, which enables the leaf to adjust its carbohydrate status in response to changes in the translocation status of the plant. The importance of leaf sucrose concentration to export is reviewed by Geiger (1979); these studies show that the sucrose concentration in the transport compartment of the leaf is one of the principal determinants of translocation.

There is compelling evidence that in photosynthesizing leaves, sucrose is synthesized in the cytoplasm from triose phosphates (dihydroxyacetone phosphate, glyceraldehyde 3-phosphate) which represent the principal photosynthetic products exported from the chloroplast (Walker, 1976; Heldt *et al.*, 1977; Robinson and Walker, 1979). As mentioned earlier, the export of triose phosphate is stoichiometrically coupled to the influx of inorganic phosphate through the "phosphate translocator" in the inner chloroplast membrane. The triose phosphates are first isomerized by cytoplasmic triose phosphate isomerase and then the isomers undergo aldol condensation to give fructose 1,6-bisphosphate (Walker, 1976). Fructose bisphosphate is hydrolyzed by fructose bisphosphatase to fructose 6-phosphate, which gives rise to hexose monophosphates and ultimately UDP glucose, the glucosyl donor for sucrose synthesis (Fig. 5). Inorganic phosphate is released to the cytoplasm at three steps: (1) the hydrolysis of fructose bisphosphate by fructose-bisphosphate phosphatase; (2) the hydrolysis of pyrophosphate by pyrophosphorylase; and (3) the conversion of sucrose phosphate to sucrose by sucrose phosphate phosphatase. Thus, the net reaction for sucrose synthesis is

$$4 \text{ Triose phosphate} + 3\ H_2O \rightarrow 1 \text{ sucrose} + 4\ P_i \qquad (9)$$

The liberation of inorganic phosphate into the cytoplasm favors continued export of triose phosphate for sucrose synthesis from the chloroplast via the phosphate translocator. Therefore, sucrose synthesis plays an important role in photosynthetic carbon metabolism, both as the principal export product from the leaves, and as a mechanism for the recycling of orthophosphate, which is needed for continued photosynthesis.

Cytoplasmic phosphate levels may play an important role in regulating both sucrose and starch synthesis, and thus export of assimilates. For instance, in species lacking mannose phosphate isomerase, mannose has been used to sequester intracellular phosphate as mannose 6-phosphate (Herold and Lewis, 1977). The mannose-induced lowering of cytoplasmic phosphate reduced photosynthesis and favored the retention of triose phosphate within the chloroplast for starch synthesis because of the unavailability of free phosphate to exchange with triose

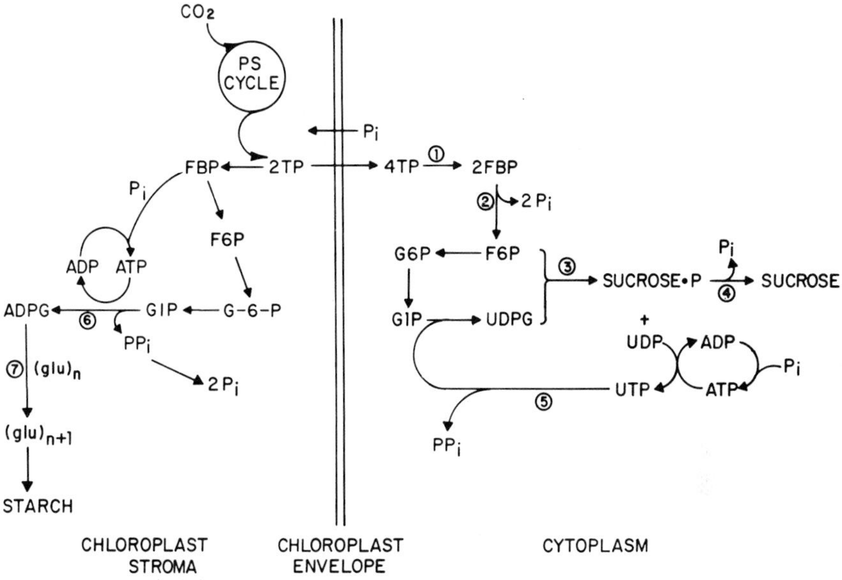

FIG. 5. Pathways of starch and sucrose biosynthesis in the mesophyll chloroplast and cytoplasm. See text for details. PS cycle, photosynthetic reduction cycle; TP, triose phosphate; FBP, fructose 1,6-bisphosphate; F6P, fructose 6-phosphate, G6P, glucose 6-phosphate; G1P, glucose 1-phosphate; ADPG; ADP glucose; UDPG, UDP glucose; PP_i, pyrophosphate; P_i, orthophosphate; $(glu)_n$, linear glucan. *Enzymes:* (1) Fructose–bisphosphate aldolase (EC 4.1.2.13); (2) Fructose–bisphosphatase (EC 3.1.3.11); (3) Sucrose–phosphate synthase (EC 2.4.1.14); (4) Sucrose–phosphatase (EC 3.1.3.24); (5) Glucose 1-phosphate uridylyltransferase (EC 2.7.7.9); (6) Glucose 1-phosphate adenylyltransferase (EC 2.7.7.27); (7) Starch synthase (EC 2.4.1.21).

phosphate from the chloroplast. Similarly, phosphate deficient plants contain substantial levels of starch (Herold and Walker, 1978) which can be decreased by increasing the phosphate status of the plant (Giaquinta and Quebedeaux, 1980). Further evidence for a possible role of P_i in carbon partitioning comes from studies correlating sucrose export with an increase in tissue phosphate level. For example, Hopkinson (1964) found that the highest rates of sucrose export from leaves during leaf development are temporally correlated with the maximum import of phosphate in the leaves. Similarly, Thorne and Koller (1974) showed that the imposition of an increased sink demand in soybeans is followed by an increase in photosynthesis, sucrose export, and a fivefold increase in leaf phosphate. These experimental observations at both the subcellular and whole-plant levels provide at least qualified support for a controlling role of orthophosphate in sucrose synthesis. Although the manner in which sucrose synthesis and degradation is regulated in leaves is not known, we shall examine, in the following sections, possible sites where control may be exerted.

1. SUCROSE PHOSPHATE SYNTHETASE AND SUCROSE PHOSPHATE PHOSPHATASE

The formation of sucrose by the coupled action of SPS and SPP is generally accepted as the predominant synthetic route in photosynthesizing leaves. The equilibrium of the SPS reaction and the hydrolysis of sucrose phosphate by the phosphatase make this an essentially irreversible reaction under *in vivo* conditions. The important role for sucrose phosphate synthetase in export is indicated by the close correlation between the onset of export capacity and the appearance of the activity of this enzyme (Giaquinta, 1978; Silvius *et al.*, 1978).

Some *in vitro* experiments suggest that SPS may be subject to regulatory control by carbon metabolism intermediates. Sucrose phosphate synthetase, unlike SS, shows absolute specificity for the glucosyl donor, UDPG, and fructose 6-phosphate. Preiss and Greenberg (1969) found that SPS activity had sigmoidal kinetics with respect to UDPG and fructose 6-phosphate concentrations, and that Mg^{2+} stimulated the maximum velocity of SPS and decreased the affinity of the enzyme to UDPG. Nucleotides, such as UDP, were found to inhibit SPS activity (Whittingham *et al.*, 1979). In addition to possible allosteric control by various photosynthetic metabolites, it has been proposed that sucrose phosphate synthetase isolated from *Vicia faba* cotyledons is associated with a "natural" activator, which can be removed by freezing and thawing (de Fekete, 1971). In the absence of this activator, the reduced enzyme activity was allosterically activated by UDPG, fructose 6-phosphate and citrate. The inhibitor effect of UDP and other free nucleotides only occurred in the presence of the proposed activator. Also, low citrate concentrations inhibited SPS activity in the presence of the activator, whereas high citrate concentration restored the enzyme activity (Davies, 1974; Whittingham *et al.*, 1979). The identity or physiological role of this activator in sucrose synthesis is not at all clear, but the concept that sucrose biosynthesis may be regulated at the level of SPS is an important aspect of carbon partitioning that warrants further attention.

Several studies at both the cellular and whole-plant levels are consistent with a regulatory role of SPS. Salerno and Pontis (1978a,b) reported that SPS activity was inhibited by sucrose, P_i, and nucleoside triphosphates. They also speculated that SPS *in vivo* existed in two forms with markedly different affinities for sucrose binding. One form was sensitive to sucrose inhibition, whereas the other was not. It was also pointed out that regulation by sucrose at the level of SPS, rather than at the level of SPP (Hawker, 1967) would be advantageous in that it would allow UDP glucose to be used for other cell reactions.

The level of SPS activity in source leaves may be related to export capacity. Silvius *et al.* (1979) found that soybean plants acclimated to high irradiance (950 $\mu E/m^2 \cdot sec$) have similar photosynthesis rates, but lower starch accumulation rates than those not acclimated to a high light intensity. Interestingly, the rate of sucrose synthesis and translocation are higher in the plants acclimated to high irradiance than in nonacclimated plants or in plants that are acclimated to moderate irradiance (600 $\mu E/m^2 \cdot sec$). The increased translocation rates in high irradiance-acclimated plants are correlated with a twofold increase in SPS activity on a protein basis, even though the sucrose pool size remains relatively constant in both acclimated and nonacclimated plants. The activity of ADPG pyrophosphorylase is not affected by irradiance treatment suggesting that the increase in SPS activity, which may result from either protein synthesis or activation of the existing enzyme, is relatively specific for SPS. These results suggest that the activity of SPS and, thus, the rate of sucrose synthesis, plays an important role in the partitioning of photosynthate between starch and sucrose and thus translocation. This view is strengthened by the recent findings of Huber (1980a) on both inter- and intraspecific variations in photosynthetic sucrose and starch formation in various leaf protoplasts. The protoplast system, which is free from the constraints of variations in sink demand and rates of product removal from the leaf, presumably allows one to assess the potential for sucrose and starch synthesis at the biochemical level. Although photosynthesis rates were generally similar, protoplasts from wheat and barley leaves (low starch formers) partitioned more carbon on a rate basis into sucrose versus starch (50% sugar versus 17% starch) than did leaf protoplasts from high starch formers such as peanuts, soybeans, and tobacco (31% sugar versus 35% starch). The sugar:starch ratio ranged from a high of 5.4 in barley to a low of 0.8 in peanuts. Differences in the partitioning of photosynthetic carbon between sucrose and starch were also observed in protoplasts derived from different varieties of wheat. Interestingly, protoplasts from two high yielding wheat varieties partitioned more carbon into sucrose and less into starch than the two lower yielding varieties. These results indicate that the partitioning of carbon into starch and sucrose and thus its availability for export may be both biochemically and genetically controlled within the mesophyll cell. In addition, Huber (1980b) found that this partitioning difference may be accounted for by the activity and regulation of

SPS. Species having a low potential for sucrose formation (i.e., high starch formers) have less SPS activity per cell than do high sucrose producers. Most interestingly, however, SPS activity from leaves having a low sucrose potential is inhibited by exogenous sucrose whereas SPS activity from species having a high potential for sucrose formation is not inhibited by sucrose. Collectively, these experiments strongly suggest that sucrose formation and, thus, the availability of translocate is controlled, in part, by SPS activity and that different forms of SPS may exist in various species.

Although there is no conclusive evidence for a direct hormonal involvement in sucrose biosynthesis, there have been reports that ABA stimulates both SPS and SS activity (Pontis, 1977). Interestingly, although gibberellin did not affect the activity of these enzymes, it did prevent the ABA-induced stimulation. At this time, though, the role of hormones on sucrose synthesis is neither clear nor convincing and thus open to much speculation. The possibility that the signals between source and sink regions in the plant are mediated by hormones may have some relevance to the sucrose biosynthesis.

Sucrose phosphate phosphatase is generally regarded as not being rate-limiting for sucrose synthesis (Whittingham *et al.,* 1979). It is, nevertheless, inhibited *in vitro* by the end product sucrose (Hawker, 1967). Much (perhaps too much) has been made about the observed *in vitro* inhibition of this enzyme by high exogenous sucrose concentrations in relation to the mechanism by which a distant sink region can alter translocation and carbon partitioning. It has been hypothesized that a feedback inhibition of SPP by sucrose (which presumably builds up in the source leaf when sink demand is diminished) would sequester intracellular phosphate in the form of sucrose phosphate and other sugar phosphates (see Herold and Lewis, 1977). The lowering of the cytoplasmic orthophosphate concentration would favor the retention of triose phosphate within the chloroplasts for starch synthesis. Alternatively, an increase in sink demand would enhance sucrose synthesis by increasing the intracellular phosphate concentration, which favors continued export of triose phosphate to the cytoplasm for additional sucrose synthesis and continued photosynthesis. Although this mechanism provides a convenient biochemical basis for sink control of export, the operation of these events in the intact, translocating plant remains to be established. The limited brief response to increased sink–source ratio should also be kept in mind. The findings of Salerno and Pontis (1978a,b) and Huber (1980a,b) on the regulation of SPS by sucrose may have more physiological relevance than the regulation of SPP by sucrose, and it deserves further attention.

Since sucrose biosynthesis is dependent upon the concentration of

UDP glucose and fructose 6-phosphate, control is also possible at the reactions governing the synthesis of these substrates, particularly at the level of fructose bisphosphate phosphatase and UDP glucose pyrophosphorylase. Fructose bisphosphate phosphatase is sensitive to changes in cytoplasmic pH, Mg^{2+}, NH_4^+, AMP, P_i and F 6P concentrations (Zimmermann *et al.*, 1978), whereas UDP glucose pyrophosphorylase is inhibited by UDP glucose and UTP (Whittingham *et al.*, 1979).

At this time, we are uninformed as to the *in vivo* regulation of sucrose biosynthesis and the relevance of *in vitro* data in relation to assimilate translocation in the intact plant.

2. SUCROSE SYNTHASE

Sucrose synthase, unlike SPS, catalyzes the readily reversible reaction of sucrose synthesis and degradation (Reaction 7). Sucrose synthase activity, however, is low in photosynthesizing leaves (Giaquinta, 1978). Although the enzyme is thought to catalyze sucrose degradation in nonphotosynthetic portions of the plant in order to generate UDP glucose necessary for various biosynthesis, its ability to function in both directions may be relevant in determining sucrose concentration *in vivo.*

Differential regulation of the synthetic and degradative reactions of this enzyme from nonphotosynthetic tissues have been reported. For example, it has been reported that NADP, pyrophosphate, iodoacetic acid, and gibberellic acid activated sucrose degradation and inhibited sucrose synthesis, whereas fructose 1-phosphate and Mg^{2+} inhibited degradation (see Davies, 1974). That the synthesis and degradation reactions of SS may be independently regulated is further indicated by the effects of limited proteolysis on the activity of this enzyme. Wolosiuk and Pontis (1974) found that trypsin treatment of the isolated enzyme markedly inhibited the sucrose degradative activity without much effect on the synthetic direction. In addition, the saturation curves of the trypsin-treated enzyme for sucrose were no longer hyperbolic but sigmoidal in shape. Pontis (1978) has reported preliminary evidence for a Ca^{2+}-dependent protease isolated from *Helianthus* tuber extracts, which acts similar to trypsin on SS. The significance of this protease in regulating sucrose concentration *in vivo* is not known, but it is interesting that the sucrose content of soybean leaves markedly increases during senescence, a time in which leaf protease activity is also markedly increased (Giaquinta and Quebedeaux, 1980).

3. INVERTASE

Invertases have been extensively studied in the actively growing regions (sinks) of many plants. Its presence in these growing tissues is correlated with the hydrolysis of translocated sucrose for growth and metabolism. Although the presence of invertases in importing regions is well documented (Giaquinta, 1980a), the role of invertase in mature, photosynthesizing leaves remains obscure. For example, Giaquinta (1978) showed that invertase activity was present in substantial and equal levels in both sugar beet sink and source leaves. In importing leaves, the invertase activity was correlated with a rapid hydrolysis of both exogenously supplied [^{14}C]sucrose and [^{14}C]sucrose, which was translocated to these leaves. In contrast, exporting leaves, which had similar levels of invertase activity, hydrolyzed both photosynthetically derived sucrose and exogenous sucrose to a much lesser extent than sink leaves. Although the most obvious explanation is that invertase in source leaves may be compartmented within the cells, perhaps within the vacuole, the expression of the activity of invertase appears to be highly regulated since there is good evidence for both vacuolar and cytoplasmic sucrose pools (Fisher and Outlaw, 1979). The observation that hexose levels increase in sugarbeet source-leaves, after disrupting the translocation pathway by heat girdling of the petiole, suggests that invertase activity can be expressed when the translocation status of the leaf is altered (Geiger and Swanson, 1965). The factors that regulate the expression of invertase activity *in vivo* are not known, but the presence of an endogenous inhibitor, membrane associations, isozymes, or control by metabolite levels such as glucose repression or substrate availability are all possibilities at this point (Giaquinta, 1980a).

D. Regulation of Assimilate Export

Since export is principally determined by the kinetics of change in sucrose in the translocation pool of the leaf, the mechanisms governing sucrose biosynthesis, degradation, and compartmentation obviously play an important role in export. As a whole, the preceding enzymes provide the basis for controlling the rate of sucrose synthesis and its cellular concentration in relation to the metabolic needs and translocation status of the plant. An important aspect of the control of export of products of photosynthesis that needs considerable study centers on how changes in the assimilate demand of distant sinks during ontogeny or in response to environment are transduced to the various biochemical mechanisms operating in the source leaves. As noted previously, the regulation of assimilate partitioning may, in part, be mediated by the intracellular phosphate concentrations and/or the regulatory properties of enzymes such as SPS and ADP glucose pyrophosphorylase. It is important to stress, though, that control of assimilate export in the intact plant is almost surely much more complex than outlined earlier in that it cannot be adequately explained solely in terms of phosphate regulation or perhaps any single biochemical mechanism. This certainly does not question the importance of the role of phosphate or enzyme regulation in carbon partitioning, but instead, emphasizes our lack of understanding of the biochemical basis for the control of assimilate partitioning in the whole plant. A case in point is the remarkable study of Chatterton and Silvius (1979) on photosynthate partitioning in the soybean plant. As mentioned earlier, they showed that reducing the photosynthetic period from 14 hours to 7 hours resulted in an increased partitioning of photosynthate into starch in the leaves at the end of the respective photosynthetic periods (10 versus 15% of the leaf dry weight after 14 hours and 7 hours, respectively). Leaves from 14-hour photosynthetic period plants partitioned 60% of the total accumulated foliar carbohydrate into starch, whereas the 7-hour photosynthetic period plants partitioned 90% into starch, even though the carbon exchange rate on an area basis was decreased by 18% in the 7-hour plants. These authors proposed that starch synthesis is controlled independent of the photosynthesis rate, since reducing both the rate and duration of photosynthesis would be expected to decrease starch accumulation if starch synthesis resulted from a retention of a constant proportion of the carbon fixed in the chloroplast. Chatterton and Silvius (1979) suggest that starch synthesis is a programmed process possibly responding to photomorphogenic control and is influenced by the energy demands of the dark period. Although the biochemical basis for these observations is not readily apparent in light of the preceding discussions on cellular control of starch synthesis, it would be most interesting to determine the metabolic correlates of these alterations occurring at the whole-plant level.

V. Conclusions

In this chapter, we have discussed several approaches to increasing crop yield, emphasized that plants are integrated and adapted systems, discussed the relation between translocation and photosynthesis, pointed out the means of controlling the export of photosynthate from source leaves, and presented information on synthesis and degradation of sucrose and starch. We hope that this knowledge will be used by the reader in formulating ways and means to improve plant productivity by controlling translocation processes.

REFERENCES

Amir, S., and Reinhold, L. (1971). *Physiol. Plant.* **24,** 226–231.

Baker, D. A. (1978). *New Phytol.* **81,** 485–497.

Borchers-Zampini, C., Glamm, A. B., Hoddinott, J., and Swanson, C. A. (1980). *Plant Physiol.* **65,** 1116–1120.

Brovchenko, M. I., Slobodskaya, G. A., Chmora, S. N., and Lipatova, T. F. (1975). *Sov. Plant Physiol. (Engl. Transl.)* **23,** 1042–1049.
Carmi, A., and Koller, D. (1977). *Ann. Bot. (London)* [N. S.] **41,** 59–67.
Carmi, A., and Koller, D. (1978). *Photosynthetica* **12,** 178–184.
Carmi, A., and Koller, D. (1979). *Plant Physiol.* **64,** 285–288.
Challa, H. (1976). *Agric. Res. Rep. (Wageningen)* **861.**
Chatterton, N. J., and Silvius, J. E. (1979). *Plant Physiol.* **64,** 749–753.
Cho, B.-H., and Komor, E. (1980). *Plant Sci. Lett.* **17,** 425–435.
Conti, T. R., and Geiger, D. R. (1982). *Plant Physiol.* **70,** 168–172.
Coster, H. G. L., Steudle, E., and Zimmermann, U. (1977). *Plant Physiol.* **58,** 636–643.
Cram, W. J. (1976). *Encycl. Plant Physiol., New Ser.* **2,** Part A, 284–316.
Davies, D. R. (1974). *Ann. Proc. Phytochem. Soc. (London)* **10,** 61–81.
de Fekete, M. A. R. (1971). *Eur. J. Biochem.* **19,** 73–80.
Doman, D. C., and Geiger, D. R. (1979). *Plant Physiol.* **64,** 528–533.
Donzou, P., and Maurel, P. (1977). *Trends Biochem. Sci.* **3,** 14–17.
El-Sheikh, A. M., and Ulrich, A. (1970). *Plant Physiol.* **46,** 645–649.
Evert, R. F., Eschrich, W., and Heyser, W. (1977). *Planta* **136,** 77–89.
Evert, R. F., Eschrich, W., and Heyser, W. (1978). *Planta* **138,** 279–294.
Fisher, D. B., and Outlaw, W. H. (1979). *Plant Physiol.* **64,** 481–483.
Fisher, D. B., Housley, R. L., and Christy, A. L. (1978). *Plant Physiol.* **61,** 291–295.
Fliege, R., Flugge, U.-I., Werdan, K., and Heldt, H. W. (1978). *Biochim. Biophys. Acta* **502,** 232–247.
Fondy, B. R., and Geiger, D. R. (1980). *Plant Physiol.* **66,** 945–949.
Fondy, B. R., and Geiger, D. R. (1982). *Plant Physiol.* **70,** (in press).
Fraser, D. E., and Bidwell, R. G. S. (1974). *Can. J. Bot.* **52,** 2561–2570.
Geiger, D. R. (1975). *Encycl. Plant Physiol., New Ser.* **1,** 395–431.
Geiger, D. R. (1976a). *Can. J. Bot.* **54,** 2337–2345.
Geiger, D. R. (1976b). *In* "Transport and Transfer Processes in Plants" (I. F. Wardlaw and J. B. Passioura, eds.), pp. 167–183. Academic Press, New York.
Geiger, D. R. (1979). *Bot. Gaz. (Chicago)* **140,** 241–248.
Geiger, D. R., and Batey, J. (1967). *Plant Physiol.* **42,** 1743–1749.
Geiger, D. R., and Swanson, C. A. (1965). *Plant Physiol.* **40,** 942–947.
Geiger, D. R., Giaquinta, R. T., Sovonick, S. A., and Fellows, R. J. (1973). *Plant Physiol.* **52,** 585–589.
Geiger, D. R., Sovonick, S. A., Shock, T. L., and Fellows, R. J. (1974). *Plant Physiol.* **54,** 892–898.
Giaquinta, R. T. (1976). *Plant Physiol.* **57,** 872–875.
Giaquinta, R. T. (1977). *Plant Physiol.* **59,** 750–755.
Giaquinta, R. T. (1978). *Plant Physiol.* **61,** 380–385.
Giaquinta, R. T. (1979). *Plant Physiol.* **63,** 744–748.
Giaquinta, R. T. (1980a). *In* "Biochemistry of Plants" (J. Preiss, ed.), Vol. 3, pp. 271–320. Academic Press, New York.
Giaquinta, R. T. (1980b). *In* "Plant Membrane Transport—Current Conceptual Issues" (R. M. Spanswick, W. J. Lucas, and J. Dainty, eds.), pp. 273–284. Elsevier/North-Holland, Amsterdam.
Giaquinta, R. T. (1980c). *Ber. Dtsch. Bot. Ges.* **93,** 187–201.
Giaquinta, R. T., and Quebedeaux, B. (1980). *Plant Physiol.* **65,** 119.
Glass, A. D. M., and Dunlop, J. (1979). *Planta* **145,** 395–397.
Gordon, A. J., Ryle, G. J. A., Powell, C. E., and Mitchell, D. (1980). *J. Exp. Bot.* **31,** 461–473.

Harbron, S., Foyer, C., and Walker, D. (1981). *Arch. Biochem. Biophys.* **212,** 237–246.
Hawker, J. (1967). *Biochem. J.* **102,** 401–406.
Hawker, J. S., Marschner, H., and Downton, W. J. S. (1974). *Aust. J. Plant Physiol.* **1,** 491–501.
Heldt, H. W., Chow, C. J., Maronde, D., Herold, A., Stankovic, Z. S., Walker, D. A., Kraminer, A., Kirk, M. R., and Heber, U. (1977). *Plant Physiol.* **59,** 1146–1155.
Herold, A., and Lewis, D. H. (1977). *New Phytol.* **79,** 1–40.
Herold, A., and Walker, D. A. (1978). *In* "Handbook on Transport" (G. Giebisch, D. C. Tostesen, and H. H. Ussing, eds.), Vol. 2, pp. 411–439. Springer-Verlag, Berlin and New York.
Heyser, W., Evert, R. F., Fritz, E., and Eschrich, W. (1978). *Plant Physiol.* **62,** 491–494.
Ho, L. C. (1976a). *J. Exp. Bot.* **27,** 87–97.
Ho, L. C. (1976b). *Ann. Bot. (London)* [N. S.] **40,** 1153–1162.
Ho, L. C. (1977). *Ann. Appl. Biol.* **87,** 191–200.
Ho, L. C. (1978). *Ann. Bot. (London)* [N. S.] **42,** 155–164.
Hodges, T. K. (1973). *Adv. Agron.* **25,** 163–207.
Hopkinson, J. M. (1964). *J. Exp. Bot.* **15,** 125–137.
Huber, S. (1981a). *Z. Pflanzenphysiol.* **101,** 49–54.
Huber, S. (1981b). *Z. Pflanzenphysiol.* **102,** 443–450.
Hunt, W. F., and Loomis, R. S. (1979). *Ann. Bot. (London)* [N. S.] **44,** 5–17.
Hutchings, V. M. (1978). *Planta* **138,** 237–241.
Kaiser, W. M., Paul, J. S., and Bassham, J. A. (1979). *Z. Pflanzenphysiol.* **94,** 377–385.
Kamanina, M. S., and Anisimov, A. A. (1977). *Sov. Plant Physiol. (Engl. Transl.)* **24,** 628–632.
Koller, H. R., and Thorne, J. H. (1978). *Crop Sci.* **18,** 305–307.
Komor, E. (1977). *Planta* **137,** 119–131.
Kuo, J., O'Brien, T. P., and Canny, M. J. (1974). *Planta* **121,** 97–118.
Kursanov, A. L., and Brovchenko, M. I. (1970). *Can. J. Bot.* **48,** 1243–1250.
Levi, C., and Gibbs, M. (1976). *Plant Physiol.* **57,** 933–935.
Lush, W. M. (1976). *Planta* **130,** 235–244.
Malek, F., and Baker, D. A. (1978). *Plant Sci. Lett.* **11,** 233–239.
Mengel, K., and Haeder, H.-E. (1977). *Plant Physiol.* **59,** 282–284.
Mengel, K., and Viro, M. (1974). *Physiol. Plant.* **30,** 295–300.
Milburn, J. A. (1974). *Planta* **117,** 303–319.
Milford, G. F. J., and Lenton, J. R. (1976). *Ann. Bot. (London)* [N. S.] **40,** 1309–1315.
Neales, T. F., and Incoll, L. D. (1968). *Bot. Rev.* **34,** 107–125.
Okita, T. W., Greenberg, E., Kuhn, D. N., and Preiss, J. (1979). *Plant Physiol.* **64,** 187–192.
Outlaw, W. H., and Fisher, D. B. (1975). *Aust. J. Plant Physiol.* **2,** 435–439.
Pearson, C. J. (1974). *Planta* **119,** 59–70.
Pontis, H. G. (1977). *Int. Rev. Biochem.* **13,** 79–117.
Pontis, H. G. (1978). *Trends Biochem. Sci.* **3,** 137–139.
Preiss, J., and Greenberg, E. (1969). *Biochem. Biophys. Res. Commun.* **36,** 289–295.
Preiss, J., and Kosuge, T. (1970). *Annu. Rev. Plant Physiol.* **21,** 433–466.
Preiss, J., and Levi, C. (1979). *Encycl. Plant Physiol., New Ser.* **6,** 282–312.
Preiss, J., Ghosh, H. P., and Wittkop, J. (1967). *In* "Biochemistry of Chloroplasts" (T. W. Goodwin, ed.), Vol. 2, pp. 131–153. Academic Press, New York.
Robinson, S. P., and Walker, D. A. (1979). *FEBS Lett.* **107,** 295–299.
Ryle, G. J. A., and Powell, C. E. (1976). *J. Exp. Bot.* **27,** 189–199.
Salerno, G. L., and Pontis, H. G. (1978a). *FEBS Lett.* **86,** 263–267.
Salerno, G. L., and Pontis, H. G. (1978b). *Planta* **142,** 41–48.

Santarius, K. A., and Heber, U. (1965). *Biochim. Biophys. Acta* **102,** 39–54.
Servaites, J. C., and Geiger, D. R. (1974). *Plant Physiol.* **54,** 575–578.
Setter, T. L., Brun, W. A., and Brenner, M. L. (1980a). *Plant Physiol.* **65,** 884–887.
Setter, T. L., Brun, W. A., and Brenner, M. L. (1980b). *Plant Physiol.* **65,** 1111–1115.
Silvius, J. E., Kremer, D. F., and Lee, D. R. (1978). *Plant Physiol.* **62,** 54–58.
Silvius, J. E., Chatterton, J. N., and Kremer, D. F. (1979). *Plant Physiol.* **64,** 872–875.
Smith, J. A. C., and Milburn, J. A. (1980a). *Planta* **148,** 28–34.
Smith, J. A. C., and Milburn, J. A. (1980b). *Planta* **148,** 35–41.
Smith, J. A. C., and Milburn, J. A. (1980c). *Planta* **148,** 42–48.
Snyder, F. W., and Carlson, G. E. (1978). *Crop Sci.* **18,** 657–661.
Steup, M., Peavey, D. G., and Gibbs, M. (1976). *Biochem. Biophys. Res. Commun.* **72,** 1554–1561.
Stitt, M., Bulpin, P. V., and Ap Rees, T. (1978). *Biochim. Biophys. Acta* **544,** 200–214.
Thorne, J. H., and Koller, H. R. (1974). *Plant Physiol.* **54,** 201–207.
Tyree, M. T. (1970). *J. Theor. Biol.* **26,** 181–214.
Van Bel, A. J. E., and Van Erven, A. J. (1979). *Planta* **145,** 77–82.
Walker, D. A. (1976). *Encycl. Plant Physiol., New Ser.* **3,** 85–136.
Walker, D. A., and Herold, A. (1977). *Plant Cell Physiol., Spec. Iss.* pp. 295–310.
Wardlaw, I. F., and Marshall, C. (1976). *Aust. J. Plant Physiol.* **3,** 389–400.
Wareing, P. F., Khalifa, M. M., and Treharne, K. J. (1968). *Nature (London)* **220,** 453–457.
Whittingham, C. P., Keys, A. J., and Bird, I. F. (1979). *Encycl. Plant Physiol., New Ser.* **6,** 313–326.
Wolosiuk, R. A., and Pontis, H. G. (1974). *Mol. Cell. Biochem.* **4,** 115–123.
Zimmermann, G., Kelly, G. J., and Latzko, E. (1978). *J. Biol. Chem.* **253,** 5952–5956.
Zimmermann, U., and Steudle, E. (1978). *Adv. Bot. Res.* **6,** 45–117.

11

Predicting Photosynthesis

JOHN D. HESKETH
JOSEPH T. WOOLLEY
DOYLE B. PETERS

ABBREVIATIONS

Chl	Chlorophyll
LAI	Leaf area index
NADPH	Reduced form of nicotinamide adenine dinucleotide phosphate
NAR	Net assimilation rate
PS	Photosystem
RuBP	Ribulose bisphosphate

ABSTRACT

Plant growth models, based in part upon the prediction of the photosynthetic process and its interactions with other processes, are being tested in simple integrated pest management situations. Such models are also being used to predict when to irrigate the fields. This application of photosynthetic research deserves attention. We present a general overview of the kinds of photosynthetic problems encountered.

Carbon budget methods central to photosynthetic prediction are being used to quantify translocation and aspects of nitrogen metabolism (NO_3 reduction, nodule N_2 fixation), as well as respiration requirements for plant maintenance and growth. Such applications of photosynthetic techniques and knowledge in plant physiological research resemble those being attempted in plant growth modeling research at the agro–ecosystem level.

Photosynthesis: Development, Carbon Metabolism, and Plant Productivity, Vol. II

ISBN: 0-12-294302-3

A photosynthetic model should represent a quantitative synthesis from existing literature. We have made many simple (at the elementary textbook level) quantitative statements about the photosynthetic process, which are not accepted by all factions of the photosynthetic research community. Criticisms of the scientific "soundness" of such statements are central to photosynthetic prediction and should be the focus of future reviews and research.

The complexity of the physiology of whole plant systems in field situations may shock photosynthetic specialists. We give examples where simple "textbook" physiology alone fails to describe such systems. Specialists should become more aware of how their system integrates with the whole plant and the associated scientific problems.

I. Introduction

Implicit in the agro–ecosystem approach to scientific research is the immediate application of basic information about photosynthetic processes to the solution of provincial–global problems confronting agriculturalists and mankind (see Gifford and Jenkins, Chapter 12, and Gifford, Chapter 13, this volume). We shall evaluate here the current role of photosynthetic knowledge in predicting agro–ecosystem behavior, including that of crop, forest, desert, mountainous, polar, and urban landscapes, as well as waterscapes. One typical global problem is how increases in atmospheric carbon dioxide (CO_2) from burnt fossil fuel will influence photosynthetic and general ecosystem behavior in the near future, and how such behavior will interact with the earth's climate. Another global problem is the prediction and alleviation of potential famines. At the other end of the scale are economic, energy, irrigation, land use, and pest-control problems that confront us today.

In the past, one scientific objective of related research has been to predict and understand how photosynthetic processes are controlled by climate and physiological status of the plant. One crop- and forest-related objective has been the understanding of factors limiting production of beneficial or economically rewarding plant components. We anticipated that such factors, once defined, could be manipulated genetically or culturally to increase economic returns. Such objectives have easily melded with those associated with predicting plant productivity and behavior, without disrupting ongoing research programs based upon earlier objectives.

In fact, we have been deficient under current agro–ecosystem objectives in generating quantitative summaries of past and ongoing research, which could contribute to the solution of problems under study. Such summaries need continual updating in a format that is understandable by the general audience, with tables of various kinds of data needed to quantify the behavior of whole-plant systems. Old data may need to be reanalyzed and tabulated in a more quantitative format. Such literature reviews do not come easily, but are beginning to appear. In the interest of generating public funding for photosynthetic research, scientists should attempt to meet this demand in their literature reviews and research. (Such an attempt may require surprisingly little effort.)

It has not always been clear how most efficiently to approach the problem of crop yield or ecosystem behavior prediction. The availability of tools like mathematical logic and computers for solving complex mathematical problems has dominated some efforts, to the detriment of analyzing available information and developing productive research programs. One of our conclusions here is that ongoing research is proceeding at a healthy pace, generating needed information and biophysical principles or theory. In some cases, information seems to be appearing before experiments can be planned and executed by those directly involved with the prediction problem. However, such information might be more helpful if researchers in general would participate more in the various aspects of the prediction effort.

Historically, the computer-oriented effort had its roots in systems engineering and its use in space technology and governmental management (defense spending in the United States in the early 1960s). The approach manifested itself in the late 1960s in the United States– International Biological Program, based mainly in the ecological disciplines (Hammond, 1972; Blair, 1977), as well as in a United States multiuniversity (and USDA) integrated pest management research effort, known as the "Huffaker" project (Huffaker and Messenger, 1976; Gutierrez and Wang, 1979). There was some effort within the USDA at the same time to predict plant behavior (McKinion *et al.*, 1975; Huck, 1977), and this effort interacted successfully with the other projects. It quickly became clear that other predictive efforts involving plants (pest management, global impact of increasing CO_2) needed a plant-prediction model before any progress could be made.

We have taken some space to introduce our subject. There is considerable confusion as to what is involved in the prediction effort underway, when it should be obvious that any new information about how plants behave under present global conditions, as well as postulated CO_2 levels and temperatures, is essential for progress.

We review here the information base available for photosynthesis prediction, as well as many of the ways such information may be used to develop a model. Models in use will vary considerably, depending upon their purpose, but one important purpose is the understanding of how whole plant systems behave.

II. Predicting Leaf Behavior

Interest developed in the prediction of canopy photosynthesis from leaf physiological information with the combination of Michaelis– Menten equations (cf. Thornley, 1976, p. 12), describing photosynthetic light response curves for leaves, and the Bouguer–Lambert–Beer equation (cf. Nobel, 1974, p. 20; Thornley, 1976, p. 83), describing light-interception characteristics of the canopy and light levels (irradiance densities, photosynthetic photon flux densities) at leaf surfaces. Here we will concern ourselves with predicting leaf photosynthesis; the effects of environment and physiological status of the plant on leaf expansion and duration, as covered in other sections, are also very relevant to predicting leaf behavior. Since we know so much more about these effects on leaf as compared with plant behavior, such an approach is important to the prediction effort.

A. The Light Reactions

1. ATP AND NADPH SUPPLY IN THE CHLOROPLAST

Nobel (1974) and Thornley (1976), as well as various contributors to Hesketh and Jones (1980) and Barfield and Gerber (1979), provided excellent background discussions related to much of the material discussed in this chapter. Included are rigorous derivations of fundamental equations involved.

For the reduction of a CO_2 molecule, four electrons need to pass

$$2\ H_2O \rightarrow O_2 + 4\ H \tag{1}$$

$$CO_2 + 4\ H \rightarrow [CH_2O] + H_2O \tag{2}$$

through the "Z scheme," requiring, at least, eight photons of light and yielding 2.6–4.0 ATP and 2 NADPH (see Fig. 1 in Chapter 16 by Vermaas and Govindjee, this volume; and Nobel, 1974, for a discussion of the controversies involved, which are inherent in every statement we make here and which are important for assessing how accurately one can predict). These photons are collected by about 400 light-harvesting chlorophyll (Chl) molecules (Thornber *et al.*, 1977) surrounding two reaction centers (in photosystems (PS) I and II) of the Z scheme (for a background, see Govindjee, 1975, 1982). Alberte and co-workers (Alberte *et al.*, 1977; De Jong *et al.*, 1979), using methods for measuring the components of such a photosynthetic unit, reported variations associated with genotype (including the Chl mutants), physiological status of the leaf (greening, stressing, senescing), and the environment in which the cell or leaf grows. The Chl *a/b* ratio reflects the size of the light harvesting shell made up of one part Chl *a* to one part *b*. This Chl *a/b* shell is large in shade-grown leaves and small in sun-grown leaves, with more P700s (PSI reaction center Chl) and chloroplasts per unit leaf area in sun leaves. (Also, consult Chapter 9 by Berry and Downton, this volume.) Some of these factors can change when sun leaves are exposed to shade, or shade leaves are exposed to sun (Bunce *et al.*, 1977), but associated leaf anatomical differences change little after being established.

The Chl *a/b* shell is small in some Chl mutants that grow well under natural conditions, with leaf photosynthetic CO_2 exchange rates comparable to those for normal green leaves in intense (but not weak) light. The Chl *a/b* shell was reduced in size in water-stressed and aging leaves (Alberte *et al.*, 1977). This conceptual model of a photosynthetic unit was based upon measurements of P700 and the distribution of antenna Chl–protein complexes and soluble proteins subjected to gel electrophoresis. This integration of concepts, methodology, and ecological data, which synthesize diverse sets of data to understand how cells, leaves, and plants behave, is an example of the systems approach. A study of exceptions to the underlying concepts that exist, or consideration of new concepts, will lead to new information about the system.

Assuming that chloroplasts occupy about 4% of the leaf volume, with 10^{-3} mole Chl/liter leaf volume, Nobel (1974, p. 201) estimated 25 mmoles Chl/liter chloroplast volume. On a clear day with 1.2×10^{17} photons/cm^2-sec, a 2×10^{-4}-cm-thick chloroplast absorbs about one-quarter of the incident photons in the visible wavelengths (see e.g., Nobel, 1974, p. 241). Thus, 3×10^{16} photons/cm^2 sec are absorbed by a chloroplast containing [(2×10^{-4} cm)(25×10^{-6} mole Chl/cm^3)(6.02×10^{23} molecules/mole)] 3×10^{15} Chl/cm^2 chloroplast area, or 3×10^{16} photons/3×10^{15} Chl molecules/cm^2 sec, or 10 photons/sec for each Chl molecule. For a processing time per photon per reaction center of 0.01 sec and for a 400 molecule photosynthetic unit, 40-(0.01 sec $\times$ 400 molecules $\times$ 10 photons/sec)–Chl would be excited every 0.01 sec, of which two could be processed photosynthetically at the two reaction centers. Much of the preceding material, as paraphrased from Nobel (cf. 1974, pp. 201, 241), is oversimplified, but illustrates how one might begin to develop a model for predicting the conversion of light into ATP and NADPH in the leaf. Amounts of available ADP and NADP are critical. Every assumption is important in evaluating how successful such a model might be, and the various parameters involved vary with position of the chloroplast in the leaf and the photosynthetic unit in the chloroplast, light absorption characteristics within the leaf, as well as other plant variables (such as genotype, mutations, environment, and physiological status of the leaf). Better chloroplast measurements are appearing in the literature (Table I; also see Watanabe, 1973).

B. The Dark Reactions

1. CO_2 TRANSPORT TO THE REACTION SITE

Gaastra (1959) revolutionized leaf photosynthetic CO_2 exchange research with his "electrical resistance" analog for CO_2 flux into the leaf, which was based upon CO_2- and H_2O-flux measurements. Meidner and Mansfield (1968), Nobel (1974, p. 8, Chapter 7, p. 302), and Cooke and Rand (1980) reviewed the theory and research involved; we will not duplicate this important information here. Considerable literature has accumulated on the resistance analog, and it represents a monumental contribution to photosynthetic prediction and research methodology.

Korner *et al.* (1979) listed typical resistances encountered and showed once again (cf. El-Sharkawy and Hesketh, 1965; Ludlow and Wilson, 1971a,b,c) the relative roles of stomatal and residual resistances in controlling gas exchange among genotypes. There is some evidence that the stomatal resistance, in short-term CO_2 studies, adjusts to maintain a constant CO_2 level near the cell surfaces inside the leaf (de Wit *et al.*, 1978; Goudriaan and van Laar, 1978; Wong *et al.*, 1979), but genotypic and other exceptions to such behavior have been reported. In long-term CO_2-enrichment treatments, starch and sugar levels may overwhelm the system in many environments (see Mauney *et al.*, 1978, 1979), with starch grains contributing to chloroplast resistances to CO_2 flux.

Estimates of cell wall, plasmalemma, cytoplasmic, and chloroplast membrane–stroma resistances were reported by Nobel (1974); more research is needed on this part of the analog. Sinclair *et al.* (1977) estimated such resistances (cf. Cooke and Rand, 1980) for leaves with arbitrary anatomical characteristics. Such arbitrarily defined systems are easier to handle mathematically and put limits on important aspects of leaf behavior.

Nobel and co-workers (cf. Nobel, 1977) have related photosynthetic increases to increased photosynthetic cell wall surface areas inside the leaf. They assumed that the cell-wall resistance to CO_2 flux is accordingly reduced. However, increased numbers of associated chloroplasts, photosynthetic units, and RuBP carboxylase molecules contribute to the photosynthetic activity of such surfaces.

TABLE I
Chloroplast Characteristics in *Brassica*[a]

Characteristic[b]	Units	Range of values reported: 9 *Brassica* Lines, 3 species	Range of values reported: 65 F_1,F_2 hybrids
1. Chlorophyll per unit leaf area	10^{-1} g m^{-2}	1.5–3.8	3.0–9.5
2. Chlorophyll per chloroplast	10^{-12} g	1.2–2.3	1.2–3.2
3. Chloroplasts per unit leaf area	10^{11} m^{-2}	1.3–2.1	2–3.7
4. Photosynthetic rate	10^4 μmole O_2/m^2-hr	1–2	2–4.7
5. Surface area per chloroplast	10^{-12} m^2 (μ^{2}m)	21–23.5	
6. Chloroplast area index (chloroplast area/leaf area)	(μm^2/μm^2)	2.7–4.4	
7. Chloroplast volume	10^{-18} m^3 (μm^3)	62–72	
8. Chloroplast short diameter	10^{-6} m (μm)	3–4.5	
9. Chloroplast long diameter	10^{-6} m (μm)	5.5–6.9	
10. Grana per chloroplast		28–32	
11. Area per granum	10^{-12} m^2 (μm^2)	0.38–0.45	
12. Granum area index (granum area/ chloroplast area)		0.63–0.71	
13. Chlorophyll per granum	10^{-14} g	4–7	
14. Chlorophyll per chloroplast area	10^{-2} g m^{-2}	5.3–10.4	
15. Chl *a*/*b* ratio		2.7–3.2	

[a]Such data are useful for quantifying aspects of photosynthetic behavior. (Adapted from Kariya and Tsunoda, 1972, 1973, with permission.)

[b]Correlation coefficients: characteristic 1 versus 2, 0.82, 0.74, 1 versus 3, 0.88, 0.18; 4 versus 1, 0.75, −0.13; 4 versus 2, 0.44, −0.44; 4 versus 3, 0.87, 0.45; 4 versus 6, 0.94, 0.55; and 4 versus 7, 0.95. The first correlation coefficient is for the nine lines; the second is for the hybrids.

2. ACCLIMATION TO WATER STRESS AND CHANGES IN TEMPERATURE

Commercial porometers for measuring stomatal resistance have been available for a decade or more and thus considerable information (cf. Turner, 1974) has accumulated about the magnitude of this resistance. There is, though, some confusion as to the role of stomatal guard cells in restricting photosynthesis and transpiration, as water becomes limiting in the soil–plant–atmosphere system. Such limitations occur when the plant's potential for transpiration, as controlled by exposed leaf area and atmospheric stress (irradiance load), cannot be met by water extraction from the soil by roots. Several options are available to a plant in such a situation including (cf. Passioura, 1976): (1) developing larger negative leaf water potentials to hasten water flux through the soil–plant system by enlarging the soil-to-leaf water potential gradient (assuming a water flux resistance analog), (2) increasing osmotic potentials to maintain turgor and associated growth processes (cell division and expansion), (3) slowing or stopping expansion of new leaves, (4) shedding of older leaves (McMichael *et al.*, 1973), (5) closing stomata in older (Jordan *et al.*, 1975) or all the leaves, and (6) changing leaf orientation to reduce irradiance interception. Options 4 and 5 can immediately slow transpiration; options 3 and 4 can limit subsequent transpiration.

Much of our understanding of these processes until recently has been based upon the behavior of plants growing in pots. Such plants abruptly encounter a very limited water supply when pot irrigation is stopped. In the field, or in properly stress-manipulated potted plants (cf. Brown *et al.*, 1976), young leaves of wilted or partially wilted plants frequently show little or no change in leaf–water potential or stomatal resistance, as compared with those of well-watered plants (Jordan and Ritchie, 1971). In such cases, options 3 to 6 have been shown to be important factors in slowing transpiration. Closed stomata or large stomatal resistances have been measured for field-grown plants in sandy or shallow soils, soils with sandy discontinuities (sand layers), and large well-watered and well-fertilized plants entering drought conditions for the first time. Plants in containers can behave like field-grown plants if the water supply is slowly reduced. Such plants, exposed to several cycles of water stress, can develop larger negative leaf-water potentials before stomata close, than measured in some field-grown plants (Brown *et al.*, 1976; Thomas *et al.*, 1976).

We are faced with a very dynamic system with large possible variations in atmospheric stress, transpiration surfaces, rates of root proliferation into wet soil, rates of water extraction, and soil properties. To complicate the system further, large reductions in photosynthesis have been encountered with no increase in the negative leaf water potential value or stomatal resistance. However, data are becoming available from enough conditions to develop a general model for predicting stomatal and photosynthetic behavior under water-limiting conditions. The laboratory-potted-plant response may occur under a wider range of natural conditions than suggested previously, but the model and its predicted variations in responses under natural conditions will not be as simple as the classical literature might suggest.

When one conducts a water-stress experiment, the ability of a leaf to remain photosynthetically active when "severely wilted" is dramatic. This phenomenon occurs whether potted- or field-grown plants are studied, and its lack of quantification by appropriate measurements causes much of the confusion existing in this area of research.

Also, the water potential at which stomata close in wilted plants is frequently near that maintained in fully active plants under well-watered conditions. Once stomata close, they may not reopen immediately after the plant recovers from stress, resulting in low negative water potentials and closed stomata. In the field, factors such as radiation load and humidity are quite variable. During periods of low evaporative demand, stressed plants frequently have time to recover by growing more roots into wet soil. In such a dynamic system, simple relationships between evaporative demand, stomata, leaf water potential, plant water flux, and time do not always hold. Reported high correlations between photosynthetic processes and water-stress characteristics in potted plants obtained under the so-called (but undefined) "careful conditions" can be misleading, at least until all important ecosystem factors and interactions are defined and understood in a conceptual model, and appropriate microclimatic and leaf characteristics can be measured and predicted in the field. Our goal is a quantitative dynamic model with stochastic elements predicting variations encountered in leaf populations in plant stands. Even in the field, under well-watered conditions, selecting leaves with uniform stomatal resistances for study of other variables becomes an art. The diurnal light environment of each leaf is important. Stomatal differences among treatments can easily be a technical (sampling) artifact. Such is the important role of stomata in controlling photosynthesis, not to mention other undefined green-thumb-related factors.

Stomatal guard cells also acclimate to temperature, with short- and long-term responses that are completely different. The time involved is less than 24 hours (Ludlow and Wilson, 1971b) and severely limits how one interprets a set of experimental results. Both effects have significant ecological implications and cause some of the variations in data collected from field studies. As one might expect, the situation described in this section has resulted in some interesting scientific discussions in the water relations literature and at scientific meetings. Nevertheless, progress has been dramatic in the 20 years since the original Gaastra (1959) paper was published, and resolution of all the preceding difficulties is imminent. Obviously, the laboratory must be taken to the field to obtain a data base for predictive purposes.

3. LIGHT ACCLIMATION AND GENOTYPIC–PHOTOSYNTHETIC DIFFERENCES

Photosynthesis rates can vary considerably depending upon light conditions during leaf expansion (cf. Elmore *et al.*, 1967). Much of the pre-1960 (and considerable current) photosynthetic knowledge is based upon behavior of plant material grown in dimly lit (winter sunlight), badly constructed (poor light penetration) greenhouses. Differences in photosynthetic rate (Boardman, 1977; Patterson, 1980) have been associated with differences in P700 density, chloroplast number, cell number and size, RuBP carboxylase levels, nitrogen (N) or protein levels, leaf thickness, and stomatal density and conductances. Light can vary considerably in nature, and its effects on leaf and canopy behavior need to be accounted for. The expansion rate of leaves in the canopy is an important variable in such an analysis. At some growth stages and under some growing conditions, a new layer of leaves exposed to the sun can appear in a very short time. Insect defoliation and subsequent behavior of shade leaves down in the canopy is an important factor controlling canopy photosynthesis in the agro–ecosystem.

The light-adaptation effect resembles the differences found among late and early maturing genotypes (Murata, 1961; Oritani *et al.*, 1979), and photosynthesis is correlated with many of the same leaf characteristics (cf. Kallis and Tooming, 1974). The supply of N and curtailment of leaf expansion at flowering may result in accumulation of N in young leaves, resulting in high photosynthetic rates (Dornhoff and Shibles, 1970). This balance between photosynthate and N supply and their demand for leaf expansion or vegetative growth, which we suggested caused sun and shade leaf differences, may also cause varietal differences, especially when differences in flowering dates are involved. Kallis and Tooming (1974) endeavored to model some of these effects (see Section III,B,5) in an attempt to determine the optimum leaf to dry weight (or N) ratio and expansion rate for maximum productivity. Since they ignored the floral physiology involved, their model led to confusing results. Nevertheless, the problem is central to the interpretation of results from many photosynthetic genetic studies, particularly in soybeans (Kaplan and Koller, 1977; Hesketh *et al.*, 1981); leaf expansion and flowering characteristics may be overwhelming factors in the control of photosynthesis and most certainly yield.

4. DARK BIOCHEMISTRY

A C_3-leaf in O_2-depleted air (in a suppressed photorespiration condition), utilizing the Calvin–Benson–Bassham pathway for CO_2-fixation, requires 3 ATP and 2 NADPH for each carbon (C) molecule fixed (see Bassham and Buchanan, Chapter 6, this volume). The chemistry is well known, but it is difficult to generate kinetic rate constants for a dynamic biochemical model. Still, such models with approximated rate constants do indicate how various pathways in the leaf are controlled by available ATP and NADPH.

For a leaf at 25°C, 300 ppm CO_2 and 21% O_2, with RuBP carboxylase/oxygenase fixing 4 CO_2's for every O_2 fixed, about 4.6 ATP and 2.9 NADPH are needed for every net CO_2 fixed (0.5 mole CO_2 is photorespired for every O_2 fixed and 2 ATP and 2.5 NADPH are required to restore the C_2 and C_3 acids to the C_5 CO_2-acceptor; cf Ogren, 1978a). (For a discussion of photorespiration, see Chapter 7 by Ogren and Chollet, this volume.) Leaves of C_4 plants require 5 or 6 ATP and 2 NADPH for every CO_2 fixed, assuming no photorespiration (Edwards *et al.*, 1977). This estimate does not include energy requirements for transporting C_4 acids between cells.

From a specific activity of 1 μmole CO_2/nmole-min RuBP carboxylase protein (80 μg CO_2/mg RuBP caroboxylase protein-min, 550,000 g/mole enzyme) and a V_{max} value of 150 mg CO_2/dm^2-hr or 57 μmole/dm^2-min (W. Ogren, personal communication, 1979), we obtain 57 nmole RuBP carboxylase/dm^2 (V_{max}/specific activity) or [(57×10^{-9}) (6.023×10^{23}) =] 3.4×10^{16} molecules RuBP carboxylase/dm^2. Jensen (1977) cited 50–60 nmole/dm^2. For 8 sites/molecule, we get 2.5×10^{17} sites/dm^2.

For 4.5 mg Chl/dm^2 and 893.5 g/mole Chl, we get [$(5 \times 10^{-6}$ mole/dm$^2)(6.023 \times 10^{23})$ =] 30×10^{17} molecules/dm^2. For about 300 molecules Chl/P700, we obtain 10^{16} Z scheme units/dm^2, each with 2 reaction sites. These sites can process 100 photons per second per site (see earlier). For each 2 photons processed per site, 1.33 ADP and 1 $NADP^+$ are phosphorylated and reduced, respectively, or 4 photons processed per photosynthetic unit (2 electrons flowing through the Z scheme). 10^{16} photosynthetic units per dm^2 then can process 2×10^{18} photons/dm^2-sec or 7.2×10^{21} photons/dm^2-hour, phosphorylating 2.4×10^{21} ADP-[$(7.2 \times 10^{21} \times 1.33/4)$]-and reducing 1.8×10^{21} NADP-($7.2 \times 10^{21}/4$)-molecules per dm^2-hr. A V_{max} of 150 mg CO_2/dm^2-hr is equivalent to 2×10^{21} molecules of CO_2/dm^2-hour, requiring 6×10^{21} and 4×10^{21} molecules of ATP and NADPH per dm^2-hour (photorespiration suppressed for the V_{max} determination, 3 ATP and NADPH per CO_2 fixed). The preceding calculations are for a leaf exposed to full sunlight and saturating CO_2 concentrations, and therefore set upper limits on the photosynthetic system.

These preceding examples, involving many assumptions and gross simplifications, indicate the kinds of information needed to quantify and predict the behavior of a photosynthetic system. Other numerous examples are given by Nobel (1974). Further experimental and theoretical refinement of all the earlier material might lead to models that would have an impact on scientific progress in other disciplines.

5. CO_2 CONCENTRATIONS IN THE CELL-WALL LIQUID AND AT THE REACTION SITE

Nobel (1974, p. 329) used the partition coefficient for calculating the liquid-phase CO_2 concentration at the air–liquid interface; this coefficient multiplied by the air CO_2 concentration at the interface gives the liquid CO_2 concentration. The air–liquid coefficient is a function of temperature (1.19, 10°C; 0.88, 20°C; 0.67, 30°C; and 0.53, 40°C) and pH (0.88 at pH = 4; 0.91 at 5; 1.23 at 6; 4.4 at 7; and 35 at 8).

The K_m values for CO_2 (the concentration at one-half the maximum photosynthetic rate) of 10 μM (in the liquid medium outside the chloroplast or cell) have been cited for chloroplast and cell systems ($K_m \cong$ 0.44 mg/liter, with air containing 0.54 mg/liter at 300 ppm CO_2), with pH and temperature effects (Ogren, 1978b). Lower K_m values for CO_2 have been estimated for CO_2 in the cell wall water ($\cong$ 5 μM) and much lower values have been predicted for the reaction site ($\cong 0.1 - 1$ μM) (cf. Tenhunen *et al.*, 1979).

6. NITRATE REDUCTION

Considerable nitrate may be reduced to ammonia or amino acids in the leaf:

$$2\ \text{NADH} + NO_3^- \rightarrow NO_2^- + 2\ \text{NAD} + H_2O \tag{3}$$

$$NO_2^- + 6\ \text{NADH} \rightarrow NH_4^+ + H_2O + 2\ OH^- + 6\ \text{NAD} \tag{4}$$

Nitrate reduction to NO_2 takes place in the cytoplasm (Beevers and Hageman, 1969; Menezel and Kirkby, 1978) with NO_3 reductase as the enzyme catalyst, while nitrite reduction takes place in the chloroplast via NO_2 reductase. Apparently both reactions are light-dependent (Beevers and Hageman, 1969; Nicholas *et al.*, 1976). A PGA–DHAP shuttle between the cytoplasm and chloroplasts (chloroplast membranes) may provide the NADH for NO_3 reduction in the light (Heber, 1974; Schnarrenberger and Fock, 1976; Walker, 1976). About 0.3 g N/m^{2}day as NO_3^- can be reduced in soybeans (60% of the accumulated N/day; Weber, 1966; Hanway and Weber, 1971), with 0.2–0.45 g N/m^2 day in maize (Hanway and Weber, 1971), or 0.015–0.032 mole N/m^2-day, requiring 0.12–0.26 mole NADPH/m^2-day. The maximum carbohydrate gross gain for a day is roughly 40 (soybeans) to 80 (maize) g [CH_2O]m^2-day or

1.3–2.6 mole/m^2-day, requiring the equivalent of 5–10 moles NADPH/ m^2-day (including the ATP equivalent). Thus, about 2–5% of the reductant used on a clear day might be tied up in NO_3 reduction.

The NO_3^- uptake is associated with water uptake and the transpiration stream. The NO_3^- supply then is controlled by the radiation load, with it controlling transpiration. Some NO_3^- may be reduced in roots at the expense of glucose (12 moles NADPH/mole glucose; 8.6 g glucose/g N or 1.8–3.9 g glucose/m^2-day to reduce 0.12–0.26 *M* N/m^2-day, or about 5% of the gross supply of photosynthate).

The flux of NO_3^- and reduced N in the xylem transpiration stream may give a clue as to how much N is reduced in the roots (cf. McClure and Israel, 1979). This assumes no cycling of reduced N in the plant between tops and roots, or flux rates of reduced N in the phloem would have to be accounted for.

7. LIGHT RESPIRATION IN THE LEAF

We have discussed the energetic requirements of photorespiration. Also, there is the importance of CO_2, which is evolved in the mitochondria after some metabolism of the C_2 acid in the peroxisomes (cf. Schnarrenberger and Fock, 1976). The vascular tissue is surrounded by mitochondria, suggesting a high energy cost for translocation. Penning de Vries *et al.* (1974) assumed that 5% of the translocate moved was used in respiration to provide energy (ATP) for phloem loading. For 1 ATP/mole sucrose translocated (and 76 ATPs/mole sucrose converted to CO_2), 1/76 or 1.3% of the sucrose would be used up in translocation. However, the sucrose must cross several membranes getting to the phloem, including those of the chloroplast and cell.

Nitrate reduction may also result in light respiration, if sugars are respired to generate the energy required. As discussed earlier, a shuttle system transferring reductant from the chloroplast to the cytoplasm and light activation of NO_3 reductase would diminish light respiration of new photosynthate to CO_2. As yet, there is no proof for such a shuttle.

C. *Energy Balance of the Leaf*

Gates (1968) described the energy budget of a leaf, or light absorbed as heat—reflected, transmitted, and photosynthesized—into chemical energy. Absorbed heat is convected to the surrounding air or is used up in the heat of vaporization of water. Nobel (1974) and Thornley (1976) discussed the processes involved. At high temperatures (above 36°C) or where advection is involved, the leaf temperature may be lower than air temperature in full sunlight, but the sunlit leaf is usually warmer than the air. Hence, canopy or leaf temperatures can be different from air temperature.

Leaf temperature is an important parameter when estimating H_2O and CO_2-flux resistance in and out of the leaf, particularly the temperature at the cell-wall surfaces inside the leaf where the transpiration water is changing from liquid to vapor state. Growth and associated respiration rates are affected by cell temperature, which may or may not be different from air temperature, depending upon the energy budget of the cell or organ in question.

About 55% of the energy in glucose can be converted to energy in ATP, the rest released as heat. More energy is lost as heat when the ATP is coupled with biosynthesis of molecules.

D. The Leaf Model

There are many equations that one might use to explain available experimental data describing photosynthesis (cf. Thornley, 1976; Tenhunen *et al.*, 1979) such as that at a reaction site and integrating such behavior over the many sites inside the leaf (cf. Cooke and Rand, 1980). Thornley (1976) integrated the Michaelis–Menten equation with the Bouguer–Lambert–Beer equation describing light interception by photosynthetic sites inside the leaf and showed that the resultant equation could be similar in shape to the Michaelis–Menten equation [see p. 98, Fig. 4.2 in Thornley (1976)]. The new equation had parameters for the leaf extinction coefficient and thickness.

The slopes of photosynthetic rates versus light and CO_2 at low levels of one or the other, as well as maximum rates at saturating CO_2 or light, can be used to estimate parameters in many of the various equations used (cf. Thornley, 1976). Photosynthesis versus CO_2 at the cell-wall surfaces inside the leaf would yield parameters independent of stomatal behavior. The CO_2 at the cell-wall surface inside the leaf can be estimated from the equations derived by Gaastra (1959), and the associated methodology. In such a case, one would need a model predicting stomatal behavior. Photosynthesis versus CO_2 in O_2-depleted air would provide another set of parameters, in which case one would need a photorespiration model.

Tenhunen *et al.* (1976) used a theory from Johnson *et al.* (1942) and Koffler *et al.* (1947) to describe the effects of temperature on the maximum photosynthetic rate in saturating CO_2, light, and O_2-depleted air. This theory included an energy of activation for the enzyme catalyzed reaction (RuBP carboxylase) as well as that for the denaturation equilibrium at high temperatures. There were variables in the theory involved, which could be adjusted to give a good fit of the theory to experimental results.

E. Conclusions

Tenhunen *et al.* (1980) critically analyzed several of the leaf models available in the literature. The various processes discussed earlier, combined with conceptual and mathematical theory, as well as experimental determinations of the many parameters involved, represent one way of summarizing what we know about photosynthesis, as well as a way of teaching it to students. One can also determine the importance of the various factors in controlling photosynthesis. For example, temperature affects the stomatal resistance, with short- and long-term effects, the partition coefficients for concentration of CO_2 in liquid versus that in air, the K_m values for O_2 and CO_2 for RuBP carboxylase, and the energy of activation for the RuBP carboxylase-catalyzed reaction, as well as that for carboxylase denaturation at high temperatures. Temperature during growth also affects the size of the light-harvesting Chl *a*/*b* shell of the photosynthetic unit. For a range of ecological and physiological conditions, what is the relative role of each in controlling leaf photosynthetic activity? Simple models of the system might answer such a question. The research and pedagogical potential for such an approach are obvious at this point.

III. Predicting Canopy Behavior

A. Interfacing Leaf and Canopy Light Interception Models

As mentioned earlier, leaf or cellular photosynthesis has been studied much more than canopy photosynthesis, and the possibility of putting all this information to work to predict canopy behavior has intrigued many scientists. At the same time, the approach has potential for understanding how the canopy behaves.

The objective is to describe beam and diffuse irradiance densities incident to various fractions of the canopy leaf area, taking into account leaf angle, height, and orientation, as well as time of day and sky conditions. The age of the leaf also needs to be accounted for; Norman (1975, 1979, 1980), Ross and Nilson (1975), and Thornley (1976) have reviewed various aspects of this problem. The equations are rather complex with the geometry involved. Norman (1980) compared various approaches under carefully defined conditions with leaf-light curves for two C_3 and one C_4 species. Differences in leaf rates of 1:1.7:2.9 predicted canopy differences of 1:1.3:1.8 at 400 W/m^2.

B. Growth Analysis

Growth-analysis data are important for several reasons. Gas-exchange results need to be checked against dry-matter production rates to test their validity. The ultimate test of any photosynthate predictor is how well it can predict actual biomass production rates in a natural ecosystem. Second, we know enough about respiration costs to generate estimates of gross photosynthetic rates from biomass data. Third, leaf area behavior is a very important part of any photosynthetic prediction model, and its study most appropriately comes under the heading of "growth analysis." Fourth, the approach is fundamental to the study of photosynthate partitioning.

Gregory (1917) introduced the growth-analysis equations in a paragraph in an agricultural experiment station report and thereby qualified himself as one of the fathers of the prediction effort. The equations were quickly developed by others (Blackman, 1919). Considerable information has accumulated about crop behavior, using this approach (cf. Leopold and Kriedemann, 1975). Heinicke and Childers (1937) and Thomas (1949) developed carbon budgets for whole plant systems and compared gas exchange with dry matter results. Dry matter data have also played an important role in determining translocation rates (Canny, 1973).

1. DRY MATTER PRODUCTION AND RESPIRATION

In a recent survey of the soybean literature, J. D. Hesketh, J. T. Woolley, and D. B. Peters (unpublished) found 27 sets of dry matter production data with maximum reported values ranging from 10–20 g/m^2-day. One such set was from the large scale Japanese International Biological Productivity effort, which was run for several years at many locations, with a fairly complete set of weather data (Murata, 1975). Some sets were taken for different plant populations (Weber *et al.*, 1966; Buttery, 1969); frequently dry matter was reported for various fractions of the plant, such as roots, nodules, stems, leaves, pods, and beans (Hume and Criswell, 1973). Penning de Vries (1972, 1975) and Penning de Vries *et al.* (1974) developed methods for estimating respiratory costs, based upon relevant biochemical reactions and the ATP requirement, as well as the composition of the biomass synthesized (fats, carbohydrates, and proteins). Considerable plant composition data were available for parts of the soybean plant (J. D. Hesketh, J. T. Wooley, and D. B. Peters, unpublished), and gross photosynthate requirements for plant organs and whole plants could be predicted.

It is also possible to estimate growth and maintenance respiration coefficients experimentally from measurements of dry matter and respiration (Hesketh *et al.*, 1971; Hori, 1977); available estimates (Hirota and Takeda, 1978) tallied well with the predictions from the Penning de Vries' approach. In such an analysis, one needs carbon (C) to dry matter ratios; once again experimental values agreed well with biochemical predictions (Watanabe, 1975).

2. PHOTOSYNTHATE PARTITIONING

The next logical step is to take C-budget information for various plant organs and generate numbers for the amount of the gross photosynthate required for growth of various plant parts. One can generate partition coefficients from the derivative of dry matter accumulation versus time for various plant fractions (Ojima and Fukui, 1966; Uchijima, 1975, for soybeans) or from ^{14}C analysis of various plant parts 24 hr and later after photosynthetic exposure to $^{14}CO_2$ (Hume and Criswell, 1973; Silvius *et al.*, 1977, for soybeans).

By manipulating the supply of photosynthate and N by shading, CO_2-enrichment, light enrichment, applied N (or, in the case of soybeans, utilizing non-nodulated genetic isolines), and applied organic matter high in cellulose and low in N, one can determine for plants in a specific temperature regime the C and N requirements of individual organs, the rate of production of such organs, and the priority with which such organs are allowed to develop or abscise. From such information, one can develop a dynamic model for partitioning photosynthate, depending upon the supply and demand for both photosynthate and nitrogen. Interactions between CO_2 enrichment, temperature, species, nitrogen supply, light, and vegetative dry matter or leaf area production and effects of these parameters on the partitioning of photosynthate have indeed been studied in detail (MacDowell, 1972a,b; Imai and Murata, 1976, 1977, 1978, 1979a,b). Leaf expansion rates and area per plant sometimes are important; in many cases, photosynthesis (NAR) is greatly increased by CO_2 enrichment of the surrounding atmosphere.

When photosynthesis is enhanced while plants are setting and developing fruit, the number of matured fruit typically increase, with not as great an increase in average fruit weight (Table II) (cf. Downs and Hellmers, 1975, pp. 86–90; Allen, 1979, for other reported yield responses from CO_2 enrichment and interactions involved). Such responses depend upon factors listed in the previous paragraph, as well as morphogenic options available to the plant. Detailed studies of the dy-

TABLE II
Effect of CO_2 Enrichment (or Depletion) on Dry Matter Production and Partitioning among Crops

Crop	Temperature (°C)	Light	CO_2 level (treatment/control) (ppm/ppm)	Growth stage treated	Ratio of treatment to that of controls: Nodules	Roots	Stems	Leaves	Fruit	Total	Number of fruit	Weight of fruit	Source
Soybean	28/23[a]	48 (klux)	1000/300	Vegetative						1.5			Imai and Murata (1979b)
		27								1.4			
	23/20	48								1.3			
		27								1.2			
	Less than 5°C above ambient (Minnesota)		1200/ambient	Continuous			2.1	1.7	1.35	1.5	1.75	0.8	Hardman and Brun (1971)
				Preflower			1.0	1.0	1.0	0.9	0.9	1.0	
				Flowering			1.7	1.5	1.05	1.2	1.5	0.7	
				Postflower			1.4	1.1	2.0	1.2	1.2	1.0	
	Open top		1000/ambient	Continuous	3	1.7	1.4	1.4	2.0	1.7	1.5	1.2	Hardy and Havelka (1976)
Maize	28/23	48 (klux)	1000/300	Vegetative						1.2			Imai and Murata (1979b)
		27								1.1			
	23/20	48								1.1			
		27								1.1			
Eight cool climate C_3 spp.	30	50 (klux)	1000/300–2500/300	Vegetative						2.5–3.1			Akita and Tanaka (1973)[c]
Five warm C_3 climate spp.										1.8–1.8 (2.0–2.1)			
Twelve C_4 spp.										0.9–1.1 (1.1–1.1)			
Cotton	35/21[b]	April–	630/330	Postflowering							2.0	1.0	Hesketh *et al.* (1972)

		May			1.7		1.8	0.9	Guinn *et al.* (1976)
		Arizona				2.0	1.4	1.1	Mauney *et al.* (1978)
Wheat	Less than 3°C above ambient	May–July (Minnesota)	600/ambient	Vegetative	1.0		1.0	1.0	Krenzer and Moss (1975)
				Floral bud	1.2		1.2	1.0	
				Growth					
				After anthesis	1.3		1.0	1.3	
	21/16	400 (1 year/day)	140/290	Continuous	0.56	0.57	0.57	1.0	Gifford (1977)
			490/290		1.4	1.3	1.4	1.1	
	15	290 (1 year/day)	750/ ambient	12–39 days from seeding	1.02		1.06	0.97	Fischer and M.-Aguilar (1976)
	15	320		40–67	1.06		1.1	0.95	
	16.5	360		68–88	1.13		1.2	0.94	
	16	520		89–121	1.0		0.96	1.05	
				Postflowering					
Rice	Less than 4°C above ambient		900/ambient	Vegetative	1.4		1.2	1.1	Cock and Yoshida (1973)
				Postflowering	1.2		1.1	1.1	
	27	480 (1 year/day)	900/ambient	33 days before flowering	1.3	1.0	1.2	1.1	Yoshida (1973)
				30 days after flowering	1.1	1.2	1.0	1.1	
Tomato			650/350				1.1		Madsen (1974)
			1000/350				1.4		
			1500/350				1.4		
			2200/350				1.2		
			3200/350				0.6		

[a]Light period/dark period.
[b]Maximum/minimum.
[c]See Imai and Murata (1976, 1977, 1978, 1979a,b) for other data.

namics of plant morphogenesis are needed, and computer technology will aid considerably in synthesizing a whole plant C and N budget from budgets of individual organs. Such exercises are instructive for understanding how plants behave in different environments at various stages of growth or after different prehistories. The approach is also central to studying and understanding how sink demand (the number of growing plant organs requiring photosynthate, or more specifically the number of dividing cells requiring photosynthate) can affect the ability of the whole plant to fix CO_2.

To complete such an analysis, we need to know how much photosynthate can be stored or carried over from one 24-hr period to the next and how fast it can be translocated in the plant. We also need information on how plants utilize their reserves when photosynthate is limiting, particularly during the fruit-development stages.

3. TRANSLOCATION

Canny (1973) reviewed research involving dry matter changes in leaves and fruits for the determination of translocation rates. In this method, the cross-sectional area of the phloem system must be estimated, and respiration in fruits must be accounted for. Lush (1976) compared results obtained from such methods with those with ^{14}C methods. Canny (1971, 1973) presented an analysis of predicting translocation; he recommended a return to the earlier quantitative approach involving actual measurements of translocate mass, as Lush (1976) and others have done.

4. N_2 FIXATION BY NODULES

A special case of the partitioning problem is the photosynthate requirement for N_2 fixation by nodules. The N supply from nodules and the soil plays an important role in determining RuBP carboxylase levels and leaf expansion rates. The NO_3, as pointed out earlier, can be reduced to NH_3, and, therefore, becomes part of the photosynthesis prediction problem.

From detailed dry matter, N and gas exchange (including H_2 production by nodules) analyses of nodule–root systems (Schubert and Evans, 1976; Atkins *et al.*, 1978; Mahon, 1979), it is possible to estimate, in a particular case, that 7.3 g glucose/g N fixed is required by the nodule for respiration and 8.3 g glucose/g N for the C required, or 1.56 g glucose/6.25 g amino acids synthesized. The root–nodule system, because of its relative large root biomass, requires about 15 g glucose/g N fixed for respiration, the difference between nodules and the nodule–root system being photosynthate requirements for growth and maintenance of the root biomass; the latter has other physiological functions to perform. It also seems, as a very rough estimate, that 40-g nodules (dry weight) are needed to fix 1 g N/day. This approach to the cost of N fixation by nodules is at the forefront of N-fixation research, and better estimates of the various parameters involved greatly improve our ability to predict gross photosynthesis rates and C and N budgets in the plant stand.

5. LEAF EXPANSION AND BEHAVIOR

The role of leaf area in controlling canopy photosynthesis was clearly indicated in Gregory's (1917) derivation of the growth analysis equations. It has frequently been suggested that leaf physiology should receive more emphasis in studies of whole-plant photosynthesis, which are supposed to be related to the solution of agricultural problems (Milthorpe, 1956). Hori (1977) listed leaf morphogenesis and partitioning of photosynthates as two major factors limiting successful simulations of plant growth. Maksymowych (1973), Williams (1975), and Leffler (1980) reviewed leaf-behavior research.

Leaves first appear as a few meristematic cells on the edge of a hemispherical dome at nodes on the plant where a leaf petiole meets a stem or branch, or at mainstem or branch apices. Thornley (1976) and Jones and Hesketh (1980) discussed a model for meristematic cell division and growth. We need to know the fraction of recently divided cells that are capable of further division, as well as cell division and expansion rates. There is limited information from which to quantify such a model (for exception, see Verbelen and de Greef, 1979), but, we can easily generate a logarithmic curve for leaf expansion from such a model.

We can discover the potential for a species to initiate and develop leaves by growing plants under conditions where photosynthesis and N supply are greatly enhanced (such as in enriched CO_2 atmospheres) (Hardman and Brun, 1971; Hesketh *et al.*, 1973; Hardy and Havelka, 1976) or by pruning apical meristems and watching axillary meristematic growth (Mauney, 1968). Most leaf initials remain quiescent after limited development, depending upon the supply of photosynthate and nitrogenous compounds. Photosynthate and N levels interact somehow with growth regulators. One theory is that vascular tissue somehow controls the leaf initial and that a translocation resistance to sucrose flux must be overcome before additional cell divisions can recommence. Such a leaf must supply some of its own photosynthate to overcome such a resistance. These aspects of leaf development are not well defined or quantified.

Leaves expand faster, to ever larger final areas, as one progresses up the soybean main stem, node-by-node (Kumura and Naniwa, 1965; Kumura, 1969; Hofstra *et al.*, 1977). Floral initiation, either by generation of hormones or by competition, reverses this correlation. In soybeans, floral buds quickly become much larger than nearby leaf primordia at the microscopic level, suggesting competition for nutrients. The prevention of flowering will reveal the potential size of leaves at various nodes on the plant. However, in some genotypes, enhanced vegetative branching seems to result in smaller main-stem leaves.

Temperature and N supply (Hesketh *et al.*, 1973; Miyasaka *et al.*, 1975; Tanaka, 1975; Hofstra *et al.*, 1977) affect the rate of appearance of new leaf initials and subsequent expansion. In soybeans, 130 degree days (7°C base, preferably soil degree days) are needed from planting to emergence (Major *et al.*, 1975). From planting to the unifoliolate, 140 to 170 degree days are needed, and 70–90 degree days are needed from the unifoliolate to the first trifoliolate stage (Hesketh *et al.*, 1973; Hofstra *et al.*, 1977). Three sets of soybean data for plants growing in a range of temperatures (Hesketh *et al.*, 1973; Shibles *et al.*, 1975; Sato, 1976) suggested that 50 degree days are needed for the appearance of a new leaf with roughly 10-cm^2 area; field results suggest 53 days (Sivakumar *et al.*, 1977; J. D. Hesketh and D. Peters, unpublished). For expansion from 10 cm^2 to full size, 200 degree days are required. Leaves endure about 400–850 degree days in the field (Table III). Maximum photosynthetic rates at full expansion can last about 300 degree days before the rates decline (Woodward and Rawson, 1976). Such reductions in leaf photosynthesis with age have been associated with decreases in N, P, K, and Chl (Murata, 1961, 1975; Boote *et al.*, 1978) and increases in stomatal resistance (Ludlow and Wilson, 1971c). Hackett and Rawson (1974) analyzed the significance of leaf aging in a C-budget analysis of a growing tobacco plant. We can construct a leaf area model from such information, with additional parameters needed for a model for branch development (cf. Kumura, 1969; Hofstra *et al.*, 1977); also, we can directly measure the leaf area (LAI) versus degree days for plants growing under different conditions. The degree-day concept is fundamental to predicting growth (Jones and Hesketh, 1980), but interactions with other variables, such as stress, are not understood or have not been quantified.

Elmore (1980), in a survey of some 82 studies of genotypic differences in leaf photosynthesis, found eight reports of good correlations between photosynthesis and leaf N content and seven reports of good correlations between photosynthesis and specific leaf weight (g/cm^2). Researchers have selected wheat cultivars that have rapid leaf expansion rates and large leaves containing less dry matter and N per unit leaf area (Evans and Dunstone, 1970; Khan and Tsunoda, 1970). The same cor-

TABLE III
Days or Degree Days from Leaf Appearance to Yellowing and Abscission for Soybeans

	76 cm row width		51[b] cm row width;	25[b] cm row width;	Kumura and Naniwa (1965);		Jones and Hesketh (1980); day/night temperature (°C)	
	27 plants per m	3.5 plants per m	27 plants per m	27 plants per m	potted plants	Kumura (1969)	32/26	26/20
				Days duration				
Cotyledons	14	14	14					
Unifoliolates	29	29	29			29		
2[a]	28	33	28	25	30	24		43
4	37	36	37	44	37	21	37	46
6	53	49	52	55	50	17	47	61
8	51	56	56	58	56	38	53	66
10	49	53	56	51	49	48	56	68
12	47	46	48	35		58		
14	40	43						
				Degree days duration ($\times 10^{-1}$)				
2	60[c]	60	47	40				
4	55	57	42	42			79	
6	82	65	73	73			97	75
8	78	84	78	84			111	99
10	82	82	86	78			138	102
12	55	70	51	36				
14		56						

[a]Trifoliolate node on the main stem counting up from the unifoliolates.
[b]Field experiment, Urbana South Farm, University of Illinois at Champaign-Urbana, 1978.
[c]600 degree days, 7°C base. Degree days equals sum of [(maximum 24-hr temperature plus the minimum 24-hr temperature, divided by 2) minus 7°C] for each accumulated 24-hr period.

relations have been reported for rice cultivars with photosynthesis negatively correlated with area per leaf (−0.87) and positively correlated with N (0.75), soluble protein (0.85), and Chl (0.64) (Oritani *et al.*, 1979). Such considerations would affect C and N needs for expanding leaves. Presumably, some combination of expansion and photosynthetic rate is optimal for growth and yield. Kallis and Tooming (1974) tested such a hypothesis in a model, but their model lacked a flowering subsystem and associated inhibition of leaf expansion, with excess leaf production in some examples.

The leaf area:dry weight or N ratio does change with photosynthate supply, so that leaf expansion rates in some species tend to remain constant over a wide range of supply values. In such cases, the photosynthetic rate is affected (Patterson, 1980) in the manner described earlier. Leaf expansion is also controlled by leaf turgor.

Hori (1977) presented an interesting discussion of the important role of the area:weight ratio in developing simulation models, after developing a sunflower model. We recommend Hori and Udagaway (1971) and Hori (1977) for a good "state-of-the-art" example of model development for predicting photosynthesis.

Despite Hori's misgivings, an integrated modeling-experimental effort might efficiently lead to an understanding of the physiology of variations in area:weight ratios and photosynthate partitioning (see Sections I and V).

C. Gas Exchange of Plant Stands in Controlled Environments

Heinicke and Childers (1937) and Thomas and Hill (see Thomas, 1949) measured the gas exchange of apple trees, alfalfa, and other crop and horticultural stands of plants growing under natural conditions in the 1930s. Both groups developed C budgets for their test plants from simultaneous growth analyses and respiration measurements. Their sets of data are still valid for generating logic and parameters for predicting photosynthesis for the various test species used.

Gas-exchange techniques, as well as methods for predicting respiration, have improved since the early work of Heinicke and Childers. This approach can be used to predict dry matter production, but the latter technology has not improved much and sampling errors result in data unfit for rigorous model testing. Tanaka (1972), Hori (1977), de Wit *et al.* (1978), Takeda (1978) and others have used canopy gas exchange data to test models built from information about photosynthetic and respiratory behavior of plant parts.

Both $^{14}CO_2$ exposure and subsequent sampling of leaves from defined positions in the canopy revealed activities of various photosynthetic organs within the canopy (Angus *et al.*, 1972; Puckridge, 1972). Such information has great potential in rigorous testing of leaf-light interception models.

Hesketh (1980) summarized some of the information available about photosynthesis of plant stands. Some crop stands characteristically have different light-response curves at different times of the day (Sakamoto and Shaw, 1967; Kumura, 1968; Puckridge and Ratkowsky, 1971). Such summaries need to be updated frequently because of the role of these kinds of data in developing crop growth models (cf. Kanemasu and Hiebsch, 1975; McKinion *et al.*, 1975; Brown and Trlica, 1977; Hodges and Kanemasu, 1977). Expressing canopy photosynthetic CO_2-exchange versus intercepted light removes some of the seasonal variability in responses to light caused by varying leaf area per unit ground area (Baker and Musgrave, 1964). Such an approach requires prediction of growth of the canopy leaf area and its light-intercepting characteristics. Photosynthesis response to light is greater as the diffuse fraction of incident light increases (Tanaka, 1972).

The plant chamber is usually placed over a crop for several days or more, allowing for modification (CO_2, temperature) of the canopy environment. Peters *et al.* (1974) developed mobile chambers that enclosed a plot for about 3 min, during which they monitored changes in CO_2 and H_2O. This system has been used on eight plots with a run every 0.5 hr—day and night. Canopy responses to light with different varieties, plant spacings, fertility levels, and plant water stress levels have been measured using this approach, with three replications (cf. Larson *et al.*, 1981). Considerable information has accumulated for maize and soybeans, using six mobile chambers. The photosynthetic, respiratory, and transpiration data will be used to construct better maize and soybean models, with consideration of the genotype in use. In addition to these objectives, Monsanto Company researchers at St. Louis have used the same system for screening effects of chemical applications on photosynthesis (see Christy, Chapter 14, this volume).

D. Micrometeorological Methods

Legg and Monteith (1975), Kanemasu *et al.* (1979), and Norman and Hesketh (1980) have reviewed the status of micrometeorological research and theory for predicting photosynthesis. The methodology was first developed for studying and predicting evapotranspiration, an example of an ecosystem-level effort supported by considerable resources for many years, long before the agro–ecosystem approach was conceived. Water is a limiting resource in many ecosystems and must be managed carefully when supplied by irrigation. Empirical models have been in use for some time to predict when to irrigate (Jensen and Haise, 1963). The evapotranspiration experience may be useful to those involved in the present effort to predict crop yields and results from various management practices.

Inoue *et al.* (1958) and Matsushima *et al.* (1958) first adapted the water-vapor-flux theory for predicting CO_2 flux into the crop. Biscoe *et al.* (1975) used existing methodology to develop a C budget for a barley stand, and successfully matched budgets generated from respiration–dry weight accumulation and leaf–light interception models. The variation they obtained when plotting photosynthesis versus light intensity indicated some of the problems with the technique. However, since they made such measurements in an undisturbed canopy over a short period, their value for predicting seasonal behavior is limited. Improvements in techniques for measuring water vapor and CO_2 exchange, based upon laser technology and new developments in theory, offer promise for the method.

Inoue (1974, 1977) developed a three-dimensional model for predicting CO_2 escape into the atmosphere from artificial releases into the crop. He predicted that little of the released CO_2 was retained in the crop from enhanced photosynthesis. Most results suggest rapid mixing of CO_2 from the atmosphere into the crop under natural conditions, even at low wind speeds.

Independent measurements of evapotranspiration (weighing lysimeters), release of tracer gases and measurements of the flux of thoron from the soil (cf. Legg and Monteith, 1975; Inoue, 1977) have been used to test various methods. The theory can be easily adapted for predicting chemical losses (insecticides, herbicides, fungicides) (Parmele *et al.*, 1972) and spore (Legg and Powell, 1979) or pollen dispersal within the ecosystem.

IV. Biophysical Principles and General Philosophy

The electrical-resistance analog is used over and over for describing flux processes in the soil–plant–atmosphere system. Enzyme kinetic theory is also frequently used to describe chemical reaction rates or transport rates across membranes. The Bouguer–Lambert–Beer equation for light absorption is frequently used as an analog for light interception by plant stands. A knowledge of the theory and assumptions involved in these general processes will form a good base for understanding the biophysical principles involved in predicting photosynthesis. The methodology for estimating parameters and manipulations necessary to make these theories useful is another matter.

Bookkeeping methods are used frequently to generate water, C, energy, and N budgets for the system, as well as for keeping track of the age of various plant parts. Degree-day concepts are used to account for temperature effects on the aging process. Nobel (1974) discussed much of the preceding material; if his treatment of this subject matter is too complicated for some biologists, then simpler presentations should be prepared. As we strive to become a generalist to study whole plant or ecosystem behavior, we should not be embarrassed if we do not understand the complexity of each specialization. If the agro–ecosystem or whole plant approach is to succeed, we must learn to communicate relevant information with specialists in different disciplines. We consider it important that granting agencies and administrators realize that funds provided for scientific writing at the general (but quantitative) level (or textbook level) are as necessary as funds for new research.

V. Conclusions

Much of the material presented here has been discussed in greater detail elsewhere (cf. Hesketh and Jones, 1980). Nevertheless, general overviews of the state-of-the-art are necessary for progress.

There have been two stages of progress in model development. The first one involved model construction, based upon leaf behavior (cf. Norman, 1980) with some effort to predict results using other methodology, including micrometeorology (cf. Norman and Hesketh, 1980) and growth analysis. In a second stage of progress, attempts have been made to improve such models with additional leaf results, including leaf-aging effects (Angus and Wilson, 1976). This second stage also involved use of several methods and models in research projects at one site and time (Hori and Udagawa, 1971; Biscoe *et al.*, 1975; Kanemasu and Hiebsch, 1975; Hodges and Kanemasu, 1977; de Wit *et al.*, 1978; Takeda, 1978; Hodges *et al.*, 1979). There had been considerable controversy while this effort evolved concerning the significance of some of the approaches used. The main concern was the frequent lack of associated experimentation to test aspects of the model; this no longer is a problem.

The potential for the application of such models to improve plant productivity justifies the effort. Some of the controversy previously discussed was, and is, associated with lack of sufficient information about crop-canopy physiology to predict behavior for a wide range of conditions. The technology and resources seem to be now at hand for the necessary exploratory research to overcome this last obstacle to progress.

REFERENCES

Akita, S., and Tanaka, I. (1973). *Proc. Crop Sci. Soc. Jpn.* **42,** 288–295.

Alberte, R. S., Thornber, J. P., and Fiscus, E. L. (1977). *Plant Physiol.* **59,** 351–353.

Allen, L. H. (1979). *In* "Modification of the Aerial Environment of Crops" (B. J. Barfield and J. F. Gerber, eds.), ASAE Monogr., pp. 500–519. Am. Soc. Agric. Eng., St. Joseph, Michigan.

Angus, J. F., and Wilson, J. H. (1976). *Photosynthetica* **4,** 367–377.

Angus, J. F., Jones, R., and Wilson, J. H. (1972). *Aust. J. Agric. Res.* **23,** 945.

Atkins, C. A., Herridge, D. F., and Pate, J. S. (1978). *In* "Isotopes in Biological Dinitrogen Fixation," pp. 211–242. IAEA, Vienna.

Baker, D. N., and Musgrave, R. B. (1964). *Crop Sci.* **4,** 127–131.

Barfield, B. J., and Gerber, J. F., eds. (1979). "Modification of the Aerial Environment of Crops." ASAE Monogr., Am. Soc. Agric. Eng., St. Joseph, Michigan.

Beevers, L., and Hageman, R. H. (1969). *Annu. Rev. Plant Physiol.* **20,** 495–522.

Biscoe, P. V., Scott, R. K., and Monteith, J. L. (1975). *J. Appl. Ecol.* **12,** 269–291.

Blackman, V. H. (1919). *Ann. Bot. (London)* **33,** 353–360.

Blair, W. F. (1977). "Big Biology: The US/IBP." Dowden, Hutchinson & Ross, Stroudsburg, Pennsylvania.

Boardman, N. K. (1977). *Annu. Rev. Plant Physiol.* **28,** 355–377.

Boote, K. J., Gallaher, R. N., Robertson, W. K., Hinson, K., and Hammond, L. C. (1978). *Agron. J.* **70,** 787–791.

Brown, K. W., Jordan, W. R., and Thomas, J. C. (1976). *Physiol. Plant.* **37,** 1–5.

Brown, L. F., and Trlica, M. J. (1977). *J. Appl. Ecol.* **14,** 215–224.

Bunce, J. A., Patterson, D. T., Peet, M., and Alberte, R. S. (1977). *Plant Physiol.* **60,** 255–258.

Buttery, B. R. (1969). *Can. J. Plant Sci.* **49,** 675–684.

Canny, M. J. (1971). *Annu. Rev. Plant Physiol.* **22,** 237–260.

Canny, M. J. (1973). "Phloem Translocation." Cambridge Univ. Press, London and New York.

Cock, J. H., and Yoshida, S. (1973). *Soil Sci. Plant Nutr.* **19,** 229–234.

Cooke, J. R., and Rand, R. H. (1980). *In* "Predicting Photosynthesis for Ecosystem Models" (J. D. Hesketh and J. W. Jones, eds.), Vol. I, pp. 93–121. CRC Press, Boca Raton, Florida.

De Jong, D. W., Alberte, R. S., Hesketh, J. D., and Woodlief, W. G. (1979). *Tob. Sci.* **23,** 78–82; *Tob. Int.* **181,** 178–183.

de Wit, C. T., Goudriaan, J., van Laar, H. H., Penning de Vries, F. W. T., Rabbinge, R., van Keulen, H., Louwerse, W., Sibma, L., and de Jonge, C. (1978). "Simulation of Assimilation, Respiration and Transpiration of Crops." Wiley (Halsted), New York.

Dornhoff, G. M., and Shibles, R. M. (1970). *Crop Sci.* **10,** 42–45.

Downs, R. J., and Hellmers, H. (1975). "Environment and the Experimental Control of Plant Growth." Academic Press, New York.

Edwards, G. E., Huber, S. C., Ku, S. B., Rathnam, C. K. M., Gutierrez, M., and Mayne, B. C. (1976). *In* "CO_2 Metabolism and Plant Productivity" (R. H. Burris and C. C. Black, eds.), pp. 83–112. University Park Press, Baltimore, Maryland.

Elmore, C. D. (1980). *In* "Predicting Photosynthesis for Ecosystem Models" (J. D. Hesketh and J. W. Jones, eds.), Vol. II, pp. 155–168. CRC Press, Boca Raton, Florida.
Elmore, C. D., Hesketh, J. D., and Muramoto, H. (1967). *J. Ariz. Acad. Sci.* **4,** 215–219.
El Sharkawy, M., and Hesketh, J. (1965). *Crop Sci.* **5,** 517–521.
Evans, L. T., and Dunstone, R. L. (1970). *Aust. J. Biol. Sci.* **23,** 725–741.
Fischer, R. A., and M.-Aguilar, I. (1976). *Agron. J.* **68,** 749–752.
Gaastra, P. (1959). *Meded. Landbouwhogesch. Wageningen* **59,** 1–68.
Gates, D. M. (1968). *Annu. Rev. Plant Physiol.* **19,** 211–238.
Gifford, R. M. (1977). *Aust. J. Plant Physiol.* **4,** 99–110.
Goudriaan, J., and van Laar, H. H. (1978). *Photosynthetica* **12,** 241–249.
Govindjee, ed. (1975). "Bioenergetics of Photosynthesis." Academic Press, New York.
Govindjee, ed. (1982). "Photosynthesis: Conversion of Energy in Plants and Bacteria" Vol. I. Academic Press, New York.
Gregory, F. G. (1917). *3rd Annu. Rep., Exp. Stn., Cheshunt,* p. 19.
Guinn, G., Hesketh, J. D., Fry, D. E., Mauney, J. R., and Radin, J. W. (1976). *Proc. Beltwide Cotton Prod. Res. Conf., 1976,* pp. 60–61.
Gutierrez, A. P., and Wang, Y. (1979). *In* "Pest Management" (G. A. Norton and C. S. Holling, eds.), Vol. 4, pp. 255–280. Pergamon, Oxford.
Hackett, C., and Rawson, H. M. (1974). *Bull.—R. Soc. N. Z.* **12,** 269–276.
Hammond, L. L. (1972). *Science* **175,** 46–48.
Hanway, J. J., and Weber, C. R. (1971). *Agron. J.* **63,** 406–408.
Hardman, L. L., and Brun, W. A. (1971). *Crop Sci.* **11,** 886–888.
Hardy, R. W. F., and Havelka, U. D. (1976). *In* "Symbiotic Nitrogen Fixation in Plants" (P. S. Nutman, ed.), IBP Vol. 7, pp. 421–439. Cambridge Univ. Press, London and New York.
Heber, U. (1974). *Annu. Rev. Plant Physiol.* **25,** 393–421.
Heinicke, A. J., and Childers, N. F. (1937). *Mem.—N.Y., Agric. Exp. Stn. (Ithaca)* **201,** 3–52.
Hesketh, J. D. (1980). *In* "Predicting Photosynthesis for Ecosystem Models" (J. D. Hesketh and J. W. Jones, eds.), Vol. I, pp. 37–50. CRC Press, Boca Raton, Florida.
Hesketh, J. D., and Jones, J. W., eds. (1980). "Predicting Photosynthesis for Ecosystem Models" Vols. I and II. CRC Press, Boca Raton, Florida.
Hesketh, J. D., Baker, D. N., and Duncan, W. G. (1971). *Crop Sci.* **11,** 394–398.
Hesketh, J. D., Fry, K. O., Guinn, G., and Mauney, J. R. (1972). *In* "Modeling the Growth of Trees" (C. E. Murphy, J. D. Hesketh, and B. R. Strain, eds.), EDFB-IBP-72-11, pp. 123–217. Oak Ridge Natl. Lab., Oak Ridge, Tennessee.
Hesketh, J. D., Myhre, D. L., and Willey, C. R. (1973). *Crop Sci.* **13,** 250–254.
Hesketh, J. D., Ogren, W. L., Hageman, E., and Peters, D. B. (1981). *Photosynth. Res.* **2,** 21–29.
Hirota, O., and Takeda, T. (1978). *Jpn. J. Crop Sci.* **47,** 336–343.
Hodges, T., and Kanemasu, E. T. (1977). *Agron. J.* **69,** 974–978.
Hodges, T., Kanemasu, E. T., and Teare, I. D. (1979). *Can. J. Pl. Sci.* **59,** 803–818.
Hofstra, G., Hesketh, J. D., and Myhre, D. L. (1977). *Can. J. Plant Sci.* **57,** 165–175.
Hori, T. (1977). *Bull. Natl. Inst. Agric. Sci., Japan, Ser. A: Physics, Statistics* **24,** 45–70.
Hori, T., and Udagawa, T. (1971). *Bull. Natl. Inst. Agric. Sci., Japan, Ser. A: Physics, Statistics* **18,** 1–56.
Huck, M. (1977). *Range Sci. Dep. Sci. Ser. (Colo. State Univ.)* **26,** 215–226.
Huffaker, C., and Messenger, P. S., eds. (1976). "Theory and Practice of Biological Control." Academic Press, New York.
Hume, D. J., and Criswell, J. G. (1973). *Crop Sci.* **13,** 519–524.
Imai, K., and Murata, Y. (1976). *Proc. Crop Sci. Soc. Jpn.* **45,** 598–606.

Imai, K., and Murata, Y. (1977). *Jpn. J. Crop Sci.* **46,** 291–297.

Imai, K., and Murata, Y. (1978). *Jpn. J. Crop Sci.* **47,** 118–123.

Imai, K., and Murata, Y. (1979a). *Jpn. J. Crop Sci.* **48,** 58–65.

Imai, K., and Murata, Y. (1979b). *Jpn. J. Crop Sci.* **48,** 409–417.

Inoue, E., Tani, N., Imai, K., and Isobe, S. (1958). *J. Agric. Meteorol. (Tokyo)* **14,** 45–53.

Inoue, K. (1974). *Bull. Natl. Inst. Agric. Sci., Japan, Ser. A: Physics, Statistics* **21,** 1–26.

Inoue, K. (1977). *Bull. Natl. Inst. Agric. Sci., Japan, Ser. A: Physics, Statistics* **24,** 19–44.

Jensen, M. E., and Haise, H. R. (1963). *Am. Soc. Civ. Eng., Drainage Div. J.* **89,** (IR 4), 15–41.

Jensen, R. G. (1977). *Annu. Rev. Plant Physiol.* **28,** 379–400.

Johnson, F., Eyring, H., and Williams, R. (1942). *J. Cell. Comp. Physiol.* **20,** 247–268.

Jones, J. W., and Hesketh, J. D. (1980). *In* "Predicting Photosynthesis for Ecosystem Models" (J. D. Hesketh and J. W. Jones, eds.), Vol. II, pp. 85–122. CRC Press, Boca Raton, Florida.

Jordan, W. R., and Ritchie, J. T. (1971). *Plant Physiol.* **48,** 783–788.

Jordan, W. R., Brown, K. W., and Thomas, J. C. (1975). *Plant Physiol.* **56,** 595–599.

Kallis, A., and Tooming, H. (1974). *Photosynthetica* **8,** 91–103.

Kanemasu, E. T., and Hiebsch, C. K. (1975). *Can. J. Bot.* **53,** 382–389.

Kanemasu, E. T., Wesely, M. L., Hicks, B. B., and Heilman, J. L. (1979). *In* "Modification of the Aerial Environment of Crops" (B. J. Barfield and J. F. Gerber, eds.), ASAE Monogr. pp. 156–182. Am. Soc. Agric. Eng., St. Joseph, Michigan.

Kaplan, S. L., and Koller, H. R. (1977). *Crop Sci.* **17,** 35–38.

Kariya, K., and Tsunoda, S. (1972). *Tohoku J. Agric. Res.* **23,** 1–14.

Kariya, K., and Tsunoda, S. (1973). *Tohoku J. Agric. Res.* **24,** 1–13.

Khan, M. A., and Tsunoda, S. (1970). *Jpn. J. Breed.* **20,** 133–140.

Koffler, H., Johnson, F., and Wilson, P. (1947). *J. Am. Chem. Soc.* **69,** 1113–1117.

Korner, C., Scheel, J. A., and Bauer, H. (1979). *Photosynthetica* **13,** 45–82.

Krenzer, E. G., and Moss, D. N. (1975). *Crop Sci.* **15,** 71–74.

Kumura, A. (1968). *Proc. Crop Sci. Soc. Jpn.* **37,** 570–582.

Kumura, A. (1969). *Proc. Crop Sci. Soc. Jpn.* **38,** 74–90.

Kumura, A., and Naniwa, I. (1965). *Proc. Crop Sci. Soc. Jpn.* **33,** 467–472.

Larson, E. M., Hesketh, J. D., Woolley, J. T., and Peters, D. B. (1981). *Photosynth. Res.* **2,** 3–20.

Leffler, H. (1980). *In* "Predicting Photosynthesis for Ecosystem Models" (J. D. Hesketh and J. W. Jones, eds.), Vol. II, pp. 133–142. CRC Press, Boca Raton, Florida.

Legg, B., and Monteith, G. (1975). *In* "Heat and Mass Transfer in the Biosphere. I. Transfer Processes in Plant Environment" (D. A. de Vries and N. H. Afgan, eds.), pp. 167–186. Scripta, Washington, D.C.

Legg, E. J., and Powell, F. A. (1979). *Agric. Meteorol.* **20,** 47–67.

Leopold, A. C., and Kriedemann, P. E. (1975). "Plant Growth and Development." McGraw-Hill, New York.

Ludlow, M. M., and Wilson, G. L. (1971a). *Aust. J. Biol. Sci.* **24,** 449–470.

Ludlow, M. M., and Wilson, G. L. (1971b). *Aust. J. Biol. Sci.* **24,** 1065–1075.

Ludlow, M. M., and Wilson, G. L. (1971c). *Aust. J. Biol. Sci.* **24,** 1077–1087.

Lush, W. M. (1976). *Planta* **130,** 235–244.

McClure, P. R., and Israel, D. W. (1979). *Plant Physiol.* **64,** 411–416.

MacDowell, F. D. H. (1972a). *Can. J. Bot.* **50,** 89–99.

MacDowell, F. D. H. (1972b). *Can. J. Bot.* **50,** 883–889.

McKinion, J. M., Baker, D. N., Hesketh, J. D., and Jones, J. W. (1975). *U.S. Dept. Agric., Agric. Res. Serv. South. Ser. Bull* **ARS S-52,** 27–82.

McMichael, B. L., Jordan, W. R., and Powell, R. D. (1973). *Agron. J.* **65,** 202–204.
Madsen, E. (1974). *Acta Agric. Scand.* **24,** 242–246.
Mahon, J. D. (1979). *Plant Physiol.* **63,** 892–897.
Major, D. J., Johnson, D. R., and Luedders, V. D. (1975). *Crop Sci.* **15,** 172–174.
Maksymowych, R. (1973). "Analysis of Leaf Development." Cambridge Univ. Press, London and New York.
Matsushima, S., Okabe, T., and Wada, G. (1958). *Proc. Crop Sci. Soc. Jpn.* **26,** 195–197.
Mauney, J. R. (1968). *In* "Advances in Production and Utilization of Quality Cotton: Principles and Practices" (F. C. Elliott, M. Hoover, and W. K. Porter, eds.), pp. 23–40. Iowa Univ. Press, Ames.
Mauney, J. R., Fry, K. E., and Guinn, G. (1978). *Crop Sci.* **18,** 259–263.
Mauney, J. R., Guinn, G., Fry, K. E., and Hesketh, J. D. (1979). *Photosynthetica* **13,** 260–266.
Meidner, H., and Mansfield, T. A. (1968). "Physiology of Stomata." McGraw-Hill, New York.
Menezel, K., and Kirkby, E. A. (1978). "Principles of Plant Nutrition." Der Bund, Bern.
Milthorpe, F. L., ed. (1956). *Proc. 3rd. Easter Sch. Agric. Sci.,* Univ. Nottingham. Butterworth, London.
Miyasaka, A., Murata, Y., and Iwata, T. (1975). *JIBP Synth.* **2,** 72–85.
Murata, Y. (1961). *Bull. Natl. Inst. Agric. Sci., Japan, Ser. D: Plant Physiol., Genet., Crops Gen.* **9,** 1–169.
Murata, Y., ed. (1975). "JIBP Synthesis: Crop Productivity and Solar Energy Utilization in Various Climates in Japan," Vol. 2. Univ. of Tokyo Press, Tokyo.
Nicholas, J. C., Harper, J. E., and Hageman, R. H. (1976). *Plant Physiol.* **58,** 731–735.
Nobel, P. S. (1974). "Introduction to Biophysical Plant Physiology." Freeman, San Francisco, California.
Nobel, P. S. (1977). *Physiol. Plant.* **40,** 127–144.
Norman, J. M. (1975). *In* "Heat and Mass Transfer in the Biosphere. I. Transfer Processes in Plant Environment" (D. A. de Vries and N. H. Afgan, eds.), pp. 187–205. Scripta, Washington, D.C.
Norman, J. M. (1979). *In* "Modification of the Aerial Environment of Plants" (B. J. Barkfield and J. F. Gerber, eds.), ASAE Monogr. pp. 249–277. Am. Soc. Agric. Eng., St. Joseph, Michigan.
Norman, J. M. (1980). *In* "Predicting Photosynthesis for Ecosystem Models" (J. D. Hesketh and J. W. Jones, eds.), Vol. II, pp. 49–68. CRC Press, Boca Raton, Florida.
Norman, J. M., and Hesketh, J. D. (1980). *In* "Predicting Photosynthesis for Ecosystem Models" (J. D. Hesketh and J. W. Jones, eds.), Vol. I, pp. 9–35. CRC Press, Boca Raton, Florida.
Ogren, W. L. (1978a). *Proc. Int. Congr. Photosynth., 4th, 1977,* pp. 721–733.
Ogren, W. L. (1978b). *In* "Photosynthetic Carbon Assimilation" (H. W. Siegelman and G. Hind, eds.), pp. 127–138. Plenum, New York.
Ojima, M., and Fukui, J. (1966). *Proc. Crop Sci. Soc. Jpn.* **34,** 448–452.
Oritani, R., Enbutsu, T., and Yoshida, R. (1979). *Jpn. J. Crop Sci.* **48,** 10–16.
Parmele, L. H., Lemon, E. R., and Taylor, A. W. (1972). *Water, Air, Soil Pollut.* **1,** 433–451.
Passioura, J. B. (1976). *In* "Transport and Transfer Processes in Plants" (I. F. Wardlaw and J. B. Passioura, eds.), pp. 373–380. Academic Press, New York.
Patterson, D. (1980). *In* "Predicting Photosynthesis for Ecosystem Models" (J. D. Hesketh and J. W. Jones, eds.), Vol. I, pp. 205–235. CRC Press, Boca Raton, Florida.
Penning de Vries, F. W. T. (1972). *In* "Crop Processes in Controlled Environments" (A. R. Rees, K. E. Cockshull, D. W. Hand, and R. G. Hurd, eds.), pp. 327–347. Academic Press, New York.

Penning de Vries, F. W. T. (1975). *In* "Photosynthesis and Productivity in Different Environments" (J. P. Cooper, ed.), pp. 459–480. Cambridge Univ. Press, London and New York.

Penning de Vries, F. W. T., Brunsting, A. H. M., and van Laar, H. H. (1974). *J. Theor. Biol.* **45,** 339–377.

Peters, D. B., Clough, B. F., Garves, R. A., and Stahl, G. R. (1974). *Agron. J.* **66,** 460–462.

Puckridge, D. W. (1972). *Aust. J. Agric. Res.* **23,** 397–404.

Puckridge, D. W., and Ratkowsky, D. A. (1971). *Aust. J. Agric. Res.* **22,** 11–20.

Ross, J., and Nilson, T. (1975). *In* "Heat and Mass Transfer in the Biosphere. I. Transfer Processes in Plant Environment" (D. A. de Vries and N. H. Afgan, eds.), pp. 327–336. Scripta, Washington, D.C.

Sakamoto, C. M., and Shaw, R. H. (1967). *Agron. J.* **59,** 73–75.

Sato, K. (1976). *Proc. Crop Sci. Soc. Jpn.* **45,** 443–449.

Schnarrenberger, C., and Fock, H. (1976). *Encycl. Plant Physiol., New Ser.* **3,** 185–234.

Schubert, K. R., and Evans, H. J. (1976). *Proc. Natl. Acad. Sci. U.S.A.* **73,** 1207–1211.

Shibles, R., Anderson, I. C., and Gibson, A. C. (1975). *In* "Crop Physiology: Some Case Histories" (L. T. Evans, ed.), pp. 151–189. Cambridge Univ. Press, London and New York.

Silvius, J. E., Johnson, R. R., and Peters, D. B. (1977). *Crop Sci.* **17,** 713–716.

Sinclair, T. R., Goudriaan, J., and de Wit, C. T. (1977). *Photosynthetica* **11,** 56–65.

Sivakumar, M. V. K., Taylor, H. M., and Shaw, R. H. (1977). *Agron. J.* **69,** 470–473.

Takeda, G. (1978). *Bull. Natl. Inst. Agric. Sci., Ser. D* **29,** 1–65.

Tanaka, I. (1975). *JIBP Synth.* **2,** 37–48.

Tanaka, T. (1972). *Bull. Natl. Inst. Agric. Sci., Japan, Ser. A: Physics, Statistics* **19,** 1–100.

Tenhunen, J. D., Yocum, C. S., and Gates, D. M. (1976). *Oecologia* **26,** 89–100.

Tenhunen, J. D., Weber, J. A., Yocum, C. S., and Gates, D. M. (1979). *Plant Physiol.* **63,** 916–923.

Tenhunen, J. D., Hesketh, J. D., and Gates, D. M. (1980). *In* "Predicting Photosynthesis for Ecosystem Models" (J. D. Hesketh and J. W. Jones, eds.), Vol. I, pp. 123–181. CRC Press, Boca Raton, Florida.

Thomas, J. C., Brown, K. W., and Jordan, W. R. (1976). *Agron. J.* **68,** 706–708.

Thomas, M. D. (1949). *In* "Photosynthesis in Plants" (J. Franck and W. Loomis, eds.), pp. 19–52. Iowa State Coll. Press, Ames.

Thornber, J. P., Alberte, R. S., Hunter, F. A., Shiozawa, J. A., and Kan, K. S. (1977). *Brookhaven Symp. Biol.* **28,** 132–158.

Thornley, J. H. M. (1976). "Mathematical Models in Plant Physiology: A Quantitative Approach to Problems in Plant and Crop Physiology." Academic Press, New York.

Turner, N. C. (1974). *Bull.—R. Soc. N.Z.* **12,** 423–432.

Uchijima, Z. (1975). *JIBP Synth.* **2,** 86–104.

Verbelen, J. P., and de Greef, J. A. (1979). *Am. J. Bot.* **66,** 970–976.

Walker, D. A. (1976). *Encycl. Plant Physiol., New Ser.* **3,** 85–136.

Watanabe, I. (1973). *Proc. Crop Sci. Soc. Jpn.* **42,** 377–386.

Watanabe, I. (1975). *Proc. Crop Sci. Soc. Jpn.* **44,** 68–73.

Weber, C. R. (1966). *Agron. J.* **58,** 46–49.

Weber, C. R., Shibles, R. M., and Byth, D. E. (1966). *Agron. J.* **58,** 99–102.

Williams, R. F. (1975). "The Shoot Apex and Leaf Growth." Cambridge Univ. Press, London and New York.

Wong, S. G., Cowan, I. R., and Farquhar, G. D. (1979). *Nature (London)* **282,** 424–426.

Woodward, R. G., and Rawson, H. M. (1976). *Aust. J. Plant Physiol.* **3,** 257–267.

Yoshida, S. (1973). *Soil Sci. Plant Nutr.* **19,** 311–316.

12

Prospects of Applying Knowledge of Photosynthesis toward Improving Crop Production*

ROGER M. GIFFORD
COLIN L. D. JENKINS

ABBREVIATIONS

C_3 Plants	Plants in which initial carboxylation reaction leads to a three-carbon intermediate (phosphoglyceric acid); Calvin cycle plants
C_4 Plants	Plants in which initial carboxylation reaction leads to a four-carbon intermediate (oxaloacetic acid); Hatch–Slack pathway plants

*To obtain a complete view on this topic, the reader is encouraged to read this chapter in conjunction with those by Portis (Chapter 1), Ogren and Chollet (Chapter 7), Berry and Downton (Chapter 9), Hesketh *et al.* (Chapter 11) and by Christy Porter (Chapter 14) (this volume). With the exception of minor updating, this chapter is based on literature available to the authors up to May 1980—Editor.

Photosynthesis: Development, Carbon Metabolism, and Plant Productivity, Vol. II

ISBN: 0-12-294302-3

CER	Carbon dioxide exchange rate per unit leaf area
CGR	Crop growth rate per unit ground area
Chl	Chlorophyll
DW	Dry weight
LAD	Leaf area duration, meaning the area under a plot of LAI against time
LAI	Leaf area index, meaning the leaf area per unit ground area
SLW	Specific leaf weight, meaning leaf DW per unit leaf area
RuBP C'ase	Ribulosebisphosphate carboxylase
HPMS	α-Hydroxypyridinemethanesulfonic acid
HBA	2-Hydroxy-3-butynoic acid
MeHBA	Methyl 2-hydroxy-3-butynoate
BuHBA	Butyl 2-hydroxy-3-butynoate
INH	Isonicotinyl hydrazide
PAL-P	Pyridoxal phosphate
PCO	Photosynthetic carbon oxidation
PCR	Photosynthetic carbon reduction

ABSTRACT

Crop photosynthesis is a hierarchical process, and its rate can be separated into "capacity" and "intensity" components. Crop photosynthetic capacity can be improved by management strategies, which extend the persistence of the leaf cover for as much of the year as possible, and by ensuring rapid attainment of full interception of the incoming solar radiation by leaves after leaf canopy development starts. Although there are circumstances where crop growth rate can suffer due to too high a leaf area, the concept of "optimum leaf area index" has not yet proved to be a useful management criterion. For C_3-crops in tropical environments, there may be some scope to improve productivity by having upright-leaved canopies, but for most situations modification of canopy architecture has little to offer. Breeding for high rates of leaf net photosynthetic CO_2 exchange and its components has not yet yielded any success in improving crop yield. Reduction in dark respiration rate, however, is meeting with some success, although the range for improvement there is limited. Several reasons for the failure of attempts to breed for high photosynthesis rate can be identified; prospects for success in the near future are not obvious. Reduction of photorespiration by metabolic inhibitors has not yet succeeded either, but some avenues remain to be explored. Inhibitors constructed to sabotage specific steps in the carbon pathway of C_4 species may provide an opportunity for herbicidal control of C_4 weeds in C_3 crops.

In general, advancement in this area seems to be hindered more by a lack of understanding of how photosynthesis is controlled and integrated into growth of the intact crop canopy than by deficiencies of knowledge of the operation of components of the process at lower levels of organization.

I. Introduction

A. A Hierarchical View of Photosynthesis

Photosynthesis is a hierarchical process operating over scales of organization ranging from 10^{-27} m^3 and 10^{-15} sec for primary photoacts to 10^5 m^3 and 10^6 sec for primary productivity of a field crop. Modellers and managers of such complex hierarchical systems often find them showing counter-intuitive behavior (Forrester, 1971)—a manipulation at one level may not lead to the intuitively expected response by the entire system after interactions and feedbacks have occurred. It is not always possible to interpret behavior at one level in terms of detailed information from two or more levels of organization below. To make progress, gross but pragmatic simplifications of information from lower levels of organization may be necessary. At the field level, simplification of low level information has to be so gross that knowledge of molecular details may not be of much help.

At the level of primary quantum absorption, the maximum potential efficiency of solar energy conversion is 45–50%, since this is the proportion of the solar irradiance that is at wavelengths absorbable by the photosynthetic pigments of higher plants. As one ascends the hierarchical ladder, more constraints come into play, and potential efficiency diminishes. At the highest level—full season net photosynthesis by a crop canopy—potential photosynthetic efficiency is about 2.5% as Hudson (1975) found for Hawaiian sugar-cane. In more typical agricultural environments with less favorable water, nutrient, and temperature regimes, the potential efficiency can still be lower by one or two orders of magnitude.

Thus, as more elements of the whole system are introduced, there is less room for improvements at low levels of organization to be manifest in terms of increased porduction. There seems, therefore, for the most part to be a greater chance of success at improving crop productivity by working on restraints that occur at high levels than at low. For example, to take extremes, maintaining an effective light-harvesting array by breeding for lodging resistance is a proven means of crop improvement (Donald, 1968; Hargrove and Cabanilla, 1979). But working at the sub-chloroplastic level to introduce further accessory pigments with wider absorption spectra* (de Kouchkovsky, 1977) is unlikely to be an effective approach to crop improvement in our view.

Despite this, there is one type of manipulation based on low level information that can be effective at the crop level, namely, sabotage (Passioura, 1979). Metabolic inhibitors and hormonal overdoses can debilitate an organism. This has been used to great effect in weed control, although herbicides have usually been discovered empirically with little or no prior knowledge of the biochemistry or physiology involved.†

*This should, however, in principle, improve photosynthesis in "shade plants," and, therefore, has the potential of increasing plant productivity under shade conditions.

†However, biochemical and biophysical studies may indeed lead to the discovery of new and more potent herbicides as discussed in Section IV.

B. Approaches to Crop Improvement

Most improvement in crop yield until now has been due to increasing the supply of water and minerals, decreasing losses to pests and pathogens, and improving the partitioning of photosynthetic assimilate into economic products both at the canopy level (by weed control) and at the plant level (by increasing the harvest index). Some of these mechanisms of improvement could operate through manipulation of photosynthesis, e.g., control of weeds by chemical inhibition of their photosynthetic processes; this is discussed in Section IV,C. The objective of Sections II and III is to explore how details of photosynthesis may be changed so as to increase yield directly. It can be treated in two parts: *capacity* (light interception by the photosynthetic system) and *intensity* (efficiency of light conversion).

II. Crop Capacity to Intercept Light

A. Introduction

The most important feature of cropping systems determining annual photosynthesis per unit of land area is the maintenance of a photosynthetic surface over the land for as much of the year as possible to maximize light interception by the crop. Though a simple concept, further application will continue to improve production. Next in importance is an adequate area of leaf surface per unit area of ground at each stage in the crop cycle. Third, the structure of the leaf array may have some influence on the canopy photosynthesis. These three aspects of *capacity* are considered in this section, the *intensity* of photosynthesis by leaves is discussed in Section III.

B. Persistence of the Chlorophyllous Layer

Leaves are usually the dominant photosynthetic organs, but green stems, floral parts, and petioles can play a significant role, as in wheat under drought (Kriedemann, 1966; Evans *et al.*, 1972). Most crop physiological research focuses on leaf *area* as the parameter to describe the extent of the photosynthetic light-harvesting system. Occasional attempts to use other criteria such as Chl per unit land area as a basis for analysis of crop production (e.g., Nishimura *et al.*, 1966) have not been developed thus far.

Evergreens have a distinct photosynthetic advantage over part-season crops in most productive environments. They can fix solar energy all year whenever other factors such as drought or extreme temperatures are not limiting. For deciduous perennial species such as fruit trees in temperate environments, the rapid springtime development of leaves from stored material allows the crop to take early advantage of sunlight. In the humid tropics, the all-year growing season can be taken advantage of by semiperennial crops like sugarcane. In the semiarid tropics, where the growing season is determined by rainfall rather than low temperature, a crop such as cassava with a large storage tuber can redevelop a large leaf canopy rapidly after rain. However, man's diet is largely of annual cereal grains. Cereal cultivation traditionally involves fallow for much of the year. Fallows have their uses, but they waste potential photosynthetic irradiance. There are several ways of minimizing this period of poor or zero light interception by foliage. In moist temperate climates, fall-planted cereals (e.g., winter wheat) generally outyield spring cultivars because they slowly develop leaves through winter and quickly take advantage of the spring light when temperatures increase. Early frosts and midseason drought of Mediterranean climates, such as in the Australian wheat belt, restrict productive photosynthesis by cereals to a brief growing period (Single, 1975). Most of the annual irradiance is therefore lost to photosynthesis. So the simple principle of maximizing the interception of the annual flow of light can be a difficult one to apply. There may be scope for further betterment in some situations through adjustment of life-cycle timing and management.

There is renewed enthusiasm for the age-old practices of multiple cropping, relay cropping, intercropping, and agro–forestry, especially in the tropics. Multiple cropping involves avoiding fallows by planting straight after harvesting year-round. Breeding for short life-cycle rice is making three or even four crops a year possible. Transplanting rice seedlings from nursery beds to multiple-cropped paddies also maximizes the time of full leaf cover. Relay cropping involves overlapping of crops on the same land, e.g., in temperate environments undersowing a cereal crop with a pasture species, which grows away after grain harvest. Intercropping, where two or more compatible crops grow in a mixture, tends to be labor intensive. In China, intercropping paddy rice with the water fern *Azolla pinnata* makes good use of solar energy to fix nitrogen by the symbiotic blue–green alga *Anabaena azollae* (Food and Agriculture Organization, 1977). One type of intercropping, agro-forestry, is more amenable to mechanical methods. Crops and/or pastures are maintained below a sparsely planted forest. Whether it is an advantageous management strategy is contentious, but it exists and is being adopted in a wide range of climates as diverse as those of India, Peru (Spurgeon, 1979), and New Zealand (Borough, 1979).

C. Leaf Area Index (LAI)

The proportion of the incoming solar flux, which is intercepted at any time, is rarely 100% the year round if there is a harvest or if unwarranted allocation of photosynthate to leaf formation is avoided. The term *leaf area index* (LAI) refers to the area of leaf lamina in a plant canopy per unit area of ground.

1. CROP GROWTH RATE (CGR) IN RELATION TO LIGHT INTERCEPTION AND LEAF AREA

The short-term crop growth rate of a species is usually linearly related to the incoming radiation, which is intercepted by foliage (Shibles and Weber, 1965, 1966; Williams *et al.,* 1965a), unless developing water stress causes denser (high interception) stands to become stressed sooner. LAI is the most important determinant of light interception, the leaf arrangement being of secondary importance (see Section II,E). Brougham (1956) found that for practical purposes growth rate was at a maximum when about 95% of the irradiance was intercepted by the plant canopy. He called this the "critical leaf area index" (Brougham, 1958a). Critical LAI is typically in the range 3–6 (ryegrass–clover pasture: Brougham, 1956, 1958a; cotton: Ludwig *et al.,* 1965; orchard grass: Brown *et al.,* 1966; Nishimura *et al.,* 1966; potato: Bremner *et al.,* 1967; soybeans: Shibles *et al.,* 1975). But there have been reports, for grass species especially, where the critical LAI exceeded 6 as in perennial ryegrass (Brougham, 1958a; Anslow, 1966), and even 8–9 for maize (Williams *et al.,* 1965a). The reasons for these high numbers are dealt with in Section II, E.

Crop growth rate (CGR), then, shows a saturating response with respect to LAI up to about 6 in typical field situations. Below an LAI of 3, an approximately linear relationship between CGR (and yield) and LAI usually holds (e.g., Watson, 1947b; Black 1957; Williams *et al.,* 1965a; Bremner and Radley, 1966; Rawson *et al.,* 1980), the slope being mainly a function of daily irradiance above the canopy (Black, 1963) and plant species (Watson, 1947a). At low irradiance the transition from linearity to a plateau occurs at a lower LAI than at high irradiance (McCloud, 1966).

2. THE CONCEPT OF AN OPTIMUM LEAF AREA INDEX

In fertile environments, LAI of field crops can exceed the critical value, particularly in forage swards (Brougham, 1958a; Anslow, 1966; Vickery *et al.,* 1971). Kasanaga and Monsi (1954) advanced the idea of an optimum LAI above which crop growth rate declines. The decline became attributed to "parasitic respiration" by lower leaves. The lower shaded leaves, surviving below the light compensation point, were supposed to import assimilate to maintain normal respiration (Nichiporovich, 1956; Davidson and Philip, 1958). The idea of optimum LAI gained support conceptually (Donald, 1961; Stern and Donald, 1961; Black, 1963; Verhagen *et al.*, 1963) and experimentally (Watson, 1958, kale; Davidson and Donald, 1958; Black, 1963, subterranean clover; Brougham, 1961, white clover; Takeda, 1961, rice; McCloud, 1965, pearl millet; Ojima and Fukui, 1966, soybean; Nishimura *et al.*, 1966, orchard grass–white clover pastures). In these studies, there was an optimum LAI; CGR or canopy CO_2 exchange rate declined when LAI exceeded this optimum value. In other studies, the curve simply plateaued, there being no disadvantage in too high an LAI [e.g., Brougham (1956) for a ryegrass–clover pasture; Wang and Wei (1964), wheat, rice, *Perilla,* and sunflower; Ludwig *et al.* (1965), cotton; Shibles and Weber (1965), soybean; Stoy (1965), wheat; Williams *et al.* (1965b), maize; Bremner *et al.* (1967), potato; Wilfong *et al.* (1967) and King and Evans (1967), alfalfa and subterranean clover; Vickery *et al.* (1971), orchard grass]. Consequently there was much controversy in the mid1960s whether optimum leaf area index was real.

It is not necessary to invoke parasitic respiration by lower leaves to define an optimum LAI for growth. It only requires that the weight of actively metabolizing plant tissue shows a positive relationship with LAI. Once full light interception is reached, further increase in plant size will cause a roughly proportional increase in whole plant respiration for tissue maintenance but no further increase in canopy net photosynthesis. There are several reasons suggested as to why an optimum LAI can sometimes be demonstrated but other times not.

In discussing optimum LAI, the distinction between CGR and instantaneous canopy net photosynthesis rate is not always made. Although the concept is plausible for CGR, it is not so easily explained for canopy net photosynthesis rate unless parasitic respiration by lower leaves is a reality. Parasitic respiration is not widely supported because the compensation point of constantly shaded leaves tends to adapt downwards over a few days thereby enabling lower canopy leaves to stay in positive carbon balance (McCree and Troughton, 1966a; King and Evans, 1967). An alternative idea (Verhagen *et al.*, 1963), that lower leaves stay above the light compensation point because canopy structure automatically adjusts to keep lower canopy illumination high enough, has not been validated.

Different experimental methods, where LAI is not truly an independent variable, have given conflicting results. Following changes in both

LAI and CGR with time after cutting a pasture sward could give spurious interpretations if rapid shoot growth from underground reserves quickly generates a large leaf area. A distinct optimum LAI was found for potato CGR by following ontogenetic changes, but this was because tuber growth slowed towards maturity, whereas LAI was still reaching its peak value (Bremner and El-Saeed, 1963). Varying LAI by whole-plant thinning or crowding can be invalid too if the CGR or canopy photosynthesis determinations are made before lower leaves, adapted to one irradiance, have adjusted metabolism to a different irradiance (Brown *et al.,* 1966). The problem can be overcome by defoliation from the bottom up without thinning (Ludwig *et al.,* 1965) or by flooding the lower canopy while determining instantaneous CO_2-exchange rate (McCloud, 1966). But such procedures are the opposite of practice where grazing is involved. Even when LAI is adjusted from the top down, the system of defoliation can affect whether an optimum LAI is observed (Vickery *et al.,* 1971).

A species can show an optimum LAI in one environment, but a plateau response, over the range of LAI examined, in another. For example, the value of the optimum is a positive function of irradiance (Brougham, 1961; Takeda, 1961; Stern and Donald, 1961; Black, 1963; Fukai and Silsbury, 1977b). Also there is evidence that the optimum band of LAIs becomes broader with lower temperatures (Ludwig *et al.,* 1965; Fukai and Silsbury, 1976).

Leaf death, abscission, and decay may have generated an apparent optimum LAI in some studies where shed leaves were not accounted for in assessing CGR (McCree and Troughton, 1966b; Hunt, 1970). Subterranean clover, much studied in relation to optimum LAI, has a rapid cycle of leaf production and decay, the more so with denser canopy and higher temperatures. Thus the higher the LAI, the greater the relative underestimation of CGR could be. Other species such as maize, which have not been described as having an optimum LAI, do not have high leaf turnover either. Although an optimum LAI generated by failure to account for leaf decay may be physiologically invalid, it may still be a managerially useful concept for the crop *system* in the field (bacteria and insects included).

The ratio of growth respiration (proportional to photosynthesis) and maintenance respiration (proportional to plant weight) may influence the result. White clover with only a weak optimum LAI for crop growth seemed to have a very low maintenance respiration (McCree and Troughton, 1966b); subterranean clover with a pronounced optimum LAI had a much larger maintenance respiration relative to growth respiration than did the white clover (Fukai and Silsbury, 1977a).

Species with a pronounced optimum LAI include some clovers that produce new leaves at the base of the canopy in the shade, whereas graminaceous species, which expand new leaves closer to the top of the canopy especially after transition to flowering (Sheehy *et al.*, 1979), usually exhibit a plateau-type response. For the former growth habit, once the canopy reaches full light interception, new leaves cannot contribute to further growth of the stands unless they are in fact "parasitic" on the rest of the plant, not for maintenance but to climb to daylight. For clovers, each successive cohort of leaves extends petioles further than the previous set thereby overtopping them (Brougham, 1958b). Thus, as LAI increases, the investmant in petiole per unit increment of leaf area also increases.

For droughted crops the picture is further confounded. If LAI increases too soon, stored water is used early and LAI declines prematurely. Therefore, it is preferable to keep LAI well below the critical value early in the season to conserve water for later. When the water supply is limited, an optimum in the peak LAI during development is likely for maximum yield, and this optimum would probably not be great enough to intercept all the light.

Discussion of optimum LAI became unfashionable after the demise of the notion of parasitic respiration. The complexity of the factors that generate or conceal it have made it an unuseful concept for crop or pasture improvement. One might expect the critical LAI (95% interception) to be a simpler tool for crop and pasture management in moist environments. A photosynthetically ideal arable crop would develop a critical LAI rapidly and then switch to storage at constant LAI (Nichiporovich, 1956). In practice, a flat-topped curve of LAI versus time does not occur for most annual crops, a relatively sharp-topped peak being more usual. As a result, if the LAI for a crop only just reaches this critical value before the decline, yield will not necessarily be maximized. In potato, Bremner and Radley (1966) found that for maximum yields it was necessary for LAI to peak one or two LAI units above the critical value of three.

It was thought that in pastures, where the commercial organ of yield is mostly leaf itself, the LAI might be a useful management criterion for grazing or controlled cutting (Davidson and Donald, 1958); defoliation whenever LAI exceeds the critical (or optimum) value might increase annual production. Similarly the height of cutting could be important; too low a cut can cause too low an LAI during the recovery period. However, often this does not in practice seem to work. For example, there was no relationship between CGR and LAI for a perennial ryegrass sward through most of the year except during the spring flush of growth (Anslow, 1966). The reason why there is suppression of temperate grass growth in summer has still not been fully established, though the degree of shading during leaf development may be involved (Woledge, 1979). In a mixed tropical pasture of C_4 grass and C_3 legume, there was no relationship between the LAI of the stand left after cutting and the yield obtained at the next cut, this being due largely to growth being more related to the sporadic rainfall (Santhirasegaram *et al.*, 1966). Similarly for pure stands of C_4 *Panicum maximum* var. *trichoglume* (green panic) or *Cenchrus ciliaris* (buffel grass), cutting to a low LAI, did not reduce yield in the subsequent growth period for several suggested reasons (Humphreys and Robinson, 1966). It was partly due to intermittent moisture shortages (wilting being positively related to LAI), because if the pasture were not cut short enough the species went to flower, and this inhibited vegetative growth—though this is contentious (see Silsbury, 1965; Deinum, 1976)—and partly because the photosynthesis rate of the remaining leaves exhibited compensatory photosynthetic stimulation, the source–sink ratio being small. This latter phenomenon has been documented by gas analysis of single leaves for several pasture species such as *Medicago sativa* (alfalfa) (Hodgkinson, 1974) and *Lolium multiflorum* (Italian ryegrass) (Gifford and Marshall, 1973) after defoliation. As a result of these opposing effects, greater seasonal forage yield may be obtained by a variable intensity of defoliation (Ollerenshaw and Incoll, 1979).

There are other reasons why it may be difficult to use LAI as a straightforward photosynthesis management tool for herbage. The compensating effect of mobilizable reserves in roots and stem bases is amplified by the fact that the new canopy, which develops so rapidly, is of young leaves close to the peak of their ontogenetic cycle of photosynthetic capacity rather than of old partially senescent leaves (Brown and Blaser, 1968). Also, the peak photosynthetic capacity of successive leaves falls due to the progressively increasing level of shade, which they experience during expansion (Woledge, 1973). Even for grass species where new leaves are well illuminated during expansion, net photosynthetic capacity per unit leaf area and apparent quantum efficiency declines with each successive leaf produced in a developing canopy (Woledge, 1973; Sheehy, 1977).

D. Leaf Area Duration (LAD)

Combining the concepts of leaf canopy persistence and leaf area index, *leaf area duration* (LAD) is defined as the integral of leaf area index over time. Growth and yield is often closely related to LAD although the parameter suffers from several of the limitations that apply to LAI. Although about one-half the variation in wheat yield across a wide range of conditions was "explained" by a linear correlation with LAD between ear emergence and maturity (Evans *et al.*, 1975), this is as likely to be partly a result of covariation between potential sink size (florets m^{-2}) and LAI at ear emergence as it is to be a causal relation between grain growth and LAD (Gifford, 1974a). That is, postanthesis LAD is not an independent variable; it is determined in part, as is potential sink size, by environmental conditions before anthesis. Correlating the two does not help in assessing sink and source control over grain yield.

Similarly correlations over entire crop cycles between yield and LAD fail to separate the causal effect of a variation in size of the photosynthetic system on growth and the allometric relations of growth. That is, large plants tend to have large everything, including leaves and fruits, and this does not necessarily originate from photosynthesis. One way to gain insight into the role of variation in canopy photosynthesis (associated with variation in LAD) in determining yield variation is to calculate a truncated LAD in which any LAI exceeding the critical value is assigned just the critical value. Thus, Bremner and Radley (1966) found for potato crops that truncating LAD above an LAI of 2.8 improved the yield versus LAD correlation suggesting that yield variation was indeed partly attributable to canopy photosynthesis variation.

E. Canopy Architecture

Species in the C_3 group experience light saturation at one-third to one-half full sunlight intensity, but even when top leaves are greatly supersaturated with light, lower leaves may be considerably below saturation. Upright leaves intercept less light from an overhead sun than the same area of horizontal ones, so it might be optimal to have erect leaves at the top of the canopy grading into horizontal ones at the bottom (e.g., Warren-Wilson, 1960). This idea, together with the experimental finding that the critical LAI varied between species, led to a great deal of literature on the geometry of light interception in relation to leaf display (e.g., de Wit, 1965; Anderson, 1966; Duncan *et al.*, 1967; Cowan, 1968; McCree and Keener, 1974).

Ideal vertical-leaved canopies (with leaves of infinitessimal thickness) would intercept no light from an overhead sun; ideal horizontal-leaved canopies would intercept no light from a sun on the horizon. When the sun is at an elevation of 45°, the difference in interception between ideal horizontal and vertical canopies is small but varies with LAI. At low LAI, the horizontal array has the advantage; at high LAI, the vertical array is preferable for the 45° solar elevation; the crossover point, where both horizontal and vertical arrays intercept light equally, is at an LAI of 2–3 (Ross, 1970). Given that an LAI of 3 is a typical value for many crops, that the difference between real-world prostrate and erect crop canopies is much less distinct than between idealized horizontal and vertical arrays, and that for any canopy the advantages of its particular geometry at one time of day or season are partly balanced by the disadvantages at another, one can appreciate that canopy architecture is not in practice very important to canopy photosynthesis. Further blurring of any distinction by fluttering of the leaves appears not to have been investigated. For C_4 species, having a more linear light response curve than C_3 species up to full sunlight, one would not expect a significant effect of leaf inclination in any environment.

Overall, in midlatitudes, where most industralized crop production is, at typical LAIs (2–5), the structure of the canopy is a tertiary consideration for yield. In low latitudes having high solar elevation, erect foliage may be of advantage to C_3 crops with abundant nutrient and water causing dense tropical growth, as has indeed been found for rice (Tanaka, 1972; Yoshida, 1979). At higher latitudes, erect canopies may be advantageous only in high input infrequently cut swards of herbage (Rhodes, 1973) for which the LAI exceeds 8.

The theoretical expectation of canopy geometry having only a marginal influence on growth and yield in most situations is borne out by the conflicting field evidence on the advantage of erectness (Trenbath and Angus, 1975; Sheehy and Peacock, 1977). Unfortunately though, most work on leaf inclination has been with the Gramineae, which is perhaps a restricted base for wide generalization. Nevertheless, offsetting the marginal influence of canopy structure may be the simple inheritance of major aspects of plant architecture such as flag leaf angle in wheat (de Carvalho and Qualset, 1978). So even though a gain from an erect canopy may be small, there might nevertheless be a worthwhile return on breeding effort.

The most important control of canopy structure in agricultural practice is probably via control of lodging. Sheehy and Peacock (1977) found no difference in CGR between grass species with contrasting leaf angle; however, when canopies lodged, efficiency of light conversion fell by about 24%. This effect was reversible by "re-erecting" the lodged canopies (Sheehy, 1977).

Other facets of canopy structure, which have been examined for possible influences on variations in crop growth and yield, include: genotypic differences in light reflection and transmission by leaves (Sheehy *et al.,* 1977); the frequency distribution in time of sunfleck flicker within the canopy (McCree and Loomis, 1969; Gross and Chabot, 1979); the penumbra effect in leaf shadows due to the discrete width of the solar disc (Philip, 1969) and the related question of the height of the leaf canopy in relation to LAI and leaf size; the different attenuation profiles of diffuse radiation and direct radiation by a given canopy; the photosynthetic response by a leaf having illumination on the underside instead of the adaxial surface or partly on both surfaces (Syvertsen and Cunningham, 1979); heliotropic movements of leaves to face or avoid the sun (Ehleringer and Forseth, 1980); and the influence of row orientation (Reddy and Prasad, 1979). As with leaf angle, these features are of minor significance individually. Perhaps if all facets of canopy structure were optimized simultaneously for a given crop-environment situation, the overall advantage compared with a standard crop might be appreciable. Indeed the bringing together of many small features into one package is the idea behind the plant-type (Jennings, 1964) or ideotype (Donald, 1979) concept, but it is hard to generalize because the pros and cons are often highly situation specific, as for example with the frequency of defoliation and level of fertility of a pasture (Rhodes, 1973). The cost of breeding numerous small physiological features into a genotype destined for one management strategy in a single region may be prohibitive.

III. Breeding for Photosynthesis Rate

A. Introduction

As well as leaf photosynthesis rate *per se,* various features of the light-harvesting system within the leaf, the CO_2-fixation system, and the CO_2-delivery system have been examined as possible manipulable aspects of the intensity element of crop canopy photosynthesis. Ideally, there should be some evidence available that the parameter examined or selected for is in some way limiting photosynthesis or yield before embarking on the investigation. In practice, researchers have usually been guided by faith in their intuition or by the capabilities of techniques and instruments available to them at the time, as is evident in the following discussion.

B. Selection for Leaf Net Photosynthesis Rate and Its Components

1. CHLOROPHYLL, ELECTRON TRANSPORT, AND PHOTOPHOSPHORYLATION

From the early days (Willstätter and Stoll, 1918), no relationship was found between CO_2-exchange rate of leaves and the amount of chlorophyll (Chl). Even at low limiting irradiance, leaf photosynthesis was saturated with respect to Chl above about 20 μg (Chl) cm^{-2} (leaf) (Gabrielsen, 1948). Since field crops grow well above the light compensation point and have leaves with more than 20 μg (Chl) cm^{-2}, it is unlikely that increasing Chl concentration could have any impact on field photosynthesis or yield. In cereals, there is no correlation between leaf CO_2-exchange rate per unit leaf area (CER) and Chl content either between modern cultivars (Ferguson *et al.*, 1973, for barley; Murthy and Singh, 1979, for wheat) or generally among the various progenitors, of different ploidy, to modern hexaploid wheat (Planchon, 1974). Even for Chl deficient mutants of barley (McCashin and Canvin, 1979), cotton (Benedict *et al.*, 1972), tobacco (Schmid and Gaffron, 1967), and ryegrass (Wilson and Cooper, 1969b), low Chl content had no, or minimal, impact on CER. Parallel changes in leaf CER and Chl content with leaf age (Saeki and Nomoto, 1958; Sestak and Catsky, 1962; Haraguchi and Shimizu, 1970) do not prove causality; other features of the leaf also rise and fall during leaf ontogeny (e.g., ribulose bisphosphate carboxylase) and such changes are not necessarily tightly in parallel (Simpson, 1978). Two studies, both with soybean, did reveal a relationship however; Watanabe (1973a) comparing five cultivars, and Buttery and Buzzell (1977) comparing 48 normal and 20 Chl deficient mutants in the field found good relationships between light-saturated leaf photosynthesis and leaf Chl content up to 35 μg cm^{-2}, above which the curves plateaued. Most normal soybean strains have at least 35 μg (Chl) cm^{-2}. Furthermore, in a study of 373 soybean genotypes, Lugg and Sinclair (1979b) found that Chl-deficient mutants also had low specific leaf weight, and this could have been the reason for any low CER in such leaves (see Section III,B,6).

Other aspects of the photochemical apparatus, in relation to CER, have received less attention than has Chl. Although genotypic differences in Hill activity and photophosphorylation have been reported (e.g., Kleese, 1966; Miflin and Hageman, 1966), the data are generally uninterpretable being expressed per unit Chl—a highly variable parameter unrelated to leaf photosynthesis rate. Maize genotypes specifically selected for high and low vegetative growth rates had chloroplasts with activities in the opposite sense (i.e., low and high respectively) with respect to phosphorylating, nonphosphorylating, and uncoupled ferricyanide reduction (Hanson and Grier, 1973). For a range of ryegrass genotypes exhibiting a wide range of leaf photosynthesis rates, there was no relationship between *in vitro* chloroplast Hill activity or PMS-mediated photophosphorylation (expressed either per unit leaf area or per unit Chl) and leaf CER (Treharne, 1972). Although Watanabe (1973a) found a linear correlation between light-saturated CER and Hill activity per unit leaf area, he concluded that it was not a causal relationship (Watanabe, 1973a,b).

2. CARBOXYLATION AND THE DARK REACTIONS OF PHOTOSYNTHESIS

Good positive correlations between light-saturated leaf CER and ribulosebisphosphate carboxylase (RuBP C'ase) activity have often been made between leaves of different ontogenetic stages and environmental histories (e.g., Smillie, 1962; Bjorkman, 1968; Bowes *et al.,* 1972; Singh *et al.,* 1974). Correlation coefficients of about 0.9 were obtained for the comparisons across a range of ryegrass genotypes and small-grain cereal cultivars. Other enzymes of the dark reactions were not significantly related to CER (Treharne, 1972). Similarly Sirohi and Ghildiyal (1975) reported a correlation between leaf CER and carboxylase activity, Murthy and Singh (1979) found a correlation between flag leaf RuBP C'ase per unit area and grain yield for a range of wheat cultivars, and Peet *et al.* (1977) related RuBP C'ase per unit area at the time of pod set to dry bean yield. However, whether such correlations have physiological significance or represent causality remains an open question. Activity of the enzyme is under endogenous control by the concentration of CO_2, pH, Mg^{2+}, certain sugar phosphates, and light. The level of enzyme protein may also be regulated according to sink demand (Jensen *et al.,* 1978). So the *in vitro* activity under standard conditions may be of doubtful comparative value.

Much of the variation in RuBP C'ase activity per unit leaf area may simply reflect leaf thickness as Dornhoff and Shibles (1970) found for soybean and Frey and Moss (1976) for barley cultivars (see also section III,B,6). But, the latter also found variation in RuBP C'ase activity per unit leaf weight and suggested this as a selection criterion. Although one might expect such selection to cause a detrimental drop in activity of other enzymes, since RuBP C'ase represents up to 70% of leaf soluble protein, data of Randall *et al.* (1977) are encouraging. They discovered a decaploid tall fescue, which not only had higher RuBP C'ase activity and lower levels of other leaf enzymes per unit of leaf protein, but also had almost twice the CER per unit leaf area and per unit leaf dry weight. High ploidy as such does not, however, incur high CER (Evans and Dunstone, 1970; Setter *et al.*, 1978; Lush and Rawson, 1979).

Selection of carboxylase of greater specific activity per unit carboxylase enzyme protein would be more desirable especially, since it is a high molecular weight enzyme with low turnover number (Hardy *et al.*, 1978). Although Anderson *et al.* (1970) found both high and low specific activity forms of RuBP C'ase spanning a threefold range among tomato mutants, and Singh and Wildman (1974) found a *Nicotiana* sp. with atypically high RuBP C'ase specific activity, the significance of this is unclear given that the way to assay the fully activated enzyme was not described until 1974 (Bahr and Jensen, 1974). Using modern methodology, Yeoh *et al.* (1980) found that $K_m(CO_2)$ of RuBP carboxylase ranged from 13 to 26 μM for the enzyme from C_3 grasses. In C_4 grasses and algae, which also possess a CO_2-concentrating mechanism, the affinity of RuBP carboxylase for CO_2 was less than for C_3 grasses indicating that a wide range in $K_m(CO_2)$ is possible. Whether it is possible to further reduce $K_m(CO_2)$ of the C_3 carboxylase by breeding is unknown, but the desirability of that objective is tempered by the observation that as the K_m of RuBP carboxylase is reduced so is the V_{max} reduced. So the question of improving the specific activity of RuBP C'ase also remains an open question which is further complicated by the oxygenase activity of that enzyme protein. This activity, necessitating the existence of the glycolate pathway of photorespiration, may be unavoidable (Lorimer and Andrews, 1973) except by using the C_4-species trick of maintaining a high CO_2:O_2 ratio in the vicinity of the enzyme.

3. PHOTORESPIRATION AND THE CO_2 COMPENSATION POINT

Early doubt was expressed (see Zelitch, 1975a) as to whether photorespiration is a necessary concomitant to the photosynthetic carbon reduction (PCR) cycle in C_3 species due to the inseparability of the oxygenase and carboxylase activities of fraction I protein. A consensus has not yet been reached (e.g., contrast the viewpoints of Oliver, 1979, and Heber and Krause, 1980). In the mid-1960s, when the markedly different values of carbon dioxide compensation points and light-saturated leaf CER between C_3 and C_4 species were first being ascribed to the lack of photorespiration in C_4 species, there was a mood of optimism that a low photorespiratory C_3-type might be bred (e.g., Menz *et al.*, 1969; Widholm and Ogren 1969). The discovery of C_3 and C_4 species within single genera (Downton *et al.*, 1969; Krenzer and Moss, 1969) and the fact that C_3 species placed in 1–2% oxygen atmospheres grow much faster than normal (Bjorkman *et al.*, 1968; Parkinson *et al.*, 1974) fueled the enthusiasm. The potential scope for improvement seems substantial. Estimates of photorespiratory CO_2 loss by C_3-species vary from 15–40% of gross photosynthesis depending on methodology (e.g., Zelitch, 1975a; Lloyd and Canvin, 1977). A method using $^{18}O_2$ (Gerbaud and Andre, 1979) gave a value of 50% for whole wheat plants in ambient CO_2 and high light.

Massive screening programs for low CO_2 compensation point genotypes and mutants have been without success. Species covered include soybeans (Cannell *et al.*, 1969; Chollet and Ogren, 1975), barley, oats, and wheat (Moss, 1976; Apel, 1979), tall fescue (Nelson *et al.*, 1975b), and potatoes (Moss, 1976). It can be argued, however, that compensation point is a poor measure, even a nonmeasure, of photorespiration (Zelitch, 1975b; Wallace *et al.*, 1976; and see material following). In a case where variation in compensation point *was* found and it correlated with photorespiration, it did not correlate with growth rate (Wilson, 1975b).

Despite the disappointments it was believed in several quarters that "the control of this process . . . . has emerged as representing one of the most promising avenues for dramatically increasing the world supply of food . . . [Chollet and Ogren, 1975, p. 167]." (See Ogren and Chollet, Chapter 7, this volume, for their current views.) The discovery of species with photosynthetic properties intermediate between C_3 and C_4, such as *Panicum milioides* (Quebedeaux and Chollet, 1977) and *Mollugo verticillata* (Sayre *et al.*, 1979), has provided further encouragement for the approach (Apel, 1979) because it is easier to make some progress by selecting for a continuously variable trait than for a discontinuous character.

Using more direct photorespiratory selection criteria than compensation point, Zelitch and Day (1968, 1973) obtained tobacco plants with both low photorespiration and high leaf CER, but it was not possible to fix the trait into a population by further cycles of selfing and selection. At an operational level, it seems likely that if photorespiration were an inevitable but wasteful process, then CER and photorespiration would be uncorrelated for intervarietal comparisons. Were photorespiration a carbon recovery system associated with an inevitable oxygenation by RuBP carboxylase/oxygenase in aerobic conditions (as suggested Lorimer and Andrews, 1973; Andrews and Lorimer, 1978; but disputed by Zelitch, 1975b), or should it serve another useful purpose, then it would be expected to be positively correlated with CER. Despite Zelitch's tobacco selections, most results show a *positive* correlation. For example, intergenotypic comparisons among orchard grass (*Dactylis glomerata*)

(Carlson *et al.*, 1971), wheat (Dunstone *et al.*, 1973), barley (McCashin and Canvin, 1979), sunflower (Lloyd and Canvin, 1977), tobacco, sugarbeet, beans (Chmora *et al.*, 1975), ryegrass (Wilson, 1975b), and tall fescue (Nelson *et al.*, 1975b) all gave a strong positive relationship between leaf CER and photorespiration rate measured by various techniques. Moreover, because of this coupling, variation in photorespiration rate either was not accompanied by variation in compensation point (Carlson *et al.*, 1971) or correlated *negatively* with compensation point (Martin *et al.*, 1972, Chmora *et al.*, 1975).

However, a ryegrass tetraploid gave more hope (Garrett, 1978; Rathnam and Chollet, 1980): it had a different RuBP carboxylase isozyme, with a higher affinity for CO_2, but the same affinity for oxygen as the normal enzyme. Unfortunately this result could not be reproduced on Garrett's genotypes by McNeil *et al.* (1981), who paid particular attention to ensuring full activation of the enzyme during assay. There is ontogenetic and developmental evidence that the ratio of photorespiration to photosynthesis is not immutable. In *Citrus madurensis,* the presence of growing fruit stimulated CER (sink stimulation, see Section III,C) and also decreased photorespiration measured using an oxygen-free atmosphere method (Lenz, 1979). However, this might have been an indirect effect if the prime result of fruiting were extra stomatal opening, resulting in increased substomatal CO_2 concentration causing altered carboxylation to oxygenation balance. In wheat, an ontogenetic (Thomas *et al.*, 1978) and diurnal (di Marco *et al.*, 1979) rise and fall in the carboxylase/oxygenase ratio of leaf extract has also been reported. Unfortunately, valid determination of oxygenase/carboxylase ratios *in vitro* are very difficult because of susceptibility to activation and deactivation (Ogren and Hunt, 1978). Thus, such findings, although encouraging, need to be viewed with caution.

Overall, the results on photorespiration are conflicting and inconclusive in terms of predicting the prospects of breeding it out. The role of the oxygenase activity and photorespiration, if any, remains undetermined. Space limitations preclude discussion here of the possible role of photorespiration in amino acid metabolism. The possibilities for its manipulation by inhibitors are discussed in Section IV,B.

4. DARK RESPIRATION

In contrast to photorespiration, the function of true respiration is clear: It mobilizes stored energy for biological maintenance and for fueling new synthesis of structure. The associated oxygen and carbon dioxide exchange is confounded with photosynthetic functions in gas exchange studies of leaves or whole plants. For that reason, it is included in this review.

Although it seems that growth respiration usually operates close to theoretical efficiency for actively growing young plants (Apel, 1979), maintenance respiration may not always be as efficient. Both Heichel (1971) and Hanson (1971) found that genotypes of maize having faster vegetative growth rate had *lower* respiration rate [either per g (DW) or per unit leaf area]. In tall fescue, Jones and Nelson (1979) found that a high growth rate selection had a dark respiration rate per g (DW) of mature leaves of only 25% of the rate for a low growth rate selection. Similarly, Wilson (1975b) selected ryegrass lines for high and low maintenance respiration rates of mature leaves and found a strong negative correlation between the rate of regrowth of a simulated sward and the rate of leaf dark respiration. This was repeatable over several years, but only a few cycles of selection were needed to exhaust the potential to a point at which heritability of any further variation of the trait had no breeding significance (Wilson, 1978). When the trait was incorporated into other ryegrass populations (Wilson, 1979), it was expressed in simulated swards in terms of reduced canopy respiration and a 10–15% increased growth rate (Robson, 1979).

There is also developmental and metabolic evidence of ineffective respiratory CO_2 loss. During wheat growth, Pearman and Thorne (1979) found that removal of growing grains not only reduced the photosynthesis rate but also caused a sustained increase in dark respiratory CO_2 release by stems until the end of growth. This is suggestive that respiratory metabolism in wheat may be able to 'burn off' unwanted carbohydrate. Cyanide-insensitive respiration in mitochondria, which is not accompanied by phosphorylation (Solomos, 1977), provides a possible way for plants to balance their carbon budgets by acting as a release valve. There also may be another "overflow" form of respiration (Lambers, 1979). If these overflow mechanisms can be shown to be activated from time to time during ontogeny, then a coordinated effort both to eliminate the machinery for the wasteful respiration and to provide alternative desirable sinks may be fruitful. It may also provide the explanation of the inverse correlation between growth rate and mature leaf respiration rate found in several Gramineae discussed earlier.

5. LEAF CO_2 EXCHANGE RATE (CER) AND COMPONENT "CONDUCTANCES"

Frequently reported cultivar differences in leaf CO_2 exchange rate per unit leaf area (CER) (usually measured at near-ambient CO_2 concentration, saturating light and optimum temperature) have been reviewed (e.g., Moss and Musgrave, 1971; Wallace *et al.*, 1972). Intervarietal differences can range as high as threefold. Broad-sense heritability (indicating the proportion of variation in a segregating population which is of genetic origin) can be quite high: for example 60% for dry beans (Wallace *et al.*, 1976) and over 90% for *Lolium perenne* (Wilson and Cooper, 1969c). Actual selection advances have been achieved in some species such as maize (Crosbie *et al.*, 1977). But, probably because of the multigene basis of this quantitative character, progress in selection can be slow (Wallace *et al.*, 1976), and low CER may show dominance over high CER as Ojima *et al.* (1969) concluded for soybeans.

Heterosis for maize leaf CER (Fousova and Avratovscukova, 1967; Heichel and Musgrave, 1969; Crosbie *et al.*, 1978a) seemed encouraging, but Ariyanayagam (1974) reported such heterosis with respect to the midparent to be positive only when the parents had low CER. For high CER parents, heterosis was negative. Generally, where heterosis for such a multicomponent character as CER is observed, it is the exception rather than the rule (Sinha and Khanna, 1975). Ozbun (1978) recommended selecting for more simply inherited subcomponents to CER, such as those discussed earlier in this section.

Nevertheless, crop lines have been successfully selected for CER with the "high" lines exceeding those of the "low" lines by 50–100% [e.g., maize (Moss and Musgrave, 1971), and tall fescue (Nelson and Wilhelm, 1976)]. Unfortunately, although the intervarietal differences were shown to be under genetic control and to persist from young vegetative stages to grain filling in maize, there was no correlation between crop yield and CER (Ariyanayagam, 1974; Crosbie *et al.*, 1978b). For tall fescue, there was either no relationship (Sheehy and Cooper, 1973; Nelson *et al.*, 1975a) or a suggestion of a negative relationship (Wilhelm and Nelson, 1978; Cohen *et al.*, 1979) between forage yield and leaf CER. Moreover, it was possible to select tall fescues with all four combinations of high or low CER with high or low growth rate (Wilhelm and Nelson, 1978). Similarly no relationship was found between crop growth or yield and leaf CER in photosynthetically contrasting genotypes of wheat (Bingham, 1967; Dantuma, 1973; Murthy and Singh, 1979), barley (Dantuma, 1973; Grafius and Barnard, 1976), alfalfa (Delaney and Dobrenz, 1974), orchard grass and timothy grass (Sheehy and Cooper, 1973), soybeans (Curtis *et al.*, 1969), or sugarcane (Irvine, 1975). There was not even a relationship of crop growth rate with short-term canopy gross photosynthesis (Sheehy and Peacock, 1975, for temperate forage grasses) or with canopy daily net CO_2 exchange (Vietor and Musgrave, 1979 for maize.

Examples of positive evidence for a yield/CER relationship often seem weak because of questionable techniques for determining leaf photosynthesis and because of the restricted number of genotypes examined. In some early-season oat isolines, supposedly differing only in crown rust resistance genes, there were several high yielding rust-resistance lines that also had high flag leaf CER. Causality cannot be assumed, however, since in another series of oat isolines for rust-resistance derived from another parent, there was again a yield advantage of the rust-resistance genes in a disease-free environment but no correlated change in flag leaf CER (Brinkman and Frey, 1978). In poplar (Gordon and Promnitz, 1976), a fast-growing clone had not only the highest peak CER but also the highest activity of other functions like peroxidase and nitrate reductase.

In several leaf gas-exchange studies relating to yield, control of CER has been separated into two components: gas-phase diffusional control (i.e., stomatal plus boundary layer conductance) and the composite of all intracellular limitations (i.e., 'mesophyll' or 'residual' conductance). Other studies have specifically examined variability in stomatal conductance or its components (stomatal frequency, length, aperture). In some, genotypic variation in CER was attributable to both stomatal and residual conductances (e.g., Criswell and Shibles, 1970, for oats; Dornhoff and Shibles, 1970, for soybeans; Dunstone *et al.*, 1973, for wheat). Other work has found either residual conductance (e.g., Wilson and Cooper, 1969a, for perennial ryegrass; Brinkman and Frey, 1978, for oats) or stomatal conductance (e.g., Gifford and Musgrave, 1973, for maize inbreds; Shimshi and Ephrat, 1975, for wheat) to be the source of variation in CER. In the latter wheat study, grain yield was correlated with stomatal conductance. For rubber, residual conductance seemed to be the major source of clone-to-clone variation in one study (Samsuddin and Impens, 1978) and stomatal conductance in another (Samsuddin and Impens, 1979).

Attempts to relate photosynthesis and yield to stomatal frequency and length have met with only limited success. Wilson (1975a) found that *Lolium* selections having few or short stomata did prove to have lower stomatal conductance under dry soil conditions, but there was no impact on photosynthesis of well-watered plants. It was suggested therefore that low stomatal frequency would be an advantage for ryegrass grown under drought conditions. Similar conclusions were obtained for barley (Miskin *et al.*, 1972). One problem is that there can be a strong negative correlation between stomatal frequency and aperture length (Jones, 1977), as Miskin and Rasmussen (1970) found in a survey of 649 barley cultivars. Also enormous variations in stomatal frequency within the field crop canopy creates a burdensome sampling problem (Lugg and Sinclair, 1979a).

6. SPECIFIC LEAF WEIGHT (SLW)

Intervarietal variability in CER is often related to leaf dry weight per unit area (SLW) such that CO_2-exchange rate per unit leaf dry weight is relatively constant (e.g., Barnes *et al.*, 1969, for alfalfa; Dornhoff and Shibles, 1970, 1976, for soybeans). Presumably SLW is really a measure of leaf thickness. Charles-Edwards (1978) suggests that most environmental, ontogenetic, and intraspecies genetic variation in CER is attributable to variation in the photosynthetic machinery per unit leaf area, i.e., to thickness. SLW is heritable (Song and Walton, 1975) and has been suggested as a selection criterion (Barnes *et al.*, 1969; Dornhoff and Shibles, 1970; Kallis *et al.*, 1974).

If, for a given amount of growth, the proportion of dry matter allocated to leaf production is constant, SLW and LAI must be inversely related. Which then is best, a large area of thin leaves or a small area of thick leaves? For a small-leaved bean mutant, crop growth rate was not significantly lower than that for the control genotype, because partitioning of dry matter to leaf was unchanged: SLW increased 30–40%, compensating for reduced LAI (Motto *et al.*, 1979). For sparse canopies, a large area of thin leaves is preferable to maximize light interception (Tsunoda, 1959). For canopies at the optimum or critical LAI, an increase in SLW could be advantageous (Kallis and Tooming, 1974) if it ensures that no photons hitting a leaf are transmitted or fruitlessly reflected. So whether selection for high SLW is worthwhile depends on the environment in which the crop is to be grown. Simultaneous increase in SLW and the light-saturated value of CER does not necessarily result in higher crop yields according to a theoretical analysis by Kallis and Tooming (1974). Furthermore, many studies show little or no correlation between SLW and CER (e.g., Heichel and Musgrave, 1969, and Crosbie *et al.*, 1977, for maize; Dunstone *et al.*, 1973, for *Triticum* spp.; Watanabe and Tabuchi, 1973, for soybean primary leaves; Pallas and Samish, 1974, for peanuts; and Brinkman and Frey, 1978, for oat isolines) or even a negative correlation (Wilson and Cooper, 1969c; Fischer *et al.*, 1981). Whether these noncorrelations were due to varying nonphotosynthetic structural or storage material per unit leaf thickness was not established.

7. APPRAISAL

Genetic improvements to many subcomponents of leaf photosynthesis rate have been attempted. In terms of increased crop productivity, none has yet been successful. In general, this may be because for crops, which are reasonably well adapted to their environments, as most are in established agrosystems, no single physiological attribute is outstandingly limiting. In the next section, some of the problems that mitigate against success in breeding for photosynthesis are discussed.

C. Reasons for Failure of Breeding for Photosynthesis Rate

Possible reasons for the disappointing record are numerous. Some have been alluded to earlier. Only a cursory discussion can be made here. Methodological problems rank high, both in terms of the techniques used and the condition and history of plant material. There are several ways, both *in vivo* and *in vitro,* to determine leaf photosynthesis and photorespiration, and their results may not be in agreement. Comparing five common methods for estimating photorespiration, Martin *et al.* (1972) concluded that none could be validly used to rank the magnitude of photorespiration—only its presence or absence. Until 1974, methodology for RuBP carboxylase activity gave dubious results because the enzyme was not activated (Bahr and Jensen, 1974). Choice of conditions for the plant before and during measurement can generate enormous variation in measured photosynthetic attributes. In some circumstances leaf photosynthesis exhibits cyclic fluctuations often with about 30–40-min periods (Cowan, 1972). Similarly, some species may be more prone than others to diurnal fluctuations in a constant environment, particularly when grown in the field. Failure to recognize these factors may cause any genetic variation to be masked, and by taking appropriate precautions, an adequately sized screening program may become practically impossible.

During screening, a well-defined irradiance in the measurement chamber is essential. Light saturation is usually chosen though this may be difficult for C_4 species. Where a contrast in light-saturated CER is obtained between genotypes, there is usually no difference in quantum yield at low light, these two criteria correlating poorly (Wilson and Cooper, 1969c). They may even be inversely correlated (Akiyama and Takeda, 1975). Thus an intergenotypic contrast in canopy CER must be less than the contrast in light-saturated leaf CER because of the lower average irradiance (Charles-Edwards, 1978). A 20% advantage for light-saturated wheat leaf CER became attenuated to a theoretical 5% advantage for the light regime of northern Europe (de Vos, 1979). Thus large differences in measured CER must be bred to have a small influence in the field.

Choosing the leaf to be measured gives problems because CER

changes with leaf age and position on the plant (e.g., Vaclavik, 1975) for many species. The period of maximum photosynthesis by a leaf may only be a few days in clovers and C_3 forage grasses, for example. Identifying when that time has come is not easy. It does not necessarily coincide with the time of full expansion of the leaf. In tobacco, Rawson and Hackett (1974) found it occurred when the leaf was only one-third expanded. To measure anything other than the peak CER for a leaf gives even worse difficulties in defining the state of the leaves being screened. Even if the time of peaking can be identified, genotypic differences in the rate of senescence may render the criterion useless. High *life–time* photosynthate production by a leaf is more important than the peak value, and a high peak rate for a genotype may be correlated with its rapid senescence, as Gordon and Promnitz (1976) found for poplar clones.

In selecting the leaf position to be measured two approaches are common; either the leaf that contributes most to a sink organ of interest is used, or attributes of a seedling leaf or whole seedling are selected. Use of seedlings makes it easier to provide uniform conditions and to screen large numbers of plants. The presumption that a photosynthetically vigorous seedling gives rise to a similar adult seems invalid, however. Lawes and Treharne (1971) found no relationship between the ranking of wheat genotypes for seedling and for flag leaf photosynthesis rates. Similarly in comparing *Triticums* (Dunstone *et al.,* 1973) and cowpeas (Lush and Rawson, 1979), the ranking of genotypes for CER changed with ontogeny. On the other hand, Crosbie *et al.* (1981) found that a maize inbred selected for high CER during the vegetative stage also had high CER at mid-grain filling.

It can be hard to separate such ontogenetic effects from environmental influences. Both the environment of a plant or leaf during growth and its recent environment, like recent temperature (Crookston *et al.,* 1974; Thiagarajah *et al.,* 1979) can profoundly influence photosynthesis rate. The nature of the genotype by environment interaction can be perplexing. In a field maize-screening program in the Philippines, Heichel and Musgrave (1969) found inbred Pa83 to have a CER of 236 ng $(CO_2)cm^{-2}$ sec^{-1} and inbred Wf9 a value of only 78 ng cm^{-2} sec^{-1}. When self-pollinated progeny from those inbreds were grown and measured in upstate New York, there was no great difference in the photosynthetic performance of the two lines, each having a CER of about 150 ng cm^{-2} sec^{-1} (Gifford and Musgrave, 1970). There are examples of differential adaptation such that the photosynthetic ranking was a function of the light environment during growth (Downes, 1971, for *Sorghum* spp.; Dunstone *et al.,* 1973, for *Triticum* spp.; Eagles and Treharne,

1969, for *Dactylis glomerata* ecotypes) or of the prior temperature environment. Thus if selection for photosynthetic attributes is ever to be effective at increasing productivity, it may be that the breeding effort will need to be highly environment specific.

Another type of problem with selecting for photosynthesis may be the root of several of the problems mentioned earlier. It relates to the control of photosynthesis by growth, i.e., sink-limited photosynthesis. Not only does photosynthesis drive growth, but growth "pulls" photosynthesis by controlling in some way leaf development, stomatal conductance, carboxylation, and leaf senescence. In response to improvement in one part of the photosynthetic system, this can lead to compensation elsewhere: selection for high leaf CER can lead to reduced leaf area (e.g., Gaskell and Pearce, 1979, for maize) and increased SLW as discussed earlier, or a good canopy structure can lead to a decline in CER (Rhodes, 1972; Fischer *et al.*, 1981). Wheat cultivars with small flag leaves can compensate with high CER (Gale *et al.*, 1974) and RuBP carboxylase (Sirohi and Ghildiyal, 1975). It is as if well-adapted crop species already make as good use of the photosynthetic resources available to them as is possible but that there are many ways to achieve optimum adaptation. Manipulation of one element simply leads to adjustments elsewhere so that either the resources are still used equally efficiently, or "selection for high photosynthesis rates could actually result in loss of productivity" if resources are not allocated optimally (Hanson, 1971, p. 338). If, despite all, the photosynthetic system does respond favorably to selection, there is still the nonphosphorylating respiratory hurdle to jump.

Finally, it is no use improving photosynthesis if the active sinks for the assimilate are unresponsive to increased supply—if they are not, then photosynthesis must somehow counter-adapt or the assimilate must be dissipated. A prerequisite for success in breeding for photosynthesis is a crop system in which the active sinks are highly source limited. But the argument becomes circular because enhanced supply of photosynthetic assimilate can generate a large sink later (Gifford and Evans, 1981).

IV. Chemical Manipulation of Photosynthesis and Photorespiration

A. Introduction

One type of manipulation, based on information from low levels of organization, which has been highly effective at the crop level, is treatment of plants with exogenous inhibitors. In addition to herbicides, applied chemicals are used to improve productivity in several ways (Nickell, 1979). Treatment with synthetic growth regulants can, for example, promote rooting, influence time of flowering and fruit set, increase organ size, and enhance resistance to environmental stresses and insect pests. This section considers two areas where knowledge of the biochemistry of photosynthesis is being utilized in efforts to improve productivity. These are the chemical control of photorespiration as a means of enhancing net photosynthesis in C_3 species (Section IV,B) and the development of inhibitors of C_4 photosynthesis as possible selective herbicides (Section IV,C).

B. Chemical Inhibition of Photorespiration

The concept of reducing photorespiration by chemical manipulation, as an alternative method to breeding it out (Section III,B,3), developed in parallel with the elucidation of its metabolic pathway (also see Ogren and Chollet, Chapter 7, this volume). Following the discovery that glycolate is the metabolic substrate for photorespiration, an early idea was developed that inhibitors of enzymes of the pathway of glycolate oxidation could enhance net photosynthesis. Compounds most frequently tried have been aldehyde bisulfite addition products, particularly α-hydroxypyridinemethanesulfonic acid (HPMS), which inhibit glycolate oxidase (Zelitch, 1959) (see Fig. 1). Treatment of leaf disks with HPMS increased their short-term net photosynthetic CO_2 uptake for tobacco (C_3) but not for maize (C_4) (Zelitch, 1966). This increase was ascribed to an inhibition of photorespiration, perhaps causing an increased CO_2 concentration gradient between the atmosphere and the chloroplast. The increase could not be sustained, however. Although CO_2 evolution into CO_2-free air was reduced after HPMS treatment of excised tobacco leaves (Moss, 1968), it also decreased the CER (e.g., Khavari-Nejad, 1977, for tomato leaves; Kaminski *et al.*, 1979, for bean leaves; Servaites and Ogren, 1977, for soybean mesophyll cells; Inoue *et al.*, 1978, for spinach leaf protoplasts). HPMS treatment is accompanied by an accumulation of glycolate at the expense of other photorespiratory intermediates but interpretation of results is complicated by other nonspecific effects of the inhibitor. Inhibition of several other enzymes by HPMS have been noted (Osmond and Avadhani, 1970), and Murray and Bradbeer (1971) indicated that CO_2 fixation by isolated spinach chloroplasts, which were free of glycolate oxidase activity, was decreased by HPMS. After longer periods the compound also induces stomatal closure (Zelitch and Walker, 1964; Meidner and Mansfield, 1966; Khavari-Nejad, 1977).

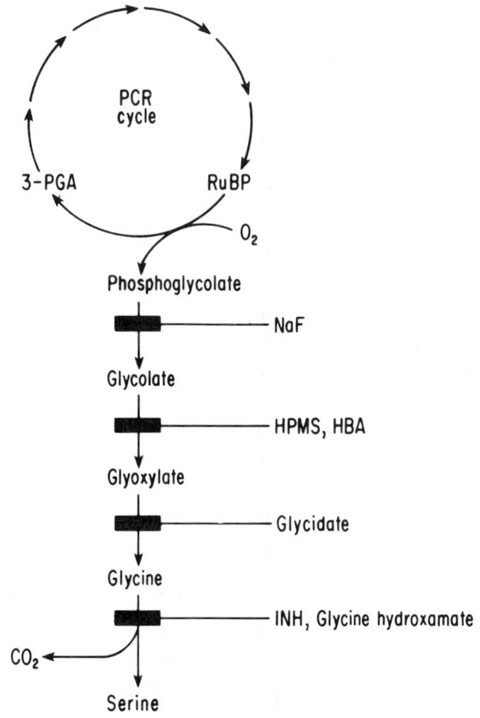

FIG. 1. A schematic diagram showing sites of action of inhibitors of the photorespiratory pathway.

In 1975, a specific inhibitor of glycolate oxidase was introduced—the acetylenic substrate analogue 2-hydroxy-3-butynoic acid (HBA). This is an enzyme-activated irreversible inhibitor ("suicide inhibitor"), which should be highly specific due to its mechanism of action. The methyl ester (MeHBA) of this compound inhibited glycolate oxidase in pea leaf disks causing glycolate accumulation (Jewess *et al.*, 1975). Net photosynthesis did not increase, however, when either MeHBA (unpublished observations of the authors) or when butyl 2-hydroxy-3-butynoate (BuHBA) (Doravari and Canvin, 1977) was applied to a variety of plant species at several concentrations. Rather, a decrease in net photosynthesis and an increased CO_2 compensation point resulted. An inhibition of photosynthesis by HBA and related esters has also been observed with wheat, barley, and maize leaf sections (Kumarasinghe *et al.*, 1977; Servaites *et al.*, 1978; Jenkins, 1979) and soybean mesophyll cells (Servaites and Ogren, 1977).

Other inhibitors tried include isonicotinyl hydrazide (INH) and 2,3-

epoxypropionate (glycidate). Glycidate, an inhibitor of glutamate: glyoxylate aminotransferase is discussed later. INH is an inhibitor of pyridoxal phosphate (PAL-P)-dependent enzymes through reaction with the cofactor PAL-P (e.g., Gore *et al.,* 1974). It has been used principally to inhibit the conversion of glycine to serine in the glycolate metabolizing pathway. This reaction, catalyzed by glycine decarboxylase and serine hydroxymethyltransferase, is considered to be the major origin of photorespired CO_2. However, INH treatment of wheat or rice leaf segments (Kumarasinghe *et al.,* 1977; Yun *et al.,* 1979) or soybean mesophyll cells (Servaites and Ogren, 1977) resulted in a decrease, rather than an increase in photosynthetic CO_2 assimilation.

Lawyer and Zelitch (1979) reported that glycine hydroxamate is a reversible competitive inhibitor that blocks the conversion of glycine to serine and CO_2. In tobacco callus tissue, the metabolism of glycolate, glyoxylate, or glycine to serine and CO_2 was prevented, but no results concerning effects on CER have yet appeared. Ishibashi *et al.* (1978) have tested many compounds for inhibition of photorespiration. Excised rice leaves treated with several of the compounds produced a decreased evolution of CO_2 into CO_2-free air, but most had either no effect or decreased net CO_2 assimilation. However, two compounds, potassium α-methylglycidate and aminoacetonitrile, show promise in that they caused an increase in net photosynthesis over the 40-min period of the experiment.

So far, then, no increased net photosynthesis due to inhibition of glycolate metabolism has yet been convincingly demonstrated. Although other side-effects cannot be ruled out with the relatively nonspecific inhibitors, such as HPMS and INH, the relationship between the specific inhibition of glycolate oxidase by HBA accompanied by an accumulation of glycolate and a subsequent inhibition of net photosynthesis is firmly established. It appears, then, that once glycolate is formed it must be further metabolized. From studies of the effects of these inhibitors on isolated soybean leaf cells, it has been suggested that it may not be feasible to improve photosynthetic production by chemically blocking glycolate metabolism (Servaites and Ogren, 1977). It is postulated that at least part of the carbon in glycolate must be returned to the chloroplast via the photosynthetic carbon oxidation (PCO) cycle (Andrews and Lorimer, 1978; Lorimer *et al.,* 1978; Lorimer and Andrews, 1981) to prevent a gradual depletion of photosynthetic carbon reduction (PCR) cycle intermediates, and subsequent inhibition of CO_2 fixation. In this case "chemicals which inhibit glycolate metabolism do not reduce photorespiration and increase photosynthetic efficiency, but rather exacerbate the problem of photorespiration [Servaites and Ogren, 1977, p.

461]." This proposal, although consistent with the data, rests on the view that the prime function of the PCO cycle is to salvage carbon lost from the chloroplast as glycolate. However, other physiological roles for the photorespiratory pathway are also possible (see Lorimer and Andrews, 1981). The possibility remains that the inhibition of net photosynthesis observed when glycolate metabolism is blocked is at least partly due to the prevention of an alternative (perhaps as yet unrecognized) function. In this respect, further studies with highly specific inhibitors of later reactions would be of interest.

A potentially more effective site for chemical inhibition is the synthesis of glycolate itself. The large accumulation of glycolate in the presence of PCO-cycle inhibitors indicates that there is no feedback inhibition by glycolate on its formation, but that glycolate synthesis may be under alternative metabolic control (Zelitch, 1979). Glycolate accumulation in the presence of HPMS was decreased in tobacco leaf disks previously floated on solutions of L-glutamate, L-aspartate, phosphoenolpyruvate, or glyoxylate (Oliver and Zelitch, 1977a,b). Reports that glycidate inhibits glycolate formation and stimulates net photosynthesis in tobacco leaf disks and isolated chloroplasts (Zelitch, 1974a,b) were later attributed to the accumulation of glutamate and glyoxylate caused by the glycidate inhibition of glutamate:glyoxylate aminotransferase (Lawyer and Zelitch, 1978; Zelitch, 1978). At variance with this, however, Chollet (1976) indicated that the effect of glycidate on photosynthesis by spinach chloroplasts was not due to an inhibition of glycolate synthesis; furthermore, Chollet (1978) showed that neither glycidate nor glyoxylate reduced photorespiration or increased net photosynthesis in tobacco leaf disks under physiological conditions when measured by other parameters. Kumarasinghe *et al.* (1977) also failed to demonstrate a stimulation of photosynthesis by glycidate.

Oliver (1978) reported that L-glutamate or L-aspartate increased net photosynthesis in isolated maize bundle sheath cells, and reduced glycolate formation and photorespiration; he suggested this may account for the low photorespiration in intact maize leaves. However, the primary site of stimulation by aspartate is on malate decarboxylation rather than CO_2 fixation (Chapman and Hatch, 1979; Chapman *et al.*, 1980). These results, in not supporting a role for the control of photorespiration by metabolites, do not favor the idea that photorespiration may be reduced through manipulation of its metabolic regulation. However, the issue remains contentious (Zelitch, 1979).

There are two steps in the major known pathway of glycolate formation—the oxygenase activity of RuBP carboxylase, which is responsible for phosphoglycolate formation, and the specific phosphatase, which is responsible for its subsequent hydrolysis to glycolate. Phosphoglycolate phosphatase was inhibited in isolated spinach chloroplasts by sodium fluoride, and though a decreased glycolate formation ensued, net assimilation was also reduced (Larsson, 1975); effects on RuBP carboxylase and perhaps other enzymes of the PCR cycle were suspected. Accumulation of phosphoglycolate following phosphoglycolate phosphatase inhibition might reduce photosynthesis as the enzyme triosephosphate isomerase is susceptible to phosphoglycolate inhibition at low concentrations (Tolbert and Ryan, 1976). As yet, no inhibitors specific for phosphoglycolate phosphatase have been described.

Since the oxygenase activity of RuBP carboxylase may be inevitable in atmospheres containing oxygen (Andrews and Lorimer, 1978), the prospect of chemical alteration of this enzyme to favor carboxylation has appeared unlikely. The relative affinities of the isolated enzyme for CO_2 and O_2 can be experimentally manipulated by changes in temperature or pH, though the ratio of maximum velocities for oxygenase and carboxylase activity remains constant (Ogren, 1978), or by replacement of Mg^{2+} by Mn^{2+} for enzyme activation and assay (Wildner and Henkel, 1978). Oxygenase activity of RuBP carboxylase from spinach leaf (Bhagwat *et al.*, 1978) and *Anabaena cylindrica* (Okabe *et al.*, 1979) may be specifically inhibited by hydroxylamine, with a stimulation of carboxylase activity in the latter case. Also, oxygenase activity of the isolated spinach leaf enzyme may be lost, whereas carboxylase activity could be preserved during anaerobiosis (Wildner and Henkel, 1980). These preliminary findings perhaps give a glimmer of hope. However, see Ogren and Chollet, Chapter 7, this volume.

The physiological role of photorespiration remains a matter of dispute, and the potential for enhanced growth by inhibition of photorespiration depends on the unresolved question (Lorimer and Andrews, 1981) as to whether it is essential or merely unavoidable. Chemical manipulation of photorespiration has yet to be practically achieved and shown effective at the plant level, but some avenues remain to be explored.

C. Development of C_4-Specific Herbicides

Examination of lists of the world's most serious weed species (Black *et al.*, 1969; Holm, 1969; Holm *et al.*, 1977) shows that they are predominated by C_4 plants. These include nutgrass (*Cyperus rotundus*), couch grass (*Cynodon dactylon*), barnyard grass (*Echinochloa spp.*) johnson grass (*Sorghum halepense*), and goosegrass (*Eleusine indica*). With the exception of sugarcane, maize, millets, and sorghum, all the major crops are C_3 species, and competition by C_4 weeds considerably reduces productivity.

Ideally, a selective herbicide should affect some metabolic process specific to the weed. Herbicide development has traditionally entailed the screening of large numbers of chemicals against several plant species until some biological activity is found, which may then be further investigated and enhanced by structural variations of the active compound (Saggers, 1976). This method requires no prior knowledge of the plant processes being affected. Although herbicides showing varying degrees of selectivity have been produced, in many cases the primary site and mode of action remains unknown. Knowledge of photosynthesis might be used both to define target processes for specific herbicidal activity in C_4 weed species and also as a guide to the chemical structures required. Potential targets in C_4 plants include C_4-specific enzymes, such as pyruvate phosphate dikinase, C_4 pathway-specific regulation processes, such as the activation of C_4 pathway enzymes by light (Hatch and Slack, 1969; Hatch, 1977), and metabolite transport processes unique to the C_4 pathway, such as the fluxes of pyruvate and phosphoenolpyruvate through the C_4 mesophyll chloroplast envelope (Huber and Edwards, 1977). Chemicals that inhibit or interfere with any of these essential processes, if they do not affect other reactions in C_3 plants, have potential as C_4-specific herbicides. This approach to the development of new selective herbicides is currently under investigation in several laboratories.

V. Conclusions

Plant scientists over the years have tried to improve individual subcomponents of crop photosynthetic systems with little or no success in terms of increased yield. This is because, in a complex system of multiple-nested subsystems, alteration of one part often results in compensation or even overcompensation elsewhere. Substantial differences at low levels of organization can become greatly attenuated at higher levels (Gifford, 1974b). The character of the feedbacks is likely to be different in different environments. Thus, no universal simple recipe for a photosynthetic route to crop improvement can be given. The experiences reviewed lend support to the idea that "the plant community is a self-regulating system, in which adaptation processes take place resulting in the maximum productivity of photosynthesis possible under the given environmental conditions [Tooming, 1967, p. 234]." It does not take a very complex model of the productive process to predict circumstances in which an increase of light-saturated photosynthesis rate per unit leaf DW and at constant SLW leads to a *decrease* in yield (Kallis and Tooming, 1974). In general, the message is that the greatest probability of success in crop improvement through photosynthesis is in seeking to make manipulations at a level of organization as close as possible to the level of organization where the problem lies. It is possible to enhance aspects of the light reactions by selection, but it does not necessarily improve leaf CER. It is possible to select for leaf CER, but it does not necessarily increase crop growth rate. It may be possible to select for high crop growth rate, but it does not necessarily increase mature crop yield. With the wisdom of hindsight, it is not surprising that most crop physiologists are circumspect about the enthusiastic hopes for productivity gains of some genetic engineers who are trying to incorporate modified packages of genes such as carbon dioxide fixation (*cfx*) or water-splitting (*lit*) genes into plants (Andersen *et al.,* 1980). Such simple recipes do not seem to reflect the hard-won lessons of crop physiology over the last several decades.

Killing C_4 weeds by inhibitors specific to C_4 metabolism may prove possible. But to improve yield by enhancing photosynthetic attributes by breeding of crops looks more problematical. If a forward way exists, it is more likely to lie, in our opinion, in the elucidation of the character of the interplay between levels of organization involved, than in the even more detailed description of, and fiddling with the constituent bits and pieces in isolation.

Acknowledgments

We are grateful to several colleagues for criticism of the manuscript and to Val Viviani for typing.

REFERENCES

Akiyama, T., and Takeda, T. (1975). *Proc. Crop Sci. Soc. Jpn.* **44,** 269–274.

Andersen, K., Shanmugam, K. T., Lim, S. T., Csonka, L. N., Tait, R., Hennecke, H., Scott, D. B., Hom, S. S. M., Haury, J. F., Valentine, A., and Valentine, R. C. (1980). *Trends Biochem. Sci.* **5,** 35–39.

Anderson, M. C. (1966). *J. Appl. Ecol.* **3,** 41–54.

Anderson, W. R., Wildner, G. F., and Criddle, R. S. (1970). *Arch. Biochem. Biophys.* **137,** 84–90.

Andrews, T. J., and Lorimer, G. H. (1978). *FEBS Lett.* **90,** 1–9

Anslow, R. C. (1966). *Proc. Int. Grassl. Congr., 9th, 1965,* Vol. 1, pp. 403–405.

Apel, P. (1979). *In* "Crop Physiology and Cereal Breeding" (J. H. J. Spiertz and T. Kramer, eds.), pp. 102–105. Centre for Agricultural Publishing and Documentation, Wageningen.

Aryiyanayagam, R. P. (1974). Ph.D. Thesis, Cornell University, Ithaca, New York.

Bahr, J. T., and Jensen, R. G. (1974). *Plant Physiol.* **53,** 39–44.

Barnes, D. K., Pearce, R. B., Carlson, G. E., Hart, R. H., and Hanson, C. H. (1969). *Crop Sci.* **9,** 421–423.
Benedict, C. R., McCree, K. J., and Kohel, R. J. (1972). *Plant Physiol.* **49,** 968–971.
Bhagwat, A. S., Ramakrishna, J., and Sane, P. V. (1978). *Biochem. Biophys. Res. Commun.* **83,** 954–962.
Bingham, J. (1967). *J. Agric. Sci.* **68,** 411–422.
Bjorkman, O. (1968). *Physiol. Plant.* **21,** 1–10.
Bjorkman, O., Hiesey, W. M., Nobs, M., Nicholson, F., and Hart, R. W. (1968). *Year Book—Carneigie Inst. Washington* **66,** 228–232.
Black, C. C., Chen, T. M., and Brown, R. H. (1969). *Weed Sci.* **17,** 338–344.
Black, J. N. (1957). *Aust. J. Agric. Res.* **8,** 335–351.
Black, J. N. (1963). *Aust. J. Agric. Res.* **14,** 20–38.
Borough, C. J. (1979). *Aust. For.* **42,** 23–29.
Bowes, G., Ogren, W. L., and Hageman, R. H. (1972). *Crop Sci.* **12,** 77–79.
Bremner, P. M., and El Saeed, A. K. (1963). *Proc. Easter Sch. Agric. Sci. (Univ. Nottingham)* **10,** 267–280.
Bremner, P. M., and Radley, R. W. (1966). *J. Agric. Sci.* **66,** 253–262.
Bremner, P. M., El Saeed, E. A. K., and Scott, R. K. (1967). *J. Agric. Sci.* **69,** 238–290.
Brinkman, M. A., and Frey, K. J. (1978). *Crop Sci.* **18,** 69–73.
Brougham, R. W. (1956). *Aust. J. Agric. Res.* **7,** 377–387.
Brougham, R. W. (1958a). *Aust. J. Agric. Res.* **9,** 39–52.
Brougham, R. W. (1958b). *N. Z. J. Agric. Res.* **1,** 707–718.
Brougham, R. W. (1961). *Proc. N. Z. Soc. Anim. Prod.* **21,** 33–46.
Brown, R. H., and Blaser, R. E. (1968). *Herb. Abstr.* **38,** 1–9.
Brown, R. H., Blaser, R. E., and Dunton, H. L. (1966). *Proc. Int. Grassl. Congr., 10th, 1966,* pp. 108–113.
Buttery, B. R., and Buzzell, R. I. (1977). *Can. J. Plant Sci.* **57,** 1–5.
Cannell, R. Q., Brun, W. A., and Moss, D. N. (1969). *Crop Sci.* **9,** 840–841.
Carlson, G. E., Pearce, R. B., Lee, D. R., and Hart, R. H. (1971). *Crop Sci.* **11,** 35–37.
Chapman, K. S. R., and Hatch, M. D. (1979). *Biochem. Biophys. Res. Commun.* **86,** 1274–1280.
Chapman, K. S. R., Berry, J. A., and Hatch, M. D. (1980). *Arch. Biochem. Biophys.* **202,** 330–341.
Charles-Edwards, D. A. (1978). *Ann. Bot. (London)* [N.S.] **42,** 717–732.
Chmora, S. N., Slobodokaya, G. A., and Nichiporovich, A. A. (1975). *Sov. Plant Physiol. (Engle. Transl.)* **22,** 963–968.
Chollet, R. (1976). *Plant Physiol.* **57,** 237–240.
Chollet, R. (1978). *Plant Physiol.* **61,** 929–932.
Chollet, R., and Ogren, W. L. (1975). *Bot. Rev.* **41,** 137–139.
Cohen, C. J., Chilcote, D. O., and Frakes, R. V. (1979). *Agron. Abstr.* **71,** 85.
Cowan, I. R. (1968). *J. Appl. Ecol.* **5,** 367–379.
Cowan, I. R. (1972). *Planta* **106,** 185–219.
Criswell, J. G., and Shibles, R. M. (1970). *Agron. Abstr.* **62,** 30.
Crookston, R. K., Lee, R., Ozbun, J. L., and Wallace, D. H. (1974). *Crop Sci.* **14,** 457–64.
Crosbie, T. M., Mock, J. J., and Pearce, R. B. (1977). *Crop Sci.* **17,** 511–514.
Crosbie, T. M., Mock, J. J., and Pearce, R. B. (1978a). *Euphytica* **27,** 657–664.
Crosbie, T. M., Mock, J. J., and Pearce, R. B. (1978b), *Crop Sci.* **18,** 87–90.
Crosbie, T. M., Pearce, R. B., and Mock, J. J. (1981). *Crop Sci.* **21,** 629–631.
Curtis, P. E., Ogren, W. L., and Hageman, R. H. (1969). *Crop Sci.* **9,** 323–327.
Dantuma, G. (1973). *Neth. J. Agric. Sci.* **21,** 188–198.

Davidson, J. L., and Donald, C. M. (1958). *Aust. J. Agric. Res.* **9,** 53–72.
Davidson, J. L., and Philip, J. R. (1958). *Arid Zone Res.* **11,** 181–187.
de Carvalho, F. I. F., and Qualset, C. O. (1978). *Crop Sci.* **18,** 561–567.
Deinum, B. (1976). *Neth. J. Agric. Sci.* **24,** 238–246.
de Kouchkovsky, Y. (1977). *Proc. Eur. Semin. Biol. Sol. Energy Conversion Syst., 1977,* pp. 3–6.
Delaney, R. H., and Dobrenz, A. K. (1974). *Agron. J.* **66,** 498–500.
de Vos, N. M. (1979). *In* "Crop Physiology and Cereal Breeding" (J. H. J. Spiertz and T. Kramer, eds.), pp. 80–89. Centre for Agricultural Publishing and Documentation, Wageningen.
de Wit, C. T. (1965). "Photosynthesis of Leaf Canopies." Centre for Agricultural Publications and Documentation, Wageningen.
di Marco, G., Grego, S., and Tricoli, D. (1979). *J. Exp. Bot.* **30,** 851–861.
Donald, C. M. (1961). *Symp. Soc. Exp. Biol.* **15,** 282–313.
Donald, C. M. (1968). *Euphytica* **17,** 385–403.
Donald, C. M. (1979). *J. Agric. Sci.* **93,** 261–270.
Doravari, S., and Canvin, D. T. (1977). *Plant Physiol.* **59,** Suppl., 42.
Dornhoff, G. M., and Shibles, R. M. (1970). *Crop Sci.* **10,** 42–45.
Dornhoff, G. M., and Shibles, R. M. (1976). *Crop Sci.* **16,** 377–381.
Downes, R. (1971). *In* "Photosynthesis and Photorespiration" (M. D. Hatch, C. B. Osmond, and R. O. Slatyer, eds.), pp. 57–62. Wiley (Interscience), New York.
Downton, J., Berry, J., and Tregunna, E. B. (1969). *Science* **163,** 78–79.
Duncan, W. G., Loomis, R. S., Williams, W. A., and Hanau, R. (1967). *Hilgardia* **38,** 181–205.
Dunstone, R. L., Gifford, R. M., and Evans, L. T. (1973). *Aust. J. Biol. Sci.* **26,** 295–307.
Eagles, C. F., and Treharne, K. J. (1969). *Photosynthetica* **3,** 29–38.
Ehleringer, T., and Forseth, I. (1980). *Science* **210,** 1094–1098.
Evans, L. T., and Dunstone, R. L. (1970). *Aust. J. Biol. Sci.* **23,** 725–741.
Evans, L. T., Bingham, J., Jackson, P. J., and Sutherland, J. (1972). *Ann. Appl. Biol.* **70,** 67–76.
Evans, L. T., Wardlaw, I. F., and Fischer, R. A. (1975). *In* "Crop Physiology: Some Case Histories" (L. T. Evans, ed.), pp. 101–149. Cambridge Univ. Press, London and New York.
Ferguson, H., Eslick, R. F., and Aase, J. R. (1973). *Agron. J.* **65,** 425–428.
Fischer, R. A., Bidinger, F., Syme, J. R., and Wall, P. C. (1981). *Crop Sci.* **21,** 367–73.
Food and Agriculture Organization (1977). *FAO Soils Bull.* **40,** p. 107.
Forrester, J. W. (1971). *Technol. Rev.* **73,** 53–68.
Fousova, S., and Avratovscukova, N. (1967). *Photosynthetica* **1,** 3–12.
Frey, N. M., and Moss, D. N. (1976). *Crop Sci.* **16,** 209–213.
Fukai, S., and Silsbury, J. H. (1976). *Aust. J. Plant Physiol.* **3,** 527–543.
Fukai, S., and Silsbury, J. H. (1977a). *Aust. J. Plant Physiol.* **4,** 159–167.
Fukai, S., and Silbury, J. H. (1977b). *Aust. J. Plant Physiol.* **4,** 485–497.
Gabrielsen, E. K. (1948). *Physiol. Plant.* **1,** 5–37.
Gale, M. D., Edrich, J., and Lupton, F. G. H. (1974). *J. Agric, Sci.* **83,** 43–46.
Garrett, M. K. (1978). *Nature (London)* **274,** 913–915.
Gaskell, M. L., and Pearce, R. B. (1979). *Agron. Abstr.* **71,** 88.
Gerbaud, A., and Andre, M. (1979). *Plant Physiol.* **64,** 735–738.
Gifford, R. M. (1974a). *Bull.—R. Soc. N. Z.* **12,** 887–893.
Gifford, R. M. (1974b). *Aust. J. Plant Physiol.* **1,** 107–117.

Gifford, R. M., and Evans, L. T. (1981). *Annu. Rev. Plant Physiol.* **32,** 485–509.
Gifford, R. M., and Marshall, C. (1973). *Aust. J. Biol. Sci.* **26,** 517–526.
Gifford, R. M., and Musgrave, R. B. (1970). *Physiol. Plant.* **23,** 1048–56.
Gifford, R. M., and Musgrave, R. B. (1973). *Aust. J. Biol. Sci.* **26,** 35–44.
Gordon, J. C., and Promnitz, L. C. (1976). *In* "Tree Physiology and Yield Improvement" (M. G. R. Cannell and F. T. Last, eds.), pp. 79–97. Academic Press, London.
Gore, M. G., Hill, H. M., Evans, R. B., and Rogers, L. J. (1974). *Phytochemistry* **13,** 1657–1665.
Grafius, J. E., and Barnard, J. (1976). *Agron. J.* **68,** 398–402.
Gross, L. J., and Chabot, B. F. (1979). *Plant Physiol.* **63,** 1033–1038.
Hanson, W. D. (1971). *Crop Sci.* **11,** 334–339.
Hanson, W. D., and Grier, R. E. (1973). *Genetics* **75,** 247–257.
Haraguchi, N., and Shimizu, S. (1970). *Bot. Mag. (Tokyo)* **83,** 411–418.
Hardy, R. W. F., Havelka, U. D., and Quebedeaux, B. (1978). *In* "Photosynthetic Carbon Assimilation" (H. W. Siegelman and G. Hind, eds.), pp. 165–178. Plenum, New York.
Hargrove, T. R., and Cabanilla, V. L. (1979). *BioScience* **29,** 731–735.
Hatch, M. D. (1977). *Plant Cell Physiol.* **3,** Spec. Issue, 311–314.
Hatch, M. D., and Slack, C. R. (1969). *Biochem. J.* **112,** 549–558.
Heber, U., and Krause, G. H. (1980). *Trends Biochem. Sci.* **5,** 32–34.
Heichel, G. H. (1971). *Photosynthetica* **5,** 93–98.
Heichel, G. H., and Musgrave, R. B. (1969). *Crop Sci.* **9,** 483–486.
Hodgkinson, K. C. (1974). *Aust. J. Plant Physiol.* **1,** 561–578.
Holm, L. (1969). *Weed Sci.* **17,** 113–118.
Holm, L., Plucknett, D. L., Pancho, J. V., and Herberger, J. P. (1977). "The World's Worst Weeds, Distribution and Biology." University Press of Hawaii, Honolulu.
Huber, S. C., and Edwards, G. E. (1977). *Biochim. Biophys. Acta* **462,** 583–602.
Hudson, J. C. (1975). *Span* **18,** 12–14.
Humphreys, L. R., and Robinson, A. R. (1966). *Proc. Int. Grassl. Congr., 10th, 1966,* pp. 113–116.
Hunt, W. F. (1970). *J. Appl. Ecol.* **7,** 41–50.
Inoue, K., Nishimura, M., and Akazawa, T. (1978). *Plant Cell Physiol.* **19,** 317–325.
Irvine, J. E. (1975). *Crop Sci.* **15,** 671–676.
Ishibashi, H., Yun, S.-J., Hyeon, S.-B., Suzuki, A., and Tamura, S. (1978). *Agric. Biol. Chem.* **42,** 1807–1809.
Jenkins, C. L. D. (1979). Ph.D. Thesis, University of Wales.
Jennings, P. R. (1964). *Crop Sci.* **4,** 13–15.
Jensen, R. G., Sicher, R. C., Jr., and Bahr, J. T. (1978). *In* "Photosynthetic Carbon Assimilation" (H. W. Siegelman and G. Hind, eds.), pp. 95–112. Plenum, New York.
Jewess, P. J., Kerr, M. W., and Whittaker, D. P. (1975). *FEBS Lett.* **53,** 292–296.
Jones, H. G. (1977). *J. Exp. Bot.* **28,** 162–168.
Jones, R. J., and Nelson, C. J. (1979). *Crop. Sci.* **19,** 367–372.
Kallis, A., and Tooming, H. (1974). *Photosynthetica* **8,** 91–103.
Kallis, A., Syber, A., and Tooming, K. L. (1974). *Sov. J. Ecol. (Engl. Transl.)* **5,** 101–106.
Kaminski, Z., Gutkowski, R., and Maleszewski, S. (1979). *Z. Pflanzenphysiol.* **91,** 17–24.
Kasanaga, H., and Monsi, M. (1954). *Jpn. J. Bot.* **14,** 304–324.
Khavari-Nejad, R. A. (1977). *Plant Physiol.* **60,** 44–46.
King, R. W., and Evans, L. T. (1967). *Aust. J. Biol. Sci.* **20,** 623–635.
Kleese, R. A. (1966). *Crop Sci.* **6,** 524–527.
Krenzer, E. G., and Moss, D. N. (1969). *Crop Sci.* **9,** 619–621.

Kriedemann, P. (1966). *Ann. Bot. (London)* [N.S.] **30,** 349–363.
Kumarasinghe, K. S., Keys, A. J., and Whittingham, C. P. (1977). *J. Exp. Bot.* **28,** 1163–1168.
Lambers, H. (1979). *Physiol. Plant.* **46,** 194–202.
Larsson, C. (1975). *Proc. Int. Congr. Photosynth., 3rd, 1974,* pp. 1321–1328.
Lawes, D. A., and Treharne, K. J. (1971). *Euphytica* **20,** 86–92.
Lawyer, A. L., and Zelitch, I. (1978). *Plant Physiol.* **61,** 242–247.
Lawyer, A. L., and Zelitch, I. (1979). *Plant Physiol.* **64,** 706–711.
Lenz, F. (1979). *In* "Photosynthesis and Plant Development" (R. Marcelle, H. Clijsters, and M. van Poucke, eds.), pp. 271–282. Junk, The Hague.
Lloyd, N. D., and Canvin, D. T. (1977). *Can. J. Bot.* **55,** 3006–3012.
Lorimer, G. H., and Andrews, T. J. (1973). *Nature (London)* **243,** 359–360.
Lorimer, G. H., and Andrews, T. J. (1981). *In* "The Biochemistry of Plants: A Comprehensive Treatise" (M. D. Hatch and N. K. Boardman, eds.), Vol. 8, pp. 329–374. Academic Press, New York.
Lorimer, G. H., Woo, K. C., Berry, J. A., and Osmond, C. B. (1978). *Proc. Int. Congr. Photosynth., 4th, 1977,* pp. 311–322.
Ludwig, L. J., Saeki, T., and Evans, L. T. (1965). *Aust. J. Biol. Sci.* **18,** 1103–1118.
Lugg, D. G., and Sinclair, T. R. (1979a). *Crop Sci.* **19,** 407–409.
Lugg, D. G., and Sinclair, T. R. (1979b). *Crop Sci.* **19,** 887–892.
Lush, W. M., and Rawson, H. M. (1979). *Photosynthetica* **13,** 419–427.
McCashin, B. G., and Canvin, D. T. (1979). *Plant Physiol.* **64,** 354–360.
McCloud, D. E. (1966). *Proc. Int. Grassl. Congr., 9th, 1965,* Vol. 1, pp. 511–517.
McCree, K. J., and Keener, M. E. (1974). *Crop Sci.* **14,** 584–587.
McCree, K. J., and Loomis, R. S. (1969). *Ecology* **50,** 422–428.
McCree, K. J., and Troughton, J. H. (1966a). *Plant Physiol.* **41,** 559–566.
McCree, K. J., and Troughton, J. H. (1966b). *Plant Physiol.* **41,** 1615–1621.
McNeil, P. H., Foyer, C. H., Walker, D. A., Bird, I. F., Cornelius, M. J., and Keyes, A. J. (1981). *Plant Physiol.* **69,** 530–34.
Martin, F. A., Ozbun, J. L., and Wallace, D. H. (1972). *Plant Physiol.* **49,** 764–768.
Meidner, H., and Mansfield, T. A. (1966). *J. Exp. Bot.* **17,** 502–509.
Menz, K. M., Moss, D. N., Cannell, R. O., and Brun, W. A. (1969). *Crop Sci.* **9,** 692–694.
Miflin, B. J., and Hageman, R. H. (1966). *Crop Sci.* **6,** 185–187.
Miskin, K., and Rasmussen, D. C. (1970). *Crop Sci.* **10,** 575–578.
Miskin, K. E., Rasmussen, D. C., and Moss, D. N. (1972). *Crop Sci.* **12,** 780–783.
Moss, D. N. (1968). *Crop Sci.* **8,** 71–76.
Moss, D. N. (1976). *In* "CO_2 Metabolism and Plant Productivity" (R. H. Burris and C. C. Black, eds.), pp. 31–41. University Park Press, Baltimore, Maryland.
Moss, D. N., and Musgrave, R. B. (1971). *Adv. Agron.* **23,** 317–336.
Motto, M., Soressi, G. P., and Salamini, F. (1979). *Euphytica* **28,** 593–600.
Murray, D. R., and Bradbeer, J. W. (1971). *Phytochemistry* **10,** 1999–2003.
Murthy, K. K., and Singh, M. (1979). *J. Agric. Sci.* **93,** 7–11.
Nelson, C. J., and Wilhelm, W. W. (1976). *Agron. Abstr.* **68,** 74.
Nelson, C. J., Asay, K. H., and Horst, G. L. (1975a). *Crop Sci.* **15,** 476–478.
Nelson, C. J., Asay, K. H., and Patton, L. D. (1975b). *Crop Sci.* **15,** 629–633.
Nichiporovich, A. A. (1956). "15th Timiryazev Lecture, 1954." Izd. Akad. Nauk SSSR, Moscow.
Nickell, L. G. (1979). *In* "Plant Growth Substances" (N. B. Mandava, ed.), pp. 263–279. Am. Chem. Soc., Washington, D.C.

Nishimura, S., Okubo, T., and Hoshino, M. (1966). *Proc. Int. Grassl. Congr., 10th, 1966,* pp. 117–120.
Ogren, W. L. (1978). *Proc. Int. Congr. Photosynth., 4th, 1977,* pp. 721–733.
Ogren, W. L., and Hunt. L. D. (1978). *In* "Photosynthetic Carbon Assimilation" (H. W. Siegelman and G. Hind, eds.), pp. 127–138. Plenum, New York.
Ojima, M., and Fukui, J. (1966). *Proc. Crop Sci. Soc. Jpn.* **34,** 448–452.
Ojima, M., Kawashima, R., and Mikoshiba, R. (1969). *Proc. Crop Sci. Soc. Jpn.* **38,** 693–699.
Okabe, K.-I., Codd, G. A., and Stewart, D. P. (1979). *Nature (London)* **279,** 525–527.
Oliver, D. J. (1978). *Plant Physiol.* **62,** 690–692.
Oliver, D. J. (1979). *Plant Sci. Lett.* **15,** 35–40.
Oliver, D. J., and Zelitch, I. (1977a). *Plant Physiol.* **59,** 688–694.
Oliver, D. J., and Zelitch, I. (1977b). *Science* **196,** 1450–1451.
Ollerenshaw, J. H., and Incoll, L. D. (1979). *Ann. App. Biol.* **92,** 133–142.
Osmond, C. B., and Avadhani, P. N. (1970). *Plant Physiol.* **45,** 228–230.
Ozbun, J. L. (1978). *HortScience* **13,** 678–679.
Pallas, J. E., Jr., and Samish, Y. B. (1974). *Crop Sci.* **14,** 478–482.
Parkinson, K. J., Penman, H. L., and Tregunna, E. B. (1974). *J. Exp. Bot.* **25,** 132–144.
Passioura, J. B. (1979). *Search* **10,** 347–50.
Pearman, I., and Thorne, G. N. (1979). *Rothamsted Exp. Stn., Annu. Rep. 1978,* Vol. 1, p. 39.
Peet, M. M., Bravo, A., Wallace, D. H., and Ozbun, J. L. (1977). *Crop Sci.* **17,** 287–293.
Philip, J. R. (1969). *In* "Physiological Aspects of Crop Yield" (J. D. Eastin, F. A. Haskins, C. Y. Sullivan, and C. H. M. van Bavel, eds.), pp. 113–115. Am. Soc. Agron., Madison, Wisconsin.
Planchon, C. (1974). *Ann. Amelior. Plant.* **24,** 201–207.
Quebedeaux, B., and Chollet, R. (1977). *Plant Physiol.* **59,** 42–44.
Randall, D. D., Nelson, C. J., and Asay, K. H. (1977). *Plant Physiol.* **59,** 38–41.
Rathnam, C. K. M., and Chollet, R. (1980). *Plant Physiol.* **65,** 489–494.
Rawson, H. M., and Hackett, C. (1974). *Aust. J. Plant Physiol.* **1,** 551–560.
Rawson, H. M., Constable, G. A., and Howe, G. N. (1980). *Aust. J. Plant Physiol.* **7,** 575–586.
Reddy, R. R., and Prasad, R. (1979). *Biol. Plant.* **21,** 85–91.
Rhodes, I. (1972). *J. Agric. Sci.* **78,** 509–511.
Rhodes, I. (1973). *Herb. Abstr.* **43,** 129–133.
Robson, M. J. (1979). *Grassl. Res. Inst., Annu. Rep. 1978,* pp. 60–61.
Ross, J. (1970). *Pred. Meas. Photosynth. Prod., IBP/PP (Int. Biol. Programme/Prod. Processes) Tech. Meet., 1969,* pp. 29–45.
Saeki, T., and Nomoto, N. (1958). *Bot. Mag.* **71,** 235–241.
Saggers, D. T. (1976). *In* "Herbicides, Physiology, Biochemistry, Ecology" (L. J. Audus, ed.), Vol. 2, pp. 447–473. Academic Press, London.
Samsuddin, Z., and Impens, I. (1978). *Exp. Agric.* **14,** 173–177.
Samsuddin, Z., and Impens, I. (1979). *Biol. Plant.* **21,** 154–156.
Santhirasegaram, K., Coaldrake, J. E., and Salih, M. H. M. (1966). *Proc. Int. Grassl. Congr., 10th, 1966,* pp. 125–129.
Sayre, R. T., Kennedy, R. A., and Pringnitz, D. J. (1979). *Plant Physiol.* **64,** 293–299.
Schmid, G. H., and Gaffron, H. (1967). *J. Gen. Physiol.* **50,** 563–582.
Servaites, J. C., and Ogren, W. L. (1977). *Plant Physiol.* **60,** 461–466.
Servaites, J. C., Schrader, L. E., and Edwards, G. E. (1978). *Plant Cell Physiol.* **19,** 1399–1405.
Sestak, Z., and Catsky, J. (1962). *Biol. Plant.* (Praha) **4,** 131–140.
Setter, T. L., Schrader, L. E., and Bingham, E. T. (1978). *Crop Sci.* **18,** 327–332.

Sheehy, J. E. (1977). *Ann. Bot. (London)* [N.S.] **41,** 593–604.
Sheehy, J. E., and Cooper, J. P. (1973). *J. Appl. Ecol.* **10,** 239–250.
Sheehy, J. E., and Peacock, J. M. (1975). *J. Exp. Bot.* **26,** 679–691.
Sheehy, J. E., and Peacock, J. M. (1977). *Ann. Bot. (London)[N.S.]* **41,** 567–578.
Sheehy, J. E., Windram, A., and Peacock, J. M. (1977). *Ann. Bot. (London)* [N.S.] **41,** 579–592.
Sheehy, J. E., Cobby, J. M., and Ryle, G. J. R. (1979). *Ann. Bot. (London)* [N.S.] **43,** 335–362.
Shibles, R. M., and Weber, C. R. (1965). *Crop Sci.* **5,** 575–577.
Shibles, R. M., and Weber, C. R. (1966). *Crop Sci.* **6,** 55–59.
Shibles, R. M., Anderson, I. C., and Gibson, A. H. (1975). *In* "Crop Physiology: Some Case Histories" (L. T. Evans, ed.), pp. 151–190. Cambridge Univ. Press, London and New York.
Shimshi, D., and Ephrat, J. (1975). *Agron. J.* **67,** 326–330.
Silsbury, J. H. (1965). *Aust. J. Agric. Res.* **16,** 903–913.
Simpson, E. (1978). *In* "Photosynthetic Carbon Assimilation" (H. W. Siegelman and G. Hind, eds.), pp. 113–125. Plenum, New York.
Singh, M., Ogren, W. L., and Widholm, J. M. (1974). *Crop Sci.* **14,** 563–566.
Singh, S., and Wildman, S. G. (1974). *Plant Cell Physiol.* **15,** 38–41.
Single, W. V. (1975). *In* "Australia Field Crops" (A. Lazenby and E. M. Matheson, eds.), Vol. 1, pp. 364–383. Angus & Robertson, Sydney, Australia.
Sinha, S. K., and Khanna, R. (1975). *Adv. Agron.* **27,** 123–174.
Sirohi, G. S., and Ghildiyal, M. C. (1975). *Indian J. Exp. Biol.* **13,** 42–44.
Smillie, R. M. (1962). *Plant Physiol.* **37,** 716–721.
Solomos, T. (1977). *Annu. Rev. Plant Physiol.* **28,** 279–297.
Song, S. P., and Walton, P. D. (1975). *Crop Sci.* **15,** 649–652.
Spurgeon, D. (1979). *Nature (London)* **280,** 533–534.
Stern, W. R., and Donald, C. M. (1961). *Nature (London)* **189,** 597–598.
Stoy, V. (1965). *Physiol. Plant., Suppl.* **4,** 1–125.
Syvertsen, J. P., and Cunningham, G. L. (1979). *Photosynthetica* **13,** 287–293.
Takeda, T. (1961). *Jpn. J. Bot.* **17,** 403–437.
Tanaka, T. (1972). *Bull. Natl. Inst. Agric. Sci., Ser. A* **19,** 1–100.
Thiagarajah, M. R., Hunt, L. A., and Hunter, R. B. (1979). *Can. J. Bot.* **57,** 2387–2393.
Thomas, S. M., Hall, N. P., and Merrett, M. J. (1978). *J. Exp. Bot.* **29,** 1161–1168.
Tolbert, N. E., and Ryan, F. J. (1976). *In* "CO_2 Metabolism and Plant Productivity" (R. H. Burris and C. C. Black, eds.), pp. 141–159. University Park Press, Baltimore, Maryland.
Tooming, H. (1967). *Photosynthetica* **1,** 233–240.
Treharne, K. J. (1972). *In* "Crop Processes in Controlled Environments" (A. R. Rees, ed.), pp. 285–303. Academic Press, New York.
Trenbath, B. R., and Angus, J. F. (1975). *Field Crop Abstr.* **28,** 231–244.
Tsunoda, S. (1959). *Jpn. J. Breed.* **9,** 161–168.
Vaclavik, J. (1975). *Biol. Plant.* **17,** 411–415.
Verhagen, A. M. W., Wilson, J. H., and Britten, E. J. (1963). *Ann. Bot. (London)* [N.S.] **27,** 627–640.
Vickery, P. J., Brink, V. C., and Ormond, D. P. (1971). *J. Br. Grassl. Soc.* **26,** 85–90.
Vietor, D. M., and Musgrave, R. B. (1979). *Crop Sci.* **19,** 70–75.
Wallace, D. H., Ozbun, J. L., and Munger, H. M. (1972). *Adv. Agron.* **24,** 97–145.
Wallace, D. H., Peet, M. M., and Ozbun, J. L. (1976). *In* "CO_2 Metabolism and Plant Productivity" (R. H. Burris and C. C. Black, eds.), pp. 43–58. University Park Press, Baltimore, Maryland.
Wang, T. D., and Wei, J. (1964). *Acta Bot. Sin.* **12,** 154–158.

Warren-Wilson, J. (1960). *Proc. Int. Grassl. Congr., 8th, 1960,* pp. 275–279.
Watanabe, I. (1973a). *Proc. Crop Sci. Soc. Jpn.* **42,** 377–386.
Watanabe, I. (1973b). *Proc. Crop Sci. Soc. Jpn.* **42,** 428–436.
Watanabe, I., and Tabuchi, K. (1973). *Proc. Crop Sci. Soc. Jpn.* **42,** 437–441.
Watson, D. J. (1947a). *Ann. Bot. (London)* [N.S.] **11,** 41–76.
Watson, D. J. (1947b). *Ann. Bot. (London)* [N.S.] **11,** 375–407.
Watson, D. J. (1958). *Ann. Bot. (London)* [N.S.] **22,** 37–55.
Widholm, J. M., and Ogren, W. L. (1969). *Proc. Natl. Acad. Sci. U.S.A.* **63,** 668–675.
Wildner, G. F., and Henkel, J. (1978). *FEBS Lett.* **91,** 99–103.
Wildner, G. F., and Henkel, J. (1980). *FEBS Lett.* **113,** 81–84.
Wilfong, R. T., Brown, R. H., and Blaser, R. E. (1967). *Crop Sci.* **7,** 27–30.
Wilhelm, W. W., and Nelson, C. J. (1978). *Crop Sci.* **18,** 405–408.
Williams, W. A., Loomis, R. S., and Lepley, C. R. (1965a). *Crop Sci.* **5,** 211–215.
Williams, W. A., Loomis, R. S., and Lepley, C. R. (1965b). *Crop Sci.* **5,** 215–219.
Willstätter, R., and Stoll, A. (1918). "Untersuchungen über die Assimilation der Kohlensäure." Springer-Verlag, Berlin.
Wilson, D. (1975a). *Ann. Appl. Biol.* **79,** 67–82.
Wilson, D. (1975b). *Ann. Appl. Biol.* **80,** 323–338.
Wilson, D. (1978). *Annu. Rep.—Welsh Plant Breed. Stn., Univ. Coll., Wales, 1977,* pp. 157–158.
Wilson, D. (1979). *Annu. Rep.—Welsh Plant Breed. Stn., Univ. Coll., Wales, 1978,* p. 152.
Wilson, D., and Cooper, J. P. (1969a). *New Phytol.* **68,** 627–644.
Wilson, D., and Cooper, J. P. (1969b). *New Phytol.* **68,** 645–655.
Wilson, D., and Cooper, J. P. (1969c). *Heredity* **24,** 633–649.
Woledge, J. (1973). *Ann. Appl. Biol.* **73,** 229–237.
Woledge, J. (1979). *Ann. Bot. (London)* [N.S.] **44,** 197–207.
Yeoh, H.-H., Badger, M. R., and Watson, L. (1980). *Plant Physiol.* **66,** 1110–1112.
Yoshida, S. (1979). *Int. Rice Comm. Newsl.* **24**(1), 5–16.
Yun, S.-J., Ishii, R., Hyeon, S.-B., Suzuki, A., and Murata, Y. (1979). *Agric. Biol. Chem.* **43,** 2207–2209.
Zelitch, I. (1959). *J. Biol. Chem.* **234,** 3077–81.
Zelitch, I. (1966). *Plant Physiol.* **41,** 1623–1631.
Zelitch, I. (1974a). *Arch. Biochem. Biophys.* **163,** 367–377.
Zelitch, I. (1974b). *Plant Physiol.* **53,** Suppl., 29.
Zelitch, I. (1975a). *Science* **188,** 626–632.
Zelitch, I. (1975b). *Annu. Rev. Biochem.* **44,** 123–145.
Zelitch, I. (1978). *Plant Physiol.* **61,** 236–241.
Zelitch, I. (1979). *Encyl. Plant Physiol., New Ser.* **6,** 353–367.
Zelitch, I., and Day, P. R., (1968). *Plant Physiol.* **43,** 1838–1844.
Zelitch, I., and Day, P. R. (1973). *Plant Physiol.* **52,** 33–37.
Zelitch, I., and Walker, D. A. (1964). *Plant Physiol.* **39,** 856–862.

13

Global Photosynthesis in Relation to Our Food and Energy Needs*

ROGER M. GIFFORD

ABBREVIATIONS

β	Biotic growth factor
DW	Dry weight
ER	Energy ratio
NEP	Net ecosystem productivity
NPP	Net primary productivity

ABSTRACT

Man is sustained by two sources of photosynthetically fixed energy—biomass and fossil fuels. Global net photosynthetic productivity is estimated to be 78×10^9t (carbon) year^{-1} of which about 7% is harvested directly or through animals for food, biomass fuel, and timber. Fossil fuel consumption has grown exponentially at 4.3% year^{-1} for over a century

*With the exception of minor updating, this chapter is based on literature available to the author up to May 1980.

Photosynthesis: Development, Carbon Metabolism, and Plant Productivity, Vol. II

ISBN: 0-12-294302-3

and is now about equal to the gross harvest of biomass. With world population size unlikely to be stabilized below 10 billion and with the huge inequalities of fuel distribution, the potential energy demand is likely to grow to a scale exceeding that of global primary productivity. If the per capita fossil fuel demand of 10 billion people grew to a maximum of one-third that of the average United States citizen in the 1970s, then the fossil fuel era would peak about 2100 A.D. and finish about 2400 A.D., but then atmospheric carbon dioxide concentration would peak at six times the preindustrial value before gradually declining over the following millenia. Although agriculture could benefit during the period of rising CO_2, the indirect effects on agriculture and society via climate change may be unacceptably disruptive. So it is likely to be inadvisable for the fossil fuel resource to be used on such a scale. However, the scale of current uncommitted photosynthetic productivity is too small to act as a substantial alternative source of fuel. It can be expected, therefore, that problems in managing immediate and fossil photosynthetic resources will become even greater sources of disruption and conflict in society than they now are.

I. The Argument

Chapter 12 by Gifford and Jenkins illustrated how detailed understanding of photosynthetic phenomena at low levels of organization is alone insufficient for understanding the determination of photosynthetic productivity at the crop canopy level. In this chapter, we shall follow the flow of photosynthetically fixed solar energy through levels above that of canopy productivity to elucidate further features of critical importance to man's future.

There are two sources of photosynthetically fixed energy-driving agriculture and the evolution of human society—recently formed biomass and fossilized biomass. The first, if properly managed, is renewable; the second is a diminishing resource. Current and fossilized biomass are each used as raw materials for food, fiber, manufacturing feedstock, and fuel. The relative importance of each for these various end uses has undergone considerable change and varies widely from country to country. There are also several interactions, the properties of which are ill-understood, but which need to be understood if man is to manage his affairs soundly.

The main interactions are as follow. The high level of solar energy fixation by modern agricultural systems is supported by fossil fuels not only directly and indirectly through farm inputs of fuel and materials, but more subtly in the cities in maintaining a technological infrastructure within which advanced agricultural technology is nurtured. In the less-developed Third World, where fossil fuels have not had such an impact, power and heat as well as food come largely from current biomass through human and animal muscle and the wood- or dung-burning fireplace. The scale of exploitation of both types of fuel is leading to two global difficulties—depletion and pollution. The excessive harvest of fuel and food crops and overgrazing of the land, stimulated in one way or another by energy consumption, is leading to deforestation, desertification, and loss of fertility that undermines future photosynthetic capabilities and population-carrying capacity of the Earth. The economic inflation generated by extraordinary leaps in fossil fuel price over the last decade is undermining the stability of the industrialized world. Moreover the high price of fossilized photosynthate is causing industrialized countries to look to current photosynthate, derived from an already battered biosphere, as an alternative fuel. The most crucial pollution aspect in the long term is that the high level of fossil fuel used together with the destruction and oxidation of large tracts of the biosphere is leading to the buildup of atmospheric carbon dioxide. This may have partially self-corrective properties in the biosphere. It may also cause incalculable disruption via plant responses to climatic change. In this chapter details of this argument are developed and, where possible, quantified.

II. Photosynthetic Productivity of the Biosphere

A. Global Primary Productivity

At the ecosystem level, photosynthetic carbon fixation is approximated by annual "net primary productivity" (NPP). This is the integral of net photosynthetic carbon fixation minus carbon loss due to the respiratory activities of plant growth and tissue maintenance (plus the gain in weight due to uptake of mineral nutrients, when NPP is expressed in terms of dry weight). There is continual litter fall and death of plants in both deciduous and evergreen ecosystems, as well as heterotrophic organisms feeding on the plants. The accumulation of living biomass from year to year is the net ecosystem productivity (NEP). NPP exceeds NEP by the weight of plant matter, which died or was eaten, less the increase in mass of heterotrophic organisms. For a mature ecosystem, NEP is zero, whereas NPP may be high (Whittaker and Marks, 1975).

Several estimates of global net primary production suggest a terrestrial value of about 120×10^9 t (DW) year^{-1} from the land area of 149×10^6 km^2 and a marine value of 55×10^9 t (DW) year^{-1} from the sea area of about 361×10^6 km^2 (Box, 1975; Leith, 1975a). A detailed assessment from the U.S.S.R. (Bazilevich *et al.*, 1971; Rodin *et al.*, 1975) gave a figure about 40% higher for the land—172 × 10^9 t (DW) year^{-1}—but the same for the ocean.

The basis of the difference has not been fully resolved. One difficulty is that the scale of man's modifications of the biosphere due to agriculture, fuel gathering, mining, grazing, forestry, chemical defoliation, damming, urbanization, and pollution has now reached such a large scale that it is unclear to which year any estimate applies. The experimental and survey data on which the estimates are based have been gathered over many years. In one of Leith's assessments (1975a), it was suggested that it pertained to 1950 inasmuch as the scale of man's influence over the last three decades had not been assessed. Possible past changes in the size of the biosphere are discussed further in Section IV,B,3.

Whittaker and Likens' (1975) global primary production estimate is presented in Table I. "Standing biomass" does not include soil organic matter or organic detritus but does include the dead heartwood of standing trees. Tropical forest is the largest contributor to terrestrial NPP (29%); all types of forest contribute 68%. Between 80 and 90% of all standing biomass is in forest (Table I, Rodin *et al.*, 1974). In contrast to the view of Bazilevich *et al.* (1971), Loomis (1979) suspects that the biomass value for tropical forest contributing to Table I (1025 Gt) may be too high by a factor of 2. He feels the figures may be biased toward the most dense patches and that the designated area may embrace non-forested patches embedded in forest zones.

Net primary production of the oceans is only between one-half and one-third of that of the land, despite its 2.5-fold greater area, mainly because sedimentation of dead marine organisms to the ocean floor continuously removes mineral nutrients from the illuminated surface zone of the sea. Moreover, the rate of turnover of marine phytomass is over 200 times more rapid than for terrestrial phytomass overall (0.07 year instead of 15 years).

Whereas nutrient concentrations (particularly nitrogen and phosphorus) determine productivity in the oceans, temperature not being at all important (Bunt, 1975), primary productivity in terrestrial ecosystems seems to be largely determined by temperature and rainfall. Global productivity models based on these (Leith, 1972) or on "actual evapotranspiration," a composite of temperature and precipitation (Leith and Box, 1972), gave values of terrestrial primary production of 124 and 119 Gt (DW) year^{-1}. These agree with the estimates based on ecosystem compilations, except for those of Bazilevich *et al.* (1971).

Given the adequacy of temperature and precipitation as descriptors of terrestrial NPP (Rosenzweig, 1968), what is the role of light and

TABLE I
Global Net Primary Productivity and Net Primary Production[a]

			Net primary production			Standing biomass			
Vegetation type	Area (10^6 km^2)	Mean *NPP* (g(DW) m^{-2} year^{-1})	Total DW (Gt year^{-1})	Total carbon (Gt year^{-1})	Total enthalpy (*EJ* year^{-1})	Mean (kg m^{-2}) (DW)	Total DW (Gt)	Total C (Gt)	Total (*EJ*)
Tropical forest	24.5	2016	49.4	22.2	939	42	1025	461	19,480
Temperate forest	12	1242	14.9	6.7	283	32	385	173	7320
Boreal forest	12	800	9.6	4.3	182	20	240	108	4565
Woodland	8.5	706	6	2.7	114	6	50	22	950
Savanna	15	900	13.5	6.1	257	4	60	27	1140
Grassland	9	600	5.4	2.4	103	1.6	14	6	266
Tundra + alpine	8	140	1.1	0.5	21	0.6	5	2	95
Desert	42	40	1.7	0.8	32	0.3	13	6	247
Cultivated	14	650	9.1	4.1	173	1	14	6	266
Swamp + freshwater	4	1700	6.8	3.1	129	7.5	30	14	570
Total continental	149	782	117.5	52.9	2233	12.3	1837	826	34,900
Total marine	361	155	55	24.8	1045	0.01	4	2	76
Global total	510	338	172.5	77.7	3278	3.6	1841	828	34,976

[a]Expressed in terms of plant dry weight, carbon content [assuming 0.45 g (*C*) per g (DW)], and gross heat of combustion [assuming 19 kJ/g (DW); based on Whittaker and Likens, 1975] ($G = 10^9$; $E = 10^{18}$).

nitrogen in determining productivity? These factors are at the heart of photosynthetic carbon fixation, one as the energy source, the other as a key element in the most abundant protein on earth—ribulosebisphosphate carboxylase. Moreover one should wonder, if the C_4 syndrome is regarded as an adaptation to conserve water, how important that photosynthetic system is as a constituent of this water-limited biosphere.

With regard to the last question, C_4 species contribute insignificantly to the global phytomass, there being no C_4 tree species and most phytomass being trees. Even for arid environments, doubt exists that the C_4 syndrome confers an advantage for productivity at a community level (Caldwell, 1974, 1975). One possible reason that it is not necessary to include light as a parameter in broad models to estimate primary productivity may be that there is sufficient correlation between photosynthetic irradiance and temperature that only one of these variables needs to be used in the empirical equations to obtain adequate prediction. There can be little doubt that, if the nitrogen status of the world's soils were all enhanced, the NPP would also increase (Postgate, 1978) and probably the standing biomass too. The reason that adequate prediction of global NPP could be obtained without explicit reference to nitrogen may be that there is too little variation between ecosystems in their nitrogen status to have appreciable impact on the overall conclusions. The nitrogen effect may be implicitly embedded in the equations used.

Similarly, most vegetation would grow faster in an atmosphere enriched with carbon dioxide (see Section V,C). But CO_2 does not need to enter models to predict NPP at different localities. Carbon dioxide concentration is, however, undergoing change with time and this is discussed in Section IV,B.

B. Man's Harvest

Despite numerous uncertainties in estimating the proportion of the global primary production, which man harvests directly or indirectly, a first order estimate is made in Table II. The columns labeled "Harvested" are relatively straightforward; they were derived from the Food and Agricultural Organization commodity statistics and reasonable assumptions for moisture, bone, and enthalpy content. The primary production or "whole plant" equivalent is very approximate as indicated in the footnote. Moreover no allowance was made for either potential human food or crop residue, which was fed to animals since no data are available. Similarly no allowance is made for unrecorded noncommercial domestic food production, which does not enter FAO statistics. Most of

TABLE II
Global Biomass Harvested in 1976[a]

	Gt (DW)/year		EJ/yr		
Commodity	Whole plant equivalent[b]	Harvested[c]	Whole plant equivalent	Harvested	Available[d] to eat
Crops	5.41	1.893	103	36	13
Meat	1.3	0.051	24	1.6	2.8 (Meat, Dairy, Fish combined)
Dairy	0.78	0.059	14	1.4	
Fish	0.14	0.012	2.6	0.3	
Fiber[e]	0.043	0.015	1	0.4	—
Timber	3.91	1.369	77	27	—
Fuelwood	0.71	0.71	14	14	—
Total	12.33	4.11	236	81	15.8

[a]Approximate estimates (G = 10^9; E = 10^{18}).
[b]For crops, fiber, and timber, this is calculated from [c] assuming simply that the harvested commodity is 0.35 of the whole plant grown. For fuelwood, it is the same as the harvested amount. For meat and dairy, various plant-to-animal energy conversion ratios (range 0.03–0.10) have been assumed. For fish, an energy conversion ratio of 0.1 was taken.
[c]Based on Food and Agriculture Organization (1977a,b, 1978a).
[d]Based on Food and Agriculture Organization compilation of food energy taken into households.
[e]Includes tobacco, rubber, and greasy wool.

the fuelwood harvest, however, is by informal noncommercial collection, and a controversy exists over the FAO estimate as discussed in Section IV,B,3. The substantial discrepancy between the harvested food products and the food documented as being available to households is not altogether understood, but may be largely accounted for by nonedible components of products, food fed to animals, storage, and transport and processing losses. Overall the total direct and indirect harvest by man of net primary production approximates 12 Gt (DW) $year^{-1}$ or 0.24×10^{21} J $year^{-1}$. However it must be recognized that a further quantity, not accounted for here, is not in fact a renewable milking of NPP but is a permanent destruction of the biomass stock through desertification, urbanization, water reservoirs, etc. (Brown, 1978, see also Section IV,B,3).

III. Fossilized Photosynthesis for Fuel

A. The Growth of the Fossil Fuel Era

Only a century ago, man's harvest from recent photosynthesis provided virtually all the energy needs of cooking, heating, lighting, transport, and manufacturing. The unleashing of the nonrenewable fossilized photosynthesis of aeons as coal, petroleum, and natural gas altered cultural evolution completely even in countries where current photosynthetic fuel sources still predominate. The dominance of rural life gave way or is giving way to urban life for the majority (Gifford, 1978a). The direct and indirect impact on agriculture is treated in Section V,A.

Figure 1 documents on a logarithmic scale consumption of fossil fuels over the last century. Except for periods during the two world wars and the Great Depression, growth in fossil fuel use has been consistently exponential at a rate of 4.35% year^{-1}. The sudden surge of price increases starting in 1973 had not had much sustained impact on the trend by 1980, although a slow-down in growth is now expected by many. If, however, the major reaction is to substitute electricity and liquid fuels made from coal for petroleum (see Section III,C,2), the opposite response may occur; the rate of fossil fuel consumption may *increase*, for the efficiency of obtaining electricity and refined liquid fuel from coal is much less than the 89% efficiency of refined products from petroleum

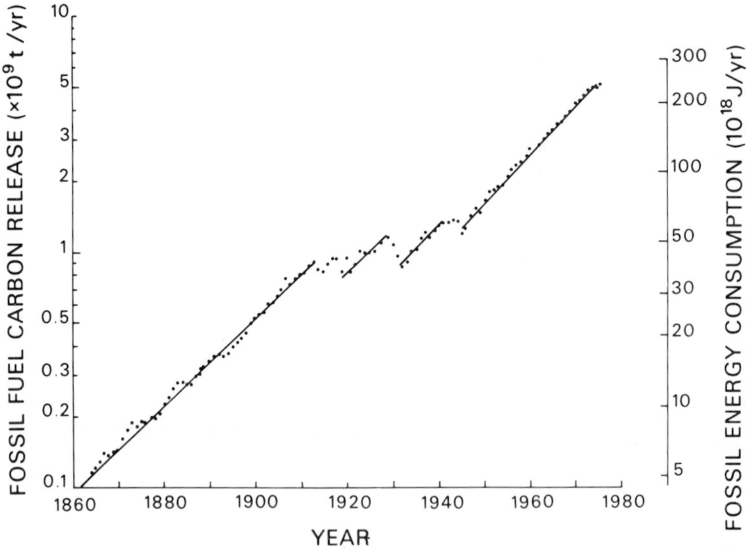

FIG. 1. The history of fossil fuel consumption. Data are based on the carbon released to the atmosphere from coal, lignite, gas, petroleum consumed, gas flared, and also a minute contribution from cement manufacture (Rotty, 1977). The energy scale, being based on equating the 1970 global fossil fuel consumption (Australian Institute of Petroleum, 1979) with Rotty's 1970 carbon release, is slightly distorted from reality, because of changes in the component fuels in the total, but is close relative to the resolution of the figure.

[e.g., 42% efficiency for methanol from coal (Stewart *et al.*, 1979), 35% for liquid hydrocarbon fuel from coal by the Fischer–Tropsch synthesis route (Rousseau, 1978) and about 25–30% for distributed electricity—see Section V,A]. There is much talk of substituting alternative energy sources such as nuclear fission, breeder and fusion, and solar energy-based alternatives such as reverting to increased reliance on biomass harvesting. The scope for the latter is discussed in Section V,B. In general, it must be recognized that historically it has taken 60–160 years for a new energy source to penetrate 50% of the market (Häfele, 1978).

B. Scale of Fossil Fuel Use Relative to Primary Production

Table III summarizes the scale of the different energy flows involving photosynthetically fixed carbon relative to current solar influx. Of the photosynthetically active part of the spectrum (about 44%), 0.28% is fixed by the biosphere as NPP, this being a weighted average between 0.13% for the ocean and 0.58% for land surface. Man's net harvest as food, fuel, wood, and fiber in 1976 was about 2.5% of the global NPP. However, both the gross harvest of biomass and the rate of combustion of fossil fuels is about three times man's net harvest of primary production, each being equivalent to almost 7% of global NPP. If the historic growth rate of fossil fuels consumption were to continue at 4.3% per annum (Fig. 1) starting from 0.224×10^{21}J in 1975, then by 2037 the rate of use would equal the rate of net solar fixation by the biosphere. Although the environmental implications of such large-scale heat and carbon release would be staggering, one needs to ask first what the prospects are for energy demand rising to such levels on the one hand, and fossil fuel supply holding out that long on the other.

TABLE III
Some Key Energy Flows on Earth[a]

	$\times 10^3$EJ/year
Solar receipt on earth	2500
Photosynthetically active solar receipt on earth	1100
Global net primary production	3.3
Gross plant harvest by man (1976)	0.24
Net biomass harvest by man (1976)	0.081
Food available to eat by man (1976)	0.016
Fossil fuels burned (1976)	0.23

[a] ($E = 10^{18}$).

C. Future Fossil Fuel Supply and Demand

1. DEMAND

One element of increasing demand derives from population growth; the other from an increase in per capita consumption. Given a world population in 1980 of 4.4×10^9 (United Nations, 1977) growing at 1.8% per year and its youthful age profile, it is unlikely that population size could stabilize below 10 billion without a sharp and undesirable rise in mortality. This is despite the fact that the average birth rate started to fall during the 1970s and that many countries now regard demographic planning as a proper function for government (United Nations, 1979a). After the birth rate stabilizes, such that each couple on average just replaces itself, it would take over 50 years for the size of the population to stabilize.

Taking the energy consumption of the United States (346 GJ/person · year in 1971) as an indication of people's potential aspirations for energy, 10 billion people would aspire to have 3.5×10^{21} J year^{-1}, although Weinberg and Hammond (1972) thought annual per capita consumption of the world might stabilize at twice the present United States consumption. It is quite possible then that unrestrained "demand" or aspiration for fossil fuel could even exceed global NPP of 3×10^{21} J year^{-1} (or 78 Gt(C) year^{-1}; see Table I).

2. SUPPLY

On the supply side, the global stock of fossilized photosynthetically reduced carbon is vast (see Section IV,A), but it may be a difficult task to ultimately recover as fuel large portions of this stock because there is no clear cutoff in accessibility. Although in the past the industry has not seen it this way, the definition of "accessibility" must relate to the net energy return from the energy extraction procedure (see Section V,A). This changes with emerging technology and in assessing potential resources, those in the industry are implicitly making judgments about the net energy balance of future fuel-gathering technology. To the extent that those involved have not seemed to think in terms of net energy return, the estimate of ultimately recoverable resources is uncertain. Nevertheless, with successive reappraisals, the estimates have become larger not smaller. Table IV summarizes the most recent appraisal for the 1977 World Energy Conference. The remaining "resource" is about 13 times greater than the current "economically recoverable reserves," 50 times greater than the total fossil fuel consumed so far, and 1360 times greater than the 1978 consumption of primary energy by the world (World Energy Conference, 1978). The potential resource is 10

TABLE IV
World Resources of Fossil Fuels Expected to Become Economically Extractable Ultimately[a]

Fuel	Already consumed to 1976		Remaining resource		Total original resource	
	(X 10^{21} J)	[Gt(C)]	(X 10^{21} J)	[Gt(C)]	(X 10^{21} J)	[Gt(C)]
Black coal	3.5	84	226	5408	230	5492
Lignite	0.4	10	70	1680	70	1690
Petroleum	2.2	38	11	218	13	257
Natural gas	0.8	11	10	153	11	164
Shale oil, tar sands, unconventional petroleum	~0	~0	26	515	26	515
Total	6.9	143	343	7974	350	8118

[a]*Sources:* Keeling, 1973; Rotty, 1973; Desprairies, 1978; Gifford, 1978b; McCormick *et al.*, 1978; Peters and Schilling, 1978. Assumptions: Black coal is 70% C; crude petroleum 84% C; 1 tonne coal equivalent = 29.3 GJ, 1 tonne lignite = 9.7 GJ, 1 tonne crude = 44 GJ, 1 m^3 natural gas = 37 MJ and weighs 0.54×10^{-3}t.

times greater than the global standing biomass (cf. Table I) and also about 10 times greater than the size of the atomosphere in terms of carbon content (see Section IV,A).

To explore how high the annual rate of fossil fuels consumption might rise, it is futile simply to extrapolate the historic 4.3% per annum to exhaustion of the presumed ultimately recoverable resource. A nonrenewable resource undergoes a depletion cycle with a more or less bell-shaped profile (Hubbert, 1969) as indicated for "ultimately discoverable" crude petroleum of the world in Fig. 2a. The traumas of rapidly increasing oil price in the 1970s occurred when the inflection point on the rising limb was reached: Exponential growth in consumption has ceased now that about one-half of the estimated ultimately discoverable resource has been discovered and about one-fifth of the ultimate resource is gone. The world must face up to decreasing annual growth rates of oil supply due to the ever greater difficulty of discovery and extraction until there is no growth around the turn of the century followed by negative growth in annual consumption thereafter. This smooth oil depletion cycle (Fig. 2a) is likely to be flattened because of political decisions to keep production below the technical maximum capacity (e.g., United States, Central Intelligence Agency, 1977; Beijdorff, 1979). But for global fossil fuel resources in toto, there is less likelihood

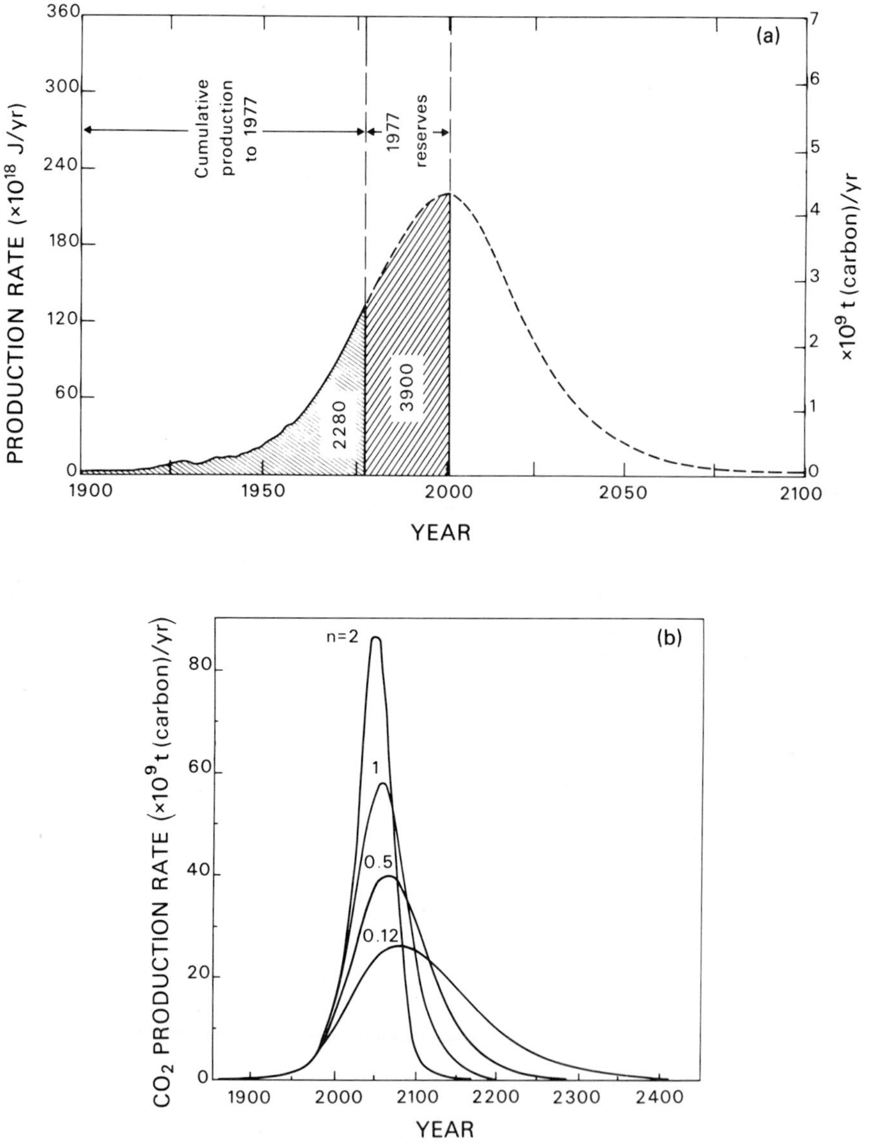

FIG. 2. Idealized smooth depletion patterns for global petroleum (a), and all fossil fuel (b) based on estimates of the ultimately discoverable recoverable resources. Curves (b) follow the equation $dC/dt = C.a(1\text{-}(C/C_\infty)^n)$ as discussed in text section III,C,2. (World Energy Conference, 1978; Keeling and Bacastow, 1977.)

of the bell-shaped depletion curve being flattened simply for political reasons, because the resource is distributed more evenly around the World (World Energy Conference, 1978).

Assuming then a smooth depletion cycle without a flat top, Keeling and Bacastow (1977) have used for heuristic purposes the convenient function $dC/dt = C \cdot a(1-(C/C_{\infty})^n)$, where C is the cumulative fossil fuel that has been consumed, C_{∞} is the ultimately exploited resource, a is a constant that is the relative growth rate during the early apparently exponential stages of the cycle, and n is a constant. Figure 2b depicts a family of such curves, for various assigned values of n, fitting the actual historic figures up to 1977. The area under each of the curves corresponds to an ultimate total for fossil fuel resources of 220×10^{21}J, rather lower than the more recent estimate in Table IV. The most free-wheeling depletion cycle ($n = 2$) would reach a total carbon consumption of over 80×10^9 $t(C)$/year, equal to global NPP, by 2050 A.D. Such a rapid buildup seems inconceivable, partly because the peak would precede the plateau in world population size if it were to stabilize at 10 billion people. More plausible is the curve with $n = 0.12$ for which per capita fossil fuel consumption by 10 billion people would peak in 2080 A.D. at one-third of present average United States consumption. Then the fossil fuel era would not finish until about 2300–2400 A.D. The peak rate would correspond to 35% of present global NPP.

In conclusion, it is possible, but not probable, that the rate of man's use of fossilized photosynthesis will reach a level commensurate with the current rate of energy fixation by the biosphere. If a cheap and abundant alternative source became available, the potential demand for such high energy use would exist. However, even the lowest rate of fossil fuel depletion depicted in Fig. 2b would probably have substantial impact on the carbon and energy balances of the Earth.

IV. The Global Carbon Cycle

A. Pool Sizes and Flows

Figure 3 indicates the estimated pool sizes of carbon in the global carbon cycle and the flow between them. Not included is the 95% of the total carbon content of earth below the lithosphere which cannot be regarded as a component of the *cycle*. It is however the source of the carbon in circulation and it continues to leak slowly via volcanoes and fumaroles into the atmosphere. Of the $60–70 \times 10^6$Gt(C), which is already part of the cycle, virtually all is locked in sedimentary rocks, and

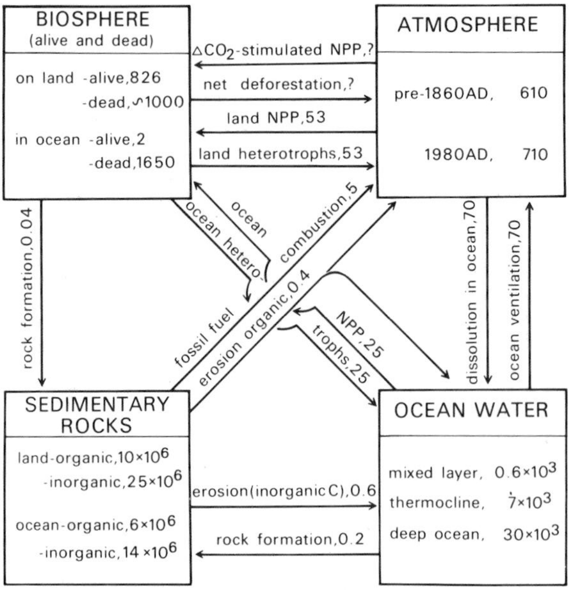

FIG. 3. Pool sizes (Gt(C) in the four boxes) and flows (Gt(C)/year on the arrows) in the global C cycle. Approximations derived from Whittaker and Likens, 1975; Baes *et al.*, 1976; Bolin, 1978; Oeschger and Siegenthaler, 1978; Broecker *et al.*, 1979. For discussion of the nonquantified 'deforestation' and 'CO_2-stimulated NPP' see Section IV,B,3 and V,C.

most of that is in oxidized form as limestone. This has been largely produced by marine organisms making their shells by the reactions of the type:

$$\underset{\text{in solution}}{Ca(HCO_3)_2} \rightleftharpoons \underset{\text{shell}}{CaCO_3} + CO_2 + H_2O \qquad (1)$$

Bicarbonate dissolved in the ocean got there by the weathering action of carbon dioxide on silicate rocks by reactions of the type (Broecker, 1973):

$$\underset{\text{metasilicate}}{CaSiO_3} + 2CO_2 + 3H_2O \rightleftharpoons Ca(HCO_3)_2 + \underset{\text{orthosilicate}}{H_4SiO_4} \qquad (2)$$

By now, redissolution of limestone by the reversal of Eq. (1) is a far greater contributor to ocean bicarbonate than is the silicate reaction (Garrels *et al.*, 1976).

The carbon pool in oceanic carbonate and bicarbonate is about 60 times larger than in the atmosphere. The surface ocean is over-saturated with calcium carbonate (Broecker, 1979). The ready reversibility of Eq.

(1) and solubility of CO_2 in water might suggest that any perturbation of the level of atmospheric carbon dioxide would in time be distributed between the oceanic and atmospheric pools in approximately the same proportions as they now exist. That is, the atmosphere should be well buffered against change, by the ocean (Garrels *et al.*, 1976). However, because the concentration of even a super-saturated calcium carbonate solution is very low and the exchange between the well-mixed surface ocean (70–100m thick) and the deep ocean through the stable unmixed thermocline is weak (Broecker *et al.*, 1979), the time for re-equilibration with the deep ocean after an atmospheric perturbation is slow (Baes *et al.*, 1976)—of the order of several thousand years (Garrels *et al.*, 1976). Moreover, because the uptake of CO_2 by the ocean as a whole exhibits a saturation response with respect to atmospheric content (its buffer capacity declines) atmospheric level might never re-equilibrate completely back to preindustrial content after the fossil fuel is all gone (Oeschger and Siegenthaler, 1978).

The fourth pool of carbon—namely hydrocarbon and coal—was produced by photosynthesis in reduced organic form (rather than oxidized inorganic, like limestone). This pool is thinly dispersed throughout sedimentary rocks and is less than one-half the size of the inorganic carbon reservoir in sedimentary rocks. A small fraction (0.1%) of it is sufficiently concentrated to be regarded as ultimately recoverable as fossil fuel (see Section III,C).

Against this background of the immensity of the C cycle relative to global biomass and the tendency for the distribution between pools to be stable in the very long term, it is necessary to examine the flows between pools to appreciate the scope for disturbance by man. Each of the net flows in Fig. 3 could be split into several components. For example, the flows into and out of the "biosphere" reservoir (here used broadly to include soil organic matter) have been separated into slowly and rapidly exchanging reservoirs associated with annual growth of easily decomposed tissues and with perennial accumulation of wood (Study of Critical Environmental Problems, 1970). Each major biome could also be treated separately.

Overall the NPP of 78 Gt(*C*)/year (Table I) is equivalent to 10–13% of the atmospheric C content. Thus the turnover time of the atmosphere by the biosphere is 7–10 years. In fact, it is faster than that, when gross carbon fixation is considered: Leaf net CO_2 exchange is 55% or higher than is implied by DW increase because of respiratory carbon losses (Yamaguchi, 1978); gross CO_2 exchange exceeds net CO_2 exchange by about 40% due to photorespiration. Thus gross photosynthesis is over two times the NPP, implying an atmosphere-to-biosphere turnover time of only 3–5 years. Thus if vegetation can respond to a change in atmospheric CO_2, the time course of its attenuation of the perturbation is one-half a decade. Figure 3 depicts the outflow and inflow to the biosphere pool as being balanced (to two significant figures). Whether or not they are exactly balanced over the annual cycle is discussed in Section IV,B,3.

Exchange between the atmosphere and surface ocean is also rapid, having a turnover time similar to that with the biosphere—about 6–7 years—but as discussed earlier the surface ocean to deep ocean exchange is very slow indeed.

The release of carbon from the reduced carbon pool by man's burning of fossil fuels is relatively small, but it is not balanced by the equivalent formation of new deposits of reduced carbon on the sea floor.

B. The Changing Atmospheric Content and Its Causes

Accurate records of atmospheric CO_2 started in 1958, and long runs are available for air over Hawaii, the South Pole, Scandinavia, Alaska (Machta, 1979), the Tasman Sea (Pearman, 1977), and New Zealand (Lowe *et al.,* 1979). Many new monitoring stations are being built (Dahlman *et al.,* 1980). Superimposed on an annual fluctuation, all records exhibit a secular trend upwards with the annual increase being between 0.5 vpm (volumes per million) and 2.2 vpm averaging, since 1958, 0.9 vpm/year. There is indication of an acceleration of the average annual rate of increase from about 0.75 vpm/year to a little over 1 vpm/year. The 1979 concentration was 335 vpm. The annual fluctuation is considered to be due largely to the seasonal cycle of terrestrial photosynthesis (Hall *et al.,* 1975; Machta, 1979). The scatter of available values suggest 290 vpm as most probable for the last decades of the nineteenth century (Callendar, 1938; Revelle and Suess, 1957; Bray, 1959). There is indirect evidence that before 1850 A.D. the atmospheric content may have been as low as 270 vpm (Stuiver, 1978; Wilson, 1978).

Four possible causes of the observed change in atmospheric CO_2 can be identified.

1. FOSSIL FUEL USE

Figure 1 indicates the history of use of fossil fuel carbon released as CO_2 into the atmosphere, totalling from 1860–1979 A.D. 140 Gt(*C*) (Bolin, 1978). Had it all remained airborne it would be equivalent to an increase in atmospheric content of almost 70 vpm CO_2. In fact, however, this rate of release has been about twice the observed rate of build-up of CO_2 in the atmosphere. Until around 1975, it was generally accepted that fossil fuel combustion was the sole cause of the atmospheric rise, the residual of CO_2 released taken up most likely by the oceans and the biosphere.

2. OCEANIC EXCHANGE

With uptake and release of CO_2 by the surface ocean being so large (70 Gt/year) (Fig. 3), it would take an imbalance of only 2–3% per year to account for the observed atmospheric rise. The only reason one can suggest for such a consistent imbalance is a change in ocean temperature. From about 1880 to 1945 mean global surface temperature increased by about 0.5°C, but has fallen since then (Lamb, 1974) by about 0.1°–0.2°C per decade, at least in the northern hemisphere where most of the recording stations are located (Kukla *et al.*, 1977). These changes seem to be part of a long-term systematic oscillation reflected in the $^{18}O/^{16}O$ ratio of a millenial record of snowfall on the Greenland ice cap (Broecker, 1975). Since CO_2 is more soluble in cooler water, such a cooling in recent decades would be expected to lead to decreasing, not increasing, atmospheric CO_2. Ericksson (1963) estimated that a 1°C change in ocean temperature would lead to a 6% change in atmospheric CO_2 at equilibrium. Unfortunately, there are no sea–surface temperature records over most of the southern hemisphere (where most of the ocean is) especially between latitudes 30° and 60°S (Kukla *et al.*, 1977). However in New Zealand (Salinger, 1975), in higher latitudes of the southern hemisphere (Damon and Kunen, 1976), and in Australia (Tucker, 1975), the sparse evidence suggests that a slight warming rather than a cooling trend is occurring. So although there is no clear evidence that the current rise in atmospheric CO_2 can be partly attributed to net release from the oceans, it cannot be wholly ruled out as a partial contributor.

3. THE LIVING AND DEAD BIOSPHERE

In the mid-1970s, a popular idea spread that the net rate of destruction of tropical forest may be as high as 1.5–2%/year or even higher due to lumbering, clearing for agricultural land, development, informal fuel-wood gathering by subsistence communities, and forest fires. Assuming this destruction rate is world-wide and that it is all burned, it can be seen (from Table 1) that it would release around 10 Gt(*C*)/year into the atmosphere, plus any releases due to enhanced oxidation of soil organic matter after its exposure (Woodwell, 1978; Woodwell *et al.*, 1978). This is more than the release from fossil fuels. Others, while accepting that official deforestation figures may be conservative, doubt that net deforestation is as high as 1.5–2%/year (Food and Agriculture

Organization, 1978b, Broecker *et al.*, 1979; Loomis, 1979; Ralston, 1979). Other broad estimates of the current annual net release of CO_2 from the biota due to man's utilization are lower than Woodwell's, being in the range 1–4 Gt(*C*)/year (Adams *et al.*, 1977; Bolin, 1977; Kohlmaier *et al.* 1978; Wong 1978). None of the estimates include the possible stimulation of biosphere photosynthesis by the higher CO_2 itself. This is discussed in Section V,C. Studies with an indirect method using carbon isotope discrimination by photosynthetic enzymes have not yet been very conclusive (Freyer, 1978; Stuiver, 1978; Wilson, 1978). They agree that there were substantial releases of carbon from the biosphere in the second half of the last and the first part of this century, but difficulties in the method preclude fine resolution between fossil fuel and biotic contributions over the last 2 or 3 decades.

Concerning oxidation of soil organic matter following cultivation, Bohn (1978) estimated that from 1870 to 1980 the release from that source averaged 1–2 Gt(*C*)/year, but that it would be partially offset by the formation of new peat at the rate of 0.4 Gt(*C*)/year. Another countering effect may be that as the mean surface temperature of the northern hemisphere has declined by 0.5°C since 1945 (Lamb, 1974) the amount of soil organic matter would increase. Using Jenny's (1930) classical soil formation equation this might amount to a sequestering of 155 Gt(*C*) since 1945 (Loomis, 1979)—if real, it is a very large sink, which will become a large additional *source* when the temperature cycle turns upwards again as it is expected to do in the mid-1980s (Broecker, 1975).

Application of industrially fixed nitrogen and other inorganic fertilizers to agricultural lands may stimulate accumulation of carbon as standing biomass and dead organic matter. The global application of N at the end of the 1970s was about 50 Mt/year (United Nations, 1979b) to 60 Mt/year (Söderlund and Svensson, 1976). Further, man-induced nitrogen fixations by grain legume crops (40 Mt), managed hay and pasture legumes (28 Mt), and emissions from fossil fuel combustion (20 Mt) (Delwiche, 1977) bring the total man-induced N fixation to about 148 Mt/year. This is twice as great as the natural preindustrial biological and atmospheric fixation of nitrogen. As an upper limit, if all the 148 Mt of nitrogen were to be locked up in terrestrial and aquatic organic matter at a C:N ratio of 10, a further 1.5 Gt(*C*)/year would be fixed. However, it is probable that much of this "artificial" nitrogen fixation escapes into the atmosphere as nitrous oxide and nitrogen (Söderland and Svensson, 1976). Photosynthesis and nitrogen-cycling on the continental shelves, however, provide a possible nitrogen-primed carbon pump delivering carbon onto the continental slopes. The mechanism proposed (Walsh *et al.*, 1981) is that organic matter produced on the shelf with a C:N ratio of

5 moves onto the slopes and is further processed biologically such that one-half the nitrogen is released to the water column being recycled perhaps 10 times per year, leaving a residue of material with C:N ratio of 10 to continue to move down the slopes in the sediment. This mechanism could lead to a further sequestering of up to $0.7 \times$ Gt(*C*)/year of which about one-half might be due to man-made nitrogen. Nitrogen, which vents to atmosphere as nitrous oxide, contributes to the "greenhouse effect" in the same way as CO_2 does (Flohn, 1978; Schneider, 1978). Furthermore, nitrous oxide, which reaches the stratosphere, may have feedback effects on terrestrial productivity via interactions with the protective ozone layer (Pratt, 1977) as indeed might the buildup of stratospheric carbon dioxide (Groves *et al.*, 1978) and man-generated chlorine compounds (Whitten *et al.*, 1980).

Most of the phosphorus fertilizer used is immobilized in the top few centimeters of soil. Organic matter formed in association with phosphorus enrichment of fresh and sea water corresponds to only about 2% of the global fossil-carbon release (Broecker *et al.*, 1979).

In summary, although quantitative assessment eludes us, it is plausible that positive and negative aspects balance, leaving the dead and alive biosphere as an approximately constant-sized pool at present.

4. GEOLOGICAL ORGANICS IN SEDIMENTARY ROCKS: EROSION

The global rate of continental erosion increased about threefold after the appearance of *Homo sapiens* on earth (Degens *et al.*, 1976) from a long term value of 9 Gt of sedimentary rock per year before man to about 24 Gt/year (Judson, 1968). Taking the average organic carbon content of sedimentary rocks to be 1.8% (Bolin, 1978), the reduced carbon becomes available for oxidation from sedimentary rocks at the rate of about 0.4 Gt/year. But some of this would wash directly into the oceans. Garrels *et al.* (1976) used a model of the global carbon/oxygen sediment cycles to suggest that a (geologically) abrupt threefold increase in erosion rate would cause over the following 10 million years a decreased atmospheric oxygen content of 15% and increased atmospheric carbon dioxide content of 2.5-fold. This corresponds to a very low annual increment in atmospheric carbon dioxide.

5. CONCLUSION

The only firmly established substantial net efflux of carbon dioxide into the atmosphere derives from fossil fuel burning. Although it seems likely that opening up of agricultural lands in the last century has contributed to the buildup of atmospheric CO_2 from 270 vpm, it is uncertain whether or not the current clearing involves a net CO_2 release.

World forests may now even be a net sink for CO_2 (Machta and Elliott, 1980). Moreover, the combined effects of man's introduction of N fertilizer into the environment, of enhanced net primary productivity due to the CO_2 enrichment itself and of a buildup of soil organic matter due to slowly falling northern hemisphere temperature, could counterbalance any deforestation effect.

It happens that the best oceanographic mixing models can only just account for the apparent partitioning between ocean uptake and atmospheric retention of fossil fuel-derived CO_2 providing that the biosphere is a small net sink for the excess CO_2 (Ekdahl and Keeling, 1973). If the biosphere were in fact a substantial net source of CO_2, as Woodwell and others have proposed (Section IV,B,3), then the disappearance of CO_2 from the atmosphere cannot be accounted for by supposing that oceanic mixing must be faster than already assumed (Siegenthaler and Oeschger, 1978; Broecker *et al.*, 1979). The patterns of future fossil fuel consumption hypothesized in Fig. 2b would be expected to lead to the time courses of atmospheric carbon dioxide shown in Fig. 4. This presumes that the increase observed over the last two decades is attributable solely to fossil fuel use. The peak CO_2 level would be something like six to eight times the preindustrial CO_2 concentration. For the lowest curve ($n = 0.12$), a doubling to 600 vpm would occur, according to the model,

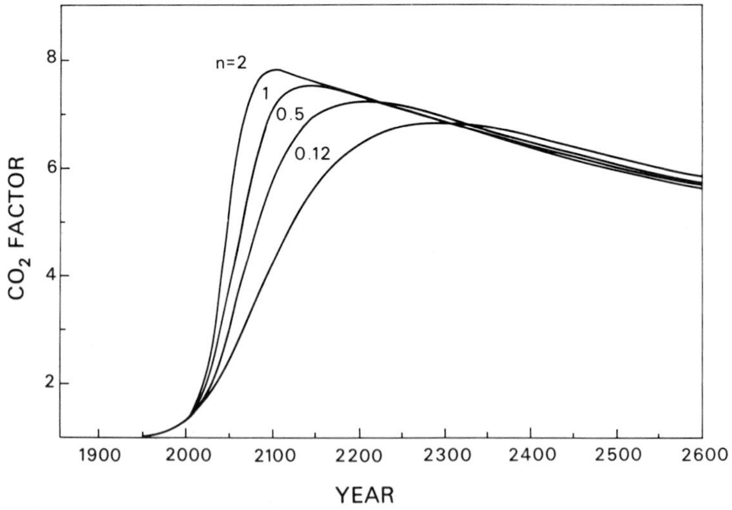

FIG. 4. Possible time course of atmospheric carbon dioxide concentration according to the hypothetical time-courses of fossil fuel depletion in Fig. 2b and a global carbon-cycle model. (From Keeling and Bacastow, 1977.)

by 2035 A.D. Such change might have a considerable impact on global photosynthetic productivity, not only directly (see Section V,C) but also via climate change.

C. Possible Climatic Impact

Global surface temperature would on average increase by 2°–3°C for each doubling of atmospheric CO_2 concentration according to results of the most comprehensive three-dimensional general circulation models published (Manabe and Wetherald, 1975, 1980; Manabe and Stouffler, 1979). The temperature increase would be maximum at high latitudes (8°–11°C) and minimal in the tropics. This may be sufficient to cause the breakup of the unstable and already thinning (Thomas, 1979) ice cap of West Antarctica leading to a 5 m sea-level rise over about 10^2 years (Mercer, 1978). There may, however, be a substantial lag in surface temperature rise due to the thermal capacity of the oceans (Mason, 1979; Newell and Dopplick, 1979). Idso (1980) believes that the temperature effect may be 10 times lower than predicted by Manabe's models. Idso, however, does not appear to have included the positive feedback through the water vapor greenhouse effect which contributes considerably to the effect in the Manabe model.

Besides general geographic heterogeneity, there is likely to be seasonal and localized geographic heterogeneity in the warming and associated rainfall effects (Smagorinsky, 1977; Chowdhury and Kukla, 1979). These effects may lead to regional changes in the length of the agricultural growing seasons insofar as they are determined by timing of first and last frosts and seasonal rainfall distribution. This could have substantial social impact (Gifford, 1979c). The local picture is made more complex by the observation that deforestation also has more direct effects on local climate; it may reduce rainfall in a deforested region (Sands, 1979).

V. Interactions between Photosynthesis, Food, and Fuels

Having described the scale of managed and "wild" global photosynthesis, of the store of "fossilized photosynthesis" and its rate of utilization and the interaction of these with the global carbon cycle, I will now consider how fossil fuel use interacts with agricultural productivity, how agricultural and biospheric production might be substituted for fossil fuels and how the modified global carbon cycle might influence it all.

A. Support Energy for Food and Fuels

Biomass becomes food only when effort is expended to channel it into a form suitable for eating. Energy is required for such an effort and this "support energy" (Leach, 1975) can be viewed, in the language of control systems, as an "energy gate" controlling the flow of solar energy into processed food (Odum, 1971). The food system does not finish at the farm gate; it includes also the factory and domestic processing stages.

Several forms of energy are used to support the photosynthetic food system—muscle energy (in labor and animal power), wood and dung especially for cooking, solar energy for field drying of hay and grain, hydropower for electricity and hydraulic rams, wind for water pumps and electricity, and fossilized photosynthesis for liquid fuels and electricity. These are very different types of energy in operational terms. A joule of fossil fuel cannot be readily compared, in terms of its utility, to a joule of muscle power, although in practice, fossil fuels are used to substitute machines for labor (de Wit, 1975). Nevertheless, in comparing diverse agricultural systems in terms of their support energy to food energy balance, analysts at first lumped all support energy inputs together to obtain a single value. This was the aggregate support enthalpy. Recognizing that enthalpy content does not reflect the useful work potential of an energy form, free energy and "availability" were considered as alternative criteria for energy analysis (International Federalion of Institutes for Advanced Study, 1974). However, in general, these parameters are not readily determinable and, since in modern systems virtually all the support energy used is from fossil fuels, use of enthalpy (heat of combustion) seemed adequate.

Much of the early work on energy analysis of food systems was carried out from the viewpoint of the ecologists interested in natural populations of animals for which the metabolic energy derived from food must exceed the energy expended in getting and preparing it (Lawton, 1973). So emphasis was placed on output:input ratios of agriculture and food systems. Economists were critical, pointing out that to suggest making management decisions on such energy efficiency criteria was to hypothesize an "energy theory of value" (Edwards, 1976; Lloyd, 1978) which is greatly inferior to the economists' techniques for analyzing the way scarce resources are allocated (International Federation of Institutes for Advanced Study, 1975).

Economists argue that it does not matter if more support energy is invested in food production than solar energy is fixed by photosynthesis into food so long as people profit by it; after all, fossil fuel is better used to aid agriculture than for many other uses. It is not the net energetics of the *food* system of a community which must be in positive balance, but the net energetics of the *total harnessed energy system* that must be in positive balance for system survival, as long as the harnessable energy resource persists. Society can subsidize the food energy system, if it wishes, as long as any net energy loss is balanced by net energy gain in another energy acquisition system. As Costanza (1980) points out, the adequacy of embodied energy as a basis of economic valuation depends on where you draw the system boundary.

The net energy gained in food gathering can be considerable for wild animals and subsistence agriculture having no nonfood energy source. For hummingbirds, the "energy ratio" [*ER* = energy gained/energy expended (Leach, 1975)] is in the range of 4–70 (Lawton, 1973). Similarly for subsistence hunter gatherers like the !Kung Bushmen of the inhospitable Kalahari Desert the *ER* is a substantial 8 (Leach, 1975). When another energy source is tapped, say to increase food production, it is necessary that the metabolic energy invested in mobilizing the second energy source does not exceed the metabolic energy provided by its use. For example, when fuelwood is harvested to improve food digestibility by cooking, the metabolic energy expended in gathering fuel should not exceed the improvement in food digestibility fostered by the cooking. Or at least, if it does so exceed, it would not be for nutritional reasons and it must not place the ratio (of total metabolic energy expended in the food system:metabolic energy derived from food) below unity. In some subsistence regions where wood is now particularly scarce due to overharvesting (e.g., in Ethiopia and Nepal), it takes a day's trek or more to acquire fuel because nearby land has been deforested. Such systems are on the borderline of viability, and the fuelwood crisis is as serious as the food crisis in many areas of the Third World (Eckholm, 1975; Food and Agriculture Organization, 1978b). Clearly such systems, which developed by depleting a store of energy recently fixed by photosynthesis, cannot persist on the same scale unless a new energy source is tapped.

In the industrialized world, the larger store of fossilized photosynthesis meets this need. Here, it is not the net energy of the food system that is of overriding importance, since current photosynthesis is not a major input, but the net energy of the major energy sources. For planning purposes, it is the net energy return on newly developed energy sources that is critical. For example, the net energy returned as refined products from Middle East crude petroleum is about 89% (Chapman *et al.*, 1974; Gifford and Millington, 1975) of its energy content. That is, output:input ratio (energy ratio) is about 89/11 = 8. For refined liquid fuels from synthetic crude oil obtained from shale yielding 36 liter (oil) per ton (shale), the ratio is likely to be about 2.3 (Chapman and Hemming, 1976). Clearly there must exist relatively inaccessible fossil hydrocarbon pools that would take more energy to extract than would be yielded. In making the assessments of the potentially discoverable fossil fuels of the world (see Section III,C,2), it is hard to know to what extent the net energy question has been considered. Ambitious ideas about returning to photosynthetically derived biomass to substitute for petroleum are sometimes espoused in ignorance of the net energy situation. This will be taken up in Section V,B after looking at the energetics of biomass production through existing agriculture.

The support energetics of many agricultural systems have now been assessed at national, regional, and farm and crop levels in terms of the energy ratio. Details vary widely but certain generalizations can be made. For crops grown in subsistence systems using muscle power and virtually no fossil fuel support energy, the energy ratio (nutritional energy output:total support energy input) can reach very high values up to 30, if the climatic and soil environment is favorable. But the ratio can also get close to the "break-even" value of unity in unfavorable environments (Gifford, 1980a). In modern systems, fossil fuel replaces muscle power and also provides the necessary means for yield-boosting inputs unavailable to subsistence farmers (fertilizer, agro-chemicals, certain types of irrigation, greenhouse protection). Yields per hectare can be high, but the energy ratio is never high (though usually in excess of unity). This is because, no matter how high the support energy input goes, the crop output can only approach asymptotically the photosynthetic ceiling imposed by annual solar irradiance and maximum canopy photosynthetic efficiency. At the level of national agriculture, a comparison of data from United States, United Kingdom, Holland, and Isreal reveals a remarkable similarity in the energy ratio of nutritional product to fossil fuel input (Gifford, 1976); each was about 0.5–0.7. By contrast, Australian agriculture had an equivalent ratio of 2.8. This discrepancy is attributed to Australia's reliance on legume nitrogen rather than bag nitrogen and to the low level of all inputs, resulting in low yields. This approach is necessitated in Australia by the economics of exporting two-thirds of the produce to distant markets.

The proportion of a nation's fuel and power supply that is devoted to agriculture is generally in the range of 3–6% and about one-half of this is energy used directly on the farm rather than for other inputs. Transport, storage, and processing of food downstream from the farm requires about five times as much energy as does the agricultural sector, such that national food systems as a whole require 10–25% of the national primary energy demand (Gifford, 1980a).

B. Biomass as an Alternative to Petroleum

In the late 1970s, there was a surge of popular enthusiasm in the industrialized world to use biomass as a source of fuel to substitute for petroleum as it becomes less available (e.g., Wade, 1979). In the light of the problems of deforestation being faced in the Third World due in part to excessive fuelwood harvesting and of the fact that petroleum was used in the industrialized world to foster growth when biomass (through fuelwood and animal fodder) would no longer suffice, this seems an unlikely substitution on a significant scale.

Knowledge of the size of global NPP relative to actual and potential energy demand (Section III,B) leads to skepticism as to the viability of substantial replacement of petroleum by biomass especially given the expanding food needs of the burgeoning population. Careful examination of the net energy recovered as useable liquid fuel from biomass (Chambers *et al.*, 1979; Stewart *et al.*, 1979) leads to further caution. Where crop residues and manure are considered, there are further questions about exacerbating the problem of soil deterioration and erosion. Suggested fuels include methane via anaerobic bacterial fermentation of high nitrogen containing organic matter like manure, ethanol from yeast fermentation of sugar or starch crops, methanol from lignocellulosic materials, natural hydrocarbons from *Euphorbia spp.* (Calvin, 1979), vegetable oils, pyrolytic oils, and char and gas from any organic matter (Tatom *et al.*, 1975).

There are several estimates of the potential scale of net production of fuel from biomass in different parts of the world. It is hard to summarize them briefly because the criteria used vary, and there are so many different options available [see Chambers *et al.* (1979) in relation to assessing whether gasohol in the United States saves fossil fuels or wastes them]. The multiplicity of options is greater when integrated fuel/food systems are considered (Lipinsky, 1978). Certainly, energy analyses of dryland crops has shown that up to the farm gate a ratio of combustible output:high grade energy input can frequently be in the range 3–6 as long as highly intensive methods are not used. But, as for the food system downstream, processing energy costs can be higher than on-farm costs. In general, to obtain satisfactory net energy yields, it is necessary to use some of the biomass feedstock as a process fuel. Since most industrial processes during the 1960s and 1970s have become geared to using high grade energy like fuel oil and natural gas for process heat, a turn-around in attitude is needed to use awkward low grade biomass fuels for the process heat needed in high-grade renewable liquid fuel production.

Vergara and Pimentel (1978) made a comparative assessment of the potential for energy from biomass in five countries. The results are summarized in Table V. This is a broad-scale study, and if each country were examined in detail region by region, it is likely that the estimated renewable fuel potential would drop because other land-use conflicts become apparent. Nevertheless, the table provides an approximate guide. The biomass-derived fuel potential for less-developed Sudan is five times the 1976 primary energy consumption in that country. But in the United States, which has a per capita energy consumption about 100 times greater than in the Sudan, the biomass–energy potential is only 2.6% of its 1976 primary energy consumption. Looked at another way, if the present population of the Sudan were to increase its primary energy demand to the United States level it could satisfy only 17% of this demand by its biomass–fuel potential.

Brazil has already embarked on a vigorous program of fuels from biomass (Hammond, 1977; Stumpf, 1978) with a goal of replacing one-half its gasoline needs by 1985 (Bazin, 1979). Special circumstances facilitate this. First, the unused arable potential, presuming stable production systems can be created, is very large and being in the tropics NPP is high; second, per capita energy consumption is low being only 5.4% of United States per capita consumption. Third, the distribution of energy

TABLE V
Commercial Energy Consumption in 1976, Noncommercial Energy Consumption in 1976, and the Potential for Fuels from Biomass Sources in Five Countries[a]

Country	Per capita commercial energy consumption (GJ/year)	National commercial energy consumption (EJ/year)	National noncommercial energy consumption (EJ/year)	Potential fuels from biomass (EJ/year)	1976 Total energy consumption potentially from biomass (%)
United States	340	72.5	0.62	1.9	2.6
Sweden	178	1.5	0.015	0.21	14
Brazil	18	2	0.87	4.4	153
India	5.7	3.5	2.3	2.1	36
Sudan	3.7	0.067	0.12	1	535

[a]See Vergara and Pimentel, 1978. Biomass sources include fuelwood, manure, sugar crops, urban refuse, sewage, food processing waste, industrial waste, and forest and sawmill residue.

consumption through the population is very uneven. A large rural labor force, otherwise unemployed, works the fuel plantations to produce fuel for the minority of fuel consumers in the cities. Brazil's National Alcohol Program cannot be seen in a straightforward manner as a model for other nations.

Australia is a large continent with few people (14 million in 1979); development of a biomass–fuel system might seem a good option. A detailed study of the potential for ethanol production from fermentable products (grain and sugar) and methanol production from lignocellulosic plant components was made for Australia (Stewart *et al.*, 1979). Agro-climatic and edaphic considerations indicated that biomass production of almost 50 Mt(DW)/year of lignocellulosic material and 17 Mt(DW)/year of fermentables was possible without encroaching on crop production. After the direct and indirect liquid fuel costs of growing, harvesting, transporting, and converting to alcohol were accounted for, this biomass would produce a net 290×10^{15}J (methanol) and 134×10^{15}J (ethanol) per year, equivalent to 60% of the demand in Australia for liquid transport fuels and 13% of national primary energy demand. The scale of the biomass production system would be about equal to the scale of the existing cropping and forestry system in Australia during the 1970s. The net production of these alcohols was 67% of the gross alcohol production, a seemingly high figure because wherever possible lignocellulose rather than liquid fuel was used as an energy source in the notional production system. This contrasts with the recent situation in the United States where, if standard maize production methods and conventional ethanol distilleries were used, the net energy of ethanol production for gasohol (10% ethanol, 90% gasoline), would be negative (Chambers *et al.*, 1979). As in Australia, options do exist in the United States to adjust the technology to give a positive net energy balance.

Alcohol fuels from biomass are expensive (Stewart *et al.*, 1979; Weisz and Marshall, 1979), being two to five times more costly than refined petroleum products from petroleum at 1979 world prices or from methanol with coal as feedstock.

In conclusion, biomass fuels are likely to play an increasing role in the fuel supply of industrialized nations, but will meet only a small fraction of the uses to which fossil fuels are now put. Moreover, except in small areas, which have no alternatives, they are likely to be less attractive economically than coal-derived fuels. This suggests that fossil fuels will continue to be exploited, perhaps on an expanding scale for a long time yet. Hence the atmospheric carbon dioxide question will loom larger. How will this affect the biosphere and agriculture?

C. Atmospheric Carbon Dioxide, Agricultural Production, and Biosphere Productivity

1. BIOSPHERE

The biosphere's photosynthetic response to rising atmospheric carbon dioxide concentrations is one of the major uncertainties in models of the global carbon cycle (Section IV,B). Countering carbon release by net deforestation, there may be enhanced fixation by the remaining vegetation due to CO_2-stimulated photosynthesis. This could explain why the amplitude of the annual cycle of atmospheric CO_2 concentration had not changed since continual monitoring began (Hall *et al.*, 1975). Although at the single leaf level detailed models of photosynthesis with respect to carbon dioxide exist, at the level of the global photosynthetic mantle a simple empirical relationship is more appropriate, as a submodel to global carbon-cycle models. The equation commonly used (e.g., Bacastow and Keeling, 1973) is $NPP = NPP_o (1 + \beta \ln(C/C_o))$, where NPP_o is the net primary productivity at a reference or baseline atmospheric carbon dioxide concentration, C_o, and β is the "biotic growth factor," which can be interpreted as the percentage increase in NPP following a 1% increase in atmospheric carbon dioxide. Most carbon-cycle modellers have set low values of β (around 0.2–0.3) on the grounds that other environmental factors like water, temperature, light, and nutrients limit global productivity so much (see Section II,A) that CO_2 enrichment would have little effect. Lemon (1977) reasoned that where environmental conditions should be most favorable for a CO_2 response (high light and high temperature) the greater water stress associated with such conditions would dampen a CO_2 response. However, Gifford (1979a) presents data showing that β for wheat canopy growth increases, rather than decreases, to values as high as 0.8 when the water supply and light environment become more limiting. High values of β were also found for nitrogen-limited cotton by Wong (1980b). The result suggests that the global carbon-cycle models may need to allow for greater biospheric response to atmospheric enrichment than before.

In assessing the role of biosphere in absorbing excess anthropogenic carbon dioxide, however, it is not the change in *NPP* but the *store* of living and dead biomass which counts. This is determined by the way ecosystems partition incremental *NPP* between wood and ephemeral material oxidized within the year by heterotrophs, on the one hand, and the turnover of the woody material on the other. It is conceivable that despite an enhancement of tree growth rate, total biomass may not change if the death and decay rates increase to an equivalent degree

(Botkin, 1977; Lemon, 1977). Interpretation is made more difficult by the time-frame discrepancy between the turnover of forest ecosystems ($> 10^2$ year) and the monitored changes in atmospheric CO_2 ($\sim 10^1$ year) and by the possibility of differential responses of different species (even within the C_3 group of plants) to CO_2 enrichment (Lemon, 1977) leading ultimately to changes in species composition of plant communities (Wright, 1974). Nevertheless, Gifford (1980b) assessed that global net storage of carbon, due to CO_2 enrichment effects on the biosphere, may be 1-1.6 Gt(*C*)/year higher in 1979 than it would have been with the 1958 atmosphere. However, the biosphere is not amenable to the controlled experiment, so accurate information will never be available. Monitoring and survey faces daunting sampling problems, and the "experiment" is already underway without adequate baseline data.

2. CROPS

Despite much horticultural work on CO_2 enrichment of greenhouse crops, information on the response of field crop growth and yield to continuous exposure to modest CO_2 enrichment is sparse. The shape of CO_2 response curves of leaf photosynthesis (Gifford and Musgrave, 1970) suggests that growth of C_4 species like maize might show little response to CO_2 enrichment above ambient levels when grown with abundant water supply. This prediction was borne out in a comparison between wheat and sorghum and between cotton and maize under conditions of abundant water and nutrient, and moderate light (R. M. Gifford, unpublished; Wong, 1980a).

A further distinction between C_3 and C_4 species may be that stomatal closure in C_4 species is reputed to be more sensitive to carbon dioxide (Akita and Moss, 1972). Thus, although growth of C_4 species may not respond to CO_2 enrichment under abundant watering, lower stomatal conductance may lead to greater water-use efficiency and hence improved yield under water-limited conditions (Morison and Gifford, in preparation). However, the two major C_4 crop species, maize and sugarcane, are not generally grown under water-limited conditions. So we cannot expect an appreciable increase in the global yield of the major C_4-species products due to increased atmospheric carbon dioxide. Despite the lesser sensitivity of wheat (C_3) stomata to CO_2 concentration, grain yield of that crop also benefits relatively more from a given CO_2 enrichment under water-limited than under well-watered conditions; wheat yield increase corresponding to an annual CO_2 increment of 1.2 vpm/year ranged from 5 to 13 kg/ha·year (Gifford, 1979b).

It is not yet possible to attempt general prediction of field crop responses to designated increases in atmospheric carbon dioxide. One of the difficulties for predictive models is the tendency for photosynthesis rate per unit leaf area to counteradapt in some species following a change in atmospheric carbon dioxide concentration. Although this was not found in wheat when flag leaf photosynthesis was studied on lifelong CO_2-enriched plants (Gifford, 1977), Raper and Peedin (1978) found no lasting effect of CO_2 enrichment in young tobacco, and Aoki and Yabuki (1977) found that the enhancement of net photosynthesis rate of cucumber due to CO_2 enrichment declined over a period of about 2 weeks to a value equal to or even less than that in ambient atmosphere when high levels of CO_2 (> 1200 vpm) were used and Wong (1980a,b) also found some counter responses. This aspect may relate to an inability of plant growth to respond to a greater photosynthetic supply. For example, Neales and Nicholls (1978) found no appreciable CO_2 enhancement of wheat-seedling growth during the first 4 weeks, but Gifford (1977) found considerable enhancement of growth of wheat over the full life cycle despite a nonsignificant effect during the first few weeks. Growth did not respond until tillering (branching) started. Once the ontogenetic stage for creation of numerous meristems was reached, the crop could respond to CO_2 enrichment and the apparent feedback inhibition of photosynthesis ceased.

A further difficulty relates to secondary effects of CO_2 increase on agriculture (and the biosphere as a whole). If a significant rise in global surface temperature occurs as a result of atmospheric CO_2 build up, then poleward shifts in agro-climatic zones may follow as the intensity of the atmospheric and oceanic circulations weaken (Section IV,C; Bach, 1978).

VI. Conclusions: Photosynthetic Resources and Man's Future

The best-recognized route through which man depends on photosynthesis is through CO_2 assimilation by crops, and failure to match food production and population size does indeed pose a threat. There are, however, other dependences on photosynthetic resources that pose equally serious threats to mankind's social stability.

Application of agro-climatic relationships to the world's soil regions, bearing in mind possibilities for irrigation, indicates that, all else equal, the agricultural system could be expanded and intensified to feed much more than the lowest plausible stable level of world population of 10–12 billion (Section III,C). In fact Buringh *et al.* (1975) concluded that the absolute maximum crop production could be over 20 times the present amount. This would require modern methods relying on energy-intensive, low-labor procedures and, by implication, the majority of people would be city dwellers. Potential photosynthetic productivity of food crops is not, then, a critical problem in feeding the world, presuming population can be stabilized at the demographically earliest time possible. However, all else would not remain equal and the implications of attempting to extend the application of our knowledge of photosynthesis and crop production to such a scale are serious for many reasons some of which involve photosynthetic resources themselves. It is sobering to note that the annual production of plant biomass associated with Buringh's absolute maximum food production estimate from land would be about 150 Gt—greater than current biosphere net primary production (Table I). As discussed in Sections III,C and V,A, the petroleum demands of such a system, both for agricultural production and for sustenance of an urban infrastructure and life-style compatible with an effective mechanized agriculture, cannot be met by the anticipated ultimately discoverable crude petroleum resource. If it implied, for example, a straight transfer of the American way to 10 billion people, then annual fossil fuel consumption would exceed the energy now fixed photosynthetically on Earth and would be equivalent to over one-half the fossil fuels cumulatively consumed by mankind so far, and to over 1% of fossil fuels anticipated ever to be discovered as potentially extractable (Table IV). So at best the system would be only a brief one. However, the implications of such a rapid rate of fossil fuel depletion on the climate-linked global carbon cycle (Section IV) are so serious that international institutions should be set up now to find ways to avoid consuming fossil fuels on such a scale (Gifford, 1979c).

One alternative to fossil fuels is to burn biomass. Section V,B shows how this is only a partial solution at best. Another extreme, a return to the land to practice labor-intensive food production, is also sometimes suggested. In a parallel study to the one on the absolute maximum food production of the world, Buringh and van Heemst (1977) estimated food production from a labor-oriented agriculture involving no machines or chemical fertilizers. They concluded that it is impossible to feed even the present world population by such a system alone. Moreover they suggest that land-use conflicts and land degradation through soil erosion and desertification would be even more disastrous than now.

The conclusion must be that, if a route exists at all, there is only a rather narrow middle course that can be safely negotiated to insure an untraumatic transition from the era of exponential growth in demand for the World's renewable and nonrenewable photosynthetic resources to one of stabilized overall demand equitably distributed at a level of consumption modest enough to remain within the buffering capacity of the global carbon cycle. The chances that the collective helmsman will aim the ship safely seem rather slight partly because there is no agreement over which channels reach dead ends and which penetrate right through to the other side.

Acknowledgments

I thank Robert B. Bacastow for providing data for Fig. 2b and 4, colleagues for discussion and criticism of the manuscript, and Val Viviani and Fleur Kelleher for typing this manuscript.

REFERENCES

Adams, J. A. S., Mantovani, M. S. M., and Lundell, L. L. (1977). *Science* **196,** 54–56.

Akita, S., and Moss, D. N. (1972). *Crop Sci.* **12,** 789–793.

Aoki, M., and Yabuki, K. (1977). *Agric. Meteorol.* **18,** 475–485.

Australian Institute of Petroleum (1979). "Oil and Australia 1979." AIP, Melbourne.

Bacastow, R. B., and Keeling, C. D. (1973). *In* "Carbon and the Biosphere" (G. M. Woodwell and E. V. Pecan, eds.), CONF-720510, pp. 86–135. U.S. Atomic Energy Commission, Washington, D.C.

Bach, W. (1978). *In* "Carbon Dioxide, Climate and Society" (J. Williams, ed.), pp. 141–167. Pergamon, Oxford.

Baes, C. F., Jr., Goeller, H. E., Olson, J. S., and Rotty, R. M. (1976). "The Global Carbon Dioxide Problem," ORNL-5194. Oak Ridge Natl. Lab., Oak Ridge, Tennessee.

Bazilevich, N. I., Rodin, L. E., and Rozov, N. N. (1971). *Sov. Geogr.* (*Engl. Transl.*) **12,** 293–317.

Bazin, M. (1979). *Nature* (*London*) **282,** 550–551.

Beijdorff, A. F. (1979). "Energy Efficiency." Shell International Petroleum Company Limited, London.

Bohn, H. L. (1978). *Tellus* **30,** 472–475.

Bolin, B. (1977). *Science* **196,** 613–615.

Bolin, B. (1978). *In* "Carbon Dioxide, Climate and Society" (J. Williams, ed.), pp. 41–43. Pergamon, Oxford.

Botkin, D. B. (1977). *BioScience* **27,** 325–331.

Box, E. (1975). *In* "Primary Productivity of the Biosphere" (H. Leith and R. H. Whittacker, eds.), pp. 265–83. Springer-Verlag, Berlin and New York.

Bray, J. R. (1959). *Tellus* **11,** 220–230.

Broecker, W. S. (1973). *In* "Carbon and the Biosphere" (G. M. Woodwell and E. V. Pecan, eds.), CONF-720510; pp. 32–50. U.S. Atomic Energy Commission, Washington, D.C.

Broecker, W. S. (1975). *Science* **189,** 460–463.

Broecker, W. S. (1979). *In* "Proceedings of the Workshop on the Global Effects of Carbon Dioxide from Fossil Fuels, Miami Beach Florida 1977" (W. P. Elliott and L. Machta, eds.), CONF-770385, pp. 18–23. U.S. Dept. of Energy, Washington, D.C.

Broecker, W. S., Takahaski, T., Simpson, H. J., and Peng, T. H. (1979). *Science* **206,** 409–418.

Brown, L. (1978). *Environment* **20,** 6–16.
Bunt, J. S. (1975). *In* "Primary Productivity of the Biosphere" (H. Leith and R. H. Whittaker, eds.), pp. 169–183. Springer-Verlag, Berlin and New York.
Buringh, P., and van Heemst, H. D. J. (1977). "An Estimation of World Food Production Based on Labour-oriented Agriculture." Centre for World Market Research, Amsterdam.
Buringh, P., van Heemst, H. D. J., and Staring, G. J. (1975). "Computation of the Absolute Maximum Food Production of the World." Agricultural University, Wageningen, Netherlands.
Caldwell, M. M. (1974). *Proc. Int. Congr. Ecol. 1st, 1974,* pp. 52–56.
Caldwell, M. M. (1975). *In* "Photosynthesis and Production in Different Environments" (J. P. Cooper, ed.), pp. 41–73. Cambridge Univ. Press, London and New York.
Callender, G. S. (1938). *Q. J. R. Meteorol. Soc.* **64,** 223–227.
Calvin, M. (1979). *BioScience* **29,** 533–538.
Chambers, R. S., Herendeen, R. A., Joyce, J. J., and Penner, P. S. (1979). *Science* **206,** 789–795.
Chapman, P. F., and Hemming, D. F. (1976). *Proc. Int. T.N.O. Conf., 9th, 1976,* pp. 119–140.
Chapman, P. F., Leach, G., and Slesser, M. (1974). *Energy Policy* **2,** 231–43.
Choudhury, B., and Kukla, G. (1977). *Nature (London)* **280,** 668–671.
Costanza, R. (1980). *Science* **210,** 1219.
Dahlman, R. C., Machta, L., Elliott, W., and MacCracken, M. (1980). "Carbon Dioxide Research Progress Report. Fiscal Year 1979." DOE/EV-0071. U.S. Dept. of Energy, Washington, D.C.
Damon, P. E., and Kunen, S. M. (1976). *Science* **193,** 447–453.
Degens, E. T., Paluska, A., and Eriksson, E. (1976). *Ecol. Bull.* **22,** 185–191.
Delwiche, C. C. (1977). *Ambio* **6,** 106–111.
Desprairies, P. (1978). *In* "World Energy Resources 1985–2020" (World Energy Conference, Conservation Commission, ed.), pp. 2–47. IPC Science and Technology Press, Guildford.
de Wit, C. T. (1975). *Neth. J. Agric. Sci.* **23,** 145–162.
Eckholm, E. (1975). "The Other Energy Crisis—Firewood." Worldwatch Pap. No. 1. Worldwatch Institute, Washington, D.C.
Edwards, G. E. (1976). *Aust. J. Agric. Econ.* **20,** 179–191.
Ekdahl, C. A., and Keeling, C. D. (1973). *In* "Carbon and the Biosphere" (G. M. Woodwell and E. V. Pecan, eds.), CONF-720510, pp. 51–85. U.S. Atomic Energy, Washington, D.C.
Ericksson, E. (1963). *JGR, J. Geophys. Res.* **68,** 3871–3876.
Flohn, H. (1978). *In* "Carbon Dioxide, Climate and Society" (J. Williams, ed.), pp. 227–237. Pergamon, Oxford.
Food and Agriculture Organization (1977a). "Production Yearbook." FAO/United Nations, Rome.
Food and Agriculture Organization (1977b). "The State of Food and Agriculture 1976." F.A.O. Agric. Ser. No. 4. FAO/United Nations, Rome.
Food and Agriculture Organization. (1978a). "Yearbook of Fishery Statistics." FAO/United Nations, Rome.
Food and Agriculture Organization (1978b). "The State of Food and Agriculture 1977." F.A.O. Agric. Ser. No. 8. FAO/United Nations, Rome.
Freyer, H. D. (1978). *In* "Carbon Dioxide, Climate, and Society" (J. Williams, ed.), pp. 69–77. Pergamon, Oxford.

Garrels, R. M., Lerman, A., and MacKenzie, F. T. (1976). *Am. Sci.* **64,** 306–15.
Gifford, R. M. (1976). Search **7,** 412–417.
Gifford, R. M. (1977). *Aust. J. Plant Physiol.* **4,** 99–110.
Gifford, R. M. (1978a). *In* "Energy, Agriculture and the Built Environment" (R. King, ed.), pp. 91–102. Centre for Environmental Studies, University of Melbourne, Australia.
Gifford, R. M. (1978b). *Aust. J. Public Admin.* **37,** 69–83.
Gifford, R. M. (1979a). *Search* **10,** 316–318.
Gifford, R. M. (1979b). *Aust. J. Plant Physiol.* **6,** 367–378.
Gifford, R. M. (1979c). *In* "Energy and People: Social Implications of Different Energy Futures" (M. Diesendorf, ed.), pp. 147–150. Society for Social Responsibility in Science (ACT)., Canberra, Australia.
Gifford, R. M. (1980a). *In* "Food Chains and Human Nutrition" (K. Blaxter, ed.), pp. 341–362. Applied Science Publishers Ltd., London.
Gifford, R. M. (1980b). *In* "Carbon Dioxide and Climate: Australian Research" (G. I. Pearman, ed.), pp. 167–81. Aust. Acad. Sci., Canberra.
Gifford, R. M., and Millington, R. J. (1975). "Energetics of Agriculture and Food Production." Bull. No. 288. Commonwealth Scientific and Industrial Research Organization, Australia.
Gifford, R. M., and Musgrave, R. B. (1970). *Physiol. Plant.* **23,** 1048–1056.
Groves, K. S., Mattingly, S. R., and Tuck, A. F. (1978). *Nature (London)* **273,** 711–715.
Häfele, W. (1978). *In* "Carbon Dioxide, Climate and Society" (J. Williams, ed.), pp. 21–34. Pergamon, Oxford.
Hall, C. A. S., Ekdahl, C. A., and Wartenberg, D. E. (1975). *Nature (London)* **255,** 136–138.
Hammond, A. L. (1977). *Science* **195,** 564–566.
Hubbert, M. K. (1969). *In* "Resources and Man" (Committee on Resources and Man, NAS/NRC, eds.), pp. 157–242. Freeman, San Francisco, California.
Idso, S. B. (1980). *Science* **207,** 1462–1463.
International Federation of Institutes for Advanced Study (1974). "Energy Analysis Workshop on Methodology and Conventions." Workshop Rep. No. 6. IFIAS, Stockholm.
International Federation of Institutes for Advanced Study (1975). "Workshop on Energy Analysis and Economics." Workshop Rep. No. 9. IFIAS, Stockholm.
Jenny, H. (1930). *Res. Bull.*—Mo., Agric. Exp. Stn. 152, 66 pp.
Judson, S. (1968). *Am. Sci.* **56,** 356–374.
Keeling, C. D. (1973). *Tellus* **25,** 174–199.
Keeling, C. D., and Bacastow, R. B. (1977). *In* "Energy and Climate" (R. Revelle, ed.), pp. 110–60. Natl. Res. Counc./Nat. Acad. Sci., Washington, D.C.
Kohlmaier, G. H., Fischbach, U., Kratz, G., and Sire, E. O. (1978). *In* "Carbon Dioxide, Climate and Society" (J. Williams, ed.), pp. 69–71. Pergamon, Oxford.
Kukla, G. J., Angell, J. K., Korshover, J., Dronia, H., Hoshiai, M., Namias, J., Rodewald, M., Yamamoto, R., and Iwashima, T. (1977). *Nature (London)* **270,** 573–580.
Lamb, H. (1974). *Ecologist* **4,** 10–15.
Lawton, J. H. (1973). *In* "Resources and Population" (B. Benjamin, P. R. Cox, and J. Peel, eds.), pp. 59–76. Academic Press, New York.
Leach, G. (1975). "Energy and Food Production." International Institute for Environment and Development, London.
Leith, H. (1972). *Nat. Resour.* **8** (2), 5–10.
Leith, H. (1975a). *Ecol. Stud.* **14,** 203–215.
Leith, H. (1975b). *In* "Unifying Concepts in Ecology" (W. H. van Dobben and R. H. Lowe-McConnell, eds.), pp. 67–88. Junk, The Hague.

Leith, H., and Box, E. (1972). *Publ. Climatol.* **25**(2), 37–46.
Lemon, E. R. (1977). *In* "The Fate of Fossil Fuel CO_2 in the Oceans" (N. R. Anderson and A. Malahoff, eds.), pp. 97–130. Plenum, New York.
Lipinsky, E. S. (1978). *Science* **199,** 644–651.
Lloyd, A. G. (1978). *In* "Energy, Agriculture and the Built Environment" (R. King, ed.), pp. 25–36. Centre for Environmental Studies, University of Melbourne, Australia.
Loomis, R. S. (1979). *In* "Proceedings of the Workshop on the Global Effects of Carbon Dioxide from Fossil Fuels, Miami Beach, Florida 1977" (W. P. Elliott and L. Machta, eds.), CONF-770385, pp. 51–62. U.S. Dept. of Energy, Washington, D.C.
Lowe, D. C., Guenther, P. R., and Keeling, C. D. (1979). *Tellus* **31,** 58–67.
McCormick, W. T., Jr., Fish, L. W., Kalisch, R. B., and Wander, T. J. (1978). *In* "World Energy Resources 1985–2020" (World Energy Conference, Conservation Commission, ed.), pp. 50–56. IPC Science and Technology Press, Guildford.
Machta, L. (1979). *In* "Proceedings of the Workshop on the Global Effects of Carbon Dioxide from Fossil Fuels, Miami Beach, Florida 1977" (W. P. Elliott and L. Machta, eds.), CONF-770385, pp. 44–50. U.S. Dept. of Energy, Washington, D.C.
Machta, L., and Elliott, W. (1980). *In* "Carbon Dioxide Research Progress Report: Fiscal Year 1979" (R. C. Dahlman *et al.*, eds.), DOE/EV-0071, pp. 11–32. U.S. Dept. of Energy, Washington, D.C.
Manabe, S., and Stouffer, R. J. (1979). *Nature (London)* **282,** 491–493.
Manabe, S., and Wetherald, R. T. (1975). *J. Atmos. Sci.* **32,** 3–15.
Manabe, S., and Wetherald, R. T. (1980). *J. Atmos. Sci.* **37,** 99–118.
Mason, J. (1979). *New Sci.* **82,** 196–198.
Mercer, J. H. (1978). *Nature (London)* **271,** 321–325.
Neales, T. F., and Nicholls, A. O. (1978). *Aust. J. Plant Physiol.* **5,** 45–59.
Newell, R. E., and Dopplick, T. G. (1979). *J. Appl. Meteorol.* **18,** 822–825.
Odum, H. T. (1971). "Environment, Power and Society." Wiley, New York.
Oeschger, H., and Siegenthaler, U. (1978). *In* "Carbon Dioxide, Climate and Society" (J. Williams, ed.), pp. 45–61. Pergamon, Oxford.
Pearman, G. I. (1977). *Clean Air* **11,** 21–26.
Peters, W., and Schilling H.-D. (1978). *In* "World Energy Resources, 1985–2020" (World Energy Conference, Conservation Commission, ed.), pp. 58–86. IPC Science and Technology Press, Guildford.
Postgate, J. R. (1978). "Nitrogen Fixation." Inst. Biol. Stud. Biol. No. 92. Arnold, London.
Pratt, P. F. (1977). *Clim. Change* **1,** 109–135.
Ralston, C. W. (1979). *Science* **204,** 1345–1346.
Raper, C. D., and Peedin, G. F. (1978). *Bot. Gaz. (Chicago)* **139,** 147–149.
Revelle, R., and Suess, H. E. (1957). *Tellus* **9,** 18–27.
Rodin, L. E., Bazilevich, N. I., and Rozov, N. N. (1974). *Proc. Int. Congr. Ecol. 1st, 1974,* pp. 176–181.
Rodin, L. E., Bazilevich, N. I., and Rozov, N. N. (1975). *In* "Productivity of the World's Main Ecosystems" (D. E. Reichle, J. F. Franklin, and D. W. Goodall, eds.), pp. 15–17, 20, 22. Natl. Acad. Sci., Washington, D.C.
Rosenzweig, M. L. (1968). *Am. Nat.* **102,** 67–74.
Rotty, R. M. (1973). *Tellus* **25,** 508–517.
Rotty, R. M. (1977). "Present and Future Production of CO_2 from Fossil Fuels—A Global Appraisal." ORAU/IEA - 77-15. Institute for Energy Analysis, Oak Ridge, Tennessee.
Rousseau, P. E. (1978). *In* "Energy 1977—Australia" (I. E. Newnham and Woodcock, J. T., eds.), pp. 57–76. Australian Academy of Technological Sciences, Melbourne.

Salinger, M. J. (1975). *Nature (London)* **256,** 397–398.

Sands, J. (1979). *J. Environ. Syst.* **8,** 99–110.

Schneider, S. H. (1978). *In* "Carbon Dioxide, Climate and Society" (J. Williams, ed.), pp. 219–225. Pergamon, Oxford.

Siegenthaler, U., and Oeschger, H. (1978). *Science* **199,** 388–395.

Smagorinsky, J. (1977). *In* "Energy and Climate" (R. Revelle, ed.), pp. 229–242. Nat. Res. Counc./Nat. Acad. Sci., Washington, D.C.

Söderland, R., and Svensson, B. H. (1976). *Ecol. Bull.* **22,** 23–73.

Stewart, G. A., Gartside, G., Gifford, R. M., Nix, H. A., Rawlins, W. H. M. and Siemon, J. R. (1979). "The Potential for Liquid Fuels from Agriculture and Forestry in Australia." Commonwealth Scientific and Industrial Research Organization, Australia.

Study of Critical Environmental Problems (SCEP) (1970). "Man's Impact on the Global Environment: Assessment and Recommendations for Action." MIT Press, Cambridge, Massachusetts.

Stuiver, M. (1978). *Science* **199,** 253–258.

Stumpf, U. E. (1978). *In* "Alcohol Fuels." pp. 2.20–2.24. Institute of Chemical Engineers, New South Wales Group, Australia.

Tatom, J. W., Colcord, A. R., Knight, J. A., Elston, L. W., and Har-Oz, P. H. (1975). *In* "Energy, Agriculture and Waste Management" (W. J. Jewell, ed.), pp. 271–288. Ann Arbor Sci. Publ., Ann Arbor, Michigan.

Thomas, R. H. (1979). *Science* **205,** 1257–1258.

Tucker, G. B. (1975). *Search* **6,** 323–328.

United Nations (1977). "Demographic Yearbook." United Nations, New York.

United Nations (1979a). "The World Population Situation in 1977." Popul. Stud. No. 63. Department of International Economic and Social Affairs, United Nations, New York.

United Nations (1979b). "Statistical Yearbook 1978." United Nations, Rome.

United States, Central Intelligence Agency (1977). "The International Energy Situation: Outlook to 1985." ER77 - 1024OU. U.S., C.I.A., Washington, D.C.

United States Department of Energy (1980). "Carbon Dioxide Research Progress Report Fiscal Year 1979." DOE/EV-0071, UC-11. U.S. Department of Energy, Washington, D.C.

Vergara, W., and Pimentel, D. (1978). *Adv. Energy Syst. Technol.* **1,** 125–173.

Wade, N. (1979). *Science* **204,** 928–929.

Walsh, J. J., Rowe, G. T., Iverson, R. L., and McRoy, C. P. (1981). *Nature (London)* **291,** 196–201.

Weinberg, A. M., and Hammond, R. P. (1972). *Peaceful Uses At. Energy, Proc. Int. Conf., 4th 1971,* Vol. 1. 171–178.

Weisz, P. B., and Marshall, J. F. (1979). *Science* **206,** 24–29.

Whittaker, R. H., and Likens, G. E. (1975). *In* "Primary Productivity of the Biosphere" (H. Leith and R. H. Whittaker, eds.), pp. 305–308. Springer-Verlag, Berlin and New York.

Whittaker, R. H., and Marks, P. L. (1975). *In* "Primary Productivity of the Biosphere" (H. Leith and R. H. Whittaker, eds.); pp. 55–118. Springer-Verlag, Berlin and New York.

Whitten, R. C., Borucki, W. J., Capone, L. A., Riegel, C. A., and Turco, R. P. (1980). *Nature (London)* **283,** 191–192.

Wilson, A. T. (1978). *Nature (London)* **273,** 40–41.

Wong, C. S. (1978). *Science* **200,** 197–200.

Wong, C. S. (1980a). *Oecologia* **44,** 68–74.

Wong, S. C. (1980b). *In* "Carbon Dioxide and Climate: Australian Research" (G. I. Pearman, ed.), pp. 159–166. Aust. Acad. Sci., Canberra.

Woodwell, G. M. (1978). *Sci. Am.* **238,** 34–43.

Woodwell, G. M., Whittaker, R. H., Reiners, W. A., Likens, G. E., Delwiche, C. C., and Botkin, D. B. (1978). *Science* **199,** 141–146.
World Energy Conference (1978). "World Energy Resources 1985–2020." IPC Science and Technology Press, Guildford.
Wright, R. D. (1974). *Am. Midl. Nat.* **91,** 360–370.
Yamaguchi, J. (1978). *J. Fac. Agric., Hokkaido Univ.* **59,** 59–129.

Special Topics

14

Canopy Photosynthesis and Yield in Soybean

A. LAWRENCE CHRISTY
CLARK A. PORTER

ABSTRACT

Data demonstrating a strong correlation between photosynthesis and yield have been very limited. Many of these studies have attempted to relate yield and point-in-time measurements of photosynthesis or yield and photosynthesis in different cultivars. Within soybean cultivars, yield appears to be strongly dependent on the total amount of photosynthesis carried on by the crop during the growing season and may not always be correlated to a point-in-time measurement of photosynthetic rate. However, when comparing cultivars, differences in their photosynthetic conversion efficiency (PCE) are important in determining the relationship between photosynthesis and yield.

The data reviewed here indicate that the yield-regulating mechanisms of soybeans, and perhaps other crops as well, are conservative in their response to prevailing growing conditions. Soybeans appear to continually readjust their sink load to concurrent photosynthesis throughout the reproductive stages of growth. This continual readjustment enables the crop to compensate for changes in the growing conditions and maximize yield.

I. Introduction

Crop yield is obviously dependent upon photosynthesis, but evidence for a direct relationship between photosynthesis and economic yield is lacking, and genetic selection for higher photosynthetic rates has not produced increases in yield. Several recent reviews have addressed various aspects of the subject of photosynthesis and plant productivity (Evans, 1975; Moss, 1976; Brun, 1978; Good and Bell, 1980). This chapter is not intended as a complete review of the subject, but it will address

Photosynthesis: Development, Carbon Metabolism, and Plant Productivity, Vol. II

ISBN: 0-12-294302-3

attention to events occurring in the crop canopy and concentrate on studies that we have completed on the relationship of canopy photosynthesis and grain yield in soybeans (also see Hesketh *et al.*, Chapter 11, this volume).

Several investigators have used field-grown soybeans to study photosynthetic rates in leaves (Beverlein and Pendleton, 1971) and canopies (Sakamoto and Shaw, 1967; Jeffers and Shibles, 1969; Egli *et al.*, 1970; Hansen, 1972) and the relationship of light interception, dry matter production, and yield (Shibles and Weber, 1966). In addition, shade and CO_2 enhancement treatments have been used as indirect means of studying the link between photosynthesis and yield in a number of crops (Moss, 1976), including atmospheric CO_2 enrichment in field-grown soybeans (Hardman and Brun, 1971). These studies have provided only limited evidence for any direct relationship between photosynthesis and economic yield. There are several factors that may account for this apparent lack of correlation between photosynthetic rates and yield. Much of the photosynthetic data is collected on individual leaves, whereas yield is measured on the entire plant or canopy. Many photosynthetic measurements are point-in-time determinations made at varying developmental stages of the plants and do not take into consideration the entire growing season. Finally, the relationship between photosynthesis and yield may be complex and differ for every crop or even cultivars of a crop. This relationship could be masked by any of a number of the biochemical and physiological events that occur between the production of photosynthate and its utilization in the accumulation of final yield.

During the past several years, we have studied the relationship between canopy photosynthesis and yield in field-grown soybeans. Photosynthetic and evapotranspiration rates are measured throughout the growing season on 2.8 m^2 soybean plots using chambers described by Peters *et al.* (1974) and Christy *et al.* (1982a). Environmental conditions are not regulated in these chambers; we attempt to measure photosynthesis under the field microclimate. Each measurement requires a plot to be enclosed by a chamber for approximately 2 min. Thermocouple measurements of leaf temperatures indicate only a 1°C rise during this time.

II. Changes in Photosynthesis during the Day

Figure 1 shows the changes that occur in several parameters in a soybean plot on an "average" day (Christy *et al.*, 1982a). However, it should be pointed out that an "average" day does not occur in the field nor does an "average" growing season occur. Each day differs with respect to temperature, humidity, cloud cover, soil moisture, and other climatic parameters. Each season differs relative to the time during the growing season that the crop "sees" warm or cold temperatures, adequate or limited rainfall, etc.

During the day, all changes are driven by the diurnal time course of light intensity (Fig. 1). The soybean canopies do not appear to light saturate during vegetative growth. However, soybeans do light saturate after the start of flowering (Christy *et al.*, 1979, 1982b). As light intensity increases, the net photosynthetic rate increases faster than the evapotranspiration rate due to less evaporative demand during cooler morning hours. This results in a high evapotranspiration ratio, i.e., water use efficiency, during the morning. After solar noon, as light intensity de-

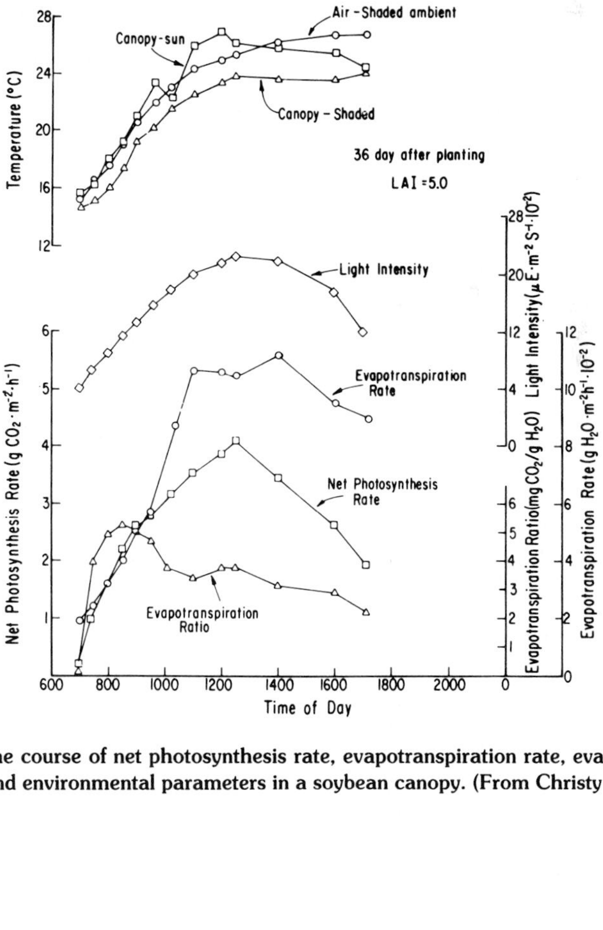

FIG. 1. **Time course of net photosynthesis rate, evapotranspiration rate, evapotranspiration ratio, and environmental parameters in a soybean canopy. (From Christy *et al.*, 1979, 1981a.)**

creases, canopy net photosynthesis decreases faster than evapotranspiration due to the higher temperature during this period of the day.

III. Effect of Plant Population

Plant population has been shown to influence the yield of crops and can be useful in studying the relationship between photosynthesis and yield (Christy *et al.*, 1976, 1979, 1982a). The time course of canopy photosynthesis during the growing season for four soybean populations grown at equidistant spacings are shown in Fig. 2. The photosynthetic advantage of the high population plots during the early part of the growing season is obvious. However, there were no significant differences in net photosynthetic rates at the three highest populations after Day 70. Egli *et al.* (1970), Beverlein and Pendleton (1971), and Hansen (1972) have reported similar maximum canopy photosynthetic rates for field-grown soybeans.

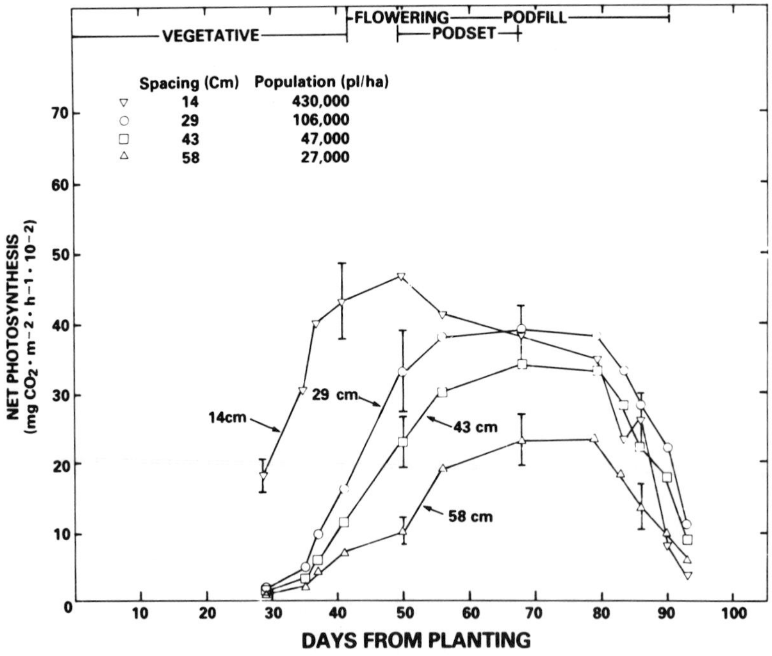

FIG. 2. Time course of canopy net photosynthesis rate of soybeans grown at four populations at equidistant spacings. All values are means of six plots, and error bars indicate one standard deviation. (From Christy *et al.*, 1982a.)

The yield for these equidistant spacing plots are listed in Table I. Yield/plot decreased with the decrease in population, and there was no significant difference in seed size indicating the observed differences in yield/plot were due to differences in the number of pods and seeds/plot. Although there was a 54% difference in yield between the 43- and the 14-cm plots, there were no significant differences in photosynthetic rate in these plots during the bean filling stage. It is obvious from Fig. 2 that major differences in photosynthesis occurred during vegetative growth, and these differences were reflected in total photosynthesis for the entire growing season. An estimate of seasonal photosynthesis may be obtained by integrating the area under the canopy net photosynthetic rate curves in Fig. 2.

Christy *et al.* (1976, 1979, 1982a) reported that grain yield is strongly dependent on seasonal photosynthesis (Fig. 3). However, it may be possible to show a relationship between yield and point-in-time measurements of photosynthesis made during the growing season (Christy *et al.*, 1976, 1982a). This may be due to the time of canopy closure, i.e., 100% light interception, with the different populations, whereas a different variable may make photosynthesis at some other time in the growing season appear indicative of yield.

Schulze (1978) reported that yield was related to photosynthesis at early bean fill. The relationship may be due to the length of the growing season for the different maturity groups under study. Although these results and those of Christy *et al.* (1976, 1982a) indicate that point-in-time measurements of photosynthesis, made at different times in the growing season, may be indicative of yield, yet these results demonstrate why this type of measurement is unsatisfactory for studies of yield and photosynthesis. The stage of development or point-in-time when a photosynthetic rate measurement can be correlated with yield may be

TABLE I
Soybean Grain Yields and Yield Components for Equidistant Spacing Plots[a]

Spacing (cm)	Plant population (plant/ha)	Yield (g/plot)	Seed number (number/plot)	Seed size (g/100 seed)
14	430,000	1533 ± 44	10725 ± 321	14.3 ± 0.6
29	106,000	1005 ± 48	7499 ± 299	13.4 ± 0.5
43	47,000	883 ± 87	6310 ± 315	14.0 ± 0.5
58	27,000	544 ± 77	3943 ± 216	13.8 ± 0.9

[a]For 1976 as reported by Christy *et al.* (1982a). All values are means of six plots.

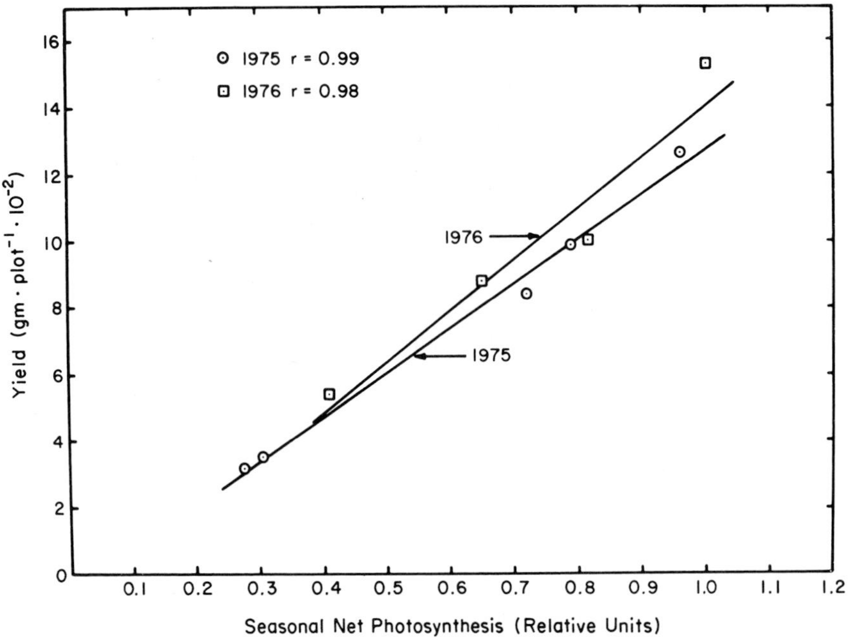

FIG. 3. Yield of soybeans (Wells) as a function of canopy seasonal photosynthesis for 2 years. Population was varied from 17,000 to 430,000 plants/ha by equidistant spacing. (From Christy *et al.*, 1982a.)

strongly influenced by the experimental design. This may be the reason for some of the confusion concerning the relationship between photosynthesis and yield.

Schulze (1978) compared canopy photosynthesis and yield in a number of soybean cultivars from four maturity groups. He reported that yield was a function of canopy photosynthesis during bean fill and yield increased with longer bean fill duration. The longer bean fill duration may be expected to lead to increased seasonal photosynthesis, and these results would be consistent with the findings of Christy *et al.* (1982a).

IV. Effect of Shading

Shading has been used to study the effect of reduced photosynthesis on yield in several crops. However, yield and yield components are usually the only data collected, and data on photosynthesis and yield from the same shading study are limited. Christy *et al.* (1982c) have studied the effect of shade on photosynthesis and yield in field-grown soybeans. Treatments using 50% shade included continuous shade and shading during vegetative growth, flowering-podset, and bean fill (Fig. 4). The effect of shade on photosynthesis was immediate, and recovery to control rates after shade removal occurred within 30 min (Christy *et al.*, 1982b). This was true even for the plots shaded during vegetative growth, although the top two-thirds of the canopy had never been exposed to full sunlight. Canopy net photosynthetic rates expressed as percentage of control (unshaded) are listed in Table II. Although 50% shade was being used, the shade treatments reduced yield only 30–40% during the treatment period. Soybean canopies light saturate at 75% full sunlight during reproductive growth (Christy *et al.*, 1976, 1979, 1982b), so 50% shade does not decrease photosynthesis 50%. In fact, preliminary studies with 25% shade during reproductive growth did not significantly affect yield or photosynthesis.

As expected, continuous 50% shade reduced yield only 25% to 35%

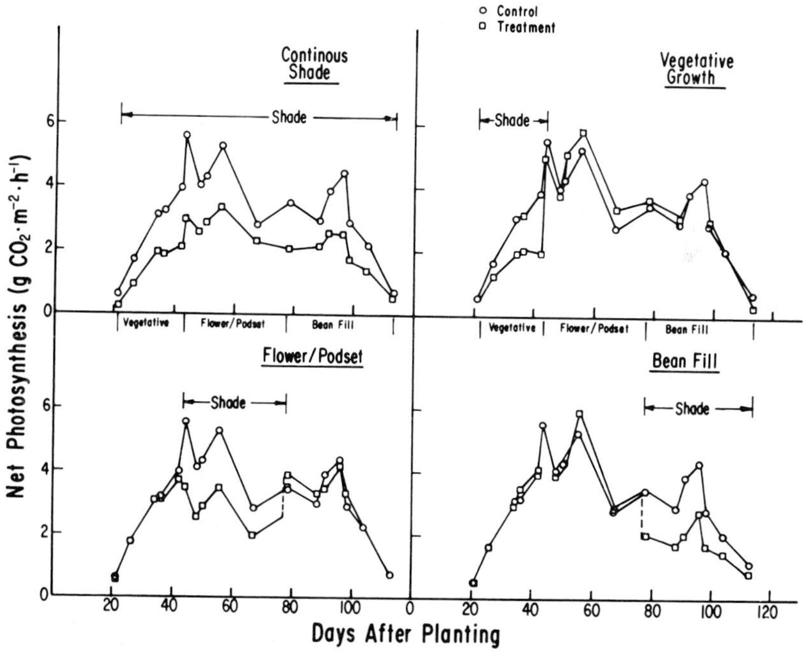

FIG. 4. Effect of shade on canopy net photosynthesis rate during the entire growing season and three developmental stages. All points are means of six plots. (From Christy *et al.*, 1982c.)

TABLE II
Effect of Shading on Total Photosynthesis during Treatment Period and Yield Components in Field Grown Soybeans[a]

	Estimated total photosynthesis during treatment period				Yield components		
Stage of shade treatment	Season	Vegetative	Flow/pod-set	Bean fill	Yield/plot	Seed size	Seed number
1977							
Continuous	66.2[b]	55.6	66.5	71.7	75	102.8	72.8
Vegetative	97.7	60.9	110.8	101.2	110.7	96.4	114.7
Flower/podset	86	96.7	66	106.1	85.7	110.9	77.4
Bean fill	88.9	100.4	102.6	63.9	82	97.9	85.4
1978							
Continuous	68.8	81.5	69.1	64	65.7	102.3	66
Vegetative	87.8	63.9	97.9	88.4	90.6	96.8	93.5
Flower/podset	81.4	97.1	63.6	89.2	79.1	103.7	76.3
Bean fill	87.4	116.6	108.5	61.3	66.4	93.4	71.2

[a]Christy *et al.*, 1982c.
[b]All values are percentages of control and means of six plots:

	1977	1978
Yield of control plots (g/plot) =	1471.8	1471.1
Estimated seasonal photosynthesis = of control plots	1.23	1.21

and shading during various stages of growth reduced yield up to 35% or in the case of vegetative shade, there was no significant reduction (Table II). Decreased photosynthesis during flowering-podset reduced seed number, but the return to full photosynthesis during bean filling allowed the plants to compensate somewhat by making larger seed. However, there was still a 15–20% reduction in yield. Reduced photosynthesis during filling resulted in only a slight reduction in seed size, but significantly reduced seed number (Table II). Yield was again dependent on estimated seasonal photosynthesis with correlation coefficients of 0.83 and 0.80 for the 2 years studied.

These results demonstrate the soybean plant's strategy for dealing with stress during the growing season. Stress during vegetative growth does not appear to be very critical, whereas stress during reproductive growth results in lower seed number but not more than a 10% reduction in seed size. Walker (1974) has reported that dry beans pass through a critical short period of development about 20 days after anthesis (full bloom). If the seed is removed from the plant before that period, it will not germinate, but it will germinate if removed after that period. Perhaps this is also the decision point for the plant. Before this point it will abort the seed if under stress, but if the seed passes this point it is retained and a less mature seed is aborted. In this way, the plant can adjust its seed load even during filling to compensate for changes in photosynthate supply due to stress and still produce viable seed capable of producing a new generation of plants.

V. Photosynthetic Conversion Efficiency

We have found that yield is strongly dependent on estimated seasonal photosynthesis across years, the cultivars and various treatments including population, row spacing, shading, and date of planting (Fig. 5). The slope of the line in Fig. 5 is essentially the harvest index or, perhaps in this case, a better term may be *photosynthetic conversion efficiency* (PCE). This is the efficiency with which the crop converts photosynthesis into grain yield. PCE is related to photosynthesis and yield as follows:

$$\text{Seasonal photosynthesis} \times \text{photosynthetic conversion efficiency} = \text{crop yield}$$

PCE differs from *harvest index* in that it is based on seasonal photosynthesis, whereas harvest index is a point-in-time measurement. PCE represents the many biochemical and physiological processes that occur between the initial photosynthetic reaction and final grain yield. The importance of the efficient conversion of photosynthate to yield or har-

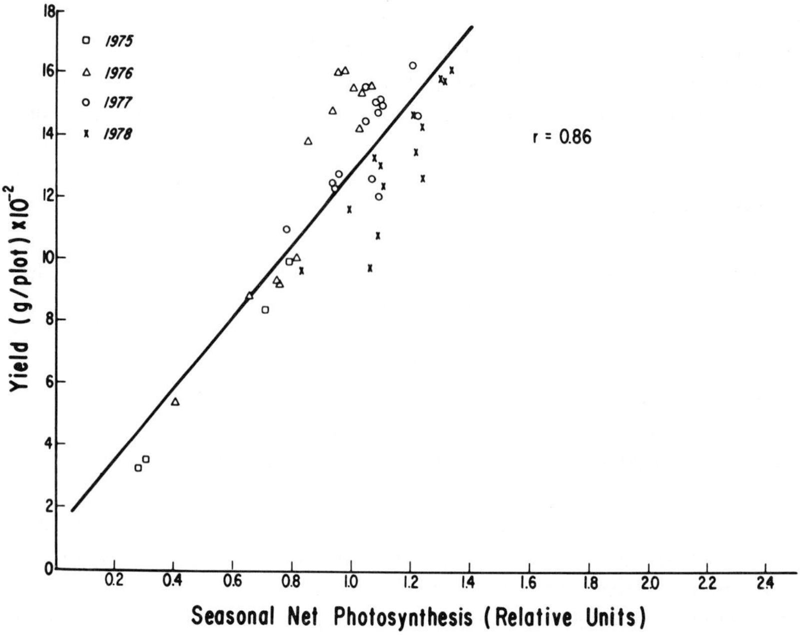

FIG. 5. Soybean grain yield as a function of seasonal photosynthesis estimated by integrating the area under curves shown in Figs. 2 and 4. These data include two cultivars, four growing seasons, and several treatments. All points are means of six plots. (From Christy *et al.*, 1979.)

vest index has been reported many times, and for an excellent review of this subject see Donald and Hamblin (1976).

Table III compares the PCE, yield components, and photosynthesis for the control plots for two soybean cultivars over two growing seasons. The PCE remained constant for each cultivar over the 2 years even though the seed number and seed size varied each year and yield differed for Wells. The yields for Williams were remarkably similar, yet the way those yields were achieved was different. In 1977, there were more seeds but they were smaller than those in 1978. There was essentially no difference in estimated seasonal photosynthesis for the 2 years, but changes did occur in total photosynthesis during the three stages of growth. In 1977, with the higher seed number, there was more photosynthate available during flowering-podset than in 1978. In 1978, when larger seeds were produced, there was more photosynthate available during filling than in 1977. The higher total photosynthetic values during these growth stages appeared to be due to the crop remaining in that stage of development longer and not due to higher photosynthetic rates

In 1977, the crop experienced above average temperatures in June, resulting in early onset of flowering and a longer flower-podset period. These data demonstrate the flexibility of the soybean crop and its capacity to compensate for changes in the environment.

VI. Discussion

Yield appears to be strongly dependent on the total amount of photosynthesis carried on by the crop during the growing season. For individual soybean cultivars, yield is strongly dependent on seasonal photosynthesis, but when comparing cultivars, differences in their PCE are important in determining the relationship of photosynthesis to their yields. The significant effect of PCE on yield may explain to some extent the confusion concerning the relationship between photosynthesis and yield.

The importance of early canopy closure to maximize seasonal photosynthesis and yield is obvious. This is reflected in the current trend toward narrow row spacings by soybean farmers. However, other agronomic characteristics must be taken into consideration when population is increased to lessen the time to canopy closure. These include increased lodging at high population and weed control. The grower must make a compromise between optimum seasonal photosynthesis and many other

TABLE III
Comparison of Canopy Photosynthesis and Yield Components for Two Varieties of Soybean over Two Growing Seasons[a]

	1975 (Wells)	1976 (Wells)	1977 (Williams)	1978 (Williams)
Yield (g/plot)	1234	1564.4	1471.8	1471.1
Seed number (number/plot)	11,017	10,729	8564	8179
Seed size (g/100)	11.2	14.6	17.21	17.99
Estimated total photosynthesis during				
Entire season	0.88	1.07	1.23	1.21
Vegetative growth	—	0.14	0.24	0.20
Flowering/podset	—	0.58	0.56	0.43
Bean fill	—	0.34	0.43	0.58
Photosynthetic conversion efficiency (PCE)	1.41	1.47	1.20	1.21

[a] All values are means of six plots grown in 24 cm rows at 494,000 plants/ha. Data taken from Christy *et al.*, 1982a,c.

factors to achieve maximum yield. Increasing photosynthesis is not the only parameter that should be considered by both the farmer and the researcher.

Several possible reasons were given in the introduction for a lack of evidence relating photosynthesis and yield. Correlation of yield with point-in-time measurements of photosynthesis appears to be somewhat fortuitous and depends to a great extent on the experimental design, prevailing weather conditions, and cultivars. Point-in-time measurements do not reflect changes in weather and other yield determining factors that occur after the measurements are made, but they may affect photosynthesis, yield, or both. Results from studies using different cultivars or hybrids could be very confusing because of differences in PCE from one cultivar to another. It should be pointed out that the conclusions drawn here may apply only to soybean and do not include other crops.

The data reviewed here indicate that soybeans, and perhaps other crops as well, have a conservative approach to reproductive growth and yield. Most of the environmental stress experienced by the crop during the growing season affects photosynthesis. Soybean yield is the result of a continual readjustment of sink load—seed number and, to some extent, seed size—to concurrent photosynthesis throughout the reproductive stages of growth. This continual readjustment enables the crop to compensate for changes in the growing conditions and produce the maximum yield that the prevailing weather conditions over the entire growing season will allow.

REFERENCES

Beverlein, J. E., and Pendleton, J. W. (1971). *Crop Sci.* **11,** 217–219.

Brun, W. A. (1978). *In* "Soybean Physiology, Agronomy, and Utilization" (A. G. Norman, ed.), pp. 45–76. Am. Soc. Agron., Madison, Wisconsin.

Christy, A. L., Westgate, M. E., and Porter, C. A. (1976). *Agron. Abstr.* p. 70.

Christy, A. L., Westgate, M. E., and Porter, C. A. (1979). *Proc. Plant Growth Regul. Work. Group* **6,** 75.

Christy, A. L., Westgate, M. E., Wideman, A. S., and Porter, C. A. (1982a). In press.

Christy, A. L., Westgate, M. E., and Porter, C. A. (1982b). In preparation.

Christy, A. L., Westgate, M. E., and Porter, C. A. (1982c). In preparation.

Donald, C. M., and Hamblin, J. (1976). *Adv. Agron.* **28,** 361–405.

Egli, D. B., Pendleton, J. W., and Peters, D. B. (1970). *Agron. J.* **62,** 411–414.

Evans, L. T. (1975). *In* "Crop Physiology: Some Case Histories" (L. T. Evans, ed.), pp. 327–355. Cambridge Univ. Press, London and New York.

Good, N. E., and Bell, D. H. (1980). *In* "The Biology of Crop Productivity" (P. E. Carlson, ed.), pp. 3–51. Academic Press, New York.

Hansen, W. (1972). Ph.D. Thesis, Iowa State University, Ames.
Hardman, L. L., and Brun, W. A. (1971). *Crop Sci.* **11,** 886–888.
Jeffers, D. L., and Shibles, R. M. (1969). *Crop Sci.* **9,** 762–764.
Moss, D. N. (1976). *In* "CO_2 Metabolism and Plant Productivity" (R. H. Burris and C. C. Black, eds.), pp. 31–41. University Park Press, Baltimore, Maryland.
Peters, D. B., Clough, B. F., Garves, R. A., and Stahl, G. R. (1974). *Agron. J.* **66,** 460–462.
Sakamoto, C. M., and Shaw, R. H. (1967). *Agron. J.* **59,** 73–74.
Schulze, L. L. (1978). Ph.D. Thesis, University of Georgia, Athens.
Shibles, R. M., and Weber, C. R. (1966). *Crop Sci.* **6,** 55–59.
Walker, K. A. (1974). Ph.D. Thesis, Yale University, New Haven, Connecticut.

15

The Functional Role of Bicarbonate in Photosynthetic Light Reaction II†

ALAN STEMLER

ABBREVIATIONS

B	A 2-electron acceptor quinone molecule that accepts electrons from Q
Chl	Chlorophyll
CO_2^*	Denotes dissolved inorganic carbon when the form CO_2, H_2CO_3, HCO_3^-, or CO_3^{2-} is left unspecified
DCMU	3-(3,4-Dichlorophenyl)-1,1 dimethylurea
DLE	Delayed light emission
P680	Reaction center Chl of PSII with one of its absorption bands at 680 nm
PSII	Photosystem II
Q	A 1-electron acceptor quinone molecule on the reducing side of PSII
S-state	A state of the oxygen evolving complex

ABSTRACT

A molecule of CO_2^* appears to be bound to each photosystem II (PSII) complex. CO_2^* depletion appears to result in two effects—a complete inactivation of some PSII complexes

†This chapter should be read in conjunction with the chapter by Vermaas and Govindjee—Editor.

Photosynthesis: Development, Carbon Metabolism, and Plant Productivity, Vol. II

ISBN: 0-12-294302-3

and slow turnover rates following light excitation in others. The effects of CO_2^* depletion are anomalously general, appearing as rate limitations on both the oxidizing and reducing sides of the reaction center II, depending on conditions and type of measurement. Three possible roles for CO_2^* in PSII have been suggested. CO_2^* may act as a controlling intermediate to balance electron transport with carbon metabolism. It may be involved in the energy conserving steps associated with PSII. Hydrated CO_2 may be the immediate source of photosynthetic O_2. Other hypotheses may yet arise, and all will require testing.

I. Introduction

The role of CO_2 as the terminal electron acceptor in the "dark" reactions of photosynthesis has been the subject of intense study for several decades. Much is now understood (see Bassham and Buchanan, Chapter 6, this volume). Less understood is the need for CO_2^* in the "light" reactions of photosynthesis. (For a discussion of the "light" reactions, see Volume I, edited by Govindjee [1982].) More specifically, PSII requires catalytic amounts of inorganic carbon in order to function. This "HCO_3^- effect" was discovered by Warburg and Krippahl (1960), but the discovery was received with great skepticism at first. Now, however, it seems probable that CO_2^* plays a more important role than previously supposed.

Despite recent advances in our knowledge of the subject, the exact role of CO_2^* in PSII remains obscure. Such a situation has led to disagreement among workers active in the area, and the reader is cautioned that the following discussion also includes the author's current thoughts and opinions on the subject, and represents a point of view not commonly taken. Along with other ideas, it will be suggested that hydrated CO_2 is most probably the immediate source of photosynthetic O_2. This "minority opinion" is offered to stimulate further thought, discussion, and experimentation in the area, and not as the definitive answer to this complex problem. For other discussions on the role of CO_2^* in PSII, which do not include a possible role for CO_2^* in the oxygen-evolving process, the reader is encouraged to consult Radmer and Cheniae (1977) and more detailed treatments by Govindjee and van Rensen (1978) and Vermaas and Govindjee (1981).

Early work on the CO_2^* problem consisted mostly of attempts, not always successful, to repeat the observations of Warburg and Krippahl (1960) and to confirm the requirement. An account of the early history of the problem has already been presented (Stemler, 1974; Govindjee and van Rensen, 1978) and will not be repeated here. Only those works that led to new discoveries about the CO_2^* requirement and that must be integrated into a reasonable explanation of the phenomenon will be mentioned.

Stern and Vennesland (1960) were first to screen a large number of chloroplast sources and to observe a CO_2^* requirement in all of them. Chloroplast grana from both C_3 and C_4 plants show the requirement (Stemler and Govindjee, 1974b). Apparently, the CO_2^* requirement is present in all higher plants. Photosynthetic bacteria, on the other hand, have not yet been shown to need CO_2^* directly for light-driven electron transport. Van Rensen and Vermaas (1981a) reported a lack of a bicarbonate effect in thylakoids isolated from the blue-green alga *Synechococcus leopoliensis*. This may indicate that CO_2^* is not an absolute requirement in all oxygen-evolving organisms or simply that the slightly modified HCO_3^- depletion procedure used by these workers failed to remove endogenous CO_2. A need for a more detailed comparative study is implied.

Good (1963) documented three important observations on the CO_2^* effect. First, he screened a large number of organic and inorganic substances, but he was unable to find a single substitute for CO_2^*. The requirement, therefore, appears specific. In addition, Good showed that the CO_2^* effect was always much greater in the presence of high NaCl and Na acetate concentrations. There appeared to be some sort of competition between CO_2^* and ions, presumably for some active site on or within the grana membranes. Later formate began to be used routinely to enhance the CO_2^* effect. Good, and also Izawa (1962), showed that added CO_2^* increased the rate of the Hill reaction much more at high light intensity than at low intensity. They concluded that CO_2^* was involved in "dark," probably enzymatic, reactions. West and Hill (1967), however, offered what appeared to be conflicting data on this last point. With 2,6-dichlorophenolindophenol as an electron acceptor, they showed that the effect of CO_2^* was independent of light intensity, which indicated an involvement of CO_2 in photochemical reactions. This apparent conflict was, in fact, the first indication that CO_2^* depletion could have two effects. Removal of CO_2^* from chloroplast grana reduces the number of PSII reaction centers able to evolve oxygen, which gives a low intensity effect, and reduces the rate of electron flow through still active centers, which gives an effect most noticeable at high light intensity. This "dual effect" of CO_2^* depletion was later confirmed by Stemler *et al.* (1974), Jursinic *et al.* (1976), and Siggel *et al.* (1977).

Despite the observations of N. Good (1963) and others, the CO_2^* effect appears not to have been taken seriously, even by those who observed it. There were probably two reasons for this; the effect was not consistently observable by everyone and it was seldom dramatic. The ability to observe routinely a large (often 10-fold) stimulation of oxygen evolution with added CO_2^* became possible with the finding (Stemler and Govindjee, 1973) that low pH—along with high concentrations of

NaCl and Na acetate (or formate) as discovered by Good, 1963—was needed to remove endogenous CO_2^* from chloroplasts. With this reliable method, the CO_2^* effect became thoroughly reproducible and could then be studied in detail.

II. Binding of CO_2^* to the Photosystem II Complex

The CO_2^* that allows electron flow through PSII is not normally free in solution. Rather, it is bound, apparently quite tightly, to the PSII complex (Stemler, 1977). This was shown by competitive-binding experiments in which chloroplasts were depleted of endogenous CO_2^*, then given ^{14}C-labeled CO_2^* along with increased concentrations of unlabeled material. The results of such experiments indicated the presence of two pools of bound CO_2^*—a small pool of high-affinity CO_2^* and a large pool of loosely bound CO_2^*. The large pool appears to consist of CO_2^* that is either dissolved in the membrane or occupies the inner thylakoid solution. This CO_2^* can be removed, apparently without any physiological effect, simply by washing the chloroplasts. The small, high-affinity pool, on the other hand, appears to control PSII activity. It is with this pool that the remaining discussion will deal.

One CO_2^* is bound to chloroplasts for every 380–400 chlorophyll (Chl) molecules (Stemler, 1977), which implies that each PSII complex has a tightly bound CO_2^*. This ligand is insensitive to washing with ordinary buffered solutions at neutral pH, but it can be removed by washing in CO_2^* depletion medium (0.1 *M* Na phosphate, pH 5, 0.175 *M* NaCl and 0.1 *M* Na formate). Under certain conditions, bound CO_2^* can also be removed by illumination (see Section III,C). It appears that removal of tightly bound CO_2^* inhibits O_2 evolution in continuous saturating light by more than 90%. It seems, therefore, that it is the characteristics of the high-affinity bound CO_2^* that must be known in order to understand the role of CO_2 in PSII.

III. Dynamic Aspects of CO_2^* Binding

Although it appears that CO_2^* is permanently bound to the PSII reaction center, the binding is actually dynamic. There are at least three important and interrelated factors that can determine the amount and form of bound CO_2^*, the rate of binding, and the reversal of binding. These factors are pH, formate concentration, and light. The rate of binding of exogenous CO_2can be measured by first depleting chloroplasts of endogenous CO_2^*, then adding back $^{14}CO_2^*$ (see Stemler,

1980a, for detailed methods). The binding of label can be stopped at any time by dilution with a high concentration of unlabeled CO_2^*. Removal of CO_2^* can be determined either by measuring the loss of previously bound $^{14}CO_2$ or by measuring a decrease in oxygen-evolving ability (Stemler, 1977, 1980a).

A. The Influence of pH

The rate of binding of exogenous CO_2^* to PSII reaction centers decreases with increasing pH in the order 6.0 > 6.8 ≫ 7.8 (Stemler, 1980a). Moreover, it is the internal thylakoid pH and formate concentration that determines the rate of binding. This suggests that the ligand binds near the inside surface and the binding rate increases as CO_2 predominates over HCO_3^-. The extent of CO_2^* binding, as well as the rate, is influenced by pH. When thylakoids were given subsaturating amounts of $^{14}CO_2$ and allowed to incubate for an extended period of time in buffered solutions which contained 0.4 *M* sucrose, $^{14}CO_2^*$ binding was maximal between pH 6.4 and 6.8 (Fig. 1, top curve), declining more sharply below than above the maximum. A possible reason for the decline in the extent of binding below pH 6.4 (where the concentration of CO_2 increases) will be discussed in Section IV.

B. The Influence of Formate

High (100 m*M*) concentration of Na formate slows the rate of binding of exogenous $^{14}CO_2^*$ to chloroplasts (Stemler, 1980a). Formate also extends the time required for HCO_3^- to reactivate the Hill reaction, according to Vermaas and van Rensen (1981); these authors also propose that formate and CO_2^* compete for the same binding site. For a model that incorporates this idea, see Vermaas and van Rensen (1981). The influence of formate, however, is not independent of pH. Formate has a much greater ability to decrease the extent of binding of added $^{14}CO_2^*$ at pH below 7.0 than at alkaline pH (Fig. 1, lower curve). The pH optimum for $^{14}CO_2^*$ binding in the presence of NaCl plus Na formate is pH 8.0 or higher compared to about 6.6 in sucrose-containing solution (upper curve). Note, however, that formate decreases the extent of $^{14}CO_2^*$ binding at every pH, including 8.0.

C. The Influence of Light

The rate of binding of added $^{14}CO_2^*$ to thylakoid membranes is slower in illuminated than in dark-adapted chloroplasts. A single saturating flash of light is enough to retard binding (Stemler, 1979). Continuous

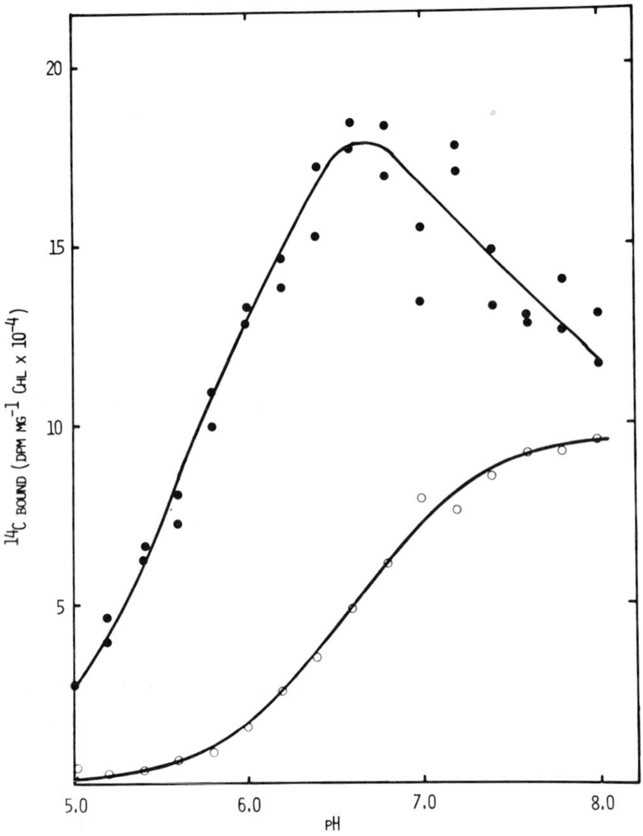

FIG. 1. The amount of $^{14}CO_2^*$ bound to chloroplasts after a 12-min incubation time as a function of pH. Chloroplasts were depleted of CO_2^*, collected by centrifugation, and resuspended in a small amount of depletion medium. Twenty μl of the chloroplast suspension was injected into the reaction mixture at 28°C. The binding of $^{14}CO_2^*$ was stopped after 12 min. The reaction mixture contained, in addition to 0.1 *M* Na phosphate, 0.33 m*M* $NaH^{14}CO_3$ (10 μCi), 0.1 m*M* $K_3Fe(CN)_6$, either (top curve) 0.01 *M* NaCl, plus 0.4 *M* sucrose or (bottom curve) 0.175 *M* NaCl plus 0.1 *M* Na formate. (See Stemler, 1980a, for experimental details. Reprinted by permission of the American Society of Plant Physiologists.)

light also prevents reactivation of the Hill reaction when CO_2^* is given back to depleted chloroplasts (Stemler, 1977; Vermaas and van Rensen, 1981). The latter workers showed, in addition, that the presence of formate is necessary for light to block reactivation of PSII. When formate is present, a dark period is necessary for added CO_2^* to act. Without formate, CO_2^* reactivates the Hill reaction even in the light. Not only can light prevent rebinding of CO_2^*, but light can cause a removal of endogenous CO_2^* so that the rate of O_2 evolution becomes progressively more dependent on added CO_2^* (Stemler, 1979, 1980a). The conditions needed for this second light effect are (1) the presence of a high formate concentration; (2) the presence of an electron acceptor such as ferricyanide; and (3) continuous light of saturating intensity. Apparently, as PSII cycles during a Hill reaction, CO_2^* is momentarily released. If formate is present, rebinding of the released ligand can be blocked. The system then becomes dependent on added CO_2^*.

The effects of pH, formate, and light, if examined individually, are difficult to explain. Taken together a dynamic picture of CO_2^* binding and release begins to emerge. This still tentative picture will be developed in the following sections. What will be discussed in Section IV represents mainly the author's interpretation of the combined influence of light, formate, and pH on CO_2^* binding. Since much of the data is relatively recent, alternate hypotheses by others have not yet had time to appear in the literature. Thus, the following section presents only a beginning attempt at a comprehensive hypothesis by the author. As mentioned in the introduction, this is offered not as the final word, but as a possible starting point for future analysis.

IV. The Active Forms of CO_2^*

Early experiments led to the conclusion that the important or "active" form of CO_2^* was HCO_3^-, not CO_2. This now appears to be an oversimplification. Stemler and Govindjee (1973) found a given amount of CO_2^* produced greater stimulation of Hill reaction at pH 6.8 than at 5.8, concluding, thereby, that HCO_3^- was active. Later, Khanna *et al.* (1977) did a more complete study of the effect of pH and found a subsaturating concentration of CO_2^* had the greatest stimulating effect between pH 6 and 7, suggesting that HCO_3^- was the active form. Why the effect decreased above pH 7.0, and where HCO_3^- concentration increased, was not explained. At the time of these experiments, however, it was not known that before CO_2^* can become active, it must bind to the membrane. Once the binding property of CO_2^* became known, we could then ask which form is bound, which is "used," and which form is momentarily released in the light.

Since formate appears to inhibit the rebinding of CO_2^* released in the light, it was supposed that a high concentration of added HCO_3^- should reverse the effect of formate, if indeed HCO_3^- was initially bound to the reaction center. Stemler (1980a) found that at pH 8.0, where > 97% of CO_2^* exists as HCO_3^-, the presence of 50 m*M* CO_2^* not only failed to reverse the effect of formate, but actually inhibited O_2 evolution even further. When the pH was lowered to 7.3, then 50 mM CO_2^* could at least partially reverse the effect of formate. In changing the pH from 8.0 to 7.3, the concentration of HCO_3^- changes very little, a drop from 97.6% to 89%. However, the concentration of CO_2 increases more than fourfold, from 2.4% to 11%. Apparently, it is the increase in CO_2, not HCO_3^-, that can reverse the effect of formate. This suggests that CO_2 is the form of CO_2^* that is initially bound to PSII. This conclusion is consistent with the finding mentioned earlier that the rate of binding of added $^{14}CO_2^*$ is 10-fold greater at pH 6.0 than at 7.8 (Stemler, 1980a). Recently, Vermaas and van Rensen (1981) confirmed the pH studies of Khanna *et al.* (1977) and proposed that CO_2^* must approach its binding site as CO_2 in order to penetrate a barrier presented by negative charges on the membrane surface.

If CO_2 is initially bound to PSII, it does not necessarily follow that CO_2 is the form "used." Earlier data, which indicated that HCO_3^- is the active form, must still be explained. The dynamics of the situation are implied in Fig. 1. In the sucrose solution (top curve), $^{14}CO_2^*$ binds maximally between pH 6.4 and 6.8. Less binding above pH 6.8 can be attributed to a lack of CO_2 (the pK_a of H_2CO_3 is about 6.4) such that even after a long incubation period, not all binding sites obtain a ligand. Below pH 6.4, however, binding also declines even though the concentration of CO_2 increases. This suggests that binding is reversible below pH of about 6.4 What appears to happen is that while CO_2 is initially bound, it is probably converted to $HCO_3^- + H^+$ after binding. The ligand seems to be stabilized as HCO_3^-; it apparently cannot exchange with free HCO_3^- (Stemler, 1977) nor can it be removed by washing. It seems that only as CO_2 can the ligand be exchanged or lost, thus, accounting for the requirement for low pH in the depletion medium.

Since PSII can operate at high pH where, if our previous inferences are correct, bound CO_2^* would be in the form of HCO_3^-, it is probable that this form is, in fact, active. From their pH studies, Khanna *et al.* (1977) and Vermaas and van Rensen (1981) also suggest that HCO_3^- is involved in the activation of PSII. The last question, then, is which form of CO_2^* is momentarily released in the light. An argument can be made that it is CO_2, not HCO_3^-. If HCO_3^- is released, it must become protonated and dissociate to $H_2O + CO_2$ in order that CO_2 can rebind; PSII recognizes only the form CO_2. However, the conversion of free HCO_3^- to CO_2 is too slow (10–100 msec; Rabinowitch, 1945) especially at high pH, relative to the rebinding rate. The rebinding time, as we deduced earlier, must be, in normal (undepleted) chloroplasts, within about 1 msec, the known recovery times of PSII following a photoact (Bouges-

Bocquet, 1973b). This time limitation means that the form of CO_2 released must be the same as that which is initially bound. By implication, HCO_3^- must be dehydroxylated in a light-driven release of CO_2.

A working hypothesis depicting the dynamic nature of the CO_2^*–PSII interaction is proposed as follows (Stemler, 1980a):

$$PSII + CO_2 \rightleftharpoons [PSII - CO_2] \quad (1)$$

$$[PSII - CO_2] + H_2O \underset{H^+}{\overset{OH^-}{\rightleftharpoons}} [PSII - HCO_3^-] + H^+ \quad (2)$$

$$[PSII - HCO_3^-] \xrightarrow{h\nu} PSII + CO_2 + OH^- \quad (3)$$

The reactions are consistent with the known effects of pH, formate, and light. Briefly, CO_2 forms a complex with the PSII, is hydrated, then dehydrated in the light. According to the scheme, both forms of dissolved CO_2^* play a role in PSII activity. Possible reasons for such cyclic behavior will be discussed in Section VII.

An apparent inconsistency in the foregoing discussion must be resolved. It is proposed that endogenous CO_2 is released and rebound to PSII cyclically in the light. This must occur quite rapidly, that is, within the turnover times of PSII. Turnover times in normal chloroplasts are about 1 msec as shown by Bouges-Bocquet (1973b). The rate of binding of exogenous $^{14}CO_2$, on the other hand, or the rate of reactivation of the Hill reaction by added CO_2^*, is measured in the time range of seconds to minutes (Stemler, 1980a; Vermaas and van Rensen, 1981). This striking difference in the binding rates of endogenous versus exogenous CO_2^* probably means that the location of the binding site is internal and inaccessible. In agreement with this idea, Vermaas and van Rensen (1981) suggested that the binding of exogenous CO_2^* is diffusion limited. After CO_2^* reaches its binding site, however, the release and rebinding in the light may well occur within 1 msec.

V. Location of the CO_2^* Binding Site

The physical location of the CO_2^* binding site has not yet been determined. The evidence pertaining to this question appears conflicting in some respects. There is fair agreement, at least, that the binding site is somewhat inaccessible, as discussed earlier. It is probably not on the external thylakoid surface. There is less agreement on the location of an internal site, however. Stemler (1980a) argued that because the binding of CO_2 was influenced by the pH and formate concentration of the internal thylakoid space that the binding site is near, or on, the inside surface of the membrane. Other views arise mainly from the fact that CO_2^* binding influences, and is influenced by, 3-(3,4-dichlorophenyl)-1,1-dimethylurea (DCMU)-type PSII herbicides. Interest in PSII herbicides and the HCO_3^- effect converged with the observation that DCMU could, in the dark, slow the removal of $^{14}CO_2^*$ from the PSII complex either by silicomolybdate treatment (Stemler, 1977) or by the HCO_3^--depletion procedure (Stemler, 1978). These initial reports were followed by more detailed studies in several laboratories. Khanna *et al.* (1981) showed that the affinity of [^{14}C]atrazine for its binding site was reduced by HCO_3^- depletion, though the number of binding sites remained unchanged. Van Rensen and Vermaas (1981b) have shown that 4,6-dinitro-*o*-cresol (DNOC) competitively inhibits restoration of Hill reaction rates when CO_2^* is given to CO_2^* depleted chloroplasts. Subsaturating amounts of DCMU and simeton also inhibited restoration, but not competitively. The latter results are confirmed by the author's unpublished data; DCMU significantly reduces the binding rate of exogenous $^{14}CO_2^*$ to thylakoid membranes.

The PSII herbicides previously mentioned are proposed to bind to the "protein shield" or "B-protein" toward the outer surface of the membrane. By implication, CO_2 is proposed to bind there also. Consistent with this idea, Vermaas and van Rensen (1981) proposed a model whereby CO_2^* binds to this protein at the bottom of a channel which opens to the external membrane surface. This location for the CO_2^* binding site would seem to be at variance with that proposed by Stemler. A possible compromise would be that the channel proposed by van Rensen and Vermaas opens to the inside thylakoid space rather than to the external medium. Clearly more work is needed to resolve this question. Perhaps in time it will also be possible to isolate and purify the CO_2^* binding component.

VI. Dual Effects of CO_2^* Depletion

CO_2^* depletion of chloroplasts results in two effects. With the removal of CO_2^*, a large fraction of the PSII units become inactive. This is indicated by reduced steady state flash yields of oxygen (Stemler *et al.*, 1974), reduced P680 absorbance change (Jursinic *et al.*, 1976), and reduced 334 nm absorbance change (X-320; Siggel *et al.*, 1977). Khanna *et al.* (1981) argued, based on reduced affinity of CO_2^*-depleted membranes for [^{14}C]-atrazine, that depletion results in a complete inactivation of a part of the total number of electron transport chains. The remaining PSII units, which remain active, are slow to recover following a photoact. This is shown by extended S-state transition times (Stemler *et al.*, 1974), slow flash-induced variable Chl *a* fluorescence decay (Jursinic *et al.*, 1976), and slow decay of flash-induced 334 nm absorbance change (Siggel *et al.*, 1977). (For definition and discussion of S-states, see Radmer and Cheniae, 1977; Wydrznski, Chapter 10, Vol. I, 1982.) Extension of S-state transition times in CO_2^*-depleted chloroplasts at pH 6.8 is represented in Fig. 2; the transition $S_1' \rightarrow S_2$ is shown. It is measured by varying the time between the first and second flash of a series given to dark-adapted chloroplasts and noting the effect in the O_2 yield of the third flash. We see that in CO_2^*-depleted chloroplasts (lower curve, Fig. 2) about 70% of the still-active PSII units complete an $S_1' \rightarrow S_2$ transition in $t_{1/2}$, 4–5 msec. However, recovery is biphasic. The remaining 30% of the units complete the transition much more slowly, such that even after 100 msec, less than

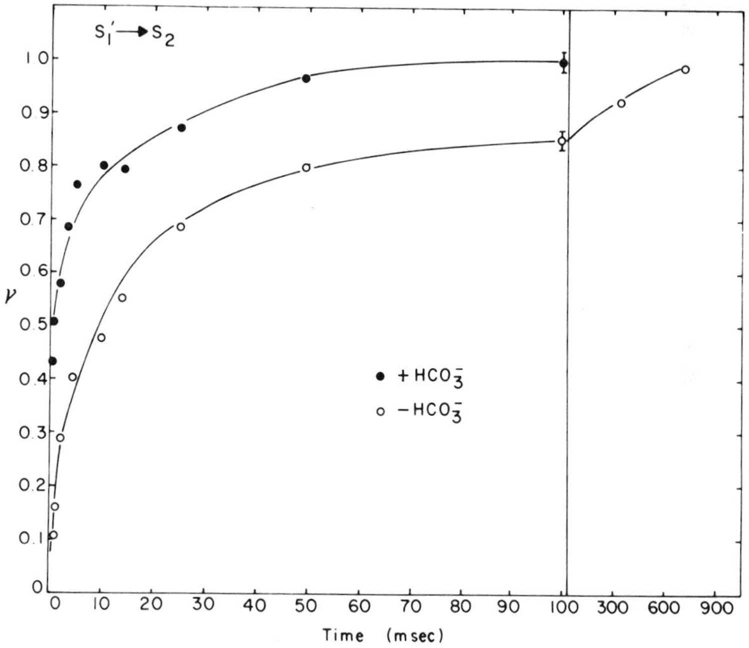

FIG. 2. PSII turnover rates (transition $S_1' \rightarrow S_2$) in CO_2^*-depleted and reconstituted maize chloroplasts after the first flash of a series. Broken chloroplasts were depleted of endogenous CO_2, then resuspended in reaction mixture which contained 0.25 mg Chl ml^{-1}, 0.05 *M* Na phosphate, 0.1 *M* NaCl, 0.1 *M* Na formate ± 0.01 *M* $NaHCO_3$, final pH 6.8. The suspension was placed on a Joliot-type electrode for O_2 measurements. Protocol for measurements and calculations was as described by Bouges-Bocquet (1973b). $\gamma \times 100$ is the percentage of reaction centers that completed a turnover in time *t* compared to the total number able to make the transition given unlimited or "infinite" time (2 sec).

one-half have done so. In reconstituted chloroplasts, nearly all units complete the transition with a half-time between 1 and 2 msec. A slow-recovering component, if present, is very small. These results are similar to those published by Stemler *et al.* (1974), only here, formate is used in the reaction mixture instead of acetate. The data as shown do not indicate differences in steady state flash yields $\pm CO_2^*$. Actually, the steady state yield was about 50% higher in the reconstituted samples. This means that added CO_2^* reactivates additional PSII units as well as shortens recovery times in all units. An alternate interpretation of these data, is that the apparent inactive PSII units are not, in fact, totally inactive, but that they recover so slowly after a photoact that recovery is not complete even between flashes spaced several seconds apart. Thus instead of a population of inactive PSII complexes, we may be observing a population which recovers extremely slowly. The resulting effect—a lowering of quantum yields of oxygen—would be the same in both cases, but the implications regarding mechanism may be very different.

The fact that CO_2^* exerts two effects on PSII makes determining the role of the ligand more complicated. On the one hand, CO_2^* appears to act as an on–off switch, which suggests that the requirement is absolute. Such a requirement would be expected if CO_2^* were a substrate or necessary catalyst. On the other hand, by regulating turnover times, CO_2^* appears to act as a rate control. This suggests that CO_2^* has only a stimulatory function; all PSII reactions can take place without CO_2^*, but at reduced rates. Such a role is characteristic of an allosteric effector, for example. Of the two effects, the rate control is most obvious. Under high intensity continuous light, the rate of electron flow can be reversibly suppressed more than 90% in CO_2^*-depleted chloroplasts. The on–off function is more subtle; it is only observed under low intensity continuous light or under a flash regime where the number of active PSII units can limit the formation of a measurable product. Often 50% or more of the PSII units are inactivated by CO_2^* depletion. In the past, greater importance has been attributed to the rate control by CO_2^*, since the effect is larger and easier to observe.

Regardless of which of the effects of CO_2^* may be termed *major* and which *minor,* relating them to the binding of CO_2^* presents a conceptual dilemma. Why, we may ask, does CO_2^* depletion not result in either complete inhibition of all PSII units, or simply slow recovery in all units. Why, in other words, are two distinct populations formed. Proposing that depletion is incomplete cannot alone explain this result. If depletion is incomplete and some PSII units remain active because they still retain bound CO_2^*, why do these same units show extended turnover times? We would expect such units to recover normally, whereas the data show (Fig. 2) that they do not.

There cannot be a simple solution to the problem of relating CO_2^* binding to the two effects it exerts. At this point, speculations must assume a certain complexity. For example, there may be two binding sites for CO_2^*, each having a unique function. There may be two populations of PSII units, each one unique in the effect CO_2^* exerts upon it. These two possibilities are by no means exhaustive, neither do they explain the CO_2^* effect. They do, however, incorporate the idea of two effects and can, again, be viewed simply as possible starting points in further analysis. This problem requires much more attention, thought, and experimentation.

VII. Possible Roles for CO_2^* in Photosystem II Chemistry

It is possible to formulate several hypotheses to explain the CO_2^* requirement, based on what is now known about it. It was suggested by Stemler and Govindjee (1973) and by Vermaas and van Rensen (1981) that the requirement represented a control mechanism to balance the reducing power and ATP output of the light reactions with the need for these substances in CO_2 fixation. This hypothesis, although reasonable, has found little supporting evidence. Because CO_2^* is tightly bound to the membranes, the Hill reactions normally show no sensitivity whatsoever to ambient CO_2^* concentrations. It may be, however, that *in vivo* conditions are such that internal thylakoid CO_2 concentration can regulate PSII and balance the light and dark reactions of photosynthesis. One wonders, though, why the levels of other carbon metabolism intermediates, such as phosphoglyceric acid (which, unlike CO_2, is the immediate reducing power and ATP-requiring metabolite) have never been shown to directly regulate PSII activity. CO_2^* would seem to play a unique role in this respect. The dynamic nature of CO_2^* binding in the light and the proposed changes in form experienced by the ligand also suggest a more complex role for CO_2^* than simple control. Thus, while this hypothesis is not entirely without merit, it cannot, at this time, be accepted as the explanation for the CO_2^* effect.

The scheme presented in Section IV suggests two possible roles for CO_2^*. Reaction (2) of the scheme releases a proton, whereas reaction (3) may release an hydroxyl ion. If the two events occurred on opposite sides of the thylakoid membrane, a pH gradient would be established to energize ATP formation. In this way, CO_2^* could play a central role in an energy transducing mechanism. Since the ligand is bound within the thylakoid, however, such a scheme raises the question of how OH^- tunnels through the membrane to make the stroma basic. Although attractive in many ways, the details of this hypothesis are difficult to imagine at present.

In the author's opinion, a more plausible hypothesis, and one consistent with current evidence, is that hydrated CO_2 is the immediate source of photosynthetic O_2. The claim that O_2 is derived from splitting CO_2 rather than H_2O has long been associated with Otto Warburg (1964), although it was a widely held belief before the acceptance of van Niel's hypothesis. Metzner (1966) proposed that HCO_3^-, not CO_2, is the source of photosynthetic O_2; this fundamentally different suggestion from that of Warburg is consistent with van Niel's hypothesis. In this case, water is the ultimate source of electrons, protons, and O_2, but CO_2 is a necessary catalyst. The abridged scheme can be written as:

$$2\,HCO_3^- + 2H^+ \xrightarrow{4\,h\nu} 2CO_2 + O_2 + 4H^+ + 4e^-$$

$$\text{(}2CO_2 \xrightarrow{+2\,H_2O} 2\,HCO_3^- \text{ + } 2H^+\text{, returned to the start)}$$

Here CO_2 is hydrated in a dark reaction, then dehydrated in light-driven reactions. This hypothesis, like the previous one, is consistent with the dynamic interaction between CO_2 and PSII that seems to occur.

At first examination, it would seem easy to eliminate at least one of the last two hypotheses mentioned. If it could be shown that CO_2^* binds and operates strictly on the reducing side of PSII, the last hypothesis, that CO_2^* is the immediate source of O_2, can be dismissed. However, this does not seem to be the case.

VIII. Sites of Action of CO_2^*

Attempts to determine the primary site of action of CO_2^* have produced equivocal results. This is partly because nearly every measurement made of PSII activity, whether it probes the donor or acceptor side of the reaction center, shows a CO_2^* effect. The picture is made more complicated by the dual effect of CO_2^* depletion discussed in Section VI. It is necessary to consider the site of action of CO_2^* that leads to activation of PSII and the site of action that increases the rate of turnovers of the unit. These sites may or may not be the same.

We do not know in detail what happens when entire PSII units are turned off completely by CO_2^* removal, that is, where the block is initially generated. These inactive complexes can account for the near 40% reduction in O_2 flash yields and in the amplitude of the oxidized form of reaction center of PSII 680^+ observed by Döring (quoted by Jursinic *et al.*, 1976), the reduced amplitude of the ESR signal II_{vf} (Jursinic *et al.*, 1976), and the reduced X-320 signal as observed by Siggel *et al.* (1977). All of these measurements were done with repetitive-flash technique, which cannot pinpoint the exact site of a block on the electron transport chain through PSII. We cannot tell, for example, if charge separation fails to take place at all in the reaction center II of CO_2^*-depleted grana or if secondary electron transport reactions fail to allow repeated activity of the centers. What happens in blocked PSII units during and after a single flash would be of interest, but this is not well described in the literature. From other data, only tentative inferences can be made. In the presence of DCMU, the variable Chl *a* fluorescence yield induced by low intensity continuous light is greater in CO_2^*-depleted chloroplasts as compared to reconstituted grana (Stemler and Govindjee, 1974a). If high fluorescence yield reflects the population Q^- (where Q is an electron acceptor of PSII; see Duysens and Sweers, 1963) then it appears that at least as much charge separation can, in fact, take place in CO_2^*-depleted chloroplasts as in reconstituted ones. If so, a direct site of action of CO_2^* on the reaction center P680 itself seems unlikely; failure of a CO_2^*-depleted PSII unit to evolve oxygen, among other things, must be due to the failure of a secondary reaction, either on the oxidizing or reducing side of the reaction center. The location of that reaction is problematic, relevant evidence is still too meager. One additional point should be made, however. There is a marked difference in the inactivation of PSII units by CO_2^* depletion compared to inactivation by DCMU. Kok *et al.* (1970) showed that addition of subsaturating amounts of DCMU lowered the flash yields of oxygen from chloroplasts, but did not change the initial oscillatory pattern. DCMU appears to permanently remove activity in some fraction of the PSII units. CO_2^* depletion, in contrast, not only lowers flash yields of oxygen, but induces strong damping of the initial oscillatory pattern (Stemler *et al.*, 1974). It appears that under a flash regime, CO_2^*-depleted PSII units take turns, to some extent, being inactive. Perhaps this means that residual CO_2^* can move from one PSII complex to another, but many other explanations are possible.

Not all PSII units are inactivated by CO_2^* depletion. Those that show extended turnover rates instead, also raise the question of a primary site of action. Although some evidence suggests that it lies on the acceptor side of PSII, other evidence is not easily explained by an exclusive location there.

Wydrzynski and Govindjee (1975) provided the first evidence that

CO_2^* was acting on the reducing side of PSII. They used Tris-washed chloroplasts (unable to evolve O_2) and artificial electron donors to PSII. When they observed Chl *a* fluorescence transients in these chloroplasts, they found that CO_2^* could still control the electron flow rate. They proposed that, since CO_2^* was acting between the site of electron donation and intersystem electron carriers, the site of action was on the reducing side of PSII. Further evidence was offered by Jursinic *et al.* (1976). These authors measured the rise and decay of variable fluorescence during and after flashes from a xenon lamp. The kinetics of the rise in variable fluorescence during the flash was the same in CO_2^*-depleted and reconstituted chloroplasts. The data were normalized, and possible amplitude changes (which might indicate the relative number of active PSII reaction centers) were not described. The decay of variable fluorescence, however, was much slower in CO_2^*-depleted chloroplasts compared to reconstituted samples. This was interpreted as evidence of a slow reaction, $Q^- \rightarrow B(B^-)$, on the reducing side of the reaction center in CO_2^*-depleted grana. In confirmation, Siggel *et al.* (1977) measured the absorbance change at 334 nm (reflecting Q/Q^-) and found the decay of this signal to be fivefold longer in CO_2^*-depleted chloroplasts as compared to reconstituted samples.

A different way of observing the rate-limiting reaction in CO_2^*-depleted chloroplasts is to measure turnover times in a PSII unit as it undergoes an S-state transition. Such a measurement was already shown in Fig. 2, Section VI. Comparison of these data to the results of Jursinic *et al.* (1976) and Siggel *et al.* (1977) presents a problem, however. As shown in Fig. 2, CO_2^* removal imposes a rate-limiting step, which extends the turnover time of the PSII complex. It is not possible, from these data alone, to tell if the limiting reaction is on the donor or the acceptor side of the reaction center. However, since the depicted $S_1' \rightarrow S_2$ transition is measured after the first flash given to dark adapted samples the only acceptor-side reaction which could be rate-limiting is electron flow from Q^- to B. According to Jursinic *et al.* (1976) the half-time of this reaction, as measured by variable fluorescence decay, was 2.6 msec in CO_2-depleted chloroplasts. Siggel *et al.* (1977) measured the same reaction as an absorbance change at 334 nm and found a half-time of 7 ± 3 msec. Even considering slightly different reaction conditions, these values cannot alone account for the slow turnover rates in CO_2^*-depleted chloroplasts shown in Fig. 2. The turnover half-time is, in fact, about 5–6 msec (in fair agreement with the values of Jursinic *et al.* and Siggel *et al.*), but this time is for the fast component only, the values do not account for the very slow recovery ($t_{1/2}$ 100–200 msec) seen in nearly 30% of the units. We must conclude that either (1) the PSII turnover rates in depleted grana are limited by a reaction on the donor side or (2) turnover rates are limited on the acceptor side, but measurements taken thus far have not looked for a slow component of the $Q^- \rightarrow B(B^-)$ reaction. Judging from the S-state recovery rates shown in Fig. 2, however, a large component of the Q^- population must be oxidized with a half-time of > 100 msec in CO_2^*-depleted chloroplasts.

Govindjee *et al.* (1976) proposed that the rate-limiting reaction in CO_2^*-depleted chloroplasts given continuous saturating light is electron transfer from B^{2-} to PQ. Arguments to this effect take up a large portion of a review article already published (Govindjee and van Rensen, 1978) and will not be repeated here. The authors, however, did not discuss the possibility of a slow component of the $Q^- \rightarrow B(B^-)$ reaction or the relative contribution such a component might make to rate-limitation under continuous saturating light. Additional experiments are needed to clarify this point.

Many experimental results on the CO_2^* effect are difficult to explain simply by a site of action on the reducing side of PSII between Q^- and $B(B^-)$ or beyond. In the presence of DCMU, for example, the rise in variable fluorescence caused by weak continuous light is more rapid in CO_2^*-depleted chloroplasts compared to reconstituted ones (Stemler and Govindjee, 1974a). When the actinic light is terminated, reconstituted samples emit more long term (0.1–10 sec) delayed light emission (DLE) than do CO_2^*-depleted chloroplasts. The difference in DLE is more striking when DCMU is present during illumination. If DCMU blocks the reoxidation of Q^-, these results indicate that CO_2^* has a site of action before Q. A slower rise in variable fluorescence in the CO_2^*-reconstituted samples may mean that back reactions are taking place in these samples, keeping the population of Q^- lower. This is consistent with the larger amount of long-term DLE by CO_2^*-reconstituted grana. Back reactions in PSII leading to long-term DLE are thought to be a function of the oxidation states of components on both sides of the PSII reaction center. In DCMU-treated chloroplasts, the necessary component on the reducing side of PSII can only be Q^-. Since Q^- appears to be formed in both CO_2^*-depleted and reconstituted chloroplasts (to account for the fluorescence rise), differences in back reactions leading to DLE suggest that there is a difference in the two types of chloroplasts on their oxidizing side.

The interpretation of DLE data is often difficult; conclusions, therefore, are usually tentative. Less ambiguous evidence showing an effect of CO_2^* on the oxidizing side of PSII resulted from measurements of turnover times of PSII in the presence and absence of high (100 m*M*) formate concentrations. There is a considerable amount of evidence indicating an interaction between formate ions and membrane-bound CO_2^*. Formate is needed in CO_2^*-depletion medium for maximal removal of endogenous CO_2^* (Stemler, 1977). It also retards the binding of exogenous $^{14}CO_2^*$ to thylakoid membranes. Formate is required in a Hill reaction medium in order for light to induce dependence on exogenous CO_2^* (Stemler, 1979, 1980a). Khanna *et al.* (1977) and Vermaas and van Rensen (1981) showed that a greater concentration of CO_2^* is needed to restore Hill reaction rates when formate is present in the reaction medium. The latter authors postulate a competitive interaction between formate and CO_2 for the CO_2 binding site. Given the ability of formate to interact with endogenous CO_2, turnover rates of the PSII units were measured in the presence and absence of formate in normal (undepleted) chloroplasts following saturating light flashes. Stemler (1980b) found that, under a condition of high pH, formate extended turnover rates only in those PSII units undergoing S-state transitions $S_2' \rightarrow S_3$ and $S_3' \rightarrow S_0$. The rates of transitions $S_0' \rightarrow S_1$ and $S_1' \rightarrow S_2$ were unaffected by the presence of formate. Most remarkably, the transitions extended by formate were already the slowest in the formate-free controls. This suggests that formate acts directly on the O_2-evolving side of PSII. If it were imposing a rate-limiting step on the reducing side, all S-state transitions should be affected equally, or at least the effect should be most noticeable in those transitions that were fastest in the formate-free controls. If formate acts by retarding the rebinding of CO_2, it was reasoned that CO_2 must be released by reactions that occur on the oxidizing side of PSII. The experiments just described were done at a high pH (8.2). It was not possible to show reversibility of formate effects by exogenous CO_2^* because, at that pH, it is not possible to add CO_2, only HCO_3^-. CO_2 is what binds initially to the PSII complex (see Section IV). The author repeated parts of these experiments at low pH (5.3). At this pH, the binding of endogenous CO_2^* (90% CO_2, 10% HCO_3^-) is reversible in the dark. S-state transitions $S_0' \rightarrow S_1$ and $S_1' \rightarrow S_2$ were measured in the presence and absence of 100 m*M* formate. To some of the formate-containing samples, 5 m*M* $NaHCO_3$ was also added; the pH was adjusted accordingly. The results are shown in Fig. 3. At pH 5.3, S-state transitions are quite slow ($t_{1/2}$, 4 msec) even in formate-free controls (top curve). Addition of formate dramatically extended the turnover times of the transitions; however, not all to the same degree. The half-time of the $S_1' \rightarrow S_2$ transition (second from bottom curve) was 17 msec, whereas that for the $S_0' \rightarrow S_1$ (bottom curve) was more than 30 msec. This difference is of particular interest because both transitions are measured simultaneously, on the same sample, by varying the time between the first and second flash given to a dark-adapted sample. If the presence of formate

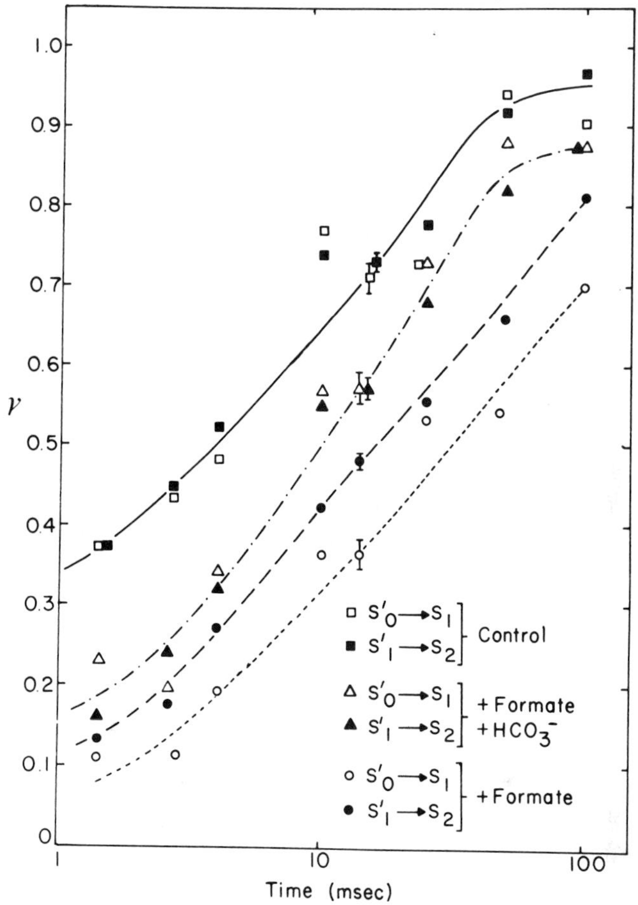

FIG. 3. Formate-induced decrease of PSII turnover rates at low pH in maize chloroplasts and partial reversibility of the formate effect by added CO_2^*. Undepleted broken chloroplasts were suspended in reaction mixture and placed on a Joliot-type O_2 electrode. The reaction mixture for the controls contained 0.25 mg Chl ml^{-1}, 0.05 *M* Na phosphate, and 0.1 *M* NaCl. Experimental samples contained, in addition, 0.1 *M* Na formate ± 5 m*M* $NaHCO_3$. The pH was brought to 5.3 in all samples. Protocol for measurements and calculations was as described by Bouges-Bocquet (1973b). Temperature, 20°C. Standard error bars at the 14 msec points are each derived from five separate measurements. Both S-state transitions $S_0' \rightarrow S_1$ and $S_1' \rightarrow S_2$ are measured after the first flash of a series given to dark-adapted chloroplasts. Note that formate slows the rate of transition $S_0' \rightarrow S_1$ more than $S_1' \rightarrow S_2$.

induced a rate-limiting step restricted to the reducing side of PSII, that step could only be the reaction $Q^- \rightarrow B(B^-)$ after the first flash. The two S-state transitions occurring after the first flash, $S_0' \rightarrow S_1$ and $S_1' \rightarrow S_2$ should be equally limited by this slow $Q^- \rightarrow B(B^-)$ reaction. What is observed, however, is that the $S'_0 \rightarrow S_1$ is nearly twice as slow as the $S'_1 \rightarrow S_2$. These results strongly suggest that formate acts directly on the oxidizing side of PSII. At pH 5.3, however, in contrast to pH 8.2, the action of formate is to a large extent reversed by added CO_2^*. Samples, which contained formate plus 5 m*M* $NaHCO_3$, showed turnover rates for both transitions $S'_0 \rightarrow S_1$ and $S'_1 \rightarrow S_2$ of 10 msec half-time (second-from-top curve). Addition of 5 m*M* $NaHCO_3$ was evidently not sufficient to completely overcome the formate effect. This was perhaps because of the limited solubility of CO_2 and the reduced affinity for the ligand at this pH (Fig. 1).

Additional indication that CO_2^* has a site of action on the oxidizing side of PSII comes from measurements of the flash-induced release rate of O_2 from the PSII complex (Stemler, 1981). The rate of O_2 release from PSII is usually impossible to measure with an unmodulated Joliot-type O_2 electrode (Joliot and Joliot, 1968) because the response time of the instrument is too long (3–4 msec). O_2 is released from PSII in 1–2 msec (Joliot *et al.*, 1966; Bouges-Bocquet, 1973a). However, if chloroplasts are suspended in reaction mixture with low pH (5.3) and a high formate concentration, i.e., CO_2^*-depletion medium, the release rate of O_2 from the PSII complex can be delayed beyond the response time of the instrument to 5.5 ± 0.27 msec (for the standard deviation, $n = 12$). Addition of 2 m*M* CO_2^* partially reverses the effects of formate, and O_2 is then released with a half-time of 4.93 ± 0.18 ms. The release rate of O_2 from PSII in response to widely spaced flashes should not be influenced by limiting reactions on the reducing side of the reaction center.

To summarize this section, it seems that with CO_2^* depletion, a rate-limiting step can appear on either the oxidizing or reducing side of PSII, depending on the conditions and on the type of measurement. An explanation for such anomalous results is not obvious, but one possible explanation will be developed in the next section.

IX. A Working Hypothesis: Hydrated CO_2 as the Immediate Source of Photosynthetic O_2

We cannot, at this time, eliminate any of the possible roles for CO_2^* in PSII chemistry that were offered in Section VII. Nevertheless, in the opinion of the author, the most plausible hypothesis is that HCO_3^- is the immediate source of photosynthetic O_2. There is certainly no proof of this hypothesis in the form of direct evidence, yet there is circumstantial evidence and other reasons to take the hypothesis seriously. The following discussion will review and elaborate those reasons. Model reactions will then be proposed showing how CO_2^* could be the immediate source of O_2.

CO_2 appears to undergo cyclic "dark" hydration and light-driven dehydration (Stemler, 1980a). Such behavior has an alternate explanation (Section VII), but is consistent with the idea that CO_2^* could be the immediate source of O_2.

The binding of CO_2^* is influenced by the internal thylakoid pH and formate concentration (Stemler, 1980a). A corresponding effect of internal pH on O_2 evolution (Harth *et al.*, 1974; Reimer and Trebst, 1975) would seem, at least, to place the two processes on the same side of the membrane.

Differential effects of formate on certain S-state transitions indicate that this ion inhibits reactions on the oxidizing side of PSII (Stemler, 1980b). Reversibility of formate effects by CO_2^* at low pH, a condition that allows the form CO_2 to be present, suggests that CO_2^* is also involved in reactions occurring on the O_2-evolving side of the reaction center.

The rate of release of O_2 from the O_2-evolving mechanism during an $S_3' \rightarrow S_0$ transition is slowed at low pH by formate (Stemler, 1981). Reversibility by CO_2^* implies that this substance is somehow needed in the final O_2-releasing reaction.

A word can also be said about the lack of evidence that H_2O, not HCO_3^-, is the immediate source of O_2. Although it is assumed that H_2O binds directly to the O_2-evolving enzyme (perhaps to complexed manganese; Wydrzynski and Sauer, 1980), there is, as yet, no proof of this. Experiments with deuterated H_2O by Arnason and Sinclair (1976) failed to show that the cleavage of an O—H bond is a rate-limiting step in O_2 evolution. (This cannot be taken to mean, however, that it does not occur; it is simply not rate-limiting.) Other stable isotope experiments discussed by Metzner (1975; see also Metzner *et al.*, 1979) have led to apparent conflicts with the idea that O_2 comes directly from H_2O. These experiments are, unfortunately, inconclusive; they do not prove that either H_2O or HCO_3^- (or some other intermediate) is the immediate source of O_2. This is, in fact, an extremely difficult question to answer, for reasons that will be discussed in Section X. Given this lack of conclusive evidence, we are permitted to keep an open mind with respect to the chemistry of oxygen evolution and the role of various possible chemical intermediates.

Considering the preceding set of circumstances, there exists the possibility that hydrated CO_2 is the immediate source of photosynthetic O_2. I have therefore proposed (Stemler, 1980b) a set of reactions as a working model for O_2 evolution (Fig. 4). The model elaborates the four-step

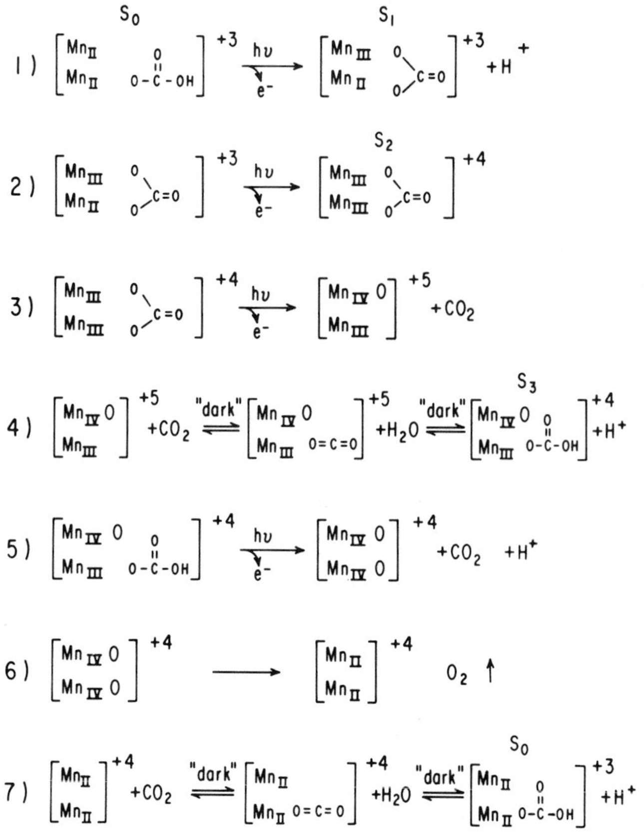

FIG. 4. A working hypotheses for photosynthetic O_2 evolution showing the role of hydrated CO_2. Positive charges on the manganese-containing complex, which are not neutralized by HCO_3^- or CO_3^{2-}, may be balanced by anions such as Cl^- in the medium. (Reprinted by permission of Elsevier Publishing Company; see Stemler, 1980b.)

mechanism of Forbush *et al.* (1971) and also includes, with some modification, the model recently developed by Wydrzynski and Sauer (1980), which suggests a likely role for manganese in O_2 evolution (see Chapter 10 by T. Wydrzynski in Vol. I, Govindjee, 1982).

Attached to the PSII reaction center complex are two Mn ions, which can change oxidation states, and a CO_2^* ligand, which can convert sequentially from CO_2 to HCO_3^- to CO_3^{2-}. Light-driven reactions (1) and (2) increase the oxidation states of the two Mn(II) ions to Mn(III) and release a proton from HCO_3^-. Reactions (3) and (4) show S-state transition $S_2 \rightarrow S_3$. Here, when an electron is removed, Mn(IV) is produced. Mn(IV) then extracts O^{2-} from CO_3^{2-}, releases CO_2, and produces $[Mn(IV)O]^{+2}$.

The CO_2 recombines with the Mn(III) still present, becomes hydrated and forms HCO_3^-, releasing a proton. Reactions (5), (6), and (7) show transition $S_3 \rightarrow S_0$. In Reaction (5), an electron is removed, as Mn(III) is oxidized to Mn(IV). Mn(IV) extracts O^{2-} from HCO_3^- (or CO_3^{2-}), again releasing CO_2 and a proton, and results in the formation of a second $[Mn(IV)O]^{+2}$. In (6), O_2 is evolved as the 2 $[Mn(IV)O]^{+2}$ intermediates react to form 2 Mn(II). In (7), the Mn complex prepares for the next cycle. CO_2 rebinds, becomes hydrated, and forms $HCO_3^- + H^+$. The S_0 state is again operative.

The reactions shown in (4) and (7) are inhibited by formate, which competes with CO_2 for the CO_2 binding site. This could explain why, at high pH, formate extends the turnover times of only the S-state transitions $S_2' \rightarrow S_3$ and $S_3' \rightarrow S_0$.

The chemistry of Mn proposed here is not without precedent. Latimer (1938) suggested that $[Mn(IV)O]^{+2}$ is an intermediate in the reduction of permanganate to manganous ions. Furthermore, Harriman *et al.* (1978) proposed, as a possible intermediate in photosynthetic O_2 evolution, a complex that contains two Mn(IV) ions bound to oxygen ligands.

In the model, CO_2 is released during transitions $S_2' \rightarrow S_3$ and $S_3' \rightarrow S_0$. Proton evolution is shown here to follow S-state transitions in a 1,0,1,2 sequence. However, the actual sequence of proton release is still not certain (Fowler, 1977; Saphon and Crofts, 1977; Junge and Ausländer, 1978; Junge and Jackson, Chapter 13, Vol. I, 1982). The proposed model would predict that the actual sequence could be, to some extent at least, a function of pH and reflect the pK of carbonic acid. In any case, the model is somewhat flexible with respect to proton evolution.

If the model is to provide a comprehensive explanation of the CO_2^* effect, it must deal with the undeniable evidence that CO_2^* has a site of action on the reducing side of PSII between Q and B. It may be that CO_2^*-mediated reactions, which occur on the oxidizing side of the PSII reaction center, can influence the $Q^- \rightarrow B$ reaction indirectly. After a charge separation occurs in the reaction center, Q^- may need to dissociate from P680 momentarily, move a short distance to contact B, and allow electron transfer. The dissociation of the anion, Q^-, may be influenced by the positive charges remaining on the reaction center complex, not necessarily on P680, which will be reduced by secondary electron donors, but on other components of the oxidizing-side complex. By allowing the release of protons, as in Reaction 1, H_2CO_3 can neutralize positive charges on the complex. This may facilitate the dissociation of Q^- from P680 and hence speed electron transfer in the forward direction. An effect of donor-side positive charges on the $Q^- \rightarrow B(B^-)$ reaction has already been suggested by van Gorkom and Donze (1973). Such an effect was proposed to account for the S-state dependence of the decay rate of fluorescence in the 3–200 msec time range in *Chlorella* as observed by Joliot *et al.* (1971). Thus CO_2^* may have two distinct functions in PSII chemistry. First, it may react with states S_2' and S_3' to transfer oxygen and electrons to manganese. Second, it may neutralize positive charges on the oxidizing side of the reaction center and allow Q^- to dissociate from P680. This second proposed function could explain most of the evidence, which shows that CO_2 controls electron transfer rates on the reducing side of PSII. It may even come into play when artificial electron donors supply electrons to PSII (Wydrzynski and Govindjee, 1975). It may also be expected that if a strong electron acceptor such as silicomolybdate could be positioned next to Q^-, dissociation of Q^- from the reaction center complex may not be necessary to bring about electron transfer. This may be why Khanna *et al.* (1977) and Stemler (1977) found nearly normal electron flow rates in CO_2^*-depleted chloroplasts when silicomolybdate was used as a PSII electron acceptor. (The effect of HCO_3^- on the silicomolybdate-mediated Hill reaction is complex, however. Crane and Barr (1977) and also van Rensen and Vermaas (1981b) reported that HCO_3^- inhibits, to some extent, silicomolybdate reduction. The latter workers suggested that silicomolybdate and HCO_3^- compete for the same binding site, but more work is needed to test this, and other, hypotheses.)

The model presented here is not the only one that could conceivably be developed using hydrated CO_2 as a "catalytic" intermediate. As mentioned earlier, in fact, a less detailed scheme has been proposed by Metzner (1978). According to this scheme, two bicarbonate ions (2 HCO_3^-) act as electron donors to intermediates (perhaps manganese), which supply electrons to $P680^+$. The resulting bicarbonate radicals ($2HCO_3$) combine to form peroxidicarbonic acid, $H_2C_2O_6$. This ultimately decomposes to yield 2 CO_2, H_2O, and O. For a more complete explanation of this scheme and further rationale, see Metzner (1978). Although the schemes for O_2 evolution suggested here do explain much of the available evidence on the CO_2^* effect, much more testing is required.

X. Labeling the O_2-Evolving Precursor

The arguments presented thus far, that hydrated CO_2 may be the immediate source of photosynthetic O_2, are based on indirect or circumstantial evidence. A number of attempts have been made to determine directly the immediate source of O_2 by labeling the O_2-evolving precursor with stable isotopes, particularly ^{18}O (Ruben *et al.*, 1941; Stemler and Radmer, 1975; Radmer and Ollinger, 1980). All such studies have invariably produced results in which the isotopic composition of evolved O_2 was very near, though not exactly (Metzner, 1975), the isotopic composition of the medium H_2O. Although indicating that H_2O is the ultimate source of photosynthetic O_2, these studies have revealed very little about the mechanism of H_2O oxidation. Attempts with mass spectrometry and isotope labels to prove that HCO_3^- is, or is not, the immediate source of photosynthetic oxygen arrive at a dilemma. If for the moment one supposes that O_2 arises from HCO_3^-, the resulting CO_2 must be hydrated again to $H^+ + HCO_3^-$ before the reaction can proceed a second time. The spontaneous hydration of CO_2 is a fairly slow process, requiring several seconds for completion. In full light, this would impose an intolerable rate limitation for the PSII complex. Therefore, one must suppose, in addition, that to overcome this rate limitation, the complex must itself catalyze the rehydration of CO_2. But if the PSII complex can catalyze the hydration of CO_2, it can catalyze isotopic exchange between bound CO_2^* and medium H_2O. We are left with the dilemma. If HCO_3^- is the immediate source of O_2, it should be difficult, perhaps impossible, to label it and keep it labeled at the PSII complex long enough to do an experiment. Considering that the binding site for CO_2^* is fairly inaccessible and that once bound a ligand is not easily replaced by another (obscuring any possible carbonic anhydrase activity detectable in the external medium), to arrive at an experimental protocol yielding conclusive results is a formidable challenge.

A method for circumventing the isotopic exchange problem is being developed by Metzner *et al.* (1979). This method makes use of the natural abundance of various oxygen isotopes and the fact that slight partitioning effects occur such that dissolved CO_2 tends to be slightly enriched in the heavier ^{18}O isotope compared to the solution H_2O. Theoretically then, evolved O_2 could be slightly enriched in ^{18}O if it were coming from CO_2^*. Since only a very small degree of enrichment is expected, this method requires very accurate measurement, elimination or measurement of O_2 uptake reactions (which could also discriminate among oxygen isotopes), careful avoidance of atmospheric contamination (O_2 in the air is enriched in ^{18}O compared to sea water; Dole, 1935), and assumptions as to what sort of isotopic partitioning and discrimination will occur at the reaction center. So far the results are not conclusive but, with further refinement, this method may lead to a direct determination of the O_2-evolving precursor.

XI. Conclusions

At the present time, there is little that can be said with certainty about the role of bicarbonate in PSII activity. Several hypotheses have been discussed. One in particular, that HCO_3^- could play a direct role in oxygen evolution, has been emphasized. One reason is that this hypothesis has been almost completely neglected in past reviews; see, for example, Vermaas and Govindjee (1981). But rapid progress is taking place in this area so that it is wise to consider all ideas on this topic as extremely tentative. Assertions to the effect that there is only one way to interpret the evidence related to the role of bicarbonate in PSII activity should be viewed with skepticism until a larger body of experimental results can be assembled and critically evaluated.

REFERENCES

Arnason, T., and Sinclair, J. (1976). *Biochim. Biophys. Acta* **449,** 581–586.
Bouges-Bocquet, B. (1973a). *Biochim. Biophys. Acta* **292,** 772–785.
Bouges-Bocquet, B. (1973b). *Biochim. Biophys. Acta* **314,** 250–256.
Crane, F. L., and Barr, R. (1977). *Biochem. Biophys. Res. Commun.* **74,** 1362–1368.
Dole, M. (1935). *J. Am. Chem. Soc.* **57,** 2731.
Duysens, L. N. M., and Sweers, H. E. (1963). *In* "Studies on Microalgae and Photosynthetic Bacteria" (J. Ashida, ed.), pp. 353–372. Univ. of Tokyo Press, Tokyo.
Forbush, B., Kok, B., and McGloin, M. P. (1971). *Photochem. Photobiol.* **14,** 307–321.
Fowler, C. F. (1977). *Biochim. Biophys. Acta* **462,** 414–421.
Good, N. E. (1963). *Plant Physiol.* **38,** 298–304.
Govindjee, ed. (1982). "Photosynthesis: Energy Conversion by Plants and Bacteria," Vol. I. Academic Press, New York.
Govindjee, and van Rensen, J. J. S. (1978). *Biochim. Biophys. Acta* **505,** 183–213.
Govindjee, Pulles, M. P. J., Govindjee, R., Van Gorkom, H. J., and Duysens, L. N. M. (1976). *Biochim. Biophys. Acta* **449,** 602–605.
Harriman, A., Porter, G., and Duncan, I. (1978). *In* "Photosynthetic Oxygen Evolution" (H. Metzner, ed.), pp. 393–403. Academic Press, New York.
Harth, E., Reimer, S., and Trebst, A. (1974). *FEBS Lett.* **42,** 165–168.
Izawa, S. (1962). *Plant Cell Physiol.* **3,** 221–227.
Joliot, P., and Joliot, A. (1968). *Biochim. Biophys. Acta* **153,** 625–634.
Joliot, P., Hoffnung, M., and Chabaud, R. (1966). *J. Chem. Phys.* **10,** 1423–1441.
Joliot, P., Joliot, A., Bouges, B., and Barbieri, G. (1971). *Photochem. Photobiol.* **14,** 287–305.
Junge, W., and Ausländer, W. (1978). *In* "Photosynthetic Oxygen Evolution" (H. Metzner, ed.), pp. 213–228. Academic Press, New York.
Junge, W., and Jackson, J. Baz. (1982). *In* "Photosynthesis: Energy Conversion by Plants and Bacteria." (Govindjee, ed.), Vol. I, pp. 589–646. Academic Press, New York.
Jursinic, P., Warden, J., and Govindjee (1976). *Biochim. Biophys. Acta* **440,** 322–330.
Khanna, R., Govindjee, and Wydrzynski, T. (1977). *Biochim. Biophys. Acta* **462,** 208–214.
Khanna, R., Pfister, K., Keresztes, A., van Rensen, J. J. S., and Govindjee (1981). *Biochim. Biophys. Acta* **634,** 105–116.

Kok, B., Forbush, B., and McGloin, M. (1970). *Photochem. Photobiol.* **11,** 457–475.
Latimer, W. M. (1938). "Oxidation Potentials." p. 224. Prentice-Hall, Englewood Cliffs, New Jersey.
Metzner, H. (1966). *Naturwissenschaften* **53,** 141–150.
Metzner, H. (1975). *J. Theor. Biol.* **51,** 201–231.
Metzner, H. (1978). *In* "Photosynthetic Oxygen Evolution" (H. Metzner, ed.), pp. 59–76. Academic Press, New York.
Metzner, H., Fischer, K., and Bazlen, O. (1979). *Biochim. Biophys. Acta* **548,** 287–295.
Rabinowitch, E. I. (1945). "Photosynthesis and Related Processes." Vol. 1, p. 175. Wiley (Interscience), New York.
Radmer, R., and Cheniae, G. (1977). *In* "Primary Processes in Photosynthesis" (J. Barber, ed.), pp. 339–341. Elsevier, Amsterdam.
Radmer, R., and Ollinger, O. (1980). *FEBS Lett.* **110,** 57–61.
Reimer, S., and Trebst, A. (1975). *Biochem. Physiol. Pflanz.* **168,** 225–232.
Ruben, S., Randall, M., Kamen, M., and Hyde, J. L. (1941). *J. Am. Chem. Soc.* **61,** 877–879.
Saphon, S., and Crofts, A. R. (1977). *Z. Naturforsch., C: Biosci.* **32C,** 617–626.
Siggel, U., Khanna, R., Renger, G., and Govindjee (1977). *Biochim. Biophys. Acta* **462,** 196–207.
Stemler, A. (1974). Ph.D. Thesis, University of Illinois, Urbana.
Stemler, A. (1977). *Biochim. Biophys. Acta* **460,** 511–522.
Stemler, A. (1978). *In* "Photosynthetic Oxygen Evolution" (H. Metzner, ed.), pp. 393–403. Academic Press, New York.
Stemler, A. (1979). *Biochim. Biophys. Acta* **545,** 36–45.
Stemler, A. (1980a). *Plant Physiol.* **65,** 1160–1165.
Stemler, A. (1980b). *Biochim. Biophys. Acta* **593,** 103–112.
Stemler, A. (1981). *Proc. Int. Congr. Photosynth., 5th, 1980,* **II,** 389–394.
Stemler, A., and Govindjee (1973). *Plant Physiol.* **52,** 119–123.
Stemler, A., and Govindjee (1974a). *Photochem. Photobiol.* **19,** 227–232.
Stemler, A., and Govindjee (1974b). *Plant Cell Physiol.* **38,** 298–304.
Stemler, A., and Radmer, R. (1975). *Science* **190,** 457–458.
Stemler, A., Babcock, G. T., and Govindjee (1974). *Proc. Natl. Acad. Sci. U.S.A.* **71,** 4679–4683.
Stern, B. K., and Vennesland, B. (1960). *J. Biol. Chem.* **235,** PC51–PC53.
van Gorkom, H. J., and Donze, M. (1973). *Photochem. Photobiol.* **17,** 333–342.
van Rensen, J. J. S., and Vermaas, W. F. J. (1981a). *Physiol. Plant* **51,** 106–110.
van Rensen, J. J. S., and Vermaas, W. F. J. (1981b). *Proc. Int. Congr. Photosynth., 5th, 1980,* **II,** 151–156.
Vermaas, W. F. J., and Govindjee (1981). *Proc. Indian Natl. Sci. Acad.* Biological Sciences Series B, **4,** 581–605.
Vermaas, W. F. J., and van Rensen, J. J. S. (1981). *Proc. Int. Congr. Photosynth., 5th, 1980,* **II,** 157–165.
Warburg, O. (1964). *Annu. Rev. Biochem.* **33,** 1–14.
Warburg, O., and Krippahl, G. (1960). *Z. Naturforsch., B: Anorg. Chem., Org. Chem., Biochem., Biophys., Biol.* **15,** 367–369.
West, J., and Hill, R. (1967). *Plant Physiol.* **42,** 819–826.
Wydrzynski, T. (1982). *In* "Photosynthesis: Energy Conversion by Plants and Bacteria" (Govindjee, ed.), Vol. I, pp. 469–506. Academic Press, New York.
Wydrzynski, T., and Govindjee (1975). *Biochim. Biophys. Acta* **387,** 403–408.
Wydrzynski, T., and Sauer, K. (1980). *Biochim. Biophys. Acta* **589,** 56–70.

16

Bicarbonate or Carbon Dioxide as a Requirement for Efficient Electron Transport on the Acceptor Side of Photosystem II*

WIM F. J. VERMAAS
GOVINDJEE

ABBREVIATIONS

B (or R)	Second quinone-type PSII electron acceptor
Chl	Chlorophyll
DAD	2,3,5,6-Tetramethylphenylenediamine
DBMIB	2,5-Dibromo-3-methyl-6-isopropyl-*p*-benzoquinone
DCMU	3-(3,4-Dichlorophenyl)-1,1-dimethylurea
DCPIP	2,6-Dichlorophenolindophenol
DPC	Diphenylcarbazide
EPR	Electron paramagnetic resonance
FeCy	Ferricyanide [$Fe(CN)_6^{3-}$]
HCO_3^-*	Species (CO_2 or HCO_3^-) that binds to a specific binding site and allows efficient electron transport (≡ CO_2^* in Stemler, Chapter 15)

*This chapter should be read in conjunction with the chapter by A. Stemler—Editor.

Photosynthesis: Development, Carbon Metabolism, and Plant Productivity, Vol. II

ISBN: 0-12-294302-3

M	Charge accumulator in the oxygen-evolving system
MV	Methyl viologen
Pheo	Pheophytin
PQ	Plastoquinone
PSII	Photosystem II
Q	First quinone-type PSII electron acceptor
S	State of the oxygen-evolving system
SiMo	Silicomolybdate ($SiMo_{12}O_{40}{}^{4-}$)
$Z_{(1,2)}$	Donor(s) to P680

ABSTRACT

In this chapter, we reviewed observations showing that HCO_3^-* (CO_2 or HCO_3^-) greatly affects the electron transport rates between the quinone-type intermediates [Q, B, and the PQ (plastoquinone) pool] on the acceptor side of photosystem II (PSII) without directly influencing the donor side of PSII.* The major observations against an important role of HCO_3^-* on the donor (oxidizing) side and in favor of its role on the acceptor (reducing) side of PSII are

1. There is no HCO_3^-* effect on the electron transport from H_2O to Q as measured with silicomolybdate as electron acceptor in the presence of DCMU or with ferricyanide in trypsin-treated chloroplasts (Section II,A).
2. There is no HCO_3^-* effect on the kinetics of electron flow from the oxygen evolving system to the electron donor Z to P680, and from Z to P680 (Section II,E).
3. There is a large HCO_3^-* effect on electron transport from Q^- to B, as measured by both fluorescence and absorption methods (Sections II,B, and II,C); an additional, but smaller, effect is observed between P680 and Q.
4. The Chl *a* fluorescence transient in CO_2-depleted samples is similar, but not identical, to that in the presence of DCMU; the latter blocks electron flow beyond Q. Detailed analysis suggests that the absence of HCO_3^-* causes a major block after B (Section II). A large effect on the electron transport from B^{2-} to PQ is also confirmed by measurements of (a) Chl *a* fluorescence after single flashes of light; and (b) absorption changes at 265 nm (Sections II,B and C).
5. The location of the binding site of HCO_3^-* is close to that of the herbicides that bind near Q and B (Section II,F).

The bottle-neck reaction in CO_2-depleted chloroplasts seems to be electron transport from B^{2-}† to PQ, with a $t_{1/2}$ of at least 100 msec; in thoroughly depleted samples, this reaction takes seconds suggesting a "complete" block. After HCO_3^- addition, the PQH_2 oxidation ($t_{1/2} \sim 25$ msec) becomes rate-limiting, like in control chloroplasts. Most of the Q^- reoxidation by $B^{(-)}$ is faster than the PQ reduction in CO_2-depleted chloroplasts. No experimental data, thus far, point unequivocally to any CO_2/HCO_3^- effect on the donor side of PSII.

This chapter does not offer a complete picture of the HCO_3^- or CO_2 action in thylakoid membranes; it only deals with the site of action of the "bicarbonate effect" on photosynthetic electron transport. Therefore, in order to obtain an "overview" of this subject, this chapter should be read in conjunction with the cited reviews and with Chapter 15.

†In this chapter, the description "B^{2-}" is used to denote fully reduced B in general, and the use of "B^{2-}" does not exclude protonation of the reduced form of this secondary quinone.

I. Introduction

A crucial role of CO_2 or bicarbonate on the oxygen evolving mechanism of photosynthesis, as presented in Chapter 15 of this volume, is a point of view that is interesting but speculative inasmuch as it does not yet have a firm experimental backing. The present chapter is written to provide the reader with information on the action of CO_2 or HCO_3^- on the acceptor side of PSII, for which, contrary to a bicarbonate effect on the donor side of PSII, strong evidence exists. The absence of HCO_3^-* results in a severe slowing down of electron transport between Q, the first quinone-type PSII acceptor, and the PQ pool, without directly affecting electron transport on the donor side of PSII (see Fig. 1). This effect of HCO_3^-* has been established by many types of experiments in several laboratories (see, e.g., Wydrzynski and Govindjee, 1975; Govindjee *et al.*, 1976; Jursinic *et al.*, 1976; Khanna *et al.*, 1977, 1980, 1981; Siggel *et al.*, 1977; Stemler, 1977, 1979; van Rensen and Vermaas, 1981a; Vermaas and van Rensen, 1981). Some of these experiments will be summarized later. It is our hope that the reader after having read this and Chapter 15 may have a better appreciation of the complicated, but interesting and important, bicarbonate problem.

In addition to its effect on electron transport, bicarbonate has been shown to influence other photosynthetic processes in the thylakoid membrane, including enhancement of photophosphorylation at pH 7.0–7.5 (Punnett and Iyer, 1964). However, at pH 8.0, the pH optimum for photophosphorylation, no stimulation of phosphorylation is observed. Nelson *et al.* (1972) concluded from their experiments with isolated coupling factor protein that HCO_3^- might cause a conformational change in this protein. This suggestion has been confirmed by Cohen and MacPeek (1980).

In this chapter, we will concentrate on the HCO_3^-* effects on photosynthetic electron transport. For a background on electron transport, the reader is referred to Vermaas and Govindjee (1981b) and Cramer and Crofts, Chapter 9, Vol. I (1982). No attempt is made to speculate on the "active species," i.e., the species (CO_2, HCO_3^- or CO_3^-) that is responsible for restoration of electron transport or about the molecular mechanism of the specific HCO_3^-* binding to the thylakoid membrane. For publications that cover these questions, see Govindjee and van Rensen (1978), Stemler (Chapter 15, this volume), Vermaas and Govindjee

Since it is unknown whether CO_2 or HCO_3^- is necessary for efficient electron transport, we will use the description HCO_3^- to denote the CO_2 or HCO_3^- that is specifically bound to the membrane and that exerts its effect on electron transport.

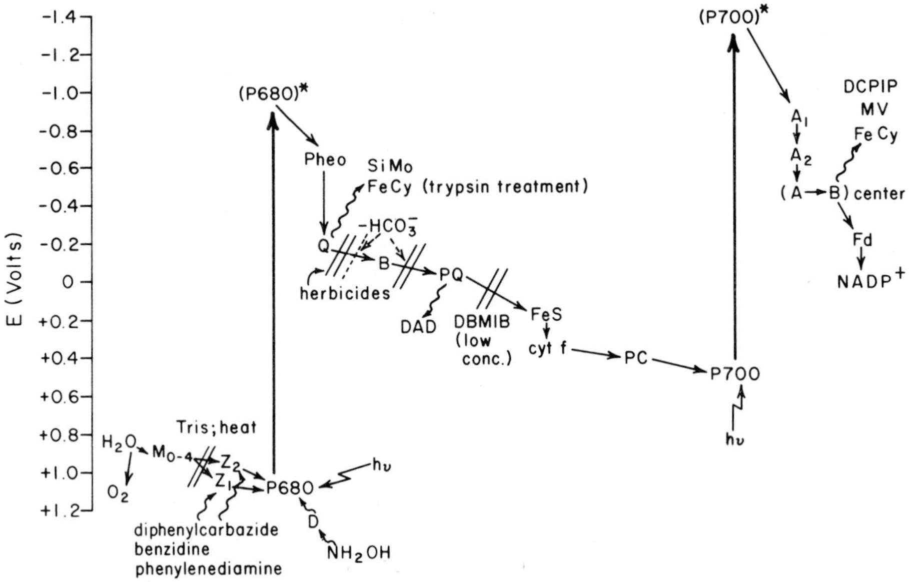

FIG. 1. Pathway of noncyclic electron flow from H_2O, the electron donor in photosynthesis, to $NADP^+$ (nicotinamide adenine dinucleotide phosphate), the "physiological" electron acceptor. E on the ordinate stands for midpoint redox potential. Light quanta ($h\nu$) are absorbed in two sets of antenna Chl molecules; the excitation energy is transferred to the reaction center Chl *a* molecules of PSII (P680) and of PSI (P700), which forms (P680)* and (P700)*, respectively. The latter two initiate electron transport. Inhibitors of photosynthetic electron transport are known to work at specific sites in the electron transport chain. Inhibition of electron transport by a certain treatment or molecule is indicated as ⇸, slowing down of electron transport as ⇢, and electron donation or acceptance by an artificial donor or acceptor is drawn as a curved line. M_{0-4}, the charge accumulator of the oxygen-evolving system that can exist in several charged states; D, an unknown electron donor to P680; FeS, the Rieske iron–sulfur center; cyt *f*, cytochrome *f*; PC, plastocyanin; A_1, the primary PSI acceptor, is suggested to be a Chl molecule; A_2, an iron containing protein, which is considered equivalent to the so-called X; (A → B) center, iron–sulfur centers observed in EPR spectra; and Fd, ferredoxin. (Also see the List of Abbreviations.) (For further details, see Volume I, also edited by Govindjee, 1982.)

(1981a), Sarojini and Govindjee (1981), and Vermaas and van Rensen (1981).

II. A Site of HCO_3^-* Action on the Acceptor Side but Not on the Donor Side of Photosystem II

Warburg and Krippahl (1958, 1960) showed that HCO_3^-* is necessary for efficient electron transport in thylakoids or intact *Chlorella* cells, using quinone as an artificial electron acceptor. After this discovery, several groups investigated this "bicarbonate effect" further (see, e.g., Abeles *et al.*, 1961; Stern and Vennesland, 1962; Good, 1963; West and Hill, 1967). These initial studies were limited due to variability in the CO_2-depleted material. Once a reproducible method for CO_2 depletion was developed (Stemler and Govindjee, 1973), a determination of the site of HCO_3^-* action was possible. Although at first, Stemler and Govindjee (1973) observed no bicarbonate effect (i.e., stimulation of electron transport by HCO_3^- addition) using DPC (diphenylcarbazide) and DCPIP (2,6-dichlorophenolindophenol) as artificial donor and acceptor, respectively, experiments performed later by Wydrzynski and Govindjee (1975) showed a significant bicarbonate effect on the DPC → DCPIP reaction, but the effect was lower than in the H_2O to DCPIP Hill reaction. These observations could indicate an action of HCO_3^-* both on the acceptor and on the donor side of PSII. However, care should be taken in interpreting these data because DPC might have effects on the electron transport chain other than electron donation to the donor Z of the PSII reaction center (P680), e.g., an increase in the efficiency of PSII (Harnischfeger, 1974). Fischer and Metzner (1981) showed that the methyl viologen-mediated Mehler reaction (monitoring noncyclic electron transfer through PSII and PSI) in thylakoids treated with hydroxylamine (blocking $P680^+$ reduction by the physiological donor Z and reducing $P680^+$ via an unknown donor D) is relatively insensitive to HCO_3^-*. However, Wydrzynski and Govindjee (1975) observed a large HCO_3^-* effect using NH_2OH as an electron donor. In order to resolve this uncertainty, the induction kinetics of the variable Chl *a* fluorescence after CO_2 depletion may be examined. The variable Chl *a* fluorescence monitors, among other things, the redox state of Q; Q is a quencher of Chl *a* fluorescence, whereas Q^- is not (see, e.g., a review by Lavorel and Etienne, 1977). Therefore, a rapid accumulation of Q^- due to an inhibition or slowing down of electron transport beyond Q is easily detected by fluorescence measurements. The CO_2 depletion causes a fast increase in the variable fluorescence yield (Stemler and Govindjee, 1974; Wydrzynski and Govindjee, 1975), similar to that observed in the presence of DCMU [3-(3,4-dichlorophenyl)-1,1-dimethylurea] (Wydrzynski and Govindjee, 1975); the latter blocks reoxidation of Q^- by $B^{(-)}$, the second quinone-type PSII acceptor. Blockage of the donor side by heat treatment (Katoh and San Pietro, 1967; Homann, 1968) causes a much lower yield of the variable fluorescence (Wydrzynski and Govindjee, 1975). Since fluorescence induction traces of CO_2-depleted chloroplasts resemble, on a short time scale, traces of normal chloroplasts in the presence of DCMU, but do *not* resemble traces in which the donor side is blocked, it may be concluded that the primary site of HCO_3^-* action is not on the oxygen evolving system (Wydrzynski and Govindjee, 1975). Furthermore, after alkaline Tris-washing (which deactivates the oxygen-evolving system (Yamashita and Butler, 1968)) and CO_2 depletion of chloroplasts, the addition of artificial PSII electron donors (e.g., DPC and NH_2OH) is unable to change $-HCO_3^-*$ fluorescence characteristics into $+HCO_3^-*$ characteristics, which points to a bicarbonate effect between the site of donation by these external donors and PQ (Wydrzynski and Govindjee, 1975). If thoroughly CO_2-depleted chloroplasts are used, a major fast rise without any significant slow phase in the fluorescence transient is observed. This points to a (nearly) complete blockage of electron transport on the acceptor side of PSII (Vermaas and Govindjee, 1982). Since the area over the fluorescence induction curve in CO_2-depleted chloroplasts is about twice as large without DCMU than with it a major block must occur beyond B (Vermaas and Govindjee, 1982). If the CO_2 depletion is not complete, then a fast rise in the fluorescence induction curve followed by a slow rise to the maximum fluorescence level is observed (Stemler and Govindjee, 1974). (For a detailed explanation of the latter data, see Vermaas and Govindjee, 1981a.)

A. *Electron Transport Rates in the Presence of Artificial Electron Acceptors*

In the presence of SiMo (silicomolybdate; $SiMo_{12}O_{40}^{4-}$) and DCMU, no significant bicarbonate effect is observed in CO_2-depleted chloroplasts (Khanna *et al.*, 1977; van Rensen and Vermaas, 1981a). [SiMo is an artificial acceptor known to accept electrons from Q (Giaquinta and Dilley, 1975; Zilinskas and Govindjee, 1975).] It is difficult to imagine a significant CO_2-depletion effect on the donor side of PSII that would not inhibit electron flow from H_2O through P680 to SiMo. However, the addition of HCO_3^- causes a dramatic increase in the Hill reaction with oxidized DAD (2,3,5,6-tetramethylphenylenediamine) as acceptor when DBMIB (2,5-dibromo-3-methyl-6-isopropyl-*p*-benzoquinone) is present to inhibit electron flow beyond the PQ pool (Khanna *et al.*, 1977; Sarojini *et al.*, 1981; see also Fig. 1). No measurable effect of HCO_3^-* is observed in a PSI reaction as monitored by electron flow from reduced DAD to MV (methyl viologen) in the presence of DCMU (Khanna *et al.*, 1977). These data indicate that the most important effect of HCO_3^-* is between Q and PQ and not on the donor side of PSII, since the reaction $H_2O \rightarrow$ SiMo appears to be almost insensitive to 10 m*M* bicarbonate. Further support against an effect of HCO_3^-* on the donor side of PSII is provided by data on trypsin-treated chloroplasts. Ferricyanide (FeCy) is known to be able to accept electrons directly from Q after trypsin treatment (Renger, 1976; van Rensen and Kramer, 1979); after trypsin incubation, no bicarbonate effect on the FeCy Hill reaction is observed (Khanna *et al.*, 1981; van Rensen and Vermaas, 1981b), suggesting that HCO_3^-* does not influence the donor side of PSII.

B. Absorption Changes Due to Quinones

An independent although somewhat more complicated way to monitor electron transport is by spectrophotometry. Flash-induced absorption changes in the 320–334-nm region are ascribed to the oxidation or reduction of Q (Stiehl and Witt, 1968; Witt, 1973; van Gorkom, 1974; Renger, 1976; Siggel *et al.*, 1977), whereas changes in absorption at 265 nm are correlated with changes mainly in the B and PQ pool (Stiehl and Witt, 1969; Haehnel, 1976; Siggel *et al.*, 1977).

Using repetitive flashes (with 250 msec dark time between the flashes), a part of the Q^- decay, as measured by 334-nm absorbance changes, is slowed down significantly in CO_2-depleted chloroplasts [450 μsec (in control) → 6 msec ($-HCO_3^-$*)]. After HCO_3^- addition, this slow phase nearly disappears, and the kinetics become similar to that in the control. The slowing down of the Q^- decay in the absence of HCO_3^-* suggests a decrease in the rate of Q^- oxidation by B. Furthermore, the amplitude of the absorbance change at 334 nm is smaller before than after HCO_3^- addition (Siggel *et al.*, 1977). This was interpreted as a (reversible) inactivation of P680 (Siggel *et al.*, 1977). However, no bicarbonate effect on steady state electron transport from H_2O to Q has been detected (Khanna *et al.*, 1977; van Rensen and Vermaas, 1981a); so, the explanation of a reversible P680 inactivation by CO_2 depletion is not supported by other data. A more reasonable explanation for this phenomenon is that part of Q^- is reoxidized very slowly in the absence of HCO_3^-* (~ 1 sec) (Jursinic and Stemler, 1981). Therefore, part of Q^- formed after one flash is not yet reoxidized when the next flash arrives at the sample; this results in a lower [Q] just before a flash and in a lower amplitude of the absorbance change at 334 nm in the absence of HCO_3^-*. It is not yet known if this very slow component of Q^- reoxidation is due to electron transfer to $B^{(-)}$ or to another electron acceptor. This means that certain PSII reaction centers are not completely "turned off" by CO_2 depletion, but rather have had no time to recover between flashes. Thus, absence of HCO_3^-* does not decrease the number of active reaction centers (P680).

When absorption changes at 265 nm that are induced by 85 msec flashes (spaced 5 sec apart) are compared in CO_2-depleted chloroplasts with and without HCO_3^-* addition, it is observed that CO_2 depletion reduces the amplitude of the 265 nm signal and slows down its decay. The kinetics of formation of the signal are determined by the opening time of the shutter (~ 10 msec) and are not significantly influenced by HCO_3^-*. To explain these data, it was suggested that in the absence of HCO_3^- the reduction of the PQ pool is slow compared to its oxidation, causing a low steady state [PQH_2] during illumination and, therefore, a small absorbance change. The observed absorbance change was interpreted to be due mainly to the formation of B^{2-}.† Then, the slow decay has to be ascribed to a slow electron transport from B^{2-} to PQ ($t_{1/2}$ ~ 100 msec), followed by a relatively fast reoxidation of PQH_2 ($t_{1/2}$ ~ 25 msec) (Siggel *et al.*, 1977). Further experiments are needed to confirm these interpretations.

C. *Chlorophyll* a *Fluorescence Yields after Light Flashes*

Experiments pointing to a significant slowing down of the oxidation of B^{2-} by PQ in CO_2-depleted chloroplasts were presented by Govindjee *et al.* (1976). When CO_2-depleted chloroplasts are subjected to one or two saturating flashes, spaced approximately 30 msec apart, the Chl *a* fluorescence yield 160 msec after the flash(es) are as low as in CO_2-depleted chloroplasts after HCO_3^- addition (Fig. 2). This observation indicates a reasonably fast (< 30 msec) reoxidation of Q^- after the first two flashes in CO_2-depleted chloroplasts. However, when three or more flashes are applied the fluorescence yield (and, thus, the Q^- concentration) 160 msec after the last flash is much higher in these chloroplasts than in CO_2-depleted chloroplasts to which HCO_3^- is added (Fig. 2). These data show a relatively efficient electron transport from Q^- to $B^{(-)}$ of only two electrons in CO_2-depleted chloroplasts. This is the number of electrons that can be accepted by B after dark adaptation when there is no appreciable electron transport from B to PQ. Although the dark time between the flashes in these experiments was somewhat short (not all Q^- after the first flashes is oxidized before the next flash reaches the sample; see Jursinic and Stemler, 1981), these data strongly imply the existence of the main site of HCO_3^-* action on the oxidation of B^{2-} instead of on the oxidation of Q^-. If the oxidation of Q^- were to be slowed down by CO_2 depletion into the 100 msec range, the fluorescence 160 msec after the first two flashes would be as high as after the other flashes. This was not observed (Fig. 2). Therefore, these data point to a lack of reoxidation of B^{2-} and, thus, to a severe slowing down of electron transport from B to PQ. However, quantitative analyses require further experiments.

† See footnote on p. 542.

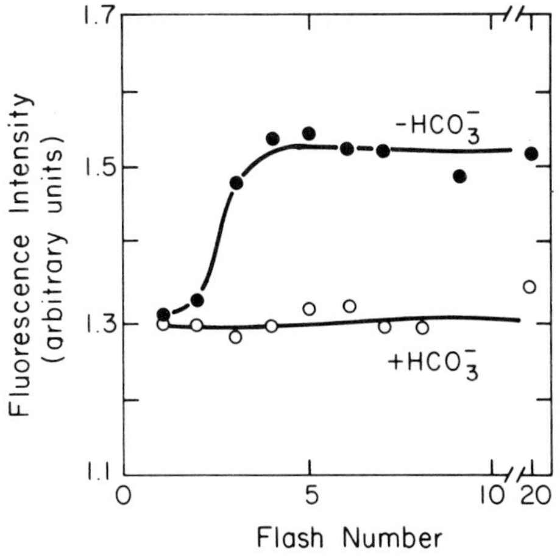

FIG. 2. Intensity of Chl *a* fluorescence 160 msec after the last of a series of 3 μsec saturating flashes, spaced approximately 30 msec apart, as a function of the number of flashes in CO_2-depleted chloroplasts with (○) or without (●) 20 m*M* $NaHCO_3$; 20 μg Chl ml^{-1}. The reaction mixture contained 50 m*M* Na phosphate, 100 m*M* NaCl, 100 m*M* Na formate (pH, 6.8). (From Govindjee *et al.*, 1976.)

The preceding conclusion is confirmed by experiments in which DCMU is added to a sample after a certain number of flashes. DCMU addition is known to shift the $Q^-B \rightleftharpoons QB^-$ equilibrium to the left (Velthuys and Amesz, 1974) and is a useful probe to monitor the redox state of B by means of fluorescence, which is sensitive to the redox state of Q. In CO_2-depleted chloroplasts to which HCO_3^- is added, a damped oscillation of fluorescence as a function of flash number (with a period of 2) is observed when DCMU is added before fluorescence is measured (Fig. 3). This observation is expected if electron transport between B^{2-} and PQ is relatively fast. B is mainly in the oxidized form after dark adaptation, and DCMU addition does not cause a large back reaction (i.e., oxidation of B^- or, perhaps, B^{2-} by Q) because $[B^-]$ is low and $[B^{2-}]$ negligible. After one flash, B is mainly in the B^- form just before DCMU is added (B^- is relatively stable), resulting in a high fluorescence yield. After two flashes, B is mainly fully oxidized (the B^{2-} oxidation by PQ is much faster than the DCMU addition). The third and fourth flashes give results similar to the first and second flashes, respectively. However, in CO_2-depleted chloroplasts such an oscillation with a period of 2 is not observed (Fig. 3); after dark adaptation the fluorescence yield is low, whereas after one flash a high fluorescence intensity is observed. The fluorescence yield after DCMU addition remains high after the subsequent flashes, pointing to a high relative concentration of B^- and/or B^{2-} in these samples after one or more flashes. A reasonable explanation for this observation is to assume that most of the reduced B formed by a flash is still present when DCMU is added. This indicates a prolonged lifetime of reduced B in CO_2-depleted chloroplasts (Govindjee *et al.*, 1976).

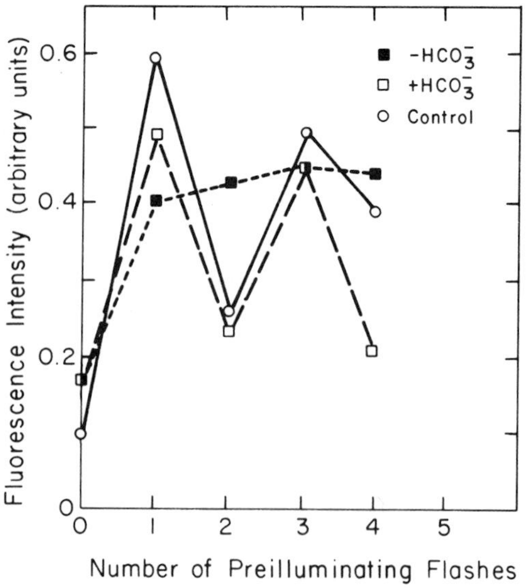

FIG. 3. DCMU-induced increase in Chl *a* fluorescence as a function of the number of preilluminating flashes. 20 μg Chl ml^{-1}; ○, nondepleted (control) chloroplasts; ■, CO_2-depleted chloroplasts; □, CO_2-depleted chloroplasts + 20 m*M* $NaHCO_3$. Reaction medium as described in the legend of Fig. 2. (From Govindjee *et al.*, 1976.)

D. Light-Induced Proton Uptake and Release by the Thylakoid Membrane

Proton uptake and release by thylakoid membranes can be monitored spectrophotometrically using indicator dyes and appropriate buffers in an accurate and elegant manner (Junge and Ausländer, 1973; Junge *et al.*, 1979). When flash-induced pH changes in the internal space of the thylakoid are measured in CO_2-depleted chloroplasts, the amplitude of the rapid pH change due to proton production by the oxygen-evolving system is decreased compared to control chloroplasts, while a slow component ($t_{1/2}$ ~ 100 msec), present in control chloroplasts, disappears (Khanna *et al.*, 1980). This slow component is related to the oxidation of PQH_2, which causes proton movement into the inside (Ausländer and Junge, 1975). The absence of this component indicates that PQH_2 is formed only very slowly in CO_2-depleted chloroplasts (Khanna *et al.*, 1980). The decrease in amplitude of the rapid phase (due to H^+ release into the intrathylakoid space) was interpreted by Khanna *et al.* (1980) as indicating an inactivation of PSII. This is probably due to an irreversible inactivation by the depletion procedure, and not to a reversible inactivation by a HCO_3^-* effect on the Q^- oxidation as the dark time between the flashes (10 sec) was long enough to obtain full recovery of Q. Khanna *et al.* (1980) could not determine the reversibility of these effects because readdition of HCO_3^- caused an increase in the buffering capacity of the system, resulting in a dramatic decrease of the pH changes. Further experiments are needed to test the reversibility of these effects. Our tentative interpretation is that CO_2 depletion affects the PQH_2 formation much more than the O_2 evolution. (The changes in the PQH_2 formation rate are expected to be reversible by HCO_3^- addition, whereas the change in O_2 evolution is probably an irreversible "artifact" of the depletion procedure.)

E. Reduction of the Electron Donor Z and of the Photosystem II Reaction Center

To investigate if HCO_3^-* causes changes in $P680^+$ and Z^+ reduction kinetics, which would be expected if HCO_3^-* had any effect on the donor side of PSII, fluorescence measurements in the μsec range were performed, and the decay kinetics of the EPR signal II_{vf} were recorded (Jursinic *et al.*, 1976). If repetitive flashes are used and, thus, the $P680^+$ reduction by Z slowed down into the (sub) μsec range (e.g., Sonneveld *et al.*, 1979), the rise in Chl *a* fluorescence yield in the μsec range is indicative, among other things, of the rate of electron flow from Z to $P680^+$ because $P680^+$ is known to be a quencher of fluorescence (Okayama and Butler, 1972; Butler *et al.*, 1973). Therefore, $P680^+$ reduction results in an increase in fluorescence yield. These rise kinetics are not influenced by HCO_3^-* (Jursinic *et al.*, 1976), indicating that the $P680^+$ reduction by Z is insensitive to HCO_3^-*. The EPR signal II_{vf} kinetics, monitoring the reduction of Z^+ by the oxygen-evolving system (Blankenship *et al.*, 1975), which were measured in an experiment using repetitive flashes, are also unchanged by CO_2 depletion (Jursinic *et al.*, 1976). However, the amplitude of the signal is decreased by 40% in the absence of HCO_3^-*. One explanation for this decrease is that the dark time between the flashes (1 sec) was not long enough to reoxidize Q^- completely (the Q^- reoxidation in CO_2-depleted chloroplasts has a very slow component as mentioned previously). The P680* (i.e., P680 in the excited state) cannot transfer an electron to Q^-, and there is no stable charge separation. For this reason, a non-negligible portion of Z is not oxidized after a flash, and the EPR II_{vf} signal is expected to be lower in CO_2-depleted chloroplasts.

In our opinion, the absence of a significant change in the EPR II_{vf} decay kinetics in CO_2-depleted chloroplasts, monitoring the reduction of Z^+ by the oxygen-evolving system, rules out under these experimental conditions any significant role of HCO_3^-* on the oxygen-evolving system, whereas the unchanged μsec fluorescence kinetics indicate that the electron donation to $P680^+$ is not influenced by HCO_3^-*.

F. *Relationship between Herbicides and HCO_3^-**

Experiments in which $H^{14}CO_3^-$ is added to CO_2-depleted chloroplasts and the $H^{14}CO_3^-$* binding monitored as a function of the $H^{14}CO_3^-$* concentration suggest that there is only one specific HCO_3^-* binding site per PSII reaction center (Stemler, 1977). Furthermore, it has been shown that the DCMU and HCO_3^-* binding sites are close together because $H^{14}CO_3^-$* removal by SiMo washing is greatly decreased by prior addition of DCMU (Stemler, 1977). A close spatial relationship between the HCO_3^-* and herbicide binding sites has now been described (Khanna *et al.*, 1981; van Rensen and Vermaas, 1981b; Vermaas *et al.*, 1982). Since herbicides are known to bind to a surface-exposed protein, very close to Q and B, and to be accessible to degradation by trypsin (Renger, 1976; Trebst, 1979; Arntzen *et al.*, 1981), it is reasonable to assume that HCO_3^-* binds to the same or to an adjacent protein molecule. This explains why a HCO_3^-* interaction with the oxygen-evolving system or even with the donor side of PSII, although possible, is unlikely.

III. Postulated Effects of HCO_3^-* on the Donor Side of Photosystem II

There are some HCO_3^-* effects that are easily explained by assuming HCO_3^-* action on the water splitting system directly (see Stemler, Chapter 15, this volume). (For a recent review on water oxidation, S-states, and O_2 evolution, see Wydrzynski, Chapter 10, Vol. I, 1982.) However, here we argue that these data do not necessarily indicate a HCO_3^-* influence even on the donor side of PSII.

A. *Effects on the S-States*

When oxygen evolution in CO_2-depleted chloroplasts is measured as a function of flash number, the damping of the oscillation with a period of 4 (Kok *et al.*, 1970) is faster without HCO_3^- addition than with it (Stemler *et al.*, 1974). This indicates that the miss parameter α, the probability of not undergoing a net change in the charge accumulator M after a flash, is higher in the absence of HCO_3^-*. This might be due to a HCO_3^-* effect on the donor side, but is equally well explained by an effect on the acceptor side. If Q^- is not fully reoxidized in the dark time between the flashes (1 sec), which is probably the case (Jursinic and Stemler, 1981), then a certain number of PSII centers (those of the type

P680.Q^-) cannot transfer an electron from P680* to the first quinone acceptor as it is in Q^- form and, therefore, no change is observed in those centers. Assuming that the slowly oxidizing Q^- does not belong to a special type of PSII center, it is obvious that the high α value in CO_2-depleted chloroplasts can readily be explained by a HCO_3^-* effect on the Q^- reoxidation.

Furthermore, it is known that the time necessary to allow another turnover of the charge accumulator M (often denoted as the $S'_n \rightarrow S_{n+1}$ reaction rate) is extended by a factor of 10 or more by CO_2 depletion (600 μsec → 10 msec) (Stemler *et al.*, 1974). This time necessary for "relaxation" (i.e., the $S'_n \rightarrow S_{n+1}$ reaction) is the time that it takes to open the PSII trap, i.e., to oxidize Q^- or to reduce $P680^+$, whichever is slower. Since the Q^- oxidation is slowed down by CO_2 depletion to a value of at least 4–10 msec (Jursinic *et al.*, 1976; Siggel *et al.*, 1977), it is probable that the slowing down of the $S'_n \rightarrow S_{n+1}$ reaction rate is caused by an effect of HCO_3^-* on the reoxidation of Q^-. Under these circumstances, the very slow component of Q^- reoxidation does not seem to show up in the rate of the $S'_n \rightarrow S_{n+1}$ reaction.

The kinetics of oxygen evolution ($S_4 \rightarrow S_0$) as measured by an unmodulated Joliot electrode are not affected by HCO_3^-* at pH 6.8 (Stemler *et al.*, 1974). However, at pH 5.3, small differences in the kinetics of O_2 evolution were detected as stated in Chapter 15 ($t_{1/2}$ = 4.93 ± 0.18 msec, +HCO_3^-*; $t_{1/2}$ = 5.52 ± 0.27 msec, −HCO_3^-*) (Stemler, 1981); these differences are very small as compared to other HCO_3^-* effects, and they could not be confirmed by the authors (Fig. 4).

Sodium formate ($NaHCO_2$), which seems to compete with HCO_3^-* (N. Good, unpublished; Khanna *et al.*, 1977; Vermaas and van Rensen, 1981), was shown to lengthen the relaxation time of $S'_2 \rightarrow S_3$ and $S'_3 \rightarrow S_0$ at pH 8.2, without affecting $S'_0 \rightarrow S_1$, $S'_1 \rightarrow S_2$, α and the "double hit" parameter β (Stemler, 1980a). Stemler proposed that those S-state transitions that show an extended relaxation time in the presence of formate must result in momentary release and rebinding of CO_2 (see Stemler, Chapter 15, this volume). If formate under these conditions would be able to replace HCO_3^-*, then a higher α value should be obtained in the presence of formate than in the absence of formate, whereas $S'_0 \rightarrow S_1$ and $S'_1 \rightarrow S_2$ also should be slowed down because of the slowing down of Q^- decay in the absence of HCO_3^-*. This was not observed. Therefore, we suggest that under these conditions (pH 8.2) formate may not be able to remove HCO_3^-* from its binding site. The differences that were observed in the $S'_2 \rightarrow S_3$ and $S'_3 \rightarrow S_0$ reactions in the presence and absence of formate could also be explained by a formate effect, other than HCO_3^-* removal, which affects S_2 and S_3 specifically. Bouges-

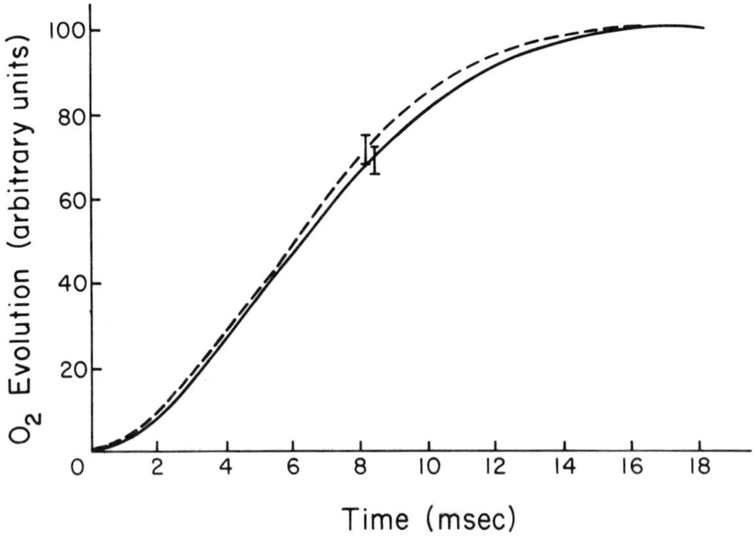

FIG. 4. Oxygen evolution, after the third flash, as a function of time in the presence (——) and the absence (- - -) of 10 m*M* $NaHCO_3$. Measurements were made with an unmodulated Joliot electrode. CO_2-free medium consisted of 50 m*M* Na phosphate, 100 m*M* $NaHCO_2$, 100 m*M* NaCl, 5 m*M* $MgCl_2$, pH, 5.3, and CO_2-sufficient medium was 50 m*M* Na phosphate, 100 m*M* $NaHCO_2$, 10 m*M* $NaHCO_3$, 90 m*M* NaCl, 5 m*M* $MgCl_2$, pH, 5.3. Xenon flashes, $\leq$ 10 μsec; number of experiments, 5. Note the absence of decrease in $t_{1/2}$ of the signal rise upon addition of CO_2. (Data of W. F. J. Vermaas and Govindjee, unpublished observations, 1980–1981.)

Bocquet (1980) suggested that two different electron donors, Z_1 and Z_2, donate electrons to P680 in parallel. Both of them are related to transitions in two S-states. Z_1 is connected to S_0 and S_1, and Z_2 is connected to S_2 and S_3. Thus, formate may slow down the Z_2 reduction but not the Z_1 reduction.

It is difficult to understand why at pH 8.2 formate does not affect $S_0' \rightarrow S_1$ and $S_1' \rightarrow S_2$ transitions, whereas at low pH those transitions are affected, but differentially (see Chapter 15). In Stemler's hypothetical scheme of CO_2 involvement in oxygen evolution, $S_0' \rightarrow S_1$ and $S_1' \rightarrow S_2$ would not be influenced by HCO_3^-* (Stemler, 1980a), and, therefore, this scheme does not explain his $S_0' \rightarrow S_1$ and $S_1' \rightarrow S_2$ data at pH 5.3 either. A possible explanation is that the Q^- reoxidation is S-state dependent. It is known that the rate of Q^- reoxidation by a back reaction is dependent on the S-state (Joliot *et al.*, 1971; Lavergne and Etienne, 1982), but more experiments are needed to check if such a statement can explain the observed difference under these low pH conditions.

B. $H^{14}CO_3^-$* Binding Studies

Stemler (1980b) showed that the rate of $H^{14}CO_3^-$* binding to CO_2-depleted chloroplasts is pH independent in the first 2 mins. However, when the thylakoids are equilibrated for 5 min followed by $H^{14}CO_3^-$* addition, then the binding is faster at low pH than at high pH (pH 6.0 versus 7.8). These results were interpreted by Stemler (1980b) to suggest that the internal pH (which was assumed to be equal to the external pH after 5 min equilibration but not in the first 2 mins) rather than the external pH governs HCO_3^-* binding. Since the oxygen-evolving site is located on the inner side of the thylakoid membrane, an interaction of HCO_3^-* with the oxygen-evolving system was suggested (see Stemler, Chapter 15, this volume). However, CO_2-depleted chloroplasts are uncoupled by the depletion procedure (Khanna *et al.*, 1977) and, thus, a fast pH equilibration (in the second range) between the inside and the outside of the thylakoid vesicle may be obtained. Furthermore, even in nondepleted (control) chloroplasts, the pH equilibration over the thylakoid membrane is rather rapid (rate constant of proton leakage: 0.73 sec^{-1}) (Khanna *et al.*, 1980). This means that if the internal pH is important, differences in $H^{14}CO_3^-$* binding are expected to be lost within a couple of seconds. This was not observed. Further theoretical and experimental analyses are required, and, in our opinion, these data cannot yet be used to support HCO_3^-* binding to the inside of the thylakoid membrane.

IV. Conclusions

Although it has been assumed that HCO_3^-* may act on the donor side of PSII, perhaps even as a direct source of O_2 produced in the oxygen-evolving system (Warburg and Krippahl, 1958, 1960; Stemler, 1980a), there is an overwhelming amount of experimental observations now that support a major HCO_3^-* function between Q and PQ on the acceptor side of PSII (for earlier reviews, see Govindjee and van Rensen, 1978; Jordan and Govindjee, 1980; Vermaas and Govindjee, 1981a).

We believe that there are, as yet, no clear experimental results that point to any HCO_3^-* function on the donor side of PSII (Section III). In our opinion, the hypothetical scheme for a HCO_3^-* function in oxygen evolution as presented by Stemler (Chapter 15, this volume) is not based on convincing experimental data. We prefer a scheme in which HCO_3^-* binds to a site on or close to the herbicide binding protein, causing a conformational change in that protein that results in an allowance of efficient electron transfer between Q and PQ (e.g., see Khanna *et al.*, 1981; Vermaas and van Rensen, 1981). Clearly, more experiments are needed to understand the molecular mode of action and the function of the bicarbonate effect in intact systems.

Acknowledgments

The authors thank John Whitmarsh for critical reading of the manuscript. We thank the National Science Foundation Grant PCM 24532 for financial support during the preparation of this chapter.

REFERENCES

Abeles, B., Brown, A. H., and Mayne, B. C. (1961). *Plant Physiol.* **36,** 202–207.

Arntzen, C. J., Pfister, K., and Steinback, K. E. (1981). *In* "Herbicide Resistance in Plants" (H. LeBaron and J. Gressel, eds.), Wiley, New York (in press).

Ausländer, W., and Junge, W. (1975). *FEBS Lett.* **59,** 310–315.

Blankenship, R. E., Babcock, G. T., Warden, J. T., and Sauer, K. (1975). *FEBS Lett.* **51,** 287–293.

Bouges-Bocquet, B. (1980). *Biochim. Biophys. Acta* **594,** 85–104.

Butler, W. L., Visser, J. W. M., and Simons, H. L. (1973). *Biochim. Biophys. Acta* **292,** 140–151.

Cohen, W. S., and MacPeek, W. A. (1980). *Plant Physiol.* **66,** 242–245.

Cramer, W., and Crofts, A. R. (1982). *In* "Photosynthesis: Energy Conversion by Plants and Bacteria" (Govindjee, ed.), Vol. I, Academic Press, New York.

Fischer, K., and Metzner, H. (1981). *Photobiochem. Photobiophys.* **2,** 133–140.

Giaquinta, R. T., and Dilley, R. A. (1975). *Biochim. Biophys. Acta* **387,** 288–305.

Good, N. E. (1963). *Plant Physiol.* **38,** 298–304.

Govindjee ed. (1982). "Photosynthesis: Energy Conversion by Plants and Bacteria" Vol. I. Academic Press, New York.

Govindjee, and van Rensen, J. J. S. (1978). *Biochim. Biophys. Acta* **505,** 183–213.

Govindjee, Pulles, M. P. J., Govindjee, R., van Gorkom, H. J., and Duysens, L. N. M. (1976). *Biochim. Biophys. Acta* **449,** 602–605.

Haehnel, W. (1976). *Biochim. Biophys. Acta* **440,** 506–521.

Harnischfeger, G. (1974). *Z. Naturforsch., C: Biosci.* **29C,** 705–709.

Homann, P. H. (1968). *Biochem. Biophys. Res. Commun.* **33,** 229–234.

Joliot, P., Joliot, A., Bouges, B., and Barbieri, G. (1971). *Photochem. Photobiol.* **14,** 287–305.

Jordan, D., and Govindjee (1980). *Natl. Acad. Sci. (India), Golden Jubilee Commemoration Vol.* pp. 369–378, National Academy of Science, India, Allahabad.

Junge, W., and Ausländer, W. (1973). *Biochim. Biophys. Acta* **333,** 59–70.

Junge, W., Ausländer, W., McGeer, A. J., and Runge, T. (1979). *Biochim. Biophys. Acta* **546,** 121–141.

Jursinic, P., and Stemler, A. (1981). *Abstr., 9th Annu. Meet. Am. Soc. Photobiol.* **WAM-C5,** 136.

Jursinic, P., Warden, J., and Govindjee (1976). *Biochim. Biophys. Acta* **440,** 322–330.

Katoh, S., and San Pietro, A. (1967). *Arch. Biochem. Biophys.* **122,** 144–152.

Khanna, R., Govindjee, and Wydrzynski, T. (1977). *Biochim. Biophys. Acta* **462,** 208–214.

Khanna, R., Wagner, R., Junge, W., and Govindjee (1980). *FEBS Lett.* **121,** 222–224.
Khanna, R., Pfister, K., Keresztes, A., van Rensen, J. J. S., and Govindjee (1981). *Biochim. Biophys. Acta* **634,** 105–116.
Kok, B., Forbush, B., and McGloin, M. (1970). *Photochem. Photobiol.* **11,** 457–475.
Lavergne, J., and Etienne, A. L. (1982). *Proc. Int. Congr. Photosynth., 5th, 1980,* **III,** pp. 759–771.
Lavorel, J., and Etienne, A. L. (1977). *In* "Primary Processes of Photosynthesis" (J. Barber, ed.), pp. 203–268. Elsevier, Amsterdam.
Nelson, N., Nelson, H., and Racker, E. (1972). *J. Biol. Chem.* **247,** 6506–6510.
Okayama, S., and Butler, W. L. (1972). *Biochim. Biophys. Acta* **267,** 523–527.
Punnett, T., and Iyer, R. V. (1964). *J. Biol. Chem.* **239,** 2335–2339.
Renger, G. (1976). *Biochim. Biophys. Acta* **440,** 287–300.
Sarojini, G., and Govindjee (1981). *Biochim. Biophys. Acta* **634,** 340–343.
Sarojini, G., Daniell, H., and Vermaas, W. F. J. (1981). *Biochem. Biophys. Res. Commun.* **102,** 944–951.
Siggel, U., Khanna, R., Renger, G., and Govindjee (1977): *Biochim. Biophys. Acta* **462,** 196–207.
Sonneveld, A., Rademaker, H., and Duysens, L. N. M. (1979). *Biochim. Biophys. Acta* **548,** 536–551.
Stemler, A. (1977). *Biochim. Biophys. Acta* **460,** 511–522.
Stemler, A. (1979). *Biochim. Biophys. Acta* **545,** 36–45.
Stemler, A. (1980a). *Biochim. Biophys. Acta* **593,** 103–112.
Stemler, A. (1980b). *Plant Physiol.* **65,** 1160–1165.
Stemler, A. (1981). *Proc. Int. Congr. Photosynth., 5th, 1980,* **II,** pp. 389–394.
Stemler, A., and Govindjee (1973). *Plant Physiol.* **52,** 119–123.
Stemler, A., and Govindjee (1974). *Photochem. Photobiol.* **19,** 227–232.
Stemler, A., Babcock, G. T., and Govindjee (1974). *Proc. Natl. Acad. Sci. U.S.A.* **71,** 4679–4683.
Stern, B. K., and Vennesland, B. (1962). *J. Biol. Chem.* **237,** 596–602.
Stiehl, H. H., and Witt, H. T. (1968). *Z. Naturforsch., B: Anorg. Chem., Org. Chem., Biochem., Biophys., Biol.* **23B,** 220–224.
Stiehl, H.-H., and Witt, H. T. (1969). *Z. Naturforsch., B: Anorg. Chem., Org. Chem., Biochem., Biophys., Biol.* **24B,** 1588–1598.
Trebst, A. (1979). *Z. Naturforsch., C: Biosci.* **34C,** 986–991.
van Gorkom, H. J. (1974). *Biochim. Biophys. Acta* **347,** 439–442.
van Rensen, J. J. S., and Kramer, H. J. M. (1979). *Plant Sci. Lett.* **17,** 21–27.
van Rensen, J. J. S., and Vermaas, W. F. J. (1981a). *Proc. Int. Congr. Photosynth., 5th, 1980,* **II,** pp. 151–156.
van Rensen, J. J. S., and Vermaas, W. F. J. (1981b). *Physiol. Plant.* **51,** 106–110.
Velthuys, B. R., and Amesz, J. (1974). *Biochim. Biophys. Acta* **333,** 85–94.
Vermaas, W. F. J., and Govindjee (1981a). *Proc. Indian Natl. Sci. Acad., Part B* **47,** 581–605.
Vermaas, W. F. J., and Govindjee (1981b). *Photochem. Photobiol.* **34,** 775–793.
Vermaas, W. F. J., and Govindjee (1982). *Biochim. Biophys. Acta.* **680,** 202–209.
Vermaas, W. F. J., and van Rensen, J. J. S. (1981). *Biochim. Biophys. Acta.* **636,** 168–174.
Vermaas, W. F. J., van Rensen, J. J. S., and Govindjee (1982). *Biochim. Biophys. Acta.* **681,** 242–247.
Warburg, O., and Krippahl, G. (1958). *Z. Naturforsch., B: Anorg. Chem., Org. Chem., Biochem., Biophys., Biol.* **13B,** 509–514.
Warburg, O., and Krippahl, G. (1960). *Z. Naturforsch., B: Anorg. Chem., Org. Chem., Biochem., Biophys., Biol.* **15B,** 367–369.
West, J., and Hill, R. (1967). *Plant Physiol.* **42,** 819–826.

Witt, K. (1973). *FEBS Lett.* **38,** 116–118.
Wydrzynski, T. (1982). *In* "Photosynthesis: Energy Conversion by Plants and Bacteria" (Govindjee, ed.), Vol. I, Academic Press, New York.
Wydrzynski, T., and Govindjee (1975). *Biochim. Biophys. Acta* **387,** 403–408.
Yamashita, T., and Butler, W. L. (1968). *Plant Physiol.* **43,** 1978–1986.
Zilinskas, B., and Govindjee (1975). *Biochim. Biophys. Acta* **387,** 306–319.

Index

S

Stuart Coward (editor). DEVELOPMENTAL REGULATION: Aspects of Cell Differentiation, 1973

I. L. Cameron and J. R. Jeter, Jr. (editors). ACIDIC PROTEINS OF THE NUCLEUS, 1974

Govindjee (editor). BIOENERGETICS OF PHOTOSYNTHESIS, 1975

James R. Jeter, Jr., Ivan L. Cameron, George M. Padilla, and Arthur M. Zimmerman (editors). CELL CYCLE REGULATION, 1978

Gary L. Whitson (editor). NUCLEAR–CYTOPLASMIC INTERACTIONS IN THE CELL CYCLE, 1980

Danton H. O'Day and Paul A. Horgen (editors). SEXUAL INTERACTIONS IN EUKARYOTIC MICROBES, 1981

Ivan L. Cameron and Thomas B. Pool (editors). THE TRANSFORMED CELL, 1981

Arthur M. Zimmerman and Arthur Forer (editors). MITOSIS/CYTOKINESIS, 1981

Ian R. Brown (editor). MOLECULAR APPROACHES TO NEUROBIOLOGY, 1982

Henry C. Aldrich and John W. Daniel (editors). CELL BIOLOGY OF *PHYSARUM* AND *DIDYMIUM*, Volume I: Organisms, Nucleus, and Cell Cycle, 1982; Volume II: Differentiation, Metabolism, and Methodology, 1982

John A. Heddle (editor). MUTAGENICITY: New Horizons in Genetic Toxicology, 1982

Potu N. Rao, Robert T. Johnson, and Karl Sperling (editors). PREMATURE CHROMOSOME CONDENSATION: Application in Basic, Clinical, and Mutation Research, 1982

George M. Padilla and Kenneth S. McCarty, Sr. (editors). GENETIC EXPRESSION IN THE CELL CYCLE, 1982

David S. McDevitt (editor). CELL BIOLOGY OF THE EYE, 1982

P. Michael Conn (editor). CELLULAR REGULATION OF SECRETION AND RELEASE, 1982

Govindjee (editor). PHOTOSYNTHESIS, Volume I: Energy Conversion by Plants and Bacteria, 1982; Volume II: Development, Carbon Metabolism, and Plant Productivity, 1982

In preparation

John Morrow. EUKARYOTIC CELL GENETICS, 1983

John F. Hartmann (editor). MECHANISM AND CONTROL OF ANIMAL FERTILIZATION, 1983